S0-BSV-537

MOLECULAR MECHANISMS IN BIOENERGETICS

New Comprehensive Biochemistry

Volume 23

General Editors

A. NEUBERGER
London

L.L.M. van DEENEN
Utrecht

ELSEVIER
Amsterdam · London · New York · Tokyo

Molecular Mechanisms in Bioenergetics

Editor

LARS ERNSTER

Department of Biochemistry, Arrhenius Laboratories for Natural Sciences, Stockholm University, S-106 91 Stockholm, Sweden

1992
ELSEVIER
Amsterdam · London · New York · Tokyo

ELSEVIER SCIENCE B.V.
Sara Burgerhartstraat 25
P.O. Box 211, 1000 AE Amsterdam, The Netherlands

First edition 1992
Second impression 1994

Paperback edition 1994

ISBN 0 444 81912 6 (Paperback)
ISBN 0 444 89553 1 (Hardbound)
ISBN 0 444 80303 3 (Series)

© 1992, 1994 ELSEVIER SCIENCE B.V. All rights reserved.

No part of this publication may be reproduced, stored in a retrieval system or transmitted in any form or by any means, electronic, mechanical, photocopying, recording or otherwise, without the prior written permission of the publisher, Elsevier Science B.V., Copyright & Permissions Department, P.O. Box 521, 1000 AM Amsterdam, The Netherlands.

Special regulations for readers in the U.S.A.-This publication has been registered with the Copyright Clearance Center Inc. (CCC), Salem, Massachusetts. Information can be obtained from the CCC about conditions under which photocopies of parts of this publication may be made in the U.S.A. All other copyright questions, including photocopying outside of the U.S.A., should be referred to the copyright owner, Elsevier Science B.V., unless otherwise specified.

No responsibility is assumed by the publisher for any injury and/or damage to persons or property as a matter of products liability, negligence or otherwise, or from any use or operation of any methods, products, instructions or ideas contained in the material herein.

This book is printed on acid-free paper

Printed in The Netherlands

Introduction

'Research is to see what everybody has seen and think what nobody has thought'

Albert Szent-Györgyi: Bioenergetics
(Academic Press, New York, 1957)

Bioenergetics is the study of energy transformations in living matter. It is now well established that the cell is the smallest biological entity capable of handling energy. Every living cell has the ability, by means of suitable catalysts, to derive energy from its environment, to convert it into a biologically useful form, and to utilize it for driving life processes that require energy. In recent years, research in bioenergetics has increasingly been focused on the first two of these three aspects, i.e., the reactions involved in the capture and conversion of energy by living cells, in particular those taking place in the energy-transducing membranes of mitochondria, chloroplasts and bacteria. This area, often referred to as *membrane bioenergetics*, has been the topic of the volume on Bioenergetics published within this series in 1984 (volume 9). As pointed out in the Introduction of that volume, important progress had just begun towards a merger between bioenergetics and molecular biology, i.e., a transition of *membrane bioenenergetics* into *molecular bioenergetics*. This progress has now reached the stage where the publication of a volume on *Molecular Mechanisms in Bioenergetics* was felt to be timely. As in the previous volume, the purpose of this Introduction is to put these developments into a historical perspective. For details, the reader is referred to the large number of historical reviews on bioenergetics that have appeared over the past years, a selection of which is listed after this Introduction.

Bioenergetics as a scientific discipline began a little over 200 years ago, with the discovery of oxygen. Priestley's classical observation that green plants produce and animals consume oxygen, and Lavoisier's demonstration that oxygen consumption by animals leads to heat production, are generally regarded as the first scientific experiments in bioenergetics. At about the same time Scheele, who discovered oxygen independently of Priestley, isolated the first organic compounds from living organisms. These developments, together with the subsequent discovery by Ingen-Housz, Senebier and de Saussure that green plants under the influence of sunlight take up carbon dioxide from the atmosphere in exchange for oxygen and convert it into organic material, played an important role in the development of concepts leading to the enunciation of the First Law of Thermodynamics by Mayer in 1842.

A recurrent theme in the history of bioenergetics is vitalism, i.e., the reference to 'vital forces', beyond the reach of physics and chemistry, to explain the mechanism of life processes. For about half a century following Scheele's first isolation of organic material from animals and plants it was believed that these compounds, which all contained carbon, could only be formed by living organisms – hence the name organic – a view which, however, was not shared by some chemists, e.g., Liebig and Wöhler. Indeed, in 1828 Wöhler succeeded for the first time in synthesizing an organic compound, urea, in the laboratory. This breakthrough was soon followed by other organic syntheses. Thus, the concept that only living organisms can produce organic compounds could not be maintained.

At the same time, however, it became increasingly evident that living organisms could produce these compounds better, more rapidly and with greater specificity, than could the chemist in his test tube. The idea, first proposed by Berzelius in 1835, that living organisms contained catalysts for carrying out their reactions, received increasing experimental support. Especially the work of Pasteur in the 1860s on fermentation by brewer's yeast provided firm experimental basis for the concept of biocatalysis. Pasteur's work was also fundamental in showing that fermentation was regulated by the accessibility of oxygen – the 'Pasteur effect' – which was the first demonstration of the regulation of energy metabolism in a living organism. In attempting to explain this phenomenon Pasteur was strongly influenced by the cell theory developed in the 1830s by Schleiden and Schwann, according to which the cell is the common unit of life in plants and animals. Pasteur postulated that fermentation by yeast required, in addition to a complement of active catalysts – 'ferments' – also a *force vitale* that was provided by, and dependent on, an intact cell structure. This 'vitalistic' view was again strongly opposed by Liebig, who maintained that it should be possible to obtain fermentation in a cell-free system. This indeed was achieved in 1897 by Büchner, using a press-juice of yeast cells.

In the early 1900s important progress was made toward the understanding of the role of phosphate in cellular energy metabolism. Following Büchner's demonstration of cell-free fermentation, Harden and Young discovered that this process required the presence of inorganic phosphate and a soluble, heat-stable cofactor which they called cozymase (later identified as the coenzyme nicotinamide adenine dinucleotide). These discoveries opened the way to the elucidation of the individual enzyme reactions and intermediates of glycolysis. The identification of various sugar phosphates by Harden and Young, Robison, Neuberg, Embden, Meyerhof, von Euler and others, and the clarification of the role of cozymase in the oxidation of 3-phosphoglyceraldehyde by Warburg are the most important landmarks of this development.

A milestone in the history of bioenergetics was the discovery of ATP and creatine phosphate by Lohmann and by Fiske and Subbarow in 1929. Their pioneering findings that working muscle splits creatine phosphate and that the creatine so formed can be rephosphorylated by ATP, were followed in the late 1930s by Engelhardt's and Szent-Györgyi's fundamental discoveries concerning the role of ATP in muscle contraction. At about the same time Warburg demonstrated that the oxidation of 3-phosphoglyceraldehyde is coupled to ATP synthesis and Lipmann identified acetyl phosphate as the product of pyruvate oxidation in bacteria. In 1941, Lipmann developed the concept of 'phosphate-bond energy' as a general principle for energy transfer between energy-generating and energy-utilizing cellular processes. It seemed that it was only a question of time until most of these processes could be reproduced and investigated using isolated enzymes.

Parallel to these developments, however, vitalism re-entered the stage in connection with studies of cell respiration. In 1912 Warburg reported that the respiratory activity of tissue extracts was associated with insoluble cellular structures. He called these structures 'grana' and suggested that their role is to enhance the activity of the iron-containing respiratory enzyme, *Atmungsferment*. Shorty thereafter Wieland, extending earlier observations by Battelli and Stern, reached a similar conclusion regarding cellular dehydrogenases. Despite diverging views concerning the nature of cell respiration – involving an activation of oxygen according to Warburg and an activation of hydrogen according to Wieland – they both agreed that the role of the cellular structure may be to enlarge the catalytic surface. Warburg referred to the 'charcoal model' and Wieland to the 'platinum model' in attempting to explain how this may be achieved.

In 1925 Keilin described the cytochromes, a discovery that led the way to the definition of the respiratory chain as a sequence of redox catalysts comprising the dehydrogenases at one end and *Atmungsferment* at the other, thereby bridging the gap in opinion between Warburg and Wieland. Using a particulate preparation from mammalian heart muscle, Keilin and Hartree subsequently showed that Warburg's *Atmungsferment* was identical to Keilin's cytochrome a_3. They recognized the need for a cellular structure for cytochrome activity, but visualized that this structure may not be necessary for the activity of the individual catalysts, but rather for facilitating their mutual accessibility and thereby the rates of interaction between the different components of the respiratory chain. Such a function, according to Keilin and Hartree, could be achieved by 'unspecific colloidal surfaces'. Interestingly, the possible role of phospholipids was not considered in these early studies and it was not until the 1950s that the membranous nature of the Keilin-Hartree heart-muscle preparation and its mitochondrial origin were recognized.

During the second half of the 1930s important progress was made in elucidating the reaction pathways and energetics of aerobic metabolism. In 1937 Krebs formulated the citric acid cycle, and the same year Kalckar presented his first observations leading to the demonstration of aerobic phosphorylation, using a particulate system derived from kidney homogenates. Earlier, Engelhardt had obtained similar indications with intact pigeon erythrocytes. Extending these observations, Belitser and Tsybakova concluded from experiments with minced muscle in 1939 that at least two molecules of ATP are formed per atom of oxygen consumed. These results suggested that phosphorylation probably occurs coupled to the respiratory chain. That this was the case was further suggested by measurements reported in 1943 by Ochoa, who deduced a P/O ratio of 3 for the aerobic oxidation of pyruvate in heart and brain homogenates. In 1945 Lehninger demonstrated that a particulate fraction from rat liver catalyzed the oxidation of fatty acids, and in 1948–1949 Friedkin and Lehninger provided conclusive evidence for the occurrence of respiratory chain-linked phosphorylation in this system using β-hydroxybutyrate or reduced nicotinamide adenine dinucleotide as substrate.

Although mitochondria had been observed by cytologists since the 1840s, the elucidation of their function had to await the availability of a method for their isolation. Such a method, based on fractionation of tissue homogenates by differential centrifugation, was developed by Claude in the early 1940s. Using this method, Claude, Hogeboom and Hotchkiss concluded in 1946 that the mitochondrion is the exclusive site of cell respiration. Two years later this conclusion was further substantiated by Hogeboom, Schneider and Palade with well-preserved mitochondria isolated in a sucrose medium and identified by Janus Green staining. In 1949 Kennedy and Lehninger demonstrated that mitochondria are the site of the citric acid cycle, fatty acid oxidation and oxidative phosphorylation.

In 1952–1953 Palade and Sjöstrand presented the first high-resolution electron micrographs of mitochondria. These micrographs served as the basis for the now generally accepted notion that mitochondria are surrounded by two membranes, a smooth outer membrane and a folded inner membrane giving rise to the *cristae*. In the early 1950s evidence also began to accumulate indicating that the inner membrane is the site of the respiratory-chain catalysts and the ATP-synthesizing system. In the following years research in many laboratories was focussed on the mechanism of electron transport and oxidative phosphorylation, using both intact mitochondria and 'submitochondrial particles' consisting of vesiculated inner-membrane fragments.

Studies with intact mitochondria, performed in the laboratories of Boyer, Chance, Cohn, Green, Hunter, Kielley, Klingenberg, Lardy, Lehninger, Lindberg, Lipmann, Racker, Slater and others, provided information on problems such as the composition, kinetics and the localization of energy-coupling sites of the respiratory chain, the control of respiration by ATP synthesis and its abolition by 'uncouplers', and various partial reactions of oxidative phosphorylation. Most of the results could be explained in terms of the occurrence of non-phosphorylated high-energy compounds as intermediates between electron transport and ATP synthesis, a chemical coupling mechanism envisaged by several laboratories and first formulated in general tems by Slater. However, intensive efforts to demonstrate the existence of such intermediates proved unsuccessful.

Studies with beef-heart submitochondrial particles initiated in Green's laboratory in the mid-1950s resulted in the demonstration of ubiquinone and of non-heme iron proteins as components of the electron-transport system, and the separation, characterisation and reconstitution of the four oxidoreductase complexes of the respiratory chain. In 1960 Racker and his associates succeeded in isolating an ATPase from submitochondrial particles and demonstrated that this ATPase, called F_1, could serve as a coupling factor capable of restoring oxidative phosphorylation to F_1-depleted particles. These preparations subsequently played an important role in elucidating the role of the membrane in energy transduction between electron transport and ATP synthesis.

A somewhat similar development took place concerning studies of the mechanism of photosynthesis. Although the existence of chloroplasts and their association with chlorophyll had been known since the 1830s, and their identity as the site of carbon dioxide assimilation was established in 1881 by Engelmann using isolated chloroplasts, it was not until the 1930s that the mechanism of photosynthesis began to be clarified. In 1938 Hill demonstrated that isolated chloroplasts evolve oxygen upon illumination, and beginning in 1945 Calvin and his associates elucidated the pathways of the dark reactions of photosynthesis leading to the conversion of carbon dioxide to carbohydrate.

The latter process was shown to require ATP, but the source of this ATP was unclear and a matter of considerable dispute. The breakthrough came in 1954 when Arnon and his colleagues demonstrated light-induced ATP synthesis in isolated chloroplasts. The same year Frenkel described photophosphorylation in cell-free preparations of bacteria. Photophosphorylation in both chloroplasts and bacteria was found to be associated with membranes, in the former case with the thylakoid membrane and in the latter with structures derived from the plasma membrane, called chromatophores. In the following years work in a number of laboratories, including those of Arnon, Avron, Chance, Duysens, Hill, Jagendorf, Joliot, Kamen, Kok, San Pietro, Trebst, Witt and others, resulted in the identification and characterization of various catalytic components of photosynthetic electron transport. Chloroplasts and bacteria were also shown to contain ATPases similar to the F_1-ATPase of mitochondria.

By the beginning of the 1960s it was evident that both oxidative and photosynthetic phosphorylation were dependent on an intact membrane structure, and that this requirement probably was related to the interaction of the electron-transport and ATP-synthesizing systems rather than the activity of the individual catalysts. However, contemporary thinking concerning the mechanism of ATP synthesis was dominated by the chemical coupling hypothesis and did not readily envision a role for the membrane. This impasse was broken in 1961 when Mitchell first presented his chemiosmotic hypothesis, according to which energy transfer between electron transport and ATP synthesis takes place by way of a transmembrane proton gradient.

Mitchell's hypothesis was first received with skepticism, but in the mid-1960s evidence began to accumulate in favour of the chemiosmotic coupling mechanism. It was shown that electron-transport complexes and ATPases, when present in either native or artificial membranes, are capable of generating a transmembrane proton gradient and that this gradient can serve as the driving force for electron transport-linked ATP synthesis. Agents that abolished the proton gradient uncoupled electron transport from phosphorylation. Proton gradients were also shown to be involved in various other membrane-associated energy-transfer reactions, such as the energy-linked nicotinamide nucleotide transhydrogenase, the synthesis of inorganic pyrophosphate, the active transport of ions and metabolites, mitochondrial thermogenesis in brown adipose tissue, and light-driven ATP synthesis and ion transport in *Halobacteria*. In recent years it has also been demonstrated that in several instances a sodium ion gradient, rather than a proton-motive force, can serve as the electrochemical device in membrane-associated energy transduction in connection with both electron transport and ATP synthesis. The chapters of this volume give an overview of our present state of knowledge concerning these processes.

The major problems in this field that remain to be solved concern, on the one hand, the topologic and dynamic aspects of membrane-associated energy-transducing catalysts at the molecular level; and, on the other hand, the mechanisms responsible for the biosynthesis and regulation of these catalysts in the intact cell and organism.

Although there is a great deal of information available today about the primary structure and subunit composition of the various catalysts, knowledge of their tertiary structure and membrane topology is still rather limited; in fact, the only example of a membrane-associated energy-transducing enzyme complex whose structure is known at the atomic level of resolution is the photosynthetic reaction center of purple bacteria. Also, relatively little is known about the conformational events – active-site rearrangements, protein-subunit and lipid-protein interactions – that take place during catalysis, and about the mechanisms by which these events are linked to the translocation of protons or other charged species that are instrumental in establishing the electrochemical gradients mediating energy transfer between the various catalysts. Indeed, there is not a single instance of precise knowledge about the mode of operation of a protein involved in the translocation of protons or any other ions across a biological membrane.

Regarding the biosynthesis and regulation of energy-transducing catalysts, important progress has been made over the last few years especially in the understanding of the role of the mitochondrial and chloroplast genomes in the synthesis of various subunits of the energy-transducing electron-transfer complexes and ATP synthase. Questions of great current interest are the mechanisms by which the synthesis of these subunits is coordinated with that of their nuclear-encoded counterparts and with the transport of the latter into the organelles. Another still poorly understood problem is the function of the noncatalytic subunits of various energy-transducing enzyme complexes, their variation in number and structure from one species or organ to another, and their possible role in the regulation of the biosynthesis and assembly of these complexes.

The chapters of this volume deal with some of the above problems, describing progress that has been made during the last few years due to the development of new methods and concepts within various disciplines, including biophysics, biochemistry, molecular and cell biology, genetics and pathophysiology. Due to these developments, we can foresee a continued rapid progress in understanding the molecular details of cellular energy transduction. At the same time, this progress has widened our perspective of bioenergetics, from molecules, membranes, organelles and cells back to the organism as a whole, i.e. where the whole story began over two centuries ago.

Before terminating this introduction it is a true pleasure to express my thanks to the authors of the various chapters for having accepted the invitation to contribute to this volume and, in particular, for their efforts to submit their manuscripts in time which has made it possible to publish this volume while its contents are still reasonably up-to-date. I was deeply shocked and saddened by the death of Peter Mitchell on April 10th, 1992. He had agreed to write a chapter on 'Chemiosmotic Molecular Mechanisms' but was unable to complete it because of illness. I am greatly indebted to Vladimir Skulachev for his willingness to extend his already submitted chapter on 'Na$^+$ Bioenergetics' by including an introductory section on chemiosmotic systems in general.

Finally I wish to thank my colleague Kerstin Nordenbrand at the Arrhenius Laboratories for her valuable help with the editorial work, and the staff of Elsevier Science Publishers B.V., in particular the Acquisition Editor Amanda Shipperbottom, the Desk Editor Dirk de Heer and the Promotion Manager Anthony Newman, for friendly and efficient cooperation.

Lars Ernster
Department of Biochemistry
Arrhenius Laboratories
Stockholm University
S-106 91 Stockholm
Sweden

Some reviews on topics related to the history of bioenergetics

Rabinowich, E.I. (1945) Photosynthesis and Related Processes. Interscience, New York.

Lindberg, O. and Ernster, L. (1954) Chemistry and Physiology of Mitochondria and Microsomes. Springer, Vienna.

Krebs, H.A. and Kornberg, H.L. (1957) A survey of the energy transformations in living matter. Ergeb. Physiol. 49, 212–298.

Novikoff, A.B. (1961) Mitochondria (Chondriosomes). In: The Cell, Vol. II, pp. 299–421. Brachet, J. and Mirsky, A.E (eds.) Academic Press, New York.

Lehninger. A.L. (1964) The Mitochondrion. Benjamin, New York.

Keilin, D. (1966) The History of Cell Respiration and Cytochrome. Cambridge University Press, Cambridge.

Slater, E.C. (1966) Oxidative Phosphorylation, Comprehensive Biochemistry, Vol. 14, pp. 327–396. Elsevier, Amsterdam.

Kalckar, H.M. (1969) Biological Phosphorylations, Development of Concepts. Prentice-Hall, Englewood, NJ.

Krebs, H.A. (1970) The history of the tricarboxylic acid cycle. Perspect. Biol. Med. 14, 154–170.

Wainio, W.W. (1970) The Mammalian Mitochondrial Respiratory Chain. Academic Press, New York.

Lipmann, F. (1971) Wonderings of a Biochemist. Wiley-Interscience, New York.

Fruton, J.S. (1972) Molecules and Life. Wiley-Interscience, New York.

Arnon, D.I. (1977) Photosynthesis 1950–1975, Changing concepts and perspectives. In: Photosynthesis I, Trebst, A. and Avron, M. (eds.) Encyclopedia of Plant Physiology, New Series, Vol. 5, pp. 7–56. Springer, Heidelberg.

Boyer, P.D., Chance, B., Ernster, L., Mitchell, P., Racker, E. and Slater, E.C. (1977) Oxidative phosphorylation and photophosphorylation. Annu. Rev. Biochem. 46, 955–1026.

Racker, E. (1980) From Pasteur to Mitchell: A hundred years of bioenergetics. Fed. Proc. 39, 210–215.

Bogorad, L. (1981) Chloroplasts. J. Cell Biol. 91, 256s–270s.

Ernster, L. and Schatz, G. (1981) Mitochondria: A historical review. J. Cell Biol. 91, 227s–255s.

Skulachev, V.P. (1981) The proton cycle: History and problems of the membrane-linked energy transduction, transmission, and buffering. In: Chemiosmotic Proton Circuits in Biological Membranes, pp. 3–46. Skulachev, V.P. and Hinkle, P.C. (eds.) Addison-Wesley, Reading, MA.

Slater, E.C. (1981) A short history of the biochemistry of mitochondria. In: Mitochondria and Microsomes, pp. 15–43. Lee. C.P., Schatz G. and Dallner, G. (eds.) Addison-Wesley, Reading, MA.

Tzagoloff, A. (1982) Mitochondria. Plenum Press, New York.

Hoober, J.K. (1984) Chloroplasts. Plenum Press, New York.

Ernster, L., ed. (1984) Bioenergetics. New Comprehensive Biochemistry, Vol. 9. Elsevier, Amsterdam.

Lee, C.P., ed. (1984, 1985, 1987, 1991) Current Topics in Bioenergetics, Vols. 13–16. Academic Press, New York.

Quagliariello, E., Slater, E.C., Palmieri, F., Saccone, C. and Kroon, A.M., eds. (1985) Achievements and Perspectives of Mitochondrial Research. Elsevier, Amsterdam.

Slater, E.C. (1987) Cytochrome systems: From discovery to present developments. In: Cytochrome Systems: Methods, Molecular Biology and Bioenergetics, pp. 3–11. Papa, S., Chance, B. and Ernster, L. (eds.) Plenum Press, New York.

Ernster, L. and Lee, C.P. (1990) Thirty years of mitochondrial pathophysiology: From Luft's disease to oxygen toxicity. In: Bioenergetics: Biochemistry, Molecular Biology, and Pathology, pp. 451–465. Kim, C.H. and Ozawa, T. (eds.) Plenum Press, New York.

Barber, J., ed. (1992) The Photosystems: Structure, Function, Molecular Biology. Topics in Photosynthesis, Vol. 11. Elsevier, Amsterdam.

List of contributors

B. Andersson, 121
Department of Biochemistry, Arrhenius Laboratories for Natural Sciences, Stockholm University, S-106 91 Stockholm, Sweden.
G. Attardi, 483
Division of Biology, California Institute of Technology, Pasadena, PA 91125, U.S.A.
H. Baltscheffsky, 331
Department of Biochemistry, Arrhenius Laboratories for Natural Sciences, Stockholm University, S-106 91 Stockholm, Sweden.
M. Baltscheffsky, 331
Department of Biochemistry, Arrhenius Laboratories for Natural Sciences, Stockholm University, S-106 91 Stockholm, Sweden.
G. Bechmann, 199
Institut für Biochemie, Heinrich-Heine-Universität Düsseldorf, Universitätsstraße 1, W-4000 Düsseldorf 1, Germany.
B. Cannon, 385
The Wenner-Gren Institute, Arrhenius Laboratories for Natural Sciences F3, Stockholm University, S-106 91 Stockholm, Sweden.
A. Chomyn, 483
Division of Biology, California Institute of Technology, Passadena, CA 91125, USA.
G.B. Cox, 283
Membrane Biochemistry Group, Division of Biochemistry and Molecular Biology, John Curtin School of Medical Research, Australian National University, Canberra, A.C.T. 2601, Australia.
R.L. Cross, 317
Department of Biochemistry & Molecular Biology, State University of New York, College of Medicine, Health Science Center, 750 East Adams Street, Syracuse, NY 13210, U.S.A.
J. Deisenhofer, 103
Howard Hughes Medical Institute Research Laboratories, University of Texas, Southwestern Medical Center at Dallas, 5323 Harry Hines Blvd., Room Y4-206, Dallas, TX 75235–9050, U.S.A.
R.J. Devenish, 283
Centre for Molecular Biology and Medicine, Department of Biochemistry, Monash University, Clayton, Vic. 3168, Australia.
L. Ernster, v
Department of Biochemistry, Arrhenius Laboratories for Natural Sciences, Stockholm University, S-106 91 Stockholm, Sweden.
L.-G. Franzén, 121
Department of Biochemistry, Arrhenius Laboratories for Natural Sciences, Stockholm University, S-106 91 Stockholm, Sweden.
F. Gibson, 283
Membrane Biochemistry Group, Division of Biochemistry and Molecular Biology, John Curtin School of Medical Research, Australian National University, Canberra, A.C.T. 2601, Australia.
T. Haltia, 217
Helsinki Bioenergetics Group, Department of Medical Chemistry, University of Helsinki, Siltavuorenpenger 10A, SF-001 70 Helsinki, Finland.

Y. Hatefi, 265
Division of Biochemistry, Department of Molecular and Experimental Medicine, Research Institute of Scripps Clinic, 10686 North Torrey Pines Road, La Jolla, CA 92037, U.S.A.
L. Hederstedt, 163
Department of Microbiology, University of Lund, Sölvegatan 21, S-223 62 Lund, Sweden.
J.B. Hoek, 421
Department of Pathology and Cell Biology, Thomas Jefferson University, Rm. 271 JAH, 1020 Locust Street, Philadelphia, PA 19107, U.S.A.
S.M. Howitt, 283
Membrane Biochemistry Group, Division of Biochemistry and Molecular Biology, John Curtin School of Medical Research, Australian National University, Canberra, A.C.T. 2601, Australia.
B. Kadenbach, 241
Fachbereich Chemie, Biochemie der Philipps-Universität, Hans-Meerwein-Straße, 3550 Marburg, Germany.
R. Krämer, 359
Institut für Biotechnologie I, Forschungszentrum Juelich, Juelich, Germany.
H. Michel, 103
Abteilung Molekulare Membranbiochemie, Max-Planck-Institut für Biophysik, Frankfurt am Main, Germany.
P. Nagley, 283
Centre for Molecular Biology and Medicine, Department of Biochemistry, Monash University, Clayton, Vic. 3168, Australia.
J. Nedergaard, 385
The Wenner-Gren Institute, Arrhenius Laboratories for Natural Sciences F3, Stockholm University, S-106 91 Stockholm, Sweden.
T. Ohnishi, 163
Department of Biochemistry and Biophysics, University of Pennsylvania, A606 Richards Building, Philadelphia, PA 19104-6089, U.S.A.
F. Palmieri, 359
Department of Pharmaco-Biology, Laboratory of Biochemistry & Molecular Biology, University of Bari, Trav. 200 Re David 4, I-70125 Bari, Italy.
G.K. Radda, 463
MRC Biochemical and Clinical Magnetic Research Unit, Department of Biochemistry, University of Oxford, Oxford, OX1 3QU, United Kingdom.
R.R. Ramsay, 145
Molecular Biology Division, Veterans Affairs Medical Center, San Francisco, CA 94121, U.S.A.
A. Reimann, 241
Fachbereich Chemie, Biochemie der Philipps-Universität, Hans-Meerwein-Straße, 3550 Marburg, Germany.
R. Renthal, 75
Division of Earth & Physical Sciences, The University of Texas, San Antonio, TX 78249, U.S.A.
C. Richter, 349
Laboratory of Biochemistry I, Swiss Federal Institute of Technology, Universitätstraße 16, CH-8092 Zürich, Switzerland.
U. Schulte, 199
Institut für Biochemie, Heinrich-Heine-Universität Düsseldorf, Universitätsstraße 1, W-4000 Düsseldorf 1, Germany.

T.P. Singer, 145
Molecular Biology Division, 151-S, Vetarans Administration Medical Center, 4150 Clement Street, San Francisco, CA 94121, U.S.A.
V.P. Skulachev, 37
Department of Bioenergetics, A.N. Belozersky Institute of Physico-Chemical Biology, Moscow State University, Moscow 119899, Russia.
D.J. Taylor, 463
MRC Biochemical and Clinical Magnetic Research Unit, Department of Biochemistry, University of Oxford, Oxford, OX1 3QU, United Kingdom.
K. van Dam, 1
E.C. Slater Institute for Biochemical Research, University of Amsterdam, Plantage Muidergracht 12, 1018 TV Amsterdam, The Netherlands.
H. Weiss, 199
Institut für Biochemie, Heinrich-Heine-Universität Düsseldorf, Universitätsstraße 1, W-4000 Düsseldorf 1, Germany.
H.V. Westerhoff, 1
E.C. Slater Institute for Biochemical Research, University of Amsterdam, Plantage Muidergracht 12, 1018 TV Amsterdam, The Netherlands.
M. Wikström, 217
Helsinski Bioenergetics Group, Department of Medical Chemistry, University of Helsinski, Siltavuorenpenger 10, SF-001 70 Helsinski, Finland.
M. Yamaguchi, 265
Division of Biochemistry, Department of Molecular and Experimental Medicine, Research Institute of Scripps Clinic, 10686 North Torrey Pines Road, La Jolla, CA 92037, U.S.A.

Contents

Introduction, by L. Ernster v
Some reviews on topics related to the history of bioenergetics xi
List of Contributors xiii
Contents xvii
Non-conventional abbreviations xix

1. *Thermodynamics and the regulation of cell functions*
H.V. Westerhoff and K. van Dam 1
2. *Chemiosmotic systems and the basic principles of cell energetics*
V.P. Skulachev 37
3. *Bacteriorhodopsin*
R. Renthal 75
4. *High-resolution crystal structures of bacterial photosynthetic reaction centers*
J. Deisenhofer and H. Michel 103
5. *The two photosystems of oxygenic photosynthesis*
B. Andersson and L.-G. Franzén 121
6. *NADH –ubiquinone oxidoreductase*
T.P. Singer and R.R. Ramsay 145
7. *Progress in succinate: quinone oxidoreductase research*
L. Hederstedt and T. Ohnishi 163
8. *Mitochondrial ubiquinol–cytochrome* c *oxidoreductase*
G. Bechmann, U. Schulte and H. Weiss 199
9. *Cytochrome oxidase: notes on structure and mechanism*
T. Haltia and M. Wikström 217
10. *Cytochrome* c *oxidase: tissue-specific expression of isoforms and regulation of activity*
B. Kadenbach and A. Reimann 241
11. *The energy-transducing nicotinamide nucleotide transhydrogenase*
Y. Hatefi and M. Yamaguchi 265
12. *The structure and assembly of ATP synthase*
G.B. Cox, R.J. Devenish, F. Gibson, S.M. Howitt and P. Nagley 283
13. *The reaction mechanism of* F_0F_1*-ATP synthases*
R.L. Cross 317
14. *Inorganic pyrophosphate and inorganic pyrophosphatases*
M. Baltscheffsky and H. Baltscheffsky 331
15. *Mitochondrial calcium transport*
C. Richter 349
16. *Metabolite carriers in mitochondria*
F. Palmieri and R. Krämer 359
17. *The uncoupling protein thermogenin and mitochondrial thermogenesis*
J. Nedergaard and B. Cannon 385
18. *Hormonal regulation of cellular energy metabolism*
J.B. Hoek 421
19. *The study of bioenergetics in vivo using nuclear magnetic resonance*
G.K. Radda and D.J. Taylor 463
20. *Recent advances on mitochondrial biogenesis*
A. Chomyn and G. Attardi 483

Index 511

Non-conventional abbrevations

AAC	ADP/ATP carrier (adenine nucleotide translocator)
AcPyAD	3-acetylpyridine adenine dinucleotide
AcPyADP	3-acetylpyridine adenine dinucleotide phosphate
AIB	α-aminoisobutyrate
APS	adenosine-5′-phosphosulphate
BChl	bacteriochlorophyll
BFM	brown fat mitochondria
BHM	beef heart mitochondria
BLM	beef liver mitochondria
BPh	bacteriopheophytin
bR	bacteriorhodopsin
BSA	bovine serum albumin
BzATP	3′-*o*-(4-benzoyl)benzoyl adenosine 5′-triphosphate
cAMP	cyclic adenosine monophosphate
CCCP	carbonylcyanide *p*-chlorophenylhydrazone
CoQ	coenzyme Q (ubiquinone)
COX	cytochrome oxidase (cytochrome *c* oxidase)
CRP	cAMP receptor protein
cyt	cytochrome
DABS	*p*-diazobenzene sulfonate
DAN-ATP	1,5-dimethylaminonaphtoyl 3′-*o*-ATP
DCCD	*N,N*′-dicyclohexylcarbodiimide
DCMU	dichloromethylurea (diurone)
DEA	diethylammonium acetate
DMBIB	2,5-dibromo-3-methyl-6-isopropylbenzoquinone
DMSO	dimethylsulfoxide
DTNB	5,5′-dithio-*bis*-(2-nitrobenzoate)
EDTA	ethylenediamine tetraacetate
EEDQ	*N*-(ethoxycarbonyl)-2-ethoxy-1,2-dihydroquinoline
EGTA	ethylene-*bis*(oxyethylenenitrilo)-tetraacetate
ENDOR	electron nuclear double resonance
ER	endoplasmic reticulum
ETHI-57	*N,N*′-dibenzyl-*N,N*′-diphenyl-1,2-phenylene-diacetamide
ETP	electron-transport particles
EXAFS	extended X-ray absorption fine structure
FBPase	fructose-1,6-biphosphatase
FCCP	carbonylcyanide *p*-fluoromethoxyphenylhydrazone
Fd	ferredoxin
FRD	soluble subcomplex of quinol-fumarate reductase
FSBA	*p*-sulfonylbenzyl-5′-adenosine
FSH	follicle stimulating hormone
HA	hydroxyapatite
HMG	high mobility group
HMG-CoA	hydroxymethyl-glutaryl-coenzyme A
HMHQQ	7-(*n*-heptadecyl)mercapto-6-hydroxy-5,8-quinolinequinone
HOQNO	2-*n*-heptyl-1,4-hydroxyquinoline-*N*-oxide
HQNO	hydroxyquinoline-*N*-oxide
hR	halorhodopsin
hsp	heat shock protein
IDH	isocitrate dehydrogenase
IMAC	inner membrane anion carrier
INS	insulin-degrading enzyme (insulinase)
Ins-1,4,5-P_3	inositol-1,4,5-triphosphate
KGDH	α-ketoglutarate dehydrogenase
LDAO	*N,N*′-dimethyl-dodecylamine-*N*-oxide
LEFE	linear electric field effect
LHCI	light-harvesting complex I
MCD	mutagenic circular dichroism
MIGB	*m*-iodobenzylguanidine
MNET	mosaic non-equilibrium thermodynamics
MOA	β-methoxyacrylates
MP	matrix protease
MPP	matrix processing peptidase
MPP^+	*N*-methyl-4-phenylpyridinium
MPTP	*N*-methyl-4-phenyl-tetrahydropyridine
m-TERF	mitochondrial termination factor
m-TF	mitochondrial transcription factor

mTF	mitochondrial transcription factor
MVC	maximal volume contraction
NA	nicotinamide
Nbf-Cl	4-chloro-7-nitrobenzofurazan
NEM	*N*-ethylmaleimide
NET	non-equilibrium thermodynamics
OGC	oxoglutarate carrier
ORF	open reading frame
OSCP	oligomycin sensitivity conferring protein
PBF	presquence binding factor
PC	plastocyanine
PCr	phosphocreatine
PDE	phosphodiesters
PDH	pyruvate dehydrogenase
PEP	phosphoenolpyruvate
PEP	processing-enhancing protein
PFK	phosphofructokinase
PHM	pig heart mitochondria
PIC	phosphate carrier
PM	purple membrane
PME	phosphomonoesterase
PP_i	inorganic pyrophosphate
PPase	inorganic pyrophosphatase
PQ	plastoquinone
prot K	proteinase K
PSI	photosystem I
PSII	photosystem II
PTU	propylthiouracil
Q	coenzyme Q (ubiquinone)
QFR	quinol-fumarate reductase
RC	reaction center
RCR	respiratory control ratio
RLM	rat liver mitochondria
RR	resonance Raman
S-13	5-chloro-3-*t*-butyl-2′-chloro-4′-nitrosalicylanilide
SDH	succinate dehydrogenase (soluble subcomplex of SQR)
SMP	submitochondrial particles
SQR	succinate-quinone reductase (Complex II)
SR	sarcoplasmic reticulum
SR	sensory rhodopsin
tbh	*tert*-butylhydroperoxide
TCA cycle	tricarboxylic acid cycle
TMPD	tetramethyl-*p*-phenylenediamine
TNBS	2,4,6-trinitrobenzene sulfonate
TNM	tetranitromethane
TPB^-	tetraphenylboron anion
TTFA	2-thenoyltrifluoroacetone
TUTase	terminal uridyl transferase
UCP	uncoupling protein
UHDBT	5-*n*-undecyl-6-hydroxy-4,7-dioxobenzothiazol
UHNQ	3-*n*-undecyl-2-hydroxy-1,4-naphthoquinone

L. Ernster (Ed.) *Molecular Mechanisms in Bioenergetics*
© 1992 Elsevier Science Publishers B.V. All rights reserved

CHAPTER 1

Thermodynamics and the regulation of cell functions

HANS V. WESTERHOFF[1,2] and KAREL van DAM[1]

[1]E.C. Slater Institute for Biochemical Research, University of Amsterdam, Plantage Muidergracht 12, NL-1018 TV Amsterdam, The Netherlands and
[2]Division of Molecular Biology, The Netherlands Cancer Institute, Plesmanlaan 121, NL-1066 CX Amsterdam, The Netherlands

Contents

1. Principles 2
1.1. Free-energy transduction is essential for life 2
1.2. The energetics do not necessarily control cell function 3
1.3. Kinetic versus thermodynamic (energetic) control 4
1.4. Linearity and control 6
1.5. Precise analyses of control 7
1.6. Thermodynamic control analysis 8
1.7. Hierarchies in the control of energy metabolism 9
1.8. Signal transduction 14
1.9. Mosaic non-equilibrium thermodynamics and the central role of free energies in control and regulation 14
2. Control of free-energy metabolism 17
2.1. Control analyses 17
2.2. Is the control kinetic or thermodynamic? 18
2.3. Is $\Delta\mu_H$ *the* energetic intermediate? 19
3. Energetics of microbial growth 19
3.1. Description of growth by MNET 23
3.2. Control of microbial metabolism 24
3.2.1. Control by substrates 24
3.2.2. Control by enzymes 25
3.2.3. Control by intermediates? 26
3.2.4. Control by free energy 26
3.2.5. Control of metabolism by energetics; the yeast case 27
3.2.5.1. Glycolysis 27
3.2.5.2. Role of mitochondria 28
4. Energetics and the control of gene expression 28
4.1. The intracellular need for ATP 28
4.2. Do the energetic properties of cells ever change? 30
4.3. Does gene expression change when the energetics change? 30
5. Concluding remarks 31

Acknowledgement 31
Symbols 31
References 32

1. Principles

1.1. Free-energy transduction is essential for life

The paradox noted by Schrödinger ([1], see also ref. [2]) that whilst the universe progresses towards maximum chaos, living systems seem to do the opposite, illustrates the special importance biology has for thermodynamics and vice versa. Because living systems tend to operate in isothermal, isobaric conditions more than in conditions of heat isolation and constant volume, the same paradox is more properly formulated as the question why living systems seem to increase in free energy, whilst experience and thermodynamics tell us that processes tend towards minimal free energy [2,3]. Free energy here represents the Gibbs free energy G ($= U + pV - TS$). The tendency of G towards a decrease comprises the tendency of the true energy U towards a decrease, the tendency of volume V to decrease and the tendency of entropy S to increase.

The resolution of Schrödinger's paradox has many aspects, the ultimate of which is one of the fundamental problems of biology. The first aspect is that many living systems operate at steady state and therefore do not increase in free-energy content. This point does not quite resolve the problem, however, because an essential aspect of living systems is that processes occur in them. For processes to occur with net effect, they need to be driven by a free-energy difference. Consequently, the occurrence of processes is necessarily accompanied by the destruction (dissipation) of free energy (production of entropy). Just to maintain the steady-state processes in living systems, free energy must be dissipated. Living systems are able to destroy free energy without decreasing in free-energy content due to their ability to import free energy [4]. Often, this free energy has the form of the import of chemical compounds that are richer in free energy than the products that are excreted.

When looking at a living system as a thermodynamic black box, this import of free energy solves the paradox. However, inspection of the molecular contents of a living system reveals another paradox. Like 'dead' systems, living systems house many processes that run downhill in the thermodynamic sense; the free-energy content of their substrates exceeds that of their products, hence they dissipate free energy. Here, enzymes serve to enhance the rate of a process that would otherwise also occur, be it at a much slower a rate. What is more characteristic of (though not unique to) living systems is that they also house reactions that run uphill in the thermodynamic sense, i.e. reactions in which free energy is increased. Examples are the synthesis of complex molecules or new cells, the synthesis of ATP for use in muscle contraction and the uptake of food from the cellular environment.

Special biological machines are needed to accomplish such tasks. They do this by coupling the reaction that is thermodynamically uphill to a reaction that is thermodynamically downhill, such that the overall process always dissipates free energy whilst free energy is being stored in the uphill process. These biological machines may be proteins, or entire metabolic networks.

These biological free-energy transducers can only support the uphill reaction when their inner workings are in good order, i.e., when they do not 'slip' too much, and when the downhill reaction that is driving them has sufficient input free energy. Clearly, input free energy and the free-energy transducing machinery are essential for the function of living systems.

1.2. The energetics do not necessarily control cell function

The fact that something is essential for some function does not imply that it also controls or is even involved in the regulation of the latter. Indeed, a free-energy transducer may have excess input free-energy available to it. An example is that of a proton pumping ATPase in the presence of 10 mM ATP, whilst the K_m of the enzyme is 0.1 mM. On the other hand, under conditions of extreme starvation, or in the presence of agents that compromise the mechanism of the free-energy transducer, the free-energy transduction must become 'limiting'. It is clear that the question to what extent the energetics control and perhaps even regulate cellular functions is legitimate and should allow for an answer that depends on the conditions under which the cell operates. What then would be a sensible definition?

It seems fairly straightforward to translate the question whether a given free energy, or a given free-energy transducer controls a function, say a flux J, into: What is the effect on J if I modulate the magnitude of the free energy, or the activity of the transducer, respectively? However, this question is not precise enough, because its answer depends on the magnitude of the modulation. When the modulation is taken as a complete knock out, the question really asks for what is essential for the cellular function. As indicated above, the answer will then be that both the input free-energy and the transducer are essential.

The knock-out definition and particularly the parallel experiment is nonphysiological in that the information it yields derives in part from a condition that is very far removed from the physiological condition of the cell. It seems more natural to modulate the activity of the transducer, or the magnitude of the free energy more subtly and then measure the effect on flux J. The effect on J relative to the magnitude of the modulation of the free energy or the transducer may then be taken as a measure of the control exerted by either of the latter on the flux. When the modulation is made small, this measure of the control becomes independent of the magnitude of the modulation. To detach this measure of control of the dimensions (units) of the two compared properties, the modulation and the change in flux may be expressed in relative terms. Consequently, a coefficient serving to quantify the control of a transducer on a flux J may be defined as the percentage change in the flux J resulting from a 1% increase in the activity of the transducer.

When one substitutes p% for the 1% in the above definition and takes the limit to infinitely small p, then the above definition becomes a differential, which brings the bonus that it can become the subject of mathematical analyses. It is this definition that is called the control coefficient in Metabolic Control Analysis (MCA) [5,3,6]. This refinement need not bother us at this moment, however.

What is more important at this point is that, with this definition of control, one has a way to establish unequivocally to what extent the free energy or a transducer controls a flux or any other functional property of a living system. This extent (e.g., the percentage change in J) may turn out to be very close to zero, in which case one has a rationale to state that that particular free energy or transducer does not control the functional property of interest under the condition of interest, even though it is beyond doubt that the free energy and the transducer are both essential for cell function. Of course, if that percentage change (i.e., the control coefficient) is close to one, then the functional property is proportional to the activity of the free-energy transducer, so that it is reasonable to summarize the situation by stating that the transducer exerts full control on the functional property.

Interestingly, this definition of control allows for subtlety in the analysis of control of cell physiology that has turned out to be important for the understanding thereof. For instance, a control coefficient may turn out to be 0.5, implying that a 1% change in the transducer affects the functional property by 0.5%. Without this subtlety, one would have had to conclude that the transducer is without control or fully in control of the functional property, since more subtle categories would not exist. Also, it allows the scientific analysis of the intuitive notion that the control of a cellular function may be partly in one factor, partly in a second factor and for the rest in a third factor. The definition of control coefficients provides for a method to demonstrate that three factors are indeed important in such a case and what their relative importance is.

Let us suppose that for a particular function, say growth rate, a free-energy transducer has a control coefficient of 0.9. The transducer may then be said to be virtually in control of growth ('growth limiting'). Is this sufficient reason to conclude that the transducer is also involved in the regulation of growth? The answer to this question is no. One speaks of regulation when growth rate is changed as a response to a certain change in environment [7,8] (or [9] when it is buffered against such environmental changes). Such regulation may occur without involving a change in the activity of the transducer; it may proceed through activation of a different reaction in the system, which has perhaps a control coefficient of only 0.1 with respect to growth rate, but is activated to a large extent. To quantify regulation one refers to the magnitude and effects of actual modulations that occur in a system that is being regulated. Control coefficients refer to the fact whether an experimental modulation of a reaction would affect the functional property of interest without asking whether or not such a modulation occurs in a physiological transition.

1.3. Kinetic versus thermodynamic (energetic) control

In this section on principles, it may be appropriate to discuss the distinction that has been made in the literature [e.g. 10,11] between thermodynamic and kinetic control. If one discusses the control of a concentration, there is a relatively straightforward definition of thermodynamic control of a concentration [X] by another substance S: [X] is controlled thermodynamically by S if the reaction converting S to X is and remains at equilibrium. In this case, [X] equals the equilibrium constant times the concentration of S.

However, often the term thermodynamic control is used with respect to a flux; this pertains to a non-equilibrium situation. Let us consider a conversion of a substrate S to a product P, which has a free-energy difference $\Delta G = \mu_S - \mu_P$ as its driving force. μ_S and μ_P refer to the chemical potentials of S and P, respectively ($\mu_S = \mu_S^0 + RT\ln[S]$). At equilibrium, the rate of the process is solely determined by the driving force; independent of any parameter value, whenever the driving force is zero, the flux is zero; purely thermodynamic control therefore, since the driving force is a thermodynamic property. Away from equilibrium, however, the reaction rate may be determined independently by both [S] and [P], hence not by the driving force alone. The special case that the reaction rate is determined by the driving force alone, may sensibly be called the case of 'thermodynamic control'.

As an aside, we may mention that thermodynamic control is somewhat of a misnomer. For, concentration and chemical potential of a substance are related by a simple exponential function. And, more strictly speaking, solution kinetics tends to use activities rather than concentrations in its rate equations, where activities are purely thermodynamic properties. Indeed the rate of conversion of S to P can be written as a function of the chemical potentials of S and P [3,12].

Should one ever expect to encounter thermodynamic control away from equilibrium? There are two cases where one should, one of which is of theoretical interest, and the other of biochemical interest. There is a unique relationship between the activities of substances S and P and their concentrations. Consequently, 'kinetic' control is defined as the case where the rate of the reaction is controlled by μ_S and μ_P independently rather than by $\Delta G = \mu_S - \mu_P$. In other words thermodynamic control is obtained when the dependence of the rate on μ_S equals minus its dependence on μ_P. Onsager ([13], cf., ref. [3]) showed that near equilibrium the latter situation must occur, hence that there must be thermodynamic control.

Thermodynamic control may also be anticipated in enzyme-catalyzed reactions when the enzyme is saturated with substrate and/or product. In that case, only the ratio of the concentrations of the two determines which will be bound to the catalytic site and from there the rate [14,12,3].

Let us elaborate this point in more detail for the simplest reversible enzyme-catalyzed reaction, which has the following rate equation:

$$v = [\mathrm{e}]\{(SV_S/K_S) - (PV_P/K_P)/\Sigma.$$

Here V_S and V_P represent the forward and reverse maximum turnover numbers, respectively. K_S and K_P represent the corresponding Michaelis constants. [e] represents the total concentration of the enzyme that catalyses the reaction. Σ measures the extent of saturation of the enzyme with its substrate and product:

$$\Sigma = 1 + (S/K_S) + (P/K_P).$$

These equations may serve to illustrate that in general a reaction rate is controlled by the properties of the enzyme that catalyzes it (through [e], V_S, V_P, K_S, and K_P), as well as by the concentrations of its substrate and product. The latter may then be called kinetic control. To illustrate the discussion around kinetic control, the above equation may be rewritten as [14,3,12]:

$$v = [\mathrm{e}][\mathrm{C}](V_P/K_P)\{(\Gamma/K_{eq}) - 1\}/\Sigma',$$

$$\Sigma' = 1 + \Gamma + [\mathrm{C}]\ \{(\Gamma/K_S) + (1/K_P)\},$$

$$[\mathrm{C}] = [\mathrm{S}] + [\mathrm{P}].$$

Since

$$\Gamma/K_{eq} = \mathrm{e}^{\Delta G/RT},$$

Γ represents the thermodynamic driving force, hence thermodynamic control. [C] represents the total concentration of substrate plus product. When it significantly exceeds both Michaelis constants, the reaction rate only becomes a function of the kinetic constants of the enzyme and Γ/K_{eq}; a case of thermodynamic control. When the sum concentration of substrate and product is low, the reaction rate becomes dependent on the kinetic properties of the enzyme, on the driving force and on the sum concentration; then control is not only thermodynamic but also by the total concentration of S and P, i.e., kinetic.

Gnaiger [15] has pointed out that there exist analogies between the term [C] $\{(\Gamma/K_{eq}) - 1\}$

and a pressure; he calls this term the reaction pressure to indicate that it is more than just a driving force. Indeed, at low substrate concentrations, both the driving force and the concentration of reactants determine the reaction rate.

Are there also cases where control is purely kinetic, i.e., where a change in the driving force does not affect the reaction rate, whereas changes in substrate and product concentrations do affect the rate? The above equation shows that this can indeed be the case; the dependence of the reaction rate on the driving force tends to be sigmoidal, with a minimum rate at low driving forces and a maximum rate at high driving forces. The maximum rate still depends hyperbolically on [C], the total concentration of substrate and product. In a sense, thermodynamic and purely kinetic control are reciprocal: if the enzyme is saturated either with low or high free energy, control becomes purely kinetic; if the enzyme is saturated with substrate plus product, control becomes thermodynamic.

It is noteworthy that whether control becomes purely kinetic or not, depends on the choice one makes for describing the reaction rate. Here we have chosen to discuss the rate as a function of the driving force and the sum concentration of substrate and product. However, one can also make the choice to write the rate as a function of driving force and product concentration [16]. In that case, at high driving forces, there is no kinetic control (i.e., by [P], although at low (highly negative) driving forces, there is).

In this section we have mostly illustrated the distinctions between kinetic and thermodynamic control by referring to single reactions. We have therefore mainly addressed control at the local level, i.e., 'elasticity' (see below). However, the same discussion holds for control at the global level.

1.4. Linearity and control

In physics many phenomena may be described by linear relationships between cause and effect. Ohms law, an example of a linearly proportional relationship between the rate of a process and its driving force, is extremely accurate for many cases. This inherent simplicity of the physics of dead matter should, however, not lead one uncritically to expect linear relations between cause and effect in biology. Indeed the very complexity of the catalysts of biological reactions should greatly diminish any expectation of linear flow–force relationships.

Consequently, if the relationship between a process and the thermodynamic force that drives it is nonlinear, this should not be taken as evidence that an extra control mechanism is operative. Examples are the relationships between rates of substrate uptake and transmembrane electric potential in various free-energy transducing membranes. Occasionally the nonlinearity has been interpreted as 'gating'.

On the basis of enzyme kinetics flow–force relationships have been calculated and plotted [14, 16–18]. Generally they are sigmoidal, exhibiting strong nonlinearity ('gating') and saturation, without any actual gating or control mechanism being present.

Why then are linear flow–force relations observed fairly frequently in bioenergetics? Near equilibrium there is no region of exceptional linearity (contrary to what is often suggested) [3]. However, almost of necessity, flow–force relations of enzyme-catalyzed reactions are sigmoidal. Hence in the middle of their range of flows, they are quasi-linear (they exhibit an inflection point), which is a linearity that may experimentally not be distinguishable from true linearity over a tenfold range of reaction rates [14].

Although the relationship between nonlinearity and control is not as obvious as sometimes suggested, there is a rather sophisticated relationship, which we shall detail below. Here we only indicate the relationship qualitatively. Let us consider two reaction rates that depend on the same metabolic variable, such as the concentration of a shared metabolite or an electric potential. If one reaction is nonlinearly dependent on that metabolic variable in the sense of a higher order dependence, then a change in the activity in that reaction leading to a change in that metabolic variable, will be virtually reversed by the response of that reaction rate to that change in metabolic variable. As a result, an enzyme with a strong dependence on the metabolic variable will exert little control on the pathway. Consequently, nonlinear relations that lead to strong dependencies tend to reduce the control exerted by the enzyme that exhibits the nonlinear property. On the other hand, strong nonlinearity that entails a very weak dependence on metabolic variables (such as is the case with 'saturation') imparts a strong control on the enzyme in question.

1.5. Precise analyses of control

Discussing metabolic control with fellow biochemists in the lab's corridor can be like discussing soccer in the Sunday afternoon pub: the discussion is vivid, partly because everyone is interested, partly because everyone thinks he/she is an expert and because discussion can remain vague. In fact the subject of metabolic control is prone to vagueness, because it is much more complex than it would seem at first sight. Let us consider the example of a microbe growing on glucose and let us discuss what controls growth. Will it be the external glucose concentration, the internal glucose concentration, catabolism, anabolism, the enzymes that catalyse catabolism? Is it an energy limitation? Does that mean that the concentration of ATP is limiting, or that the activity of the H^+-ATPase is limiting? Part of the confusion resides in the point that one often considers the option that something which itself is a variable, controls something else. For instance, in the above example, it is confusing to suggest that [ATP] or intracellular glucose may be controlling growth, because these two concentrations are themselves controlled by other factors and as a consequence there is no operational definition of what their control would mean. It is, for instance, impossible to change [ATP] and see what the effect is on growth, because, after the addition of ATP, the system will respond by taking away the added ATP and returning to the old ATP concentration (if the ATP is a true variable, cf. ref. [3]). Below (Section 3) we shall discuss possible ways to analyze the involvement of metabolic variables in control and regulation.

The systematic approach to metabolic control (for review see refs. [3,6,19]) starts by defining what in the system is variable and what is fixed. The fixed properties include properties set from the outside (temperature, pressure, the concentration of an external substrate for a metabolic pathway) and immutable properties on the inside (V_m's, K_m's, enzyme concentrations in systems with constant gene expression). The variables include the concentrations of metabolites, free-energy differences across reactions, transmembrane potentials, reaction rates. Legitimate control questions ask to what extent any of the variables is controlled by any of the fixed properties. The magnitude of that control may be denoted in terms of the magnitude of the corresponding control coefficient, defined in Section 1.2.

Returning to the question what controls the growth of the microbe, we may now specify that with growth we refer to growth rate and that the candidates to be considered are any of the enzyme activities in the microbial cell (we here assume that all reactions are enzyme-catalyzed), the corresponding kinetic properties, and the concentration of glucose in the environment. An important relationship exists between the controls exerted by the enzymes: the sum of the control coefficients of the enzymes must equal 1 when they refer to a flux, and 0, when they refer to a concentration or free-energy difference. These are the summation laws of metabolic control theory, for review, see refs. [3,6,19].

The control exerted by pathway substrate and pathway product are not limited by a summation law [20]. They are however related to the control exerted by the enzyme they affect through

$$C^J_{[S]} = C^J_{[e_1]} \cdot \varepsilon^1_{[S]}.$$

$\varepsilon^1_{[S]}$ is the so-called elasticity coefficient of enzyme 1 with respect to S, defined by the percentage change in reaction rate of enzyme 1 caused by a 1% increase in concentration of S, at constant magnitude of all other factors that might affect the rate.

A similar situation occurs in the case of oxidative phosphorylation by isolated mitochondria incubated at a fixed concentration ratio of succinate/fumarate × $[O_2]$. One may then ask to what extent mitochondrial respiration is controlled by the redox potential difference, and this will be given by

$$C^J_{\Delta G_{ox}} = C^J_o \cdot \varepsilon^o_{\Delta G_{ox}}.$$

If J refers to the rate at which ADP is being phosphorylated, this relationship demonstrates that the question whether the redox side controls oxidative phosphorylation must be split into two questions: to what extent does the externally clamped redox potential control the rate of phosphorylation (C^J_o and to what extent does the respiratory chain (referred to by the index o) control the process ($C^J_{\Delta G_{ox}}$).

The first reaction in a metabolic sequence may also be sensitive to the concentration of its product and this may be characterized by its elasticity coefficient towards that product, defined in the same way as the elasticity coefficient for its substrate (see above). In fact for any enzyme in the system elasticity coefficients can be defined with respect to any of the metabolite concentrations and free-energy differences to which they respond. It turns out that metabolic control is largely determined by these elasticity coefficients. This is because of the following connectivity law:

$$\Sigma\, C^Y_{e_i} \cdot \varepsilon^{e_i}_X = -1 \text{ for } X = Y, \text{ and } 0 \text{ in any other case.}$$

The summation is over all enzymes in the system. Y may be a flux, X and Y may be metabolite concentrations, or free-energy differences, such as the free energy of hydrolysis of ATP.

1.6. Thermodynamic control analysis

At first, the metabolic control theory has been devised with respect to intermediary metabolism. More recently however, the theory was elaborated for thermodynamic properties [21,3]. Often this does not involve more than a nomenclature change. One may ask for instance to what extent the growth of a microbe is controlled by the free energy of its substrate. In fact, this question is virtually identical to the question concerning the control by the concentration of the substrate, for the latter may be written as

$$C^{J_a}_{[S]} = (dJ/J)/(d[S]/[S])\,,$$

i.e., the % change in growth rate resulting from a 1% increase in substrate concentration. Because (μ_S symbolising the chemical potential of growth substrate S; *not* the growth rate on S):

$$d[S]/[S] = d\ \ln[S] = d(\mu_S/RT),$$

one finds the identity (J_a symbolizing growth rate):

$$C^{J_a}_{[S]} = C^{J_a}_{\mu_S}.$$

This identity holds for as long as one may equate concentration to thermodynamic activity. We note that C^Y_X is interpreted as $(dY/Y)/(dX/X)$ whenever X and Y refer to concentrations or fluxes. Because of the correspondence between free energies and logarithm of concentrations, whenever Y or X is a free energy, the derivative with respect to that free energy normalized by RT (hence *not* by the free energy itself) is taken.

One may also ask to what extent a chemical potential (such as pH) is controlled by a parameter (such as the activity of the mitochondrial respiratory chain). This question is identical to asking to what extent the activity ('free concentration') of protons in the mitochondrial matrix is controlled by mitochondrial respiration:

$$C^{[H^+]}_{[o]} = d[H^+]/[H^+]/(do/o) = d\{\mu_H/(RT)\}/(do/o) = -2.3\ d\ pH/(do/o).$$

Here o represents the activity of the mitochondrial respiratory chain. The above equations suggest the generalization to the definition of the coefficient of control of the proton motive force ($\Delta\mu_H$) by respiration:

$$C^{\Delta\mu_H}_{[o]} = d[H^+]/[H^+]/(do/o) = d(\Delta\mu_H/(RT)/(do/o).$$

Similarly there are control coefficients quantifying the control of the phosphorylation potential (when it is variable) by various cellular processes such as respiration, proton permeability of the membrane and H^+-ATPase activity.

The summation and connectivity theorems have been shown to remain valid when translated into the thermodynamic properties [21]. This thermodynamic control analysis has been amply applied to mitochondrial oxidative phosphorylation. Results have included the demonstration that the strong respiratory control by the proton motive force (i.e., the high elasticity of mitochondrial respiration with respect to the electrochemical potential difference for protons) is responsible for the phenomenon that cytochrome oxidase exerts rather little control on mitochondrial respiration and the adenine nucleotide translocator (although closer to equilibrium) exerts comparatively much control [22]. Also, the control exerted by the mitochondria on the extracellular phosphorylation potential turned out to depend strongly on the elasticity of the extramitochondrial enzyme that consumes most of the ATP [23].

1.7. Hierarchies in the control of energy metabolism

Up to this point energy metabolism and its control have been discussed largely with respect to processes that directly transduce free energy. Gibbs free energies may affect such processes in the same sense as substrate and product concentrations affect reaction rates. Near equilibrium, such an effect is given by [3]:

$$J = L \cdot \Delta G.$$

Notably, in this equation there is a second factor, L; the flux is not only a function of ΔG as driving force, but also of L. L comprises the 'conductance' of the enzyme catalyzing the process [15]. It is proportional to the activity of that enzyme. Consequently, the flow can also be modified in the absence of changes in ΔG by processes that change enzyme activities [12,3]. And, when there are changes in ΔG, these may affect the flux through their action as a driving force at constant enzyme activity or by changing the activity of the enzyme.

Of course, this phenomenon does not depend on our use of the near equilibrium flow–force relation and is also valid far away from equilibrium. The rate of an enzyme-catalyzed reaction does not only depend on the concentrations of its substrates and products but also on the V's and K_m's of the enzyme catalyzing the reaction for the substrates and products. It is useful to distinguish three ways in which such enzyme kinetic parameters may change. The first is by allosteric effects of other metabolites or free energies that are involved in the same pathway in the cell. These may lead to 'apparent' changes in thermokinetic parameters. Typically these effects disappear as the enzyme is extracted from the cell.

A second class of alterations of thermokinetic parameters of enzymes results from more permanent changes, often covalent modifications that are stable upon isolation. Typically, these alterations change the catalytic effectiveness of the enzyme, without being able to affect the position of the equilibrium; this is what distinguishes them from a second, free-energy dissipating reaction coupled to the reaction under consideration.

A third class of alterations in thermokinetic parameters involves changes in gene expression. Here the simplest case is that in which the concentration of the enzyme in the cell changes. This simply leads to an apparent increase in the V's, hence in the L parameter of the NET descriptions. In more complicated cases, a mutated form may arise, or a gene encoding an isozyme may become expressed to a higher extent. Then changes in other parameters, such as K_m's and $\Delta G^{\#}$ (cf. ref. [3]), may arise as well.

Lately, the essence of the difference between the first class and the other two classes of alterations in thermokinetic parameters has been stressed ([24,25], for earlier work see for instance refs. [26,27]). In the latter case, the metabolism of the property affecting the enzyme is *not* involved in the same metabolic network. Consequently, in the latter case, it becomes useful to consider the total of chemical conversions in the cells not as a single, horribly complex, network, but as a constellation of a number of 'modules': unconnected networks that influence each other in the absence of metabolic conversions between them.

An example is the case of a pathway where the enzyme concentrations are subject to changes through regulated gene expression. This may be conceived of as a hierarchy, with the metabolic pathway at the bottom [24]. One layer up is the level of metabolism of each of the enzymes involved in the catalysis of the metabolic pathway, typically involving protein synthesis by the ribosomes, protein transport and protein degradation. The next level up is that of the metabolism of the mRNA encoding each particular protein. The three levels are independent of one another to the extent that the nucleotides and amino acids involved in protein and RNA metabolism are not involved in the metabolic pathway under study. When in this three-level hierarchical system there is no feedback from the lowest level to one of the higher levels, it may be called a 'dictatorial hierarchy' [24,28]. In many cases there are some feedback effects from the metabolic level to the gene-expression level, e.g., because the concentration of a metabolite inhibits the transcription of the gene that encodes the protein that synthesizes it. Then one may speak of a democratic hierarchy. In a democratic hierarchy, it may be difficult to discern which should be called the upper and which the lower levels; every level may control and be controlled by every other level [25,29]. In fact, as we shall discuss below, it becomes subtle to define control.

The possibility is open that the expression of genes encoding the enzymes involved in cellular

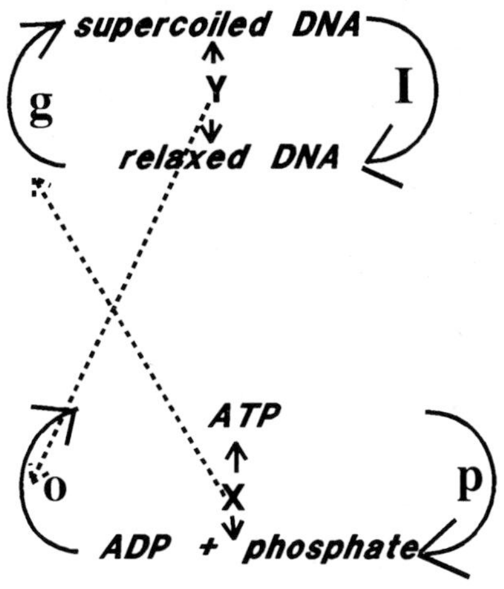

Fig. 1. Scheme of possible reciprocal regulation of the intracellular free energy state (in the form of the free energy of hydrolysis of ATP) and the energetics of DNA. g, and I represent DNA gyrase and topoisomerase I, respectively. o and p refer to ATP synthesizing enzymes (oxidative phosphorylation for instance, or the *atp* genes) and ATP consuming enzymes respectively. X and Y refer to the hydrolytic free energy of ATP and the free energy of DNA supercoiling, respectively. Full lines refer to chemical conversions, dashed lines refer to influences: ATP/ADP affecting DNA gyrase, as quantified by an elasticity coefficient $\varepsilon^{g}_{\Delta G_p} = \varepsilon^{g}_{X}$ and degree of supercoiling affecting the intracellular concentration of enzymes synthesizing ATP (as quantified by the elasticity coefficient ε^{o}_{Y}). The amount of ATP consumed by DNA gyrase is neglected in this scheme.

free-energy transduction is affected by the cellular energy state. To the extent that that is the case (cf. ref. [30]), cellular free-energy transduction is a democratic hierarchy in terms of its regulation.

To a large extent biochemistry has been successful because it dissected the complex cellular system into smaller parts that could then be analyzed and understood. However, the way back, from understood elemental process (e.g., the reaction catalyzed by a single enzyme) to an understanding of the entire cellular system, has remained almost untrodden [31]. The custom to study intermediary metabolism, separate from the study of gene expression is an example. The concept that, essentially, cell physiology is modular (hierarchical), is in fact a reflection of the biochemical approach. Importantly, modern, 'hierarchical' control theory has demonstrated that a modular approach is legitimate. In terms of control coefficients, the control of cell physiology should be understandable by (i) first analyzing the control characteristics of each of its modules ('levels' in the hierarchy), (ii) then defining the influences of the modules on one another (in terms of elasticity coefficients), and finally (iii) just calculating the implications for the control of the entire physiology [25,32]. The fact that the last step can be a calculation, legitimizes the biochemical approach of dissecting cell physiology into modules for separate experimental study.

We shall now present an example of the approach. Figure 1 shows the case where oxidative phosphorylation is in part regulated through variable expression of the *atp* genes. In this scheme the possibility is considered that the intracellular level of ATP, through DNA gyrase, affects transcription. For simplicity we have taken transcription and translation as a single module and the work-floor level of ATP metabolism as the second module. It is assumed that the ATP consumption at the level of regulated gene expression is negligible compared to work-floor turnover; this makes for metabolically unconnected modules. To allow us to illustrate the essence of the method, each module has been kept as simple as possible, containing a single energetic intermediate. This is the hydrolytic free energy of ATP at the work floor and the energy of DNA supercoiling at the gene expression level.

First we may analyze the work-floor level separately. We focus on the control of the phosphorylation potential. 'o' and 'p' referring to the enzyme (systems) catalyzing the synthesis and the consumption of ATP, respectively, the summation and connectivity theorems read, respectively [3,33]:

$$C_{\mathrm{o}}^{X} + C_{\mathrm{p}}^{X} = 0,$$

$$C_{\mathrm{o}}^{X} \cdot \varepsilon_{X}^{\mathrm{o}} + C_{\mathrm{p}}^{X} \cdot \varepsilon_{X}^{\mathrm{p}} = -1.$$

X refers to the phosphorylation potential. $\varepsilon_{X}^{\mathrm{o}}$ is the elasticity of the enzyme systems that make ATP with respect to the phosphorylation potential. When $-\varepsilon_{X}^{\mathrm{o}}$ is high, the systems that make ATP are greatly accelerated by a drop in the phosphorylation potential (a drop in [ATP]). The control exerted by the phosphorylating and the ATP consuming systems, C_{o}^{X} and C_{p}^{X}, respectively, on X is then given by:

$$1/C_{\mathrm{o}}^{X} = -1/C_{\mathrm{p}}^{X} = -\,\varepsilon_{X}^{\mathrm{o}} + \varepsilon_{X}^{\mathrm{p}}.$$

Referring to the degree of supercoiling by 'Y', to DNA gyrase by 'g', and to the counteracting topoisomerase I by 'I', one may deal analogously with the upper module:

$$1/C_{\mathrm{g}}^{Y} = -1/C_{\mathrm{I}}^{Y} = -\varepsilon_{Y}^{\mathrm{g}} + \varepsilon_{Y}^{\mathrm{I}}.$$

Experimentally this corresponds to the determination of the elasticity coefficients of gyrase ($\varepsilon_{Y}^{\mathrm{g}}$) and topoisomerase I ($\varepsilon_{Y}^{\mathrm{I}}$) with respect to the degree of supercoiling of the DNA and the subsequent evaluation of the control properties (C_{g}^{Y} and C_{I}^{Y}) of the gene expression module.

The second stage of the analysis would be the determination of the ways in and extents to which the two modules affect each other. Among the known effects are the effect of the intracellular phosphorylation potential on DNA gyrase, describable by the elasticity coefficient $\varepsilon_{X}^{\mathrm{g}}$. We shall take the sensitivity of topoisomerase to changes in the cellular phosphorylation potential to equal zero. If the system is a dictatorial hierarchy, then the work-floor level is not affected by the gene-expression level. In that case, the control of DNA supercoiling by the enzymes catalyzing oxidative phosphorylation is given by:

$$G_{\mathrm{o}}^{Y} = C_{\mathrm{g}}^{Y} \cdot \varepsilon_{X}^{\mathrm{g}} \cdot C_{\mathrm{o}}^{X}.$$

Here the control coefficient involving the entire (two-module) system is indicated by a G. The C control coefficients refer to the control as defined within a module; they follow the expression into the local elasticity coefficients given above. The latter equation represents the third phase in the control analysis; it calculates control properties of the entire system in terms of the control properties of the component parts and the effects the modules have on one another.

More interesting perhaps is the case of the democratic hierarchy. To illustrate this, we shall assume that the upper module affects the lower module by an effect of DNA supercoiling on the transcription of the gene encoding the enzymes catalyzing oxidative phosphorylation. For simplicity we shall describe this by an elasticity coefficient $\varepsilon_Y^{\mathrm{o}}$ (which is legitimate; it plays the role of an overall elasticity coefficient [21,3], but now in an hierarchical context). The control of oxidative phosphorylation on the degree of supercoiling is modified due to this feedback:

$$G_{k_{\mathrm{o}}}^{Y} = G_{\mathrm{o}}^{Y}/(1-R^{\mathrm{O}}).$$

Here $G_{k_{\mathrm{o}}}^{Y}$ refers to the effect of a change in the turnover rate constant of the oxidative phosphorylation system on DNA supercoiling and R^{O} to the circular regulation coefficient (the supercript 'O' is a circle referring to this circular aspect) (cf. ref. [25]) defined by:

$$R^{\mathrm{O}} = \varepsilon_Y^{\mathrm{o}}\, C_{\mathrm{g}}^{Y}\, \varepsilon_X^{\mathrm{g}}\, C_{\mathrm{o}}^{X},$$

all coefficients referring to (C's) intramodule control or (ε's) to intermodule effects. It is gratifying to recognize the intuitive notion that negative feedback regulation (negative $\varepsilon_X^{\mathrm{o}}$) reduces the control of any component of the system; it stabilizes the system. Conversely, positive feedback, leading to negative values of R^{O} may destabilize the system (when the denominator in the expression for $G_{k_{\mathrm{o}}}^{Y}$ becomes zero or negative).

We stress that by using this method recursively [25], control of the physiology of a large chunk of the cell, as denoted by control coefficients of the type G, can be evaluated from the control properties of the separate modules. Interestingly, this allows for the analysis of the involvement of energetic control at various levels in the cell.

It is of interest also to evaluate the control exerted by an enzyme on a concentration in its own pathway. For instance, in the democratic case:

$$1/G_{\mathrm{o}}^{X} = -\varepsilon_X^{\mathrm{o}} - \varepsilon_X^{\mathrm{o}\mathrm{O}} + \varepsilon_X^{\mathrm{p}}.$$

with

$$\varepsilon_X^{\mathrm{o}\mathrm{O}} = \varepsilon_Y^{\mathrm{o}} \cdot C_{\mathrm{g}}^{X} \cdot \varepsilon_X^{\mathrm{g}}.$$

The latter pseudo-elasticity coefficient measures the regulation of o by X through the upper level in the hierarchy and becomes zero in the case of a dictatorial hierarchy. The equation for G_{o}^{X} shows that the indirect regulation through the upper level acts in just the same way as the internal regulation given by the elasticity coefficient of o with respect to X, $\varepsilon_X^{\mathrm{o}}$. Clearly, regulation involving other modules can substitute for direct regulation. For any real system this suggests the question how much of the regulation is direct and how much involves other modules. For a case with regulated gene expression this may lead to the question how much of the control of the intracellular phosphate potential occurs directly at the level of oxidative phosphorylation itself and how much of it involves regulated gene expression.

There are two ways of distinguishing the direct from the indirect regulatory route. One involves the measurement of the separate control and elasticity coefficients. Although the most systematic method, it is difficult in practice. The second method makes use of the possibility that although two regulatory routes have similar effects at steady state, the one route may act more quickly than the other. Typically regulation involving altered gene expression reaches a new steady state more slowly than does metabolic regulation.

1.8. Signal transduction

The collection of phenomena often called 'cellular signal transduction' [34,35] constitutes an example of a hierarchical (modular) system. Typically an extracellular signaller binds to a membrane receptor and may, for instance, cause dimerization of the latter [34]. This is a pathway of chemical processes and constitutes one level in the hierarchy. The receptor molecule may be a kinase that phosphorylates an intracellular protein, 'P'. Its kinase activity may depend on whether the receptor is a dimer or a monomer. In some cases the receptor molecule is a bifunctional enzyme with both kinase and phosphatase activity. In other cases another protein phosphatase is able to dephosphorylate P. The reactions of phosphorylation and dephosphorylation of P constitute the second level in the hierarchy. P may also have a kinase/ phosphatase activity, hence affect the phosphorylation state of a second protein, 'Q'. These phosphorylation/dephosphorylation reactions then constitute a third level in the hierarchy. Q may be a DNA binding protein. The association/dissociation reaction of DNA and Q then constitutes the fourth level in the hierarchy. As the binding of Q to the DNA may affect the transcription of a local gene, this level may again affect the processes determining the level of mRNA corresponding to that gene. Alternatively Q may be a Ca^{2+} channel in the plasma membrane, which is gated by phosphorylation.

Signal transduction has bioenergetic aspects to it. One such aspect is that in many cases the signal is the presence versus absence of phosphorylation of a protein, or a phospholipid, in a sense an E ~ P complex. A second aspect is that some of the signals correspond to non- equilibrium states. That is, they arise because of one reaction, disappear because of a second one and the sum of the two reactions constitutes a process of free-energy dissipation. An example is the kinase–phosphatase signalling; when both reactions occur, hydrolysis of ATP occurs. It has been shown [36,37,38] that such free-energy dissipating signalling cycles can amplify signals and this may be one of the reasons for their existence.

1.9. Mosaic non-equilibrium thermodynamics and the central role of free energies in control and regulation.

Metabolic and hierarchic control analyses discuss the magnitude of control coefficients. These are defined as the effect of very small changes in parameters on system properties. In the reality of biological regulation, changes are often not very small (i.e., 10% or smaller). They may amount to changes by factors of 10.

The reason why the standard control analysis is not directly applicable to larger changes is that the effects are not usually proportional to the magnitude of the perturbation (where the latter may even be taken to any constant power). Consequently, the value of the control coefficient will drift away as the magnitude of the perturbation is increased. Also, in the case of simultaneous changes in two parameters, the effect arises that the result of a change in one parameter will change as the other parameter is changed; unless limited to very small changes, a control coefficient will change as an unrelated parameter is altered. A rather obvious corollary is the phenomenon that as one enzyme is more and more inhibited, its control on the pathway flux tends to increase, whereas the control of the other enzymes tends to decrease.

To date, the only exact description of control and regulation in terms of large changes requires the knowledge of the kinetic equations for all processes in the system, combined with a computer solution of the steady states attained before and after the regulatory event. This computer integration of the entire system is an important tool as illustrated by the pioneering examples from the groups of Garfinkel [39] and Wright [40,41]. When used alone, however, the method has at least three shortcomings. One is that the programming languages used to program metabolic and cellular systems are phrased in terms of concepts and operations endogenous to mathematical rather than biological analysis. Only recently, object-oriented programming is beginning to allow removal of this limitation; the simulation programs themselves may now be formulated in terms of biological objects and operations [42]. The second disadvantage is that quite a lot of detailed knowledge is required of every single enzyme in the system. Although the kinetic properties of many enzymes have been studied, rarely has this been done for all enzymes in a pathway under identical conditions. For glycolysis, for instance, no such complete data set is available for conditions relevant to the in vivo situation (see Section 3.2.5). Indeed, the work of Garfinkel and colleagues [39] and of Wright and colleagues [40,41], who have come close to this aim, have been little short of heroic. The latter has analyzed the detailed kinetics of Krebs cycle enzymes in *D. discoideum*. Such analyses may not always be feasible for an organism under study.

The third disadvantage is related to the more philosophical question, as to what constitutes 'understanding'. Most of us have lived through discussions between experts in intermediary metabolism, knowing how to evaluate the simultaneous effects of, say, five factors. To most of us, such discussions soon become hard to follow. Rather we listen to a lecture in which the lecturer has abstracted from the many small effects and highlights the major regulatory influences in the system. Clearly the understanding attainable by human beings is itself subject to major limitations; contrary to computers, we are not fit to understand problems with many components in terms of all their details in quantitative precision. In keeping with our evolutionary origin, our understanding has been optimized towards allowing relatively quick decisions on the basis of imprecise and often insufficient qualitative information. The aim of quantitative analysis of cell physiology may then not be to calculate how the cell behaves on the basis of all its kinetic details, but just to help the human mind a little in understanding the major factors involved in control, energetics and regulation of cell physiology.

It is with this insight that many scientists in many sciences (e.g., Newton in mechanics) have proposed to describe reality in a somewhat simplified but not too simplified manner (e.g., by assuming mass to be independent of velocity, or by assuming Michaelis Menten kinetics for enzymes). The idea was that the end result of such an analysis should be an approximate understanding of how reality works. The understanding should be approximate in that it would not be precise to the third digit after the decimal point, but it would be manageable for the human mind, because nonlinear equations would not be involved.

The same strategy has been applied to cell biology by various schools. One of the main problems is, which simple equations serve best to (i) approximate the actual kinetic behaviour of the individual processes of the cell and (ii) be integrated so as to allow understanding of the essence of cell physiology. Initially the approximation that all metabolite concentrations would be below the corresponding Michaelis constants, was used. This made most kinetics linear and led to simple equations describing fluxes and concentrations in metabolic pathways as a function of the properties of the enzymes. Although lacking much of what we would consider typical for biochemistry (enzyme saturation for instance), this approach could help in understanding some phenomena of control and regulation, as reviewed by Heinrich and colleagues [43]. A somewhat more sophisticated approach was developed by Savageau [26,44]. He approximated the rate laws of individual processes by power laws, e.g., $v = k\,[S]^a$ for the dependence of a rate on the concentration of the substrate. This allows for a more accurate description of enzyme kinetics than the linear approximation did [45]. Yet, by then working in terms of logarithms of rates and concentrations, the equations become linear and can be integrated on paper by pencil [26]. The method works fairly nicely when there are only unidirectional reactions in the system. However, when reversible reactions occur or branches in pathways, the coefficients k and a become complex phenomenological mixtures of kinetic properties of many enzymes. As a consequence one may not be able to understand system behaviour as a function of any one of the components of the system after all (although progress is being booked along these lines [45]).

In this book the emphasis is on free-energy transduction, hence on thermodynamics. Therefore, it is more than appropriate to focus further on two related approaches that were aimed at understanding free-energy transduction in biology, though the methods could in fact serve the more general purpose of semi-qualitative understanding of other physiological processes as well. The one approach started from the non-equilibrium thermodynamics developed for near-equilibrium processes [46,47]. As we discussed above, in that approach the rates of processes are formulated as linear functions of their thermodynamic driving forces. Near equilibrium the relationships are Onsager reciprocal [13,3]. This original non-equilibrium thermodynamic method has improved understanding of a number of principles relevant for bioenergetics, such as the phenomenon of coupling between processes, the thermodynamic efficiency of free-energy transduction (and the ways in which this may be evaluated) and the point that the optimum state for a biological system is not necessarily that of complete coupling and 100% thermodynamic efficiency [48,49,3].

Disadvantages in this method of phenomenological non-equilibrium thermodynamics were: (i) The coefficients relating the flows to the forces were phenomenological, i.e., they bore no reference to the mechanism. (ii) The method was legitimated only for processes that are close to equilibrium, whereas most biological processes of interest for free-energy transduction are far from equilibrium. (iii) The relations were too dogmatically linear in the sense that they suggested that flows become infinite at infinite values of the forces, whereas a key property of enzymes is that the flow is limited between the V_m's in the forward and the reverse direction. (iv) The relations were assumed to be reciprocal, although in many cases of interest clear deviations from reciprocity are observed [12].

The mosaic non-equilibrium thermodynamic method (e.g., refs. 14,17,3]) is an extension of non-equilibrium thermodynamics that seeks to avoid these disadvantages of the earlier method. It is based on translation of standard enzyme kinetic rate equations into flow–force relationships which exhibit large linear regions that are however different from the near-equilibrium thermodynamic ones in that they do not extrapolate through the zero force-zero flow point. Also, these flow–force relationships allow for the absence of reciprocity and are not limited therefore to the near-equilibrium domain. Enzyme saturation is reflected by the fact that the flow–force relations used are piecewise linear with both an upper and a lower bound to a central linear piece. In the previous book on bioenergetics in this series we have given a more extensive overview of this method and its applications [12] and more details and applications may be found in ref. [3].

Certainly in the sense of enhancing understanding without leading to a completely accurate description, MNET has been applied successfully to a number of cases of biological free-energy transduction, including mitochondrial oxidative phosphorylation, light driven free-energy transduction in bacteriorhodopsin liposomes and microbial growth (review in refs. [3,12]). All these cases were taken to consist of an input reaction coupled to the generation of a high intermediary free energy (ΔG_X in Fig. 2), an output reaction driven by this intermediate free energy and a leak reaction dissipating this free energy. In principle this may also be done for the living cell more in general: write the free energy of hydrolysis of ATP as the central free-energy intermediate through which various processes in the cell communicate with one another. This view is depicted in Fig. 2. The question now is warranted as to the extent to which intracellular

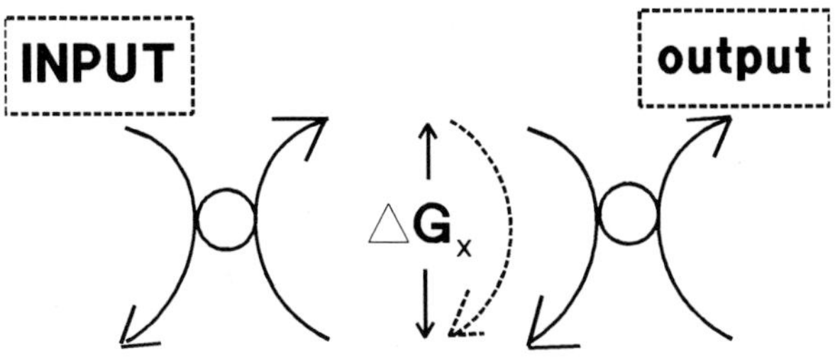

Fig. 2. General scheme for cell physiology where a central free energy (ΔG_X) is involved in the regulation. In practice, ΔG_X may be the phosphorylation potential ΔG_p, or the proton motive force $\Delta \mu_H$. The dashed arrow refers to reactions dissipating the central free energy without coupling to biosynthetic processes ('leakage').

free energies such as the phosphorylation potential and the redox potential of NAD(H) do indeed function as signals between the various processes. Clearly an alternative is that signal transduction pathways take care of the communication. For the hydrolytic free energy of ATP to function as a means through which processes are mutually adjusted to one another, it should (i) change as cell physiology is regulated, and (ii) affect the processes that are to be regulated. As will be discussed more extensively below, it is at present not clear if the hydrolytic free energy of ATP does play such a role in cell physiology. What is clear is that there are many additional routes of signal transduction.

2. *Control of free-energy metabolism.*

2.1. *Control analyses*

The emphasis of this review is on the question if free energies may mediate control of cell physiology. The reciprocal question if and how free-energy metabolism is itself controlled, is important too. In bioenergetics only a subset of the many processes of biological free-energy transduction, i.e., oxidative and photophosphorylation, traditionally received much of the attention, even to the extent that their study is occasionally considered as defining the topic of bioenergetics. Also in this section (but see the subsequent section) we will hardly tread beyond the more specific question how oxidative phosphorylation is controlled. However, we consider it likely that, as so often in the past, the conceptual shifts that have occurred in this field, are also relevant for the other systems of biological free-energy transduction.

The question which reaction determines the rate of mitochondrial oxidative phosphorylation, had been discussed and studied for a long time when conceptual frameshifts led to the resolution that the question itself was improper. Briefly, (i) there is no single such step, (ii) the distribution of control over the various steps varies between conditions, and (iii) control is not directly related to position in the pathway or distance from equilibrium. In isolated rat liver mitochondria in State 3, control turned out to be distributed over the translocators of succinate and ATP, and cytochrome oxidase. In the absence of excess work load (less added hexokinase), there was more control in the proton leak across the membrane and in hexokinase [50,51,52,53]. The kinetic (non-equilibrium thermodynamic) properties of the work load also determine the extent of control exerted by the mitochondria themselves on oxidative phosphorylation [23]. Meanwhile it has been shown that in yeast mitochondria [54] the control is distributed differently. In studies with isolated mitochondria one tends to provide the mitochondria with excess redox equivalents. Mitochondria in situ may witness less redox substrate, such that control may be shifted towards the input of redox equivalents. For heart mitochondria in situ, it has indeed been proposed that control of their respiration lies in the redox input rather than in the phosphorylative systems (e.g., refs. [55,56]). As evidence the phenomenon is quoted that with increased work load, the intracellular phosphorylation potential hardly decreases. Clearly, this reduces the potential role of the intracellular free energy of hydrolysis of ATP as a regulator. The control exerted on mitochondrial respiration by the translocator of ATP across the inner mitochondrial membrane, as measured in isolated liver cells, was rather similar to the control measured in isolated rat liver mitochondria [57].

That it has become possible to assess the distribution of control of mitochondrial free-energy transduction over the component processes, shifts the interest to what is the basis for the observed control distribution. Metabolic control theory demonstrates that that basis should lie in the elasticity coefficients of the component processes with respect to the variable intermediates [58,59,60,61,62]. In this case the variable intermediates would include the intramitochondrial redox potential, the electrochemical potential difference for protons across the inner mitochondrial membrane ($\Delta\mu_H$) and the intra- and extramitochondrial phosphorylation potentials [22,3]. The complete analysis of this has proved too difficult, but considerable insight has been gained by conceptually dividing the overall process of mitochondrial oxidative phosphorylation into three subprocesses (in line with Fig. 2). The first is the respiratory part, driving proton pumping. The second is the subsystem that phosphorylates extramitochondrial ADP (it comprises the adenine nucleotide translocator, the phosphate translocator and the H^+-ATPase). The third is the proton leak. The overall system may then be treated as if consisting of three simple subsystems and the extent to which the respiratory subsystem controls respiration is given by $1/(1-{}^*\varepsilon^{o}_{\Delta\mu_H}/{}^*\varepsilon^{p}_{\Delta\mu_H})$ [22]. The $^*\varepsilon$'s are so-called 'overall' elasticity coefficients. Westerhoff and colleagues have thus shown that the limited degree to which the respiratory subsystem controls mitochondrial oxidative phosphorylation, is due to the high magnitude of the ratio of overall elasticities in the above expression, i.e., due to the high respiratory control by $\Delta\mu_H$ [22,3].

This method of dividing a complex metabolic system into smaller, but still composite parts, has been further elaborated by Westerhoff and colleagues [21,3, Schuster, Kahn and Westerhoff, in preparation] and by Brand and colleagues (who renamed the method the 'top–down approach') [63]. The latter group has also greatly extended the corresponding experimental analyses.

2.2. Is the control kinetic or thermodynamic?

For quite some time, there have been confusing discussions in which it was stated that the control of mitochondrial free-energy transduction is kinetic, implying 'rather than thermodynamic'. Rarely, it was defined what was precisely meant. In line with Section 1.3 of the present paper, we would conclude the following with respect to the control of mitochondrial free-energy transduction:

(i)There is control in the participating enzymes (potentially summing up to at least 60% even halfway between States 3 and 4) [50].

(ii) There is control in external substrates such as ADP, ATP [14,65], phosphate [54] and, in vivo [56], the redox substrates.

(iii) Below a sum concentration of some 1 mM, the mitochondrial respiration is a function of both [ADP] and [ATP] [14,11]. This function can of course be written as a function of log{[ATP/[ADP]}, hence phosphorylation potential, and some other variable (e.g., [ATP] + [ADP]). Thus one may understand the control as just kinetic or as thermodynamic plus kinetic (the latter case). In the latter case, because the sum concentration of ATP and ADP is often constant (in the more extended analysis, AMP is also accounted for [3]), the control may just be understood as thermodynamic. At adenine nucleotide concentrations above 1 mM, the adenine nucleotide translocator tends to become saturated with adenine nucleotides, such that only the ratio [ATP]/[ADP] matters [14]; then the control is automatically thermodynamic (but can of course also be phrased kinetically).

(iv) Of course, in none of these cases control is thermodynamic in the sense of equilibrium thermodynamics; it is all non-equilibrium thermodynamic; the process is always far from equilibrium.

(v) To illustrate the latter: the relationship between mitochondrial respiration and extramitochondrial phosphorylation potential is sigmoidal rather than linear and the sigmoidicity resides largely in the adenine nucleotide translocator [65].

It is clear that in as far as mitochondrial free-energy transduction goes, the phosphorylation potentials and/or its components are important factors in control. That their changes do not always correlate with the changes in fluxes, may simply reflect that control is distributed also over other components (cf. refs. [55,56]).

2.3. Is $\Delta\mu_H$ the energetic intermediate?

Mitchell [66] has proposed that the electrochemical potential difference for protons across the inner mitochondrial membrane is *the* free-energy intermediate between mitochondrial respiration and the phosphorylation of ADP. This proposal has been supported by experimental evidence, but it is too much to be reviewed here. As always however, it is hard to test whether the proposed mechanism functions in all its details. For one, the proton pumps involved may have more special properties than anticipated; they may slip or have variable stoichiometries (e.g., ref. [67]). This has consequences for the control properties [3].

It has been a long standing issue whether the actual intermediate did indeed behave as a macroscopic electrochemical potential difference such as 'the' $\Delta\mu_H$ [68]. Rather, a different $\Delta\mu_H$ might be the intermediate (in or close to the membrane), or the energetic intermediates might be so divided that the small number of high-energy protons did not allow treatment in terms of such a thermodynamic quantity [69]. More recently, it has been pointed out that the various proton pumps in mitochondrial membranes that are so close together should rather be expected to exchange free energy through dynamic electric interactions, in addition to through proton transfer [70].

At present the experimental evidence for rat liver mitochondria is largely compatible with just a single, classical $\Delta\mu_H$, although, so should be noted, all the evidence for slip and non-Ohmic conductance could be equally well explained by local protonic coupling. At present for us the null hypothesis is delocalized chemiosmotic coupling with slipping proton pumps that may occasionally exchange extra free energy.

3. Energetics of microbial growth

The effectiveness of microbial growth can be expressed in different terms. Classically, one considers for each of the elements that are contained in the nutrients, its 'conversion efficiency' into biomass (review in ref. [71]). Thus, the 'carbon conversion efficiency' is called 50% if half of the carbon in the carbon-containing substrate is ultimately found in the cells. Similarly, one can define a 'nitrogen conversion efficiency'. Provided one counts all influx of carbon, this efficiency must always be lower than one. This is because of mass (better: element) balance; this type of efficiency has no thermodynamic bounds [3].

A thermodynamic efficiency should reflect how close a process has come to achieving the maximum possible in thermodynamic terms. It should represent the rate at which free energy (which in isothermal, isobaric systems corresponds to the amount of work that could maximally be done) leaves the system, divided by the rate at which free energy flows into the system. However, although this prescribes what 100% efficiency should mean, it does not define smaller efficiencies. Consequently, there are a number of non-equivalent definitions of thermodynamic efficiency of microbial growth in the literature. Part of the problem may be illustrated by considering (Fig. 3) the conversion of a substance S to a substance P and by attempting to define the thermodynamic efficiency of that process. According to the above definitions, the efficiency η equals:

$$\eta = \mu_P/\mu_S.$$

The problem with this definition derives from the fact that the free-energy contents (μ's) of S and P are each only relative quantities. According to a chemical convention they are taken with reference to the independent elements in their most stable form. The latter reference state

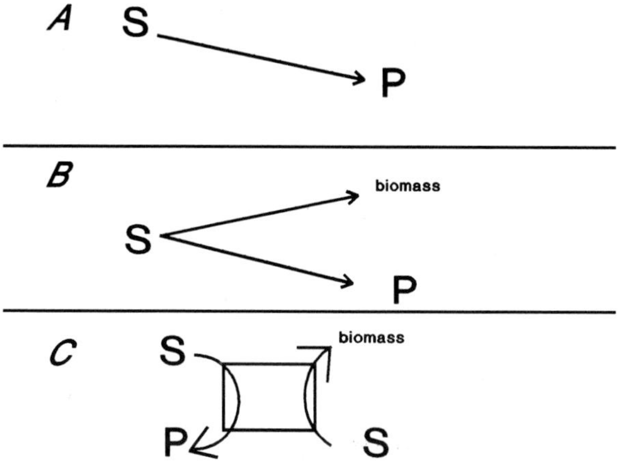

Fig. 3. Three ways of viewing processes or microbial growth, each leading to its own definition of the thermodynamic efficiency thereof. (A), a single process downhill in the thermodynamic sense, the efficiency may be defined as the ratio of the free energy of the product to that of the substrate. This ratio will depend on the reference state for either. (B), microbial growth seen as a conversion of substrate into biomass, with some waste going to product; efficiency may be defined as the free energy production in the form of biomass divided by the free-energy consumption in the form of substrate, forgetting about the metabolic products altogether, or using the state of the metabolic products as the reference state. (C), microbial growth seen as a free-energy transduction between two processes; the input is a conversion of some of the substrate to product; the output is the conversion of the rest of the substrate to new biomass; efficiency is defined as the free-energy gain in the latter reaction divided by the free-energy loss in the former reaction. See also the text.

makes no biological sense. However, switching to another reference state for free energies would alter the magnitude of any value of η differing from one (relative to the elements, μ_S and μ_P might be 90 and 100 kJ/mol, respectively; taken relative to a 'standard state' of complete oxidation at pH = 7, they might be 390 and 400 kJ/mol, respectively, leading to calculated efficiencies of 90 and 97.5% for the same process, just depending on the thermodynamic state used for reference).

As illustrated by Fig. 3B microbial growth on a single substrate may be conceived of as a process leading from a single substrate to two products, i.e., new biomass and catabolic products such as CO_2. A second problem with defining the thermodynamic efficiency arises when there is more than one substrate or more than one product: one needs to decide for each of these, whether it is to be considered part of the input or part of the output. For this case of microbial growth, the efficiency could be defined as

$$\eta = (J_b \cdot \mu_b + J_{Pr} \cdot \mu_{Pr}) / \{J_S \cdot \mu_S\},$$

or as

$$\eta = (J_b \cdot \mu_b) / \{J_S \cdot (\mu_S - \mu_{Pr})\}.$$

In the latter definition it is stressed that only that part of the free energy of the substrate is used that exceeds that of the products. In either case the calculated value for the efficiency still depends on the reference state used. A current definition of the thermodynamic efficiency of microbial growth, introduced by Roels [71], does indeed depend on the reference state used. It takes μ_{Pr} as the reference state. Comparing the free energy of combustion of the used nutrients with that of the generated biomass, the standard state chosen is that of complete oxidation of all compounds:

$$\eta = J_b \cdot (\mu_b - \mu_{Pr})/\{J_S \cdot (\mu_S - \mu_{Pr})\}.$$

For aerobic microbial growth this may seem fair; the efficiency is always smaller than one and positive. On the other hand,the definition seems not to be quite as fair in that it overestimates the work done at the output side of the free-energy transduction process: the biomass really is not made by starting from the level of the catabolic products, but starting from the substrate level.

Following this line, one would see as the input process (cf. Fig. 3C) the combustion of part of the substrate and as the output process the conversion of another part of the substrate into biomass. This then suggests for the definition of the efficiency [72,3]:

$$\eta = J_b \cdot (\mu_b - \mu_S)/\{ J_{Pr} \cdot (\mu_S - \mu_{Pr}) \}.$$

With all these definitions the efficiency is always smaller than one. Yet the calculated values for a particular case of microbial growth differ between the definitions.

Kedem and Caplan [73] have shown that thermodynamic efficiencies can be defined in a manner that is independent of the thermodynamic reference state for cases where there is free-energy transduction between *processes* (rather than between substances). Ideally, there is one clearly recognizable 'input' process (serving no other function than to drive the output process) and a chemically unconnected 'output' process. Usually, the latter runs opposite its own free-energy difference and must therefore be driven by free energy transduced from the input process. Westerhoff and colleagues [72,74] have shown that the latter definition of the thermodynamic efficiency of microbial growth corresponds to the efficiencies defined by this method. Consequently, the efficiency defined in this manner can be compared to all other free-energy transducers for which the efficiency is defined in this manner, and the calculated efficiency does not depend on the standard state chosen for the chemical potentials.

This then may be considered a truly thermodynamic definition of the efficiency of microbial growth [3]. With such a definition the efficiency must be smaller than one, according to the second law, but it may also become negative. The latter occurs when the substrate for growth contains more free energy per molecule of carbon than the produced biomass does; then growth is downhill in the thermodynamic sense and efficiency should indeed be negative (just like riding a bike downhill may be nice but is negatively efficient in the thermodynamic sense).

Quite importantly, Rutgers [75] and Heijnen [76] have pointed out that, even though this way of defining the thermodynamic efficiency of microbial growth on a single substrate is independent of the reference state, it still depends rather strongly on the precise definition of both the anabolic (output) and the catabolic reaction [114]. For instance, the catabolic reaction may be written as

$$C_6H_{12}O_6 + 6\ O_2 \leftrightarrow 6\ CO_2 + 6\ H_2O,$$

or as

$$C_6H_{12}O_6 + 6\ H_2O \leftrightarrow 6\ CO_2 + 12\ H_2.$$

The Gibbs free-energy differences of these reactions are different. Hence calculated efficiencies depend on which definition one chooses. Of course, for aerobic growth, the former definition is the more realistic one. For anaerobic growth it may not be. Moreover, for the anabolic reaction, a problem occurs rather readily whenever the substrate has a higher degree of oxidation that the biomass has. If then molecular oxygen is used as the compound closing the redox balance, one will find that molecular oxygen is assumed to be produced in the assimilation reaction. Such an assimilation reaction is not realistic biologically. Comparison of the efficiency of the process with the efficiency of other free-energy transducers may still be useful then, but it must be made clear that the comparison is made for the specified anabolic reaction.

What is crucial for any definition of thermodynamic efficiency is that one painstakingly defines what is considered the input process and what is considered the output process [76]. The calculated efficiency is only valid for that combination of processes and when one defines input and output process slightly differently, different values for the thermodynamic efficiency may be obtained. In most cases a natural and physiologically relevant choice of input and output is available however.

Whatever definition we choose, the possibilities for microbial growth to be improved are limited by the second law of thermodynamics *and* by the available biochemical mechanisms. This has led Stouthamer [77] to introduce a definition based on knowledge of the biochemical pathways. He reasoned that the generation of the constituents of biomass would be accompanied by utilization of a definable number of molecules of ATP; knowing the biochemistry of the anabolic pathways that number may be calculated. Stouthamer called this the maximum yield of ATP (Y_{ATP}^{max}). It turned out, however, that in those cases where the number of molecules of ATP generated by catabolism could be calculated, the yield of biomass was less than half of that expected [77]. Thus, an unexplained gap remained between theory and experiment.

3.1. Description of growth by MNET

When one wants to discuss the concepts of efficiency and energetics of biological processes and their relationships with some of the relevant mechanisms, the MNET model is very appropriate, because it makes use of the quantitative relation between rates of processes and free-energy changes [72,3]. This of course should not take away the consciousness that MNET is only an approximation of a completely precise description (see above; yet to understand, one usually does not require complete, but just sufficient precision).

In the simplest MNET scheme of microbial growth (cf. Fig. 2) we include (i) catabolism, (ii) anabolism and (iii) a generalized ATP leak. In a first approximation, the first two processes are fully coupled reactions between a driving reaction and a driven reaction, i.e. (i) substrate breakdown coupled to ATP synthesis and (ii) ATP hydrolysis coupled to biomass synthesis. The scheme is completed by addition of a separate possibility of ATP breakdown (which may be catalyzed via any subreaction, as long as it is not coupled to catabolism or anabolism). Formally, the whole arrangement may be compared to the scheme of oxidative phosphorylation, where we have two coupled proton pumps and a proton leak across the membrane; the electrochemical potential difference for protons then takes the place of the phosphorylation potential in the case of microbial growth. In each case a reaction has a rate that is related to the free-energy change connected with that elemental reaction.

Starting from this simple MNET scheme, one can derive a relation between substrate utilization and growth and predict how this relation will depend on the factor that limits the growth [72]. Many of the fundamental observations on microbial growth can already be accommodated by this description when relations between flows and forces are taken to be strictly proportional and Onsager reciprocal.

A more realistic description, however, takes into account the fact that there is in general no proportionality between forces and fluxes in enzyme-catalyzed reactions. This can be accounted for by including an extra constant in the equations and allowing asymmetry in the dependence on the forces [72,20,3]. Although with such a more refined description a realistic relation between substrate utilization and growth is available, its experimental verification remains problematic: the constants depend on the cell composition and it is very difficult to keep that constant while varying the external conditions.

It is, however, possible to measure actual thermodynamic efficiencies, as defined in the previous paragraph. To do this, we have to measure the concentration of substrates and products (including biomass) as a function of time. Such measurements were available from the literature, but under non-standardized conditions. Rutgers and colleagues [78] repeated those experiments using one microorganism (*Pseudomonas oxalaticus*), that is capable of growing on a large range of substrates. They further substantiated the earlier conclusion that the highest thermodynamic efficiency of growth is reached with the most oxidized growth substrates (cf. refs. [71,72] for review) and that it approaches a value of 24% [74]. This value coincides with that calculated by Stucki for an energy converter that is optimized towards maximal output rate at optimal efficiency [49].

The actual way in which this optimization of growth towards a certain parameter has been achieved remains a matter of speculation. Variation of coupling can be achieved either by introducing slips in the primary reactions or leak of ATP. Slip in a primary catabolic reaction may for instance be visualized by considering electron transport chains with different stoichiometries. Leak of ATP can occur if there are futile cycles involving ATP. It can be calculated, using the MNET formalism, that slips or leaks have a different effect on the force ratio at which maximal efficiency is reached [3]. Moreover, the order in which the different parameters are optimized, has an effect on the outcome of the optimization procedure [3]. For various other modes of optimization we refer to a recent review by Heinrich et al. [79].

3.2. Control of microbial metabolism

The rate of microbial metabolism is of course determined by the concentrations of substrates and products and by the enzymatic makeup of the cells. In principle, it is possible to determine the control coefficient of each of these factors on overall metabolism of the cells. It is instructive to see what are the relevant measurements to make when determining control in a quantitative sense. To begin with, we will simplify matters by only looking at steady states in which the composition of the cells is not changing. In a more refined model such changes, for instance by induction of expression of enzymes can be accommodated (in a hierarchical series of steady states, [25]).

3.2.1. Control by substrates

When one wants to determine the control of microbial metabolism (growth) by growth substrates, an experimental problem arises because of the high affinity of most substrate uptake systems for the substrate. The range in which the rate of substrate uptake changes with its concentration is at such low concentrations that negligible changes in biomass accompany large changes in substrate concentration, if one works in a batch culture.

A solution to this problem is afforded by the so-called 'chemostat' [80]. This device allows a steady state of growth by continuous dilution of a microbial culture with fresh growth medium, while the total volume is kept constant by efflux of culture fluid. In the steady state, biomass and all substrate concentrations are constant. Usually one of the growth substrates is 'limiting'; its concentration then changes strongly with the dilution (=growth) rate. To determine the control of a particular substrate on growth rate, one experimentally sets the growth rate and measures which substrate concentration is accompanying it in the steady state. For small changes, the relative change in dilution rate divided by the relative change in substrate concentration equals the coefficient for the control of growth by that substrate [29]. Thus, the control by substrate is determined indirectly.

By varying the relative concentrations of substrates in the input medium, Rutgers et al. [81] were able to construct plots from which the control by more than one substrate could be simultaneously read. This latter experiment actually proved that simultaneous growth control by more than one substrate can occur.

It would, however, be more satisfying if one could set the substrate concentration and measure the ensuing growth rate. Recently, a practical modification of the chemostat was reported, in which this can be done by monitoring the substrate concentration in the effluent continuously [82]. By a feed back mechanism, the medium flux is adjusted to keep that concentration constant, thus practically dictating the growth rate by setting the substrate concentration.

Microbiologists consider different possible 'limitations' in microbial growth. Thus, they distinguish between catabolite-, anabolite- and energy-limited cultures. In principle, one might consider the former two as special cases of 'control by substrate' if it would be possible to define purely catabolic or anabolic substrates. Unfortunately, almost any substrate for growth is used both in catabolism and anabolism. This is certainly true for those cases where the carbon substrate is 'limiting', because it will be required both for generation of ATP and for generation of cellular building blocks. One possible exception may be found in photosynthetic organisms, where light reactions may serve as the free-energy providing 'catabolic' reaction (especially in those microorganisms that do not fix CO_2). In other organisms, control theory may be used to do what experiment cannot do: split the single substrate for catabolism and anabolism in two conceptual substrates and calculate the relative control exerted by the two.

Otherwise, the definition of 'anabolite-limited' and of 'catabolite-limited' is rather arbitrary. The definition of 'energy-limited' is even more ambiguous. It seems to be usually implied that under those conditions in some way the supply of ATP would limit the rate of growth. However, a quantitative measurement of ATP concentration (or its turnover) in the cells is seldom reported. This point will be discussed further in Sections 3.2.3 and 3.2.4.

3.2.2. Control by enzymes

The control of enzymes on microbial metabolism (growth) is simpler to measure than the control of substrates. However, also in this case pitfalls arise in the interpretation of experiments.

For one thing, one has to be careful to restrict the analysis to comparable conditions with small relative changes. This is difficult to do in many cases. The simplest approach would be to vary a specific enzyme activity with a specific inhibitor, just like in the case of analysis of control in oxidative phosphorylation (see Section 2). Unfortunately, in the complex mixture of enzymes that constitutes a microbial cell, it is very hard to find an absolutely specific inhibitor. Nevertheless, this approach has been used to some extent and conclusions about the control of, for instance, enzymes of glycolysis on the growth of cells have been drawn [83,84,85, review: 86]. It may be illustrative that these conclusions are sometimes contradictory.

A more specific way of determining the control by an enzyme is by varying its activity through genetic means. One elegant procedure has been developed by Walsh and Koshland [87], by making use of introduction of the relevant gene on an inducible plasmid. In this way, activities of the enzyme to be investigated can be modulated over any desired range and its control can be determined from its effect on the physiological activity.

Some more recent examples of this approach are the following. Birkenhead et al. determined the control by the dicarboxylate translocator on growth and metabolism of *Bradyrhizobium japonicum* [88]. They measured the change in growth rate and the activity of the dicarboxylate translocator in these cells. From the results one can conclude that the control of the translocator on growth is small. This is not unexpected, since the uptake step is only one in a large pathway. Because of the summation theorem, one can expect a small control by each of the constituent steps.

Schaaff et al. looked at the effect of overproduction of each of the enzymes of glycolysis on ethanol production and growth rate in *Saccharomyces cerevisiae* [89]. From their results, one has to conclude that none of the glycolytic enzymes has significant control on the glycolytic flux in yeast. This might again be the consequence of the large number of steps involved. However, another reason may be that most of the control resides in the uptake step ([90], Van Dam and Oehlen, in preparation). Bailey and Galazzo [91] have calculated the control distribution for yeast glycolysis from elasticity coefficients.

Jensen determined the control of ATP synthase on fluxes in *E.coli* [30; P.R. Jensen, H.V. Westerhoff and O. Michelsen, in preparation]. Surprisingly, the control of this crucial enzyme on growth rate was virtually zero. Furthermore, its control on respiration was negative. Perhaps this reflects homeostatic control mechanisms involving higher levels in the control hierarchy, that enable *E. coli* to activate substrate level phosphorylation when oxidative phosphorylation is compromised.

The control of enzyme II of the PEP:glucose phosphotransferase system (PTS) on glucose oxidation, growth or uptake was studied by Ruijter et al. [92]. The control turned out to be high on the isolated uptake step, but low on the more extensive pathways of oxidation or growth. Again: the magnitude of control by enzymes depends on the pathway considered and on external conditions.

3.2.3. Control by intermediates?

It is one of the tenets of metabolic control theory, that only parameters (i.e., properties that may be set but are otherwise constant) may control variables (properties that always vary between steady states and that fluctuate [59,3]). As a consequence intermediates in pathways of metabolism, free-energy transduction and microbial growth cannot control other properties such as growth rate or pathway flux.

Because intermediates are still involved in control, attempts have been made to quantify their contribution to control. In two cases [93,8], this addressed the involvement of an intermediate in the regulation of a pathway flux by an external effector. In another case [7,9] it did involve some sort of 'control' by the intermediate itself. The latter control (or even regulation) was zero, for reasons of stability of the system [59]. The total of zero could however be split up into terms that corresponds to different routes, each of which would not be zero [9].

As an alternative to these earlier works, let us consider what one might mean with control by intracellular ATP, thinking of that ATP as a settable parameter using the general definition:

$$C^{J}_{\mathrm{ATP}} = \mathrm{d}(\ln J)/\mathrm{d}(\ln[\mathrm{ATP}]).$$

The problem is: how can one experimentally determine the magnitude of this control? The idea would be to clamp the intracellular ATP level at the steady-state value it has anyway and then shift that value as if it were a parameter, keeping the other parameters in the system constant. It is not simple to vary the intracellular concentration of ATP in an independent way, because the cell membrane is not permeable to this compound. On the other hand, if one would vary the steady-state concentration of ATP by changing either the rate of the metabolic reactions leading to its synthesis or of the reactions leading to its hydrolysis, one would violate the condition of constant parameters. The most satisfactory way out of this dilemma would seem to be the controlled introduction of a new 'side reaction', leading to ATP hydrolysis, which in itself constitutes a 'metabolic dead end'.

For microbial growth one may now ask to what extent such a parameterized [ATP] would control growth. The answer corresponds to the (overall) elasticity coefficient of anabolism with respect to [ATP] (cf. refs. [12,3], Schuster, Kahn and Westerhoff, in preparation). The question has not really been asked experimentally in approaches where other influences than through ATP have been quite excluded (often also the membrane potential varies). Yet, of course, protonophores do inhibit growth, at moderate decreases of intracellular phosphorylation potentials [94].

Similar considerations hold for the experimental determination of the control of metabolic fluxes by any other metabolite. In each case, one would have to find a reaction that changes the concentration of the metabolite in question, but which does not connect further with the metabolic fluxes.

3.2.4. Control by free energy

One may define the control of a free energy on microbial metabolism J (cf. ref. [3]) as

$$C^{J}_{\Delta G} = \mathrm{d}(\ln J)/\mathrm{d}(\Delta G/RT).$$

This definition is straightforward when the free energy under consideration is a parameter, such as the free energy of catabolism, or the free energy of anabolism, set by the outside world. No explicit measurement of the control of microbial growth by either external free energy has been reported to date, although control coefficients could be calculated from existing measurements of the control by growth substrates.

Analogously to what was noted in the previous section, the definition of control by a free energy is problematic when the free energy under consideration is itself a variable, such as the free energy of ATP hydrolysis (the 'phosphate potential') or the proton motive force. The definition that one may then give is that for the identical system, at the same magnitude of the free energy, where however, the free-energy difference has become a parameter, which can now be changed by interference with an added process. For the case of the proton motive force one would add a protonophoric uncoupler, which will decrease that force by allowing a controlled backflow of protons across the membrane.

For mitochondrial oxidative phosphorylation, this approach has been taken experimentally [22]. What would be defined as the coefficient of control by $\Delta\mu_H$ on ATP synthesis, corresponds to the, then defined, 'overall elasticity coefficient of ATP synthesis with respect to $\Delta\mu_H$'. The coefficient for the control exerted by $\Delta\mu_H$ on respiration corresponds to the then defined 'overall elasticity coefficient of the respiratory subsystem for $\Delta\mu_H$'. It was shown that the ratio of these two coefficients determined how the control of mitochondrial respiratory rate (or phosphorylation rate) was distributed over the respiratory and the ATP synthesizing subsystems (cf. Section 2.1). Thus, the quasi-control coefficients with respect to parameterized free energies may have some meaning for control and regulation of free-energy metabolism.

We can look at the control of ΔG's on partial reactions (this then is equivalent to the elasticity coefficients of those reactions for those ΔG's) or on overall reactions of metabolism (which is then equal to 'overall' elasticity coefficients [3]). According to the MNET description, the rate of each process depends on each of the ΔG's involved in it. If the rate were proportional to a free energy, with proportionality coefficient L, the control by that free energy on the flux would be $RT/\Delta G$. However, such a proportionality does normally not occur, not even in the elementary reactions (in contrast to the classical NET description) (see Section 1.9).

3.2.5. *Control of metabolism by energetics; the yeast case*

3.2.5.1. Glycolysis. Glycolysis was the first metabolic pathway to be studied by biochemists, but up to this day its regulation has remained unsolved. Although many regulatory aspects at the level of individual and important enzymes have been discovered and analyzed in great detail, the quantitative understanding of the extent to which the various enzymes and regulators control the overall process, is lacking. We will discuss here the control of glycolysis in yeast.

The pathway from glucose to ethanol and CO_2 in yeast starts with the uptake of the sugar across the plasma membrane. There are indications that this step may have a large control on the overall flux ([90], Van Dam and Oehlen, in preparation). In any case, overexpression of all the subsequent enzymes (or combinations of them) does not lead to a significant increase in the overall glycolytic flux [89]. This is presumably one of the reasons why the uptake has been studied extensively, but a number of unsatisfactory mysteries remain. In the uptake of glucose one sometimes distinguishes a low- and a high-affinity system [95]. Induction of the latter would occur at low glucose concentrations. A close connection between uptake and phosphorylation is indicated by the fact that the high-affinity system cannot be found in yeast cells in which the two forms of hexokinase have been deleted [96]. Indeed, it is still unclear, whether the uptake is facilitated by diffusion followed by rapid phosphorylation of the sugar via the kinases or, alternatively, there is a more direct connection between the uptake and phosphorylation steps.

Intricate regulation of glycolysis must occur, since it is a pathway with typical feedback properties: in the initial part ATP is needed to direct glucose into the pathway, in the later part of the pathway more ATP is formed. It has been shown that in such a scheme one can expect oscillations in the concentration of intermediates and overall flux after addition of glucose, provided the kinetic properties of the enzymes involved fall within a certain region [97]. For this, the enzymes do not even need to be allosterically regulated [98,99]. Of course, the fact that several of the glycolytic enzymes are subject to strong allosteric regulation can further enhance this phenomenon of oscillations. Indeed, the occurrence of oscillations in the degree of reduction of NAD has been shown, both in intact yeast cells and in extracts containing all the glycolytic enzymes [100,97].

On top of the short-term regulation of glycolysis, there is also a long-term effect of addition of glucose to cells that have been cultured aerobically on another substrate, for instance ethanol. This effect has been called the 'glucose effect' and involves the modulation of a number of enzymes via different mechanisms (for review see ref. [101]). Thus, several mitochondrial and vacuolar enzymes are repressed, whereas some enzymes (such as pyruvate decarboxylase) are induced. In these cases, presumably the regulation involves transcription factors.

An example of an enzyme of which the activity is modulated in a different way is fructose-1,6-bisphosphatase (FBPase). Apparently, there is a rapid turnover of this enzyme, involving its synthesis followed by uptake into the vacuoles and proteolysis. Upon addition of glucose the concentration of the receptor for FBPase in the vacuolar membrane is increased, so that breakdown of the latter is stimulated and as a consequence the steady-state concentration of FBPase is lowered [102]. Regulation of the activity of a number of enzymes through a cascade involving the action of cAMP and protein kinases has also been suggested (for review see ref. [103]).

3.2.5.2. Role of mitochondria. Glycolysis is a pathway that is in itself 'redox neutral', i.e. it does not lead to generation or dissipation of NADH. However, in the partial reactions of glycolysis NAD is needed. Thus, all processes that do lead to changes in the redox status of NAD will affect the flux through glycolysis. The most notable possibility for such processes is the mitochondrial respiratory chain. Oxidation of NADH by the mitochondria will in fact have a dual effect: it disturbs the redox balance and also adds to the formation of ATP. It is, therefore, not surprising that the above-mentioned phenomenon of oscillations in intracellular NADH depends on an intricate balance between the activity of mitochondrial and cytosolic proteins [104].

4. Energetics and the control of gene expression

4.1. The intracellular need for ATP

Because they are thermodynamically uphill, many intracellular processes require the input of ATP. Many processes involved in regulated gene expression are among these. Of course, both transcription and translation involve nucleoside triphosphates as substrates for reaction. Consequently, there could be a rather direct effect of changes in intracellular phosphorylation potential on transcription and translation. This potential route of energetic regulation has received little attention.

A second route by which cellular energetics may affect gene expression involves the energet-

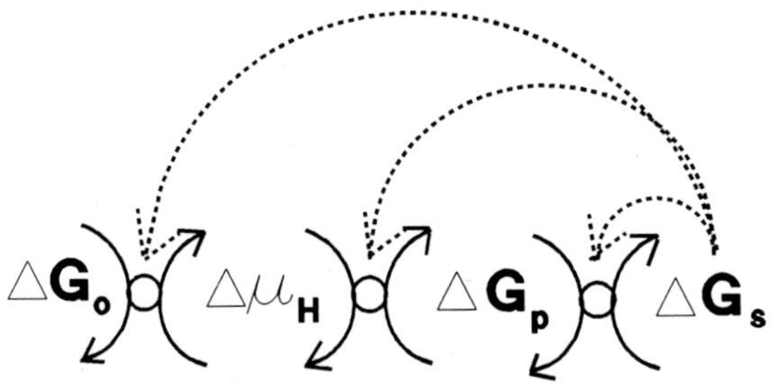

Fig. 4. Possible involvement of various free energies in the control of cell physiology in prokaryotes. ΔG_o, $\Delta \mu_H$, ΔG_p, and ΔG_s represent the (?) intracellular redox potential, the proton motive force, the phosphorylation potential and the free energy of DNA supercoiling, respectively. Full arrows refer to relations due to coupled chemical conversions, dashed arrows refer to 'allosteric' influences, notably those of DNA supercoiling through gene expression on the concentration of enzymes.

ics of the structure of DNA. In eukaryotes, the DNA is largely wound around nucleosomes, consisting of histone proteins. Taking the interaction with these and other proteins into account, the DNA is, on the average, in topological equilibrium ([105], see however, ref. [106]).

In prokaryotes however, the DNA is not in topological equilibrium [107]. In prokaryotes an ATPase, called DNA gyrase, reduces the number of times the one strand of the double helix is wound around the other in steps of two. This causes the DNA to be in isomeric forms that are of higher free-energy content than a relaxed flat circle and even than the form that would have the lowest free energy taking into account the interaction with the DNA binding proteins. The degree of supercoiling of the DNA is also determined by the activity of a second enzyme called topoisomerase I, which allows changes in linking number and acts in the direction of the lowest-free-energy molecule.

In terms of a pathway of transduction of metabolic free energy into free energy of DNA, this leads to the situation depicted in Fig. 4: intracellular redox free energy may be transduced to an electrochemical potential difference for protons across the plasma membrane of *E. coli*; this proton motive force may be transduced to the intracellular hydrolytic free energy of ATP (by the membrane bound H^+-ATPase) and the latter may be transduced into free energy of supercoiling.

For this route to be relevant for regulation of the structure of DNA, the reactions should in practice be sensitive to changes in the free-energy differences. The rate at which H^+-ATPases synthesize ATP is known to be rather sensitive to changes in the magnitude of the proton motive force, as well as to changes in the magnitude of the phosphorylation potential [108]. In vitro, the activity of DNA gyrase is known to be affected by changes in the concentrations of ATP and ADP. In fact this has been shown to be a case of so-called 'thermodynamic control', i.e. the effects could be described as a dependence on [ATP]/[ADP] ratio rather than as separate functions of [ATP] and [ADP] [109].

The possibility that DNA structure may be regulated by the energy state of the cell is of interest in itself. However, its potential interest is much wider, as the energy state of the DNA is known to affect the expression of many genes. The transcription of a large number of genes is enhanced by negative supercoiling. This may be understood as due to facilitated opening of the DNA double helix by RNA polymerase. The transcription of many other genes is, however, reduced at increased negative supercoiling, which must be due to a different mechanism. Be it as it may, the expression of only a minority of the genes of *E. coli* is unaffected by changes in supercoiling [107].

In Fig. 4, the effect of DNA supercoiling on gene expression opens up the possibility of a hierarchical control structure. Indeed, the elements for one such a hierarchical mode of regulation are known: the transcription of the genes encoding DNA gyrase is inhibited by negative supercoiling, and the expression of the gene encoding topoisomerase I is stimulated by negative supercoiling. The two gene expression routes thus constitute a hierarchical route of homeostatic control of the extent of supercoiling of the DNA [24,28].

In addition, however, there is the possibility that also the intracellular free-energy state is subject to hierarchical control involving gene expression. If the transcription of genes encoding enzymes involved in oxidative phosphorylation were sensitive to DNA supercoiling, then the intracellular free-energy state could be coregulated through gene expression. Jensen has been studying this possibility to some extent [30]. He observed that the expression of some of the promoters of the *atp* genes of *E. coli* varies with growth rate in a way consistent with an effect of DNA supercoiling. This possibility is now in further investigation.

4.2. Do the energetic properties of cells ever change?

For a property to be a mediator of regulation not only must the processes be sensitive to changes in that property, also those changes must actually occur during regulatory transitions. It has often been suggested that intracellular energetic properties hardly change. To some extent the absence of such major changes may precisely be due to a strong regulatory role of those energetic properties. To keep cellular physiology running, these properties themselves may be strongly regulated. The connectivity law (cf. Section 1.5, and ref. [33]) shows that the control exerted by an enzyme on a free energy decreases with the absolute magnitudes of the elasticities of the reactions towards that free energy. It is noteworthy that many cellular processes are highly elastic with respect to important free-energy potentials such as the free energy of hydrolysis of ATP and the proton motive force. For the latter we refer to the strong respiratory control in eukaryotic cells [3]. Indeed, the control properties of cellular free-energy transduction may have evolved so as to allow for good buffering of potential changes in these central intracellular free energies.

The case of animal heart physiology may be an illustrative case with respect to the absence of control involving the important free-energy intermediate. In principle, with enhanced work load, the enhanced mitochondrial respiration in heart cells could have been regulated through a drop in intracellular phosphorylation potential and then classical respiratory control by reduced phosphate potential in the mitochondria. However, mitochondrial respiration has been reported to be often regulated rather at the side of input of redox equivalents than at the side of the phosphorylation potential. In yeast mitochondria, increases in respiratory rates caused by decreases in the phosphorylation potential may be due to allosteric effects on cytochrome oxidase rather than to proton motive force mediated effects of phosphorylation potential on the mitochondrial respiratory chain [110].

Clearly, it seems legitimate to question whether intracellular free-energy potentials play a role in cellular regulation.

4.3. Does gene expression change when the energetics change?

The question of effects of cellular energetics on prokaryotic gene expression has recently come under investigation. First, it was shown that the extent to which, in vitro, DNA gyrase supercoils DNA is a rather strong function of the phosphorylation potential [109]. Then, protonophores were shown to reduce negative intracellular supercoiling [24]. Subsequently Drlica and co-workers ([111], see also ref. [112]) showed that in a transition from aerobic to anaerobic growth, the supercoiling of an intracellular reporter plasmid was reduced concurrently with a drop in intracellular [ATP]/[ADP] ratio.

The changes in DNA supercoiling and intracellular [ATP]/[ADP] observed by various authors [111,24,112,113] were not strictly correlated. A possible explanation for this is that there are two regulatory levels, with different relaxation times. This is presently under investigation.

Most interestingly, Jensen has measured the coefficient referring to the control of growth of *E. coli* by the genes coding for the H^+-ATPase. For the wild-type cell he found that the control exerted by these genes is close to zero, as if the expression of these genes had been optimized towards this objective [30].

5. *Concluding remarks*

Although free energy is essential for the most important biological processes, it may not actually control or regulate them or cell physiology. The question whether intracellular free energies are involved in the regulation of cell physiology, must remain unanswered at present. The question becomes especially complex because, precisely in view of their importance, these intracellular free energies are dynamically buffered by the cellular control structure. Interestingly, these homeostatic controls may involve different levels of the control hierarchy, including those of regulated gene expression. The question of control through cellular energetics therefore becomes quite subtle and may require control analytic or non-equilibrium thermodynamic analysis for its resolution. Because signal transduction pathways have much in common with free-energy transduction pathways, they may also benefit from such analyses.

Acknowledgement

We thank the Netherlands Organization for Scientific Research (NWO) for support.

Symbols

J	flux; chemical or transport flow
V_S	forward maximum velocity
K_S	Michaelis constant (K_m) for substrate
V_p	backward maximum velocity
K_p	K_m for product
e	enzyme
[X]	concentration of X
ΔG	free-energy difference of reaction
RT	gas constant times absolute temperature
C_Y^X	coefficient quantifying control of X by Y
ε_X^e	elasticity of enzyme e with respect to [X]
J_a	growth rate
μ_S	chemical potential of S

μ_S^0 standard chemical potential of S
$\delta\mu_H$ proton motive force
ΔG_p phosphorylation potential
G_Y^X control coefficient involving > 1 modules
η thermodynamic efficiency
J_S substrate consumption rate
J_b rate of biomass production
J_{Pr} rate of formation of low-molecular carbon product
$\ln X$ natural logarithm of X
o respiration, or redox
p phosphorylation
g DNA gyrase
I topoiosomerase I

References

1 Schrödinger, A. (1945) What is life?, Cambridge University Press, Cambridge.
2 Prigogine, I. (1991) in: S. Ji (Ed.) Molecular Theories of Cell Life and Death, Rutgers University Press, New Jersey, pp. 238–242
3 Westerhoff, H.V. and van Dam, K. (1987) Thermodynamics and Control of Biological Free-Energy transduction, Elsevier, Amsterdam.
4 De Groot, S.R. and Mazur, P. (1962) Non-Equilibrium Thermodynamics, North-Holland, Amsterdam.
5 Burns, J.A., Cornish-Bowden, A., Groen, A.K., Heinrich, R., Kacser, H., Porteous, J.W., Rapoport, S.M., Rapoport, T.A., Tager, J.M., Wanders, R.J.A. and Westerhoff H.V. (1985) Trends Biochem. Sci. 10, 16.
6 Cornish-Bowden, A. and Cardenas, M.L. (Eds.) (1990) Control of Metabolic Processes, Plenum Press, New York.
7 Westerhoff, H.V. (1989) Biochimie 71, 877–886.
8 Hofmeyr, J.-H.S. and Cornish-Bowden, A. (1991) Eur. J. Biochem. 200, 223–236.
9 Kahn, D. and Westerhoff, H.V. (1992) Biotheor. Acta, in press.
10 LaNoue, K.F., Feffries, F.M.H., Radda, G.K. (1986) Biochemistry 25, 7667–7675.
11 Jacobus, W.E., Moreadith, R.W. and Vandegaer, K.M. (1982) J. Biol. Chem. 257, 2397–2402.
12 Van Dam, K. and Westerhoff, H.V. (1984) in: Ernster L. (Ed.) New Comprehensive Biochemistry, Elsevier, Amsterdam, pp. 27–53.
13 Onsager, L. (1931) Phys. Rev. 37, 405–426.
14 Van der Meer, R., Westerhoff, H.V. and van Dam, K. (1980) Biochim. Biophys. Acta 591, 488–493.
15 Gnaiger, E. (1989) in: W. Wieser and E. Gnaiger (Eds.) Energy Transformations in Cells and Organisms, Georg Thieme Verlag, Stuttgart, pp. 6–17.
16 Rottenberg, H. (1973) Biophys. J. 13, 503–511.
17 Van Dam, K., Westerhoff, H.V., Krab, K., van der Meer, R. and Arents, J.C. (1980) Biochim. Biophys. Acta 591, 240–250.
18 Garlid, K.D., Beavis, A.D. and Ratkje, S.K. (1989) Biochim. Biophys. Acta 976, 109–120.
19 Kell, D.B. and Westerhoff, H.V. (1986) FEMS Microbiol. Rev. 39, 305–320.
20 Rutgers, M., van Dam, K. and Westerhoff, H.V. (1991) Crit. Rev. Biotechnol. 11, 367–395.
21 Westerhoff, H.V., Plomp, P.J.A.M., Groen, A.K. and Wanders, R.J.A. (1987) Cell Biophys. 10, 239–267.
22 Westerhoff, H.V., Plomp, P.J.A.M., Groen, A.K., Wanders, R.J.A., Bode, J.A. and Van Dam, K. (1987) Arch. Biochem. Biophys. 257, 154–169.
23 Wanders, R.J.A., Groen, A.K., van Roermund, C.W.T. and Tager, J.M. (1984) Eur. J. Biochem. 142, 417–424.

24 Westerhoff, H.V., Koster, J.G., van Workum, M. and Rudd, K.E. (1990) in: A. Cornish-Bowden and M.-L. Cardenas (Eds.) Control of Metabolic Processes, Plenum Press, New York, pp. 399–412.

25 Kahn, D. and Westerhoff, H.V. (1991) J. Theor. Biol. 153, 255–285.

26 Savageau, M.A (1976) Biochemical Systems Analysis, a Study of Function and Design in Molecular Biology, Addison-Wesley, Reading, MA.

27 Barthelmess, I.B., Curtis, C.F. and Kacser, H. (1974) J. Mol. Biol. 87, 303–316.

28 Westerhoff, H.V. (1991) in: N. Grunnet, and D. Quistorff (Eds.) Regulation of Hepatic Function. Alfred Benzon Symposium 30, Munksgaard, Copenhagen, pp. 232–242.

29 Westerhoff, H.V., van Heeswijk, W., Kahn, D. and Kell, D.B. (1991) Antonie van Leeuwenhoek J. Microbiol. Serol. 60, 193–207.

30 Jensen, P.R. (1991) PhD Thesis, Danmarks Tekniske Hojskole.

31 Kell, D.B. and Welch, G.R. (1991) The Times Higher Education Supplement 9-8-91, p. 15.

32 Westerhoff, H.V. and Kahn, D. (1992) Biotheor. Acta, in press.

33 Westerhoff, H.V., Groen, A.K. and Wanders, R.J.A. (1984) Biosc. Rep. 4, 1–22.

34 Schlessinger, J. (1988) Trends Biochem. Sci. 13, 443–447.

35 Cantley, L.C., Auger, K.R., Carpenter, C., Duckworth, B., Graziani, A., Kapeller, R., Soltoff, S. (1991) Cell 64, 281–302.

36 Goldbeter, A., Koshland, D.E., Jr, (1984) J. Biol. Chem. 259, 14441–14447.

37 Chock, P.B., Rhee, S.G. and Stadtman, E.R. (1980) Annu. Rev. Biochem. 49, 813–843.

38 Cardenas, M.L. and Cornish-Bowden, A. (1989) Biochem. J. 257, 339–345.

39 Kohn, M.C. and Garfinkel, D. (1983) Ann. Biomed. Eng. 11, 511–531.

40 Kelly, P.J., Kelleher, J.K. and Wright, B.E. (1979) Biochem. J. 184, 589–597.

41 Wright, B.E. and Kelly, P.J. (1981) Curr. Top. Cell. Regul. 19, 103.

42 Stoffers, H.J., Sonnhammer, E.L.L., Blommestijn, G.J.F., Raat, H.N.J. and Westerhoff, H.V. (1992) CABIOS, in press.

43 Heinrich, R., Rapoport, S.M. and Rapoport, T.A. (1977) Prog. Biophys. Mol. Biol. 32, 1–83.

44 Savageau, M.A. and Voit, E.O. (1982) J. Ferment. Technol. 60, 221–228.

45 Voit, E.O. and Savageau, M.A. (1987) Biochemistry 26, 6869–6880.

46 Voit, E.O. (1990) in: A. Cornish-Bowden and M.L. Cardenas (Eds.) Control of Metabolic Processes, Plenum Press, New York, pp. 89–100.

47 Katchalsky, A. and Curran, P.F. (1965) Nonequilibrium Thermodynamics in Biophysics, Harvard University Press, Cambridge, MA.

48 Caplan, S.R. and Essig, A. (1983) Bioenergetics and Linear Non-Equilibrium Thermodynamics, Harvard University Press, Cambridge, MA.

49 Stucki, J.W. (1980) Eur. J. Biochem. 109, 269–283.

50 Groen, A.K., Wanders, R.J.A., Westerhoff, H.V., van der Meer, R. and Tager, J.M. (1982) J. Biol. Chem. 257, 2754–2757.

51 Halangk, W., Dietz, H., Bohnensack, R. and Kunz, W. (1987) Biochim. Biophys. Acta 893, 100–108.

52 Brand, M.D., Hafner, R.P. and Brown, G.C. (1988) Biochem. J. 255, 535–539.

53 Moreno-Sanchez, R. (1985) J. Biol. Chem. 260, 12554–12560.

54 Mazat, J.-P., Bart, E. J.-B., Rigoulet, M. and Guerin, B. (1986) Biochim. Biophys. Acta 849, 7–15.

55 Zweier, J.L. and Jacobus, W.E. (1987) J. Biol. Chem. 262, 8015–8021.

56 Heineman, F.W. and Balaban, R.S. (1990) Annu. Rev. Physiol. 52, 523–542.

57 Duszynski, J., Groen, A.K., Wanders, R.J.A., Vervoorn, R.C. and Tager, J.M. (1982) FEBS Lett. 146, 263–266.

58 Kacser, H. and Burns, J. (1973) in: D.D. Davies (Ed.) Rate Control of Biological Processes, Cambridge University Press, pp. 65–104.

59 Westerhoff, H.V. and Chen, Y. (1984) Eur. J. Biochem. 142, 425–430.

60 Fell, D.A. and Sauro, H.M. (1985) Eur. J. Biochem. 148, 555–561.

61 Westerhoff, H.V. and Kell, D.B. (1987) Biotechnol. Bioeng. 30, 101–107.

62 Reder, C. (1988) J. Theor. Biol. 135, 175–202.

63 Brown, G.C., Hafner, R.P. and Brand, M.D. (1990) Eur. J. Biochem. 188, 321–325.

64 Hafner, R.P., Brown, G.C. and Brand, M.D. (1990) Eur. J. Biochem. 188, 313–319.

66 Mitchell, P. (1961) Nature 191, 144–148.
67 Pietrobon, D., Zoratti, M., Azzone, G.F., Stucki, J.W. and Walz, D. (1982) Eur. J. Biochem. 127, 483–494.
68 Westerhoff, H.V., Kell, D.B., Kamp, F. and van Dam, K. (1988) in: D.P. Jones (Ed.) Microcompartmentation CRC Press, Boca Raton, FL, pp. 114–154.
69 Westerhoff, H.V. and Chen, Y. (1985) Proc. Natl. Acad. Sci. USA 82, 3222–3226.
70 Westerhoff, H.V., Kell, D.B. and Astumian, R.D. (1988) J. Electrostat. 21, 257–298.
71 Roels, J.A. (1983) Energetics and Kinetics in Biotechnology, Elsevier Biomedical Press, Amsterdam, p. 330.
72 Westerhoff, H.V., Lolkema, J.S., Otto, R. and Hellingwerf, K.J. (1982) Biochim. Biophys. Acta 683, 181--220.
73 Kedem, O. and Caplan, S.R. (1965) Trans. Faraday Soc. 21, 1897–1911.
74 Westerhoff, H.V., Hellingwerf, K.J. and van Dam, K. (1983) Proc. Natl. Acad. Sci. USA 80, 305–309.
75 Rutgers, M. (1990) Control and Thermodynamics of Microbial Growth, PhD Thesis, University of Amsterdam.
76 Heijne, J. (1991) Antonie van Leeuwenhoek J. Microbiol. Serol. 60, in press.
77 Stouthamer, A.H. (1973) Antonie van Leeuwenhoek J. Microbiol. Serol. 39, 545–565.
78 Rutgers, M., van der Gulden, H.M.L. and van Dam, K. (1989) Biochim. Biophys. Acta 973, 302–307.
79 Heinrich, R., Schuster, S. and Holzhütter, H.-G. (1991) Eur. J. Biochem. 201, 1–21.
80 Monod, J. (1942) Recherches sur la Croissance des Cultures Bacteriennes, Hermann, Paris.
81 Rutgers, M., Balk, P.A. and van Dam, K. (1990) Arch. Microbiol. 153, 478–484.
82 Kleman, G.L., Chalmers, J.J., Luli, G.W. and Strohl, W.R. (1991) Appl. Environ. Microbiol. 57, 918–923.
83 Poolman, B., Bosman, B., Kiers, J. and Konings, W.N. (1987) J. Bacteriol. 169, 5887–5890.
84 Iwami, Y. and Yamada, T. (1985) Infect. Immun. 50, 89–100.
85 Walter, R.P., Kell, D.B. and Morris, J.G. (1987) J. Gen. Microbiol. 133, 259–266.
86 Van Dam, K. and Jansen, N. (1991) Antonie van Leeuwenhoek J. Microbiol. Serol. 60, in press.
87 Walsh, K. and Koshland, D. (1985) J. Biol. Chem. 82, 3577–3581.
88 Birkenhead, K., Manian, S.S. and O'Gara, F. (1988) J. Bacteriol. 170, 184–189.
89 Schaaff, I., Heinisch, J. and Zimmermann, F. (1989) Yeast 5, 285–290.
90 Lagunas, R., Dominguez, C., Busturia, A. and Sacz, M.J. (1982) J. Bacteriol. 152, 19–25.
91 Galazzo, J.L. and Bailey, J.E. (1990) Enzyme Microb. Technol. 12, 162–172, addendum in (1991) Enzyme Microb. Technol.
92 Ruijter, G., Postma, P.W. and van Dam, K. (1991) J. Bacteriol. 173, 6184–6191.
93 Sauro, H.M. (1990) in: A. Cornish-Bowden and M.-L. Cardenas (Eds.) Control of Metabolic Processes, Plenum, New York, pp. 225–230.
94 Maloney, P. (1987) in: F.C. Neidhart (Ed.) *E. coli* and *S. typhimurium* , Am. Soc. Microbiol., Washington DC, pp. 222–243.
95 Bisson, L.F., Neigeborn, L., Carlson, M. and Fraenkel, D. (1987) J. Bacteriol. 169, 1656–1662.
96 Rose, M., Albig, W. and Entian, K-D. (1991) Eur. J. Biochem. 199, 511–518.
97 Hess, B. and Boiteux, A. (1971) Annu. Rev. Biochem. 40, 237–258.
98 Sel'kov, E.E. (1975) Eur. J. Biochem. 59, 151–157.
100 Chance, B., Estabrook, R.W. and Ghosh, A. (1964) Proc. Natl. Acad. Sci. USA. 51, 1244–1251
101 Dickinson, J.R. (1990) *Saccharomyces*, Biotechnology Handbooks, Vol.4, Plenum Press, New York, pp. 59–100.
102 Chiang, H-L. and Schekman, R. (1991) Nature 350, 313–318.
103 Thevelein, J.M. (1991) Mol. Microbiol. 5, 1301–1307.
104 Aon, M.A., Cortassa, S., Westerhoff, H.V., Berden, J.A., van Spronsen, E. and van Dam, K. (1991) J. Cell Sci. 99, 325–334.
105 Sinden, R.R., Carlson, J. and Pettijohn, D.E. (1980) Cell 21, 773–783.
106 Liu, L.F. (1989) Annu. Rev. Biochem. 58, 351–375.
107 Pruss, G.J. and Drlica, K. (1989) Cell 56, 521–523.
108 Junesch, U. and Gräber, P. (1985) Biochim. Biophys. Acta 809, 429–434.

109 Westerhoff, H.V., O'Dea, M.H., Maxwell, A. and Gellert, M. (1988) Cell Biophys. 12, 157–183.
110 Rigoulet, M. (1990) Biochim. Biophys. Acta 1018, 185–189.
111 Hsieh, L-S., Burger, R.M. and Drlica, K. (1991) J. Mol. Biol. 219, 443–450.
112 Cortassa, S., Aon, M.A., van Dam, K. and Westerhoff, H.V. (1990) EBEC Rep. 6, 95.
113 Westerhoff, H.V., Aon, M.A., van Dam, K., Cortassa, S., Kahn, D. and van Workum, M. (1990) Biochim. Biophys. Acta 1018, 142–146.

L. Ernster (Ed.) *Molecular Mechanisms in Bioenergetics*
© 1992 Elsevier Science Publishers B.V. All rights reserved

CHAPTER 2

Chemiosmotic systems and the basic principles of cell energetics

VLADIMIR P. SKULACHEV

Department of Bioenergetics, A.N. Belozersky Institute of Physico-Chemical Biology, Moscow State University, Moscow 119899, Russia

Contents

1. Introduction 37
2. The H^+ cycle 38
3. The Na^+ cycle 41
3.1. Brief history 41
3.2. $\Delta\bar{\mu}_{Na^+}$-generating systems 42
3.2.1. The Na^+-motive NADH-quinone reductase 42
3.2.2. Na^+-motive terminal oxidase 45
3.2.3. Na^+-motive decarboxylases 47
3.2.4. Na^+-motive electron transfer in methanogenic and acetogenic bacteria 50
3.2.5. Animal Na^+/K^+-ATPase and Na^+-ATPase 50
3.2.6. Bacterial Na^+-ATPases 52
3.3. Utilization of $\Delta\bar{\mu}_{Na^+}$ produced by primary $\Delta\bar{\mu}_{Na^+}$ generators 53
3.3.1. Osmotic work: Na^+, solute-symports 53
3.3.2. Na^+ ions and regulation of the cytoplasmic pH 54
3.3.3. Mechanical work: the Na^+ motor 55
3.3.4. Chemical work: $\Delta\bar{\mu}_{Na^+}$-driven ATP synthesis 56
4. Interrelations of the H^+- and Na^+-cycles 63
5. Conclusion. Three basic principles of cell energetics 66
References 68

1. Introduction

In 1941 Lipmann [1] proposed that ATP is the universal compound that couples energy-supplying and energy-consuming processes. This concept was based mainly on the facts that (i) both respiration and glycolysis in eukaryotic cells generate ATP; and (ii) a number of endergonic reactions utilize ATP as the energy source. This was a highly foresighted proposal, made before photophosphorylation was discovered and at a time when little was known about the bioenergetics of prokaryotes.

During the following two decades, Lipmann's proposal was substantiated by many new pieces of evidence. However, some exceptions were also noticed, indicating that utilization of energy sources for performance of useful work by the cell may in certain cases occur without the involvement of ATP. At the same time, there was growing evidence that membranes may be important for this kind of energy transductions.

A very important breakthrough in this connection was made by Mitchell [2] who in 1961 formulated his chemiosmotic hypothesis of oxidative and photosynthetic phosphorylation. According to the hypothesis, the energy derived from respiration and photosynthetic electron transport is used primarily to create a transmembrane electrochemical proton gradient ($\Delta\bar{\mu}_{H^+}$), which is then utilized to form ATP from ADP and P_i. Moreover, it followed from the chemiosmotic hypothesis that not only ATP formation, but also some other energy-consuming processes in mitochondria (reverse electron transfer in the respiratory chain, the nicotinamide nucleotide transhydrogenase reaction, the uphill transport of Ca^{2+}) may be directly supported by $\Delta\bar{\mu}_{H^+}$. It was assumed that ATP is not necessary to energize these processes provided that the energy is supplied by respiration [3,4].

This prediction was subsequently proven experimentally. It was found that the chain of events 'energy source $\rightarrow \Delta\bar{\mu}_{H^+} \rightarrow$ useful work' is generally valid for membrane-linked energy transduction in bacteria and eukaryotic organelles, leading to the formulation of the principle that, besides ATP, there is one more, membranous form of convertible biological energy currency namely $\Delta\bar{\mu}_{H^+}$ [5]. In the 1980s, this concept was widely accepted by the bioenergetic community.

However, further studies revealed certain exceptions when neither ATP nor $\Delta\bar{\mu}_{H^+}$ mediate the utilization of cellular energy sources. Notably, there were membrane-linked energy transductions occurring under conditions where the $\Delta\bar{\mu}_{H^+}$ level was too low to serve as an intermediate, e.g., in the presence of protonophores or at alkaline pH. Careful investigation of the exceptions showed that, at least in some cases, Na^+ served as the coupling ion. This type of energy transduction is already described in very many systems, justifying the concept that a difference in electrochemical Na^+ potentials ($\Delta\bar{\mu}_{Na^+}$) is a third form of convertible energy currency in living cells [6].

The discovery of the coupling role of $\Delta\bar{\mu}_{Na^+}$ has widened the general picture of biological energy transductions, extending the basic principles of bioenergetics that are of universal applicability. These principles will be considered at the end of this review. The major part of the chapter will be focused on the mechanism of formation and utilization of $\Delta\bar{\mu}_{Na^+}$, in view of the important progress that has been made in recent years in this particular field of cellular bioenergetics.

A detailed review of $\Delta\bar{\mu}_{H^+}$ utilizing bioenergetic systems has recently been published elsewhere [7]. This problem is also treated in several chapters of this volume. Following here is only a short summary of $\Delta\bar{\mu}_{H^+}$-generating and $\Delta\bar{\mu}_{H^+}$-consuming mechanisms.

2. *The H^+ cycle*

A general pattern of the pathways involving $\Delta\bar{\mu}_{H^+}$ is shown in Fig. 1. According to the scheme, energy of light or of respiratory substrates can be utilized to form $\Delta\bar{\mu}_{H^+}$ by enzymes of the photosynthetic or respiratory redox chains or, in halobacteria, by bacteriorhodopsin. The $\Delta\bar{\mu}_{H^+}$ formed can support various types of membrane-linked processes, with ATP synthesis being the most important $\Delta\bar{\mu}_{H^+}$-linked energy utilizer. $\Delta\bar{\mu}_{H^+}$-supported synthesis of inorganic pyrophosphate and the transport of reducing equivalents in the direction of more negative redox potentials (e.g., reverse electron transport in the respiratory chain or the nicotinamide nucleotide transhydrogenase reaction) also represent $\Delta\bar{\mu}_{H^+}$-mediated chemical energy

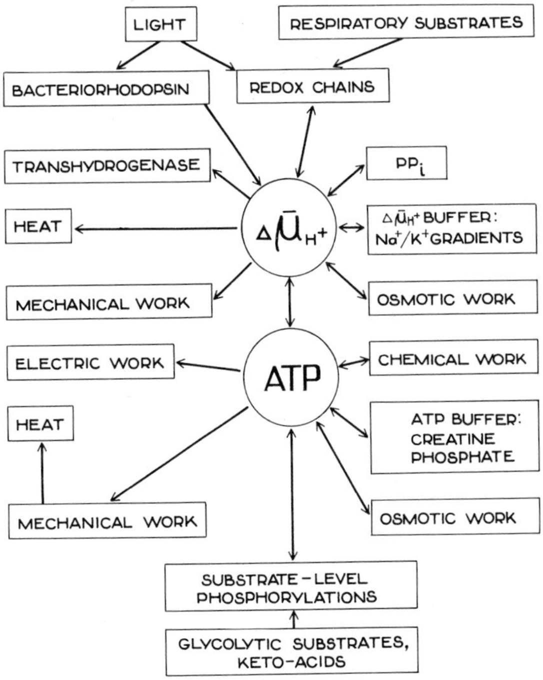

Fig. 1. Energy transductions in living systems employing $\Delta\bar{\mu}_{H^+}$ and ATP as convertible energy currencies (from ref. [8]).

transductions. The $\Delta\bar{\mu}_{H^+}$-driven uphill transport of various substances across membranes may be defined as transduction of $\Delta\bar{\mu}_{H^+}$ to osmotic energy, while rotation of the bacterial flagella exemplifies $\Delta\bar{\mu}_{H^+}$-driven mechanical work. Thermoregulatory heat formation due to $\Delta\bar{\mu}_{H^+}$ discharge by the natural protonophores may be an example of useful $\Delta\bar{\mu}_{H^+}$-mediated heat production.

Substrate-level phosphorylations serve as an alternative mechanism of ATP formation, which operates without the involvement of $\Delta\bar{\mu}_{H^+}$. Such phosphorylations occur in the glycolytic chain and in the oxidative decarboxylation of keto-acids.

Among the ATP-consuming processes there are ATP-driven endergonic reactions (chemical work) and some uphill transports (osmotic work). If the transported component is charged, a transmembrane electric potential difference is generated (electric work). ATP-linked mechanical work is performed by actomyosin-type systems. In the latter case, ATP hydrolysis may also be used as a mechanism of thermoregulatory heat production.

There are systems specialized in the buffering of $\Delta\bar{\mu}_{H^+}$ and ATP. For $\Delta\bar{\mu}_{H^+}$, such a function is performed by gradients of Na^+ and K^+; and for ATP, by creatine phosphate (for details, see refs. [7–9]).

Some $\Delta\bar{\mu}_{H^+}$ generators and consumers are listed in Table I. Among the $\Delta\bar{\mu}_{H^+}$ generators, the most wide-spread are those involved in the respiratory and photosynthetic redox chains (systems A1–A7). Systems A1–A11 can be defined as primary $\Delta\bar{\mu}_{H^+}$ generators since they directly utilize external energy sources to produce $\Delta\bar{\mu}_{H^+}$. Systems A12–A14 (H^+-ATPases) should be

TABLE I

$\Delta\bar{\mu}_{H^+}$ generators and consumers (from ref. [7])

Type of system		Distribution
A. $\Delta\bar{\mu}_{H^+}$-generators		
A1.	Reaction center complex in the cyclic bacterial photoredox chain	Purple bacteria
A2.	Reaction center complex in the non-cyclic bacterial photoredox chain	Green bacteria
A3.	Reaction center complex of photosystem I	Chloroplasts, cyanobacteria
A4.	Reaction center complex of photosystem II	Chloroplasts, cyanobacteria
A5.	NADH-Q reductase	Mitochondria, many aerobic and photosynthetic bacteria, some anaerobic bacteria
A6.	QH_2-cytochrome *c* (plastocyanin) reductase	Mitochondria, some bacteria, chloroplasts
A7.	Cytochrome *c* oxidase	Mitochondria, some aerobic bacteria
A8.	Fumarate-reducing complex	Some anaerobic bacteria
A9.	Cytochrome *o* complex	Some aerobic bacteria
A10.	Methanogenesis-linked electron transfer complex	Methanobacteria
A11.	Bacteriorhodopsin	Halobacteria
A12.	H^+-ATPase of F_0F_1-type	Some anaerobic bacteria
A13.	H^+-ATPase of vacuolar type	Tonoplast of plant and fungal vacuoles, secretory granules of animal cells, lysosomes
A14.	H^+-ATPase of E_1E_2 type	Plasma membrane of plant and fungal cells
B. $\Delta\bar{\mu}_{H^+}$-consumers		
B1.	H^+-ATP synthase of F_0F_1-type	Mitochondria, chloroplasts, eubacteria
B2.	H^+-ATP synthase similar to vacuolar H^+-ATPase	Archaebacteria
B3.	H^+-PP_i synthase	Purple bacteria, tonoplast
B4.	H^+-nicotinamide nucleotide transhydrogenase	Mitochondria, some bacteria
B5.	Reversal of NADH-Q reductase	Bacteria oxidizing substrates of positive redox potential, mitochondria
B6.	Reversal of QH_2-cytochrome *c* reductase	Bacteria oxidizing substrates of positive redox potential
B7.	H^+, solute-symporters, H^+, solute-antiporters	Mitochondria, chloroplasts, non-marine bacteria
B8.	H^+-motor	Many bacteria, some chloroplasts
B9.	Proteins mediating the fatty acid-induced uncoupling (thermogenin, ATP/ADP-antiporter)	Mitochondria

regarded as secondary $\Delta\bar{\mu}_{H^+}$ generators utilizing ATP which is formed at the expense of membrane-linked or substrate-level phosphorylations [8].

Among the $\Delta\bar{\mu}_{H^+}$ consumers, H^+-ATP-synthases (systems B1 and B2) seem to be the most important; they are responsible for ATP formation in bacteria, mitochondria and chloroplasts [7].

3. *The Na^+ cycle*

3.1. Brief history

When Mitchell [2] advanced in 1961 his chemiosmotic hypothesis proposing H^+ circulation to be the mechanism of energy coupling in biomembranes, he also suggested in the same paper a concrete chemical formulation of this general idea. The hypothesis was based on two crucial postulates.

(i) H atoms, originating from the oxidation substrates, are split to protons and electrons, the latter being translocated across the membrane by enzymes of the respiratory or photosynthetic redox chain.

(ii) The protons formed are translocated in the same direction as the electrons by another enzyme, H^+-ATP-synthase, which forms ATP and H_2O from ADP in inorganic phosphate.

According to this mechanism, H^+ is an obligatory component of the energy coupling of electron transport and ATP synthesis, which cannot be replaced by any other ion.

Over the past 30 years, the general concept of the H^+ cycle has been verified. As to the actual mechanism, a direct coupling between electron and proton transfer is fairly well established in certain instances – e.g., the ubiquinone-, menaquinone- or plastoquinone-linked H transfer in bacteria, mitochondria and chloroplasts – but not in others, e.g., the energy-linked nicotinamide nucleotide transhydrogenase. In the latter case, the reducing equivalent (H^-), removed from NADPH, is directly translocated to NAD^+, without exchange with the water phase. Thus, the redox process is coupled to the transmembrane translocation of those protons which do not take part in the redox reaction per se (for reviews, see refs. [7, 8]). Such an indirect mechanism of coupling of the redox process and H^+ translocation in transhydrogenase was recognized by Mitchell as early as 1966 [3]. An essential feature of this type of energy coupling machinery is that here the proton is no longer irreplaceable. It may, in principle, be replaced by any other ionic species provided that the energy-transducing enzymes are able to reversibly bind and translocate these species [8].

Ion transport ATPases, such as the Na^+/K^+-ATPase, Ca^{2+}-ATPase, K^+-ATPase and Cl^--ATPase, exemplify energy-transducing enzymes that translocate ion(s) other than H^+. For one of them, Na^+/K^+-ATPase of the animal cell plasma membrane, it has been shown that one of the ion gradients formed (transmembrane Na^+ concentration difference, ΔpNa) can be utilized to perform useful work. In the same membrane, Na^+, solute-symporters are present. They use ΔpNa or $\Delta\bar{\mu}_{Na^+}$ as the driving force to accumulate sugars, amino and fatty acids, as well as many other metabolites, inside the cell [10,11]. Thus, a sort of Na^+ cycle is operative in the animal plasma membrane: Na^+ is pumped uphill from the cell by the Na^+/K^+-ATPase and comes downhill back to the cytoplasm, accompanied by metabolites that move uphill.

In bacteria, the possible bioenergetic role of Na^+ was considered by Mitchell in 1968 [4]. He pointed out that the reversible exchange of inner H^+ for outer Na^+ by the Na^+/H^+ antiporter may, in principle, increase the pH buffer capacity of the bacterial cytoplasm.

Subsequently I suggested [12] that (i) it is the electrogenic K^+ influx that discharges the redox chain-produced transmembrane electric potential difference ($\Delta\Psi$), thereby converting $\Delta\Psi$ to H^+ and K^+ concentration differences (ΔpH and ΔpK, respectively) and (ii) ΔpH is then converted to ΔpNa by the Na^+/H^+-antiporter. Such a system was proposed to operate as a buffer of $\Delta\bar{\mu}_{H^+}$ [12]. According to this concept, energy is stored in ΔpK and ΔpNa when the sum of the rates of the $\Delta\bar{\mu}_{H^+}$-producing processes is higher than that of the $\Delta\bar{\mu}_{H^+}$-consuming ones. Under these conditions, K^+ is taken up and Na^+ is expelled. When the $\Delta\bar{\mu}_{H^+}$ consumption is faster than the $\Delta\bar{\mu}_{H^+}$ production, $\Delta\bar{\mu}_{H^+}$ decreases, and the direction of K^+ and Na^+ fluxes is reversed. The electrogenic efflux of K^+ generates $\Delta\Psi$, whereas the influx of Na^+ in exchange for H^+ generates ΔpH. Thus $\Delta\bar{\mu}_{H^+}$ is stabilized until the K^+ and Na^+ gradients are dissipated. The idea of K^+/Na^+ gradients as $\Delta\bar{\mu}_{H^+}$ buffer was experimentally proven by our and other groups [9,13–16].

Other studies revealed that the role of Na^+ in bacterial energetics is not restricted to $\Delta\bar{\mu}_{H^+}$ buffering. Na^+, solute-symports were described in certain bacteria [17]. These bacteria can utilize $\Delta\bar{\mu}_{Na^+}$ to perform osmotic work.

It was also found that the Na^+/H^+ antiporter is not the only $\Delta\bar{\mu}_{Na^+}$-generating mechanism found in bacterial cells. In 1980 Dimroth discovered a primary Na^+ pump in *Klebsiella pneumoniae*, namely, the Na^+-motive oxaloacetate decarboxylase [18]. In 1981–1982 Tokuda and Unemoto reported on a Na^+-motive NADH-quinone reductase in *Vibrio alginolyticus* [19,20]. The same group found that *V. alginolyticus* accumulates amino acids and sugars by means of Na^+, solute-symports [21,22]. We showed that in this bacterium (i) the flagellar motor directly utilizes $\Delta\bar{\mu}_{Na^+}$ with no $\Delta\bar{\mu}_{H^+}$ involved [23] and (ii) ATP synthesis can be driven by $\Delta\bar{\mu}_{Na^+}$ in the presence of protonophorous uncouplers [24].

Summarizing these observations in 1984, I put forward the Na^+-cycle concept, which suggests that in certain bacteria Na^+, rather than H^+, functions in membrane energization and in the performance of all three types of membrane-linked work, i.e., chemical, osmotic and mechanical [6,25].

Subsequently, several other primary Na^+ pumps were described in numerous species of marine bacteria. The great taxonomic variety of species employing Na^+ as the primary coupling ion demonstrates the ubiquitous distribution of this novel type of membrane-linked energy transduction. Obviously, besides the world of living organisms that use H^+ energetics, there is also a rather extensive area in the biosphere that may be defined as the 'sodium world'.

The discovery of this 'sodium world' has not only shown that H^+ is not unique as a coupling ion, but has also given new insight into the principles of the operation of energy-transducing mechanisms. For example, the fact that bacterial Na^+-ATP synthase is an F_0F_1-type of enzyme, which translocates H^+ when Na^+ is absent (see below, Section 3.3.4) excludes any direct involvement of translocated H^{2+} in H_2O formation during ATP synthesis.

Na^+-cycle studies may also open up new vistas for applied bioenergetics. In fact, the Na^+-cycle represents a bioenergetic mechanism vital for bacteria in their struggle for survival under unfavourable conditions. By inhibiting or stimulating Na^+ energetics, we may suppress the growth of some pathogenic bacteria and activate that of useful ones, respectively.

3.2. $\Delta\bar{\mu}_{Na^+}$-*generating systems*

3.2.1. *The Na^+-motive NADH-quinone reductase*

Tokuda and Unemoto were the first to demonstrate respiration-dependent, protonophore-resistant export of Na^+ from cells of the marine alkalotolerant *Vibrio alginolyticus* [19,20]. The same group then succeeded in isolating the *V. alginolyticus* Na^+-motive NADH-mena(ubi)-quinone reductase and in reconstituting corresponding proteoliposomes. The rate of oxidation was strongly stimulated by Na^+ [26] and specifically inhibited by very low concentrations of Ag^+ [27] or 2-heptyl-4-hydroxyquinoline *n*-oxide (HQNO) [28].

The molecular properties of H^+- and Na^+-motive NADH-quinone reductases proved to be quite different. The H^+-translocating enzyme from mitochondria or *Paracoccus denitrificans* is known to be composed of several subunits containing FMN and FeS clusters (reviewed by Yagi [29]). The Na^+-motive reductase contains only three subunits (56, 46 and 32 kDa), FAD, FMN, but no FeS clusters. The subunit stoichiometry is probably 2α, 2β and 2γ. The FAD-containing β-subunit catalyzes the oxidation of NADH and the reduction of CoQ or menaquinone to a semiquinone anion (e.g. $CoQ^{\cdot-}$), whereas the α- and γ-subunits are required for the dismutation of two molecules of $CoQ^{\cdot-}$ to $CoQH_2$ and CoQ. The latter (but not the former) process is Na^+-dependent [30–32].

Na^+-motive NADH-quinone reductases were subsequently found in *Vibrio costicola* [33], *Vibrio parahaemolyticus* [34,35], in the psychrophilic marine *Vibrio* ABE [36], the halotolerant bacterium Ba_1 [37], *Klebsiella pneumoniae* [38], in the alkalo- and halotolerant *Bacillus FTU* [39] and *E. coli* growing at alkaline pH [40], as well as in eight out of the nine strains of marine bacteria belonging to the genera *Vibrio, Alcaligenes, Alteromonas* and *Flavobacterium* tested for the presence of Na^+ pumps [35,41,42].

According to Unemoto and Hayashi [32], CoQ is necessary for reduction of FMN bound to the α-subunit of the Na^+-dependent NADH-quinone oxidoreductase, whereas the β-subunit-bound FAD can be reduced even without CoQ. Purified α-subunit shows no enzymatic activity but in combination with purified β- and γ-subunits, it can reduce CoQ. Without the γ-subunit, the reconstituted $\alpha\beta$-complex is inactive. Maximal activity is obtained when the FAD:FMN ratio in the reconstituted NADH-quinone reductase is one.

In another bacterium, *Klebsiella pneumoniae,* also possessing a Na^+-motive NADH-quinone reductase, the formation of superoxide radicals was shown to accompany the oxidation of NADH by CoQ_1 in a Na^+-free medium. The radicals disappeared upon the addition of Na^+. The Na^+ effect was abolished by HQNO. It was suggested that superoxide radicals arise from one-electron reduction of CoQ_1 [38].

The Na^+ transport by Na^+-motive NADH-quinone reductase is electrogenic, as revealed by experiments with intact cells [20], subcellular vesicles [39,40,42,43] and proteoliposomes [31]. In the latter case, CoQ_1 was used as the electron acceptor. When CoQ_1 was replaced by menadione, accepting electrons from FAD, no $\Delta\Psi$ was formed [31]. These above observations seem to indicate that Na^+ transport is coupled to the dismutation of two semiquinone molecules catalyzed by the FMN-containing α-subunit.

Two tentative schemes of energy coupling in Na^+-motive NADH-quinone reductase are shown in Fig. 2.

Scheme A, is based on the proposal by Ken-Dror et al. [37] that Na^+ combines with the semiquinone anion radical which serves as a transmembrane Na^+ carrier. It is assumed that two semiquinone molecules are formed per oxidized NADH. The process, localized close to the inner (cytoplasmic) membrane surface, is catalyzed by the FAD-containing β-subunit (scheme A, reactions 1 and 2). Two Na^+ ions, which are transported from the cytoplasm to the $Q^{\cdot-}$ binding site through a Na^+-specific channel or a Na^+-conducting pathway, are postulated to combine with two molecules of $Q^{\cdot-}$ (reaction 3). In a fourth step, $2Q^{\cdot-}Na^+$ diffuses across the membrane to FMN which is bound to the α-subunit and is localized close to the outer membrane surface. Oxidation of one of the $Q^{\cdot-}Na^+$ molecules by FMN produces Q, which contains no anion-binding groups, so that Na^+ is released to the outer medium (reaction 5). From FMN the electron passes to the second $Q^{\cdot-}Na^+$ molecule to reduce it to the quinol level (reaction 6). As a result, a monosodium salt of the quinol bianion is produced. This compound moves back to the cytoplasmic membrane surface (step 7). Here it is protonated by cytoplasmic H^+ ions and one Na^+ ion is released to the cytoplasm (step 8). Within the framework of such a mechanism, the net balance of the Na^+ export process is one Na^+ ion transported from the cytoplasm to the outer medium per oxidized NADH. To double this stoichiometry (two Na^+ per NADH), one has to assume that the monosodium salt of the quinol bianion dissociates to Na^+ and Q^{2-} close to the outer membrane surface.

The stoichiometry is not a problem for scheme B. In this case, it is proposed that semiquinone dismutation is coupled to the transport of Na^+ ions in an indirect fashion, assuming no

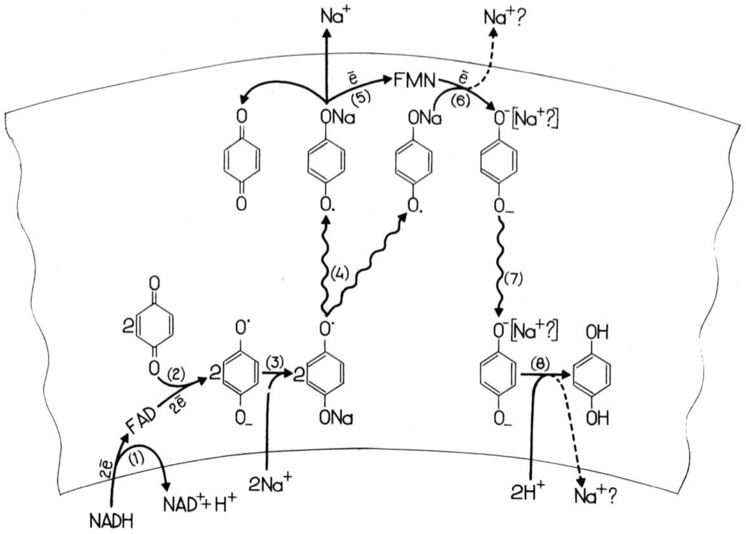

A

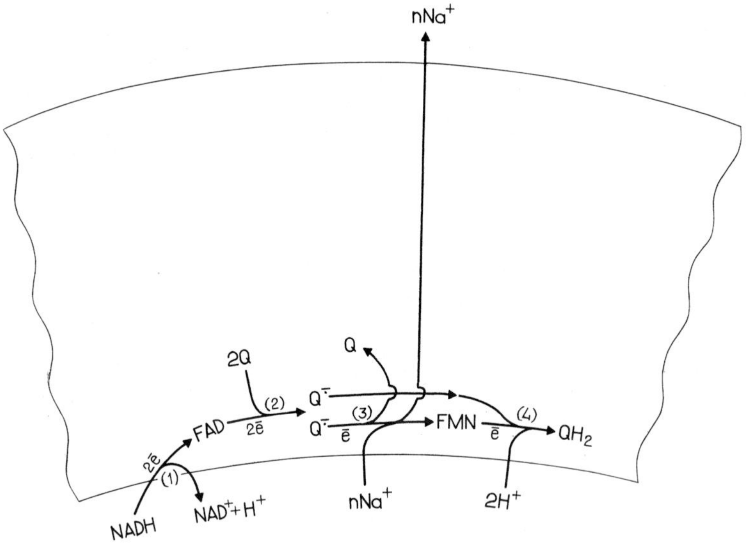

B

Fig. 2. Possible mechanisms of the Na^+-motive NADH-ubi(mena)quinone reductase. (A) Semiquinone anion ($CoQ^{\cdot-}$) is used to bind Na^+ and to translocate it across the membrane. Substituents at the benzene ring of CoQ are not shown. Wavy arrows represent the transmembrane diffusion of CoQ. (B) $CoQ^{\cdot-}$ oxidation is conformationally coupled to Na^+ translocation from the inward to the outward Na^+ channels or Na^+-conducting pathways.

interaction between Na^+ and the semiquinone anion. For instance, reduction of FMN by $Q^{\cdot-}$ may result in conformational change(s) in the protein molecule and thus cause the uphill transport of $n Na^+$ across the membrane.

An essential feature of both schemes in Fig. 2 is that they require Na^+-specific channels or pathways coupled to a redox reaction. An uncoupling of these two mechanisms was described by our group [44] in *Bacillus FTU*. It was found that submicromolar [Ag^+] not only inhibits the electron transfer reaction catalyzed by Na^+-motive NADH-quinone reductase, but also increases the passive Na^+ permeability of the membrane. This effect of Ag^+ is modulated by HQNO and by reduction of the respiration chain.

3.2.2. Na^+-motive terminal oxidase

In 1988–1991 we showed [45,46,39,40] that oxidation of ascorbate via TMPD by *Bacillus FTU* cells is coupled to Na^+ extrusion stimulated by protonophores. Inside-out *Bac. FTU* vesicles proved to be involved in the ascorbate + TMPD (or diaminodurene) oxidation-supported Na^+ uptake which was (i) stimulated by protonophorous uncouplers or valinomycin + K^+, (ii) further stimulated by a salt of a penetrating weak base and a penetrating weak acid, i.e., diethylammonium acetate (DEA) in the presence (but not in the absence) of a protonophore, (iii) inhibited by monensin, by the Na^+ ionophore *N,N′*-dibenzyl-*N,N′*-diphenyl-1,2-phenylenediacetamide (ETH157), by high cyanide concentration, by very low (10^{-8} M) concentration of Ag^+ and (iv) resistant to HQNO and amiloride. We concluded that *Bac. FTU* has a terminal oxidase carrying out electrogenic Na^+ transport. Electrogenicity explains the stimulating effect of protonophores or valinomycin. These ionophores discharge the oxidase-produced $\Delta\Psi$, allowing the large-scale transport of Na^+. Further increase in the rate of protonophore-stimulated Na^+ transport in the presence of acetic acid is accounted for by the discharge of ΔpH produced due to cooperation of the electrogenic Na^+-motive oxidase and the protonophore (Fig. 3A). The inhibitory effect of Ag^+ is explained by the enhanced Na^+ conductance (see the preceding section). It is noteworthy that the Na^+ uptake is not affected by HQNO, a specific inhibitor of the Na^+-motive NADH-quinone reductase, which is also present in *Bac. FTU*. The resistance of the process to micromolar cyanide excludes any participation of the aa_3-type oxidase, an enzyme which is also present in this microorganism, being able to perform H^+-motive terminal oxidation. Any explanation of the observed Na^+ transport by cooperation of a H^+-pump and Na^+/H^+ antiporter is excluded since (i) protonophores and the penetrating weak acid strongly stimulate, rather than inhibit, the Na^+ uptake, (ii) the added Na^+/H^+ antiporter, monensin, is inhibitory and (iii) the inhibitor of Na^+/H^+ antiporter, amiloride, is without effect.

All the above relationships were shown to be found in *Bac. FTU* grown at pH 8.6. However, if the pH of the growth medium was 7.5, then the Na^+ transport, being mediated by Na^+/H^+ antiporter, proved to be $\Delta\bar{\mu}_{H^+}$ dependent. It was still activated by valinomycin and inhibited by the Na^+ ionophore. At the same time, protonophorous uncouplers and DEA acetate were now inhibitory; Ag^+ and monensin were without effect; low cyanide, inhibiting the H^+ motive cytochrome oxidase, arrested the Na^+ transport; it was also abolished by amiloride which specifically inhibited the Na^+/H^+ antiporter (Fig. 3B).

A similar pattern of energy transduction was recently described by our group in *E. coli* as well [47,48]. In these experiments we attempted to determine which of the two *E. coli* quinol oxidases (*o* or *d*) is responsible for Na^+ transport in the cells growing at alkaline pH. For this purpose we studied Gennis' mutants lacking *o* oxidase or *d* oxidase [48]. Na^+-motive TMPD oxidation was found to occur in the *d* oxidase-containing mutant. This process was absent in the mutant with *o* oxidase. The Na^+-motive NADH-quinone oxidase was found to be present

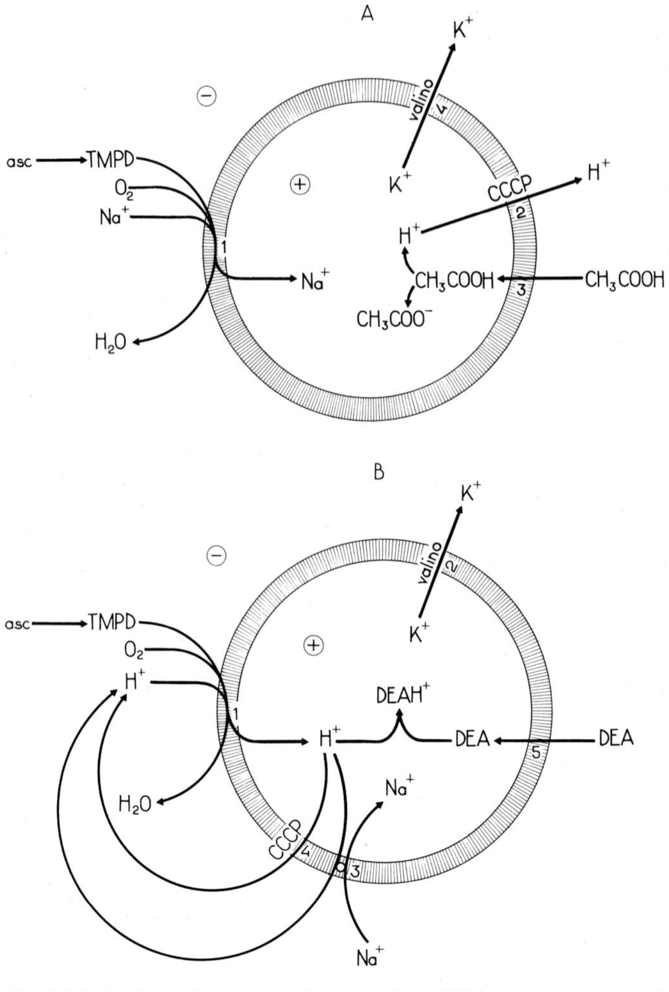

Fig. 3. Mechanisms of the terminal oxidase-linked Na^+ transport into inside-out subcellular vesicles from bacteria growing under conditions of low (A) and high (B) $\Delta\bar{\mu}_{H^+}$. (A) Primary Na^+ pump. Na^+-motive oxidase oxidizes ascorbate (asc) via tetramethyl-*p*-phenylenediamine (TMPD) and pumps Na^+ electrogenically into the vesicle. (1) Large-scale Na^+ import is limited by the oxidase-produced $\Delta\Psi$. (2) An H^+ infux mediated by the protonophorous uncoupler CCCP discharges $\Delta\Psi$. ΔpNa formation is limited by ΔpH formation (exhaustion of intravesicular H^+ pool). (3) A ΔpH discharge due to the acetic acid influx that compensates for the CCCP-induced decrease in intravesicular H^+ concentration. (4) A valinomycin-mediated K^+ efflux discharges the pump-generated $\Delta\Psi$. ΔpNa formation is limited by the size of the intravesicular K^+ pool. (B) Cooperation of the primary H^+ pump and the Na^+/H^+-antiporter. (1) H^+-motive oxidase pumps H^+ electrogenically into the vesicles. ΔpH formation is limited by $\Delta\Psi$. (2) Valinomycin-mediated K^+ efflux discharges $\Delta\Psi$ so that ΔpH is formed. (3) Na^+/H^+-antiporter utilizes ΔpH to accumulate Na^+ inside the vesicles. (4) CCCP discharges $\Delta\bar{\mu}_{H^+}$ and thus abolishes the Na^+ uptake. (5) The non-protonated form of diethylammonium (DEA) diffuses into the vesicle, discharges ΔpH and inhibits the Na^+ uptake. Valino, valinomycin. (From Kostyrko et al. [39].)

in both mutants. Further experiments showed that the d^+, o^- mutant can grow in the presence of an uncoupler at neutral pH, whereas the d^-, o^+ mutant cannot. In the wild-type *E. coli*, growth at high pH or in the presence of uncoupler was found to increase the amount of cytochrome *d* [48].

According to Green and Gennis [49], it is cytochrome *d*, not *o*, that is responsible for TMPD oxidase activity in *E. coli*. At the same time, cytochrome *o*, not *d*, is responsible for the extrusion of H^+ (oxidase-linked H^+ pump) when an electron donor is oxidized by the cytochrome *o* or *d* proteoliposomes (M. Wikström, personal communication).

Thus cytochrome *d* appears to be a good candidate for the *E. coli* Na^+-motive terminal oxidase.

Such a conclusion is, however, at variance with the proposal by Efiok and Webster [50], who have studied *Vitreoscilla*. This aerobic alkalotolerant microorganism is found mostly in benthic regions of fresh water sources and in cow dung. The authors described respiration-dependent Na^+ extrusion from the intact *Vitreoscilla* cells, which was slightly stimulated, rather than inhibited, by a protonophore, causing a transient $\Delta\Psi$ decrease [51]. In *Vitreoscilla* membrane particles or cytochrome *o* proteoliposomes, Na^+ caused a twofold increase in menadiol oxidase activity, with Li^+, K^+ and choline being ineffective. In the latter system some TMPD oxidation-supported Na^+ uptake was observed, which decreased by one-half upon the addition of 1 mM KCN [50]. The authors concluded that *Vitreoscilla* cytochrome *o* operates as a Na^+ pump [50]. However, cytochrome *d* is also present in the membrane of this bacterium [50]. As for the stimulation of menadiol oxidation by Na^+, this occurred only at rather high concentrations of this cation (the half-maximal effect at about 150 mM Na^+ in membranes and at about 50 mM Na^+ in proteoliposomes). Moreover, in contrast to the *E. coli* cytochrome *o*, the *Vitreoscilla* cytochrome *o* is active in TMPD oxidation [50] and thus resembles in this respect the *E. coli* cytochrome *d*. Also, K_d for the interaction of the oxidized enzyme and cyanide proved rather high (3.5 mM) [51a]. It is not excluded, therefore, that the *Vitreoscilla* cytochrome *o* is a functional analog of the *E. coli* cytochrome *d*. An alternative possibility is that the *Vitreoscilla* cytochrome *o* preparations were contaminated with cytochrome *d*.

3.2.3. Na^+-motive decarboxylases

As shown by Dimroth [18], decarboxylase from the anaerobically grown *Klebsiella pneumoniae* (i) converts oxaloacetate to pyruvate and CO_2 only if Na^+ is present; and (ii) is responsible for the uphill export of Na^+ from the cytoplasm to the external medium:

$$^{-}OOCCH_2COCOO^- + 2Na^+_{in} + H^+ \rightarrow CH_3COCOO^- + CO_2 + 2Na^+_{out}. \qquad (1)$$

This process could be demonstrated in intact cells and in the inside-out subcellular vesicles. The biotin-containing enzyme proved to be localized in the cytoplasmic membrane of *K. pneumoniae* [52]. The reaction sequence includes (i) transfer of the carboxyl residue from oxaloacetate to the biotin prosthetic group and (ii) release of free CO_2 from carboxylated biotin with the regeneration of the initial form of the enzyme. It is the second step that requires Na^+ [53]. Decarboxylation and Na^+ transport were sensitive to avidin, an inhibitor of biotin enzymes. The decarboxylation-dependent Na^+ uptake by everted vesicles resulted in a positive charging of the intravesicular space, which indicates that the Na^+ transport was electrogenic. The $\Delta\Psi$ and ΔpNa values were estimated by measuring the accumulation of $S^{14}CN$ and $^{22}Na^+$, repectively. $\Delta\Psi$ was about 65 mV, and ΔpNa was equivalent to 50 mV, the total $\Delta\bar{\mu}_{Na^+}$ being about 115 mV [54].

Oxaloacetate decarboxylase of *K. pneumoniae* was shown to be composed of α-, β- and γ-

subunits (64, 35 and 9 kDa, respectively). The α-subunit is a peripheral membrane protein, while the two others can be released only by means of detergent treatment. Biotin is localized exclusively in the α-subunit. The active enzyme complex can be reconstituted from isolated subunits [54].

The α-subunit was shown to be composed of carboxyltransferase (α_1) and biotin carrier (α_2) domains, which could be easily separated by trypsin treatment [55]. Electron microscopy revealed a cleft in the α-subunit, with the biotin residue localized close to the β- and γ-subunits [56]. Flip-flop movement of the biotin between the carboxyltransferase site on the α-subunit and the ligase site on the ($\beta+\gamma$)-subunits was postulated [55,57,58].

The sequence of the enzyme [59,60] shows that between residues −59 and −28 upstream of the biotin-binding lysine residue, there is a potentially very flexible region composed of 22 amino acids (7 prolines, 14 alanines and one serine). Such an alanine- and proline-rich sequence was also found in the dihydrolipoamide acetyltransferase subunit of pyruvate dehydrogenase at a similar distance from the lipoamide residue [61]. An NMR study has shown that this sequence is highly flexible [62]; a striking homology of the N-terminal region of the α_1-domain and the 5S subunit of transcarboxylase was observed, while the C-terminal region (the α_2-domain) proved to be homologous to biotin-binding sites in other biotin enzymes.

A hydropathy plot of the β- and γ-subunit sequences suggests that they form six and one α-helical transmembrane columns, respectively [60].

The passive Na^+-conductivity of decarboxylase proteoliposomes is low. It shows no significant increase upon dissociation of the α-subunit from the enzyme [58].

A mechanism for the Na^+-motive oxaloacetate decarboxylation was postulated by Dimroth [57]. It was assumed (Fig. 4) that in the first step, biotin (BH) is carboxylated by oxaloacetate, and pyruvate is formed. This event, catalyzed by the α-subunit, entails translocation of the carboxylated biotin ($BCOO^-$) to the lyase catalytic site on the β- (or γ-) subunit. In the second step, biotin is decarboxylated, CO_2 released to the cytosol, biotin returns to its initial position in the transcarboxylase site, and is protonated by the cytosolic H^+ ion. One of the events occurring in the second step is coupled to the translocation of $2Na^+$ from the cytosol to the outer medium. In the tentative scheme in Fig. 4, it is the downward movement of biotin that is Na^+-motive. However, any other partial reaction involved in the decarboxylation process in the second step can, in principle, be conformationally coupled to the Na^+ transport.

The above data were obtained in studies on the *K. pneumoniae* oxaloacetate decarboxylase. Recently, a very similar enzyme was described in *Salmonella typhimurium* [63].

Two other Na^+-motive decarboxylases were described in anaerobic bacteria. In *Acidaminococcus fermentans, Peptococcus aerogenes, Clostridium symbiosium* and *Fusobacterium nucleatum*, glutaconyl-CoA, an intermediate of the glutamate to acetate and butyrate fermentation was shown to be decarboxylated to crotonyl-CoA in a Na^+-motive fashion:

$$\text{CoA–CO–CH–CH=CH}_2\text{–COO}^- + n\text{Na}^+_{\text{in}} + \text{H}^+ \rightarrow \text{CoA–CO–CH=CH–CH}_3 + \text{CO}_2 + n\text{Na}^+_{\text{out}}. \quad (2)$$

This system was discovered by Buckel and coworkers [64–68]. The enzymes from the four above-mentioned species differ in M_r but have some common features. They contain biotin, are sensitive to avidin, require Na^+ for activity and can be reconstituted with phospholipids to form proteoliposomes competent in Na^+-motive decarboxylation. Monensin prevents the formation of a Na^+ gradient.

Independently, Dimroth's group showed that decarboxylation of methylmalonyl-CoA to propionyl-CoA is also a Na^+-motive process. This occurs in the strictly anaerobic *Veilonella*

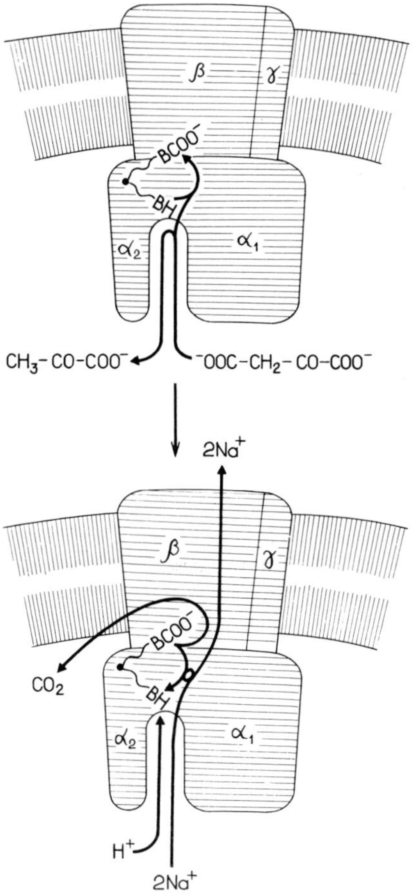

Fig. 4. A possible mechanism of Na^+-motive oxaloacetate decarboxylation. α_1 and α_2, domains of the largest (α) subunit of decarboxylase. β and γ, membraneous subunits of decarboxylase. BH and $BCOO^-$, biotin and carboxylated biotin, respectively. (Adapted from Dimroth [57]; for explanations, see text.)

alcalensis, converting lactate to acetate and propionate [69–71] and in *Propionigenium modestum*, which utilizes succinate and formes propionate [72]. Before decarboxylation, succinate was found to be converted to methylmalonyl-CoA. Decarboxylation of the latter is Na^+-motive:

$$\text{CoA–COCH(CH)}_3\text{COO}^- + 2\text{Na}^+_{\text{in}} + \text{H}^+ \rightarrow \text{CoA–COCH}_2\text{CH}_3 + \text{CO}_2 + 2\text{Na}^+_{\text{out}}. \quad (3)$$

Again, the enzyme was found to contain biotin and to require Na^+ for decarboxylation. Two Na^+ ions were transported across the membrane of the *V. alcalensis* subcellular vesicles per molecule of decarboxylated methylmalonyl-CoA [70]. Decarboxylase from *V. alcalensis* was purified, and proteoliposomes were reconstituted. The enzyme transported Na^+ from the me-

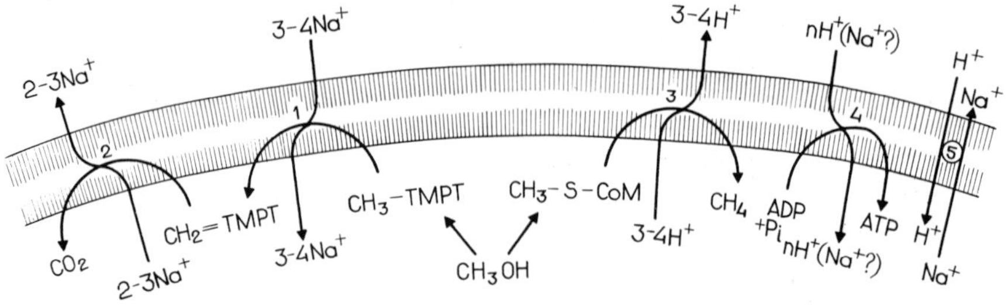

Fig. 5. A tentative scheme of the Na^+ and H^+ cycles involved in methanol disproportionation to CH_4 and CO_2 by methanogenic bacteria. (1) Endergonic oxidation of CH_3–TMPT (formed from CH_3OH) to CH_2=TMPT, coupled to the downhill import of 3 or 4 Na^+ ions. (2) Exergonic oxidation of CH_2=TMPT to CO_2, coupled to the uphill export of 2–3 Na^+ ions. (3) Conversion of CH_3–S–CoM (formed from another CH_3OH molecule) to CH_4, coupled to the uphill export of 3–4 H^+ ions. (4) ADP phosphorylation coupled to the downhill import of nH^+ or, possibly, nNa^+. (5) H^+_{out}/Na^+_{in} antiport consuming $\Delta\bar{\mu}_{H^+}$ which is produced by reaction (3). TMPT, tetrahydromethanopterin. (From Kaesler and Schönheit [79].)

dium to the proteoliposome interior. A 30-fold Na^+ gradient was formed. The rate of the Na^+ influx was accelerated by a protonophorous uncoupler, a fact indicating that the Na^+ transport process is electrogenic [70,73–75].

The enzyme was found to be composed of 60, 33, 18.5 and 14 kDa subunits. The 18.5 kDa subunit contains biotin [57,74]. The mechanism of $\Delta\bar{\mu}_{Na^+}$ formation seems to be similar to that shown in Fig. 4 (for discussion, see ref. [75]).

3.2.4. Na⁺-motive electron transfer in methanogenic and acetogenic bacteria

According to Kaesler and Schönheit [76], a partial reaction involved in the conversion of methanol to CO_2 by methanogenic bacteria is coupled to Na^+ extrusion from the bacterial cell. On the other hand, another partial reaction involved in the same process is $\Delta\bar{\mu}_{Na^+}$ driven; it is coupled to the downhill Na^+ influx into the cell [76–81] (Fig. 5).

As suggested by Heise et al. [82], a Na^+-motive oxidoreduction system resembling that in methanobacteria is also present in acetogenic bacteria. It was shown that the reduction of methylenetetrahydrofolate to methyltetrahydrofolate is activated by the addition of Na^+ and may be coupled to the uphill Na^+ efflux from the bacterial cell. In the same group, an indication was obtained that *Acetobacterium woodi* has Na^+-driven ATP-synthase [82a].

3.2.5. Animal Na⁺/K⁺-ATPase and Na⁺-ATPase

Na^+/K^+ ATPase, an enzyme described by Skou in 1957 [83], has been the object of numerous studies. Localized in the plasma membrane of animal cells, it is responsible for $\Delta\Psi$, ΔpNa and ΔpK generation across this membrane. Na^+/K^+ ATPase catalyzes the exchange of $2K^+_{out}$ for $3Na^+_{in}$ per molecule of ATP hydrolyzed. The enzyme is composed of four subunits: two α and two β, with molecular masses of about 95 and 45 kDa, respectively [84,85]. Electron microscopy of two-dimensional crystals of the enzyme (2 nm resolution) carried out by Ovchinnikov's group [86], showed the vertical dimension of the $\alpha\beta$-protomer to be about 10 nm, i.e., larger than the lipid bilayer thickness. The α-subunit was found to protrude from the inner, as well as from the outer membrane surfaces, while the β-subunit was exposed on the outer surface only. The C-terminal region of the β-subunit is glycosylated. The mass of carbohydrate moiety is 7 kDa. Chemical modification and antibody-technique data showed that almost two-third of the

α-subunit is localized outside the membrane, its cytoplasmic domain being three times larger than that located on the outer surface. The α-subunits are somewhat narrowed in the central part and contact each other in the intramembrane region. The height of the contact region is about 3 nm.

Na^+/K^+ ATPase was reconstituted into proteoliposomes which are involved in energy transduction. The freeze-etching technique showed that the proteoliposome membrane contains 9–12 nm particles, a size corresponding to the $\alpha_2\beta_2$-complex.

An analysis of the hydropathy profiles of the amino acid sequence indicates the presence of seven and one α-helical columns in α- and β-subunits, respectively [87].

The N-terminal domain of the α-subunit was found to face the cytoplasmic membrane side. The carboxyl of a dicarboxylic amino acid residue, localized in this domain, is phosphorylated by ATP so that a phosphoenzyme (EP) is formed [84].

The Na^+/K^+ ATPase mechanism is described in terms of the scheme for so-called E_1E_2 ATPases [87a]:

$$E_1 + ATP \xrightarrow{Na^+_{in}} E_1P + ADP; \tag{4}$$

$$E_1P \rightarrow E_2P; \tag{5}$$

$$E_2P \xrightarrow{K^+_{out}} E_2 + P_i; \tag{6}$$

$$E_2 \rightarrow E_1. \tag{7}$$

According to the scheme, Na^+ (presumably Na^+_{in}) is required for enzyme phosphorylation. The phosphoenzyme undergoes a conformational change ($E_1 \rightarrow E_2$ transition) and then decomposes to form E_2 and P_i in a K^+_{out}-dependent fashion. E_2 relaxes to E_1, ready to enter the next cycle. It is the E_2P state that is blocked by ouabain, a specific Na^+/K^+ ATPase inhibitor. Oligomycin was found to affect the cycle at the E_1 and E_1P steps, stimulating Na^+ binding to the enzyme (K_d for Na^+ decreases from 340 to 60 μM). K_d for K^+ was found to be the same with and without oligomycin (6 μM). The binding sites for $3Na^+$ and $2K^+$ showed positive cooperatively (reviewed in refs. [85,87a]).

Rb^+, Cs^+, NH_4^+, Tl^+, Li^+ and Na^+ were reported to substitute to some degree for K^+ at the K^+ sites. On the other hand, the Na^+-binding sites are known to have relatively high specificity. Among the above-mentioned cations, only Li^+ was found to partially substitute for Na^+. The Li^+/K^+ ATPase activity is, however, very weak and the affinity for Li^+ is much lower than for Na^+. It is interesting that H^+ can substitute for Na^+ and the affinity for H^+ is, in fact, at least two orders of magnitude higher than for Na^+. Nevertheless, at physiological pH, $[Na^+]$ is very much higher than $[H^+]$, so that Na^+, rather than H^+, is translocated by Na^+/K^+ ATPase. However, under slightly acidic conditions (pH = 5.7), ATPase activity coupled to H^+/K^+ antiport was reported in Na^+/K^+ ATPase proteoliposomes [88].

As already noted, the α-subunit is directly involved in the catalysis. As to the β-subunit, its role is still disputed. At least one of the β-subunit isoforms (β_2) was found to function in astrocytes as an adhesion molecule on glia [89,90]. Quite recently, Sverdlov and colleagues reported that the level of the β_2-subunit mRNA is strongly decreased in human tumors of the kidney, lung and liver as compared with normal tissue [91]. This observation may be important within the framework of the concept that contact inhibition of cell proliferation is a feature distinguishing normal cells from cancer ones [92].

Another Na^+-transporting ATPase was described by Del Castillo and Robinson and Proverbio and coworkers in basolateral membranes from small intestinal epithelial cells [93] and from renal proximal tubular cells [94–96] (see also ref. [97]). It differs from Na^+/K^+-ATPase in several respects. First, K^+ is not necessary for ATPase activity, which is specifically increased by Na^+. Ouabain is ineffective, while fucosemide, which does not affect Na^+/K^+-ATPase, is strongly inhibitory. Among other cations, only Li^+ activates the enzyme, though to a lesser extent than Na^+. In contrast to Na^+/K^+-ATPase, it is activated by micromolar $[Ca^{2+}]$. It shows a pH optimum and K_d for Na^+ differing from those of Na^+/K^+-ATPase. The authors concluded that the enzyme, defined as Na^+-ATPase, is responsible for the ouabain-insensitive part of Na^+ transport in the animal cells studied. Molecular characteristics of the enzyme are not yet available.

3.2.6. *Bacterial Na^+-ATPases*

In 1982 Heefner and Harold presented convincing evidence for Na^+-motive ATPase in *Streptococcus faecalis*. It was found that (i) the inside-out subcellular vesicles exhibit sodium-activated ATPase activity and (ii) $^{22}Na^+$ is accumulated in the vesicle interior in an ATPase-dependent fashion [98]. Neither DCCD nor vanadate inhibits the *S. faecalis* enzyme [99]. It was also found that (i) the Na^+ transport does not generate $\Delta\Psi$, (ii) Na^+ is absolutely necessary for K^+ accumulation by the *S. faecalis* mutant lacking H^+-ATPase and (iii) valinomycin dissipates the K^+ gradient generated by Na^+-ATPase. It was assumed that this enzyme catalyzes the electroneutral Na^+/K^+ antiport activity coupled to ATP hydrolysis [99].

As found by Kakinuma et al. [100,100a], EDTA treatment of the *S. faecalis* membranes results in the detachment of an about 400 kDa protein which is not observed in the mutant lacking Na^+-ATPase, and this detachment is increased under conditions of induction of this enzyme (see below, Section 4). The protein consisted mainly of polypeptides of 73 kDa, 52 kDa and 29 kDa.

In membrane fragments, Na^+-ATPase was sensitive to nitrate and *N*-ethylmaleimide in the same way as vacuolar H^+-ATPase [100]. The amino acid sequence of the largest subunit of Na^+-ATPase proved also to be similar to that of vacuolar H^+-ATPase [100a]. Concerning the induction of *S. faecalis* Na^+-ATPase under low $\Delta\bar{\mu}_{H^+}$ conditions and the possible physiological significance of this regulation, see Section 4.

It may be noted in this context, that Selwyn's group found indications that the facultative alkalophile *Exiguobacterium aurantiacum* has a primary Na^+ pump exporting Na^+ ions at the expense of glycolytic energy. The process was found to be $\Delta\bar{\mu}_{H^+}$ independent [101]. The simplest explanation of these data is that the pump involves a Na^+-motive ATPase.

A pathway including Na^+-motive ATPase is apparently present in *Mycoplasma* and *Acholeplasma*. According to Leblanc and colleagues [102,103], there are two mechanisms for Na^+ extrusion in *Mycoplasma mycoides*: (i) a $\Delta\bar{\mu}_{H^+}$-driven Na^+/H^+ antiport and (ii) a $\Delta\bar{\mu}_{H^+}$-independent, ATP- driven Na^+ uniport (or possibly Na^+/K^+ antiport). In the same *Mycoplasma*, a Na^+-activated ATPase was found. Vanadate and ouabain, as well as K^+, had no effect on the enzyme, which differs in this respect from mammalian Na^+/K^+ ATPase. A similar enzyme was described in *Acholeplasma laidlawii* [104–108]. The ATPase was found to contain five types of subunits with molecular masses similar to that of F_1-ATPase. Like factor F_1, the ATPase was inactivated when stored in the cold. On the other hand, it was insensitive to aurovertin, a specific inhibitor of F_1-ATPase activity [105,106].

As shown by Shirvan et al. [109,110], *Mycoplasma gallisepticum* extrudes Na^+ in a protonophore-resistant, DCCD-sensitive fashion. In the same microorganism, the authors have found an ATPase which was stimulated threefold by Na^+ at pH 8.5 but very little at pH 5.5. Apparently, it is this enzyme that is responsible for Na^+ extrusion in the presence of a protonophore.

The recent observations on Na^+-ATPase activity in *Propionigenium modestum* and methanogenic bacteria will be described in Section 3.3.4 in the context of Na^+-dependent ATP synthesis.

3.3. *Utilization of $\Delta\bar{\mu}_{Na^+}$ produced by primary $\Delta\bar{\mu}_{Na^+}$ generators*

3.3.1. *Osmotic work: Na^+, solute-symports*

It has been shown that in the alkalotolerant *Vibrio alginolyticus*, having an Na^+-motive NADH-quinone reductase, there are Na^+, solute-symporters responsible for the accumulation of 19 amino acids as well as sucrose [21,22]. The import of α-aminoisobutyrate (AIB) was studied in detail. It was found that (i) Na^+ is necessary for AIB accumulation inside the cell, (ii) respiration is competent in supporting AIB accumulation and (iii) the accumulation is protonophore resistant. It was also demonstrated that the accumulation of K^+ in *V. alginolyticus* cells at alkaline pH is driven by $\Delta\Psi$ produced by the Na^+-motive NADH-quinone reductase. At pH > 8.0, Na^+-dependent K^+ import caused a $\Delta\Psi$ decrease without any concomitant ΔpH formation.

Na^+-dependent import of nutrients into alkalophilic *Bacilli* was described by several groups [111–114].

In neutrophilic bacteria living at low or moderate NaCl concentrations, $\Delta\bar{\mu}_{H^+}$, rather than $\Delta\bar{\mu}_{Na^+}$, is used to support osmotic work. However, some exceptions to this rule have also been reported. For example, in *Mycobacterium phlei* [115], *Salmonella typhimurium* [116] and *E. coli* [117,118], a Na^+, proline-symporter is used to accumulate proline inside the cell. A significant sequence homology was found between the *E. coli* Na^+, proline-symporter and the animal plasma membrane Na^+, glucose-symporter [119].

An interesting example of a 'dualistic' mechanism of metabolite import was described in *E. coli* which was shown to use H^+ or Na^+ alternatively as coupling cation for cotransport when it takes up melibiose [120,121].

Both Na^+ and H^+ are symported with adenosine when the latter is taken up by *Vibrio parahaemolyticus* [121a]. A similar carrier is responsible for the import of citrate by *Klebsiella pneumoniae*. It symports $citrate^{3-}$, $2Na^+$ and $2H^+$. This means that $\Delta\Psi$, ΔpNa and ΔpH are the driving forces for the uphill citrate import [122]. *Bacillus stearothermophilus* was shown to transport glutamate anions with one H^+ and one Na^+ [123]. On the contrary, $succinate^{2-}$ crosses the *Selenomonas ruminantium* membrane together with $3Na^+$ [124]. The authors postulated that such a mechanism is used by the bacterium to generate $\Delta\bar{\mu}_{Na^+}$ by means of an efflux of succinate, which is known to be the endproduct of glucose fermentation by *S. ruminantium*. This system looks similar to that in *Oxalobacter formigenes* converting $oxalate^{2-}$ + H^+ to $formate^-$. An efflux of $formate^-$ in exchange for $oxalate^{2-}$ generates $\Delta\Psi$, whereas fermentation per se generates ΔpH so that $\Delta\bar{\mu}_{H^+}$ is formed [125].

Brodie and coworkers [115] succeeded in isolating a 20 kDa Na^+, proline-symporter from *Mycobacterium phlei*. The purified symporter was reconstituted with phospholipids to form proteoliposomes which were able to support proline accumulation driven by an artificially imposed $\Delta\Psi$. To generate $\Delta\Psi$, a valinomycin-induced downhill K^+ efflux was used. The K^+ efflux-driven proline accumulation required Na^+, and was lowered by a protonophorous uncoupler, discharging $\Delta\Psi$, and by sulphydryl reagents.

Partial purification and reconstitution of a Na^+, aspartate-symporting system from halophilic *H. halobium* was described by Greece and MacDonald [126].

Generally, marine and halophilic bacteria, like alkalophiles, employ Na^+, solute (not H^+, solute) symporters. This is also true for the plasma membrane of higher animals, the phenomenon being yet another testimony of the assumption that blood is a small part of the ocean trapped inside our body. In this membrane, $\Delta\bar{\mu}_{Na^+}$ is generated due to Na^+ extrusion by the Na^+,K^+-ATPase. The $\Delta\bar{\mu}_{Na^+}$ formed is utilized by numerous carriers catalyzing the symport of Na^+ and amino acids, sugars fatty acids and other compounds into the cell [10,11]. Some of these Na^+, solute-symporters were isolated and studied at the proteoliposomal level (see, e.g., ref. [127]).

In some animal tissues, H^+-ATPase is localized in the outer cell membrane. In these cells, H^+, solute-symporters have been described (see ref. [8]).

3.3.2. Na^+ ions and regulation of the cytoplasmic pH

In bacteria, the pH_{in} homeostasis mechanism operates very effectively. As shown by Slonczewski et al. [128], this homeostasis in *E. coli* is quite efficient for pH_{out} values between 5.5 and 9.0. Within this range, pH_{in} could be described by:

$$pH_{in} = 7.6 + 0.1\ (pH_{out} - 7.6), \tag{8}$$

so that pH_{in} was 7.4 and 7.8 at pH_{out} 5.5 and 9.0, respectively.

When pH_{out} decreases, acidification of the cytoplasm can be prevented by K^+ uptake, discharging the $\Delta\Psi$ that is produced by the $\Delta\bar{\mu}_{H^+}$ generators. As a result, $\Delta\Psi$ is converted to ΔpH, the cytoplasm becoming more alkaline than the outer medium (Fig. 6A).

At neutral pH_{out}, the only problem is how to return to the cytoplasm those protons that are pumped out from the cell by $\Delta\bar{\mu}_{H^+}$ generators. The problem can be solved by an electroneutral Na^+/H^+ antiporter (Fig. 6B).

The electroneutral Na^+/H^+ antiporter is effective in the maintenance of pH homeostasis only if $pH_{out} < pH_{in}$. Other systems are necessary to acidify the cytoplasm when pH_{out} is higher than pH_{in}. Bacteria face this problem when the growth medium becomes alkaline. Now, there is no reason for the electroneutral Na^+/H^+ antiporter to carry H^+ from the outside to the interior of the cell since both the Na^+ and H^+ gradients are in an unfavourable direction. One may overcome this difficulty, assuming that at alkaline pH_{out} (i) the Na^+/H^+ antiporter is electrogenic, transporting more than one H^+ per Na^+ and (ii) the $\Delta\Psi$ formed by $\Delta\bar{\mu}_{H^+}$ generators is so large that it can compensate for the unfavourable ΔpH and ΔpNa (Fig. 6C). It is noteworthy that electroneutral and electrogenic Na^+/H^+ antiporters cannot coexist in the same bacterium simultaneously, since a combination of the processes shown in the schemes in Figs. 6B and 6D discharges $\Delta\bar{\mu}_{H^+}$ [129].

An alternative possibility is shown in Fig. 6D. There is a primary Na^+ pump here which charges the membrane, the H^+ uniporter allowing H^+ ions to move electrophoretically into the cytoplasm.

Quantitative measurements carried out in *Bacillus alcalophilus* by Guffanti et al. [113] are in good agreement with the latter scheme. It was found that at pH_{out} 11.5, pH_{in} is 9.0 and $\Delta\Psi$ is about 160 mV. These values mean that ΔpH is almost equal to $\Delta\Psi$, as predicted in Fig. 6D. (Yet, it should be noted that Guffanti and his colleagues explain their results in terms of the electrogenic Na^+/nH^+ antiport scheme [130,131].)

Each of the two ion transfer systems shown in Fig. 6D was described in some types of biological membranes. Primary Na^+ pumps using energy of respiration, ATP hydrolysis or decarboxylation were found in certain prokaryotes (see above, Section 3.2). As to the H^+ uniport, also required for the scheme in Fig. 6D, it occurs in the inner mitochondrial membrane

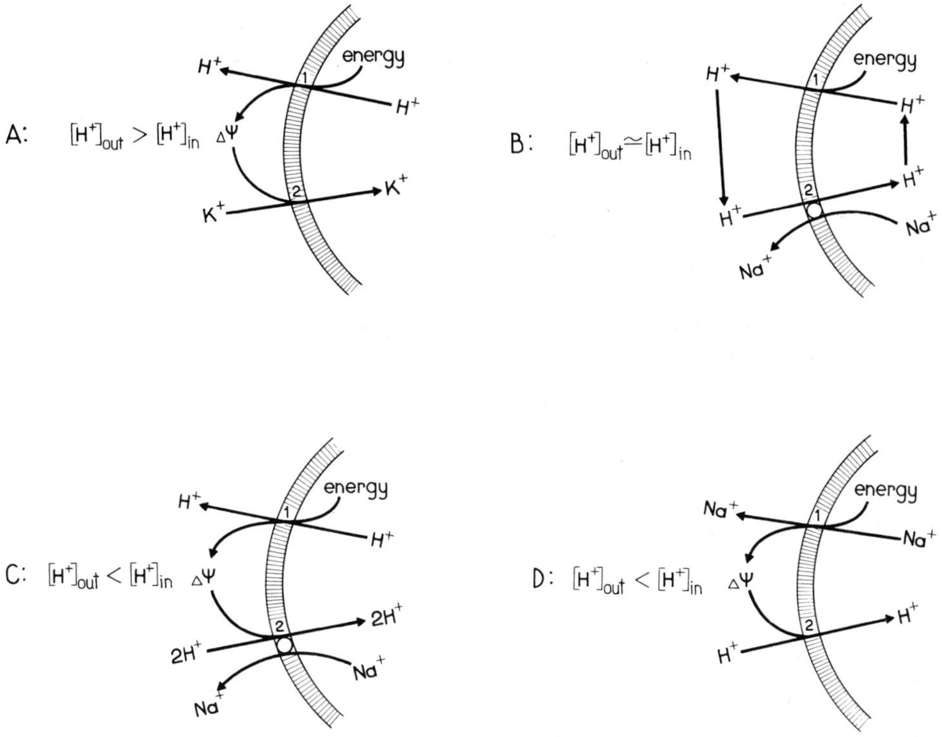

Fig. 6. Possible mechanisms for pH homeostasis in bacteria. (A) $[H^+]_{out} > [H^+]_{in}$. $\Delta\bar{\mu}_{H^+}$ generator extrudes H^+ from the cell and forms $\Delta\Psi$ (reaction 1). This $\Delta\Psi$ is discharged by the K^+ influx so that $\Delta\Psi \rightarrow \Delta pH$ transition occurs, the cell interior is more alkaline (reaction 2). (B) $[H^+_{out}] = [H^+_{in}]$ under non-energized conditions. The H^+ loss due to the activity of $\Delta\bar{\mu}_{H^+}$ generators (reaction 1) is compensated by H^+_{out} influx in exchange for Na^+_{in} (reaction 2). (C) and (D) $[H^+_{out}] < [H^+_{in}]$. (C) The H^+ pump extrudes H^+ and forms $\Delta\Psi$ (reaction 1). To discharge $\Delta\Psi$, $Na^+/2H^+$-antiporter exchanges one Na^+_{in} for two H^+_{out} (reaction 2). (D) A primary Na^+ pump charges the membrane, extruding Na^+ (reaction 1). Uniport of H^+ ions into the cell discharges $\Delta\Psi$ and acidifies the cytoplasm (reaction 2).

of brown fats and some other tissues of newborn or cold-adapted animals [8,132]. (About mechanisms of adaptation of bacteria to alkaline pH, see also Section 3.4.)

Na^+ ions are also involved in eukaryotic pH homeostasis. In animal cells, it has been firmly established that it is the electroneutral Na^+/H^+ antiporter that is in the system responsible for the maintenance of a stable pH_{in} (for discussion, see ref. [8]). In plants, the Na^+/H^+ antiporter seems to be present in the outer cell membrane and in the tonoplast [133,134]. A study of the latter system showed that the antiporter is sensitive to amiloride, just as it is in animals and bacteria [134].

3.3.3. Mechanical work: the Na^+ motor

There is at least one example of a bacterium performing mechanical work at the expense of $\Delta\bar{\mu}_{Na^+}$, generated by a primary Na^+ pump. Dibrov and coworkers in our group [23,135–139] showed that motility of *V. alginolyticus* (i) occurs only in the presence of Na^+, (ii) can be supported by an artificially imposed ΔpNa in a monensin-sensitive fashion (under the same conditions ΔpH was ineffective) and (iii) is carried out at a lowered but measurable rate in the presence of a very high concentration of a protonophore, such as 1×10^{-4} M *m*-carbonylcyanide phenylhydrazone (CCCP). A 100-fold lower protonophore concentration completely arrested the motility if the medium is supplemented with monensin. Added without CCCP, monensin decreased the respiration-supported motility only slightly. It was concluded that the flagellar motor of *V. algionolyticus* is driven by $\Delta\bar{\mu}_{Na^+}$ rather than by $\Delta\bar{\mu}_{H^+}$ [23,135–139].

It was found in our laboratory [8] that the motility of *V. cholerae* and of the marine *V. harveyi* is basically of the same nature as that of *V. alginolyticus*. Ferris et al. [140] described the structures of the basal bodies of the *V. cholerae* flagellum. A striking similarity to that of *V. alginolyticus* was revealed [136]. According to Unemoto (unpublished result), *V. cholerae* possesses Na^+-motive NADH-quinone reductase.

It is common knowledge that *V. cholerae* produces a toxin causing salt extrusion from the tissues to the intestinal lumen. This salt may be necessary for *V. cholerae* to support its Na^+ energetics [8].

$\Delta\bar{\mu}_{Na^+}$-supported motility was also demonstrated in alkalophilic *Bacilli. Bacillus sp.* YN-1 and *Bacillus firmus* were used for this purpose [141–147]. In *Bacillus sp.* YN-1, studied in much detail, an Na^+ requirement for motility and partial resistance of motility to the protonophore were shown. It was also found that ΔpNa and $\Delta\Psi$, generated enzymatically, are equivalent in supporting motility. The effect of an artificially imposed $\Delta\bar{\mu}_{Na^+}$ was not studied. Since the primary Na^+ pumps have not yet been described in these microorganisms, it is assumed that electrogenic Na^+/nH^+ antiporter is the only mechanism of $\Delta\bar{\mu}_{Na^+}$ generation [128,129]. It should be mentioned that it was in *Pseudomonas stutzeri* that an Na^+-dependent motility was observed for the first time, by Kodama and Taniguchi in 1977 [148]. Unfortunately, the problem of $\Delta\bar{\mu}_{Na^+}$ formation has never been studied in this bacterium.

3.3.4. *Chemical work: $\Delta\bar{\mu}_{Na}$-driven ATP synthesis*

The first indication that ΔpNa (together with ΔpK) can, in principle, be utilized to reverse an ion-transfer ATPase reaction was obtained in 1966 by Glynn and Garrahan [149] who described ATP synthesis by Na^+/K^+-ATPase in erythrocytes. The process was observed when $[Na^+]_{in}$ was higher than $[Na^+]_{out}$, and $[K^+]_{in}$ was lower than $[K^+]_{out}$. However, this reaction can hardly occur under natural conditions because of the opposite direction of the ion gradients.

Some evidence that $\Delta\bar{\mu}_{Na} \rightarrow$ ATP energy transduction really occurs in the living cell was obtained by Dimroth and colleagues [72] who studied *Propioniogenium modestum*, a strictly anaerobic marine bacterium discovered in 1982 by Schink and Pfennig [150]. The only biologically useful energy for *P. modestum* is gained through decarboxylation of succinate to propionate:

$$^{-}OOC{-}CH_2{-}CH_2{-}COO^- + H_2O \rightarrow CH_3{-}CH_2{-}COO^- + HCO_3^- + 5\ \text{kcal.} \qquad (9)$$

The energy yield of the reaction is approximately one-half of the energy price for ATP synthesis for ADP and inorganic phosphate under physiological conditions (this is why Schink and Pfennig called this microbe *modestum*, i.e., modest with respect to its energy requirement). It seems clear that the mechanism of substrate-level phosphorylation cannot be used by *P. modestum*, since it presumes the formation of one ATP molecule per molecule of the substrate utilized. A solution to the problem may be found if substrate utilization were coupled to the formation of a difference in electrochemical potentials of some ions, e.g., H^+ or Na^+. In this case, utilization of one substrate molecule may result in the export of nH^+ or nNa^+ ions, while the formation of one ATP molecule may require the import of, e.g., $2nH^+$ or $2nNa^+$ ions.

According to Dimroth et al. [72], decarboxylation of methyl-malonyl-CoA formed from succinyl-CoA is coupled to the Na^+ extrusion from the *P. modestum* cell (see above, Section 2.3). The formed $\Delta\bar{\mu}_{Na^+}$ is assumed to be employed to synthesize ATP by a reversal of Na^+-ATPase found in large amounts in the cytoplasmic membrane of *P. modestum*. As was reported by Dimroth, this ATPase is of the F_0F_1-type. When membrane-bound, it is sensitive to DCCD, trialkyltin and venturicidin. Detachment from the membrane results in DCCD resistance and in the disappearance of the activating effect of Na^+. Detachable and membraneous sectors showed subunit compositions similar to F_1 and F_0, respectively [151,152]. The amino acid sequence of the β-subunit of *P. modestum* resembles that of *E. coli, V. alginolyticus* and other eubacteria [153]. Also, the c-subunits of *P. modestum* and *V. alginolyticus* reveal some sequence homology [154].

When the *P. modestum* F_0F_1-proteoliposomes were partially depleted of factor F_1, ATP hydrolysis was ineffective in sustaining the Na^+ gradient. But the addition of *E. coli* F_1 made it possible to sustain the gradient [155]. It was also found that decarboxylation of malonyl-CoA or, alternatively, ATP hydrolysis, is coupled to an uphill electrogenic Na^+ uptake by the everted subcellular vesicles of *P. modestum* [72]. The ATP-dependent Na^+ uptake was inhibited by monensin and resistant to CCCP. Further experiments showed that ATP is synthesized at the expense of the $\Delta\bar{\mu}_{Na^+}$ formed by decarboxylase. Again, monensin was inhibitory. Unfortunately, the coupling of decarboxylation and phosphorylation in the vesicles was so poor that the rate of ATP synthesis was about 10^4 times slower than that of decarboxylation. Moreover, the ATPase activity of the vesicles was about 10^3 times faster than that of ATP synthase. Thus, this observation should be regarded as a qualitative indication rather than the final proof of $\Delta\bar{\mu}_{Na^+}$-driven ATP synthesis.

Nevertheless, the entire logic and some related observations are in favour of the authors' statement that decarboxylation promotes ATP synthesis, whereby $\Delta\bar{\mu}_{Na^+}$ is the intermediate of the process.

(i) *P. modestum* must have a mechanism for ATP synthesis coupled to the conversion of succinate to propionate since this conversion is found to be the only energy-releasing reaction in the studied bacterium.

(ii) It has been firmly established that the energy released by decarboxylation is utilized for $\Delta\bar{\mu}_{Na^+}$ generation.

(iii) Na^+ causes a 20-fold activation of membrane-bound ATPase.

(iv) The addition of ATP to membrane vesicles of *P. modestum* or to Na^+-ATPase proteoliposomes results in a monensin-sensitive uphill Na^+ transport; this transport is strongly increased by the protonophore CCCP as well as by valinomycin + K^+.

(v) ATP hydrolysis can be coupled to carboxylation of propionate to methylmalonyl-CoA in a $\Delta\bar{\mu}_{Na^+}$-dependent mode [72]. Dimroth's version of the Na^+ cycle in *P. modestum* is shown below in Fig. 8A.

$\Delta\bar{\mu}_{H^+}$-independent ATP synthesis supported by an artificially imposed $\Delta\bar{\mu}_{Na^+}$ has been demonstrated in another type of anaerobic microbes, namely, in methanogenic bacteria. In 1983 Schönheit and Perski [156] reported that the valinomycin-mediated K^+ efflux from *Methanobacterium thermoautotrophicum* cells results in an increase in ATP level. The effect was greatly stimulated by Na^+ [156] or Li^+ [157]. The authors suggested that one of the possible explanations of this phenomenon lay in the reversal of hypothetical Na^+-ATPase by means of the $\Delta\Psi$ generated by the K^+ efflux. However, they failed to show the presence of a sodium-stimulated ATPase in the membrane fraction of this bacterium [156].

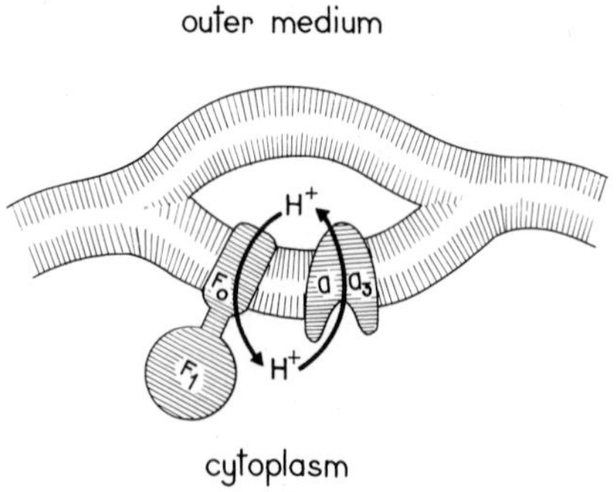

Fig. 7. Hypothetical scheme explaining the operation of the H^+ cycle in non-marine alkalophilic *Bacilli.* It is assumed that a lens in the cytoplasmic membrane allows respiratory enzymes and H^+-ATP synthase to operate without direct contact with the outer medium (From ref. [7].)

In 1985–1986 the above observation was extended by Lancaster and coworkers to *Methanococcus voltae* [158–161]. It was found that ATP synthesis can be induced by the diffusion potential of K^+ (+ valinomycin) or H^+ (+ a protonophorous uncoupler). In the latter case, a pulse of KOH was added to bacteria incubated with an uncoupler at acidic pH. The uncoupler-mediated downhill H^+ efflux from the cell could generate a $\Delta\Psi$ (negative inside). ATP was postulated to be synthesized at the expense of this $\Delta\Psi$ by a reversal of the Na^+-ATPase. In agreement with this explanation, (i) Na^+ was required for uncoupler-induced ATP synthesis and (ii) valinomycin + K^+, which discharged the H^+ efflux-generated $\Delta\Psi$, abolished the ATP synthesis [159]. The following experiments showed that the Na^+ pulse is also competent in ATP formation, with monensin and diethylstilbestrol being inhibitory [158]. The latter did not affect phosphorylation coupled to methanogenesis.

Recently, the same group showed that the relationships found in *M. voltae* were also found in *Methanobacterium thermoautotrophicum.* The only difference was that diethylstilbestrol failed to inhibit the ATP synthesis supported by the artificially imposed $\Delta\Psi$ in *M. thermoautotrophicum.* The process was specifically arrested by harmaline, an inhibitor of Na^+-dependent systems [160]. As found by Smigan et al. [162], the level of Na^+-ATPase increased when a high Na^+ concentration was present in the growth medium.

Unfortunately, it is not clear whether $\Delta\bar{\mu}_{Na^+}$-driven ATP formation in methanogenes really occurs under physiological conditions. The authors regarded their observation only as an indication of the existence of Na^+-ATPase, which is necessary for expelling Na^+ from the cell of this marine halotolerant microbe. The $\Delta\bar{\mu}_{Na^+}$ formed can be utilized to perform osmotic work. Sprott's group has found that many amino acids are cotransported with Na^+ into *Methanococcus voltae* cells [163,164].

An important problem was whether oxidative phosphorylation in the respiring bacteria could be Na^+-coupled. The evidence that respiratory phosphorylation may be Na^+-dependent was obtained in 1977, when intact cells of *Pseudomonas stitzeri* were studied by Kodama and Taniguchi [148]. The possibility that Na^+ circulation is involved in the respiratory energy coupling was not considered in that paper.

Such a possibility was mentioned for the first time in 1981 by Guffanti et al. [165] when they discussed various mechanisms of respiratory phosphorylation in soil alkalophilic bacilli. The authors, however, concluded that these micoorganisms had chosen another way to solve their problems.

Later we reinvestigated this question using the alkalotolerant marine *V. alginolyticus* [24,25,166,167]. It was shown that the addition of lactate to *V. alginolyticus* cells, exhausting the pool of endogenous substrates and ATP, resulted in a strong stimulation of oxygen consumption and in a manifold increase in the intracellular ATP level, which could reach that in the non-exhausted cells. This ATP increase was stimulated by Na^+, resistant to CCCP and sensitive to HQNO. Reverse ΔpNa (high $[Na^+]_{in}$ versus low $[Na^+]_{out}$) completely inhibited the oxidative ATP synthesis, without any decrease in the rate of lactate oxidation and coupled $\Delta\Psi$ generation. In further experiments, ATP synthesis, supported by an artificially imposed ΔpNa and ΔpK with proper directions, was demonstrated. To this end, NaCl was added to K^+-loaded cells incubated without Na^+. It was found that such an Na^+ pulse resulted in a fast [ATP] increase. In 1–2 min, the ATP concentration returned to the level close to that before NaCl addition. Subsequent addition of lactate caused a steady increase in ATP level.

Analysis of the effects of the Na^+ pulse and lactate revealed the following.

(1) The effect of the Na^+ pulse, but not of lactate, was abolished by monensin.
(2) The effect of lactate, but not of the Na^+ pulse, was arrested by HQNO.
(3) CCCP did not abolish either the Na^+-pulse- or the lactate-induced [ATP] increases.
(4) CCCP + monensin almost completely inhibited both the Na^+-pulse and lactate effects.
(5) A decrease of the added [NaCl] from 0.25 to 0.05 M strongly lowered the magnitude of the Na^+-supported [ATP] rise, but did not affect lactate-induced phosphorylation.
(6) Neither K^+ nor Li^+ could substitute for Na^+.
(7) Both the Na^+-pulse and lactate effects were inhibited by DCCD (note that DCCD also inhibits Na^+-ATP synthase of *P. modestum* [71]).

The observations listed in (1)–(6) indicate that the Na^+-pulse- and lactate-induced phosphorylations are mediated by Na^+-ATP synthase, whereas (7) suggests that this enzyme is DCCD-sensitive.

Recently Guffanti and Krulwich [168] and Tsuchiya and coworkers [168a] confirmed our finding of Na^+-pulse-supported ATP formation in *V. alginolyticus* and *V. parahaemolyticus* cells, respectively. Some indications of Na^+-driven ATP synthesis in the psychrophilic *Vibrio sp.* strain ABE-1 were published by Takada et al. [168b].

The next step in the study was to demonstrate that ATP hydrolysis is coupled to the Na^+ uptake by inside-out *V. alginolyticus* vesicles. This was accomplished in our group by Dibrov et al. [169]. The Na^+ transport was strongly stimulated by CCCP or valinomycin + K^+, and completely inhibited by DCCD and diethylstilbestrol. Moreover, it was found that the same vesicles were involved in ATP-supported H^+ uptake which was stimulated by valinomycin and inhibited by CCCP. DCCD was also inhibitory [170].

Also in our group, Dmitriev et al. succeeded in isolating ATP synthase from *V. alginolyticus* [171]. Its sensitivity to DCCD, triphenyltin and venturicidin, its subunit composition and the N-terminal sequences of the major F_1-subunits proved to be similar to those of the F_0F_1-type H^+-ATP synthase of *E. coli* and Na^+-ATP synthase of *P. modestum* [171,172]. Further indications of similarities of ATP synthases in these three species of bacteria were revealed when the complete sequence of the *V. alginolyticus* ATPase operon [173,174] and that of β- and c-units of the *P. modestum* Na^+-ATPase [153,154] were obtained.

How can similar F_0F_1 ATP synthases transport H^+ or Na^+? This question proved to be especially intriguing when it was reported [175] that in proteoliposomes one and the same *P. modestum* ATP synthase transports either Na^+ or H^+, depending on the $[Na^+]$ level in the incubation medium. It was the H^+ ion that was transported when $[Na^+]$ was < 0.2 mM. At

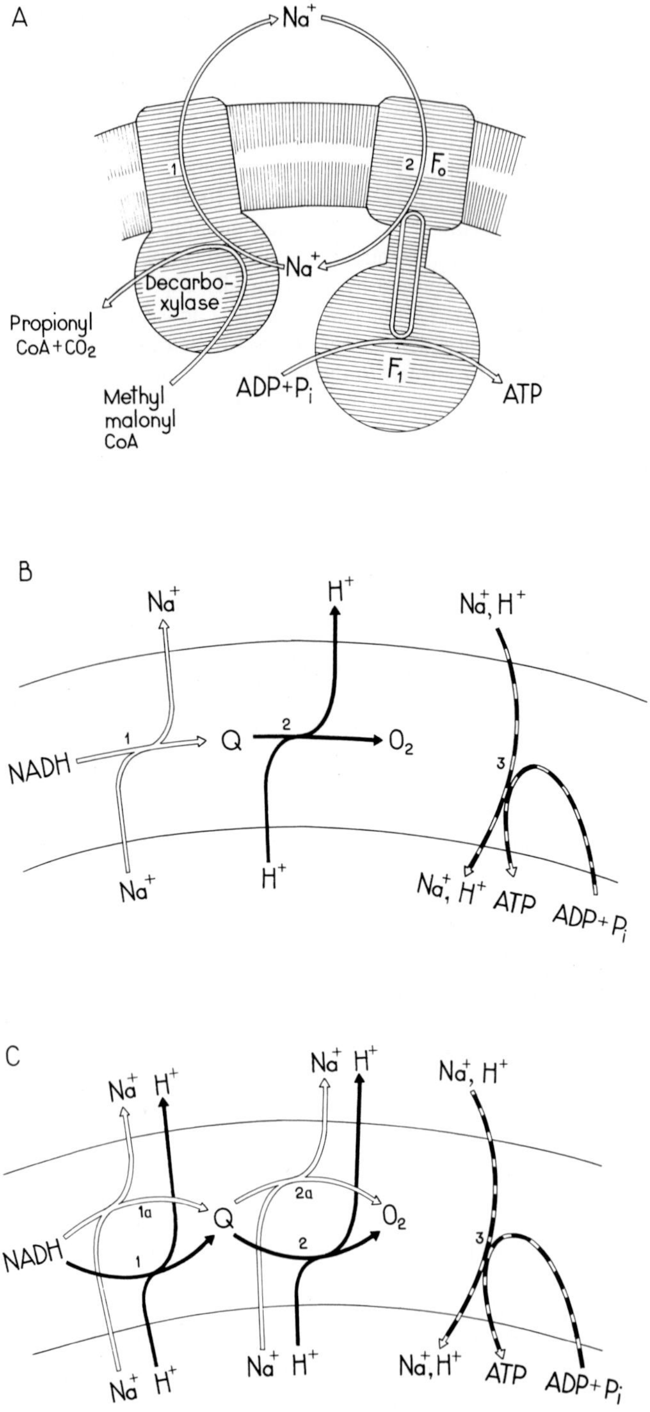
A
Na+
1
2
Fo
Na+
Decarbo-
xylase
Propionyl
CoA+CO2
Methyl
malonyl
CoA
ADP+Pi
F1
ATP
B
Na+
H+
Na+, H+
1
NADH
Q
2
O2
3
Na+
H+
Na+, H+
ATP
ADP+Pi
C
Na+ H+
Na+ H+
Na+, H+
1a
2a
Q
O2
NADH
1
2
3
Na+ H+
Na+ H+
Na+,H+
ATP
ADP+Pi

D

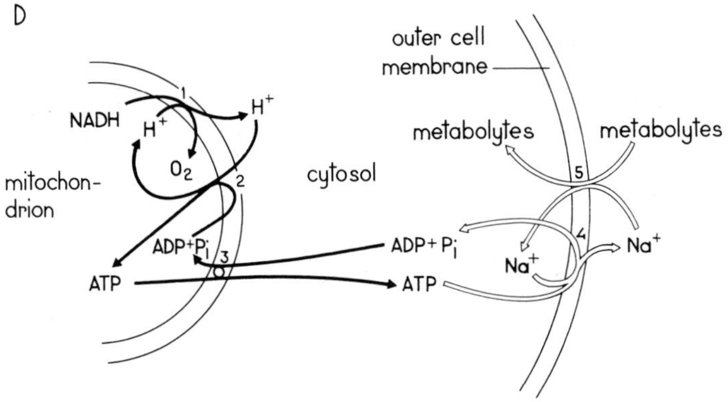

←

Fig. 8. Interrelation of Na^+- and H^+-linked energy transducers. Heavy and light arrows, systems specific for H^+ or for Na^+, respectively. Dotted arrows, systems equally effective with H^+ and Na^+. (A) The Na^+ cycle in *P. modestum*. Purely sodium energetics (no $\Delta\bar{\mu}_{H^+}$ generators and consumers are present). (1) Decarboxylase converts methylmalonyl CoA to propionyl CoA and CO_2 in an Na^+-motive fashion. (2) Downhill Na^+ influx via Na^+-ATP synthase is coupled to ATP formation. (From ref. [71].) (B) *V. alginolyticus* energetics. One and the same respiratory chain includes $\Delta\bar{\mu}_{Na^+}$ and $\Delta\bar{\mu}_{H^+}$ generators in its initial (1) and terminal (2) segments, respectively. One and the same ATP synthase (3) is postulated to consume $\Delta\bar{\mu}_{Na^+}$ or $\Delta\bar{\mu}_{H^+}$. (C) Energetics of *Bac. FTU* and *E. coli* growing at alkaline pH. In addition to the H^+-motive respiratory chain (1,2) there is the Na^+-motive one (1a,2a). Each chain includes at least two energy-coupling sites, one before quinone and another after quinone. (D) Energetics of the animal cell. The Na^+ and H^+ cycles are localized in two different membranes, i.e., the inner mitochondrial membrane and the plasma cell membrane. (1) In mitochondria, respiration generates $\Delta\bar{\mu}_{H^+}$, which is utilized by H^+-ATP synthase (2). ATP formed is exchanged for extramitochondrial ADP and P_i via the ATP/ADP-antiporter (3) and phosphate, H^+-symporter (to simplify the scheme, H^+ infux via the latter system is not shown). (4) ATP is hydrolyzed by the Na^+/K^+-ATPase in the plasma membrane and Na^+ is pumped out of the cell. (5) Downhill influx of Na^+ via Na^+, metabolite-symporters results in accumulation of these metabolites in the cytosol. (From ref. [7].)

higher [Na^+], no H^+ pumping occurred while Na^+ was pumped. Thus, under physiological conditions, the only reason for the *P. modestum* ATP synthase to operate as a Na^+ pump, rather than an H^+ pump, is higher [Na^+] than [H^+].

A similar situation was observed with relation to the animal Na^+/K^+-ATPase. As already mentioned, this enzyme, which is known to be very specific for Na^+, was shown to catalyze the H^+/K^+, instead of the Na^+/K^+, antiport at slightly acidic pH and low [Na^+] [176,177]. On the other hand, H^+/K^+-ATPase in the gastric mucosa plasma membrane exchanges Na^+ for K^+ at high [Na^+] and low [H^+] [178,179]. It is an H^+ flux that is shown to be responsible for ion conductance through the Na^+ channel in nerve fibres in the absence of Na^+ [180]. A few examples are described where an ion, metabolite-symporter recognizes H^+ and Na^+ (see refs. [175,181] and Section 3.3.1). Mutants were described in which substitution in a single amino acid residue resulted in a change of H^+ to Na^+ as the cotransport ion. H^+, melibiose-symporter, is an example of this kind [182].

It may be important for this consideration that in water solutions most of the H^+ ions combine with H_2O to form H_3O^+. As recently pointed out by Boyer [183], crown ethers of Na^+ and H_3O^+ are structurally very similar. Boyer proposed that H_3O^+ rather than H^+ interacts with ATPases, ATP synthases, porters and channels. Crown-like structures are apparently formed by peptide bond carbonyls of the α-helical columns which are so typical of membrane proteins.

At neutral pH, the concentration of H_3O^+ in sea water is 10^{-7} M, i.e., 5×10^6 times lower than that of Na^+ (5×10^{-1} M). This means that the gate of the proton channel in factor F_0 must be very specific to interact with H_3O^+ rather than with Na^+. Thus it seems probable that any change, even a small one, in the gate structure will result in a specificity decrease and, hence, in the transport of Na^+ instead of H^+ In agreement with the above reasoning, Junge and coworkers demonstrated [184] that the rearrangement or fragmentation of the chloroplast F_0 under certain conditions causes Na^+ and K^+ permeability which, like that of H^+, is venturicidin-sensitive. One may speculate that the F_0F_1-type Na^+-ATP synthase of *P. modestum* has appeared as a result of a mutation in the H^+-ATP synthase gene, which caused some decrease in its ion specificity.

At physiological $[H^+]$ and $[Na^+]$, *P. modestum* ATP synthase operates with Na^+, not H^+ [175]. As for the *V. alginolyticus* ATP synthase, one may assume that it transports either H^+ or Na^+ at neutral or alkaline pH, respectively. In this context, the following observations are warranted.

(1) The initial and terminal segments of the *V. alginolyticus* respiratory chain are H^+- and Na^+-motive, respectively [8].
(2) ATP hydrolysis by subcellular vesicles causes uphill transport of H^+ as well as Na^+ [169].
(3) Only one type of F_0F_1 ATPase can be revealed in the *V. alginolyticus* membrane [171] and only one ATPase gene is found in the genome of this bacterium [174].
(4) Both $\Delta\bar{\mu}_{H^+}$- and $\Delta\bar{\mu}_{Na^+}$-driven ATP synthase activities can be observed in *V. alginolyticus* cells [8].

Thus the *V. alginolyticus* oxidative phosphorylation system seems to be composed of (i) $\Delta\bar{\mu}_{Na^+}$- and $\Delta\bar{\mu}_{H^+}$-generators consecutively included into the respiratory chain and (ii) one and the same F_0F_1-ATP synthase which transports alternatively Na^+ or H^+ at low or high $[H^+]$, respectively (see below, Fig. 8B). Such mechanisms seem to be quite reasonable for *V. alginolyticus* cells living in mats of algae where strong pH fluctuations occur due to the photosynthetic activity. pH, which is neutral in the morning, shifts to high values in the evening. Such a shift may cause the energetics switch-over from the H^+ to the Na^+ cycle.

An alternative possibility was suggested by Tsuchiya's group [185]. The authors reported that protonophore-resistant oxidative phosphorylation and Na^+ pulse-supported ATP synthesis were found in the cells of *Vibrio parahaemolyticus*, a close relative of *V. alginolyticus*. A mutant, showing a strongly lowered level of α- and β-subunits of factor F_1, was selected. In it, the rate of ATP synthesis supported by artificially imposed $\Delta\bar{\mu}_{H^+}$ was much lower than in the wild type, whereas that supported by $\Delta\bar{\mu}_{Na^+}$ was of the same order of magnitude as in the wild type. These observations were interpreted as an indication that *V. parahaemolyticus* Na^+-ATP synthase is an enzyme other than F_0F_1. Such a conclusion seems to be very speculative since in both the wild and mutant cells, the rates of ATP synthesis driven by $\Delta\bar{\mu}_{Na^+}$ were very much lower than those driven by $\Delta\bar{\mu}_{H^+}$. In a recent publication, Tsuchiya and coworkers [168a], becoming more flexible, have mentioned that maybe their mutant 'possesses altered F_1F_0 and lost the ability to utilize H^+ as a coupling ion'.

Concluding this section, it should be stressed that under alkaline conditions, *V. alginolyticus* was shown to use respiratory $\Delta\bar{\mu}_{Na^+}$ to synthesize ATP. This fact completes the experimental verification of the sodium-cycle hypothesis proposing that Na^+ can be employed, instead of H^+, to couple energy-releasing processes to the performance of the three main types of work in the living cell: chemical, osmotic and mechanical [6]. In fact, *V. alginolyticus* proved to be the first species with a complete Na^+ cycle.

The list of $\Delta\bar{\mu}_{Na^+}$ generators and $\Delta\bar{\mu}_{Na^+}$ consumers is given in Table II.

4. Interrelations of the H^+- and Na^+-cycles

The lowering of $[H^+_{out}]$ seems to be an obvious reason for the cell to substitute Na^+ for H^+ as the coupling ion. An alkaline pH of the medium creates at least two difficulties for the use of the H^+ cycle. One is of kinetic nature. H^+ ions, extruded from the cell by, e.g, the H^+-motive respiratory chain, are immediately neutralized by the external alkaline solution, so that proton-acceptor devices of the $\Delta\bar{\mu}_{H^+}$ consumers must have a very high affinity for H^+ in order to be operative. This should create problems for the release of the H^+ ions, translocated by the consumer, to the cytoplasm where the pH is lower than outside.

An even more dramatic problem arises if we take into account the thermodynamics of the process. As long as pH_{in} is lower than pH_{out}, the formation of $\Delta\Psi$ (inside negative) by an H^+-extruding $\Delta\bar{\mu}_{H^+}$ generator is counterbalanced by a ΔpH in the opposite direction. Alkalinization of the cytoplasm is impossible because of inactivation of the intracellular enzymes. A further increase in $\Delta\Psi$ (to compensate the opposite ΔpH) is also impossible because a high $\Delta\Psi$ can cause an electric break-down of the cytoplasmic membrane. Reorientation of the $\Delta\bar{\mu}_{H^+}$ generators in the membrane can hardly solve the problems since in this case the direction of the transmembrane electric field would be opposite to the usual one (inside positive). But this makes it impossible to have a normal arrangement of all the membrane proteins that are

TABLE II

$\Delta\bar{\mu}_{Na^+}$ generators and consumers (from ref. [7])

Type of the system	Distribution
A. $\Delta\bar{\mu}_{Na^+}$-generators	
A1. Na^+-NADH-Q reductase	Many marine bacteria, *Bacillus FTU, E. coli* growing at alkaline pH
A2. Na^+-QH_2 oxidase	*Bacillus FTU, E. coli* growing at alkaline pH, *Vitreoscilla*
A3. Na^+-decarboxylases	Some anaerobic bacteria
A4. Methanogenesis-linked electron transfer	Methanobacteria
A5. Na^+-ATPase similar to vacuolar H^+-ATPase	*Streptococcus faecalis*
A6. Na^+-ATPase of methanogenic bacteria	Methanogenic bacteria
A7. Na^+/K^+-ATPase	Animal plasma membrane
B. $\Delta\bar{\mu}_{Na^+}$-consumers	
B1. Na^+-ATP synthase	Some marine bacteria
B2. Methanogenesis-linked electron transfer	Methanobacteria
B3. Na^+, solute-symporters, Na^+, solute-antiporters	Marine bacteria, animal plasma membrane
B4. Na^+-motor	Alkalophilic and alkalotolerant bacteria

electrophoretically oriented inside the membrane (negatively and positively charged groups are localized mainly on the outer and on the inner membrane sides, respectively [8]).

Substitution of Na^+ for H^+ solves all these problems. Na^+ extrusion forms a $\Delta\Psi$ in the right direction. The downhill Na^+ influx actuates $\Delta\bar{\mu}_{Na^+}$ consumers which perform various types of work in spite of the unusual direction of the pH gradient. This is why alkalophilic and alkalo-tolerant species require Na^+ to survive. The use of the sodium cycle seems especially convenient for marine bacteria since here $[Na^+]_{out}$ is always high.

In non-marine alkalophilic *Bacilli*, both osmotic and mechanical kinds of work were clearly shown to be equally $\Delta\bar{\mu}_{Na^+}$ dependent. On the other hand, the mechanisms of $\Delta\bar{\mu}_{Na^+}$ generation and ATP synthesis remain obscure. Attempts to identify a primary Na^+ pump in these microorganisms have not been successful thus far [186]. According to Krulwich and Guffanti [186], $\Delta\bar{\mu}_{Na^+}$ is formed by an electrogenic Na^+/nH^+ antiporter, which produces $\Delta\bar{\mu}_{Na^+}$ at the expense of the low $\Delta\bar{\mu}_{H^+}$ still created by the usual respiratory chain $\Delta\bar{\mu}_{H^+}$ generators found in these *Bacilli*. As to the respiratory ATP formation, the authors believe that it is catalyzed by a 'microchemiosmotic' mechanism when H^+ ions, pumped by the respiratory chain enzymes, are directly channeled to an H^+-ATP synthase molecule without being neutralized by the external alkaline medium. Such a direct transfer of protons between the respiratory and the ATP-forming enzymes, combining to a supercomplex, was postulated as early as in 1961 [187], but has never been proven experimentally, in spite of the many attempts by numerous laboratories (for reviews, see refs. [8,188]).

In this connection I would like to suggest an alternative possibility, namely, that in non-marine alkalophiles, oxidative phosphorylation is carried out by H^+-motive respiratory chain enzymes and H^+-ATP synthases localized in lenses, formed by the cytoplasmic membrane. Inside the lens, there is a 'third' water phase, separated from the alkaline outer medium by an H^+-impermeable membrane (Fig. 7). Such a scheme explains, without postulating any direct H^+ exchange between respiratory and ATP-synthase enzymes, the main paradoxes of soil alkalophilic bacilli, i.e., (i) the uncoupler-sensitive ATP formation supported by respiration at a very low bulk phase $\Delta\bar{\mu}_{H^+}$ level [186], (ii) the inefficiency of the same bulk $\Delta\bar{\mu}_{H^+}$ when it was artificially imposed across the membrane of non-respiring cells [186] and (iii) the similarity of the respiratory chain enzymes [186] or ATP-synthase enzymes [186,189–191] of these bacilli and the corresponding enzymes of neutrophilic bacteria.

Lenses may be specifically required for such a function as ATP synthesis which has a high $\Delta\bar{\mu}_{Na^+}$ threshold. In non-marine bacteri, $[Na^+_{out}]$ is apparently too low to maintain a high ΔpNa and, hence, a high $\Delta\bar{\mu}_{Na^+}$. As to the Na^+, solute-symporters and the Na^+ motor, they do not show such a high $\Delta\bar{\mu}_{Na^+}$ threshold. Therefore, they can be supported by a low $\Delta\bar{\mu}_{H^+}$ maintained by operation of H^+-motive respiratory enzymes localized in membrane regions other than lenses. Substrate-level phosphorylation is one more mechanism of ATP synthesis when $\Delta\bar{\mu}_{H^+}$ is too low to actuate H^+-ATP synthase. According to Tsuchija and coworkers [168a] this mechanism can effectively substitute for Na^+-ATP synthase. It was found that *V. parahaemolyticus*, growing aerobically at alkaline pH on lactate, responds to an NaCl pulse by an increase in the ATP level. This does not occur when the cells were grown on glucose under anaerobic conditions.

Thus, substitution of Na^+ for H^+ is not the only bioenergetic mechanism of adaptation to an alkaline medium. On the other hand, such an adaptation is not the only reason for the cell to use the Na^+ cycle. Rather, the Na^+ energetics looks like a way of surviving under certain conditions where $\Delta\bar{\mu}_{H^+}$ is low. Within the framework of this concept, a high pH is a particular case of an adaptive mechanism of more general importance. An indication that this is really the case was recently obtained by our group [192]. It was found that the Na^+-motive respiratory chain of *Bac. FTU* can be induced not only under alkaline conditions but also at neutral pH, provided that the growth medium was supplemented with a protonophore (CCCP) or a low concentration of cyanide, which specificially inhibits the H^+-motive respiration. Similarly, the Na^+ cycle could be induced, according to Avetisyan et al. [40], in *E. coli* by either alkaline conditions or by growth in the presence of an uncoupler.

Analogous relationships were revealed with respect to Na^+/K^+-ATPase of *S. faecalis* (see above, Section 3.2.6). This enzyme, according to Kinoshita et al. [193], is induced when glycolytic ATP fails to energize the membrane of this anaerobe due to mutation in H^+-ATPase. Moreover, the induction can be initiated by growth in the presence of the uncoupler CCCP [193,194] or at high pH [195]. As shown by Kakinuma and Igarashi [196], a mutant deficient in Na^+/K^+-ATPase was unable to grow under alkaline conditions.

Quite recently, Krulwich's group reported on an increase in cytochrome oxidase level in *Bac. firmus* growing at high pH or, alternatively, at neutral pH in the presence of a protonophore [197].

Adaptation to low-$\Delta\bar{\mu}_{H^+}$ conditions presumes that the bacterial cell can measure, directly or indirectly, its $\Delta\bar{\mu}_{H^+}$ level. A device monitoring $\Delta\bar{\mu}_{H^+}$ called 'protometer', was postulated some years ago in our laboratory by Glagolev and coworkers [198–201] to explain the paradoxical effect of the respiratory chain and H^+-ATPase inhibitors strongly increasing the attractant effect of long-wavelength light on halobacteria [199]. It was suggested that bacteriorhodopsin transduces the light energy to $\Delta\bar{\mu}_{H^+}$; this $\Delta\bar{\mu}_{H^+}$ is monitored by a protometer which sends an attractant signal to the flagellar motor if $\Delta\bar{\mu}_{H^+}$ increases. Such an effect should be less pronounced when alternative mechanisms of $\Delta\bar{\mu}_{H^+}$ generation, i.e., respiration and ATP hydrolysis, are operative. Quite recently this observation was confirmed and extended in our group in cooperation with Oesterhelt's laboratory [202,203].

Thus, the Na^+ cycle substitutes for the H^+ cycle in bacteria as *E. coli, Bac. FTU* (Fig. 8C) or *S. faecalis* under low $\Delta\bar{\mu}_{H^+}$ conditions, the effect being mediated, most probably, by a $\Delta\bar{\mu}_{H^+}$ receptor. In the same species growing under high $\Delta\bar{\mu}_{H^+}$ conditions, the primary Na^+ pumps are absent and Na^+ ions are extruded from the cell by cooperation of the H^+ pumps and the Na^+/H^+-antiporter. Apparently, the latter is repressed when $\Delta\bar{\mu}_{H^+}$ decreases [40].

In *V. alginolyticus*, the Na^+-motive NADH-quinone reductase was active even when the cells were growing at neutral pH. At the same time, the terminal oxidase was always H^+-motive (Fig. 8B). The *V. alginolyticus* is always well equipped for living either under neutral or alkaline conditions. This fact can easily be explained by adaptation to the niche where the pH is fluctuating between low and high values (see Section 3.4).

For *P. modestum* (Fig. 8A), always living at neutral pH, we should find another explanation why Na^+, and not H^+, is used as the coupling ion. The simplest explanation is that the Na^+-motive decarboxylases were discovered during evolution, whereas the H^+-motive ones were not. Such an explanation seems reasonable, especially if we assume, after Rosen [204], that evolutionarily the Na^+ cycle has appeared prior to the H^+ energetics, which arose as adaptation to fresh water when marine microorganisms came to colonize rivers and lakes as well.

It seems, however, more probable that the H^+ cycle was the first to appear. In the case of a direct (redox loop) chemiosmotic mechanism, an electrochemical potential difference of hydrogen ions seems to be the inevitable consequence of the chemistry of the energy-supplying oxidative reaction. The very simplicity of such a system may point to its evolutionary priority.

On the other hand, indirect chemiosmotic energy transducers, such as H^+-transhydrogenase and, most probably, H^+-ATP synthase look like the latest inventions of biological evolution. Here $\Delta\bar{\mu}_{H^+}$ formation is not a direct consequence of the energy-releasing process, and, hence, the energy coupling is organized in a more sophisticated manner. Systems of this type are not inevitably connected with the transport of a hydrogen ion. It is not surprising therefore that one and the same enzyme transports either H^+ or Na^+, as in the case in the *P. modestum* ATP synthase.

The $\Delta\bar{\mu}_{Na^+}$ generators seem to fall in a class of indirect energy transducers and thus may be regarded as the latest in the evolution of membrane bioenergetics.

It is remarkable that such an evolutionarily young membrane in the most progressive kingdom of living organisms as the animal plasma membrane employs Na^+ as the coupling ion. The only type of work performed by this membrane is osmotic, i.e., the uphill transport of metabolites. As to the H^+ cycle, it is localized in a quite different place, namely, in the inner mitochondrial membrane (Fig. 8D).

The problem of evolutionary relationships between the H^+ and Na^+ energetics is complicated by the absence of a consensus among specialists concerning the pH level and the ion composition of the early ocean. A common viewpoint is that it was was acidic and quite low in sodium (see, e.g., ref. [205]). Recently, however, the opposite ('soda ocean') concept, suggesting that both the pH and the $[Na^+]$ levels were high, has been advanced [206–208].

Assuming that 'protonists' appeared before 'sodists', one may visualize a possible way of the evolution of Na^+ bioenergetics. The use of Na^+ and K^+ gradients as $\Delta\bar{\mu}_{H^+}$ buffers could be the initial step. Direct utilization of $\Delta\bar{\mu}_{Na^+}$ formed by the Na^+/H^+-antiporter to support, e.g., osmotic work via Na^+, solute-symporters was the next step toward the sodium cycle. The cycle was completed when biological evolution invented primary $\Delta\bar{\mu}_{Na^+}$ generators operating with no $\Delta\bar{\mu}_{H^+}$ involved. Most of these events apparently took place at a rather early stage in the evolution, so that now the sodium cycle can be found in quite different taxa: *Vibrio, Bacillus, Escherichia, Salmonella, Propionigenium, Alcaligenes, Alteromonas, Flavobacterium, Klebsiella, Veilonella, Acidaminococcus, Clostridium, Fusobacterium, Streptococcus, Peptostreptococcus, Vitreoscilla,* methano- and acetogenic bacteria, cyanobacteria, *Pseudomonas, Mycoplasma, Acholeplasma,* and, finally, in animal cells [8,25,41,51,64,71,181,209–212].

Such a great taxonomic variety of organisms employing the sodium cycle points to an ubiquitous distribution of this novel type of membrane-linked energy transductions.

The actual extent of the sodium world is yet to be determined. However, already today we may assume that it occupies a fairly large region in the biosphere.

It should be stressed that the search for sodium world representatives is complicated by the fact that Na^+ and H^+ energetics can coexist in one and the same species, switching over from H^+ to Na^+ and vice versa, depending on the actual conditions.

5. *Conclusion. Three basic principles of cell energetics*

The above consideration allows the formulation of three basic principles of cell energetics:

First Principle. *The living cell avoids direct utilization of external sources in the performance of useful work. It transforms energy of these sources to a convertible biological energy currency, i.e., ATP,* $\Delta\bar{\mu}_{H^+}$ or $\Delta\bar{\mu}_{Na^+}$, *which is then spent to support various types of energy-consuming processes.*

Second Principle. *Any living cell possesses at least two energy currencies, one water- soluble (ATP) and the other membrane-linked (*$\Delta\bar{\mu}_{H^+}$ *and/or* $\Delta\bar{\mu}_{Na^+}$*).*

Third Principle. *All the energy requirements of the living cell can be satisfied if at least one of three convertible energy currencies is produced at the expense of external energy sources.*

The interrelations of three convertible energy currencies are shown in Fig. 9. In spite of the universal applicability of this scheme, some specific bioenergetic features seem to be characteristic of bacteria, animals, plants or fungi. For instance, in the majority of bacteria, both $\Delta\bar{\mu}_{H^+}$ and $\Delta\bar{\mu}_{Na^+}$ are maintained across one and the same (cytoplasmic) membrane. In animal cells, $\Delta\bar{\mu}_{H^+}$ and $\Delta\bar{\mu}_{Na^+}$ are inherent in the outer cell membrane and intracellular membranes (mitochondria, lysosomes, endosomes, secretory granules), respectively. In plants and fungi, $\Delta\bar{\mu}_{H^+}$ is used, as a rule, in both outer cell and intracellular membranes to perform membrane-linked work, whereas Na^+ ions are pumped from the cytosol to the extracellular space or to the vacuole in a $\Delta\bar{\mu}_{H^+}$-dependent fashion, i.e., by means of Na^+/H^+-antiporters.

ATP is equilibrated with $\Delta\bar{\mu}_{H^+}$ and $\Delta\bar{\mu}_{Na^+}$ due to the reversibility of the corresponding ATP synthases and ATPases. As for the $\Delta\bar{\mu}_{H^+} \leftrightarrow \Delta\bar{\mu}_{Na^+}$ interconversion that is catalyzed by the Na^+/H^+-antiporter, it should take place only when $\Delta\bar{\mu}_{H^+}$ is high, otherwise energy will be dissipated.

The most compelling evidence for the validity of the Third Principle originates from the studies of biological niches. There are several examples where a single exergonic reaction can supply all the necessary energy for the cells occupying a niche. We already mentioned *P. modestum* employing the Na^+-motive decarboxylation of methylmalonyl-CoA as the only energy source (Section 3.2.3). Fumarate reducing anaerobic bacteria and mitochondria of *Ascaris lumbricoides* exemplify systems that use the initial step of the respiratory chain as the only $\Delta\bar{\mu}_{H^+}$ generator. In *Rhizobium japonicum*, which oxidizes succinate by nitrate, $\Delta\bar{\mu}_{H^+}$ generation is localized at the level of the middle region of the respiratory chain. In *Thiobacillus*

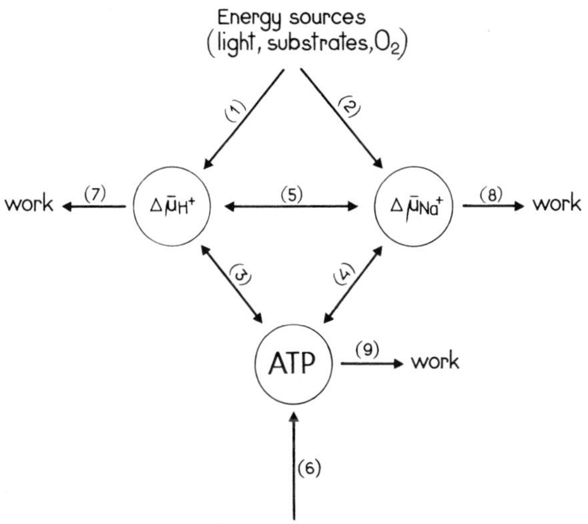

Fig. 9. Three convertible energy currencies of living systems. (1) Utilization of energy sources (light or respiratory substrates + O_2) by photosynthetic or respiratory $\Delta\bar{\mu}_{H^+}$ generators (H^+-motive photosynthetic or respiratory chain enzymes and bacteriorhodopsin). (2) Utilization of respiratory substrates + O_2 and decarboxylase substrates by $\Delta\bar{\mu}_{Na^+}$ generators (Na^+-motive respiratory chain enzymes and decarboxylases). (3) The $\Delta\bar{\mu}_{H^+} \leftrightarrow$ ATP interconversion catalyzed by H^+-ATP synthase (H^+-ATPase). (4) The $\Delta\bar{\mu}_{Na^+} \leftrightarrow$ ATP interconversion catalyzed by Na^+-ATP synthase (Na^+-ATPase. (5) $\Delta\bar{\mu}_{H^+} \leftrightarrow \Delta\bar{\mu}_{H^+}$ interconversion catalyzed by the Na^+/H^+-antiporter. (6) ATP formation by substrate-level phosphorylation enzymes (glycolysis, oxidative decarboxylations). (7), (8) and (9) Chemical, osmotic or mechanical work supported by $\Delta\bar{\mu}_{H^+}$, $\Delta\bar{\mu}_{Na^+}$ and ATP, respectively.

ferrooxidans, which oxidizes Fe^{2+} ions using O_2, the H^+-motive terminal oxidase of aa_3 type is the only mechanism of energy production (for review, see ref. [8]).

There are also examples where energy is supplied by a substrate-level phosphorylation and primary $\Delta\bar{\mu}_{H^+}$ and $\Delta\bar{\mu}_{Na^+}$ generators are absent. However, in these cases H^+-ATPase or Na^+-ATPase are always present to form $\Delta\bar{\mu}_{H^+}$ or $\Delta\bar{\mu}_{Na^+}$ at the expense of, e.g., glycolytic ATP (see above, Section 3.2.6). This fact illustrates well the validity of the Second Principle.

As to the First Principle, for some time the very fact that certain bacteria can grow on glycolytic substrates in the presence of protonophorous uncouplers has been regarded as evidence that only one energy currency, i.e., ATP, can be sufficient for life. However, more careful investigations showed that, under the conditions used, $\Delta\bar{\mu}_{Na^+}$ was formed or some $\Delta\bar{\mu}_{H^+}$ was still present [8]. The latter is also true for the soil alkalophilic bacilli, which seem to use lenses or another microchemiosmotic mechanism to form ATP at the expense of respiration. As it is repeatedly stressed by Krulwich and her colleagues, there is still some $\Delta\bar{\mu}_{H^+}$ even at the most alkaline pH at which the bacilli are still growing [186].

Perhaps tomorrow an exception to the general principles of the bioenergetics formulated above will be revealed. But today I fail to find in the available literature a single well-documented and independently confirmed observation which is at variance with the above three principles. I therefore think that it is time to include these principles in textbooks of molecular and cell biology.

References

1 Lipmann, F. (1941) Adv. Enzymol. 1, 99–107.
2 Mitchell, P. (1961) Nature 191, 144–148.
3 Mitchell, P. (1966) Biol. Rev. 41, 445–502.
4 Mitchell, P. (1968) Chemiosmotic Coupling and Energy Transduction, Glynn Research, Bodmin.
5 Skulachev, V.P. (1977) FEBS Lett. 74, 1–9.
6 Skulachev, V.P. (1984) TIBS 9, 483–485.
7 Skulachev, V.P. (1992) Biosci. Rep., accepted.
8 Skulachev, V.P. (1988) Membrane Bioenergetics, Springer, Berlin.
9 Skulachev, V.P. (1980) Can. J. Biochem. 58, 161–175.
10 Hoshi, T. and Himukai, M. (1982) in: R. Sato and Y. Kagawa (Eds.), Transport and Bioenergetics in Biomembranes, Japan Sci. Soc. Press, Tokyo, pp. 111–135.
11 West, I.C. (1983) The Biochemistry of Membrane Transport, Chapman and Hall, London.
12 Skulachev, V.P. (1978) FEBS Lett. 87, 171–179.
13 Wagner, G., Hartmann, R. and Oesterhelt, D. (1978) Eur. J. Biochem. 89, 169–179.
14 Arshavsky, V. Yu., Baryshev, V.A., Brown, I.I., Glagolev, A.N. and Skulachev, V.P. (1981) FEBS Lett. 133, 22–26.
15 Brown, I.I., Galperin, M. Yu., Glagolev, A.N. and Skulachev, V.P. (1983) Eur. J. Biochem. 134, 345–349.
16 Michels, M. and Bakker, E.P. (1985) J. Bacteriol.161, 231–237.
17 Drapeau, G.R. and MacLeod, R.A. (1963) Biochem. Biophys. Res. Commun. 12, 111–115.
18 Dimroth, P. (1980) FEBS Lett. 122, 234–236.
19 Tokuda, H. and Unemoto, T. (1981) Biochem. Biophys. Res. Commun. 102, 265–271.
20 Tokuda, H. and Unemoto, T. (1982) J. Biol. Chem. 257, 10007–10014.
21 Nakamura, T., Tokuda, H. and Unemoto, T. (1982) Biochim. Biophys. Acta 692, 389–396.
22 Tokuda, H., Sugasawa, M. and Unemoto, T. (1982) J. Biol. Chem. 257, 788–794.
23 Chernyak, B.V., Dibrov, P.A., Glagolev, A.N., Sherman, M. Yu and Skulachev, V.P. (1983) FEBS Lett. 164, 38–42.

24 Dibrov, P.A., Lazarova, R.L., Skulachev, V.P. and Verkhovskaya, M.L. (1986) Biochim. Biophys. Acta 850, 458–465.
25 Skulachev, V.P. (1985) Eur. J. Biochem. 155, 199–208.
26 Tokuda, H. (1984) FEBS Lett. 176, 1125–1128.
27 Asano, M., Hayashi, M., Unemoto, T. and Tokuda, H. (1985) Agric. Biol. Chem. 49, 2813–2817.
28 Tokuda, H., Udagawa, T. and Unemoto, T. (1985) FEBS Lett. 183, 95–98.
29 Yagi, T. (1991) J. Bioenerg. Biomembranes 23, 211–225.
30 Hayashi, M. and Unemoto, T. (1986) FEBS Lett. 202, 327–330.
31 Hayashi, M. and Unemoto, T. (1987) Biochim. Biophys. Acta 890, 47–54.
32 Unemoto, T. and Hayashi, M. (1989) J. Bioenerg. Biomembranes 21, 649–662.
33 Udagawa, T., Unemoto, T. and Tokuda, H. (1986) J. Biol. Chem. 261, 2616–2622.
34 Tsuchiya, T. and Shinoda, S. (1985) J. Bacteriol. 162, 794–798.
35 Tokuda, H. and Kogure, K. (1989) J. Gen. Microbiol. 135, 703–709.
36 Takada, Y., Fukunaga, N. and Sasaki, S. (1988) Plant Cell Physiol. 29, 207–214.
37 Ken-Dror, S., Lanyi, J.K., Schöbert, B., Silver, B. and Avi-Dor, Y. (1986) Arch. Biochem. Biophys. 244, 766–772.
38 Dimroth, P. and Thomer, A. (1989) Arch. Microbiol. 151, 439–444.
39 Kostyrko, V.A., Semeykina, A.L., Skulachev, V.P., Smirnova, I.A. Vaghina, M.L. and Verkhovskaya, M.L. (1991) Eur. J. Biochem. 198, 527–534.
40 Avetisyan, A.V., Dibrov, P.A., Semeykina, A.L., Skulachev, V.P. and Sokolov, M.V. (1991) Biochim. Biophys. Acta 1098, 95–104.
41 Tokuda, H. (1989) J. Bioenerg. Biomembranes 21, 693–704.
42 Kogure, K. and Tokuda, H. (1989) FEBS Lett. 256, 147–149.
43 Dibrov, P.A., Kostyrko, V.A., Lazarova, R.L., Skulachev, V.P. and Smirnova, I.A. (1986) Biochim. Biophys. Acta 850, 449–457.
44 Semeykina, A.L. and Skulachev, V.P. (1990) FEBS Lett. 269, 69–72.
45 Verkhovskaya, M.L., Semeykina, A.L. and Skulachev, V.P. (1988) Dokl. Acad. Nauk SSSR 303, 1501–1503.
46 Semeykina, A.L., Skulachev, V.P., Verkhovskaya, M.L., Bulygina, E.S. and Chumakov, K.M. (1989) Eur. J. Biochem. 183, 671–678.
47 Avetisyan, A.V., Dibrov, P.A., Skulachev, V.P. and Sokolov, M.V. (1989) FEBS Lett. 254, 17–21.
48 Avetisyan, A.V., Bogachev, A.V., Murtasina, R.A. and Skulachev, V.P. (1992) FEBS Lett., in press.
49 Green, G.N. and Gennis, R.B. (1983) J. Bacteriol. 154, 1269–1275.
50 Efiok, B.J.S. and Webster, D.A. (1990) Biochem. Biophys. Res. Commun. 173, 370–375.
51 Efiok, B.J.S. and Webster, D.A. (1990) Biochemistry 29, 4734–4739.
51a Georgiou, C.D. and Webster, D.A. (1987) Biochemistry 26, 6521–6526.
52 Dimroth, P. (1981) Eur. J. Biochem. 115, 353–358.
53 Dimroth, P. (1982) Eur. J. Biochem. 121, 435–441.
54 Dimroth, P. and Thomer, A. (1988) Eur. J. Biochem. 175, 175–180.
55 Dimroth, P. and Thomer, A. (1983) Eur. J. Biochem. 137, 107–112.
56 Däkena, P., Rohde, M., Dimroth, P. and Mayer, F. (1988) FEMS Microbiol. Lett. 55, 35–40.
57 Dimroth, P. (1990) in: The Molecular Basis of Bacterial Metabolism, Springer, Berlin.
58 Dimroth, P. (1990) Res. Microbiol. 141, 332–336.
59 Schwarz, E., Oesterhelt, D., Reinke, H., Beyrenther, K. and Dimroth, P. (1988) J. Biol. Chem. 263, 9640–9645.
60 Laußermair, E., Schwarz, E., Oesterhelt, D., Reinke, H., Beurenther, K. and Dimroth, P. (1989) J. Biol. Chem. 264, 14710–14715.
61 Stephens, P.E., Darlison, M.G., Lewis, H.M. and Guest, J.R. (1983) Eur. J. Biochem. 133, 481–489.
62 Radford, S.E., Lane, E.D., Perham, R.N., Miles, J.S. and Guest, J.R. (1987) Biochem. J. 247, 641–649.
63 Wifling, K. and Dimroth, P. (1989) Arch. Microbiol. 152, 584–588.
64 Buckel, W. (1986) Methods Enzymol. 125, 547–558.
65 Buckel, W. and Semmler, R. (1982) FEBS Lett. 148, 35–38.

66 Buckel, W. and Semmler, R. (1983) Eur. J. Biochem. 136, 427–434.
67 Wohlfarth, G. and Buckel, W. (1985) Arch. Microbiol. 142, 128–135.
68 Buckel, W. and Ziedtke, H. (1986) Eur. J. Biochem. 156, 251–257.
69 Hilpert, W. and Dimroth, P. (1983) Eur. J. Biochem. 132, 579–587.
70 Hilpert, W. and Dimroth, P. (1984) Eur. J. Biochem. 138, 579–583.
71 Dimroth, P. (1987) Microbiol. Rev. 51, 320–340.
72 Hilpert, W., Schink, B. and Dimroth, P. (1984) EMBO J. 3, 1665–1680.
73 Rohde, M., Dakena, P., Mayer, F. and Dimroth, P. (1986) FEBS Lett. 195, 280–284.
74 Hoffmann, A., Hilpert, W. and Dimroth, P. (1989) Eur. J. Biochem. 179, 645–650.
75 Hilpert, W. and Dimroth, P. (1991) Eur. J. Biochem. 195, 79–86.
76 Kaesler, B. and Schönheit, P. (1989) Eur. J. Biochem. 184, 309–316.
77 Blaut, M., Müller, V., Fiebig, K. and Gottschalk, G. (1985) J. Bacteriol. 164, 95–101.
78 Müller, V., Blaut, M. and Gottschalk, G. (1988) Eur. J. Biochem. 172, 601–606.
79 Kaesler, B. and Schönheit, P. (1989) Eur. J. Biochem. 186, 309–316.
80 Müller, V., Winner, C. and Gottschalk, G. (1988) Eur. J. Biochem. 178, 519–525.
81 Gottschalk, G. and Blaut, M. (1990) Biochim. Biophys. Acta 1018, 263–266.
82 Heise, R., Müller, V. and Gottschalk, G. (1989) J. Bacteriol. 171, 5473–5478.
82a Heise, R., Reidlinger, J., Müller, V. and Gottschalk, G. (1991) FEBS Lett. 295, 119–122.
83 Skou, J.C. (1957) Biochim. Biophys. Acta 23, 394–401.
84 Schwartz, A., Collins, J.H. (1982) in: A.N. Martonosi (Ed.) Membranes and Transport, Vol. 1, Plenum Press, New York, pp. 521–527.
85 Matsui, H. (1982) in: R. Sato and Y. Kagawa (Eds.) Transport and Bioenergetics in Biomembranes, Japan Sci. Soc. Press, Tokyo, pp. 165–187.
86 Ovchinnikov, Yu. A., Demin, V.V., Barnakov, A.N., Kuzin, A.P., Lunev, A.V., Modyanov, N.N. and Dzhandzhugazyan, K.N. (1985) FEBS Lett. 190, 73–76.
87 Ovchinnikov, Yu. A., Modyanov, N.N., Broude, N.E., Petrukhin, K.E., Grishin, A.V., Arzamazova, N.M., Aldanova, N.A., Monastyrskaya, G.F. and Sverdlov, E.D. (1986) FEBS Lett. 201, 237–245.
87a Nakao, M. (1991) in: Y. Mukohata (Ed.) New Era of Bioenergetics, Academic Press, New York, pp. 1–46.
88 Hara, Y., Yamada, J., Nakao, M. (1986) J. Biochem. 99, 531–539.
89 Pagliusi, S., Antonicek, H., Gloor, S., Frank, R., Moos, M. and Schachner, M. (1989) J. Neurosci. Res. 22, 113–119.
90 Gloor, S., Antonicek, H., Sweadner, K.J., Pagliusi, S., Frank, R., Moos, M. and Schachner, M. (1990) J. Cell Biol. 110, 165–174.
91 Akopyanz, N.S., Broude, N.E., Bekman, E.P., Marzen, E.O. and Sverdlov, E.D. (1991) FEBS Lett. 289, 8–10.
92 Gelfand, I.M. and Vasiliev, Yu. M. (1981) Neoplastic and Normal Cell in Culture, Cambridge Univ. Press, Cambridge.
93 Del Castillo, J.R. and Robinson, J.W.L. (1985) Biochim. Biophys. Acta. 812, 413–422.
94 Proverbio, F., Gondrescu-Guidi, D. and Whittembury, G. (1975) Biochim. Biophys. Acta 394, 281–292.
95 Proverbio, F. and Del Carillon, J.R. (1981) Biochim. Biophys. Acta 646, 99–108.
96 Marin, B., Proverbio, T. and Proverbio, F. (1985) Biochim. Biophys. Acta 817, 299–306.
97 El Mernissi, G., Barlet-Bas, C., Khadouri, C., Marsy, S., Cheval, L. and Doucet, A. (1991) Biochim. Biophys. Acta 1064, 205–211.
98 Heefner, D.L. and Harold, F.M. (1982) Proc. Natl. Acad. Sci. USA 79, 2798–2802.
99 Kakinuma, Y. and Igarashi, K. (1989) J. Bioenerg. Biomembranes 21, 679–692.
100 Kakinuma, Y. and Igarashi, K. (1990) FEBS Lett. 271, 97–101.
100a Kakinuma, Y., Igarashi, K., Konishi, K. and Yamato, I. (1991) FEBS Lett. 292, 64–68.
101 McLaggan, D., Selwyn, M.D. and Dawson, A.P. (1984) FEBS Lett. 165, 254–258.
102 Benyoucef, M., Rigaud, J.-L. and Leblanc, G. (1982) Biochem. J. 208, 529–538.
103 Benyoucef, M., Rigaud, J.-L. and Leblanc, G. (1982) Biochem. J. 208, 539–547.
104 Jinks, D.C., Silvins, J.S. and McElhaney, R.N. (1978) J. Bacteriol. 136, 1027–1036.

105 Lewis, R.N.A.H. and McElhaney, R.N. (1983) Biochim. Biophys. Acta 735, 113–122.
106 Chen, J.-W., Sun, Q. and Hwang, F. (1984) Biochim. Biophys. Acta 777, 151–154.
107 George, R. and McElhaney, R.N. (1985) Biochim. Biophys. Acta 813, 161–166.
108 Mahajan, S., Lewis, R.N.A.H., George, R., Sykes, B.D. and McElhaney, R.N. (1988) J. Bacteriol. 170, 5739–5746.
109 Shirvan, M.H., Schuldiner, S. and Rottem, S. (1989) J. Bacteriol. 171, 4410–4416.
110 Shirvan, M.H., Schuldiner, S. and Rottem, S. (1989) J. Bacteriol. 171, 4417–4424.
111 Koyama, N., Kiyomiya, A. and Nosoh, Y. (1976) FEBS Lett. 72, 77–78.
112 Kitada, M. and Horikoshi, K. (1977) J. Bacteriol. 131, 784–788.
113 Guffanti, A.A., Susman, P., Blanco, R. and Krulwich, T.A. (1978) J. Biol. Chem. 253, 708–715.
114 Krulwich, T.A., Guffanti, A.A., Bornstein, R.F. and Hoffstein, J. (1982) J. Biol. Chem. 257, 1885–1889.
115 Lee, S.-H., Cohen, N.S., Jacobs, A.J. and Bordie, A.F. (1979) Biochemistry 18, 2232–2238.
116 Cairney, J., Higgins, C.F. and Booth, I.R. (1984) J. Bacteriol. 160, 22–27.
117 Stewart, L.M.D. and Booth, I.R. (1983) FEMS Microbiol. Lett. 19, 161–164.
118 Chen, C.-C., Tsuchiya, T., Yamane, Y., Wood, J.M. and Wilson, T.H. (1985) J. Membr. Biol. 84, 157–164.
119 Hedigen, M.A., Turk, E. and Wright, E.M. (1989) Proc. Natl. Acad. Sci. USA 86, 5748–5752.
120 Bassilana, M., Diamino-Forano, E. and Leblanc, G. (1985) Biochem. Biophys. Res. Commun. 129, 626–631.
121 Tsuchiya, T. and Wilson, T.H. (1978) Membr. Biochem. 2, 63–79.
121a Okabe, Y., Sakai-Tomita, Y., Mitani, Y, Tsuda, M. and Tsuchiya, T. (1991) Biochim. Biophys. Acta 1059, 332–338.
122 Dimroth, P. and Thomer, A. (1986) Hoppe-Seyler's Z. Physiol. Chem. 367, 813–823.
123 De Vrij, W., Bulthuis, R.A., van Iwaarden, P.R. and Konings, W.N. (1989) J. Bacteriol. 171, 1118–1125.
124 Michael, T.A. and Macy, J.M. (1990) J. Bacteriol. 172, 1430–1435.
125 Anantharam, V., Allison, M.J. and Maloney, P.C. (1989) J. Biol. Chem. 264, 7244–7250.
126 Greece, R.V. and MacDonald, R.E. (1984) Arch. Biochem. Biophys. 229, 576–584.
127 Koepsell, H., Korn, K., Ferguson, D., Menuhr, H., Ollig, D. and Haase, W. (1984) J. Biol. Chem. 259, 6548–6558.
128 Slonczewski, L.L., Rosen, B.P., Algeri, J.R. and Macnab, R.M. (1981) Proc. Natl. Acad. Sci. USA 78, 6271–6275.
129 Castle, A.M., Macnab, R.M. and Shulman, R.G. (1986) J. Biol. Chem. 261, 7797–7806.
130 Krulwich, T.A., Federbush, J.G. and Guffanti, A.A. (1985) J. Biol. Chem. 260, 4055–4058.
131 Guffanti, A.A. (1983) FEMS Lett. 17, 307–310.
132 Nicholls, D.G. and Locke, R.M. (1984) Physiol. Rev. 64, 1–64.
133 Blumwald, E. and Poole, R. (1985) Plant Physiol. 78, 163–167.
134 Pole, P.J. (1978) Annu. Rev. Plant Physiol. 29, 437–460.
135 Bakeeva, L.A., Drachev, A.L., Metlina, A.L., Skulachev, V.P., Chumakov, K.M. (1987) Biokhimiya 52, 8–14.
136 Bakeeva, L.E., Chumakov, K.M., Drachev, A.L., Metlina, A.L. and Skulachev, V.P. (1986) Biochim. Biophys. Acta 850, 466–472.
137 Tokuda, H., Asano, M., Shimmer, Y., Unemoto, T., Sugiyama, S. and Imae, Y. (1988) J. Biochem. 103, 650–655.
138 Liu, J.Z., Dapice, M. and Khan, S. (1990) J. Bacteriol. 172, 5236–5244.
139 Yoshida, S., Sugiyama, S., Hojo, Y., Tokuda, H. and Imae, Y. (1990) J. Biol. Chem. 265, 20346–20350.
140 Ferris, F.G., Beveridze, T.J., Marcean-Day, M.L. and Larson, A.D. (1984) Can. J. Microbiol. 30, 322–339.
141 Hirota, N., Kitada, M. and Imae, Y. (1981) FEBS Lett. 132, 278–280.
142 Hirota, N. and Imae, Y. (1983) J. Biol. Chem. 258, 10577–10581.

143 Sugiyama, S., Matsukura, H., Koyama, N., Nosoh, Y. and Imae, Y. (1986) Biochim. Biophys. Acta 852, 38–46.
144 Kitada, M., Guffanti, A.A. and Krulwich, T.A. (1982) J. Bacteriol. 152, 1096–1104.
145 Sugiyama, S., Matsukura, H. and Imae, Y. (1985) FEBS Lett. 182, 265–268.
146 Imae, Y., Matsukura, H. and Kobayashi, S. (1986) Methods Enzymol. 125, 582–592.
147 Imae, Y. and Atsumi, T. (1989) J. Bioenerg. Biomembranes 21, 705–716.
148 Kodama, T. and Taniguchi, S. (1977) J. Gen. Microbiol. 98, 503–510.
149 Glynn, I.M. and Garrahan, P.J. (1966) Nature 211, 1414–1415.
150 Schink, B. and Pfennig, N. (1982) Arch. Microbiol. 133, 209–216.
151 Laubinger, W. and Dimroth, P. (1987) Eur. J. Biochem. 168, 475–480.
152 Laubinger, N. and Dimroth, P. (1988) Biochemistry 27, 7531–7537.
153 Amann, R., Ludwig, W., Laubinger, W., Dimroth, P. and Schleifer K.H. (1988) FEMS Microbiol. Lett. 56, 253–260.
154 Ludwig, W., Kaim, G., Laubinger, W., Dimroth, P., Hoppe, J. and Schleifer, K.H. (1990) Eur. J. Biochem. 193, 395–399.
155 Laubinger, W., Deckers-Heberstreit, G., Altendorf, K. and Dimroth, P. (1990) Biochemistry 29, 5458–5463.
156 Schönheit, P. and Perski, H.-J. (1983) FEMS Lett. 20, 263–267.
157 Schönheit, P. and Beimborn, D.B. (1985) Arch. Microbiol. 142, 354–361.
158 Carper, S.W. and Lancaster, J.R. (1986) FEBS Lett. 200, 177–180.
159 Crider, B.P., Carper, S.W. and Lancaster, J.R. (1985) Proc. Natl. Acad. Sci. USA 82, 6793–6796.
160 Al-Mahrouq, H.A., Carper, S.W. and Lancaster, J.R. (1986) FEBS Lett. 207, 262–265.
161 Lancaster, J.R. (1989) J. Bioenerg. Biomembranes 21, 717–740.
162 Smigan, P., Horovska, L. and Greksak, M. (1988) FEBS Lett. 242, 85–88.
163 Ekiel, I., Jarrell, K.F. and Sprott, G.D. (1985) Eur. J. Biochem. 149, 437–444.
164 Jarrell, K.F. and Sprott, G.D. (1985) Can. J. Microbiol. 31, 851–855.
165 Guffanti, A.A., Borstein, R.F. and Krulwich, T.A. (1981) Biochim. Biophys. Acta 635, 619–630.
166 Verkhovskaya, M.L., Dibrov, P.A., Lazarova, R.L. and Skulachev, V.P. (1987) Biokhimiya 52, 15–23.
167 Dibrov, P.A., Lazarova, R.L., Skulachev, V.P. and Verkhovskaya, M.L. (1989) J. Bioenerg. Biomembranes 21, 347–357.
168 Guffanti, A.A. and Krulwich, T.A. (1988) J. Biol. Chem. 263, 14748–14752.
168a Sakai-Tomita, Y., Tsuda, M. and Tsuchiya, T. (1991) Biochem. Biophys. Res. Commun. 179, 224–228.
168b Takada, Y., Fukunaga, N. and Sasaki, S. (1991) FEMS Microbiol. Lett. 82, 225–228.
169 Dibrov, P.A., Skulachev, V.P., Sokolov, M.V. and Verkhovskaya, M.L. (1988) FEBS Lett. 233, 355–358.
170 Smirnova, I.A., Vulfson, E.N. and Kostyrko, V.A. (1987) FEBS Lett. 214, 343–346.
171 Dmitriev, O. Yu., Grinkevich, V.A. and Skulachev, V.P. (1989) FEBS Lett. 258, 219–222.
172 Capozza, G., Dmitriev, O. Yu., Krasnoselskaya, I.A., Papa, S. and Skulachev, V.P. (1991) FEBS Lett. 280, 274–276.
173 Krumholz, L.R., Esser, U. and Simoni, R.D. (1989) Nucleic Acids. Res. 17, 7993–7994.
174 Krumholz, L.R., Esser, U. and Simoni, R.D. (1990) J. Bacteriol. 172, 6809–6817.
175 Laubinger, W. and Dimroth, P. (1989) Biochemistry 28, 7194–7198.
176 Hara, Y., Yamada, J. and Nakao, M. (1986) J. Biochem. 99, 531–539.
177 Polvani, C. and Blostein, R. (1988) J. Biol. Chem. 263, 16757–16763.
178 Polvani, C., Sachs, G. and Blostein, R. (1989) Biophys. J. 55, 337A.
179 Rabon, E.C., Bassilian, S., Sachs, G. and Karlish, S.J.D. (1990) J. Biol. Chem. 265, 19594–19599.
180 Mozhayeva, G.N. and Naumov, A.P. (1983) Pflügers Arch. 396, 163–173.
181 Skulachev, V.P. (1989) FEBS Lett. 250, 106–114.
182 Kawakami, T., Akizawa, Y., Ishikawa, T., Shimamoto, T., Tsuda, M. and Tsuchiya, T. (1988) J. Biol. Chem. 263, 14276–14280.
183 Boyer, P.D. (1988) TIBS 13, 5–7.

184 Schönknecht, G., Althoff, G., Apley, E., Wagner, R. and Junge, W. (1989) FEBS Lett. 258, 190–194.
185 Sakai, Y., Moritani, C., Tsuda, M. and Tsuchiya, T. (1989) Biochim. Biophys. Acta 973, 450–456.
186 Krulwich, T.A. and Guffanti, A.A. (1989) Annu. Rev. Microbiol. 43, 435–463.
187 Williams, R.J.P. (1961) J. Theor. Biol. 1, 1–13.
188 Skulachev, V.P. (1984) TIBS 9, 182–185.
189 Hicks, D.B. and Krulwich, T.A. (1990) J. Biol. Chem. 265, 20547–20554.
190 Hoffmann, A., Laubinger, W. and Dimroth, P. (1990) Biochim. Biophys. Acta 1018, 206–210.
191 Hoffmann, A. and Dimroth, P. (1991) Eur. J. Biochem. 196, 493–497.
192 Semeykina, A.L. and Skulachev, V.P. (1991) FEBS Lett. 296, 77–81.
193 Kinoshita, N., Unemoto, T. and Kobayashi, H. (1984) J. Bacteriol. 158, 844–848.
194 Kakinuma, Y. and Harold, F.M. (1985) J. Biol. Chem. 260, 2086–2091.
195 Kakinuma, Y. and Igarashi, K. (1990) FEBS Lett. 261, 135–138.
196 Kakinuma, Y. and Igarashi, K. (1990) J. Bacteriol. 172, 1732–1735.
197 Quirk, P.Q., Guffanti, A.A., Plass, R.J., Clejan, S. and Krulwich, T.A. (1991) Biochim. Biophys. Acta 1058, 131–140.
198 Glagolev, A.N. (1980) J. Theor. Biol. 82, 171–185.
199 Baryshev, V.A., Glagolev, A.N. and Skulachev, V.P. (1981) Nature 192, 338–340.
200 Glagolev, A.N. (1984) Trends Biochem. Sci. 9, 397–400.
201 Glagolev, A.N. (1984) Motility and Taxis in Prokaryotes, Harwood Acad. Publ., Amsterdam.
202 Bibikov, S.I. and Skulachev, V.P. (1989) FEBS Lett. 243, 303–306.
203 Bibikov, S.I., Grishanin, R.N., Marwin, W., Oesterhelt, D. and Skulachev, V.P. (1991) FEBS Lett. 295, 223–226.
204 Rosen, B.P. (1986) Annu. Rev. Microbiol. 40, 263–286.
205 Maisonneuve, J. (1982) Sediment. Geol. 31, 1–11.
206 Kempe, S. and Degens, E.T. (1985) Chem. Geol. 53, 95–108.
207 Kazmierczak, J. and Degens, E.T. (1986) Mitt. Geol.-Paläontol. Inst. Universität Hamburg, 61, 1–20.
208 Kempe, S., Kazmierczak, J. and Degens, E.T. (1990) in: R.E. Crick (Ed.) Origin, Evolution, and Modern Aspects of Biomineralization in Plants and Animals, Plenum Press, New York, pp. 29–43.
209 Skulachev, V.P. (1989) J. Bioenerg. Biomembranes 21, 635–647.
210 Dibrov, P.A. (1991) Biochim. Biophys. Acta 1056, 209–224.
211 Brown, I.I., Taranenko, V.D., Timofeyev, I.V. and Timofeyeva, E.S. (1990) Biol. Membr. 7, 1271–1274 (Russ.).
212 Brown, I.I., Fadeyev, S.I., Kirik, I.I., Severina, I.I. and Skulachev, V.P. (1990) FEBS Lett. 270, 203–206.

L. Ernster (Ed.) *Molecular Mechanisms in Bioenergetics*
© 1992 Elsevier Science Publishers B.V. All rights reserved

CHAPTER 3

Bacteriorhodopsin

ROBERT RENTHAL

Division of Earth & Physical Sciences, The University of Texas at San Antonio, San Antonio, TX 78249, U.S.A. and Department of Biochemistry, The University of Texas Health Science Center at San Antonio, San Antonio, TX 78284, U.S.A.

Contents

1. Introduction 76
2. Bacteriorhodopsin structure 77
2.1. Atomic model of bacteriorhodopsin 77
2.2. Helix prediction 78
2.3. Helix packing 79
2.4. Inside-out protein? 79
2.5. Proline and glycine in transmembrane helices 82
2.6. Retinal binding pocket 84
2.7. Proton channel 85
3. The structure of the retinal chromophore 86
3.1. Light-adapted bacteriorhodopsin 86
3.2. Other chromophore states 87
3.2.1. Dark adaption 87
3.2.2. Purple-to-blue transition 87
3.2.3. Red and pink pigments 88
3.2.4. Dehydrated PM 88
3.3. The purple membrane photocycle 88
4. Proton pump cycle 90
4.1. Stoichiometry 90
4.2. Proton pump mechanism 90
4.2.1. bR–K. *Trans–cis* isomerization; disruption of protonated Schiff base interaction with counter ion 91
4.2.2. K–L. Protein relaxation around 13-*cis* retinal 91
4.2.3. L–M_1. Deprotonation of the Schiff base; protonation of Asp^{85}; proton release from the extracellular side of bR 92
4.2.4. M_1–M_2. Protein conformational change 92
4.2.5. M_2–N. Deprotonation of Asp^{96}; reprotonation of the Schiff base 93
4.2.6. N–bR and N–O–bR. *Cis–trans* isomerization of retinal; reprotonation of Asp^{96} from the cytoplasmic side of bR; dissociation of Asp^{85} 93
4.3. Energy transformations 94
4.4. Bacteriorhodopsin and halorhodopsin 94
5. Summary 96
Acknowledgements 96
References 96

1. Introduction

Under conditions of plentiful oxygen, *Halobacterium halobium* uses oxidative metabolism and an electron transport chain [1] to generate ATP. But when oxygen is scarce, under conditions such as growth to high cell populations, *H. halobium* can utilize sunlight to meet its energy needs. A specialized membrane region, purple membrane (PM), is synthesized. PM contains only a single type of protein, bacteriorhodopsin (bR), arranged in a hexagonal lattice [2–4]. bR consists of 248 amino acids (Fig. 1) with all-*trans* retinal attached via a Schiff base to a lysine residue. Light absorption by retinal activates an outwardly directed proton pump which is coupled to ATP synthetase [5,6].

Halobacteria also contain many other rhodopsin-like proteins, including halorhodopsin (hR), a light-driven chloride pump [7–11], and sensory rhodopsins (sR), a group of pigments involved in coupling flagellar motions to phototropic responses [12–15].

Since the 1984 review in this series by Lanyi [16], the molecular basis for understanding the bacteriorhodopsin proton pump has become clear. Many experimental results appear to be consistent with a specific sequence of proton binding and dissociation steps involving particular amino acid side chains. The understanding of the bacteriorhodopsin mechanism is based largely on three types of experiment: high-resolution cryo-electron microscopy, mutagenesis, and spectroscopy. At the same time, the key steps in conversion of light energy to a proton gradient have been identified in terms of thermodynamics.

The amino acid sequence similarity found between halorhodopsin and bacteriorhodopsin indicates that the chloride pump and the proton pump must share many similarities. Recent progress in understanding both ion pumps has benefited from comparisons of bacteriorhodopsin and halorhodopsin.

This review primarily covers papers between 1984 and mid-1991 on the structure of purple

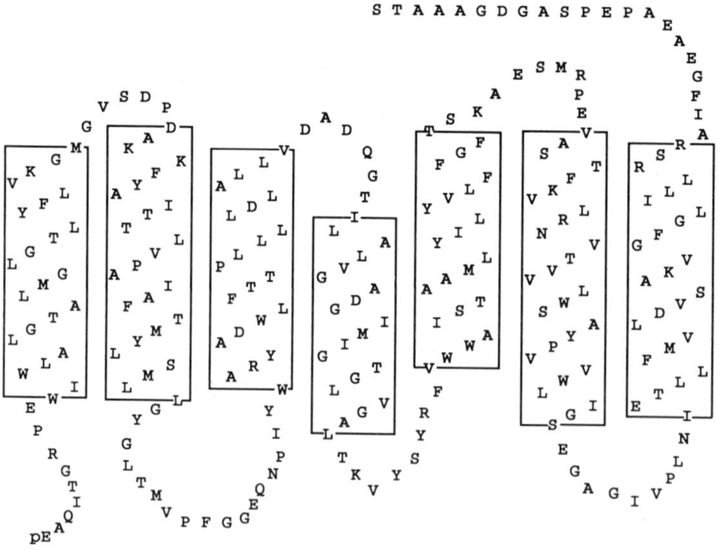

Fig. 1. Amino acid sequence of bacteriorhodopsin. Sequence from ref. [44]. Boxed sequences are transmembrane α-helices, as determined by electron microscopy [25]. N-terminal residue, pG, on the extracellular side of the membrane, is pyroglutamic acid. Helical segments are labled A–G from the left.

membrane, the proton pump mechanism studied by mutagenesis and spectroscopy, the photocycle, and comparisons between bacteriorhodopsin and halorhodopsin. Several reviews on bR have appeared during this period [17–24].

2. *Bacteriorhodopsin structure*

2.1. *Atomic model of bacteriorhodopsin*

Remarkable progress in high-resolution electron cryo-microscopy has provided sufficient information to construct an atomic resolution model of bR [25]. A diagram of the secondary structure of bR is shown in Fig. 2. There are seven transmembrane helical segments, labeled A–G from the amino terminus in Figs. 1, 2 and 5. The orientation of the figures is with the cytoplasmic side up and the extracellular side down. The retinal chromophore is attached to Lys^{216} in helix G. In addition to the transmembrane helices, Henderson et al. [25] identified some evidence for short helical segments parallel to the extracellular surface.

The current model has a resolution horizontally of 3.5 Å and vertically of 10 Å. This was sufficient for observation of patterns of bulky side chains and for fitting the amino acid sequence into the electron density with some confidence. The 1990 bR listing in the Brookhaven Protein Data Bank gives atomic coordinates for 1359 atoms of 170 amino acids. Only the coordinates of the membrane-embedded part of the protein were listed, because the surface loops were rather indistinct at the current resolution. In addition, 262 of the transmembrane atoms listed had low or ambiguous electron density. These poorly defined atoms are about equally divided between backbone, polar and nonpolar side chains. In a few cases, the missing atoms will be troublesome for those interested in deriving pump mechanisms from the present model. For example, there is uncertainty in the locations of Arg^{82}, Asp^{85}, and many of the atoms in the peptide backbone of helices F and G.

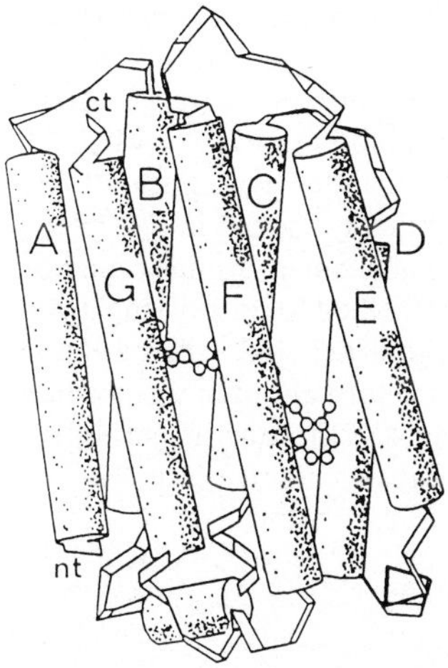

Fig. 2. Structure of bacteriorhodopsin. Spatial arrangement of helical segments of bacteriorhodopsin, as determined by electron microscopy. Position of retinal chromophore is shown. Reprinted from ref. [25], with permission. Copyright 1990, Academic Press.

Of the amino acids in the transmembrane helices, 36 (22%) are polar (D, E, K, N, Q, R, S, T), 116 (70%) are nonpolar* (A, F, I, L, M, P, V, W, Y), and 13 are glycine. The percentage of polar residues is about the same as that of the transmembrane helices of the photosynthetic reaction center (21% polar, 70% nonpolar) [26], the only other membrane protein for which atomic resolution structural information is available at this time. However, many of the 11 reaction center transmembrane helices extend one or two turns beyond the membrane surface, and most of the charged side chains are found in these extensions. The polar amino acids in the membrane-embedded sections of the reaction center helices are largely serine and threonine, with no ionizable groups found near the center of the bilayer. By contrast, the ion transport function of bR requires interior ionizable groups.

2.2. *Helix prediction*

The bR structure has been used by Argos and coworkers [27] to develop a transmembrane helix-prediction method. The basis set of helices chosen for a recent version of this method [28] are somewhat offset from the helices observed by Henderson et al. [25]. bR has also served as a testing ground for helical domain prediction algorithms for transmembrane proteins. The two most widely used methods are the hydropathy scale of Kyte and Doolittle (KD) [29] and the polarity scale of Goldman, Engelman and Steitz (GES) [30], recently reviewed by Engelman et al. [31]. The helical segments identified by Henderson et al. from the electron microscopy structure are shown in Table I along with the predictions from the KD and GES scales. The KD scale tends to overpredict bR helices by an average of 2.0 residues per helix end, and it tends to underpredict bR helices by an average of 1.2 residues per helix end (i.e. the predicted helices tend to be offset from the observed ones). Conversely, the GES scale overpredicts by 0.5 residues and underpredicts by 2.1 residues. Since the uncertainty from the structural model is ± 1 residue per helix end, obviously both scales are very satisfactory at predicting the bR helices, with the KD scale tending to slightly overpredict and the GES scale tending to slightly underpredict. The GES scale also slightly underpredicts the helix ends of reaction center helices [31]. As pointed out by Engelman et al. [31], the window used to predict the bR helices by Kyte and Doolittle was 7 amino acids. It is not clear why the more realistic window of 19 amino acids gives less satisfactory results with the KD scale.

Secondary structure analysis of bR from spectroscopic data produced a range of results. Circular dichroism spectra have been interpreted as showing appreciable β-sheet structure [32]. Raman scattering indicated 72–82% helix [33]. The actual helix composition appears to be about 70%, assuming about 10 surface residues are helical in addition to the 165 in transmembrane helices. This is very close to the 71% of very slowly exchanging amide groups found by deuterium–hydrogen exchange [34]. A mutagenesis and spin label method for identifying transmembrane helix boundaries [35] was applied to the sequence region 125–142, where helix D ends and E begins. The beginning of helix E was placed one helical turn ahead of the position found by electron microscopy. Henderson et al. placed the residues of this turn as part of the D–E connecting loop, with the helix axis parallel to the plane of the membrane (Fig. 2).

---

*Proline is considered nonpolar on this list because of the hydrocarbon side chain. However, the tertiary α-amine makes the main chain more polar than in a typical peptide bond.

2.3. *Helix packing*

Transmembrane helices are often depicted schematically as if they are exactly perpendicular to the membrane plane. However, even at low resolution [4], it was clear that four of the bR helices are tilted. Analysis of the helix packing geometry of a large number of water-soluble globular proteins shows that helices tend to form bundles that pack against each other at angles determined by the pattern of side chain ridges and backbone grooves [36–38]. For example, a relatively common four-helix bundle motif is found in proteins such as cytochrome *c′*, cytochrome b_{562} and hemerythrin. The interhelical angles between the four helical segments of these proteins average between 11° and 29° (Table II). bR may be thought of as consisting of two four-helix bundles with one helix common to both: helices A, B, C and G in one and C, D, E and F in the other. The interhelical angles are shown in Table II. The helix packing in bR is similar to the four-helix bundles of small, soluble proteins. The fact that helical domains in both globular proteins and integral membrane proteins are found to pack in similar ways implies that helix–helix interactions may be more important than protein–lipid interactions in determining integral membrane protein tertiary structure.

Attempts have been made to measure the average angle of helix tilt from the membrane normal, with mixed results. Analysis of the powder line shape by ^{13}C NMR [39] showed that the bR helices are not all perpendicular to the plane of the membrane. Ultraviolet circular dichroism indicated an angle of 0° to the membrane normal [40], whereas infrared linear dichroism gave a variety of results, dependent on the choice of helix model and solvent conditions [41–43]. Cassim and coworkers [43] have argued that the helix tilt observed by electron microscopy is an artifact of the sample preparation procedure.

2.4. *Inside-out protein?*

What can be said of the non-covalent forces that hold the bR structure together? Examination of the amino acid sequence of bR [44,45], model-building [46], and neutron diffraction [47], gave rise to the idea that bR is an 'inside-out' protein, with most of the polar groups facing inward, in contrast to the nonpolar core of water-soluble proteins. The idea is plausible, but the

TABLE I

Helix predictions compared with the bR structural model

Helix	Electron microscopy [25]	Kyte and Doolittle [29]	Engelman et al. [31]
A	10–32	10–34	10–29
B	38–62	44–68	44–63
C	80–101	78–102	82–101
D	108–127	106–130	108–127
E	136–157	134–158	135–154
F	167–193	175–199[a]	178–197
G	203–227	200[a]–224	204–223
Error range: (per helix end)	± 1	+2.0 –1.2	+0.5 – 2.1

[a]Boundary between helices F and G unresolved; arbitrarily set at 199/200.

TABLE II

Transmembrane interhelical angles of bR compared with four-helix bundle proteins

Helices	Inter-helix angle (°)
Average of 3 proteins (hemerythrin, cytochrome *c′*, cytochrome b_{562} [215])	
1,2	16 ± 3[a]
2,3	11 ± 3
3,4	29 ± 10
4,1	19 ± 8
bR bundle I	
A,B	24
B,C	5
C,G	19
G,A	10
bR bundle II	
F,C	19
C,D	11
D,E	26
E,F	11

[a]Standard deviation.

recent structural model does not seem to completely support it. Figure 3 shows two projections of the bR backbone on the membrane plane. In one, the side chains of nonpolar residues are plotted, and in the other, the side chains of the polar residues. Surprisingly, the distribution does not show a favorable inward or outward orientation for either group as a whole. Both types of side chain seem to be distributed about equally pointing toward the exterior and interior of the protein. Rees et al. [48] have devised methods for calculating the polarity of helix orientation in proteins. Their hydrophobicity scale ranges from −1.76 (Arg) to 0.73 (Ile). The average polar surface of bR helices has a hydrophobicity of 0.08 and the average nonpolar surface has a hydrophobicity of 0.33. Combining results from a wide variety of membrane proteins, they conclude that the surface hydrophobicity of membrane proteins is indeed reversed from that found in water-soluble proteins, although the interior hydrophobicity is about the same as that found in soluble proteins. However, Rees et al. used a Fourier method to identify the inside and outside of the membrane proteins, based on a number of assumptions. When the bR structural model is used to identify interior and surface amino acids, the results are somewhat different. The average hydrophobicity of the inside is about 0.15, whereas the outside is actually less hydrophobic. If several surface Lys and Arg residues that appear to be interacting with lipids are removed, the surface hydrophobicity is about 0.20. The average hydrophobicity of the interior of bR is the same as the average calculated by Rees et al. for the sides of 35 transmembrane helices that are thought to be facing the interiors of integral membrane proteins. However, the surface residues seem to be much less hydrophobic than the average calculated by Rees et al., and less hydrophobic than the surface of photosynthetic reaction centers. This lower hydrophobicity of bR may in part be due to a number of Ser and Thr side chains that appear to face the exterior of the protein.

It was initially proposed that ionizable groups buried in the interior of bR would be buried

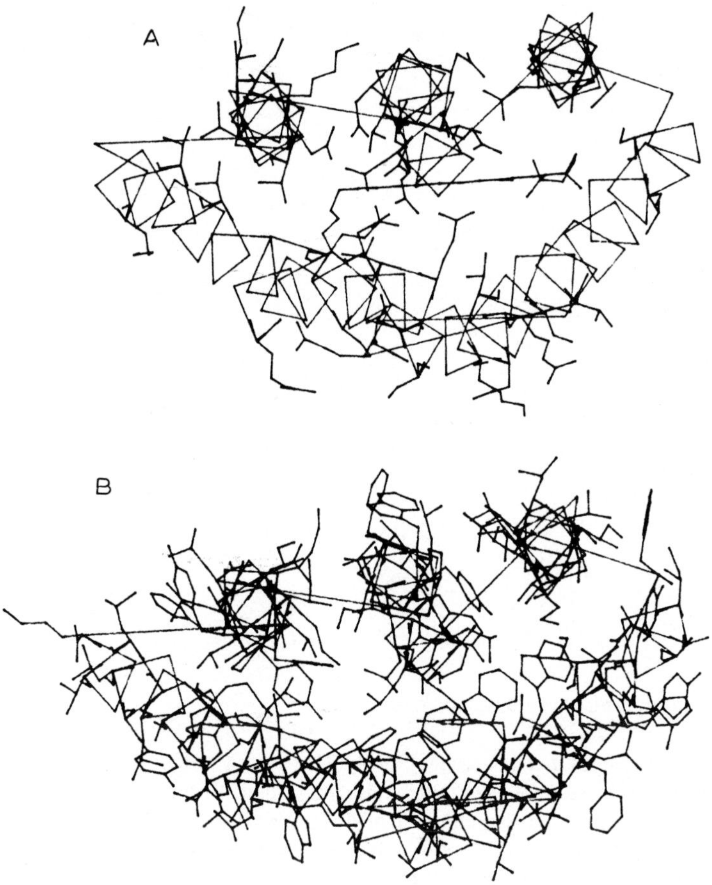

Fig. 3. Orientation of polar and nonpolar side chains. Projected structure of bR in the plane of the membrane. (A) Polar side chains. (B) Nonpolar side chains.

as ion pairs [46]. There has been some debate about the thermodynamics of embedding an ion pair inside a membrane, but the analysis by Honig and Hubbell [49] suggests that the energy barrier is not large. The bR structure shows one possible buried ion pair. The side chains of Asp^{212} and Lys^{216} have been located in the structure within 4 Å, close to the ion pairing distance, and the groups have been shown to be ionized by spectroscopic methods (see below). No other buried ion pairs have been identified. Other Lys and Arg groups are located near the ends of helices and the cations are likely to be in contact with lipid head groups or water. Three other Asp groups are buried in the transmembrane region. Asp^{85} is ionized and is probably in contact with the Schiff base, and Asp^{96} and Asp^{115} are buried in the protonated state (see below for further discussion).

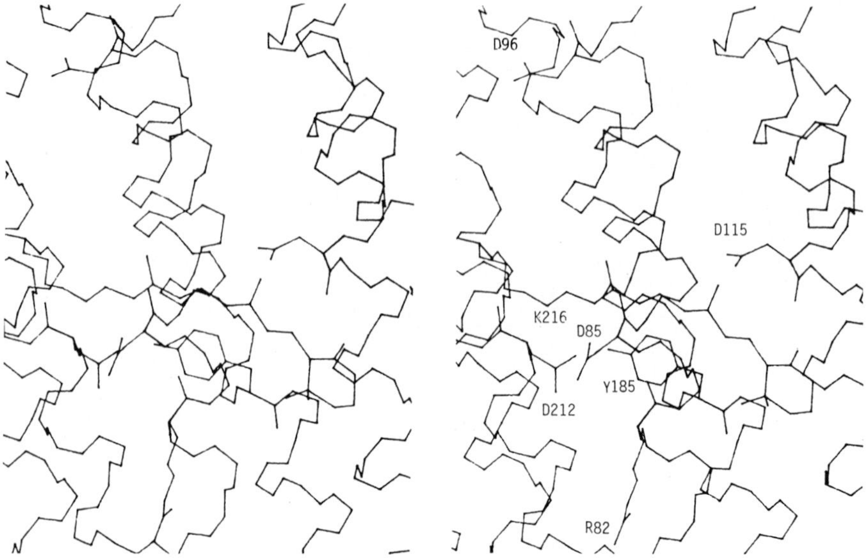

Fig. 4. Retinal binding site. Stereo view of the retinal binding pocket.

2.5. Proline and glycine in transmembrane helices

Proline and glycine are rarely found in α-helices [50]: proline because it cannot form a peptide backbone H-bond, and glycine because of the higher conformational entropy of the backbone around the α-carbon in the absence of a side chain. Yet bR contains 13 Gly and 3 Pro out of 165 transmembrane helical amino acids (10%). Photosynthetic reaction center transmembrane helices contain similar Pro + Gly content (10%). However, all of the transmembrane helix prolines of reaction centers are found within four residues of the helix ends, whereas all of the bR prolines are in the interior of the helices. Pro^{50} and Pro^{91} are near the centers of helix B and C, respectively, and Pro^{186} is about three-quarters of the way down helix F. These Pro residues produce noticeable kinks in the helices.

Proline was suggested as a possible interior site for transient proton binding in a hypothetical pump mechanism [51]. The presence of prolines has also been observed in putative transmembrane helices in a number of transport proteins, and a transport theory was proposed involving *cis–trans* isomerization of proline [52]. All three transmembrane prolines in bR have their imide nitrogens roughly facing the interior of the protein, so a role in transport is not ruled out on structural grounds alone. However, mutagenesis experiments [53] cast doubt on this idea. All prolines in light-adapted bR were shown by NMR spectroscopy to be in the *trans* conformation [54]. During proton pumping, C–N bond rotation was observed by FTIR in ^{15}N-labeled protein [55].

The glycine residues are clustered mostly in helices A and D. In both cases, they form a stripe along the surface, the interior surface in the case of helix A. It is conceivable that the glycines have a structural role in helix–helix contacts, but the pattern is not conserved in hR or sR.

TABLE III

Comparison of retinal contacts in bR with putative transmembrane sequences in the rhodopsin superfamily

Protein		Sequence range
Helix A	*	
bR	WIWLALGTALMGLGTLYFLVKGM	10–32
hR	SLWVNVALAGIAILVFVYMGRTI	11–33
sR	TAYLGGAVALIVGVAFVWLLYRS	6–28
bov rho	FSMLAAYMFLLIMLGFPINFLTL	37–59
h blu op	FYLQAAFMGTVFLIGFPLNAMVL	34–56
h red op	YHLTSVWMIFVVTASVFTNGLVL	53–75
pig α_2AR	ILTLVCLAGLLMLFTVFGNVLVI	53–75
rat β_1AR	TAGMGLLLALIVLLIVVGNVLVI	58–80
Helix B	* * *	
bR	DAKKFYAITTLVPAIAFTMYLSMLL	38–62
hR	RPRLIWGATLMIPLVSISSYLGLLS	37–61
sR	PHQSALAPLAIIPVFAGLSYVGMAY	33–57
bov rho	QHKKLRTPLNYILLNLAVADLFMVF	64–88
h blu op	RYKKLRQPLNYILVNVSFGGFLLCI	61–85
h red op	KFKKLRHPLNWILVNLAVADLAETV	80–104
pig α_2AR	TSRALKAPQNLFLVSLASADILVAT	60–84
rat β_1AR	KTPRLQTLTNLFIMSLASADLVMGL	85–109
Helix C	** ** *	
bR	WARYADWLFTTPLLLLDLALLV	80–101
hR	WGRYLTWALSTPMILLALGLLA	85–106
sR	GLRYIDWLVTTPILVGVVGYAA	71–92
bov rho	TGCNLEGFFATIGGEIALWSLV	108–129
h blu op	HVCALEGFLGTVAGLVTGWSLA	105–126
h red op	PMCVLEGYTVSLCGITGLWSLA	124–145
pig α_2AR	IYLALDVLFCTSSIVHLCAISL	108–129
rat β_1AR	LWTSVDVLCVTASIETLCVIAL	133–154
Helix D	* * *	
bR	ILALVGADGIMIGTGLVGAL	108–127
hR	LFTVIAADIGMCVTGLAAAM	113–132
sR	IIGVMVADALMIATGAGAVV	99–118
bov rho	AIMGVAFTWVMALACAAPPL	153–172
h blu op	ALTVVLATWTIGIGVSIPPF	150–169
h red op	AIVGIAFSWIWSAVWTAPPI	169–188
pig α_2AR	IKAIIVTVWVISAVISFPPL	150–169
rat β_1AR	ARALVCTVWAISALVSFLPI	175–194
Helix E	* * *	
bR	VWWAISTAAMLYILYVLFFGFT	136–157
hR	AFYAISCAFFVVVLSALVTDWA	142–163
sR	ALFGVSSIFHLSLFAYLYVIFP	126–147
bov rho	ESFVIYMFVVHFIIPLIVIFFC	201–222
h blu op	ESYTWFLFIFCFIVPLSLICFS	198–219
h red op	QSYMIVLMVTCCIIPLAIIMLC	217–238
pig α_2AR	KWYVISSCIGSFFAPCLIMILV	194–215
rat β_1AR	RAYAIASSVVSFYVPLCIMAFV	222–243

TABLE III

Protein		Sequence range
Helix F	* ** *	
bR	VASTFKVLRNVTVVLWSAYPVVWLIGS	167–193
hR	TAEIFDTLRVLTVVLWLGYPIVWAVGV	171–197
sR	QIGLFNLLKNHIGLLWLAYPLVWLFGP	156–182
bov rho	VTRMVIIMVIAFLICWLPYAGVAFYIF	250–276
h blu op	VSRMVVVMVGSFCVCYVPYAAFAMYMV	247–273
h red op	VTRMVVVMIFAYCVCWGPYTFFACFAA	266–292
pig α_2AR	FTFVLAVVIGVFVVCWFPFFFTYTLTA	372–398
rat β_1AR	ALKTLGIIMGVFTLCWLPFFLANVVKA	311–337
Helix G	* *	
bR	IETLLFMVLDVSAKVGFGLILLRSR	203–227
hR	VTSWAYSVLDVFAKYVFAFILLRWV	208–232
sR	GVALTYVFLDVLAKVPYVYFFYARR	192–216
bov rho	FGPIFMTIPAFFAKTSAVYNPVIYI	283–307
h blu op	LDLRLVTIPSFFSKSACIYNPIIYC	280–304
h red op	FHPLMAALPAYFAKSATIYNPVIYV	299–323
pig α_2AR	VPPTLFKFFFWFGYCNSSLNPVIYT	403–427
rat β_1AR	VPDRLFVFFNWLGYANSAFNPIIYC	343–367

References: bacteriorhodopsin (bR) [44], halorhodopsin (hR) [216], sensory rhodopsin (sR) [14], bovine rhodopsin (bov rho) [61], human blue cone opsin(h blu op) [217], human red cone opsin (h red op) [217], pig α_2 adrenergic receptor (pig α_2AR) [218], rat β_1 adrenergic receptor (rat β_1AR) [219].
*Sequence positions in retinal binding pocket of bR [25].

2.6. Retinal binding pocket

From the structure model, Henderson et al. [25] identified 21 amino acid side chains in contact with retinal in the center of the protein (Table III). These include some polar groups that are close to the protonated Schiff base (Asp^{85}, Thr^{89}, Thr^{90}, Asp^{212} and Lys^{216}) and other polar groups that are near the β ionone ring (Asp^{115}, Ser^{141}). Four tryptophans ($Trp^{86,138,182,\text{ and }189}$) and two tyrosines ($Tyr^{57\text{ and }185}$) are also in contact with retinal. The retinal binding pocket is shown in Fig. 4. A number of the amino acids in contact with retinal had been correctly predicted by site-directed mutagenesis [56] or spectroscopy [57].

Some similarities of the retinal binding pocket in bR to the chromophore or agonist binding sites in opsins or other receptors have been noted [56,58–60]. Sequence alignments with members of the heptahelical transmembrane protein family are shown in Table III. The alignments in the table were made to maximize the apparent similarities between the bR retinal binding pocket and sequences believed to be transmembrane helices in the opsins and receptors. Helices A and D show rather weak sequence similarity. The relationship of bR helix B to the G-protein receptor family is shown in Table III offset from the usual transmembrane position (cf. ref. [61]) in order to align the sequence KK that appears at the beginning of helix B [25] and is also found in the putative A–B connecting loop of many of the opsins. The alignment has the virtue of placing a conserved Asp of the G-protein receptor family in the Tyr^{57} retinal contact position of bR. It also aligns either Ile or Phe with bR's Val^{49} retinal contact position. This interesting alignment causes a few problems, however, such as the extremely basic cluster forced into the opsin helices and the very short A–B connecting loop.

In order to align the Glu^{113}/Asp^{113} position of the visual opsins [59] and adrenergic receptors [62] with Asp^{85} of bR the alignment for helix C in Table III is shifted 12 residues from that suggested by Mogi et al. [56] . This position is thought to be near the Schiff base in all the opsins, and near the positive charge on adrenergic receptor agonists. The helix C sequences selected in Table III also align Thr^{90}, a retinal contact point, with Thr residues in 7 of the 8 sequences. In helix E, all 8 sequences have an aromatic residue in the position of Trp^{138} in bR. The other two retinal contact points, Ser^{141} and Met^{145}, align to within one residue of the conserved Ser^{204} and Ser^{207} of the adrenergic receptor binding sites for catechol hydroxyl groups [58]. Helix F shows strong conservation of the Trp^{182}/Tyr^{185} pair as aromatic groups in all eight sequences, as pointed out by Mogi et al. [56]. Helix G contains a conserved Lys in all of the opsins.

Although the overall sequence similarities are weak, there is enough similarity to leave the question open as to the evolutionary relatedness of bR and the opsin family. This leads to the deeper question of possible mechanistic similarities between energy transduction by bR and signal transduction by rhodopsin or related receptors. Both types of proteins are believed to use their ligand to trigger a protein conformational change. It is easy to imagine how one process could evolve into the other. Through natural selection, a signal representing an energy source might be substituted for the energy source itself. Thus, bR may resemble an energy-transducing ancestor of the signal-transducing opsin family.

2.7. Proton channel

Studies of the bR structure at low resolution failed to show any water channels through the membrane [4,63]. Nagle and coworkers developed the idea of proton translocation by means of a 'proton wire' [64], which could, in principle, span the membrane, with proton transfers between protein side chains. However, at higher resolution, the bR structure shows a channel, presumably filled with water, leading from the cytoplasmic side of the membrane to the Schiff base, and leading from the Schiff base to the extracellular side of the membrane (Fig. 5). The proton release channel, on the extracellular side, is large and more hydrophilic than the proton uptake channel, on the cytoplasmic side. Proton conduction in bR-containing vesicles is considerably faster when retinal has been removed from bR [65,66], suggesting that retinal blocks the channel. The existence of this channel simplifies the proton pump mechanism to a problem involving only a few proton transfers within the protein.

Henderson et al. [25] have identified 31 amino acids that line the proton channel. Ten of these are also part of the retinal binding pocket. The channel is primarily composed of the inner surfaces of helices B, C and G. Eight amino acids in the channel have carboxyl side chains and three have amino side chains. There are also six hydroxylic groups (four Thr and two Tyr). The location of the channel has been confirmed by neutron diffraction [67]. The involvement of bound water in the proton uptake process has been demonstrated by kinetic spectroscopy of PM in solutions of diminished water activity [68]. Indicator dye studies with frozen PM [69] suggest the water in the channel is ice-like.

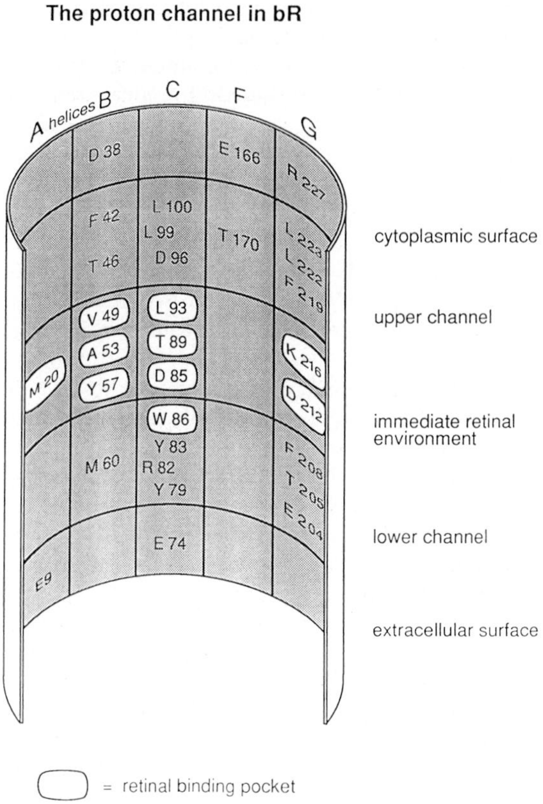

Fig. 5. Proton channel. Schematic diagram of the amino acids lining the proton channel. Reprinted from ref. [25], with permission. Copyright 1990, Academic Press.

3. *The structure of the retinal chromophore*

3.1. *Light-adapted bacteriorhodopsin*

Continuous illumination of PM produces light-adapted bR, which has an absorbance maximum at 568 nm with an extinction coefficient of 63 000 M^{-1} cm^{-1} [70]. The retinal chromophore in light-adapted bR is more than 98% all-*trans* [71]. The *trans* double bonds extend into the β ionone ring (i.e. 6 s-*trans*), in contrast to the 6 s-*cis* conformation adopted by retinals in solution [72]. The Schiff base is protonated [73] and the C=N double bond is anti [74]. The polyene is essentially planar [75]. These results, along with linear dichroism measurements [76] indicate that the Schiff base proton is pointing toward the extracellular surface and the polyene methyl groups are pointing toward the cytoplasmic surface.

The large redshift that occurs in retinals when bound to opsins (the opsin shift) is the source of their usefulness as biological light energy converters and sensors, since the absorbance maximum of free retinal is in the UV to blue spectral region. The first detailed analysis of the opsin shift used interrupted polyenes to identify the minimum chromophore, with the idea that local charge perturbations might explain the rhodopsin spectra [77]. However, recent spectral data suggest other more substantial contributions to the opsin shift of bR, particularly from the weak hydrogen bonding of the protonated Schiff base [78] and the 6 s-*trans* conformation of the polyene [72].

3.2. Other chromophore states

3.2.1. Dark adaptation

In the absence of light, the chromophore undergoes a shift in wavelength of its absorbance maximum from 568 to about 558 nm, and the extinction coefficient decreases [2]. The half-time for dark adaptation at 26°C and pH 5 is about 1 h [79]. The retinal chromophore exists as a 2:1 mixture of 13-*cis*:all-*trans* isomers in dark-adapted bR [71], although earlier workers using less reliable methods obtained various ratios near 1:1 [80]. Thus, in the dark, bR is a retinal isomerase. The absorbance maximum of the pure 13-*cis* form is calculated to be 554 nm [71]. The protonated Schiff base of the 13-*cis* isomer is C=N syn, whereas the all-*trans* isomer is the same as in light-adapted bR [81]. The combination of 13-*cis* and C=N syn produces only minimal structural perturbations in the retinal binding site [82], in contrast to the steric effects caused by the 13-*cis* C=N anti chromophore that forms during the photocycle (see below). The β ionone ring has the 6 s-*trans* conformation in both the 13-*cis* and all-*trans* isomers of dark adapted bR [72,83]. The 13-*cis* chromophore of dark adapted bR undergoes a separate photochemical reaction cycle which does not pump protons [84–86] but which can connect with the photocycle of light-adapted bR [87,88]. Dark adaptation has been studied in site-specific mutants of bR expressed in *E. coli* [89]. Mutations affecting amino acids in the region of the Schiff base (Tyr^{185}, Arg^{82}, Asp^{85}, and Asp^{212}) caused anomalies in light–dark adaptation rates, *cis-trans* isomer ratios, and absorbance maxima.

3.2.2. Purple-to-blue transition

Acidification [2,80,90,91] or deionization [92,93] of PM produces blue membrane, which has an absorbance maximum near 600 nm. Although blue membrane is photochemically active [80,94], the Schiff base does not deprotonate [95] and it fails to pump protons [96]. Blue membrane was reported to contain about a 2:3 ratio of 13-*cis* to all-*trans* retinal, insensitive to light–dark adaptation [80,97]. Substitution of Asn for Asp^{85} by site-directed mutagenesis produces a blue chromophore which is completely inactive in proton pumping, suggesting that neutralization of Asp^{85} directly results in blue membrane formation [98]. This conclusion, implicating titration of a single site in the purple-to-blue transition, is consistent with spectroscopic studies [99] and cation binding experiments [100]. Alternatively, it has been argued that the purple-to-blue transition is regulated by a conformational change which is controlled by the surface charge of PM [101–103]. There is evidence for structural changes in the purple-to-blue transition [99,104], but it remains to be shown whether these changes are the cause or the effect of the titration of Asp^{85}.

Experiments designed to measure the side of the membrane from which the blue pigment forms give conflicting results [105–107]. Vesicles in which bR orients with its cytoplasmic side facing the external medium (inside-out vesicles) pump protons inward, and eventually the interior of the vesicle becomes so acidic that blue pigment forms. This would appear to be consistent with the role of Asp^{85} in controlling the formation of blue pigment, since this group faces the extracellular side of the membrane, which is on the inside of the vesicles. But exposure of these same inside-out vesicles to external acid pulses results in rapid formation of blue pigment followed by a slower phase. Nasuda-Kouyama et al. [108] suggest that this results

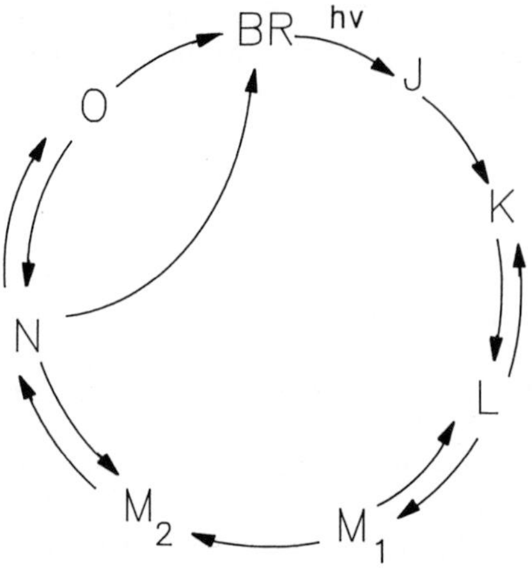

Fig. 6. Photocycle. Photointermediates of bacteriorhodopsin. Typical kinetic constants for each step are given in Table IV. Reprinted from ref. [191], with permission. Copyright 1991, American Chemical Society.

from the dark adaptation equilibrium between 13-*cis* and all-*trans* retinal: they propose that the all-*trans* pigment is accessible to H^+ only from the cytoplasmic side of the membrane (an externally applied acid pulse), while the 13-*cis* pigment is accessible to H^+ only from the extracellular side of the membrane (acidification of the interior due to proton pumping, or slow leak inside of an externally applied acid pulse).

3.2.3. *Red and pink pigments*

A pigment with an absorbance maximum at 480 nm forms when PM is exposed to organic solvents [109] or certain inhalation anaesthetics [110], or when bR is reconstituted in vesicles and raised to alkaline pH [111]. The 480 nm pigment is photochemically active [109] and contains a protonated Schiff base [112]. Pande et al. [112] showed that the formation of the 480 nm pigment induced by DMSO is strongly inhibited in PM with the 13-*cis* dark-adapted chromophore. With inside-out oriented vesicles, the alkaline 480 nm pigment forms from the extracellular side in the all-*trans* chromophore [108].

A 480 nm chromophore has also been observed as a photoproduct of blue membrane [113]. This pink membrane is different from the solvent- or alkali-red pigment, since it is photoreversible back to the blue form, unlike the other red pigments which photocycle through an M-like form [109].

3.2.4. *Dehydrated PM*

The purple chromophore of bacteriorhodopsin is perturbed by dehydration at relative humidities below 20%, resulting in a blue-shifted absorbance maximum near 530 nm [114–116]. The Schiff base is deprotonated in the 530 nm form [115]. The photoreactions of bR are significantly slowed with partial dehydration [117–119].

3.3. *The purple membrane photocycle*

The initial finding of a rhodopsin-like pigment in halobacteria stimulated a search for photo-transients analogous to those found in the animal visual pigments. Kinetic spectroscopy of light-adapted PM exposed to short flashes of light revealed a series of intermediate chromophore states that appeared to progress in a simple cycle: bR_{568} – K_{590} – L_{545} – M_{410} – O_{640} – bR_{568} (where the subscript indicates the wavelength of maximum visible absorption) [120,121]. The primary photoproduct was found to be a state called J, which forms in about 500 fs and decays to K in a few picoseconds [122–125]. An additional photointermediate, N_{560}, was established between M and O [126–128], as had been initially proposed by Lozier and Niederberger [121]. The simple sequential cycle suffers from a number of deficiencies, some of which were summarized by Nagle and coworkers [129–131]. The decay of M was found to be multi-exponential [132], suggesting multiple forms of bR with parallel photocycles [133–136] or branching before L [137]. There has been disagreement over parallel photocycle models. Several problems, including inconsistencies in amplitudes of fitted kinetic components for parallel photocycles, have been discussed [138]. Evidence for ground-state heterogeneity of bR, required by some parallel photocycle models, has been presented [139,140].

Introduction of reversible steps was found to fit the kinetics to the simple sequential scheme [130], and recent evidence has accumulated to support this idea for the reactions K to L [119], L to M [119,141–143], M to N [141,143–145], and N to O [145–147]. Intermediates N and O display a strong pH dependence [121,128]. The results of Varo et al. [146] indicated a branch point at N, with bR forming directly from N below pH 6. A number of problems remained with the L to M and M to N steps (summarized by Varo and Lanyi [138]), which were resolved by introducing two forms of M connected by an irreversible step [138]. Thus, a consensus is emerging for a model of the bR photocycle similar to that shown in Fig. 6. Properties of the photointermediates during the photocycle are summarized in Table IV.

TABLE IV

The bacteriorhodopsin photocycle

Transition	Lifetime[a]	Conformation of retinal	Schiff base[b]
bR–J	500 fs	13-*cis*	(NH^+)
J–K	3 ps	13-*cis*, twisted	NH^+
K–L	f 1.3 μs	13-*cis*, less twisted	NH^+
	r 1.8 μs		
L–M_1	f 35 μs	13-*cis*	N
	r 15 μs		
M_1–M_2	56 μs	13-*cis*	N
M_2–N	f 3.5 ms	13-*cis*	NH^+
	r 5.9 ms		
N–O	f 5.0 ms	all-*trans*, twisted	NH^+
	r 2.4 ms		
N–bR	5.0 ms	all-*trans*, planar	NH^+
O–bR	8.0 ms		

[a]References for kinetic data: bR–K [122–125]; other steps show inverse of rate constant determined at pH 7 in 0.1 M NaCl and 0.05 M phosphate [138]. f = forward; r = reverse.

[b]NH^+ = protonated; N = unprotonated.

4. *Proton pump cycle*

4.1. *Stoichiometry*

The number of protons pumped by bR per cycle has been extensively studied. The stoichiometry of the pump is the ratio of protons transported vectorially across the membrane to bR molecules cycling. Early studies measured the stoichiometry with pH indicator dyes [148,149], dilatometry [150], pH electrodes [148,151] and conductivity [152]. The quantum yield for the photocycle (i.e. the fraction of bR molecules absorbing light that proceeds through the photocycle) was found to be 0.25 to 0.30 at −30°C [153,154], whereas the quantum yield for proton pumping appeared to be dependent on ionic strength. At high ionic strength it was reported to be about 0.5. Thus the pump stoichiometry appeared to be as high as two protons per cycle. This result complicated the task of modeling a molecular mechanism of proton pumping. Although it was straightforward to explain one proton translocated through the Schiff base site during each cycle, a second proton was more difficult. An additional problem in these early studies was the relationship between the kinetics of photocycle intermediates and the release and uptake of protons. Some workers found the proton release step to coincide with the rise of M [96,155,156], while others saw a lag between M formation and proton release [120,157].

Recently, these issues have been resolved. Tittor and Oesterhelt [158] and Govindjee et al. [159] remeasured the quantum yield for the photocycle and obtained a value of 0.64 ± 0.04 at pH 7 and room temperature. Drachev et al. [160] noted that measurements showing a lag in proton release were made with indicator dyes in weakly buffered media. When stronger buffering was used, the equilibration of the released protons with the bulk media was faster and revealed the actual release kinetics to be simultaneous with M formation. CO_2 also appeared to influence the calibration of the buffering capacity of the system. Both Drachev et al. [160] and Grzesiek and Dencher [161] carefully remeasured the pump stoichiometry and found about 1 H^+ pumped per cycle, independent of ionic strength. Possible sources of error in some of the early studies include the effect of CO_2 on the buffering capacity calibration [160], and the assumption that all bR that is photocycling can be measured by the M concentration. At slightly alkaline pH and high ionic strength, appreciable concentrations of N coexist with M (see for example Ames and Mathies [143]). Underestimating the amount of photocycling bR would give a higher number of protons per cycle.

When bR is reconstituted in vesicles or dissolved in detergent, the quantum yield for proton pumping seems to decrease to about 0.06, nearly ten times lower than with an intact PM [162–165]. These low values have not been explained. Nevertheless, even with this low quantum yield, the stoichiometry is approximately one proton per cycle [103].

4.2. *Proton pump mechanism*

The bR proton pump consists of two key features. First, aqueous channels lead into the center of bR from the two membrane surfaces. Second, a light-driven molecular impeller releases a proton to the extracellular channel and picks up a proton from the cytoplasmic channel. The *trans–cis* photo-isomerization initiates the pump cycle, but spectroscopic evidence [166] suggests that the chromophore motions are not sufficient to complete the cycle. Instead, protein conformational changes are probably involved in returning the pump to the initial state. Although the details of the pump mechanism are not yet complete, the outlines are now clear (Fig. 7) and are discussed below with reference to the states of the photocycle.

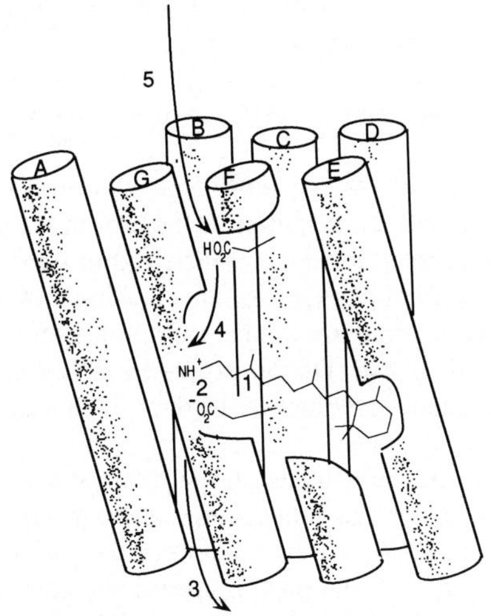

Fig. 7. Proton pump mechanism. Main steps in the proton pump cycle. (1) Light absorption causes *trans–cis* isomerization around 13–14 double bond of retinal. (2) Proton transfer from Schiff base to Asp^{85}. (3) At the same time as step 2 a proton is released from extracellular surface from an unknown group. (4) Proton transfer from Asp^{96} to Schiff base. (5) Proton uptake from cytoplasm by Asp^{96}.

4.2.1. bR– K. Trans–cis isomerization; disruption of protonated Schiff base interaction with counter ion

In bR, the Schiff base forms a weak ion pair with Asp^{212} and Asp^{85} [25,167]. The photo-induced formation of 13-*cis* retinal (the K state) moves the protonated Schiff base away from Asp^{212}/Asp^{85}. The *trans–cis* isomerization occurs in less than a picosecond [168], and electrical changes can be detected in less than 5 ps [169], but protein relaxation around the isomerized prosthetic group takes longer. The situation is very similar to ligand photolysis from hemoglobin or myoglobin, where the conformation of the prosthetic group changes in less than a picosecond, but the protein response is on a longer time scale (for a recent discussion, see ref. [170]). Some changes in protein structure have been noted in K. UV and IR difference spectra suggested the possibility of a tyrosinate existing in bR that becomes protonated in K [171]. The UV and IR changes have also been attributed to a tyrosine that undergoes changes in H-bonding between bR and K [172]. A tyrosine with an intrinsic pK of 8.3 was observed in light-adapted bR by UV absorption [139], but this was not detected by NMR [173,174] or Raman [175] methods. Changes in the conformation of the side chain of Lys^{216} in K have been proposed, based on FTIR studies [176].

*4.2.2. K–L. Protein relaxation around 13-*cis *retinal*

The details of the protein relaxation are unclear. There is some evidence for a change in H-bonding of Asp^{96} during the formation of L [175,177,178] or even deprotonation [178,179]. Also, changes in quenching of tryptophan fluorescence are observed with the formation of L [180]. The Schiff base becomes much more acidic in L, either due to its new environment, or due to bond rotation which reduces the charge delocalization along the polyene [181]. In bR, the pK of the Schiff base is 13 [161,182]. Removal of Asp^{85} by mutagenesis decreases the Schiff base pK to 7 or 8 [183], whereas removal of Asp^{212} has little effect [167]. Thus, when the Schiff base moves away from Asp^{85}, the protonated Schiff base becomes a proton donor. If its pK is below 5, it would be expected to deprotonate in tens of microseconds.

4.2.3. L–M_1. Deprotonation of the Schiff base; protonation of Asp^{85}; proton release from the extracellular side of bR

With the formation of M, the Schiff base deprotonates [73]. At the same time, a proton is released from the extracellular surface of bR [160,184]. However, it cannot be the proton from the Schiff base. Instead, it seems very likely that the Schiff base proton is transferred to Asp^{85}, which is observed to protonate by FTIR spectroscopy [177–179,185]. Mutants lacking a carboxyl group at position 85 do not form an M intermediate and are unable to pump protons [137,164,183,186,187]. In a mutant with Asp^{85} converted to Asn (D85N) expressed in *E. coli*, the Schiff base can transfer a proton directly into the aqueous medium in the dark, with a pK of 7 [183].

It is not yet clear what group releases the proton from the extracellular surface with the decay of L. Water [179] and Arg^{82} [23] have been suggested as possible proton release groups. It seems unlikely that arginine could dissociate in the hydrophilic environment of the proton release channel [25]. An additional argument against Arg^{82} as the proton release site comes from mutagenesis experiments showing that it is not essential for proton pumping [98,188].

A second carboxylate has been observed by FTIR spectroscopy to form in M [177,179,185], and this has been attributed to Asp^{212}. Mutations in Asp^{212} expressed in *E. coli* and studied in detergent show severely inhibited proton pumping [164,186], although the kinetics of proton release do not seem to be affected. Mutation of Asp^{212} to Asn in *H. halobium* (i.e. native PM sheets) gives somewhat different results [167]. Above pH 7, the chromophore is blue: no M is produced at all. Below pH 7, the chromophore is purple: diminished amounts of M are produced and proton pumping is one third of wild type. Recent pump models have suggested Asp^{212} remains deprotonated throughout the pump cycle [23,25], although it may experience an environmental change in M. Changes in tyrosine protonation or H-bonding have also been observed to coincide with formation of M [171,172,189,190].

4.2.4. M_1–M_2. Protein conformational change

This important step is necessary to move the Schiff base from connectivity with the proton release channel to connectivity with the proton uptake channel; to change the Schiff base from a proton donor to a proton acceptor; and as a source of back pressure for the proton gradient [138,191]. In principle, the change could simply be *cis–trans* re-isomerization, but this does not occur until the formation of the O intermediate. A 14 s-*cis–trans* isomerization has been suggested as a mechanism for moving the Schiff base [192], but spectroscopic evidence against this idea has been given [166], with the implication that there are no significant changes in the retinal structure between L and O. Thus, it seems most likely that the conformational change consists of movements of protein side chains. This conformational change was called the C–T model by Fodor et al. [166], postulating one protein conformation that forms in response to 13-*cis* retinal (C) and a different conformation that forms around all-*trans* retinal (T). Fodor et al. [166] initially placed the C–T transition at the M–N stage of the photocycle, but more recently Varo and Lanyi suggested that a significant conformational change must occur at the M_1–M_2 transition [138,191]. There is considerable spectroscopic evidence for protein conformational changes around the time of the M_1–M_2 transition [55,185,193,194]. A large change in reactivity of Asp^{115} toward chemical modification was observed under conditions that favor significant steady-state concentrations of M [195,196]. Asp^{115} is near the β-ionone ring of retinal [25]. Conformational changes between bR and M were directly observed at 8 Å resolution by neutron diffraction [197]. Small motions, involving a few amino acids or a slight helix tilt, were observed near the Schiff base and the β-ionone ring of retinal. Small differences between bR and M were also observed by electron cryomicroscopy at 3.3 Å resolution [198] and by time-resolved X-ray diffraction [199].

Insight into the conformational change occurring at M might be obtained from studies of stable bR states with certain features of the M intermediate. For example, dehydrated bR has an unprotonated Schiff base, and blue membrane or the D85N mutant has a protonated Schiff base and a neutral group at Asp^{85}. An argument has been given for structural similarities between blue membrane and M [200]. Szundi and Stoeckenius have proposed that a protein conformational change is coupled to the formation of blue membrane [101,102,201]. The conformations of these pigments have not yet been studied in detail.

4.2.5. M_2–N. Deprotonation of Asp^{96}; reprotonation of the Schiff base

With the appearance of N, Asp^{96} deprotonates [177,178,202] and the Schiff base reprotonates [166]. This proton transfer presumably occurs as a result of the M_1–M_2 conformational change which is proposed to place the Schiff base in an environment that increases its basicity. Asp^{96} is likely to be the proton donor to the Schiff base, because mutation of Asp^{96} to Asn (D96N) slows the proton uptake reaction [164,177,183,186,203]. The activation enthalpy for reprotonation of the Schiff base was found to be 42 kJ/mol greater in wild-type PM than in D96N [68]. This was attributed to charge separation during deprotonation of Asp^{96} during the formation of intermediate N. The inhibitory effects of the D96N mutation can be overcome by the presence of small, hydrophobic weak acids such as HN_3, which apparently can diffuse through the proton uptake channel and directly reprotonate the Schiff base [204]. The 10 Å distance between Asp^{96} and the Schiff base means that other groups must participate in the proton transfer, either protein side chains or water molecules. These proton transfer steps may resemble the 'proton wire' model of Nagle and Morowitz [64]. Cao et al. [68] have shown that the removal of water from PM by osmotically active solutes such as sucrose or ethylene glycol slows primarily the M_2–N step of the photocycle, implying that water is essential for proton transfer from Asp^{96} to the Schiff base.

4.2.6. N–bR and N–O–bR. Cis–trans *isomerization of retinal; reprotonation of Asp^{96} from the cytoplasmic side of bR; dissociation of Asp^{85}*

The retinal in the O intermediate re-isomerizes to all-*trans*. There is little information about the groups involved in the isomerase activity, other than the importance of Asp^{85}, Arg^{82}, Tyr^{185} and Asp^{212} in dark adaptation, which also involves isomerization of retinal [89]. Reprotonation of the Schiff base in N may contribute to subsequent re-isomerization by allowing a delocalized cationic intermediate at C13 which would diminish the bond order of the C13–14 double bond, facilitating rotation [181]. With the formation of bR, Asp^{96} picks up a proton from the cytoplasmic surface [175,177,179]. By comparing the rates of M–N–bR in wild-type with M–bR in D96N, Cao et al. [68] estimated that the carboxyl group on Asp^{96} accelerates proton transfer to the Schiff base by a factor of about 3000. They attributed this difference to the electrostatic effect of the negative charge that appears on Asp^{96} after it transfers its proton to the Schiff base.

The deprotonation of Asp^{85} that occurs with the formation of bR [175,177,179] cannot be a release from the membrane, since no pH increase is observed in this time frame [160]. Therefore, this must be an internal proton transfer. The identity of the proton acceptor is unknown, but it is probably the same as the group that releases a proton to the extracellular side during the formation of M.

4.3. Energy transformations

The light energy absorbed by retinal is transferred to the protein, and ultimately proton transfer reactions produce a change in the chemical potential of H^+ between the two sides of the membrane. The flow of energy can now be traced in detail. Varo and Lanyi [191] have measured the activation enthalpies, activation entropies and activation free energies for all of the steps of the photocycle (Fig. 8). For the reversible steps, this analysis also gives the relative changes in enthalpy, entropy and free energy. The interesting result is that there are no large free energy differences between K, L and M_1, or between M_2, N and O. The free energy input at K, estimated to be about 49 kJ/mol [205], must therefore be dissipated in the steps M_1–M_2, N–bR and O–bR. A pK change of the Schiff base in L from 13 to 5 would be about equal to the entire free energy input into the system from light absorption. The proton transfer in L–M_1 appears to trigger the subsequent transfer of free energy to the protein from the chromophore (i.e. retinal and the side chains in contact with it), which occurs during the M_1–M_2 conformational change.

The entropy change of the M_1–M_2 transition has been deduced by Varo and Lanyi [191] from the following data: (1) The enthalpy difference between M_2 and bR was set to the value measured by Ort and Parson [206]. (2) The free energy differences between O and bR, and between M_1 and M_2 were estimated from limits on the back reactions [138,191]. (3) The measured enthalpy input at K [205] was assumed to be equal to the free energy change from bR to K. The entropy difference between M_1 and M_2 was calculated to be 300 J/mol K, which is consistent with a protein conformational change involving as many as 20 amino acids.

The steps with large free energy decreases correspond to the irreversible steps in the proton pump. Irreversible formation of bR results in full cycling, maximizing the availability of chromophore in the initial state for utilization of high light fluxes. The irreversible M_1–M_2 step allows the proton pumping to occur in the presence of a proton gradient across the membrane. Under these conditions, M accumulates in the steady state [207,208] (presumably M_2), due to the slowing of the pH-dependent N–bR step, but intermediates before M do not build up.

4.4. Bacteriorhodopsin and halorhodopsin

The amino acid sequence similarity between hR and bR (Table III) indicates similarity in three-dimensional structure [9,25]. Many of the conserved residues between the two proteins are in the retinal binding pocket and in the protein release and uptake channels. One notable exception is Asp^{85}, which is Thr in hR. This absence explains why hR is not a proton pump (see discussion above about D85N bR). But, could bR pump chloride? Surprisingly, the answer seems to be yes. At low pH, Asp^{85} becomes protonated, resulting in blue membrane. It has been known for some time that blue membrane turns purple when titrated to pH 0 with HCl. Although Fisher and Oesterhelt suggested that this effect is due to halide binding near the Schiff base [91], the low pH halide form of PM has been thought of for some time as a consequence of the low pH rather than the halide (e.g., refs. [80,209,210]). Recent work has clarified the relationship between acid and halide in the acid halide PM, and strong evidence for halide binding has been given [78,94,211]. The chromophore changes on chloride binding are analogous to those observed on chloride binding to hR [11]. The analogy is clinched by the findings of Keszthelyi and coworkers [212–214] that acidified PM pumps Cl^- inward. Thus, much of the mechanism of proton pumping must be identical in the chloride pump.

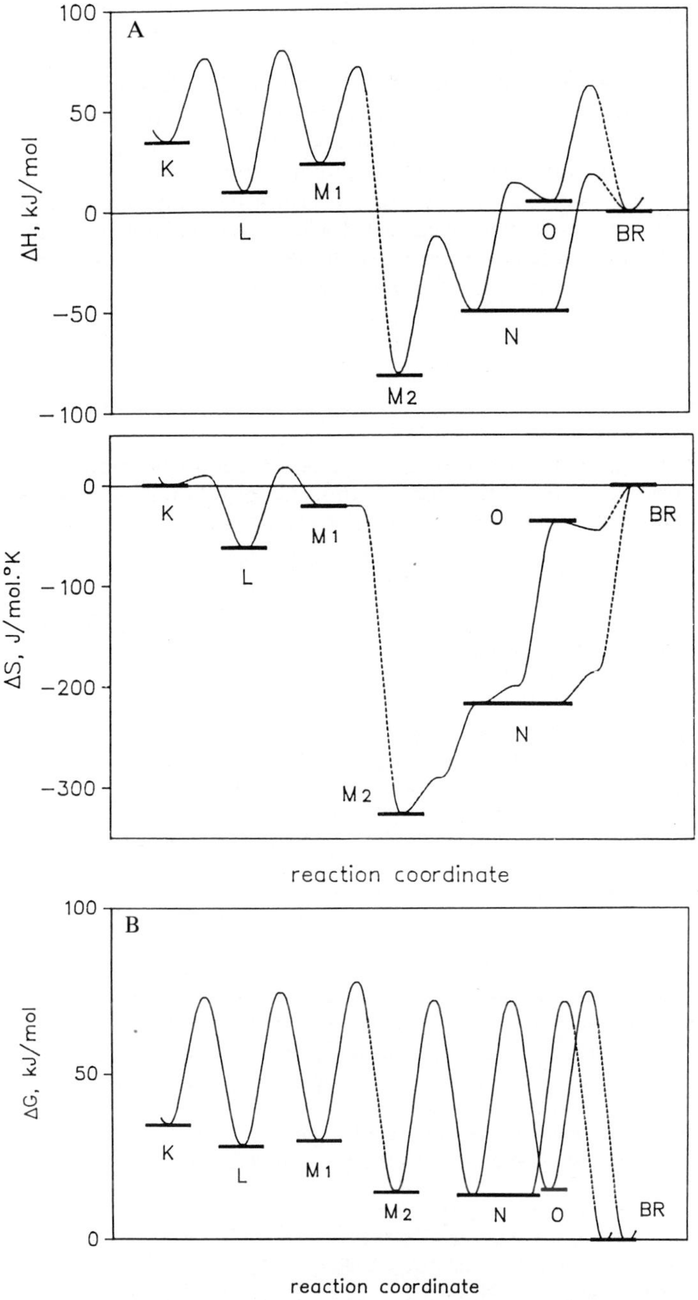

Fig. 8. Thermodynamics of the photocycle. (A) Enthalpy and entropy changes. Large decrease in entropy at M_1–M_2 suggests a protein conformational change. (B) Free energy changes. Key free energy decreases are at M_1–M_2 and N–O–bR, which are proposed to be irreversible steps. Reprinted from ref. [191], with permission. Copyright 1991, American Chemical Society.

5. *Summary*

The molecular basis of the bacteriorhodopsin proton pump is now understood in considerable detail. Light energy absorbed by the retinal chromophore is transferred to the protein through steric and electrostatic effects. As a result, the proton affinities of key ionizable protein side chains are altered in a kinetic sequence that transports protons from a cytoplasmic water channel through the protein to an extracellular water channel. Protein conformational changes that occur during the pump cycle remain to be elucidated. The structural similarity of bacteriorhodopsin to the seven-helix receptor family suggests that the bR conformational switch may turn out to be a general mechanism found in both transmembrane signaling and in energy transduction.

Acknowledgements

I thank Janos Lanyi for helpful discussions, manuscript preprints, and bibliographic assistance, and Richard Henderson for a list of the atomic coordinates of bR. Support from the U.S. Public Health Service (GM 25483, GM 08194, and EY 06324) is gratefully acknowledged.

References

1 Lanyi, J.K. (1968) Arch. Biochem. Biophys. 128, 716–724.
2 Oesterhelt, D. and Stoeckenius, W. (1973) Proc. Natl. Acad. Sci. USA 70, 2853–2857.
3 Blaurock, A.E. and Stoeckenius, W. (1971) Nature 233, 152–155.
4 Henderson, R. and Unwin, P.N. (1975) Nature 257, 28–32.
5 Racker, E. and Stoeckenius, W. (1974) J. Biol. Chem. 249, 662–663.
6 Danon, A. and Stoeckenius, W. (1974) Proc. Natl. Acad. Sci. USA 71, 1234–1238.
7 Lanyi, J.K. (1986) Annu. Rev. Biophys. Biophys. Chem. 15, 11–28.
8 Hegemann, P., Tittor, J., Blanck, A. and Oesterhelt, D. (1987) in: Yu.A. Ovchinnikov (Ed.), Retinal Proteins, VNU Science Press, Utrecht, pp. 333–352.
9 Oesterhelt, D. and Tittor, J. (1989) TIBS 14, 57–61
10 Lanyi, J.K., Duschl, A., Hatfield, G.W., May, K. and Oesterhelt, D. (1990) J. Biol. Chem. 265, 1253–1260.
11 Lanyi, J.K. (1990) Physiol. Rev. 70, 319–330.
12 Bogomolni, R.A. (1984) in: Information and Energy Transduction in Biological Membranes, A. Bolis, H. Helmreich, and H. Passow, (Eds.) Alan R. Liss, Inc., New York, pp. 5–12.
13 Spudich, J.L. and Bogomolni, R.A. (1988) Annu. Rev. Biophys. Biophys. Chem. 17, 193–215.
14 Blanck, A., Oesterhelt, D., Ferrando, E., Schegk, E.S. and Lottspeich, F. (1989) EMBO J. 8, 3963–3971.
15 Marwan, W., Schafer, W. and Oesterhelt, D. (1990) EMBO J. 9, 355–362.
16 Lanyi, J.K. (1984) in: L. Ernster (Ed.) Comparative Biochemistry: Bioenergetics, Elsevier, Amsterdam, pp. 315–350.
17 Crouch, R.K. (1986) Photochem.Photobiol. 44, 803–807.
18 Ovchinnikov, Yu.A. (1987) Photochem. Photobiol. 45, 909–914.
19 Kouyama, T., Kinosita, K. and Ikegami, A. (1988) Adv. Biophys. 24, 123–175.
20 Khorana, H.G. (1988) J. Biol. Chem. 263, 7439–7442.
21 Trissl, H.W. (1990) Photochem. Photobiol. 51, 793–818.
22 Birge, R.R. (1990) Biochim. Biophys. Acta Bio-Energetics 1016, 293–327.
23 Mathies, R., Lin, S.W., Ames, J.B. and Pollard, W.T. (1991) Annu. Rev. Biochem. 20, 491–518.
24 Lanyi, J.K. (1992) J. Bioenerg. Biomembranes, in press.

25 Henderson, R., Baldwin, J.M., Ceska, T.A., Zemlin, F., Beckmann, E. and Downing, K.H. (1990) J. Mol. Biol. 213, 899–929.
26 Rees, D., Komiya, H., Yeates, T., Allen, J. and Feher, G. (1989) Annu. Rev. Biochem. 58, 607-633.
27 Argos, P., Rao, J.K. and Hargrave, P.A. (1982) Eur. J. Biochem. 128 (2–3), 565–575.
28 Mohana Rao, J.K. and Argos, P. (1986) Biochim. Biophys. Acta 869, 197–214.
29 Kyte, J. and Doolittle, R.F. (1982) J. Mol. Biol. 157, 105–132.
30 Steitz, T.A., Goldman, A. and Engelman, D. (1982) Biophys. J. 37, 124–125.
31 Engelman, D.M., Steitz, T.A. and Goldman, A. (1986) Annu. Rev. Biophys. Biophys. Chem. 15, 321–353.
32 Jap, B.K., Maestre, M.F., Hayward, S.B. and Glaeser, R.M. (1983) Biophys. J. 43, 81–89.
33 Vogel, G. and Gartner, W. (1987) J. Biol. Chem. 262, 11464–11469.
34 Downer, N.W., Bruchman, T.J. and Hazzard, J.H. (1986) J. Biol. Chem. 261, 3640–3647.
35 Altenbach, C., Marti, T., Khorana, H.G. and Hubbell, W.L. (1990) Science 248, 1088–1092.
36 Crick, F.H.C. (1953) Acta Crystallogr. 6, 689–697.
37 Chothia, C., Levitt, M. and Richardson, D. (1971) Proc. Natl. Acad. Sci. USA 74, 4130–4134.
38 Chothia, C., Levitt, M. and Richardson, D. (1981) J. Mol. Biol. 145, 215–250.
39 Lewis, B.A., Harbison, G.S., Herzfeld, J. and Griffin, R.G. (1985) Biochemistry 24, 4671–4679.
40 Muccio, D.D. and Cassim, J.Y. (1979) Biophys. J. 26, 427–440.
41 Rothschild, K.J. and Clark, N.A. (1979) Biophys. J. 25, 473–487.
42 Nebedryk, E. and Breton, J. (1981) Biochim. Biophys. Acta 635, 515–524.
43 Draheim, J.E., Gibson, N.J. and Cassim, J.Y. (1991) Biophys. J. 60, 89–100.
44 Khorana, H.G., Gerber, G.E., Herlihy, W.C., Gray, C.P., Anderegg, R.J., Nihei, K. and Biemann, K. (1979) Proc. Natl. Acad. Sci. USA 76, 5046–5050
45 Ovchinnikov, Yu.A., Abdulaev, N.G., Feigina, M.Y., Kiselev, A.V. and Lobanov, N.A. (1979) FEBS Lett. 100, 219–224.
46 Engelman, D.M., Henderson, R., McLachlan, A.D. and Wallace, B.A. (1980) Proc. Natl. Acad. Sci. USA 77, 2023–2027.
47 Engelman, D.M. and Zaccai, G. (1980) Proc. Natl. Acad. Sci. USA 77, 5894–5898.
48 Rees, D., DeAntonio, L. and Eisenberg, D. (1989) Science 245, 510–513.
49 Honig, B. and Hubbell, W.L. (1984) Proc. Natl. Acad. Sci. USA 81, 5412–5416.
50 Chou, P.Y. and Fasman, G.D. (1974) Biochemistry 13, 211–221.
51 Dunker, A.K. (1982) J. Theor. Biol. 97, 95–127.
52 Brandl, C. and Deber, C. (1986) Proc. Natl. Acad. Sci. USA 83, 917–921.
53 Mogi, T., Stern, L.J., Chao, B.H. and Khorana, H.G. (1989) J. Biol. Chem. 264, 14192–14196.
54 Deber, C.M., Sorrell, B.J. and Xu, G.-Y. (1990) Biochem. Biophys. Res. Commun. 172, 862–869.
55 Gerwert, K., Hess, B. and Engelhard, M. (1990) FEBS Lett. 261, 449–454.
56 Mogi, T., Marti, T. and Khorana, H.G. (1989) J. Biol. Chem. 264, 14197–14201.
57 Rothschild, K.J., He, Y.-W., Mogi, T., Marti, T., Stern, L.J. and Khorana, H.G. (1990) Biochemistry 29, 5954–5960.
58 Strader, C.D., Candelore, M.R., Hill, W.S., Sigal, I.S. and Dixon, R.A.F. (1989) J. Biol. Chem. 264, 13572–13578.
59 Sakamar, T.P., Franke, R.R. and Khorana, H.G. (1989) Proc. Natl. Acad. Sci. USA 86, 8309–8313.
60 Henderson, R. and Schertler, G.F. (1990) Philos. Trans. R. Soc. London Ser. B 326, 379–389.
61 Hargrave, P.A., McDowell, J.H., Curtis, D.R., Wang, J.K., Juszczak, E., Fong, S.L., Rao, J.K.M. and Argos, P. (1983) Biophys. Struct. Mech. 9, 235–244.
62 Strader, C.D., Candelore, M.R., Hill, W.S., Dixon, R.A.F. and Sigal, I.S. (1989) J. Biol. Chem. 264, 16470–16477.
63 Zaccai, G. and Gilmore, D.J. (1979) J. Mol. Biol. 132, 181–191.
64 Nagle, J.F. and Morowitz, H.J. (1978) Proc. Natl. Acad. Sci. USA 75, 298–302.
65 Konishi, T. and Packer, L. (1978) FEBS Lett. 89, 333–336.
66 Burghaus, P.A. and Dencher, N.A. (1989) Arch. Biochem. Biophys. 275, 395–409.
67 Papadopoulos, G., Dencher, N.A., Zaccai, G. and Bldt, G. (1990) J. Mol. Biol. 214, 15–19.
68 Cao, Y., Varo, G., Chang, M., Ni, B., Needleman, R. and Lanyi, J.K. (1991) Biochemistry, in press.

69 Heberle, J. and Dencher, N.A. (1990) FEBS Lett. 277, 277–280
70 Rehorek, M. and Heyn, M.P. (1979) Biochemistry 18, 4977–4983.
71 Scherrer, P., Mathew, M.K., Sperling, W. and Stoeckenius, W. (1989) Biochemistry 28, 829–834.
72 Harbison, G.S., Smith, S.O., Pardoen, J.A., Courtin, J.M., Lugtenburg, J., Herzfeld, J., Mathies, R.A. and Griffin, R.G. (1985) Biochemistry 24, 6955–6962.
73 Lewis, A., Spoonhower, J., Bogomolni, R.A., Lozier, R.H. and Stoeckenius, W. (1974) Proc. Natl. Acad. Sci. USA 71, 4462–4466.
74 Smith, S.O., Myers, A.B., Pardoen, J.A., Winkel, C., Mulder, P.P.J., Lugtenburg, J. and Mathies, R.A. (1984) Proc. Natl. Acad. Sci. USA 81, 2055–2059.
75 Smith, S.O., Braiman, M.S., Myers, A.B., Pardoen, J.A., Courtin, J.M.L., Winkel, C., Lugtenburg, J. and Mathies, R.A. (1987) J. Am. Chem. Soc. 109, 3108–3125.
76 Lin, S.W. and Mathies, R.A. (1989) Biophys. J. 56, 653–660.
77 Nakanishi, K., Balogh-Nair, V., Arnaboldi, M., Tsujimoto, K. and Honig, B. (1980) J. Am. Chem. Soc. 102, 7945–7947.
78 De Groot, H.J.M., Smith, S.O., Courtin, J., Van den Berg, E., Winkel, C., Lugtenburg, J., Griffin, R.G. and Herzfeld, J. (1990) Biochemistry 29, 6873–6883.
79 Casadio, R., Gutowitz, H., Mowery, P., Taylor, M. and Stoeckenius, W. (1980) Biochim. Biophys. Acta 590, 13–23.
80 Mowery, P.C., Lozier, R.H., Chae, Q., Tseng, Y.W., Taylor, M. and Stoeckenius, W. (1979) Biochemistry 18, 4100–4107.
81 Harbison, G.S., Smith, S.O., Pardoen, J.A., Winkel, C., Lugtenburg, J., Herzfeld, J., Mathies, R. and Griffin, R.G. (1984) Proc. Natl. Acad. Sci. USA 81, 1706–1709.
82 Dencher, N.A., Papadopoulos, G., Dresselhaus, D. and Bldt, G. (1990) Biochim. Biophys. Acta Bio-Membr. 1026, 51–56.
83 Van der Steen, R., Biesheuvel, P.L., Mathies, R.A. and Lugtenburg, J. (1986) J. Am. Chem. Soc. 108, 6410–6411.
84 Fahr, A. and Bamberg, E. (1982) FEBS Lett. 140, 251–253.
85 Ohno, K., Takeuchi, Y. and Yoshida, M. (1977) Biochim. Biophys. Acta 462, 575–582.
86 Lozier, R.H., Niederberger, W., Ottolenghi, M., Sivorinovsky, G. and Stoeckenius, W. (1978) in: S.R. Caplan and M. Ginzburg (Eds.) Energetics and Structure of Halophilic Microorganisms, Elsevier/ North-Holland, Amsterdam, pp. 123–139.
87 Dencher, N.A., Rafferty, C.N. and Sperling, W. (1976) Ber. Kernforschungsanlage Juelich 1347, 1–42.
88 Sperling, W., Carl, P., Rafferty, C.H. and Dencher, N.A. (1977) Biophys. Struct. Mech. 3, 79–94.
89 Dunach, M., Marti, T., Khorana, H.G. and Rothschild, K.J. (1990) Proc. Natl. Acad. Sci. USA 87, 9873–9877.
90 Moore, T.A., Edgerton, M.E., Parr, G., Greenwood, C. and Perham, R.N. (1978) Biochem. J. 171, 469–476.
91 Fischer, U. and Oesterhelt, D. (1979) Biophys. J. 28, 211–230.
92 Chang, C.-H., Chen, J.-G., Govindjee, R. and Ebrey, T. (1985) Proc. Natl. Acad. Sci. USA 82, 396–400.
93 Kimura, Y., Ikegami, A. and Stoeckenius, W. (1984) Photochem. Photobiol. 40, 641–646.
94 Varo, G. and Lanyi, J.K. (1989) Biophys. J. 56, 1143–1151.
95 Chronister, E.L. and El-Sayed, M.A. (1987) Photochem. Photobiol. 45, 507–513.
96 Drachev, L.A., Kaulen, A.D. and Skulachev, V.P. (1978) FEBS Lett. 87, 161–167.
97 Pande, C., Chang, C.H. and Ebrey, T.G. (1985) Photochem. Photobiol. 42, 549–552.
98 Subramaniam, S., Marti, T. and Khorana, H.G. (1990) Proc. Natl. Acad. Sci. USA 87, 1013–1017.
99 Dunach, M., Padros, E., Seigneuret, M. and Rigaud, J.L. (1988) J. Biol. Chem. 263, 7555–7559.
100 Jonas, R. and Ebrey, T.G. (1991) Proc. Natl. Acad. Sci. USA 88, 149–153.
101 Szundi, I. and Stoeckenius, W. (1987) Proc. Natl. Acad. Sci. USA 84, 3681–3684.
102 Szundi, I. and Stoeckenius, W. (1988) Biophys. J. 54, 227–232.
103 Holz, M., Drachev, L.A., Mogi, T., Otto, H., Kaulen, A.D., Heyn, M.P., Skulachev, V.P. and Khorana, H.G. (1989) Proc. Natl. Acad. Sci. USA 86, 2167–2171.

104 Heyn, M.P., Dudda, C., Otto, H., Seiff, F. and Wallat, I. (1989) Biochemistry 28, 9166–9172.
105 Lind, C., Hojeberg, B. and Khorana, H.G. (1981) J. Biol. Chem. 256, 8298–8305.
106 Druckmann, S., Ottolenghi, M. and Korenstein, R. (1985) Biophys. J. 47, 115–118.
107 Bakker-Grunwald, T. and Hess, B. (1981) J. Membr. Biol. 60, 45–49.
108 Nasuda-Kouyama, A., Fukuda, K., Iio, T. and Kouyama, T. (1990) Biochemistry 29, 6778–6788.
109 Oesterhelt, D., Meentzen, M. and Schuhmann, L. (1973) Eur. J. Biochem. 40, 453–463.
110 Nishimura, S., Mashimo, T., Hiraki, K., Hamanaka, T., Kito, Y. and Yoshiya, I. (1985) Biochim. Biophys. Acta 818, 421–424.
111 Hwang, S.B. and Stoeckenius, W. (1977) J. Membr. Biol. 33, 325–350.
112 Pande, C., Callender, R., Henderson, R. and Pande, A. (1989) Biochemistry 28, 5971–5978.
113 Chang, C.-H., Liu, S.-Y., Jonas, R. and Govindjee, R. (1987) Biophys. J. 52, 617–623.
114 Lazarev, Y.A. and Terpugov, E.L. (1980) Biochim. Biophys. Acta 590, 324–338.
115 Hildebrandt, P. and Stockburger, M. (1984) Biochemistry 23, 5539–5548.
116 Renthal, R. and Regalado, R. (1991) Photochem. Photobiol. 54, 931–935.
117 Korenstein, R. and Hess, B. (1977) Nature 270, 184–186.
118 Varo, G. and Keszthelyi, L. (1983) Biophys. J. 43, 47–51.
119 Varo, G. and Lanyi, J.K. (1991) Biophys. J. 59, 313–322.
120 Lozier, R.H., Bogomolni, R.A. and Stoeckenius, W. (1975) Biophys. J. 15, 955–963.
121 Lozier, R.H. and Niederberger, W. (1977) Fed. Proc. 36, 1805–1809.
122 Nuss, M.C., Zinth, W., Kaiser, W., Kolling, E. and Oesterhelt, D. (1985) Chem. Phys. Lett. 117, 1–8.
123 Sharkov, A.V., Pakulev, A.V., Chekalin, S.V. and Matveetz, Y.A. (1985) Biochim. Biophys. Acta 808, 94–102.
124 Polland, H.J., Franz, M.A., Zinth, W., Kaiser, W., Kolling, E. and Oesterhelt, D. (1986) Biophys. J. 49, 651–662.
125 Petrich, J.W., Breton, J., Martin, J.L. and Antonetti, A. (1987) Chem. Phys. Lett. 137, 369–375.
126 Drachev, L.A., Kaulen, A.D., Skulachev, V.P. and Zorina, V.V. (1986) FEBS Lett. 209, 316–320.
127 Dancshazy, Zs., Govindjee, R., Nelson, B. and Ebrey, T.G. (1986) FEBS Lett. 209, 44–48.
128 Kouyama, T., Nasuda-Kouyama, A., Ikegami, A., Mathew, M.K. and Stoeckenius, W. (1988) Biochemistry 27, 5855–5863.
129 Nagle, J.F., Parodi, L.A. and Lozier, R.H. (1982) Biophys. J. 38, 161–174.
130 Parodi, L.A., Lozier, R.H., Bhattacharjee, S.M. and Nagle, J.F. (1984) Photochem. Photobiol. 40, 501–506.
131 Xie, A.H., Nagle, J.F. and Lozier, R.H. (1987) Biophys. J. 51, 627–635.
132 Eisenbach, M., Bakker, E.P., Korenstein, R. and Caplan, S.R. (1976) FEBS Lett. 71, 228–232.
133 Hanamoto, J.H., Dupuis, P. and El-Sayed, M.A. (1984) Proc. Natl. Acad. Sci. USA 81, 7083–7087.
134 Dancshazy, Zs., Govindjee, R. and Ebrey, T.G. (1988) Proc. Natl. Acad.Sci. USA 85, 6358–6361.
135 Diller, R. and Stockburger, M. (1988) Biochemistry 27, 7641–7651.
136 Bitting, H.C., Jang, D.-J. and El-Sayed, M.A. (1990) Photochem. Photobiol. 51, 593–598.
137 Butt, H.J., Fendler, K., Bamberg, E., Tittor, J. and Oesterhelt, D. (1989) EMBO J. 8, 1657–1663.
138 Varo, G. and Lanyi, J.K. (1991) Biochemistry 30, 5008–5015.
139 Balashov, S.P., Govindjee, R. and Ebrey, T.G. (1991) Biophys. J. 60, 475–490.
140 Wu, S. and El-Sayed, M.A. (1991) Biophys. J. 60, 190–197.
141 Varo, G. and Lanyi, J.K. (1990) Biochemistry 29, 2241–2250.
142 Alshuth, T. and Stockburger, M. (1986) Photochem. Photobiol. 43, 55–66.
143 Ames, J.B. and Mathies, R.A. (1990) Biochemistry 29, 7181–7190.
144 Otto, H., Marti, T., Holz, M., Mogi, T., Lindau, M., Khorana, H.G. and Heyn, M.P. (1989) Proc. Natl. Acad. Sci. USA 86, 9228–9232.
145 Gerwert, K., Souvignier, G. and Hess, B. (1990) Proc. Natl. Acad. Sci. USA 87, 9774–9778.
146 Varo, G., Duschl, A. and Lanyi, J.K. (1990) Biochemistry 29, 3798–3804.
147 Chernavskii, D.S., Chizhov, I.V., Lozier, R.H., Murina, T.M., Prokhorov, A.M. and Zubov, B.V. (1989) Photochem. Photobiol. 49, 649–653.
148 Bogomolni, R.A., Baker, R.A., Lozier, R.H. and Stoeckenius, W. (1980) Biochemistry 19, 2152–2159.
149 Govindjee, R., Ebrey, T.G. and Crofts, A.R. (1980) Biophys. J. 30, 231–242.

150 Ort, D.R. and Parson, W.W. (1979) Biophys. J. 25, 341–353.
151 Kuschmitz, D. and Hess, B. (1981) Biochemistry 20, 5950–5957.
152 Marinetti, T. and Mauzerall, D. (1983) Proc. Natl. Acad. Sci. USA 80, 178–180.
153 Becher, B. and Ebrey, T.G. (1977) Biophys. J. 17, 185–191.
154 Goldschmidt, C.R., Kalisky, O., Rosenfeld, T. and Ottolenghi, M. (1977) Biophys. J. 17, 179–183.
155 Rayfield, G.W. (1983) Biophys. J. 41, 109–117.
156 Fahr, A., Lauger, P. and Bamberg, E. (1981) J. Membr. Biol. 60, 51–62.
157 Lozier, R.H., Niederberger, W., Bogomolni, R.A., Hwang, S. and Stoeckenius, W. (1976) Biochim. Biophys. Acta 440, 545–556.
158 Tittor, J. and Oesterhelt, D. (1990) FEBS Lett. 263, 269–273.
159 Govindjee, R., Balashov, S.P. and Ebrey, T.G. (1990) Biophys. J. 58, 597–608.
160 Drachev, L.A., Kaulen, A.D. and Skulachev, V.P. (1984) FEBS Lett. 178, 331–335.
161 Grzesiek, S. and Dencher, N. (1986) FEBS Lett. 208, 337–342.
162 Marinetti, T. and Mauzerall, D. (1986) Biophys. J. 50, 405–415.
163 Marinetti, T. (1987) Biophys. J. 52, 115–121.
164 Marinetti, T., Subramaniam, S., Mogi, T., Marti, T. and Khorana, H.G. (1989) Proc. Natl. Acad. Sci. USA 86, 529–533.
165 Renthal, R. and Hollub, A. (1988) Photochem. Photobiol. 48, 219–221.
166 Fodor, S.P., Ames, J.B., Gebhard, R., van der Berg, E.M., Stoeckenius, W., Lugtenburg, J. and Mathies, R.A. (1988) Biochemistry 27, 7097–7101.
167 Needleman, R., Chang, M., Ni, B., Varo, G., Fornes, J., White, S. and Lanyi, J. (1991) J. Biol. Chem. 266, 11478–11484.
168 Mathies, R.A., Brito Cruz, C.H., Pollard, W.T. and Shank, C.V. (1988) Science 240, 777–779.
169 Simmeth, R. and Rayfield, G.W. (1990) Biophys. J. 57, 1099–1101.
170 Petrich, J.W., Lambry, J.-C., Kuczera, K., Karplus, M., Poyart, C. and Martin, J.-L. (1991) Biochemistry 30, 3975–3987.
171 Rothschild, K.J., Roepe, P., Ahl, P.L., Earnest, T.N., Bogomolni, R.A., Das Gupta, S.K., Mulliken, C.M. and Herzfeld, J. (1986) Proc. Natl. Acad. Sci. USA 83, 347–351.
172 Rothschild, K.J., Braiman, M.S., He, Y.-W., Marti, T. and Khorana, H.G. (1990) J. Biol. Chem. 265, 16985–16991.
173 Herzfeld, J., Das Gupta, S.K., Farrar, M.R., Harbison, G.S., McDermott, A.E., Pelletier, S.L., Raleigh, D.P., Smith, S.O., Winkel, C., Lugtenburg, J. and Griffin, R.G. (1990) Biochemistry 29, 5567–5574.
174 McDermott, A.E., Thompson, L.K., Winkel, C., Farrar, M.R., Pelletier, S., Lugtenburg, J., Herzfeld, J. and Griffin, R.G. (1991) Biochemistry 30, 8366–8371.
175 Gerwert, K., Souvignier, G. and Hess, B. (1991) Proc. Natl. Acad. Sci. USA 87, 9774–9778.
176 McMaster, E. and Lewis, A. (1988) Biochem. Biophys. Res. Commun. 156, 86–91.
177 Gerwert, K., Hess, B., Soppa, J. and Oesterhelt, D. (1989) Proc. Natl. Acad. Sci. USA 86, 4943–4947.
178 Braiman, M.S., Bousche, O. and Rothschild, K.J. (1991) Proc. Natl. Acad. Sci. USA 88, 2388–2392.
179 Braiman, M.S., Mogi, T., Marti, T., Stern, L.J., Khorana, H.G. and Rothschild, K.J. (1988) Biochemistry 27, 8516–8520.
180 Jang, D.-J. and El-Sayed, M.A. (1989) Proc. Natl. Acad. Sci. USA 86, 5815–5819.
181 Tavan, P., Schulten, K. and Oesterhelt, D. (1985) Biophys. J. 47, 415–430.
182 Druckmann, S., Ottolenghi, M., Pande, A., Pande, J. and Callender, R.H. (1982) Biochemistry 21, 4953–4959.
183 Otto, H., Marti, T., Holz, M., Mogi, T., Stern, L.J., Engel, F., Khorana, H.G. and Heyn, M.P. (1990) Proc. Natl. Acad. Sci. USA 87, 1018–1022.
184 Grzesiek, S. and Dencher, N.A. (1986) FEBS Lett. 208, 337–342.
185 Engelhard, M., Gerwert, K., Hess, B., Kreutz, W. and Siebert, F. (1985) Biochemistry 24, 400–407.
186 Mogi, T., Stern, L.J., Marti, T., Chao, B.H. and Khorana, H.G. (1988) Proc. Natl. Acad. Sci. USA 85, 4148–4152.
187 Stern, L.J., Ahl, P.L., Marti, T., Mogi, T., Dunach, M., Berkovitz, S., Rothschild, K.J. and Khorana, H.G. (1989) Biochemistry 28, 10035–10042.

188 Stern, L.J. and Khorana, H.G. (1989) J.Biol.Chem. 264, 14202–14208.
189 Roepe, P., Ahl, P.L., Das Gupta, S.K., Herzfeld, J. and Rothschild, K.J. (1987) Biochemistry 26, 6696–6707.
190 Roepe, P., Scherrer, P., Ahl, P.L., Das Gupta, S.K., Bogomolni, R.A., Herzfeld, J. and Rothschild, K.J. (1987) Biochemistry 26, 6708–6717.
191 Varo, G. and Lanyi, J.K. (1991) Biochemistry 30, 5016–5022.
192 Gerwert, K. and Siebert, F. (1986) EMBO J. 5, 805–811.
193 Braiman, M.S., Ahl, P.L. and Rothschild, K.J. (1987) Proc. Natl. Acad.Sci. USA 84, 5221–5225.
194 Ormos, P. (1991) Proc. Natl. Acad. Sci. USA 88, 473–477.
195 Renthal, R., Cothran, M., Espinoza, B., Wall, K.A. and Bernard, M. (1985) Biochemistry 24, 4275–4279.
196 Renthal, R., Brogley, L. and Vila, J. (1988) Biochim. Biophys. Acta 935, 109–114.
197 Dencher, N.A., Dresselhaus, D., Zaccai, G. and Bldt, G. (1989) Proc. Natl. Acad. Sci. USA 86, 7876–7879.
198 Glaeser, R.M., Baldwin, J., Ceska, T.A. and Henderson, R. (1986) Biophys. J. 50, 913–920.
199 Frankel, R.D. and Forsyth, J.M. (1985) Biophys. J. 47, 387–393.
200 Marrero, H. and Rothschild, K.J. (1987) Biophys. J. 52, 629–635.
201 Szundi, I. and Stoeckenius, W. (1989) Biophys. J. 56, 369–383.
202 Bousche, O., He, Y.-W., Marti, T., Khorana, H.G. and Rothschild, K.J. (1991) J. Biol. Chem. 266, 11063–11067.
203 Thorgeirsson, T.E., Milder, S.J., Miercke, L.J.W., Betlach, M.C., Shand, R.F., Stroud, R.M. and Kliger, D.S. (1991) Biochemistry 30, 9133–9142.
204 Tittor, J., Soell, C., Oesterhelt, D., Butt, H.-J. and Bamberg, E. (1989) EMBO J. 8, 3477–3482.
205 Birge, R.R., Cooper, T.M., Lawrence, A.F., Masthay, M.B., Zhang, C.-F. and Zidovetzki, R. (1991) J. Am. Chem. Soc. 113, 4327–4328.
206 Ort, D.R. and Parson, W.W. (1979) Biophys. J. 25, 355–364.
207 Groma, G.I., Helgerson, S.L., Wolber, P.K., Beece, D., Dancshazy, Z., Keszthelyi, L. and Stoeckenius, W. (1984) Biophys. J. 45, 985–992.
208 Helgerson, S.L., Mathew, M.K., Bivin, D.B., Wolber, P.K., Heinz, E. and Stoeckenius, W. (1985) Biophys. J. 48, 709–719.
209 Smith, S.O. and Mathies, R.A. (1985) Biophys. J. 47, 251–254.
210 Albeck, A., Friedman, N., Sheves, M. and Ottolenghi, M. (1989) Biophys. J. 56, 1259–1265.
211 Renthal, R., Shuler, K. and Regalado, R. (1990) Biochim. Biophys. Acta Bio-Energetics 1016, 378–384.
212 Der, A., Toth-Boconadi, R. and Keszthelyi, L. (1989) FEBS Lett. 259, 24–26.
213 Keszthelyi, L., Szaraz, S., Der, A. and Stoeckenius, W. (1990) Biochim. Biophys. Acta Bio-Energetics 1018, 260–262.
214 Der, A., Szaraz, S., Toth-Boconadi, R., Tokaji, Zs., Keszthelyi, L. and Stoeckenius, W. (1991) Proc. Natl. Acad. Sci. USA 88, 4751–4755.
215 Murzin, A. and Finkelstein, A.V. (1988) J. Mol. Biol. 204, 749–769.
216 Blanck, A. and Oesterhelt, D. (1987) EMBO J. 6, 265–273.
217 Nathans, J., Thomas, D. and Hogness, D.S. (1986) Science 232, 193–202
218 Guyer, C.A., Horstman, D.A., Wilson, A.L., Clark, J.D., Cragoe, E.J. and Limbird, L.E. (1990) J. Biol. Chem. 265, 17307–17317.
219 Machida, C.A., Bunzow, J.R., Searles, R.P., Van Tol, H., Tester, B., Neve, K.A., Teal, P., Nipper, V. and Civelli, O. (1990) J. Biol. Chem. 265, 12960–12965.

© 1992 Elsevier Science Publishers B.V. All rights reserved

CHAPTER 4

High-resolution crystal structures of bacterial photosynthetic reaction centers

JOHANN DEISENHOFER[1] and HARTMUT MICHEL[2]

[1]Howard Hughes Medical Institute and Department of Biochemistry, University of Texas Southwestern Medical Center, Dallas, TX 75235, USA and [2]Max-Planck-Institut für Biophysik, Abteilung Molekulare Membranbiochemie, Frankfurt/Main, Germany

Contents

1. Introduction 103
2. Chemical nature of RCs of photosynthetic purple bacteria 104
3. Crystallization and X-ray structure analysis of RCs 105
   3.1. The RC of *Rhodopseudomonas viridis* 105
   3.2. The RC of *Rhodobacter sphaeroides* 111
4. Implications 112
   4.1. Pathways and mechanism of electron transfer 112
   4.2. RC mutants 113
   4.3. Photosystem II RC: similarity to RCs of purple bacteria 114
   4.4. Membrane protein structure 114
5. Prospects 115
   5.1. Crystallography 115
   5.2. Electron crystallography 116
Acknowledgements 116
References 116

1. Introduction

Photosynthetic reaction centers (RCs) are membrane-spanning complexes of polypeptide chains and cofactors that catalyze the first steps in the conversion of light energy to chemical energy during photosynthesis. Absorption of a photon in the RC, or energy transfer from light-harvesting complexes to the RC, causes rapid, efficient electron transfer from a primary donor along a chain of acceptors leading across the photosynthetic membrane. In all types of RCs, the primary donor has been shown, or is assumed to be, a closely associated pair of chlorophyll molecules [1]. As early as 1971 the existence of such special pair was postulated based on electron spin-resonance experiments [2]. The chemical nature of the acceptor molecule at the end of the electron transfer chain within RCs depends on the type of RC: in photosynthetic purple bacteria, and in photosystem II of green plants and cyanobacteria, it is a quinone which is converted to a quinol by two successive electron transfer events and protonation. The end of the electron-transfer chain in photosystem I is an iron–sulphur center [3]. In each case, part of the energy of the photon is stored in a chemical compound; this compound is processed further to generate either an electrochemical proton gradient across the photosynthetic membrane, or the strong reductant NADPH.

Photosynthetic purple bacteria were the first organisms in which RCs were discovered [4]; their relative simplicity made them logical targets for studying photosynthetic light reactions. Reed and Clayton [5] reported the first preparation of an RC, that of *Rhodopseudomonas (Rps.) sphaeroides* (later renamed *Rhodobacter (Rb.) sphaeroides*). This pioneering work and the purification of RCs of other purple bacteria opened the way for the elucidation of the main functional and structural properties of RCs (for a review see ref. [6]). The determination of the high-resolution three-dimensional structures of RCs of two species of purple bacteria has further increased the understanding of their function. Thus, the RCs of photosynthetic purple bacteria are to date the best known mediators of primary photosynthetic processes, in spite of rapid progress with studies of the photosystems of green plants and cyanobacteria.

In this chapter, we focus on X-ray crystallographic studies on RCs from purple bacteria and describe the main features of their structures, mostly using the RC of *Rps. viridis* as an example. We then discuss implications of the structures for RC function, relations to photosystem II, and aspects of membrane protein structure. Finally, we mention new developments in structural studies of other photosynthetic membrane protein complexes.

Recent review articles discuss various aspects of RCs: structure and function [1,6–8], spectroscopy and electron transfer [9,10], site-specific mutagenesis [11], primary photochemistry [8,12], and membrane protein structure [13,14]. Useful collections of articles on these topics can also be found in conference proceedings [15,16].

2. *Chemical nature of RCs of photosynthetic purple bacteria*

RCs of photosynthetic purple bacteria consist of at least three protein subunits called L (light), M (medium), and H (heavy), after their apparent molecular weights determined with SDS gel electrophoresis [17]. Subunits L and M have very hydrophobic amino acid sequences, in which five membrane-spanning segments can be distinguished [18–22]; subunit H has a single membrane-spanning segment near its NH_2 terminus [18,23,24]. The subunits L and M bind bacteriochlorophylls (BChl), bacteriopheophytins (BPh), quinones, a ferrous iron ion, and a carotenoid as cofactors.

Species differ in the amino acid sequences of their protein subunits and in the types of cofactors bound to them. Amino acid sequence homologies between subunits of the four species *Rps. viridis, Rb. capsulatus, Rb. sphaeroides* and *Rhodospirillum rubrum* range from 60 to 78% for subunit L, and from 50 to 77% for subunit M [21]. The currently known sequences of the three H-subunits show homologies between 38 and 64% [23]. RCs of different organisms can contain different types of BChl and BPh. For example, RCs of *Rb. sphaeroides* and *Rb. capsulatus* contain BChl and BPh of type *a* (BChl-*a* and BPh-*a*); RCs of *Rps. viridis* and *Thiocapsa pfennigii* contain BChl-*b* and BPh-*b* [25]. The quinones also can be of different types: *Rb. sphaeroides* has two ubiquinone-10 molecules [7], whereas *Rps. viridis* has one menaquinone-9 and one ubiquinone-9 molecule [26]. Various types of carotenoids are used in different species; for example in *Rps. viridis* the major carotenoid species is 1,2-dihydroneurosporene (I. Sinning. and H. Michel, unpublished data), in *Rb. sphaeroides* it is spheroidene [27].

Another major difference between RCs from different species of photosynthetic purple bacteria is the presence or absence of a four-heme *c*-type cytochrome as the fourth protein subunit. Such a subunit is part of the RC of e.g. *Rps. viridis* and *Thiocapsa pfennigii*; RCs of other bacteria, e.g. *Rb. sphaeroides, Rb. capsulatus,* and *Rhodospirillum rubrum,* consist of only the three protein subunits L, M, and H.

3. Crystallization and X-ray structure analysis of RCs

To date, high-resolution structural information on RCs has been obtained only with X-ray crystallography. By 'high resolution' we mean resolution sufficient to construct an accurate atomic model of a molecule; this resolution limit depends on the quality and completeness of the measured diffraction data, and on the accuracy of the phase information available; it usually lies between 3.5 and 3.0 Å.

A prerequisite for X-ray crystallographic structure analysis is the formation of well-ordered three-dimensional crystals. With steady improvement of X-ray sources, X-ray detectors, and crystallographic methods and software over the last 30 years, crystallization has become in many cases the rate-limiting step in the high-resolution structure analysis of macromolecules. This is especially true for integral membrane proteins. Until about 1980, three-dimensional crystallization of this class of proteins was generally considered impossible. Then three-dimensional crystals were obtained for bacteriorhodopsin [28] and matrix porin from *Escherichia coli* [29] using the detergent octylglucoside to form water soluble protein-detergent complexes that could be treated like soluble proteins in vapor diffusion experiments in which ammonium sulphate or polyethylene glycol served as precipitant. These experiments were soon followed by the crystallization [30] and X-ray structure analysis [31] of the RC of *Rps. viridis*.

3.1. The RC of Rhodopseudomonas viridis

The RC from the purple bacterium *Rps. viridis* is a complex of 4 protein subunits and 14 cofactors. The protein subunits are, in order of decreasing size, cytochrome [336 amino acids (a.a.)], subunit M (323 a.a.), subunit L (273 a.a.), and subunit H (258 a.a.); the complete amino acid sequences of these subunits were derived from the sequences of the genes coding for them [22,24,32]. The cofactors are four heme groups, covalently linked to the cytochrome subunit, four BChl-*b*, two BPh-*b*, a menaquinone-9 and a ubiquinone-9, a ferrous iron ion, and the carotenoid dihydroneurosporene [26,33].

Well-ordered three-dimensional crystals of the RC of *Rps. viridis* were obtained from RCs solubilized with the detergent *N*,*N*-dimethyl-dodecylamine-*N*-oxide (LDAO); the small amphiphile heptane-1,2,3-triol was an essential additive, and ammonium sulphate was the precipitant in a vapor diffusion experiment [30,34]. The crystals have the symmetry of space group $P4_32_12$. The unit cell has dimensions of $a = b = 223.5$ Å, $c = 113.6$ Å; it contains eight RC molecules [30,31]; about 70% of the unit cell's volume is occupied by solvent. The crystals diffract X-rays to at least 2.3 Å resolution; at temperatures around 4°C they are rather insensitive to radiation damage, so that a substantial portion of a diffraction data set can be collected from one crystal [31].

The structure of the RC crystals was initially determined at 3.0 Å resolution using the method of multiple isomorphous replacement with heavy-atom compounds [31]. Heavy-atom derivatives were prepared by soaking RC crystals in solutions of heavy-atom compounds. To preserve the high crystalline order during the soaking experiments, an additional purification step for the RC molecules had to be introduced, and a special 'soak buffer' had to be developed

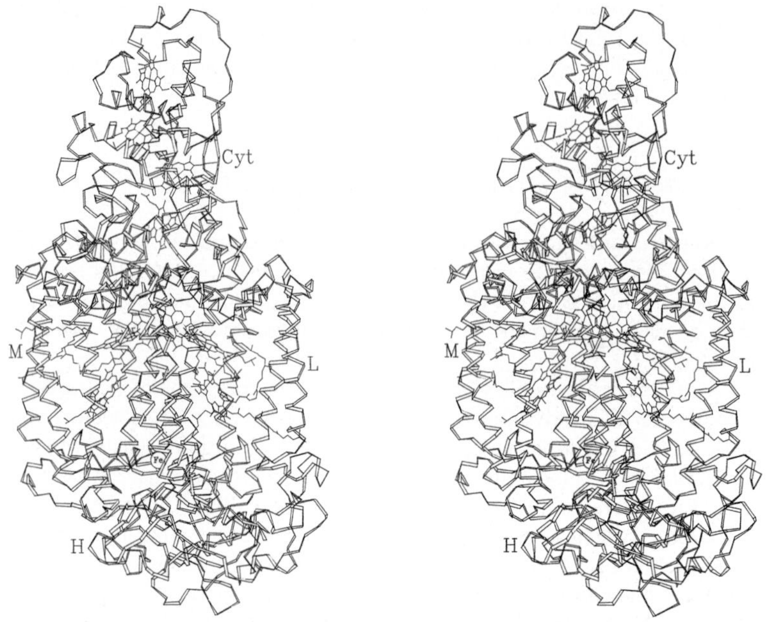

Fig. 1. (Stereo pair). Overall view of the RC of *Rps. viridis*. The polypeptide chains of the protein subunits are represented by ribbons; cofactors are drawn as wire models. The membrane plane is approximately perpendicular to the plane of projection; the approximate positions of the periplasmic (upper line) and the cytoplasmic membrane surfaces are indicated.

[31,35]. Small mercury compounds and uranyl nitrate turned out to be the most suitable for the phase determination. Phase information from five heavy-atom derivatives enabled the calculation of an electron density map, which the method of solvent flattening [36] further improved. On the basis of this electron density map, an atomic model of the RC was built for the cofactors [31] and subsequently for the protein subunits [37]. The combination of sequence information and crystallographic data allowed the description of protein–cofactor interactions [38]. Crystallographic refinement at 2.3 Å resolution (J. Deisenhofer, O. Epp, I. Sinning, H. Michel, in preparation) led to a an atomic model consisting of 10288 atoms, including e.g. 201 bound water molecules, 7 bound anions, and 1 firmly bound LDAO molecule [35,39,40]. The model has a crystallographic R-value of 0.19 for 95762 unique reflections to 2.3 Å resolution ($R = \Sigma|F_o - F_c|/\Sigma F_o$; F_o and F_c are the observed and calculated structure factor amplitudes, respectively; the summation includes all unique reflections used in the refinement); the upper limit for the average error in the atomic coordinates [41] was estimated to be 0.266 Å [35,40]. X-ray data at 2.3 Å resolution do not allow to distinguish between carbon, nitrogen, and oxygen atoms. Therefore, the orientations of the side chains of amino acids such a asparagine, glutamine, histidine, and threonine, or of acetyl groups of BChl and BPh, were chosen so that a maximum number of H-bonding interactions with suitable geometries between these groups and their environment was achieved. In a low-resolution neutron diffraction study, the distribution of the detergent in the crystal could be determined [42].

Figure 1 shows an overall view of the model of the RC of *Rps. viridis*. In its longest dimen-

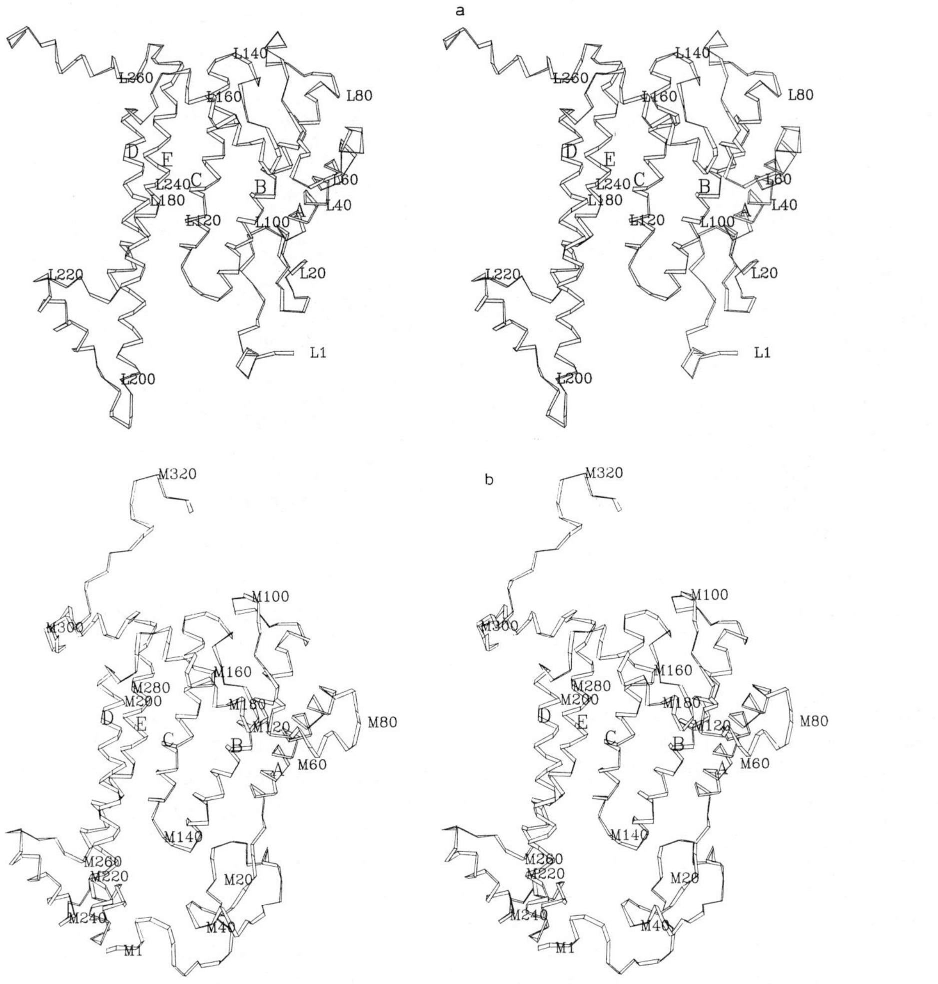

Fig. 2. (Stereo pair). L-subunit (a) and M-subunit (b) of the RC of *Rps. viridis* in ribbon representation. The transmembrane helices in each subunit are labeled A, B, C, D, and E.

sion, the complex measures ~130 Å; the maximum width perpendicular to that direction is ~70 Å. The closely associated subunits L and M, together with the bound BChl, BPh, quinones, nonheme iron, and carotenoid, form the central part of the RC. The most prominent structural features of each of the central subunits are five long hydrophobic helices that are assumed to span the bacterial membrane. This assumption is supported by the distribution of detergent in the crystal [42], by properties of the model [35], and by functional considerations. The orientation of RCs with respect to the cytoplasmic and periplasmic sides of the membrane was determined e.g. by proteolysis experiments [43–45], which showed that the NH_2 termini of subunits L and M, and subunit H are accessible at the cytoplasmic membrane surface. The polypeptide backbones of subunits L and M and the attached cofactors display a high degree

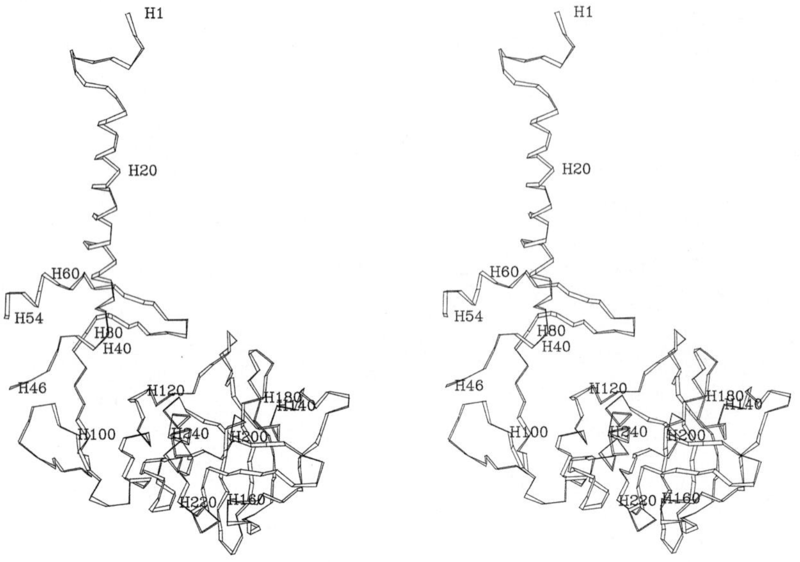

Fig. 3. (Stereo pair). The H-subunit of the RC of *Rps. viridis* in ribbon representation. Residues H47 to H53 are disordered in the crystal and not shown.

of approximate local twofold symmetry; the local twofold symmetry axis is oriented perpendicular to the membrane plane [46]. On either side of the membrane-spanning region of the L–M complex, a peripheral subunit is attached: the cytochrome with its four bound heme groups at the periplasmic side, and the globular domain of the H-subunit at the cytoplasmic side of the membrane. The H-subunit contributes a single membrane-spanning helix. Neither the cytochrome nor the H-subunit have the local symmetry found in the central part of the RC; the cytochrome has an internal local symmetry of its own.

Figure 2 shows schematic drawings of the polypeptide chain folding of the reaction center subunits L and M; the structural similarity between these subunits is obvious. Structurally similar segments in the L and M subunits include the transmembrane helices (labeled A, B, C, D, E) and a large fraction of the polypeptide chains connecting them. In total, 216 α-carbons of the M-subunit can be superimposed onto corresponding α-carbons of the L-subunit by a rotation of almost exactly 180° to a root mean square (rms) deviation of 1.22 Å. The transmembrane helices are between 21 and 28 residues long; additional, shorter helices can be found in the connecting polypeptide chains. The M-subunit is 50 residues longer than the L-subunit. Both sequence alignments [22], and structural alignments [35,37] show that the sequence insertions in M reside on either side of the transmembrane helices (20 residues at the NH_2 terminus, 7 residues in the connection between transmembrane helices MA and MB, 7 residues in the connection between MD and ME, and 16 residues at the COOH terminus). Through these insertions, the M-subunit dominates the contacts of the L–M complex to the peripheral subunits. The insertion between transmembrane helices MD and ME, containing another small helix, is of importance for binding of the nonheme iron and for the different conformations of the quinone binding sites in L and M.

The H-subunit, shown in Fig. 3, can be divided into three structural regions with different

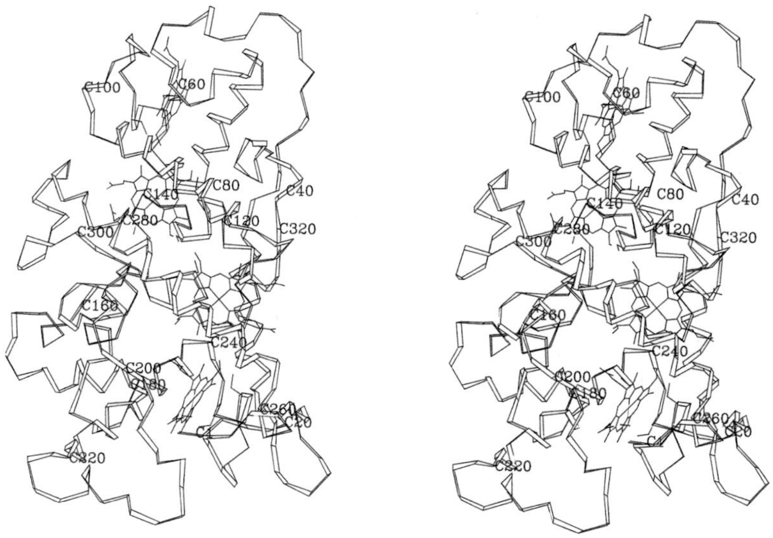

Fig. 4. (Stereo pair). The cytochrome subunit of the RC of *Rps. viridis.* The polypeptide chain is drawn as a ribbon; the hemes and the heme-binding cysteine residues are drawn as wire models. The view is approximately along the local twofold symmetry axis of the cytochrome. The hemes are numbered 1, 2, 4, 3 (top to bottom).

characteristics. The amino terminal segment, beginning with formylmethionine [24], contains the only transmembrane helix of subunit H (residues H12 to H35). Near the end of the transmembrane helix, the polypeptide chain consists of seven consecutive charged residues (H33 to H39). No significant electron density can be found for residues H47 to H53; they are disordered in the crystal. Following the disordered region, the H-chain forms an extended structure along the surface of the L–M complex, apparently deriving structural stability from that contact. The surface region contains a short helix and two two-stranded antiparallel β-sheets. The third structural segment of the H-subunit, starting at about residue H105, forms a globular domain. This domain contains an extended system of parallel and antiparallel β-sheets (between residues H134 and H203), and an α-helix (residues H232 to H248).

The cytochrome is the largest subunit in the reaction center complex [32]. Its structure, shown in Fig. 4, consists of an NH_2-terminal segment, two pairs of heme binding segments, and a segment connecting the two pairs. Each heme binding segment consists of a helix with an average length of 17 residues followed by a turn and the Cys–X–Y–Cys–His sequence typical for *c*-type cytochromes. The hemes are connected to the cysteine residues via thioether linkages in such a way that the heme planes are parallel to the helix axes. The Nε2 atoms of the histidine residues are the fifth ligands to the heme irons (the pyrrole nitrogens are ligands 1 to 4). The sixth ligands to three of the four heme irons are the sulphur atoms of the methionine residue within the helices. The exception in this scheme is heme 4, where histidine C124, located within heme binding segment 2, serves with its Nε2 atom as a sixth ligand for the iron. The two pairs of heme binding segments, containing hemes 1, 2 and 3, 4, respectively, are related by a local twofold symmetry. Within each of these regions 65 residues obey this local symmetry with an rms deviation between corresponding α-carbon atoms of 0.93 Å. The local symmetry of the

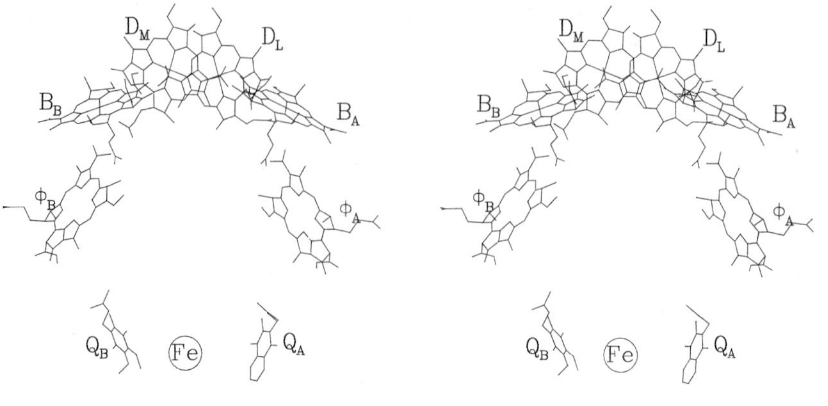

Fig. 5. (Stereo pair). Model of the cofactors associated with subunits L and M in the RC of *Rps. viridis*. The molecules are labeled according to the nomenclature used in the text. Only the head groups are shown for BChl-*b*, BPh-*b*, and quinones; the carotenoid was omitted.

cytochrome is not related to the local symmetry in the center of the RC (see above). Lack of electron density for the last four residues, C333 to C336, indicates disorder. Also disordered is the lipid moiety bound to the NH_2-terminal cysteine residue [47].

Figure 5 shows arrangement and nomenclature of the cofactors associated with protein subunits L and M, excluding phytyl side chains or isoprenoid side chains for clarity. A closely associated pair of BChl-*b*s, the special pair, resides at the origin of two branches of cofactors, each of which consists of another BChl-*b* (the accessory BChl-*b*), a BPh-*b*, and a quinone. The nonheme iron sits between the quinones and is bound to five amino acid residues, four histidines (L190, L230, M217, M264), and one glutamic acid (M232) [37,38]. The Mg^{2+} ions of the BChl-*b* are five-coordinated with two histidine residues each from the subunits L and M, forming metal–nitrogen bonds (L153 $\rightarrow$ B_A, L173 $\rightarrow$ D_L, M180 $\rightarrow$ B_B, M200 $\rightarrow$ D_M). D_L, D_M, Φ_A, Φ_B, Q_A and Q_B form hydrogen bonds with the protein [38].

A variety of nomenclatures for the cofactors have been used in publications on RC structure. The first proposal [37] was too complicated, and alternatives were discussed [48]. We use D for the special pair, B for the accessory BChl, Φ for the BPh, and Q for the quinones. The branches are denoted by subscripts A and B. Because D belongs to both branches, its two BChl, whenever they are mentioned individually, are distinguished by subscripts L and M according to the protein subunit to which their magnesium atoms are linked.

The tetrapyrrole rings of BChl-*b*, BPh-*b*, and the quinone head groups follow the same approximate local symmetry as shown by the L and M chains. This local twofold symmetry is most perfect in the arrangement of the special pair BChl-*b*s. The two molecules overlap with their pyrrole rings I. The orientation of the rings leads to a close proximity between the ring I acetyl groups and the Mg^{2+} ions, but the acetyl groups do not act as ligands to the Mg^{2+}. The pyrrole rings I of D_L and D_M are nearly parallel to each other, and to the symmetry axis; they are ~3.2 Å apart. However, the remainders of the tetrapyrrole ring systems of D_L and D_M are not parallel to each other: planes through the pyrrole nitrogens of D_L and D_M form an angle of 11.3°. Individual atoms of the rings can deviate from these planes by more than 0.5 Å. The D_M ring is considerably more deformed than the D_L ring.

The tetrapyrrole rings of the BChl-*b* and BPh-*b* of D and of the A and B branches can be superimposed using a single coordinate transformation with the reasonably low rms deviation of 0.38 Å between the positions of equivalent atoms. Close inspection, however, shows significant differences between the local symmetry operations, individually superimposing D_M to D_L (180° rotation), B_B to B_A (176° rotation), and Φ_B to Φ_A (173° rotation) [35]. Imperfect symmetry causes the interatomic distances and inter-planar angles to differ in the two branches. For example, the closest distance of atoms involved in double bonds in D and in Φ_A is 0.7 Å shorter than the corresponding distance in D and Φ_B. These structural differences, together with different environments of the cofactors [38] may account in part for the different electron transfer properties in these branches (see below).

Large deviations from local symmetry are manifest in the structures of the phytyl chains of the B and Φ cofactors, in the different chemical nature and different occupancy of the quinones (Q_B is bound to only ~30% of the RCs in the crystal), and in the presence of a carotenoid molecule near B_B only. Differences in the crystallographically refined B-values of Φ and of the phytyl chains of B indicate a higher mobility of the cofactors of the B branch.

3.2. The RC of Rhodobacter sphaeroides

The RC of *Rb. sphaeroides* consists of the three protein subunits M (307 a.a.), L (281 a.a.), and H (260 a.a.), and of ten cofactors. As mentioned above, it does not include a bound cytochrome subunit; the cofactors are 4 BChl-*a*, 2 BPh-*a*, 2 identical quinones (ubiquinone-10), and the carotenoid spheroidene. The amino acid sequences of the RC subunits M, L, and H of *Rb. sphaeroides* and *Rps. viridis* show 59, 49, and 39% homology, respectively [22–24]; these values imply very similar three-dimensional structures.

The RC of *Rb. sphaeroides* was crystallized under several conditions (see e.g. Table 1 in ref. [8]). RC crystals were reported for the carotenoidless mutant R-26 [49–52], wild-type strain 2.4.1 [53,54], and wild-type strain Y [55]. The crystals for which structures have been published so far have the symmetry of space group $P2_12_12_1$ and diffract X-rays to about 3.5 to 2.8 Å resolution [49,52]. They were grown either in the presence of β-octylglycoside and polyethylene glycol [49] or in the presence of LDAO, heptane triol, and polyethylene glycol, followed by exchange of the detergent to β-octylglucoside after completion of crystal growth [52].

The high degree of structural similarity between the RC of *Rps. viridis* (without the cytochrome subunit) and that of *Rb. sphaeroides* was demonstrated in two separate studies by the success of molecular replacement calculations: a structural model of the *Rps. viridis* RC, from which the cytochrome subunit and the animo acid side chains that are not identical in the two structures were removed, could be located in the crystals of the *Rb. sphaeroides* RC; the correctly placed model was subsequently used for the solution of the phase problem and as a starting point for crystallographic refinement [51,56]. Refinement of the *Rb. sphaeroides* RC models was more difficult than for the *Rps. viridis* RC model because the crystals were less well ordered, which led to a smaller number of measurable reflections, and, therefore, to a considerably lower ratio of the number of observations to the number of refined model parameters. Also, water molecules bound to the RC could not be located. Results of crystallographic refinement were published [52,54,57–64].

The basis of one of the studies of the RC of *Rb. sphaeroides* R-26 was a set of X-ray diffraction intensities with 2.8 Å resolution, containing 23349 reflections strong enough to be significantly measured [52]. The refined model has a crystallographic *R*-value of 0.24; the average error in the atomic coordinates in this model was estimated to ~0.5 Å [54]. As expected, the structure of the RC of *Rb. sphaeroides* is very similar to that of *Rps. viridis* (without the cytochrome subunit). The same approximate twofold symmetry is found in the L–M complex and the associated cofactors. However, significant differences in the interactions of these cofactors with the protein were described, part of which result from differences in amino acid sequence [54,59]. The Q_B binding site in *Rb. sphaeroides* appears to be fully occupied; Q_A is found in a position further apart from the nonheme iron than in the *Rps. viridis* RC [59]. Surprising differences in interactions appear to arise from altered side chain conformations of the histidine residues that bind the magnesium of D_L and B_A in *Rps. viridis*: in the *Rb. sphaeroides* RC, they seem to be engaged in intraprotein H-bonds and the magnesium ions appear only four-coordinated [54].

A second study of *Rb. sphaeroides* R-26 [63,64] used 13493 significantly measured reflections with 3.1 Å resolution; the *R*-value of the refined model was 0.22, with an estimated error in the coordinates of ~0.5 Å [63]. In general the conclusions on the structure of the RC of *Rb. sphaeroides* were very similar in both studies. However, the two studies differ in their description of the interactions between cofactors and protein, with the results of the second study being more similar to those obtained for the *Rps. viridis* RC. Clearly, more work is necessary to resolve these discrepancies.

Refinement of the RC structure of *Rb. sphaeroides* wild-type strain Y was reported in brief [62] without giving structural details.

4. *Implications*

4.1. *Pathways and mechanism of electron transfer*

The current understanding of the function of the RC was developed by combining structural information with information from other experimental techniques, notably spectroscopy, as described in recent reviews [6,12,65]. The special pair, D, is the starting point for a light-driven electron transfer reaction across the membrane. Absorption of a photon or energy transfer from light-harvesting complexes in the membrane raises D to its first excited singlet state, D*. From D*, an electron is transferred to Φ_A with a time constant of ~3 ps [66–68]. The cofactor Φ_A can be identified as the electron acceptor because Φ_A and Φ_B have absorption maxima at slightly different wavelengths; with the knowledge of the crystal structure, linear dichroism absorption experiments can distinguish between them [69–71]. The electron transfer time constant is shortened considerably when the temperature is lowered to near absolute zero. The exact nature of this electron transfer step and of the temperature dependence of the time constant is currently under experimental [72–75] and theoretical investigation [76–79]. One of the main open questions concerns the role of B_A, located between D and Φ_A: B_A was considered to be only indirectly involved in the electron transfer between D and Φ_A; very recently new spectroscopic results were interpreted as indication that B_A acts as the first electron acceptor [73,74].

From Φ_A, the electron is transferred to Q_A with a time constant of ~200 ps at room temperature [12,74,80]. This time constant shortens to ~100 ps at temperatures below 100 K [80]. In moving from D to Q_A, the electron has crossed most of the membrane. From Q_A, the electron moves on to Q_B within about 100 μs; the nonheme iron does not seem to play an essential role in this step [81]. Q_B can pick up two electrons and two protons [82]. Q_BH_2 then dissociates from the RC, and the Q_B site is refilled from a pool of quinones dissolved in the membrane. Electrons and protons on Q_BH_2, together with up to two additional protons, are transferred back through the membrane by the cytochrome bc_1 complex (ubiquinol-cytochrome *c* oxidoreductase). Cytochrome c_2 molecules deliver the electrons back to the RC. The whole process can be described as a light-driven cyclic electron flow, the net effect of which is the generation of a proton gradient across the membrane that is used to synthesize adenosine triphosphate, in accord with Mitchell's chemiosmotic theory [83].

In RCs without a bound cytochrome subunit, cytochrome c_2 directly reduces D^+. In RCs with a bound cytochrome subunit, D^+ first receives an electron from the closest heme group; the time constant for this electron transfer step in *Rps. viridis* is ~270 μs at room temperature [84]. Cytochrome c_2 in turn reduces the bound cytochrome. It is currently unknown which of the four hemes of the bound cytochrome of *Rps. viridis* accepts the electrons from cytochrome c_2. The individual redox potentials of the four hemes (listed in order of increasing distance to D) were determined in RC crystals to be +370 mV (heme 3), +10 mV (heme 4), +300 mV (heme 2), and −60 mV (heme 1) [85]; in other studies similar results were obtained [82,86,87]. Electron transfer processes between these heme groups are under investigation.

One of the major surprises from the structural work was the symmetry of the core structure, which raised questions of why only the A branch of the cofactors is used [88,89] and of the significance of the apparently unused branch. Further questions relate to electron transfer between Q_A and Q_B, the role of the nonheme iron, and the function of Q_B as a two-electron gate and proton acceptor. The atomic coordinates of RCs have been used in calculations investigating e.g. molecular dynamics or electric fields in order to answer some of these questions [90–95]. Unfortunately, atomic coordinates determined by X-ray crystallography may not be accurate enough for some of these studies (for a discussion, see ref. [10]).

4.2. *RC mutants*

RC mutants have been obtained by site-specific mutagenesis and selection of resistant strains in the presence of herbicides that block the Q_B binding site of wild-type RCs. Site-specific mutagenesis is most advanced for RCs of *Rb. capsulatus*, as reviewed recently [11]. Youvan and colleagues designed a number of interesting mutants and characterized them spectroscopically; a particularly interesting example is a species in which D is a heterodimer of BChl-*a* and BPh-*a* [96]. Unfortunately, and ironically, the RC of *Rb. capsulatus* has not yet been crystallized, so the structural effects of these mutations cannot be investigated directly.

Very recently, site-specific mutants were also reported for RCs of *Rb. sphaeroides* [97,98], but crystallographic characterizations of them have not yet been published. Due to various problems, site-specific mutagenesis of the RC of *Rps. viridis*, for which the crystal structure is known to the highest precision, has become possible only very recently; structural information on site-specific mutants has not yet been reported [133].

Spontaneous mutants that are resistant to terbutryn and atrazine, inhibitors of Q_B binding, have been selected for *Rps. viridis* [99,100] and for *Rb. sphaeroides* [97,101–103]. First results of crystallographic characterization of the two *Rps. viridis* mutants T1 (Ser L223 → Ala and Arg L217 → His) and T4 (Tyr L222 → Phe) are available [104,105]. In T1, the structural changes are restricted to the two mutated residues and Asn L212; these changes also affect the geometry of the Q_B binding. In contrast, T4 shows a shift of a 26-residue segment of the M-subunit (M25–M50). These examples indicate that mutations in the RC structure can have a wide range of structural consequences that most likely could not have been predicted with the currently available tools of macromolecular modeling.

4.3. Photosystem II RC: similarity to RCs of purple bacteria

The elucidation of the three-dimensional structure of RCs of purple bacteria significantly influenced the identification of the photosystem II (PS II) reaction center of green plants and cyanobacteria [106–108]. Weak sequence homologies between subunits L and M of purple bacteria and proteins D1 and D2 of PS II have been found [18,20,22]. The three-dimensional RC structures revealed that the sequence homologies include key amino acid residues, e.g. histidine residues that bind D_L, D_M, and the nonheme iron, and tryptophane and phenylalanine residues in contact with the quinones. These and other findings strengthen the hypothesis that the RCs of purple bacteria and of PS II are structurally related, and that proteins D1 and D2 are equivalent to the subunits L and M, respectively. Interestingly, the histidines binding B_A and B_B in the bacterial RC are not conserved in the sequences of D1 and D2, indicating the absence or a different mode of binding of the accessory chlorophylls. The preparation of a complex of D1, D2, cytochrome b^{559}, 5 chlorophylls, 2 pheophytins, and 1 β-carotene, which could perform the initial charge separation step of PS II, was strong experimental support for the PS II RC model [109]. Additional evidence comes from the study of herbicide-resistant PS II mutants, in which sequence changes are found in a region analogous to the Q_B binding site of the bacterial RC [110]. Structural information from the bacterial RC was a valuable guideline during the recent identification of two intermediate electron carriers as tyrosine residues in proteins D1 and D2 [111–114].

4.4. Membrane protein structure

Structural properties that characterize the RCs as integral membrane proteins were discussed in recent reviews [13,14] and are briefly summarized here.

The surface of the RC of *Rps. viridis* shows a sharp transition between a polar character in the peripheral subunits and a very hydrophobic character in a band of ~30 Å width around the center of the complex [35,115]. This band is covered by detergent in the crystals [42] and presumably reflects the dimensions of the hydrophobic core of the membrane. An analysis of the RC of *Rb. sphaeroides* [58] led to the same estimate for the thickness of the membrane core.

Charged amino acid residues are absent from a zone of ~25 Å thickness in the center of the RC of *Rps. viridis*; the position of this zone coincides with that of the hydrophobic surface zone. The net charges on the periplasmic and cytoplasmic side of the membrane-spanning segments of subunits L and M are negative and positive, respectively [22], thereby neutralizing at least part of the membrane potential; this neutralization may facilitate electron transfer from D to Q_A, and also promote the correct insertion of these proteins into the membrane. Other amino acid residue types (e.g. Tyr, Trp) also show characteristic distributions with respect to the membrane interior [14,35]. Firmly bound water molecules in the RC of *Rps. viridis* are primarily found outside the membrane-spanning region; only a few are located inside the membrane core [14,35].

An analysis of the RC model of *Rb. sphaeroides* and of the known amino acid sequences of related proteins showed that, as in soluble proteins, surface residues are less conserved than residues buried in the interior of the structure [13,60,116]. The study also showed that the polarity of the interior of the RC structure resembles that of soluble proteins and is intermediate between protein surfaces exposed to aqueous solution and those exposed to hydrophobic cores of membranes [116].

5. *Prospects*

The high-resolution structures of the RCs of *Rps. viridis* and of *Rb. sphaeroides* have helped to develop a deeper insight into the primary processes of photosynthesis. Further progress requires extension of the structural studies to other systems, notably photosystems I and II, and light-harvesting antenna complexes. Efforts to crystallize key photosynthetic proteins have been intensified, and new techniques for structure determination emerge; some recent results are mentioned below.

5.1. *Crystallography*

Success with crystallization experiments of photosystem I (PS I) preparations were reported recently [117–119]. A PS I preparation from the thermophilic cyanobacterium *Phormidium laminosum* was crystallized in the presence of triton X-100, sodium dodecylsulphate, or dodecylmaltoside as detergents, and polyethylene glycol as precipitant [117]. The crystals reached maximum dimensions of about 0.1 mm; X-ray diffraction experiments with them were not reported so far.

Crystals diffracting X-rays with about 4 Å resolution were obtained for the PS I reaction center of the cyanobacterium *Synechococcus* sp., using β-dodecylmaltoside as detergent, polyethylene glycol 6000 as precipitant, and various combinations of buffer and salt [118,119]. The crystal symmetry is that of space group $P6_3$ or $P6_322$; the building blocks forming the crystals are molecular aggregates with an estimated molecular weight of about 600 kDa [119]; evidence from biochemical analysis and from electron microscopy [120] supports the view that these aggregates are trimers of identical PS I reaction center complexes. Each PS I reaction center is assumed to consist of the *psa-A* and *psa-B* gene products with molecular weights of about 83 kDa each, and of four additional polypeptides with molecular weights in the range 9–18 kDa; these proteins bind a variety of cofactors, among them an as yet unknown number of chlorophyll molecules. Crystal packing considerations, assuming $P6_322$ space group symmetry, lead to a model with one PS I reaction center ($M_r \approx 200$ kDa) in the asymmetric unit [119]. The crystals are stable enough to allow collection of X-ray diffraction data. With some improvement of the crystal quality the determination of the structure at atomic resolution of this large membrane protein complex with its many bound chlorophylls should be possible.

Very recently several crystal forms of PS I RCs of the thermophilic cyanobacterium *Mastigocladus laminosus* were obtained. The best of these crystals diffract X-rays with about 5.5 Å resolution [121].

Single crystals for light-harvesting antenna complexes of several photosynthetic bacteria were grown [122–124], but in each case the crystals diffracted X-rays only with about 10 Å resolution. Crystals of a complex of B875 antenna and RCs from *Rps. palustris* [125] were reported, but apparently they were not suitable for X-ray diffraction experiments.

Recently crystallization and X-ray diffraction with 3.5 Å resolution of the B800–850 light-harvesting complex of the *Rps. acidophila* strain 10050 was published [126]. Crystallization of this complex, an integral membrane protein, was achieved in the presence of 0.9 M phosphate with β-octylglucoside as detergent, benzamidine hydrochloride as small amphiphile and ammonium sulphate as precipitant. The crystals have the symmetry of space group R32, and are assumed to contain 6 α- and 6 β-subunits, 18 BChl-*a* molecules, and 9 carotenoid molecules per asymmetric unit [126]. With the possibility of non-crystallographic symmetry averaging it should be possible to obtain an atomic model of the complex. A second crystal form of the light-harvesting complex with tetragonal symmetry diffracts only with 12 Å resolution [126].

Well-diffracting crystals were obtained also of the B800–850 light-harvesting complex of *Rhodospirillum molischianum* (H. Michel, unpublished). Their symmetry is that of space group $P42_12$, with unit cell dimensions of $a = b = 92$ Å, and $c = 208$ Å; they diffract with about 2.6 Å resolution.

5.2. Electron crystallography

Electron diffraction from two-dimensional crystals as an alternative to X-ray crystallography has recently attracted increased attention. For a long time the only structural information on membrane proteins was an electron diffraction study of bacteriorhodopsin at a resolution high enough to distinguish individual transmembrane helices, but too low to see amino acid side chains or helix connections [127]. Diffraction was observed to much higher resolution, but phasing turned out to be a very serious problem. Very recently, due to progress in methods and technology, this problem was overcome, and an atomic model of the transmembrane helices of bacteriorhodopsin was published [128,129]. Similar studies of plant light-harvesting complexes [130] and of bacterial porins [131] by electron crystallography are in progress. With a reliable method to determine three-dimensional phase sets, electron diffraction can become a most valuable tool for structural studies, especially of membrane proteins under conditions more resembling their natural environment than single crystals.

Acknowledgements

Our colleagues Otto Epp, Kunio Miki, and Irmi Sinning made significant contributions to the X-ray crystallographic studies of the RC of *Rps. viridis*. Figures 1 to 5 were prepared with a computer program by A.M. Lesk and K.D. Hardman [132].

References

1 Nitschke, W. and Rutherford, A.W. (1991) Trends Biochem. Sci. 16, 241–245.
2 Norris, J.R., Uphaus, R.A., Crespi, H.L. and Katz, J.J. (1971) Proc. Natl. Acad. Sci. USA 68, 625–628.
3 Golbeck, J.H. (1987) Biochim. Biophys. Acta 895, 167–204.
4 Duysens, L.N.M. (1989) Biochim. Biophys. Acta 1000, 395–400.
5 Reed, D.W. and Clayton, R.K. (1968) Biochem. Biophys. Res. Commun. 30, 471–475.
6 Parson, W.W. (1987) in: J. Amesz (Ed.) New Comprehensive Biochemistry: Photosynthesis, Elsevier, Amsterdam, pp. 43–61.
7 Feher, G., Allen, J.P., Okamura, M.Y. and Rees, D.C. (1989) Nature 339, 111–116.
8 Budil, D.E., Gast, P., Chang, C.-H., Schiffer, M. and Norris, J.R. (1987) Annu. Rev. Phys. Chem. 38, 561–583.
9 Friesner, R.A. and Won, Y. (1989) Biochim. Biophys. Acta 977, 99–122.
10 Boxer, S.G. (1990) Annu. Rev. Biophys. Biophys. Chem. 19, 267–299.
11 Coleman, W.J. and Youvan, D.C. (1990) Annu. Rev. Biophys. Biophys. Chem. 19, 333–367.
12 Kirmaier, C. and Holten, D. (1987) Photosynth. Res. 13, 225–260.
13 Rees, D.C., Komiya, Y., Yeates, T.O., Allen, J.P. and Feher, G. (1989) Annu. Rev. Biochem. 58, 607–633.

14 Michel, H. and Deisenhofer, J. (1990) in: T. Claudio (Ed.) Current Topics in Membranes and Transport, Vol. 36, Protein–Membrane Interactions, Academic Press, New York, pp. 53–69.
15 Michel-Beyerle, M.E. (Ed.) (1985) Antennas and Reaction Centers of Photosynthetic Bacteria: Structure, Interactions, and Dynamics, Springer, Berlin.
16 Breton, J. and Vermeglio, A. (Eds.) (1988) The Photosynthetic Bacterial Reaction Center. Structure and Dynamics, Plenum Press, New York.
17 Feher, G. and Okamura, M.Y. (1978) in: R.K. Clayton and W.R. Sistrom (Eds.) The Photosynthetic Bacteria, Plenum Press, New York.
18 Youvan, D.C., Bylina, E.J., Alberti, M., Begusch, H. and Hearst, J.E. (1984) Cell 37, 949–957.
19 Williams, J.C., Steiner, L.A., Ogden, R.C., Simon, M.I. and Feher, G. (1983) Proc. Natl. Acad. Sci. USA 80, 6505–6509.
20 Williams, J.C., Steiner, L.A., Feher, G. and Simon, M.I. (1984) Proc. Natl. Acad. Sci. USA 81, 7303–7307.
21 Bélanger, G., Bérard, J., Corriveau, P. and Gingras, G. (1988) J. Biol. Chem. 263, 7632–7638.
22 Michel, H., Weyer, K.A., Gruenberg, H., Dunger, I., Oesterhelt, D. and Lottspeich, F. (1986) EMBO J. 5, 1149–1158.
23 Williams, J.C., Steiner, L.A. and Feher, G. (1986) Proteins 1, 312–325.
24 Michel, H., Weyer, K.A., Gruenberg, H. and Lottspeich, F. (1985) EMBO J. 4, 1667–1672.
25 Seftor, R.E.B. and Thornber, J.P. (1984) Biochim. Biophys. Acta 764, 148–159.
26 Gast, P., Michalski, T.J., Hunt, J.E. and Norris, J.R. (1985) FEBS Lett. 179, 325–328.
27 Cogdell, R.J. and Frank, H.A. (1987) Biochim. Biophys. Acta 895, 63–79.
28 Michel, H. and Oesterhelt, D. (1980) Proc. Natl. Acad. Sci. USA 77, 1283–1285.
29 Garavito, R.M. and Rosenbusch, J.P. (1980) J. Cell Biol. 86, 327–329.
30 Michel, H. (1982) J. Mol. Biol. 158, 567–572.
31 Deisenhofer, J., Epp, O., Miki, K., Huber, R. and Michel, H. (1984) J. Mol. Biol. 180, 385–398.
32 Weyer, K.A., Lottspeich, F., Gruenberg, H., Lang, F.S., Oesterhelt, D. and Michel, H. (1987) EMBO J. 6, 2197–2202.
33 Thornber, J.P., Cogdell, R.J., Seftor, R.E.B. and Webster, G.D. (1980) Biochim. Biophys. Acta 593, 60–75.
34 Michel, H. (1983) Trends Biochem. Sci. 8, 56–59.
35 Deisenhofer, J. and Michel, H. (1989) EMBO J. 8, 2149–2170.
36 Wang, B.-C. (1985) Methods Enzymol. 115, 90–112.
37 Deisenhofer, J., Epp, O., Miki, K., Huber, R. and Michel, H. (1985) Nature 318, 618–624.
38 Michel, H., Epp, O. and Deisenhofer, J. (1986) EMBO J. 5, 2445–2451.
39 Deisenhofer, J. and Michel, H. (1988) in: J. Breton and A. Vermeglio (Eds.) The Photosynthetic Bacterial Reaction Center: Structure and Dynamics, Plenum Press, New York, pp. 1–3.
40 Deisenhofer, J. and Michel, H. (1989) in: T. Frangsmyr (Ed.) Les Prix Nobel 1988, Nobel Foundation, Stockholm, pp. 134–188.
41 Luzzati, P.V. (1952) Acta Crystallogr. 5, 802–810.
42 Roth, M., Lewit-Bentley, A., Michel, H., Deisenhofer, J., Huber, R. and Oesterhelt, D. (1989) Nature 340, 659–662.
43 Tadros, M.H., Frank, R., Doerge, B., Gad'on, N., Takemoto, J.Y. and Drews, G. (1987) Biochemistry 26, 7680–7687.
44 Tadros, M.H., Frank, R., Takemoto, J.Y. and Drews, G. (1988) J. Bacteriol. 170, 2758–2762.
45 Tadros, M.H., Spormann, D. and Drews, G. (1988) FEMS Microbiol. Lett. 55, 243–247.
46 Breton, J. (1985) Biochim. Biophys. Acta 810, 235–245.
47 Weyer, K.A., Schafer, W., Lottspeich, F. and Michel, H. (1987) Biochemistry 26, 2909–2914.
48 Hoff, A.J. (1988) in: J. Breton and A. Vermeglio (Eds.) The Photosynthetic Bacterial Reaction Center: Structure and Dynamics, Plenum Press, New York, pp. 98–99.
49 Chang, C.-H., Schiffer, M., Tiede, D.M., Smith, U. and Norris, J.R. (1985) J. Mol. Biol. 186, 201–203.
50 Allen, J.P. and Feher, G. (1984) Proc. Natl. Acad. Sci. USA 81, 4795–4799.

51 Allen, J.P., Feher, G., Yeates, T.O., Rees, D.C., Deisenhofer, J., Michel, H. and Huber, R. (1986) Proc. Natl. Acad. Sci. USA 83, 8589–8593.
52 Allen, J.P., Feher, G., Yeates, T.O., Komiya, H. and Rees, D.C. (1987) Proc. Natl. Acad. Sci. USA 84, 5730–5734.
53 Taremi, S.S., Violette, C.A. and Frank, H.A. (1989) Biochim. Biophys. Acta 973, 86–92.
54 Yeates, T.O., Komiya, H., Chirino, A., Rees, D.C., Allen, J.P. and Feher, G. (1988) Proc. Natl. Acad. Sci. USA 85, 7993–7997.
55 Ducruix, A.F. and Reiss-Husson, F. (1987) J. Mol. Biol. 193, 419–421.
56 Chang, C.-H., Tiede, D.M., Tang, J., Smith, U., Norris, J.R. and Schiffer, M. (1986) FEBS Lett. 205, 82–86.
57 Allen, J.P., Feher, G., Yeates, T.O., Komiya, H. and Rees, D.C. (1987) Proc. Natl. Acad. Sci. USA 84, 6162–6166.
58 Yeates, T.O., Komiya, H., Rees, D.C., Allen, J.P. and Feher, G. (1987) Proc. Natl. Acad. Sci. USA 84, 6438–6442.
59 Allen, J.P., Feher, G., Yeates, T.O., Komiya, H. and Rees, D.C. (1988) Proc. Natl. Acad. Sci. USA 85, 8487–8491.
60 Komiya, H., Yeates, T.O., Rees, D.C., Allen, J.P. and Feher, G. (1988) Proc. Natl. Acad. Sci. USA 85, 9012–9016.
61 Allen, J.P. and Feher, G. (1988) in: J. Breton and A. Vermeglio (Eds.) The Photosynthetic Bacterial Reaction Center: Structure and Dynamics, Plenum Press, New York, pp. 5–11.
62 Arnoux, B., Ducruix, A., Astier, C., Picaud, M., Roth, M. and Reiss-Husson, F. (1990) Biochimie 72, 525–530.
63 Chang, C.-H., El-Kabbani, O., Tiede, D., Norris, J. and Schiffer, M. (1991) Biochemistry 30, 5352–5360.
64 El-Kabbani, O., Chang, C.-H., Tiede, D., Norris, J. and Schiffer, M. (1991) Biochemistry 30, 5361–5369.
65 Parson, W.W. and Ke, B. (1982) in: Govindjee (Ed.) Photosynthesis: Energy Conversion by Plants and Bacteria, Vol. 1, Academic Press, New York, pp. 331–384.
66 Martin, J.-L., Breton, J., Hoff, A.J., Migus, A. and Antonetti, A. (1986) Proc. Natl. Acad. Sci. USA 83, 957–961.
67 Breton, J., Martin, J.-L., Migus, A., Antonetti, A. and Orszag, A. (1986) Proc. Natl. Acad. Sci. USA 83, 5121–5125.
68 Fleming, G.R., Martin, J.-L. and Breton, J. (1988) Nature 333, 190–192.
69 Zinth, W., Kaiser, W. and Michel, H. (1983) Biochim. Biophys. Acta 723, 128–131.
70 Zinth, W., Knapp, E.W., Fischer, S.F., Kaiser, W., Deisenhofer, J. and Michel, H. (1985) Chem. Phys. Lett. 119, 1–4.
71 Knapp, E.W., Fischer, S.F., Zinth, W., Sander, M., Kaiser, W., Deisenhofer, J. and Michel, H. (1985) Proc. Natl. Acad. Sci. USA 82, 8463–8467.
72 Kirmaier, C. and Holten, D. (1988) FEBS Lett. 239, 211–218.
73 Holzapfel, W., Finkele, U., Kaiser, W., Oesterhelt, D., Scheer, H., Stilz, H.U. and Zinth, W. (1989) Chem. Phys. Lett. 160, 1–7.
74 Holzapfel, W., Finkele, U., Kaiser, W., Oesterhelt, D., Scheer, H., Stilz, H.U. and Zinth, W. (1990) Proc. Natl. Acad. Sci. USA 87, 5168–5172.
75 Kirmaier, C. and Holten, D. (1990) Proc. Natl. Acad. Sci. USA 87, 3552–3556.
76 Marcus, R.A. (1987) Chem. Phys. Lett. 133, 471–477.
77 Plato, M., Michel-Beyerle, M.E., Bixon, M. and Jortner, J. (1989) FEBS Lett. 249, 70–74.
78 Bixon, M., Jortner, J., Michel-Beyerle, M.E. and Ogrodnik, A. (1989) Biochim. Biophys. Acta 977, 273–286.
79 Bixon, M., Jortner, J. and Michel-Beyerle, M.E. (1991) Biochim. Biophys. Acta Bio-Energetics 1056, 301–315.
80 Kirmaier, C., Holten, D. and Parson, W.W. (1985) Biochim. Biophys. Acta 810, 33–48.
81 Debus, R.J., Feher, G. and Okamura, M.Y. (1986) Biochemistry 25, 2276–2287.

82 Dracheva, S.M., Drachev, L.A., Konstantinov, A.A., Semenov, A.Y., Skulachev, V.P., Arutjunjan, A.M., Shuvalov, V.A. and Zaberezhnaya, S.M. (1988) Eur. J. Biochem. 171, 253–264.
83 Mitchell, P. (1979) Science 206, 1148–1159.
84 Holten, D., Windsor, M.W., Parson, W.W. and Thornber, J.P. (1978) Biochim. Biophys. Acta 501, 112–126.
85 Fritzsch, G., Buchanan, S. and Michel, H. (1989) Biochim. Biophys. Acta 977, 157–162.
86 Shopes, R.J., Levine, L.M.A., Holten, D. and Wraight, C.A. (1987) Photosynth. Res. 12, 165–180.
87 Vermeglio, A., Richaud, P. and Breton, J. (1989) FEBS Lett. 243, 259–263.
88 Michel-Beyerle, M.E., Plato, M., Deisenhofer, J., Michel, H., Bixon, M. and Jortner, J. (1988) Biochim. Biophys. Acta 932, 52–70.
89 McDowell, L.M., Gaul, D., Kirmaier, C., Holten, D. and Schenck, C.C. (1991) Biochemistry 30, 8315–8322.
90 Treutlein, H., Schulten, K., Deisenhofer, J., Michel, H., Brünger, A.T. and Karplus, M. (1988) in: J. Breton and A. Vermeglio (Eds.) The Photosynthetic Bacterial Reaction Center: Structure and Dynamics, Plenum Press, New York, pp. 139–150.
91 Treutlein, H., Schulten, K., Niedermeier, C., Deisenhofer, J., Michel, H. and Devault, D. (1988) in: J. Breton and A. Vermeglio (Eds.) The Photosynthetic Bacterial Reaction Center: Structure and Dynamics, Plenum Press, New York, pp. 369–377.
92 Creighton, S., Hwang, J.-K., Warshel, A., Parson, W.W. and Norris, J.R. (1988) Biochemistry 27, 774–781.
93 Treutlein, H., Niedermeier, C., Schulten, K., Deisenhofer, J., Michel, H., Brünger, A.T. and Karplus, M. (1988) in: A. Pullman, J. Jortner and B. Pullman (Eds.) Transport through Membranes: Carriers, Channels, and Pumps, Proc. 21st Jerusalem Symposium on Quantum Chemistry and Biochemistry, Kluwer Academic Publishers, Dordrecht, pp. 513–525.
94 Warshel, A., Chu, Z.T. and Parson, W.W. (1989) Science 246, 112–116.
95 Parson, W.W., Chu, Z.-T. and Warshel, A. (1990) Biochim. Biophys. Acta Bio-Energetics 1017, 251–272.
96 DiMagno, T.J., Bylina, E.J., Angerhofer, A., Youvan, D.C. and Norris, J.R. (1990) Biochemistry 29, 899–907.
97 Paddock, M.L., Rongey, S.H., Feher, G. and Okamura, M.Y. (1989) Proc. Natl. Acad. Sci. USA 86, 6602–6606.
98 Gray, K.A., Farchaus, J.W., Wachtveitl, J., Breton, J. and Oesterhelt, D. (1990) EMBO J. 9, 2061–2070.
99 Sinning, I., Michel, H., Mathis, P. and Rutherford, A.W. (1989) Biochemistry 28, 5544–5553.
100 Ewald, G., Wiessner, C. and Michel, H. (1990) Z. Naturforsch. 45c, 459–462.
101 Paddock, M.L., Williams, J.C., Rongey, S.H., Abresch, E.C., Feher, G. and Okamura, M.Y. (1987) in: J. Biggins (Ed.) Progress in Photosynthesis Research, Vol. 3, Martinus Nijhoff, Dordrecht, pp. III.11.775–III.11.778.
102 Paddock, M.L., McPherson, P.H., Feher, G. and Okamura, M.Y. (1990) Proc. Natl. Acad. Sci. USA 87, 6803–6807.
103 Paddock, M.L., Feher, G. and Okamura, M.Y. (1991) Photosynth. Res, 27, 109–119.
104 Sinning, I., Koepke, B., Schiller, B. and Michel, H. (1990) Z. Naturforsch. 45c, 455–458.
105 Sinning, I., Koepke, J. and Michel, H. (1990) in: M.E. Michel-Beyerle (Ed.) Reaction Centers of Photosynthetic Bacteria, Springer, Berlin, pp. 199–208.
106 Trebst, A. (1986) Z. Naturforsch. 41, 240–245.
107 Michel, H. and Deisenhofer, J. (1988) Biochemistry 27, 1–7.
108 Michel, H. and Deisenhofer, J. (1986) in: L.A. Staehelin and C.J. Arntzen (Eds.) Encyclopedia of Plant Physiology (New Series), Vol. 19, Photosynthesis III: Photosynthetic Membranes and Light Harvesting Systems, Springer, Berlin, pp. 371–381.
109 Nanba, O. and Satoh, K. (1987) Proc. Natl. Acad. Sci. USA 84, 109–112.
110 Trebst, A. (1987) Z. Naturforsch. 42, 742–750.
111 Barry, B.A. and Babcock, G.T. (1987) Proc. Natl. Acad. Sci. USA 84, 7099–7103.

112 Debus, R.J., Barry, B.A., Sithole, I., Babcock, G.T. and McIntosh, L. (1988) Biochemistry 27, 9071–9074.
113 Debus, R.J., Barry, B.A., Babcock, G.T. and McIntosh, L. (1988) Proc. Natl. Acad. Sci. USA 85, 427–430.
114 Babcock, G.T., Barry, B.A., Debus, R.J., Hoganson, C.W., Atamian, M., McIntosh, L., Sithole, I. and Yocum, C.F. (1989) Biochemistry 28, 9557–9565.
115 Deisenhofer, J. and Michel, H. (1989) Science 245, 1463–1473.
116 Rees, D.C., DeAntonio, L. and Eisenberg, D.S. (1989) Science 245, 510–513.
117 Ford, R.C., Picot, D. and Garavito, R.M. (1987) EMBO J. 6, 1581–1586.
118 Witt, I., Witt, H.T., Gerken, S., Saenger, W., Dekker, J.P. and Roegner, M. (1987) FEBS Lett. 221, 260–264.
119 Witt, I., Witt, H.T., di Fiore, D., Rogner, M., Hinrichs, W., Saenger, W., Granzin, J., Betzel, C. and Dauter, Z. (1988) Ber. Bunsenges. Phys. Chem. 92, 1503–1506.
120 Boekema, E.J., Dekker, J.P., Rögner, M., Witt, I., Witt, H.T. and van Heel, M. (1989) Biochim. Biophys. Acta 974, 81–87.
121 Almog, O., Shoham, G., Michaeli, D. and Nechushtai, R. (1991) Proc. Natl. Acad. Sci. USA 88, 5312–5316.
122 Allen, J.P., Theiler, R. and Feher, G. (1985) in: M.E. Michel-Beyerle (Ed.) Springer Series in Chemical Physics, Vol. 42, Springer, Berlin, pp. 82–84.
123 Cogdell, R.J., Woolley, K., Mackenzie, R.C., Lindsay, J.G., Michel, H., Dobler, J. and Zinth, W. (1985) in: M.E. Michel-Beyerle (Ed.) Springer Series in Chemical Physics, Vol. 42, Springer, Berlin, pp. 85–87.
124 Mäntele, W., Steck, K., Wacker, T. and Welte, W. (1985) in: M.E. Michel-Beyerle (Ed.) Springer Series in Chemical Physics, Vol. 42, Springer, Berlin, pp. 88–91.
125 Wacker, T., Gad'on, N., Becker, A., Mäntele, W., Kreutz, W., Drews, G. and Welte, W. (1986) FEBS Lett. 197, 267–273.
126 Papiz, M.Z., Hawthornthwaite, A.M., Cogdell, R.J., Woolley, K.J., Wightman, P.A., Ferguson, L.A. and Lindsay, J.G. (1989) J. Mol. Biol. 209, 833–835.
127 Henderson, R. and Unwin, P.N.T. (1975) Nature 257, 28–32.
128 Ceska, T.A. and Henderson, R. (1990) J. Mol. Biol. 213, 539–560.
129 Henderson, R., Baldwin, J.M., Ceska, T.A., Zemlin, F., Beckmann,E. and Downing, K.H. (1990) J. Mol. Biol. 213, 899–929.
130 Kühlbrandt, W. and Wang, D.N. (1991) Nature 350, 130–134.
131 Jap, B.K., Walian, P.J. and Gehring, K. (1991) Nature 350, 167–170.
132 Lesk, A.M. and Hardman, K.D. (1985) Methods Enzymol. 115, 381–390.
133 Laussermair, E. and Oesterhelt, D. (1992) EMBO J. 11, 777–783.

L. Ernster (Ed.) *Molecular Mechanisms in Bioenergetics*
© 1992 Elsevier Science Publishers B.V. All rights reserved

CHAPTER 5

The two photosystems of oxygenic photosynthesis

BERTIL ANDERSSON and LARS-GUNNAR FRANZÉN

Department of Biochemistry, Arrhenius Laboratories for Natural Sciences, Stockholm University, S-10691 Stockholm, Sweden

Contents

1. Introduction 122
2. Photosystem I 123
 2.1. Overview 123
 2.2. Redox components and electron transfer reactions 125
 2.3. The PSI-A and PSI-B proteins – the reaction centre heterodimer 126
 2.4. The PSI-C iron–sulphur protein 126
 2.5. The stromally exposed PSI-D and PSI-E proteins 127
 2.6. The PSI-F protein – the plastocyanin-binding subunit? 127
 2.7. The PSI-G and PSI-H proteins 128
 2.8. The low molecular weight PSI-I, PSI-J and PSI-K proteins 128
 2.9. The PSI-L protein 129
 2.10. Additional PSI subunits 129
 2.11. Organization of the photosystem I complex 129
3. Photosystem II 131
 3.1. Overview 131
 3.2. Redox components and electron-transfer reactions 132
 3.3. The D_1–D_2 protein reaction centre heterodimer 132
 3.4. Cytochrome *b*-559 135
 3.5. CP43 and CP47: chlorophyll *a* binding proteins 135
 3.6. The 9 kDa phospho-protein 135
 3.7. Low molecular weight polypeptides 136
 3.8. Extrinsic membrane proteins associated with the donor side of photosystem II 136
 3.9. The hydrophobic 22 and 10 kDa proteins 137
 3.10. Organization of the photosystem II complex 138
4. Analogies between the photosystems of oxygenic photosynthesis and photosynthetic bacteria 139
Acknowledgements 140
References 140

1. Introduction

Light-driven electron transport is nature's molecular strategy for converting solar radiant energy into chemical energy. The process of photosynthetic electron transport occurs in both prokaryotic organisms – green and purple bacteria, prochlorons and cyanobacteria – and in eukaryotic organisms – algae and green plants. These organisms all contain supramolecular membrane-bound complexes known as photosystems. The photosystems contain light-absorbing pigments, mainly chlorophylls, and a reaction centre redox couple. The light absorbed by the antenna pigments is very rapidly (<1 ps) transferred to a reaction centre chlorophyll (P) which becomes activated into an exited state (P*). The reduction potential of P* is large enough to allow electron donation to a nearby acceptor molecule. This charge separation within the reaction centre is vectorial and gives rise to a negative charge towards one side of the membrane and a positive charge towards the other side. From the reduced primary acceptor, the electron is transported via traditional biological electron transport components such as cytochromes and quinones to a terminal acceptor. The reaction centre chlorophyll (P^+) is re-reduced by some kind of electron donor. In oxygenic photosynthesis, which occurs in all eukaryotic photosynthetic organisms, in prochlorons and in cyanobacteria, the source of electrons is water whilst the terminal electron acceptor is $NADP^+$. Coupled to this electron transport is the creation of an electrochemical proton gradient across the photosynthetic membrane, which is utilized to drive ATP synthesis. In oxygenic photosynthesis the electron transport is driven by two separate complexes, photosystem I (PSI) and photosystem II (PSII). These two photosystems are connected by a quinone pool (plastoquinone-9), a cytochrome *b*/*c* type complex normally referred to as cytochrome *b*/*f*(*ronze*), and by plastocyanin or cytochrome *c*-553. The transmembrane proton gradient created by the electron transport is utilized by the CF_0/CF_1 protein complex to synthesize ATP. In many respects, photosynthetic electron transport and photophosphorylation are functionally and structurally analogous to respiratory energy conversion. However, photosystems I and II have no correspondence in respiration, but are related to the green-sulphur and purple bacterial reaction centres, respectively (see Section 4).

All the components responsible for electron transport and ATP synthesis during oxygenic photosynthesis are bound to membranes called thylakoids (Fig. 1). In eukaryotic cells the thylakoid membranes are located within the chloroplast organelle. In plants and several algae the thylakoids form a continuous folded membrane network composed of single, non-appressed membranes and stacks of tightly paired or appressed membranes (see ref. [1]). The outer thylakoid surface faces the chloroplast stroma compartment, which harbours the rubisco enzyme and the other soluble components of the Calvin cycle. This compartment is in turn separated from the cytoplasm by a double outer membrane, designated the chloroplast envelope. At the inner side, the thylakoids enclose a continuous space referred to as the thylakoid lumen.

For both photosystems the electron accepting side is located towards the stromal side of the thylakoid membrane whilst the donor side is located towards the lumenal side. This arrangement results in transmembrane electron transfer and translocation of protons from the stroma to the thylakoid lumen. Apart from this transverse asymmetry, the thylakoid membrane possesses an extreme lateral asymmetry in the distribution of components between the stacked and unstacked membrane regions (Fig. 1) [1,2]. In particular, the two photosystems are physically segregated in such a way that PSI is excluded from the tightly appressed regions of the grana stacks and confined to the non-appressed regions which are in direct contact with the surrounding stroma. In contrast, PSII is quite scarce in these regions and is mainly located within the appressed grana thylakoids.

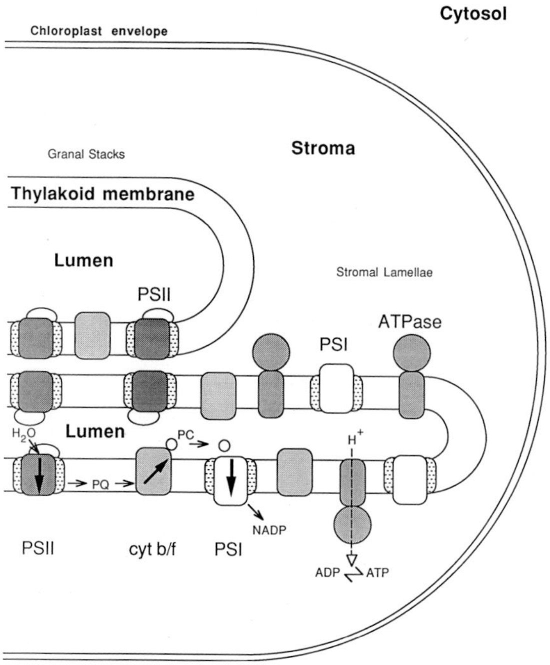

Fig. 1. Organization of the higher-plant chloroplast, including the photosynthetic electron transport components. The chloroplast is separated from the cytoplasm by a double outer membrane, designated the chloroplast envelope. The continuous network of interconnected thylakoid membranes are structurally differentiated into appressed and non-appressed regions. The outer thylakoid surface points towards the stromal compartment, while the inner surface points towards the lumenal space. The PSI complex and ATP synthase are located only in non-appressed membranes, whilst the main location of PSII is the appressed thylakoid region. The cytochrome *b/f* complex is evenly distributed with respect to these two membrane regions.

2. *Photosystem I*

2.1. Overview

PSI catalyses light-driven electron transfer from plastocyanin to ferredoxin. In higher plants and in green algae, a native PSI complex can be isolated from thylakoid membranes by detergent treatment followed by sucrose gradient centrifugation, and further fractionated into a PSI core complex and a light-harvesting complex I (LHCI) [3]. The core complex contains at least twelve different polypeptides of both plastid and nuclear origin and binds all of the electron-transfer components as well as about 100 antenna chlorophyll *a* molecules (Fig. 2, Table I). LHCI consists of several nuclear-encoded polypeptides that bind approximately 100 antenna chlorophyll *a* and *b* molecules per PSI reaction centre [4]. Cyanobacteria and red algae lack chlorophyll *a/b* binding proteins such as LHCI. Instead, they contain light-harvesting protein complexes, phycobilisomes, which are extrinsically located at the outer surface of the thylakoid membrane [5].

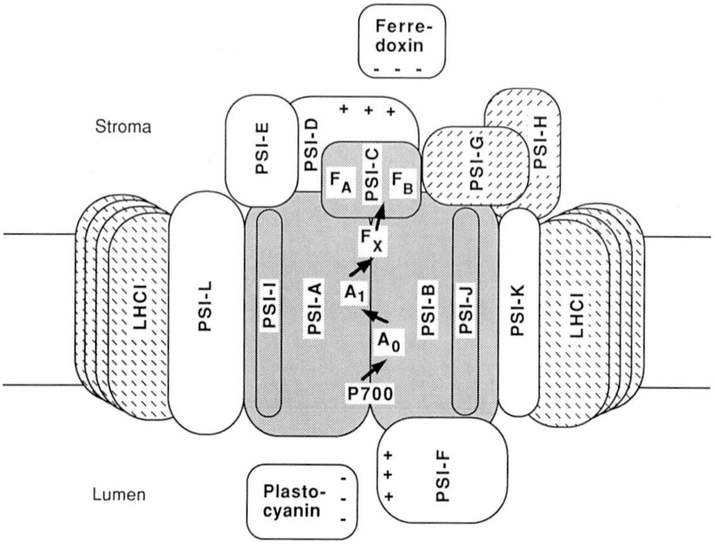

Fig. 2. A model for the organization of the photosystem I complex. The PSI-A and PSI-B subunits form a heterodimer that binds P700, A_0, A_1 and F_X. All of the redox components are located on three of the chloroplast-encoded proteins, PSI-A, PSI-B and PSI-C. For most of the subunits, the function is not known. The exact location of these subunits is arbitrary, but PSI-E, PSI-G and PSI-H appear to be located on the stromal side and PSI-I, PSI-J, PSI-K and PSI-L appear to be intrinsic membrane proteins. Chloroplast-encoded components are shaded, and components that have not been detected in cyanobacteria are hatched. Electron transfers within the PSI complex are indicated with arrows. As it is not known how the electrons are transferred between F_B and F_A, these electron transfers are not indicated (see Subsection 2.2).

The absorption of a light quantum by PSI results in electron transfer from the reaction centre chlorophyll P700 to a series of acceptors, namely A_0 (chlorophyll *a*), A_1 (phylloquinone) and several iron–sulphur centres designated F_X, F_B and F_A. The electron is then transferred to soluble ferredoxin which is loosely attached to the outer thylakoid surface. Finally there is a reduction of $NADP^+$ catalyzed by ferredoxin–NADP reductase. $P700^+$ is re-reduced by plastocyanin, a soluble type-I copper protein located at the inner thylakoid surface. In some cyanobacteria and algae, plastocyanin is replaced by a soluble cytochrome *c*-553.

In contrast to the situation in PSII (see Section 3), there are no 'trivial' names for the PSI subunits. Bengis and Nelson [6] introduced a nomenclature in which the PSI core subunits are numbered with roman numerals (I–VI) according to decreasing apparent molecular mass. However, at least ten subunits are now known in just the molecular weight range between 4 and 20 kDa and their order of electrophoretic migration on SDS-PAGE gels differs both between plant species and between gel electrophoretic systems. Sequence data are now available for all subunits. Here we define the subunits of the PSI core complex as products of their genes, *psaA–psaM*, according to the list compiled by Hallick [7]. Consequently, the product of the *psaA* gene is called the PSI-A polypeptide.

TABLE I

Photosystem I protein components[a]

Gene	Coded[b]	Mass (kDa)	Organization[c]	Function
psaA	C	82–83	intrinsic (11)	Binds redox groups and antenna Chl *a*
psaB	C	82–83	intrinsic (11)	Binds redox groups and antenna Chl *a*
psaC	C	9	extr. stromal	Binds FeS centres F_A and F_B
psaD	N	15–18	extr. stromal	Ferredoxin docking
psaE	N	8–11	extr. stromal	?
psaF	N	16–18	extr. lumenal	Plastocyanin docking
psaG	N	10–11	extr. stromal	?
psaH	N	10–11	extr. stromal	?
psaI	C	4	intrinsic (1)	?
psaJ	C	5	intrinsic (1)	?
psaK	N	8	intrinsic (1–2)	?
psaL	N	18	intrinsic (2)	?

[a] The components of the LHCI complex are not included in the Table.
[b] C = chloroplast, N = nucleus.
[c] Intrinsic (number of membrane spans) or extrinsic membrane protein. Lumenal or stromal side of the membrane.

2.2. *Redox components and electron transfer reactions*

After absorption of a light quantum by the antenna of PSI, the excitation energy is rapidly transferred to P700, the primary electron donor. Electronic excitation changes the midpoint potential of P700 from +0.43 V to approximately −1.3 V. Therefore, the excited form P700* can donate an electron to the electron acceptor A_0 (Fig. 2), which has a midpoint potential of approximately −1 V, after which $P700^+$ is re-reduced by plastocyanin, which has a midpoint potential of +0.37 V. On the acceptor side, the electron is rapidly transferred through a series of secondary acceptors, namely A_1 (midpoint potential −0.8 V), F_X (−0.7 V), F_B (−0.58 V) and F_A (−0.53 V), and then transferred to soluble ferredoxin (−0.42 V) [8].

P700 is probably a dimer of chlorophyll *a* molecules, although there are some indications that it is a chlorophyll monomer [8]. The first electron acceptors A_0 and A_1 are considered to be a chlorophyll *a* monomer and a phylloquinone (vitamin K_1), respectively. The PSI reaction centre contains two phylloquinone molecules per P700. Removal of phylloquinone by ether-extraction results in a block of room temperature electron transfer beyond A_0^- [8].

There has been a controversy whether F_X is a [2Fe–2S] centre or a [4Fe–4S] centre, but recent EXAFS and Mössbauer measurements on an isolated PSI complex devoid of F_A and F_B (cf. Subsection 2.4 below) have identified F_X as a [4Fe–4S] centre [9,10]. In addition, stoichiometric considerations also support the identification of F_X as a [4Fe–4S] centre, since the PSI-A/PSI-B heterodimer (see Subsection 2.3) contains four iron atoms and four inorganic sulphur atoms [11–13]. EPR measurements have shown that the iron–sulphur centres F_A and F_B are [4Fe–4S] centres. It is not known if the electron flow is serial through F_B and F_A or if the two centres work in parallel [8].

2.3. The PSI-A and PSI-B proteins – the reaction centre heterodimer

The primary electron donor P700 and the early electron acceptors A_0, A_1 and F_X all appear to be located on the two large, chloroplast-encoded PSI-A and PSI-B polypeptides, which are predicted to form a heterodimer known as CPI. In addition, about 100 antenna chlorophyll *a* molecules are bound to these two subunits. The *psA* and *psaB* genes have been sequenced in several organisms, including higher plants, green algae and cyanobacteria [8]. The similarity between species is very high, with 95% identity between higher plant sequences and about 80% identity between plant and cyanobacterial sequences. The PSI-A and PSI-B subunits are also homologous to one another, showing about 45% sequence identity. Therefore, it is likely that the *psaA* and *psaB* genes arose from a gene duplication of a single ancestral gene. Both polypeptides have molecular masses of 82–83 kDa according to the sequences, but migrate with apparent molecular masses of 65 kDa in SDS-PAGE. They are predicted to form as many as 11 membrane-spanning helices each.

There is presently no detailed data available on the three-dimensional structure of PSI, but recently PSI complexes from cyanobacteria have been crystallized [14–16]. Thus, PSI may very well be the next multifunctional membrane protein complex whose three-dimensional structure will be solved after the reaction centre of purple bacteria [17]. In the absence of structural information, predictions can be made from analyses of the PSI-A and PSI-B sequences and from comparisons with the D_1 and D_2 subunits of PSII (see Section 3). The PSI-A and PSI-B proteins contain a large number of conserved histidines, glutamines and asparagines that could coordinate the approximately 100 chlorophyll *a* molecules (P700, A_0 and antenna chlorophylls) bound to these two subunits. On the basis of limited sequence similarity, Robert and Moenne-Loccoz [18] suggested that transmembrane helices VI and VIII of PSI-A and PSI-B correspond to helices B and D of the D_1 and D_2 proteins of PSII. They predicted that histidine residues in helices VI and VIII coordinate the A_0 and P700 chlorophyll molecules. However, from a similar analysis, Margulies [19] suggested that helix X of PSI-B corresponds to helix D of the D_2 protein (or possibly helix D of the D_1 protein). Furthermore, helix X of PSI-B was predicted to be involved in binding P700, A_0 and A_1 in the same way as helix D of the D_2 protein interacts with P680, pheophytin and Q_A (see Section 3).

The electron acceptor F_X is a [4Fe–4S] iron–sulphur centre and requires four cysteine residues as ligands. The PSI-A and PSI-B proteins contain three and two cysteine residues, respectively, which are conserved between all species. The PSI-A/PSI-B heterodimer binds four iron atoms [11] and four acid-labile sulphur atoms per P700 [12,13]. F_X is probably bridged between the PSI-A and PSI-B proteins [20,21].

2.4. The PSI-C iron–sulphur protein

In contrast to F_X, the iron–sulphur centres of F_A and F_B are not located on the reaction centre heterodimer, but on a 9 kDa polypeptide that is encoded by the chloroplast gene *psaC* [13,22,23]. Complete sequence information is available from at least 15 different organisms, including higher plants, algae and cyanobacteria [8,24–26]. The protein encoded by the *psaC* gene contains 81 amino acids and is highly conserved. The higher-plant sequences are approximately 95% identical, and 79% of the residues are conserved between all organisms. The PSI-C protein also shows significant homology to soluble ferredoxins from bacteria [23,27]. The PSI-C protein from all sources contains nine cysteines, eight of which are expected to participate in liganding the two iron–sulphur centres F_A and F_B.

The iron–sulphur centres in the isolated PSI-C holoprotein are unstable in vitro but can be restored by incubation with $FeCl_3$, Na_2S and β-mercaptoethanol [8]. The EPR signals of F_A and F_B are much broader in the isolated PSI-C protein than the corresponding signals in the native PSI complex. Rebinding of the PSI-C holoprotein to the PSI complex causes some changes in the EPR signals, but the normal EPR signals of F_A and F_B are restored only if the PSI-D protein is also rebound to the PSI complex [28].

2.5. The stromally exposed PSI-D and PSI-E proteins

Sequence information for the PSI-D and PSI-E proteins is available from plants and cyanobacteria [8,24,26]. The eukaryotic proteins are nuclear-encoded and synthesized as large precursors with N-terminal presequences that direct the proteins into the chloroplast stroma (see ref. [29]). The PSI-D and PSI-E proteins are positively charged, hydrophilic proteins with molecular masses of 18 and 10 kDa, respectively. They do not contain hydrophobic regions that could span the thylakoid membrane. Thus, they can be classified as extrinsic membrane proteins located on the stromal side of the thylakoid membrane.

PSI-D corresponds to 'subunit II' in the nomenclature of Bengis and Nelson [6]. It is highly conserved, with about 55% identity between higher-plant and cyanobacterial proteins. It appears to be involved in the binding of ferredoxin to the thylakoid membrane. It has an extremely high isoelectric point (predicted pI = 9.6), and is consequently positively charged at physiological pH. Thus, it has the potential of binding the negatively charged ferredoxin by electrostatic interaction. Experimentally, Zanetti and Merati [30] and Zilber and Malkin [31] showed that the PSI-D protein can be chemically crosslinked to ferredoxin. Chitnis et al. [32] have specifically inactivated the *psaD* gene in *Synechocystis* 6803. The mutant cells grew very slowly under photoautotropic conditions but grew well in the presence of 5 mM glucose. The mutated PSI complex was able to perform electron transport from plastocyanin to artificial electron acceptors in vitro. Together, the in vivo and in vitro results indicate that the electron transfer from the PSI acceptor side to ferredoxin was blocked. However, it should be noted that the PSI complex isolated from the *psaD* mutant also lacked several other low molecular weight subunits, including PSI-E.

PSI-E corresponds to 'subunit IV' in the nomenclature of Bengis and Nelson [6]. Münch et al. [33] and Steppuhn et al. [34] originally designated PSI-E and PSI-F 'subunit III' and 'subunit IV', respectively. However, this has later been corrected [35]. The *psaE* gene has been specifically inactivated in the cyanobacterium *Synechocystis* 6803 [36]. The *psaE* mutant grew photoautotrophically at rates similar to those of the wild type. It appears to contain structurally and functionally normal PSI complexes, except for the absence of PSI-E. Thus, despite the fact that the protein is highly conserved, the protein is not obligatory for PSI electron transport, and its function remains to be established.

2.6. The PSI-F protein – the plastocyanin-binding subunit?

psaF cDNAs or genes have been isolated and sequenced from spinach, the green alga *Chlamydomonas reinhardtii* and the cyanobacterium *Synechocystis* 6803 [34,37,38]. The molecular mass of the protein is approximately 17 kDa according to the sequence data. The spinach and *Chlamydomonas* proteins are synthesized as larger precursors with N-terminal presequences of the lumen-targeting type (see ref. [29]), whilst the cyanobacterial protein is synthesized with a prokaryotic signal sequence. Therefore, the protein should be located in the thylakoid lumen. The sequence identity between plants and cyanobacteria is about 50%. The eukaryotic proteins have net positive charges whilst the *Synechocystis* protein is negatively charged at neutral pH.

Bengis and Nelson [6] showed that the loss of 'subunit III' (= PSI-F, cf. discussion about PSI-E above) correlated with impairment of electron donation from plastocyanin to P700. Recently, the PSI-F protein has been crosslinked to plastocyanin [39,40]. The crosslinked plastocyanin exhibited fast electron donation to P700 while steady-state PSI activity was lost. Similarly, cytochrome *c*-553 can be crosslinked to the cyanobacterial PSI-F protein [41]. These experiments identify PSI-F as the plastocyanin-docking (or cytochrome *c*-553 docking) subunit of PSI. However, proteins with N-terminal sequences identical to that of PSI-F have also been found in a preparation with ferredoxin–plastoquinone oxidoreductase activity [42], and in an LHCI preparation [43]. It should be pointed out that PSI-F can easily be lost from the PSI complex during preparations that include detergent treatments [6], and it may possibly co-purify with other components.

Chitnis et al. [38] have specifically inactivated the *psaF* gene in *Synechocystis.* This *psaF* mutant was able to grow photoautotrophically, indicating that PSI-F is not obligatory for electron donation from plastocyanin/cytochrome *c*-553 to P700. However, it is possible that this electron donation is not rate-limiting for photoautotrophic growth of *Synechocystis.* In vitro, electron donation to P700 is not completely blocked after removal of PSI-F subunit, but it is slowed down by a factor of 10^3 [44]. This slow electron donation may be sufficient to sustain photoautotrophic growth in vivo.

2.7. The PSI-G and PSI-H proteins

cDNA clones of the *psaG* and *psaH* genes from higher plants and *Chlamydomonas* have been isolated and sequenced [8,24]. The PSI-G and PSI-H proteins have so far not been found in cyanobacteria. The two proteins are 36 and 40% conserved between higher plants and *Chlamydomonas*, respectively. Both are relatively hydrophilic proteins and are synthesized as larger precursor proteins with N-terminal transit peptides of the stroma-targeting type. Consequently, they are probably located on the stromal side of the membrane. PSI-G is negatively charged (predicted pI = 4.5) in contrast to PSI-D PSI-E, PSI-F and PSI-H, which are predicted to be positively charged [8,45].

The physiological roles of the PSI-G and PSI-H proteins are not known. However, since the cyanobacterial PSI complex appears to lack these two proteins, one could conclude that their functions are not required for maintaining functional electron transport. Instead, they may have functions not required in cyanobacteria, such as the binding of LHCI to the PSI core complex. In the case of the PSI-H protein, this possibility is supported by the finding that the accumulation of *psaH* mRNA is delayed in greening experiments in a way similar to LHCI and LHCII mRNA [35].

2.8. The low molecular weight PSI-I, PSI-J and PSI-K proteins

The PSI-I and PSI-J proteins are two small chloroplast-encoded proteins with molecular masses of 4 and 5 kDa, respectively [46,47]. They have been identified and sequenced in both higher plants and cyanobacteria [8,24,48]. Both proteins are very hydrophobic and predicted to contain one membrane-spanning α-helix each.

A cDNA clone of the nuclear *psaK* gene from the green alga *Chlamydomonas* has been isolated and sequenced [45]. The molecular mass calculated from the sequence is 8.4 kDa and homologous proteins have been identified in higher plants and cyanobacteria by N-terminal sequencing [8,48]. The protein is quite hydrophobic, and predicted to have one or two membrane-spanning helices [45].

Some recent results suggest that the PSI-I, PSI-J and PSI-K proteins are tightly associated with the PSI-A/PSI-B reaction centre heterodimer [46,49,50]. Scheller et al. [46] speculated that the PSI-I protein may be involved in co-factor binding. However, Ikeuchi et al. [51] have reported that the PSI-J and PSI-K proteins are not retained in a PSI core complex prepared after treatment of PSI complex with Triton X-100. Similarly, Takabe et al. [52] have reported that the PSI-I and PSI-K polypeptides are missing in a PSI core complex that catalyzes electron transfer from plastocyanin to ferredoxin. The latter results appear to exclude an involvement of the low molecular weight proteins in co-factor binding.

2.9. The PSI-L protein

Okkels et al. [53] have isolated still another PSI subunit from barley. A cDNA clone of this subunit encodes a 22 kDa precursor protein that is probably processed to a final molecular mass of 18 kDa. The protein is hydrophobic, and is predicted to have at least two membrane-spanning helices. This protein, PSI-L has also been detected in spinach [54] and in cyanobacteria [8,47]. Based on limited sequence similarity, the cyanobacterial PSI-L sequences were until recently identified as PSI-G sequences. There are presently no experimental results that suggest any role for the PSI-L protein.

2.10. Additional PSI subunits

A distinct 9 kDa subunit of PSI from spinach and pea, co-migrating with the PSI-G and PSI-C proteins during SDS-PAGE, has been identified [54]. The protein was present in native PSI complexes but was not retained in PSI core complexes.

Koike et al. [47] have reported the N-terminal sequence of a 4.8 kDa PSI subunit from the cyanobacterium *Synechococcus vulcanus*, which does not correspond to any known PSI subunit from plants. In addition, the gene of a very small PSI subunit (31 amino acids) has been cloned from the cyanobacterium *Synechococcus* sp. and designated *psaM*. The protein is found as a component of PSI crystals from this cyanobacterium (W. Haehnel, N. Nelson and I. Witt, unpublished). The gene is homologous to ORF32 in the liverwort (*Marchantia polymorpha*) chloroplast genome [55].

2.11. Organization of the photosystem I complex

In Fig. 2, we present a current model for the organization of the PSI complex. Most of the chloroplast-encoded subunits, PSI-A, PSI-B, PSI-I and PSI-J, are hydrophobic intrinsic membrane proteins. Two of the nuclear-encoded subunits, PSI-K and PSI-L, are also predicted to be intrinsic membrane proteins. PSI-C, PSI-D, PSI-E, PSI-G, and PSI-H are extrinsic proteins located on the stromal side of the thylakoid membrane. This is known not only from sequence data (see above) but also from experiments involving salt or chaotrope washing and Triton X-114 fractionation of thylakoid membranes and isolated PSI complexes [56].

The PSI-C polypeptide and other low molecular mass subunits can be removed from the PSI complex by detergents or chaotropes. Chaotrope extraction results in a PSI core preparation that retains the redox components P700, A_0, A_1 and F_X but is depleted of the PSI-C and PSI-D and PSI-E proteins [57,58]. Isolated PSI-C holoprotein with intact iron–sulphur centres can rebind to the PSI complex, but the binding is further stabilized by rebinding of the PSI-D

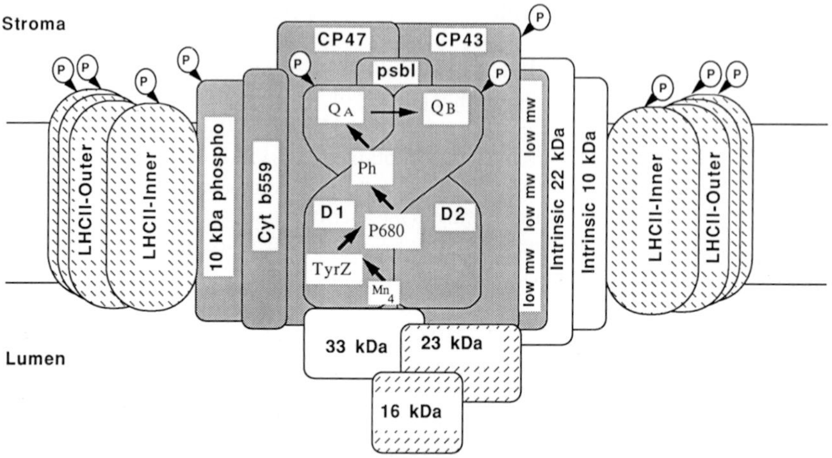

Fig. 3. A model for the organization of the photosystem II complex. The D_1 and D_2 subunits form a heterodimer that binds all the redox components of PSII mediated electron transport, including the manganese cluster involved in water oxidation. The chloroplast-encoded components are shaded and components not detected in cyanobacteria are hatched. Electron transfers within the PSII complex are indicated with arrows. Proteins undergoing light-mediated phosphorylation have been marked with a circled P.

protein. The PSI-D and PSI-E proteins, on the other hand, can only rebind to the PSI complex after rebinding of intact PSI-C protein [28,58].

As indicated above (Subsections 2.5 and 2.6), mutations in cyanobacteria and the green alga *Chlamydomonas* provide a valuable approach for studying structure–function relationships. However, the limitation of this approach is that the deletion of one particular subunit often leads to secondary effects such as partial disassembly and degradation of the protein components, and unambiguous interpretations of the role of a given subunit cannot be made. On the other hand, this phenomenon can be used to understand the assembly of the complex. Mutants in six different PSI genes have been analyzed in *Chlamydomonas* or in cyanobacteria. These include mutants in *psaA* [59,60], *psaB* [61,62], *psaC* [25,63], *psaD* [32], *psaE* [36] and *psaF* [38]. If the synthesis of one of the hydrophobic, chloroplast-encoded PSI-A or PSI-B proteins is blocked, the complete PSI complex is missing in the thylakoid membrane. The other subunits are likely to be synthesized but they degrade rapidly and do not accumulate in the membrane. In contrast, if the synthesis of one of the extrinsic, nuclear-encoded PSI-D, PSI-E or PSI-F proteins is blocked, a functional PSI complex is found in the thylakoid membrane, although its activity may not be optimal. For the PSI-C protein, which is chloroplast-encoded but extrinsic, there are conflicting results. Inactivation of the *psaC* gene in *Chlamydomonas* leads to absence of the complete PSI complex in the thylakoid membrane [25]. However, in the cyanobacterium *Anabaena variabilis* 29413, a partially functional PSI compex can assemble in the absence of the PSI-C subunit [63].

The biosynthesis of the PSI complex requires the coordinated expression of both chloroplast and nuclear genes. It appears to start with the formation of an inner core of hydrophobic, chloroplast-encoded subunits, and thereafter extrinsic, nuclear-encoded subunits can be added.

3. Photosystem II

3.1. Overview

PS I is not self-sufficient for conducting linear electron transport since P700 has to be re-reduced following primary charge separation. The evolution of PSII and its connection to PSI provided photosynthetic carbon dioxide assimilation with an unlimited source of reducing power, namely H_2O. The unique properties of PSII are that it cannot only perform light-driven charge separation but that it can also oxidize water, bringing about the extraction of electrons and protons with simultaneous release of molecular oxygen. Despite this central role, the composition and organization of PSII was virtually a black box at the end of the 1970s. However, during the last decade progress has been quite dramatic, and today we know that PSII is a multiprotein complex made up of at least 25 different subunits, including several chlorophyll *a*/*b* binding proteins (Fig. 3, Table II) (see ref. [64]). As in the case of PSI, the PSII proteins are the products of plastid or nuclear genes [65]. Despite this large number of proteins only two subunits, designated D_1 and D_2, appear to perform all the primary photochemical

TABLE II

Photosystem II protein components[a]

Protein	Gene	Coded[b]	Mass (kDa)	Organization[c]	Function
D_1	*psbA*	C	32	intrinsic (5)	Binding P680, pheophytin, Q_B, Tyr_Z, Mn-cluster
CP47	*psbB*	C	47	intrinsic (6)	Binding Chl *a* antenna
CP43	*psbC*	C	43	intrinsic (6)	Binding Chl *a* antenna
D2	*psbD*	C	34	intrinsic (5)	Binding P680, Q_A, Tyr_D, Mn cluster?
Cyt *b*-559 α	*psbE*	C	9	intrinsic (1)	Heme binding
Cyt *b*-559 β	*psbF*	C	4	intrinsic (1)	Heme binding
Phosphoprotein	*psbH*	C	9	intrinsic (1)	?
	psbI	C	4.5	intrinsic (1)	Reaction centre function?
	psbK	C	3.7	intrinsic (1)	Chl binding?
	psbL	C	4.8	intrinsic (1)	?
	psbM	C	3.7	intrinsic (1)	?
	psbN	C	4.7	intrinsic (1)	?
	psbO	N	33	extr. lumenal	Mn-stabilizing, Ca^{2+}/Cl^- sequestering
	psbP	N	23	extr. lumenal	Ca^{2+}/Cl^- sequestering
	psbQ	N	16	extr. lumenal	Ca^{2+}/Cl^- sequestering
	psbR	N	10	intrinsic (1)	Docking extrinsic subunits
22 kDa		N	22	intrinsic (1)	?
7 kDa		N	7	intrinsic?	?
5 kDa		N	5	extrinsic?	?
3.2 kDa		N	3.2	intrinsic?	?

[a] The components of the chlorophyll *a*/*b* proteins are not included.
[b] C = chloroplast, N = nucleus.
[c] Intrinsic (number of membrans spans) or extrinsic membrane proteins. Lumenal or stromal site of the membrane.

reactions [66,67] and may even harbour the manganese cluster involved in the photosynthetic water oxidation [64]. Most of this knowledge is based upon the remarkable functional and structural analogy with the L and M subunits of the reaction centre of photosynthetic purple bacteria (see Section 4), whose three-dimensional structure has been determined in the spectacular work of Deisenhofer et al. [17].

3.2. Redox components and electron transfer reactions

From the chlorophyll antennae excitation energy is rapidly transferred to the reaction centre chlorophyll *a* of PSII. This primary donor, designated P680, is excited and in 3–4 ps reduces an intermediary acceptor, a pheophytin (see ref. [68]). This primary charge separation reaction results in a radical pair that is stabilized on the reducing side of PSII by transfer (100–200 ps) of one electron to the primary plastoquinone acceptor Q_A and subsequently in a slower electron transfer reaction to the secondary plastoquinone acceptor Q_B. After double reduction and protonation of Q_B, this quinone leaves the reaction centre to equilibrate with the plastoquinone pool of the thylakoid membrane to allow for reduction of $P700^+$ via the cytochrome *b/f* complex. At the acceptor side of PSII, situated approximately midway between Q_A and Q_B, there is also an iron, whose function is not clear although a role in the binding of bicarbonate to PSII has been suggested.

On the oxidizing side $P680^+$, which has the very oxidizing potential of +1.1 V, is re-reduced in 20–200 ns by a nearby redox-active tyrosine residue (Tyr_Z) of the D_1 protein. Tyr_Z^+ is in turn reduced in the water-splitting reaction which involves a cluster of four manganese ions and possibly some organic radicals [64]. It should be noted that the PSII redox chemistry is quite complicated, involving a two-electron process in the reduction of QB and a four-electron transfer process in the oxidation of water. These events are connected by the one photon-event of the primary charge separation. Below we will outline in some detail the protein components of this light-driven water–plastoquinone oxidoreductase.

3.3. The D_1-D_2 protein reaction centre heterodimer

Since 1986 it has been generally accepted that the reaction centre of PSII is composed of a heterodimer of the two integral membrane proteins designated D_1 and D_2. This concept was based upon the functional and structural analogy between the purple bacterial reaction centre and PSII [17,67,68] and the isolation of a photochemically active PSII reaction centre particle containing the D_1 and D_2 proteins together with cytochrome *b*-559 and the *psbI* gene product [69,70].

The D_1 and D_2 proteins can be resolved by SDS-PAGE as two diffuse bands (explaining the designation D) in the 32–34 kDa region [71]. The D_1 protein is the product of the *psbA* gene of the chloroplast genome and was the first PSII chloroplast gene to be sequenced [72]. It has been shown to be highly conserved among diverse photosynthetic species [73,74]. Originally the D_1 protein was not considered as a reaction centre subunit. Instead its main function was thought to be binding of the secondary quinone and it was consequently termed the '32 kDa-Q_B protein' (see ref. [64]). The early interest in this protein was focussed on its ability to bind several kinds of herbicides targeted on PSII such as diurone (DCMU) and atrazine [75]. Later, another gene in the chloroplast genome was found that coded for a '32 kDa-like' protein. This *psbD* gene, which is also highly conserved, could be shown to code for the D_2 protein [76–78].

The D_1 and the D_2 proteins are homologous to each other and to the L and M subunits in the reaction centre of purple bacteria [66,67]. The latter suggests that the D_1 and D_2 proteins make

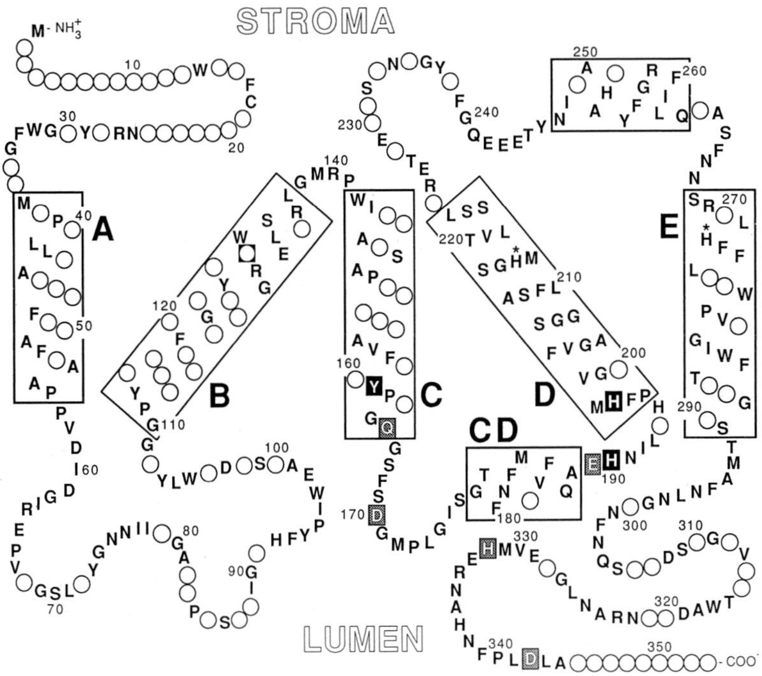

Fig. 4. Predicted folding pattern for the D_1 protein from spinach. The protein sequence from spinach was compared to 37 other sequences for the D_1 protein from other species [74]. In the figure conserved amino acids are represented by their one letter symbol, while circles represent the non-conserved residues. Shown with a white letter on a black background are amino acids with a suggested specific function (i.e. binding of redox components). The manganese-binding residues are shown with a white letter on a shaded background. The histidine ligands for the acceptor side iron are marked with asterisks.

up a heterodimer where each of the two polypeptide chains possesses five membrane-spanning α helices (Fig. 4) [66,67], a concept which has gained experimental support through membrane topological analyses [79].

From sequence alignment analyses the bacterial reaction centre proteins, two histidines D_1-His[198] and D_2-His[198] are considered to coordinate the reaction centre chlorophyll P680 [66,67]. Recently, this prediction has been supported by site-directed mutagenesis studies of these amino acids in *Synechocystis* 6803 (see ref. [80]). The active pheophytin is believed to bind to the D_1 protein only [66,67], via hydrogen bonding to Glu[130] [81,82] (Fig. 4). The primary quinone acceptor Q_A binds to the D_2 protein, probably to the peptide loop which is exposed at the outer thylakoid surface and connects trans-membrane helices D and E [66,67]. The secondary quinone acceptor Q_B is bound to the corresponding loop on the D_1 protein (Fig. 4). Experimental evidence for this assignment is that many amino acids mutated to affect herbicide binding are located in this loop [67,83]. The electron donor to P680 has been identified as Tyr[161] (Tyr_Z) of the D_1 protein (Fig. 4) again by site-directed mutagenesis studies [84,85]. The corresponding tyrosine on the D_2 protein (Tyr_D) is also redox active [86,87], but not involved in the normal steady-state transfer of electrons between the manganese cluster and the reaction centre.

From a chemical point of view, most models for the manganese cluster of the water splitting reaction propose either a distorted cubic structure or a dimer of dimers (see ref. [88]). The manganese atoms are thought to be connected by di-μ-oxo bridges and mixed μ-oxo- and carboxylato-bridging. There are also a nearby Ca^{2+} ion and a Cl^- ion which are both essential for functional water-oxidation.

There are now several lines of evidence to suggest that the four manganese atoms are tightly associated with the reaction centre heterodimer, in particular the D_1 protein (see ref. [64]). A mutant (LF-1) of the green algae *Scenedesmus* which lacks the enzyme for processing the C-terminal extension of the D_1 precursor protein is impaired in oxygen evolution and cannot bind manganese [89]. When this extension is removed by processing the enzyme in vitro, the manganese can be ligated concomitant with the restoration of water-splitting activity [90,91]. Moreover, when the D_1 protein is damaged and degraded during high-light stress (see below) four manganese atoms are released from the inner thylakoid surface [64,92]. Chemical modifications have suggested that at least one histidine in the D_1 protein is involved in manganese binding [93]. More recently, site-directed mutagenesis of *Synechocystis* has implicated Asp^{170}, His^{332} and Asp^{342} of the D_1 protein (Fig. 4) as being involved in manganese ligation (see ref. [80]). Asp^{170} seems to be of special interest since it has been shown to constitute part of a high-affinity site for one of the manganese atoms.

In contrast to the situation at the acceptor side, the donor side of PSII shows little homology to the bacterial reaction centre. Despite this, Svensson et al. [94] have made a structural prediction of the redox-active tyrosines on the donor side of PSII by computer modelling. This model was based on the structural information of bacterial reaction centre in combination with analyses of conserved amino acid sequences of the D_1 and D_2 proteins and functionally determined distances between redox components. The redox active tyrosines (Tyr_Z and Tyr_D) are suggested to be symmetrically located in hydrophobic cavities at a distance of approximately 14 Å from P680. Moreover, both tyrosines are thought to form hydrogen bonds with His^{190} from the loop connecting helices C and D in the D_1 and D_2 proteins, respectively (Fig. 4). This prediction has recently gained support from spectroscopic analysis of site-directed mutants of histidine D_2-His^{190} (see ref. [80], Styring, Svensson, Davidsson and Vermaas, unpublished results). From theoretical arguments based upon known distances, it is suggested that the manganese cluster is bound to the lumenal ends of the D_1 protein at transmembrane helices A and B and their connecting loop. In addition, a metal binding site involving the residues Gln^{165}, Asp^{170} and Glu^{189} of the D_1 protein (Fig. 4) has been predicted close to Tyr_Z [74]. According to these studies, it is not very likely that the D_2 protein contributes to ligation of the manganese cluster.

The D_1 protein, and to some extent the D_2 protein, show high turnover in contrast to the other PSII subunits (see ref. [95]). This turnover is a consequence of the photochemistry in PSII and is enhanced at high light intensities, particularly under adverse environmental conditions (see ref. [96]). The mechanisms behind this inhibition of PSII electron transport and protein turnover are currently a central area of photosynthesis research. The damage to the D_1 protein is thought to be caused by singlet oxygen formed after over-reduction of the acceptor side of PSII or the formation of long-lived oxidizing species ($P680^+$ and/or $Tyr_Z^{\dot{+}}$) at the donor side [96,97]. Once damaged, the D_1 protein has to be replaced to recover functional PSII electron transport. The damaged D_1 protein is degraded by proteolytic activities residing within the PSII complex, thereby allowing the functional incorporation of a new copy of the protein (see refs. [96,97]).

Both the D_1 and D_2 proteins are required for the assembly of the PSII complex. Deletion of either *psbA* or *psbD* genes in cyanobacteria leads to partially unassembled PSII and to lack of both the heterodimer proteins in addition of the chlorophyll-binding subunit CP47 [83,98].

3.4. *Cytochrome b-559*

Cytochrome *b*-559 is composed of two polypeptides, one subunit of 9 kDa and one subunit of 4 kDa (see ref. [99]). Both polypeptides are encoded by the chloroplast DNA and their genes are designated *psbE* and *psbF*, respectively [100]. Hydropathy predictions as well as membrane topological experiments suggest that each of the two polypeptides has one membrane span with their N-termini exposed at the outer thylakoid surface [99]. Structurally, the cytochrome is envisaged as a heterodimer in which the heme group is sandwiched between two histidines toward the outer side of the membrane. The cytochrome is closely associated with the D_1–D_2 heterodimer as revealed by its presence in isolated PSII reaction centre particles [69,70]. Deletion of the *psbE* and *psbF* genes in *Synechocystis* leads to failure to assemble the PSII reaction centre and lack of both the D_1 and D_2 proteins [101].

Cytochrome *b*-559 can attain one high-potential form (+350 mV) or several low-potential forms (5 mV–230 mV) [99]. The high-potential form is generally thought to be the active state of the cytochrome. The function of cytochrome *b*-559 is not fully understood, but it is clear that it is not involved in the major pathway of linear electron transport. However, under conditions where the water splitting system is impaired, it can donate electrons to $P680^+$ [99]. The cytochrome appears to catalyze electron flow from reduced quinones to $P680^+$, thereby mediating a cyclic electron flow around PSII [102,103]. This electron transfer possibly occurs via a chlorophyll monomer that is present in the PSII reaction centre. The physiological significance of these reactions may be to reduce strongly oxidizing $P680^+$ under conditions of inoptimal electron donation from the manganese cluster, thereby protecting the PSII reaction centre against photodamage [102,103].

3.5. *CP43 and CP47: chlorophyll a binding proteins*

Two polypeptides of PSII bind chlorophyll *a* and have a function in light harvesting and energy transfer to the reaction centre (see ref. [104]). This inner light-harvesting antenna is complemented by several chlorophyll *a*/*b* binding proteins that make up the outer antenna of PSII [4]. The two proteins referred to as CP47 and CP43 can be detected as two distinct chlorophyll *a* bands during mild SDS-PAGE [104]. Each of the chlorophyll proteins is composed of one apo-polypeptide with apparent molecular weights of 47 and 43 kDa, respectively. Both apo-polypeptides of CP47 and CP43 are encoded by the chloroplast DNA, and their genes are designated *psbB* and *psbC*, respectively [76,77,105]. Sequence analysis shows that the two proteins are related and predicted to possess six membrane spanning α-helices each [106]. These contain conserved histidines which are likely to be essential for the binding of approximately 15 chlorophyll molecules per polypeptide. Interestingly, both polypeptides have unusually large hydrophilic domains consisting of nearly 200 amino acid residues between two membrane-spanning regions. These hydrophilic loops, which most likely are exposed at the inner thylakoid surface, suggest that the two proteins may have functions other than purely light harvesting [105]. Deletion of the *psbB* gene or removal of the region Gly^{351} to Thr^{365} in the exposed loop of CP47 leads to disassembly of the PSII complex [82,83]. Deletion of the *psbC* gene leads to highly reduced levels of PSII, but residual amounts of a functional complex appear to be assembled in the absence of CP43 [83].

3.6. *The 9 kDa phospho-protein*

Several PSII proteins can be phosphorylated by a membrane-bound kinase that is under the control of the redox state of the plastoquinone pool (see ref. [107]). The physiological significance of this protein phosphorylation is not yet understood, even though functionally important subunits such as CP43 and the D_1 and D_2 proteins can be phosphorylated. The most prominent phospho-protein of PSII is a 9 kDa protein of unknown function. It is the product of the plastid *psbH* gene [108,109]. The protein is composed of 72 amino acids and is predicted to have one transmembrane α-helix. Deletion of *psbH* in *Synechocystis* does not lead to inhibition of PSII electron transport (S. Mayes and J. Barber, personal communication). However, the cell becomes more sensitive to high light intensities, suggesting some inoptimal regulation of the PSII electron flow once the 9 kDa protein is absent. In contrast to many other deletions of integral PSII membrane proteins, the deletion of the *psbH* gene does not appear to affect the functional assembly of the PSII complex.

3.7. Low molecular weight polypeptides

The presence of several low molecular weight proteins in PSII was demonstrated by Ljungberg et al. [110]. Presently at least nine different gene products in the 3–7 kDa range have been identified (see refs. [64,106]). Six of these are encoded by the chloroplast DNA, of which one is the gene product of *psbF* and associated with cytochrome *b*-559 (see Subsection 3.4). The other five plastid-encoded proteins can be assigned to open reading frames in the plastid DNA and have been designated *psbI*, *psbK*, *psbL*, *psbM* and *psbN* [106,111].

The *psbI* gene codes for a polypeptide of 4.5 kDa. Interestingly this polypeptide is present in isolated PSII reaction centre complexes together with the D_1–D_2 heterodimer and cytochrome *b*-559 [70]. Thus this polypeptide is closely associated with the reaction centre and has the potential of playing a central but as yet unidentified role in the function of PSII. It contains 36 amino acids and is predicted to possess one hydrophobic domain typical of a membrane span.

The product of the *psbK* gene corresponds to a 3.7 kDa polypeptide [112]. It contains 37 amino acids and is likely to span the membrane once. It shows slight sequence homology with the major light-harvesting polypeptide of photosynthetic bacteria, and may represent a chlorophyll-binding protein that so far has escaped detection. The *psbL* gene product is a 4.8 kDa polypeptide and is likely to possess one trans-membrane span [113]. It contains an N-terminal threonine and has the potential to represent an identified low molecular mass phosphoprotein.

The *psbM* and *psbN* proteins are two newly discovered one membrane-spanning proteins of PSII [111]. The *psbM* protein is predicted to have a molecular mass of 3.7 kDa and has significant similarity to the sequence deduced from open reading frame 34 of the plastid DNA. The *psbN* protein can be connected to open reading frame 43 and its molecular mass is estimated to be 4.7 kDa.

Three of the small polypeptides are nuclear encoded, possessing molecular masses of 7, 5 and 3.2 kDa [114]. The 7 kDa and 3.2 kDa species are not seen in all types of functional PSII complexes and are not likely to have any direct catalytic function. Both proteins show hydrophobic properties during biochemical handling and are probably integral membrane proteins [114]. The 5 kDa protein is present in all PSII core preparations analyzed so far. This protein has been purified, it is water-soluble and its amino acid composition reveals a hydrophilic character [110]. However, it cannot be released from any side of the membrane by biochemical procedures normally applied to release extrinsic membrane proteins.

3.8. Extrinsic membrane proteins associated with the donor side of photosystem II

Three extrinsic proteins influencing the oxygen-evolving reaction have been shown to be located at the inner thylakoid surface (see refs. [64,115,116]). These proteins, with apparent molecular masses of 33, 23 and 16 kDa, can be reversibly detached from and re-attached to everted thylakoid membranes, resulting in inhibition and reconstitution of water oxidation. Mutants of *Synechocystis* or *Chlamydomonas* lacking the 33 kDa protein [117–119] and a mutant of *Chlamydomonas* lacking the 23 kDa protein [120] all have reduced oxygen-evolving activities. All three proteins are highly hydrophilic and water soluble, supporting their classification as extrinsic membrane proteins.

Despite their influence on electron transport on the donor side of PSII, there is now a general consensus that these proteins have non-catalytic functions. This became clear when it was found that electron-transport activity lost by removal of the three proteins from the membrane in vitro could be restored by the addition of relatively high concentrations of Ca^{2+} and/or Cl^- ions (see refs. [115, 116]). Moreover, the gene for the 33 kDa protein has recently been deleted in *Synechocystis* and the cyanobacterial cells could still show photoautotrophic growth [117,118]. Still, the mutated cells show stronger susceptibility towards photoinhibition [118], implicating some perturbation of the electron transfer from the manganese cluster.

The 23 and 16 kDa proteins, even though present in all eukaryotic PSIIs, has not been found in any cyanobacterial species, providing direct in vivo evidence for a non-catalytic role for these two extrinsic proteins also (see ref. [116]).

Presently, it is considered that the 23 and 16 kDa proteins provide high-affinity binding sites for the Ca^{2+} and Cl^- ions which are both necessary as co-factors for photosynthetic water-splitting [115,116]. Such a function may also be considered for the 33 kDa protein, since its removal further increases the requirement of Cl^- ions for optimal water-splitting. Interestingly, it has been shown that the 33 kDa protein can bind Ca^{2+} ions in vitro, and that it shows sequence homology to a mammalian intestinal calcium binding protein [83]. However, the main role of the 33 kDa protein appears to be stabilization of the manganese cluster, although a direct metal ligation is not very likely [115,116].

The 33, 23 and 16 kDa proteins are nuclear encoded [65] by genes designated *psbO*, *psbP* and *psbQ* respectively (see refs. [7,83]). These proteins, as well as the PSI-F protein and plastocyanin, are imported as precursors with unusually long presequences (80–85 amino acids) possessing a hydrophilic chloroplast import domain and a hydrophobic thylakoid transfer domain. Notably, the proteins have to be transported through the two chloroplast outer membranes and the thylakoid membrane (Fig. 1) to reach their location at the lumenal membrane surface. The processing appears to occur in two steps, one after penetration into the chloroplast stroma and a final cut once the protein has reached the thylakoid lumen.

3.9. The hydrophobic 22 and 10 kDa proteins

The 22 and 10 kDa proteins were discovered by analysis of the nearest-neighbours of the 33 and 23 kDa proteins [121] and shown to be present in certain PSII preparations [122]. Removal of the two proteins from the thylakoid membrane, which can be achieved by treatments with chaotropic salts and low amounts of detergents, does not in itself inhibit PSII electron transport [121]. Thus, the two proteins add to the population of non-catalytic proteins in PSII whose functions remain to be established. Still, there is quite good evidence that the 10 kDa protein is involved in the docking of the extrinsic 23 kDa protein to the inner thylakoid surface [121]. The 10 kDa protein is nuclear encoded and the gene has been designated *psbR* [7,65]. The nucleotide sequence predicts one membrane span at the very C-terminal end of the protein. It is predicted that the protein is located in the lumenal space, anchored by its membrane-spanning C-terminal portion, consistent with a role as a docking protein. The presence of the 10 kDa protein in cyanobacteria has so far not been demonstrated.

Recently the *psbR* gene was inactivated by the anti-sense RNA technique in potato plants [123]. The transgenic plants showed normal levels of other PSII proteins but the steady state of oxygen evolution in vitro showed a 20–60% reduction compared to control thylakoids. The impairment was suggested to be targeted to the acceptor side of PSII, more specifically on the reoxidation of Q_A. The apparent discrepancy between the biochemical and genetic data will need further experiments.

From preliminary sequence analysis, one membrane spanning helix is also suggested for the 22 kDa protein, which is encoded by a nuclear gene (R.G. Herrmann, et al., unpublished). Both the 10 and 22 kDa proteins are synthesized as precursors, which have to be imported into the chloroplast. Most studies on the 22 kDa protein suggest some regulatory role associated with the acceptor side of PSII (see ref. [64]), but basically its function is unknown. Recently the 22 kDa protein has been shown to be present in cyanobacteria [98].

3.10. Organization of the PSII complex

As indicated above, our knowledge of the arrangement of the D_1–D_2 reaction centre heterodimer is quite advanced. In contrast, the molecular organization of the PSII complex with respect to the other subunits is less well understood. As indicated in Fig. 3, there is quite reliable knowledge of which are hydrophobic intrinsic membrane-spanning proteins and which are hydrophilic extrinsic membrane proteins. Amongst the latter category the 33, 23 and 16 kDa proteins, located at the inner thylakoid surface, are the best characterized. In contrast to PSI, there is no evidence for an extrinsic protein located at the outside of the thylakoid membrane. Possibly, there is no space for bulky extrinsic proteins in the closely appressed regions of the grana stacks where PSII has its main location. The 33, 23 and 16 kDa proteins are in close vicinity of the water oxidation site and appear to sequester it from the bulk of the lumenal space. In freeze-etch studies of the inner thylakoid surface, these proteins have been shown to make up typical tetrameric particles [124] that were once called quantasomes [125], and thought to harbour the entire photosynthetic machinery. Several studies have addressed the binding sites of the three proteins, but the picture is far from clear. For the 33 kDa protein, most experiments suggest a close connection to the D_1–D_2 reaction centre heterodimer [64], although the protein is still bound to the membrane in a cyanobacterial mutant lacking these two proteins [98]. There are data in support of the hypothesis that the 23 kDa protein is anchored to the membrane via the 10 kDa protein [121]. The 16 kDa protein in turn is bound to the 23 kDa protein.

For the integral membrane proteins, our knowledge concerning their location is largely based upon subfractionation studies. Consequently, cytochrome *b*-559 and the *psbI*-gene product are thought to be tightly associated with the D_1 and D_2 proteins due to their presence in isolated reaction centre particles [70]. Other integral subunits present in various isolated functional photosystem II preparations and hence located in the core of the complex are the two chlorophyll *a* proteins CP47 and CP43 and some low molecular weight proteins. More peripheral in the complex are the various chlorophyll *a*/*b* proteins.

It should be mentioned that there are thought to be two forms of PSII, the so-called α and β centres [126]. PSIIα is the main population, located in the appressed regions of the grana. These α centres have a full complement of light-harvesting antenna and can perform complete PSII electron transport. The minor population, PSIIβ, is located in the stroma exposed regions, being intermixed with PSI. The β centres have a smaller antenna and cannot perform electron transfer between Q_A and Q_B.

4. Analogies between the photosystems of oxygenic photosynthesis and photosynthetic bacteria

As described in Section 3, the D_1 and D_2 subunits of PSII are homologous to the L and M subunits of the purple bacterial reaction centre. Apart from this structural homology, there is also a functional homology with respect to the redox components and the electron transfer reactions. The primary electron donor (special pair) consists of (bacterio)chlorophyll. The acceptor side components are also remarkably similar: (bacterio)pheophytin, Fe^{2+} and the quinones Q_A and Q_B.

There is yet another group of photosynthetic bacteria with reaction centres similar to PSII – the green-gliding bacteria (e.g. *Chloroflexus*). However, the reaction centres from these bacteria are not as well characterized as those from purple bacteria.

Is there also homology between PSI and some bacterial reaction centre? There have been some hypotheses that PSI and the reaction centre of green-sulphur bacteria (e.g. *Chlorobium*) are analogous. Recent spectroscopic and biochemical studies have indeed shown that the PSI reaction centre is remarkably similar to the reaction centre of green-sulphur bacteria. These two reaction centres appear to contain the same type of acceptor side components: a (bacterio)chlorophyll (A_0), a quinone (A_1) and iron–sulphur centres of the F_A, F_B and F_X types [127]. Also, the reaction centre of the newly discovered heliobacteria (e.g. *Heliobacterium*) [128] has been shown to be similar to PSI [127]. Furthermore, the reaction centre of green-sulphur bacteria and of heliobacteria consists of polypeptides of apparent molecular masses comparable to those of PSI-A and PSI-B. As in PSI, these reaction centre polypeptides carry a large number of antenna pigments [127]. Thus, it is likely that the reaction centres of these bacteria contain a heterodimer of polypeptides homologous to PSI-A and PSI-B.

Consequently, photosynthetic reaction centres may be divided into two groups: PSI-type (PSI, green-sulphur bacteria and heliobacteria) and PSII-type (PSII, purple bacteria and green-gliding bacteria). However, there are several principal functional similarities between the two groups of reaction centres: (i) The primary donor consists of (bacterio)chlorophyll, most likely in the form of a dimer. (ii) The primary acceptor is a (bacterio)pheophytin or a (bacterio)chlorophyll, which are chemically similar. (iii) The secondary acceptor is a quinone molecule.

It is likely, therefore, that all photosystems originate from a common ancestor. One major difference between the two types of reaction centres is that PSI-type reaction centre polypeptides are larger and contain many antenna pigments, whilst the PSII-type reaction centre polypeptides are smaller and contain virtually no light-harvesting chlorophylls. The ancestral reaction centre polypeptides may have been PSII type, and a gene fusion might have occurred between genes for these polypeptides and the genes for light-harvesting polypeptides, creating a PSI-type reaction centre. Alternatively, gene fission may have transformed large reaction centre polypeptides into small ones and light-harvesting polypeptides [127]. In PSI type photosystems, an extra subunit carrying iron–sulphur centres (PSI-C) has been added. Probably, a soluble ferredoxin has been 'captured' by the photosystem, since the PSI-C protein shows a significant homology with soluble ferredoxins from bacteria (see Section 2.4).

Acknowledgements

This work was supported by grants from the Swedish Natural Science Research Council, Carl Trygger's Foundation, Lars Hierta's Memorial Foundation and Göran Gustafsson's Foundation. We are grateful to Drs. H. Salter and S. Styring for useful comments on the manuscript. Fig. 4 was skillfully designed by B. Svensson.

References

1 Anderson, J.M. and Andersson, B. (1982) Trends Biochem. Sci. 7, 288–292.
2 Andersson, B. and Anderson, J.M. (1980) Biochim. Biophys. Acta 593, 427–440.
3 Mullet, J.E., Burke, J.J. and Arntzen, C.J. (1980) Plant Physiol. 65, 814–822.
4 Green, B.R., Pichersky, E. and Kloppstech, K. (1991) Trends Biochem. Sci. 16, 181–186.
5 Glaser, A. and Melis, A. (1987) Annu. Rev. Plant Physiol. 38, 11–45.
6 Bengis, C. and Nelson, N. (1977) J. Biol. Chem. 252, 4564–4569.
7 Hallick, R.B. (1989) Plant Mol. Biol. Rep. 7, 266–275.
8 Golbeck, J.H. and Bryant, D.A. (1991) in: C.P. Lee, (Ed.) Current Topics Bioenergetics, Vol. 16, Academic Press, New York, pp. 83–177.
9 McDermott, A.E., Yachandra, V.K., Guiles, R.D., Sauer, K., Klein, M., Parret, K. and Golbeck, J.H. (1989) Biochemistry 28, 8056–8059.
10 Petrouleas, V., Brand, J.J., Parret, K.V. and Golbeck, J.H. (1989) Biochemistry 28, 2980–2983.
11 Høj, P.B. and Møller, B.L. (1986) J. Biol. Chem. 261, 14292–14300.
12 Golbeck, J.H. (1987) Biochim. Biophys. Acta 895, 167–204.
13 Høj, P.B., Svendsen, I., Scheller, H.V. and Møller, B.L. (1987) J. Biol. Chem. 262, 12684–12687.
14 Witt, I., Witt, H.T., Gerken, S., Saenger, W., Dekker, J.P. and Rögner, M. (1987) FEBS Lett. 221, 260–264.
15 Ford, R.C., Pauptit, R.A. and Holzenburg, A. (1988) FEBS Lett. 238, 385–389.
16 Almog, O., Shoham, G., Michaeli, D. and Nechushtai, R. (1991) Proc. Natl. Acad. Sci. USA 88, 5312–5316.
17 Deisenhofer, J., Epp, O., Miki, K., Huber, R. and Michel, H. (1985) Nature 318, 618–624.
18 Robert, B. and Moenne-Loccoz, P. (1990) in: M. Baltscheffsky (Ed.) Current Research in Photosynthesis, Vol. I, Kluwer Academic Publishers, Dordrecht, pp. 65–68.
19 Margulies, M.M. (1991) Photosynth. Res. 29, 133–147.
20 Golbeck, J.H. and Cornelius, J.M. (1986) Biochim. Biophys. Acta 849, 16–24.
21 Scheller, H.V., Svendsen, I. and Møller, B.L. (1989) J. Biol. Chem. 264, 6929–6934.
22 Hayashida, N., Matsubayashi, T., Shinosaki, K., Sugiura, M., Inoue, K. and Hiyama, T. (1987) Curr. Genet. 12, 247–250.
23 Oh-oka, H., Takahashi, Y., Kuriyama, K., Saeki, K. and Matsubara, H. (1988) J. Biochem. 103, 962–968.
24 Scheller, H.V. and Møller, B.L. (1990) Physiol. Plant. 78, 484–494.
25 Takahashi, Y., Goldschmidt-Clermont, M., Soen, S.-Y., Franzén, L.-G. and Rochaix, J.-D. (1991) EMBO J. 10, 2033–2040.
26 Mann, K., Schlenkrich, T., Bauer, M. and Huber, R. (1991) Biol. Chem. Hoppe-Seyler 372, 519–524.
27 Dunn, P.P.J. and Gray, J.C. (1988) Plant Mol. Biol. 11, 311–319.
28 Li, N., Zhao, J., Warren, P.V., Warden, J.T., Bryant, D.A. and Golbeck, J.H. (1991) Biochemistry 30, 7863–7872.
29 Von Heijne, G., Steppuhn, J., Herrmann, R.G. (1989) Eur. J. Biochem. 180, 535–545.
30 Zanetti, G. and Merati, G. (1987) Eur. J. Biochem. 169, 143–146.
31 Zilber, A.L. and Malkin, R. (1988) Plant Physiol. 88, 810–814.
32 Chitnis, P.R., Reilly, P.A. and Nelson, N. (1989) J. Biol. Chem. 264, 18381–18385.

33 Münch, S., Ljungberg, U., Steppuhn, J., Schneiderbauer, A., Nechushtai, R., Beyreuther, K.J and Herrmann, R.G. (1988) Curr. Genet. 14, 511–518.
34 Steppuhn, J., Hermans, J., Nechushtai, R., Ljungberg, U., Thümmler, F., Lottspeich, F. and Herrmann, R.G. (1988) FEBS Lett. 237, 218–224.
35 Steppuhn, J., Hermans, J., Nechushtai, R., Herrmann, G.S. and Herrmann, R.G. (1989) Curr. Genet. 16, 99–108.
36 Chitnis, P.R., Reilly, P.A., Miedel, M.C. and Nelson, N. (1989) J. Biol. Chem. 264, 18374–18380.
37 Franzén, L.-G., Frank, G., Zuber, H. and Rochaix, J.-D. (1989) Plant Mol. Biol. 12, 463–474.
38 Chitnis, P.R., Purvis, D. and Nelson, N. (1991) J. Biol. Chem. 266, 20146–20151.
39 Wynn, R.M. and Malkin, R. (1988) Biochemistry 27, 5863–5869.
40 Hippler, M., Ratajczak, R. and Haehnel, W. (1989) FEBS Lett. 250, 280–284.
41 Wynn, R.M., Omaha, J. and Malkin, R. (1989) Biochemistry 28, 5554–5560.
42 Bendall, D.S. and Davies, E.C. (1989) Physiol. Plant. 76, A87.
43 Anandan, S., Vainstein, A. and Thornber, J.P. (1989) FEBS Lett. 256, 150–154.
44 Ratajczak, R., Mitchell, R. and Haehnel, W. (1988) Biochim. Biophys. Acta 933, 306–318.
45 Franzén, L.-G., Frank, G., Zuber, H. and Rochaix, J.-D. (1989) Mol. Gen. Genet. 219, 137–144.
46 Scheller, H.V., Okkels, J.S., Høj, P.B., Svendsen, I., Roepstorff, P. and Møller, B.L. (1989) J. Biol. Chem. 264, 18402–18406.
47 Koike, H., Ikeuchi, M., Hiyama, T. and Inoue, Y. (1989) FEBS Lett. 253, 257–263.
48 Ikeuchi, M., Nyhus, K.J., Inoue, Y. and Pakrasi, H.B. (1991) FEBS Lett. 287, 5–9.
49 Hoshina, S., Sue, S., Kunishima, N., Kamide, K., Wada, K. and Itoh, S. (1989) FEBS Lett. 258, 305–308.
50 Wynn, R.M. and Malkin, R. (1990) FEBS Lett. 262, 45–48.
51 Ikeuchi, M., Hirano, A., Hiyama, T. and Inoue, Y. (1990) FEBS Lett. 263, 274–278.
52 Takabe, T., Iwaskaki, Y., Hibino, T. and Ando, T. (1991) J. Biochem. 110, 622–627.
53 Okkels, J.S., Scheller, H.V., Svendsen, I. and Møller, B.L. (1991) J. Biol. Chem. 266, 6767–6773.
54 Ikeuchi, M. and Inoue, Y. (1991) FEBS Lett. 280, 332–334.
55 Ohyama, K., Fukuzawa, H., Kohchi, T., Shirai, H., Sano, T., Sano, S., Umesono, K., Shiki, Y., Takeuchi, M., Chang, Z., Aota, S., Inokuchi, H. and Ozeki, H. (1986) Nature 322, 572–574.
56 Tjus, S.E. and Andersson, B. (1991) Photosynth. Res. 27, 209–219.
57 Golbeck, J.H., Parret, K.G., Mehari, T., Jones, K.L. and Brand, J.J. (1988) FEBS Lett. 228, 268–272.
58 Li, N., Warren, P.V., Golbeck, J.H., Frank, G., Zuber, H. and Bryant, D.A. (1991) Biochim. Biophys. Acta 1059, 215–225.
59 Choquet, Y., Goldschmidt-Clermont, M., Girard-Bascou, J., Kück, U., Bennoun, P. and Rochaix, J.-D. (1988) Cell 52, 903–913.
60 Smart, L.B., Anderson, S. and McIntosh, L. (1991) EMBO J. 10, 3289–3296.
61 Girard-Bascou, J., Choquet, Y., Schneider, M., Delosme, M. and Dron, M. (1987) Curr. Genet. 12, 489–495.
62 Toelge, M., Ziegler, K., Maldener, I. and Lockau, W. (1991) Biochim. Biophys. Acta 1060, 233–236.
63 Mannan, R.M., Whitmarsh, J., Nyman, P. and Pakrasi, H.B. (1991) Proc. Natl. Acad. Sci. USA 88, 10168–10172.
64 Andersson, B. and Styring, S. (1991) in: C.P. Lee (Ed.) Current Topics in Bioenergetics, Vol. 16, Academic Press, New York, pp. 1–82.
65 Andersson, B. and Herrmann, R.G. (1988) in: J.L. Harwood and T.J. Walton (Eds.) Plant Membranes – Structure, Assembly and Function, The Biochemical Society, London, pp. 33–45.
66 Michel, H. and Deisenhofer, J. (1986) in: A. Staehelin and C.J. Arntzen (Eds.) Encyclopedia of Plant Physiology, Vol. 19, Springer, Berlin, pp. 371–381.
67 Trebst, A. (1986) Z. Naturforsch. 41c, 240–245.
68 Rutherford, A.W. (1989) Trends Biochem. Sci. 14, 227–232.
69 Nanba, O. and Satoh, K. (1987) Proc. Natl. Acad. Sci. USA 84, 109–112.
70 Webber, A.N., Packman, L., Chapman, D.J. , Barber, J. and Gray, J.C. (1989) FEBS Lett. 242, 259–262.
71 Chua, N.H. and Gillham, N.W. (1977) J. Cell Biol. 74, 441–452.

72 Zurawiski, G., Bohnert, H.J., Whitfeldt, P.R. and Bottomley, W. (1982) Proc. Natl. Acad. Sci. USA 79, 7699–7703.
73 Erickson, J.M., Rochaix, J.P. and Delepelaire, P. (1985) in: K.E. Steinback, S. Bonitz, C.J. Arntzen and L. Bogorad (Eds.) Molecular Biology of the Photosynthetic Apparatus, Cold Spring Harbor Laboratory, Cold Spring Harbor, New York, pp. 53–65.
74 Svensson, B., Vass, I. and Styring, S. (1991) Z. Naturforsch. 46C, 765–776.
75 Trebst, A. and Draber, W. (1986) Photosynth. Res. 10, 381–392.
76 Alt, J., Morris, J., Westhoff, P. and Herrmann, R.G. (1984) Curr. Genet. 8, 597–606.
77 Holschuh, K., Bottomley, W. and Whitfeld, P.R. (1984) Nucleic Acid. Res. 12, 8819–8834.
78 Rochaix, J.-D., Dron, M., Rahire, M. and Malnoe, P. (1984) Plant Mol. Biol. 3, 363–370.
79 Sayre, R.T., Andersson, B. and Bogorad, L. (1986) Cell 47, 601–608.
80 Diner, B.A., Nixon, P.J. and Farchans, J.W. (1991) in: Current Opinions in Structural Biology, Vol. 1, Current Biology Ltd, London, pp. 546–554.
81 Moenne-Loccoz, P., Robert, B. and Lutz, M. (1990) in: M. Baltscheffsky (Ed.) Current Research in Photosynthesis, Vol. I, Kluwer Academic Publishers, Dordrecht, pp. 423–426.
82 Nabedryk, E., Andrianambinitsoa, S., Berger, G., Leonard, M., Mandele, W. and Breton, J. (1990) Biochim. Biophys. Acta 1016, 49–54.
83 Vermaas, W.F.J. (1989) in: J. Barber R. Malkin (Eds.) Techniques and New Developments in Photosynthesis Research, Plenum Press, New York, pp. 35–59.
84 Debus, R.J. Barry, B.A. Sithole, I., Babcock, G.T. and McIntosh, L. (1988) Biochemistry 27, 9071–9074.
85 Metz, J.G., Nixon, P.J., Rögner, M., Brudvig, G.W. and Diner, B.A. (1989) Biochemistry 28, 6960–6969.
86 Debus, R.J., Barry, B.A., Babcock, G.T. and McIntosh, L. (1988) Proc. Natl. Acad. Sci. USA 85, 427–430.
87 Vermaas, W.F.L., Rutherford, A.W. and Hansson, Ö. (1988) Proc. Natl. Acad. Sci. USA 85, 8477–8481.
88 Debus, R.J. (1992) Biochim. Biophys. Acta, in press.
89 Metz, J.G., Bricker, T.M. and Seibert, M. (1985) FEBS Lett. 185, 191–195.
90 Diner, B.A., Ries, D.F., Cohen, B.N. and Metz, J.G. (1988) J. Biol. Chem. 263, 8972–8980.
91 Taylor, M.A., Packer, J.C.L. and Bowyer, J.R. (1988) FEBS Lett. 237, 229–233.
92 Virgin, I., Styring, S. and Andersson, B. (1988) FEBS Lett. 233, 408–412.
93 Seibert, M., Tamura, N. and Inoue, Y. (1989) Biochim. Biophys. Acta 974, 185–191.
94 Svensson, B., Vass, I., Cedergren, E. and Styring, S. (1990) EMBO J. 7, 2051–2059.
95 Mattoo, A.K., Marder, J.B. and Edelman, M. (1989) Cell 56, 241–246.
96 Barber, J. and Andersson, B. (1992) Trends Biochem. Sci. 17, 61–66.
97 Andersson, B., Salter, A.H., Virgin, I., Vass, I. and Styring, S. (1992) J. Photochem. Photobiol., in press.
98 Nilsson, F., Andersson, B. and Jansson, C. (1990) Plant Mol. Biol. 14, 1051–1054.
99 Cramer, W.A., Furbacker, P.N., Szczepaniak, A. and Tae, G.-S. (1990) in: Current Research in: M. Baltscheffsky (Ed.) Photosynthesis Vol. 3, Kluwer Academic Publishers, Dordrecht, pp. 221–230.
100 Widger, W.R., Cramer, W.A., Hermodson, M. and Herrmann, R.G. (1985) FEBS Lett. 191, 186–190.
101 Pakrasi, H.B., Nyhus, K.J. and Granok, H. (1990) Z. Naturforsch. 45c, 423–429.
102 Thompson, L.K. and Brudvig, G.W. (1988) Biochemistry 27, 6653– 6658.
103 Telfer, A., He, W.-Z. and Barber, J. (1990) Biochim. Biophys. Acta 1017, 143–151.
104 Green, B.R. (1988) Photosynth. Res. 15, 3–32.
105 Morris, J. and Herrmann, R.G. (1984) Nucleic Acid. Res. 12, 2837–2850.
106 Gray, J.C., Hird, S.M., Wales, R., Webber, A.N. and Willey, D.L. (1989) in: J. Barber and R. Malkin (Eds.) Techniques and New Developments in Photosynthesis Research, Plenum Press, New York, pp. 423–435.
107 Bennett, J. (1983) Biochem. J. 212, 1–13.
108 Hird, S.M., Dyer, T.A. and Gray, J.C. (1986) FEBS Lett. 209, 181–186.
109 Westhoff, P., Farchaus, J. and Herrmann, R.G. (1986) Curr. Genet. 11, 165–169.

110 Ljungberg, U., Henrysson, T., Rochester, C.P., Åkerlund, H.-E. and Andersson, B. (1986) Biochim. Biophys. Acta 849, 112–120.
111 Ikeuchi, M., Koike, H. and Inoue, Y. (1989) FEBS Lett. 253, 178–182.
112 Murata, N., Miyao, M., Haayashida, N. Hidaka, T. and Suigura, M. (1988) FEBS Lett. 235, 283–288.
113 Webber, A.N., Packman, L. and Gray, J.C. (1989) FEBS Lett. 242, 435–438.
114 Schröder, W.P., Henrysson, T. and Åkerlund, H.-E. (1988) FEBS Lett. 235, 289–292.
115 Murata, N. and Miyao, M. (1985) Trends Biochem. Sci. 10, 122–124.
116 Andersson, B. and Åkerlund, H.-E. (1987) in: J. Barber (Ed.) Topics in Photosynthesis, Vol. 8, Elsevier, Amsterdam, pp. 379–420.
117 Burnarp, R.L. and Sherman, L.A. (1991) Biochemistry 30, 440–446.
118 Mayes, S.P., Cook, K.M., Self, S.J., Zhang, Z. and Barber, J. (1991) Biochim. Biophys. Acta 1060, 1–12.
119 Mayfield, S.P., Bennoun, P. and Rochaix, J.-D. (1987) EMBO J. 6, 313–318.
120 Mayfield, S.P., Rahire, M., Frank, G., Zuber, H. and Rochaix, J.-D. (1987) Proc. Natl. Acad. Sci. USA 84, 749–753.
121 Ljungberg, U., Åkerlund, H.-E. and Andersson, B. (1986) Eur. J. Biochem. 158, 477–482.
122 Ghanotakis, D.F., Demetriou, D.M. and Yocum, C.F. (1987) Biochim. Biophys. Acta 891, 15–21.
123 Stockhaus, J., Höfer, M., Renger, G., Westhoff, P., Wydrzynski, T. and Willmitzer, L. (1990) EMBO J. 9, 3013–3021.
124 Simpson, D.J. and Andersson, B. (1986) Carlsberg Res. Commun. 51, 467–474.
125 Park, R.B. and Biggins, J. (1964) Science 144, 1009–1011.
126 Melis, A. (1991) Biochim. Biophys. Acta 1058, 87–106.
127 Nitschke, W. and Rutherford, A.W. (1991) Trends Biochem. Sci. 16, 241–245.
128 Gest, H. and Favinger, J.L. (1983) Arch. Microbiol. 136, 11–16.

L. Ernster (Ed.) *Molecular Mechanisms in Bioenergetics*
© 1992 Elsevier Science Publishers B.V. All rights reserved

CHAPTER 6

NADH-ubiquinone oxidoreductase

THOMAS P. SINGER and RONA R. RAMSAY

Departments of Biochemistry and Biophysics, Pharmaceutical Chemistry and Division of Toxicology, University of California, San Francisco, CA 94143 and Molecular Biology Division, Department of Veterans Affairs Medical Center, San Francisco, CA 94121, USA

Contents

1. Introduction 145
2. Enzyme preparations and their general properties 146
3. Substrate specificity 147
4. Subunit structure 147
5. Prosthetic groups 150
6. Inhibitors 153
  6.1. Acting between NADH and FMN 153
  6.2. Acting between center 2 and Q 153
  6.3. Acting at multiple or ill-defined sites 157
7. Energy conservation site 1 157
8. Diseases related to complex I deficiency 158
Acknowledgements 159
References 159

1. Introduction

NADH-ubiquinone reductase (NADH dehydrogenase of the respiratory chain) is generally recognized as the most complex mitochondrial enzyme. This is why, 30 years after its isolation in membrane bound form ('complex I') and soluble form [1,2] many of its fundamental aspects, such as the number of subunits, the quaternary structure and the number of its iron–sulfur clusters remain controversial.

Reviews during the past decade have focused on selected aspects, such as its structure [3], its iron–sulfur clusters [4], and the molecular biology of the enzyme [5]. Since the resolution of questions concerning the enzyme often hinges on information derived from all these disciplines, as well as from the catalytic properties of the enzyme, the present review will attempt to synthesize available information from a variety of sources.

2. Enzyme preparations and their general properties

The first purified preparation derived from the respiratory chain-linked NADH dehydrogenase was the 'DPNH-cytochrome *c* reductase' of Mahler et al. [6,7]. Although it was soon recognized to be a fragment of the intact enzyme, devoid of its iron–sulfur clusters, with greatly altered catalytic properties [8,9], it retained the ability to oxidize NADH, albeit with oxidants which do not react with the native enzyme. Nevertheless, this was a significant advance since this fragment of three subunits was water-soluble, could be purified by classical procedures, and, much later served to identify the substrate and flavin binding subunits in other sources. In the ensuing years a surprising number of preparations of this three-subunit fragment were isolated by varying procedures and were given distinct names [10–13], although on critical comparison no significant difference among them could be detected [14]. The latest of this array of soluble derivatives is the 'flavoprotein' derived from complex I by chaotropic resolution [15], which has been the subject of extensive studies, but it does not show material differences from the preparations isolated in Hayaishi and Mahler's laboratories many years earlier [6,7]. The fact that a variety of rather different and relatively drastic procedures applied to mitochondria, submitochondrial particles, ill-defined heart-muscle preparations, and complex I yield the same three-subunit cluster suggests that this is indeed a preexisting domain of the large and complex NADH-ubiquinone reductase molecule. The evidence that the other two fragments derived by chaotropic resolution (the so-called 'iron protein' and 'hydrophobic protein') preexist as distinct components of the native enzyme, rather than being fortuitous products of the particular conditions applied to dissociation of the enzyme, is less persuasive.

The two preparations of the mammalian enzyme which seem to retain most of the properties seen in intact mitochondria are the membraneous 'complex I' (originally called DPNH-coenzyme Q reductase) [1,16] and the soluble NADH dehydrogenase extracted with phospholipase A [2], which has also been called the 'high molecular weight form', to distinguish it from the smaller unit dissociated from it, discussed above. Unlike the small fragment, these are not readily obtained in relatively homogeneous form, a fact which has given rise to many uncertainties and controversies. Both of these preparations retain the various Fe–S clusters seen in submitochondrial particles and full catalytic activity in the NADH–ferricyanide assay, but only the membraneous complex I catalyzes the rotenone- and piericidin-sensitive reduction of Q. One of us suggested several years ago that the loss of rotenone sensitivity and the ability to reduce Q (the two go hand-in-hand) reflects the separation of some component needed for Q binding during the extraction and purification [17]. The suggestion still seems plausible by analogy with complex II, where Q reduction has been shown to be a function of two small binding peptides removed during solubilization [18], particularly since small peptides have been identified in complex I but not in the soluble preparation [19]. This interpretation is not generally accepted, however. It is important to point out that complex I, while catalytically competent, differs significantly in sensitivity to inhibition by rotenone, piericidin A [20], and MPP^+ analogs from more intact submitochondrial particles, perhaps as a result of changes evoked by the harsh isolation procedure.

Among unicellular eukaryotic cells *Neurospora crassa* and *Candida utilis* are known to contain NADH-ubiquinone reductase closely resembling the beef heart enzyme. The dehydrogenase from *N. crassa* has been extracted with and purified in the presence of Triton X-100 in apparently monodisperse form [21]. Estimates from sedimentation equilibrium gave a molecular mass of 700 kDa, somewhat higher than a published estimate for the soluble enzyme from beef heart [2]. Both values should be considered provisional, the former because of the uncertainty of physical measurements performed in the presence of detergents, the latter because it was based on FMN content. In regard to major criteria, such as the presence of four clearly identifiable Fe–S clusters and sensitivity to rotenone and piericidin A, the *Neurospora* and mammalian enzymes behave identically. If mitochondrial protein synthesis is prevented with chloramphenicol, the fungus synthesizes a smaller form of the dehydrogenase (about 350 kDa) consisting of only 13 nuclear coded subunits. While this small form retains the Fe–S clusters of centers 1, 3, and 4, it lacks center 2 and is rotenone-insensitive [22].

Earlier studies on the yeast, *C. utilis* [23,24], revealed that this organism contained two membrane-bound NADH dehydrogenases, one present during exponential growth and when the supply of Fe or S is limited, the other in the stationary phase in the presence of adequate nutrients. The latter enzyme resembles NADH dehydrogenase from mammalian mitochondria qualitatively in all respects. It has not been isolated and studied in detail, but this is a potentially promising field for cloning its subunits and determining their function using site-directed mutagenesis approaches. In contrast, the NADH dehydrogenases of *Saccharomyces cerevisiae* mitochondria are rotenone-insensitive and bear no resemblance to the mammalian enzyme [25]. This is all the more surprising, since other known components of the respiratory chain in these two disparate sources closely parallel each other.

Mention should also be made of the fact that among the many NADH oxidizing enzymes from bacteria, those of *Paracoccus denitrificans* are of particular interest. In extensive studies, *P. denitrificans* has been shown to contain two membrane-bound NADH dehydrogenases, one of which resembles the mammalian enzyme in rotenone sensitivity, energy coupling site 1, and the nature of its Fe–S clusters, as revealed by EPR, paralleling the earlier findings on *C. utilis*. For detailed information the reader is referred to a recent review [26].

3. Substrate specificity

In addition to NADH, the purified enzyme oxidizes a variety of analogs, albeit slowly, except for the deamino NADH analog, which is a fast substrate [27]. NADPH is also oxidized by the purified enzyme but at 1/3300th the rate of NADH at pH 7.8 and 30°C [28]. That NADPH is a substrate was also shown by the fact that it bleaches both the yellow color of the enzyme and reduces center 1 completely, yielding the same intensity of EPR signal as does NADH [29]. Both NADH and NADPH induce dissociation of the subunits and inactivation on prolonged contact and protect the enzyme against thermal inactivation [28]. The slow oxidation of NADPH by the enzyme was rediscovered some ten years later by Hatefi and Hanstein, who also showed that while the rate of NADH oxidation declines with pH, that of NADPH rises sharply, reaching a maximum at pH 6.0–6.5 [30,31]. Based on their failure to observe complete reduction of center 1 by NADPH, they proposed that NADPH and NADH are oxidized at different sites and electrons originating from NADPH enter NADH-ubiquinone reductase at the level of center 3 or 4 or perhaps center 2 [31], a notion later revived with better experimental basis by Albracht's group (see Section 5). Hatefi's results contradicted the previous observations by the author [27,28,32] and were, in fact, recognized subsequently to be merely consequences of the slow oxidation of NADPH by complex I and failure to exclude O_2, resulting in auto-oxidation [33].

4. Subunit structure

NADH-Q oxidoreductase in mammalian systems is a large complex, which is estimated to be about 700 000 Da [3,5,19,35] for complex I, and about 530 000 Da for the intact high molecular weight enzyme [2]. It is a complex of 23–30 subunits which are coded for by both nuclear and mitochondrial genes and which depends for its functional integrity on the membrane in which it is located [35]. There is evidence from kinetic [36], crosslinking [37], and binding studies [31] that the complex can associate with other enzymes in the membrane and the presence of heme in complex I preparations suggests that other components of the electron transport chain also associate with the complex either artifactually or for functional reasons. Thus the isolation of 'pure' NADH-Q oxidoreductase is impossible and many spurious proteins may be found associated with any particular preparation. Whether such association could be the source of the acylcarrier protein recently reported to be present in complex I [38] or if such a protein plays a role in the complex is still unclear. The component subunits observed as bands separated electrophoretically on SDS–polyacrylamide gels in several laboratories using a variety of traditional biochemical techniques, coupled with insights from molecular biology studies, are shown in Table I. The function of only a few of these subunits is known.

As noted above, the simplest form of NADH dehydrogenase [10–15,39] consists of only three subunits and catalyzes the oxidation of NADH but with altered substrate and acceptor specificity. The 50–56 kDa subunit contains FMN [40], and has the four Cys residues required for binding an Fe–S cluster [41]. Reports [42,43] assigning center 3 to this subunit remain unconfirmed, and others have found only greatly modified EPR signals [44] and low, variable iron content in this fragment (reviewed in ref. [14]). The 24–27 kDa subunit has been reported to contain the 2Fe–2S cluster, center lb [42,43], and five Cys residues [41,45]. The third peptide is small (10–11 kDa) [39] and contains no Cys residues [41]. The function of the 50–56 kDa subunit is unambiguous. It contains the NADH binding site, identified by photoaffinity labelling with arylazido derivatives of NAD^+ [46,47], and catalyzes the oxidation of NADH to NAD^+ with reduction of FMN, as well as the FMN dependent [4B-^{3}H] NADH-H_2O exchange [48]. In complex I, several NAD^+ binding sites were identified using photoaffinity probes [46,48], the extra ones arising from impurities. In contrast to the substrate site catalyzing NADH dehydrogenation, NADH–NAD^+ transhydrogenation was found to be associated with the 42 kDa subunit and NADPH–NAD^+ transhydrogenation with a 130 kDa subunit which is not an integral part of complex I [48,49].

The subunits containing the Fe–S clusters of complex I could not be identified by EPR because of alterations caused by the chaotropic agents [43]. From sequence data likely subunits are the 51 kDa and 24 kDa subunits of the flavoprotein fragment as described above, the 75 kDa subunit of the 'iron protein' and the mitochondrially encoded 67 kDa ND5. The 'iron protein' subunit (75 kDa) fragment may contain the 4Fe–4S cluster (center 4) [5], probably associated with the CysXXCysXXCys sequence (amino acids 153–159) [50], and at least one other Fe–S cluster with two other CysXXCys sequences available [50].

The multiple 2Fe–2S clusters detected in the subunits (75, 49 and 30 kDa) of the 'iron-protein' fragment in early work [43,51] were artifacts resulting from the deleterious effects of the chaotropic agents used in the preparation. A sequence of three cysteines conserved between the bovine [52] and *Neurospora* enzymes could provide a place for another Fe–S cluster in the 49 kDa peptide but the 30 kDa one probably does not contain Fe–S clusters. All these subunits are considered peripheral and contain no membrane helices [5]. The seven subunits known to contain membrane helices (ND1–6) are all mitochondrially encoded [5,53]. ND5 (67 kDa) contains four conserved cysteines [54], probably liganding the 4Fe–4S cluster of center 2. Conserved cysteines have also been identified in ND4L (two cysteines) and in ND3 and ND6 (one each) but the function of these is unknown.

Attempts to identify the peptides involved in the interaction with Q have not yet produced

TABLE I

Subunits reported to be present in various preparations

Beef heart preparations			Fragments			Gene products		
Complex I (I)[a]	High MWT enz.[b]	Low MWT enz.[c]	Fragmented C-I[d]			Neurospora[e]		Beef heart mito[f] gene product
			fp	ip	hp	perip.	mito	
75	75			75		78		
							78	67 (ND5)
56	56	56	53			51		
53	53			53				52 (ND4)
40	40				42	40		
					39		66	39 (ND2)
33	33				33		42	36 (ND1)
30	30			30		30.4		
27	27	27	24			29.9		
23						24		
21						21.3		
						20.8		
19				18		18.3	27	19 (ND6)
16	16			15				
14	14			13			17	13 (ND3)
12								
11	11	11	10				10	11 (ND4L)
9	9							
5								

[a] Complex I [19] prepared according to Hatefi [16]. Molecular weight varies from 600–750 kDa. Some peptides are poorly resolved [3,43,51] and other groups find additional bands at 40–50 kDa (e.g., ref. 134].
[b] High molecular weight enzyme [2,19]. Molecular weight is estimated to be 550 kDa.
[c] The low molecular weight enzyme [7,10–14]. Molecular weight 70–84 kDa, is similar to the fp fragment [15].
[d] Fragments produced by chaotropic agents [15,43,51].
[e] Gene products and peptide analysis of the enzyme in *N. crassa* [5,21,61]. A small form of the enzyme without the mitochondrial subunits has a molecular weight of 330 kDa [61], whereas the whole enzyme is approximately 700 kDa [21]. Multiple smaller peptides also appear in the whole enzymes.
[f] The mitochondrial genes code for membrane spanning proteins and have been characterized for beef heart (shown in ref. [135]) and for human cells.

reliable results. Analogs of the inhibitor, rotenone, which blocks the passage of electrons from the Fe–S clusters to Q have been reported to photolabel the 30–36 kDa subunit of complex I [55,56]. However, some doubt remains since one analog [55] was azido labelled at a site remote from the groups important for binding and in the other experiment [56] binding was very low. Antisera were used to demonstrate that this subunit is the ND1 gene product [53]. Although it contains no Fe–S clusters, it may be linked by a disulfide bridge to the 75 kDa subunit [37]. Crosslinking experiments also suggest that this 30 kDa peptide is close to the 51 kDa peptide which contains center 3 and the NADH binding site. Although located in the middle of the complex [37,57], the location of rotenone binding need not be identical or even close to the Q-binding site. Crosslinking experiments with dithiobis-(succinimidyl propionate) [37] demonstrated seven peptides within 11–12 Å of each other (75,33,53,42,24,17 and 13 kDa) but in the presence of rotenone the five other subunits could not be crosslinked to the 75–33 pair [37]. Thus rotenone induces conformational changes which must (if the proposed locations of the clusters are correct) affect the proximity of center 4 on the 75 kDa subunit to the smaller peptides, including center 3 on the 53 kDa peptide and center 1b on the 24 kDa peptide. In contrast, another laboratory found no change in crosslinks involving the 30 kDa peptide after addition of rotenone, but did find that NADH abolished a 75–51 kDa link [58]. In any case, these widespread remote changes make it difficult to pinpoint the location of the rotenone binding site which interferes with Q reduction. Comparison of the deduced sequence from cDNA with other ubiquinone-binding enzymes suggests that indeed the 30–36 kDa subunit might be the site and one laboratory [59], but not another [60], demonstrated carbodiimide labelling of this peptide in parallel with the loss of activity.

It is clear that molecular biology has contributed tremendously to the recent advances in the identification of the peptides and their function. In particular, comparison of the mammalian enzyme with sequences of bacterial enzymes and with simpler forms of NADH–Q oxidoreductase, such as that in *N. crassa*, has opened up the possibility of studying the synthesis and assembly of the complex and of identifying metabolic defects in humans, particularly those of mitochondrial origin. Synthesis and assembly have been studied mainly in *N. crassa*. If mitochondrial protein synthesis is inhibited, a 'small form' of the enzyme (13 peptides) is made which lacks the hydrophobic components [22], including center 2 and the rotenone-sensitive Q site [54]. Pulse labelling demonstrated that the complete enzyme is similarly preassembled into the nuclear-encoded peripheral subunit moiety and then joined with a preassembled hydrophobic fraction [61]. Although studies of bovine complex I have suggested that the 'flavoprotein' and 'iron–sulfur protein' segments span the membrane (e.g., refs. [37,43,62,63], electron microscopic analysis of the *N. crassa* enzyme suggests that the enzyme is L-shaped with the peripheral segment projecting into the matrix and the hydrophobic moiety buried in the membrane [5]. The latter part contains the N-2 cluster and the quinone binding site which is predicted from sequence data to lie on the inner membrane surface [64].

The assembly of the mammalian enzymes has been studied by measuring the incorporation of pulse-labelled proteins into immunoprecipitable enzyme in hepatoma cell lysates [65]. Nuclear-coded proteins are imported into mitochondria and assembled into a 'scaffold' to which the mitochondrial proteins are added at different and much slower rates. The data were consistent with the model of Ragan [3] where the mitochondrial proteins sheath the hydrophilic nuclear components to anchor the complex in the membrane.

Although the *N. crassa* enzyme seems to be monomeric [5], it has been suggested that the mammalian enzyme is at least dimeric. Multiple symmetrical arrays in phase contrast electron microscopy [66] suggested a tetrameric or multiple tetrameric structure, but it must be kept in mind that highly hydrophobic molecules might well aggregate in a regular manner. The stoichiometry of center 1 to the other centers has led to the postulation of a dimer. This and kinetic evidence for a dimer will be discussed in Section 5.

5. *Prosthetic groups*

FMN was established as the flavin group in NADH dehydrogenase in 1961 [40,67] and is non-covalently bound probably to the 51 kDa subunit. It is retained during perchlorate fragmentation of the complex but in a loosely bound form, so that during further dissociation of the flavoprotein fragment (the low molecular weight enzyme) into the individual subunits, it is lost [42]. It provides a quantifiable index of the moles of dehydrogenase present in simple preparations but in ETP and intact mitochondria other systems interfere making quantitation in membranes uncertain.

A complete picture of the Fe–S clusters has still not emerged. EPR spectroscopy has been the key method for identifying and quantifying the constituent Fe–S clusters which account for 16–18 Fe/FMN [1,29,68] or even higher in less pure preparations (for a review see ref. [4]). Potential errors arise in quantifying 2Fe–2S signals in the presence of the broad underlying contribution from the 4Fe–4S clusters and from overlap with signals from other respiratory enzymes. In addition, there is the possibility of EPR-silent clusters due either to inaccessibility to reductants or to intercluster magnetic interactions. Attempts to study the Fe–S clusters by extrusion techniques [69] or by EPR on fragments [43] generated by treatment with chaotropic agents both gave ratios of 2Fe to 4Fe clusters of 2:1 but the apparently large number of 2Fe–2S clusters is now thought to result from collapse of one or more 4Fe clusters during the treatments.

There is general agreement based on EPR [4,44,70,71], magnetic circular dichroism [72], and conserved cysteines [5] that there are at least three 4Fe–4S clusters (centers 2, 3, 4) and probably one or two 2Fe–2S clusters, which would require an Fe/FMN ratio of 16, the lower end of the observed range. Additional iron in the preparations could come from 'impurities' in the sense of spurious proteins normally associated with or near to the respiratory complexes in the membrane which, once split from their own complex, do not readily dissociate into hydrophilic media, or from adventitious Fe and labile S from disintegrated clusters. The cluster initially designated center 5 (note that the designations used are those of Ohnishi [73]) may fall into the former category being found in substoichiometric amounts (<0.25 4Fe–4S/FMN) in beef heart preparations and not at all in yeast or plant mitochondria.

Of the firmly identified clusters, center 2 (4Fe–4S) has the highest potential (−20 mV) and is assumed therefore to interact with ubiquinone [74]. The other two 4Fe–4S clusters, center 3 and 4, have the same midpoint potential as center 1b (2Fe–2S). Spin coupling of the EPR signals led to the suggestion that center 3 is adjacent to the flavin [75] and is probably located in the same 51 kDa subunit. However, this does not imply that it is directly involved in the passage of electrons from the flavin. A careful study of changes in the EPR signals caused by modifying reagents has suggested rather that center 1 comes before centers 3 and 4 [76].

These three clusters presumably share electrons with relative ease due to their similar potentials and location in the same region of the enzyme, a hypothesis consistent with the irreversible disappearance of the center 4 signal without disruption of the electron flow [76]. The jump in potential between these clusters in the peripheral part and center 2 in the hydrophobic segment points to the transfer site between these being the coupling site. In *Neurospora*, the 'small' enzyme (consisting of the peripheral parts) reduces Q but is not sensitive to piericidin, leading to the suggestion of an internal Q cycle in the whole enzyme to transfer electrons to center 2 [5]. This hypothesis is further discussed in Section 7.

The stoichiometry of centers 1a plus 1b to 2 has been much debated. Most groups find slightly less than 1 : 1 stoichiometry, e.g., 0.9 [72] or 0.8 [4,77]. The latter careful study defined center 1b as a NADH-reducible 2Fe–2S cluster with a midpoint potential of −335 mV at pH 8 and measured its spin concentration as 0.8/FMN. Center 1a was reported to have a midpoint potential of <−500 mV. The combined centers la and lb accounted for 1.25 spins/FMN. In early studies on complex I, the center 1b with the higher midpoint potential was found to be very variable, depending on the concentration of the preparation. Preincubation of the preparation with NADH was also found to reduce the observed center 1 signal [76], perhaps due to the inactivation by NADH. Albracht's group consistently found a stoichiometry for 1a plus 1b of about 0.5–0.6 divided between two clusters, both of which are NADH reducible, and did not see the low-potential species [78]. They fitted the data to two axial signals in contrast to Ohnishi's report where a single rhombic spectrum fitted the signal from center 1b. The problem with the Albracht stoichiometry is that it requires postulation of a model in which there is only one center 1a and one center 1b per four centers 2, 3, and 4 and flavin, or alternatively a mixed preparation. Since other groups get higher quantitation, and early work showed variability in this center depending on the preparation, and Albracht detects no low-potential species, resolution of this discrepancy must await further characterization of complex I. In *N. crassa*, only one type of center 1 was found, in quantities equivalent to the other clusters, suggesting a simple monomeric unit which was also observed in electron microscopy [5].

Kinetic arguments have been made for a dimeric structure for beef heart NADH–Q oxidoreductase. Freeze–quench EPR was used to study the reduction of the Fe–S clusters by NADH or NADPH at pH 8, where NADPH oxidation is negligible (less than 0.1% of NADH) and at pH 6.5 where it is about 60% of the NADH rate. The study demonstrated incomplete reduction by NADPH, and biphasic reduction of the clusters by NADH. Early work by Hatefi attributed the failure to see complete reduction of the clusters by NADPH to auto-oxidation which was faster than the rate of reduction [33], but Albracht concluded that auto-oxidation was only 2% of the turnover rate [79]. At both pH 6.5 and 8.0, NADPH reduced 50% of centers 2, 3 and 4 within 30 ms but further reduction was very slow [79]. Pre-steady-state kinetics of the reoxidation of the clusters [80] likewise demonstrated two populations of clusters. Reoxidation of submitochondrial particles reduced with a pulse of NADH in the presence of excess NADPH was incomplete. At pH 8 only 50% of centers 2 and 4 and 75% of center 3 were rapidly reoxidized. The incomplete oxidation was also observed in the absence of NADPH, suggesting that half of the clusters are not involved in reoxidation of NADH. The half kept reduced by NADPH, involved in the oxidation of both NADH and NADPH, was reoxidized by ubiquinone at pH 6.5. In contrast, the oxidation of center 1 and the other half, involved only in the oxidation of NADH, was very pH sensitive, being rapid at pH 8 but slow at 6.5. It should be noted that pH 8 is the pH where most Q radical signal is observed [81]. At pH 6, only 20% of the radical signals remained and were ascribed to the neutral semiquinone radical. The role of the Q radicals in the reoxidation of the two types of clusters is not yet known.

In inhibitor studies, Albracht found that center 1 was not reduced by NADPH and that only 0.5 mol piericidin per center 2 was required for inhibition of NADPH oxidase activity [32]. However, the piericidin binding did not take into account spurious binding of the inhibitor and required the assumption that *all* the piericidin must first bind to the NADPH-reduced part of the molecule. Some differences between two putative piericidin sites have been noted [82], but in Albracht's study [32] only marginally higher amounts (1.2 versus 1 mol/mol center 2) were required in the absence of NADH, suggesting that piericidin still binds to oxidized enzyme.

Further evidence cited [32] was the observation that preincubation of particles with NADPH and 0.5 mol piericidin per mol of cluster 2 resulted in a lag phase in the rate of oxygen consumption initiated by NADH due to the slow redistribution of piericidin from the NADPH-reduced part of the dimer to both parts after reduction of the other center 2 by NADH. However, this postulated redistribution is at odds with the virtually irreversible binding of piercidin [82] and, we could not repeat the observation of the lag phase in our laboratory. Interpretation of observations after preincubation of the complex with inhibitor and reductant is further complicated by the deterioration of the lb signal under these conditions [77] and by the inhibitory effects of superoxide formed at different rates with NADH or NADPH [83,84]. At present, the dimer hypothesis, proposed only for the complex mammalian enzyme, remains an interesting possibility which is not yet widely accepted.

6. *Inhibitors*

6.1. *Acting between NADH and FMN*

The only inhibitor with a reaction site definitely localized at the substrate site of NADH dehydrogenase is rhein [85]. Although its structure (4,5-dihydroxyanthraquinone-2-carboxylic acid) does not resemble that of NADH, it is a pure competitive inhibitor of the purified 'high molecular weight' form of the enzyme (K_i = 2 μM at 30°C), and in membranes substrate competition predominates. Interestingly, the three-subunit fragment (the 'flavoprotein' of Hatefi and Stempel [15]) is activated by rhein, one of the many differences in the catalytic properties between this fragment and the intact enzyme.

The action of rhein is not entirely specific, since it also inhibits mitochondrial transhydrogenase and DT-diaphorase at low concentrations and lactate and malate dehydrogenases at higher levels [85].

6.2. *Acting between center 2 and Q*

The first class of inhibitors eventually shown to block electron transport on the oxygen side of the enzyme, i.e., between the highest potential Fe–S center and Q, were barbiturates, such as amytal [86]. The site of action was originally reported to be between substrate and flavin [87], later between flavin and iron–sulfur centers [88] on the basis of misinterpretation of the difference spectra obtained in dual wavelength spectrophotometry. However, lack of inhibition of NADH–ferricyanide activity (a measure of the substrate to center 1 segment [14]) in membranes and soluble preparations [27,89], kinetic data [90], and measurement of the rates of reduction of the various Fe–S centers of the enzyme [91] clearly placed the reaction site of barbiturates, as well as that of rotenone and piericidins, on the high-potential side of the enzyme, near the reaction site of Q. The much lower K_i value for rotenone than for barbiturates indelibly labeled the region of block as the 'rotenone site'. Later, it was shown that piericidin A and its analogs block the same site as rotenone and barbiturates but at lower concentrations [92].

Piericidin A and, to an even greater extent, rotenone, bind avidly to hydrophobic components in membranes. This has been termed 'unspecific binding' because bovine serum albumin (BSA) facilitates the dissociation of the inhibitors from such sites. In contrast, binding to the 'specific site(s)' is tighter, so that BSA dissociates rotenone from this location only slowly and piericidins not at all. Both the specific and unspecific binding sites contribute to the inhibition of the NADH dehydrogenase–Q interaction, the former far more than the latter [20]. The resultant mixed inhibition complicates determination of the stoichiometry of the binding. The figures from Scatchard plots approximate 2 per mol of enzyme [20], if one calculates the concentration of the dehydrogenase from the turnover number, determined from the FMN content of the purest preparation. Although this was the best means available at the time, it introduced uncertainties in view of the high molecular weight of the enzyme, so that even a trace of FMN-containing impurity would seriously affect the ratio. In addition, the very steep slope in the double reciprocal plots of the highly purified enzyme made accurate determination of the activity difficult. The ratio of 2 per mol of enzyme has been questioned and a 1:1 ratio proposed instead [93]. Two observations seem to lend support to the notion that more than one mole of piericidin is specifically bound to the enzyme when inhibition is nearly complete. One is that the curve relating inhibition to specifically bound piericidin is sigmoidal. The other is that treatment of inner membrane preparations with mercurials or with bile salts, as used in the preparation of complex I, nearly halves the amount of [^{14}C]piericidin bound at the specific site [22]. Indeed, the conventional complex I preparation yields a figure close to one in Scatchard plots of specific binding,in contrast to two in the intact inner membrane.

In order to arrive at a more unambiguous value for the stoichiometry of piericidin binding, van Belzen et al. [32] related the amount of added piericidin needed for complete inhibition to the center 2 content of inner membrane preparations, determined by EPR (see above). An obvious flaw of this method is that nearly half of the added piericidin gets lodged at unspecific sites, contributing little to the inhibition. While this difficulty would be easy to overcome by washing the particles with BSA, two others remain: first, precise measurement of the content of center 2 is difficult; second, the piericidin/enzyme ratio thus obtained is reported to depend on the prior treatment of the preparation [32]. At present, therefore, the question of the stoichiometry remains open.

In recent years several other types of inhibitors which block electron flux from the dehydrogenase to Q have been identified. Among these, capsaicin and its analogs have been shown [94] to act near the Q binding site in complex I preparations by the EPR techniques previously used to localize the reaction site of rotenone and piericidin [91] and by the demonstration that the inhibition is competitive with respect to Q_1 as the electron acceptor. The I_{50} value of capsaicin analogs decreases as the alkyl chain is lengthened up to 12 carbons, beyond which it rises sharply, reminiscent of the observations on MPP^+ analogs to be described. Myxalamids, with structures resembling piericidins, have also been reported to block the enzyme in this region [95]. Recent extensive studies of the inhibition of NADH-ubiquinone reductase in ETP and mitochondria by a large series of 4-hydroxypyridine and 4-hydroxyquinoline derivatives [96] arrive at conclusions fully compatible with the studies using MPP^+ analogs discussed below.

Perhaps the most interesting group of recently described inhibitors acting between NADH and Q are derivatives of *N*-methyl-4-phenylpyridinium (MPP^+), the neurotoxic product arising from the oxidation of the parent tetrahydropyridine (MPTP) by monoamine oxidases A and B in the brain. The target of the neurotoxin is the complex I region of the respiratory chain, specifically the ‘rotenone site’ [97–100]. As the alkyl chain in the 4′ position of the aromatic ring is lengthened, both the blockade of the oxidation of NAD^+-linked substrates in mitochondria and the inhibition of NADH oxidase activity increase (Table II) [100,101]. The tetraphenylboron anion (TPB^-) increases the inhibition in both systems, in mitochondria by facilitating the accumulation in response to the membrane potential and facilitating the entry into a hydrophobic site postulated to be present on the target enzyme, NADH dehydrogenase [102,103]. In inverted membranes only the latter comes into play. The action of the neutral 4-phenylpyridine analogs is not affected by TPB^-, since the latter acts by ion pairing with the cationic form to facilitate the penetration of the charged molecule into hydrophobic areas [104]. It is noteworthy in terms of the nature of their binding to the enzyme that the inhibition by highly hydrophobic MPP^+ analogs is not potentiated by TPB^- in submitochondrial preparations [101], suggesting that they are either bound differently or that their hydrophobicity overcomes the kinetic barrier that seems to be present against the penetration of short chain MPP^+ analogs to the rotenone site.

The question of how such disparate structures as barbiturates, rotenoids, piericidins, MPP^+ analogs and now capsaicin can all act in the same manner and at the same site has always been puzzling, although an important recent paper by Chung et al. [96] attempted to visualize a common hydrophobic binding site. Several years ago, evidence from binding studies clearly established that barbiturates, rotenoids, and piericidins compete for a common binding region and, hence, seem to be bound at the same site. The availability of 4′-alkyl MPP^+ analogs with I_{50} values approaching that of rotenone permitted the demonstration that these compounds

TABLE II

Inhibition of mitochondrial respiration and of NADH oxidation in ETP by 4′-alkyl-MPP^+ analogs (from ref. [101]). I_{50} values for liver mitochondria respiring on malate + glutamate were determined after 6 min preincubation at 30°C and they represent approximate values because time is a factor. Inhibition in inner membrane preparations were determined spectrophotometrically at 30°C and are independent of time

Analog	I_{50} (μM)			
	Mitochondria		ETP	
	Without TPB^-	with 10 μM TPB	without TPB^-	with 10 μM TPB
MPP^+	110	1.82	3300	825
4′-methyl-MPP^+	24	0.67	580	130
4′-propyl-MPP^+	3.8	0.23	150	30
4′-pentyl-MPP^+	1.3	0.10	18	5.6
4′-heptyl-MPP^+	0.5	0.10	3.2	2.8
4′-decyl-MPP^+	ND	ND	1.6	2.0
4-phenylpyridine	205	207	142	114
4′-methyl-4-phenylpyridine	125	129	39	36
4′-propyl-4-phenylpyridine	59	63	7.4	7.5
4′-pentyl-4-phenylpyridine	11	10	2.2	2.3
4′-heptyl-4-phenylpyridine	13	12	1.7	1.9
4′-decyl-4-phenylpyridine	10	11	4.5	5.8

compete with [^{14}C]rotenone [105] and [^{14}C]piericidin A [100] to prevent both binding of the latter and the resulting inhibition of NADH oxidase activity in submitochondrial particles. Fig. 1 illustrates these findings: when corrected for residual inhibition by the MPP^+analogs themselves, the protection from binding and from the resulting inhibition are very similar, showing that, beyond doubt, the same region of the respiratory chain is involved. Under suitable conditions even the less effective MPP^+ analogs completely prevent the interaction of the extremely tightly bound piericidin A with its inhibition site.

It is noteworthy that the most hydrophobic analogs (Fig. 1 and ref. [100]) were less effective in competing with rotenone and piericidin A, again raising the possibility that these hydrophobic MPP^+ analogs bind differently or at different sites, as also noted in extensive studies on 4-hydroxypyridine analogs [96]. Taken together with the fact (Table I) that action of the most hydrophobic MPP^+ analogs is not potentiated by TPB^-, a tentative picture emerges on the nature of the ‘rotenone site’. Recent extensive studies of the inhibition of NADH-ubiquinone reductase in ETP and mitochondria by a large series of 4-hydroxypyridine and 4-hydroxyquinoline derivatives [96] arrive at conclusions fully compatible with the studies using MPP^+ analogs.

We visualize this ‘site’ as a cleft in the complex NADH dehydrogenase molecule where tightly bound Q juxtaposes the high-potential Fe–S cluster (center 2) for its reoxidation. A number of compounds with varying degrees of hydrophobicity may enter this cleft and become bound there by hydrophobic interactions, resulting in steric hindrance to the redox reaction between center 2 and Q or to the indirect passage of electrons between them. The less hydro-

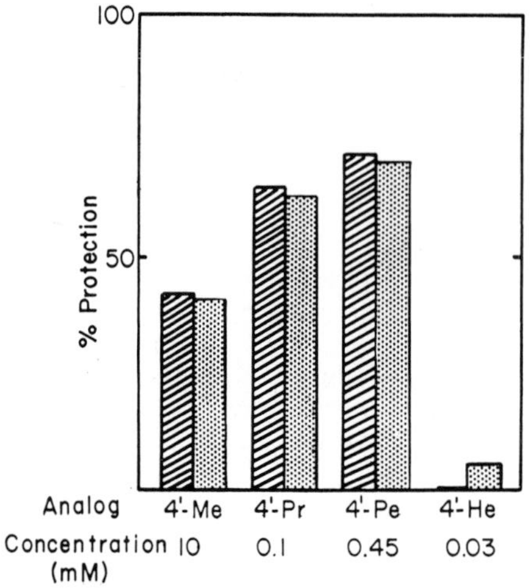

Fig. 1. Protection of ETP from binding of [^{14}C]piericidin and inhibition of NADH oxidase activity. ETP (3 mg/ml) was incubated for 5 min at 30°C with 10 μM TPB^- and the indicated concentration of MPP^+ analog. BSA was added to 2% w/v and the samples were incubated on ice for 5 min. The particles were washed with buffer containing 2% BSA by centrifugation and resuspension, once in the presence of analog to remove excess piericidin, twice in BSA-buffer only to remove the analog. NADH oxidase activity was assayed. The samples were centrifuged again and the amount of [^{14}C]piericidin bound to the pelleted particles was determined by liquid scintillation counting. In the absence of MPP^+ analog, piericidin inhibited 93% of the NADH oxidase activity after 3 washes. The hatched bars indicate the percentage of this inhibition that is relieved by including MPP^+ analog. 1.66 nmol [^{14}C]piericidin is bound per nmol NADH dehydrogenase. The stippled bars indicate what percentage of the [^{14}C]piericidin is prevented from binding by including the MPP^+ analogs.

phobic compounds, such as MPP^+ and its short alkyl chain analogs, being positively charged may be stabilized by a negatively charged side chain in this region, whereas the most hydrophobic ones are differently oriented, being bound in a manner which is not affected by a putative charged amino acid residue. Thus we are dealing with a two-stage process. The first is the penetration through a hydrophobic region, which is hindered by the positive charge but facilitated by the hydrophobicity of the compound or by ion pairing with TPB^-. The second is the orientation in the cleft which may be enhanced by an electrostatic attraction. The effect of pH on inhibition by 4′-alkyl MPP^+ is in accord with this suggestion [101]. The penetration of rotenoids, piericidins, and of barbiturates being uncharged is not affected by pH, accounting for the lack of effect of pH on their inhibition. The inhibition is regarded not as competition with Q for a specific binding site on the enzyme, but rather to come from steric hindrance or from a conformational change preventing the passage of electrons to Q. In terms of this tentative interpretation, the 'rotenone site' is not a set of amino acids in a subunit but a zone of interaction in the quaternary structure, so that its topographic identification is inherently difficult. As discussed above, experiments with photoaffinity labeled rotenone derivatives [55,56] suggested the possibility that the 33 kDa subunit is involved. While this remains a possibility, it is no more than that, in part because the azido label on the rotenone was not on the part of the rotenone molecule believed to be involved in binding, raising the possibility of spurious interaction with adjacent regions, in part because several subunits may converge in the region of interest. Evidence has also been presented [37] that rotenone may induce conformation changes in the native enzyme, moving several of the subunits away from each other. This expansion is suggested to result in preventing electron transfer between center 2 and bound ubiquinone. Ultimate resolution of the problem will have to await elucidation of the spatial interrelation of the subunits.

6.3. *Acting at multiple or ill-defined sites*

Among the numerous other compounds reported to inhibit NADH-ubiquinone reductase, a few are of special interest. Dequalinium chloride and related compounds containing two quinolinium rings separated by alkyl chains of various length inhibit NADH-ubiquinone reductase in membrane preparations from mammalian heart and *P. denitrificans* at very low concentrations. These compounds seem to act at (or near) the 'rotenone site', as well as at other, ill-defined sites [106]. The antibiotic adriamycin and other anthracyclines have also been reported to inhibit the enzyme in mammalian submitochondrial particles [107].

The potentially most interesting effects have been observed with –SH inhibitors, such as mercurials and alkylating agents on preparations of different complexity [89,108–112]. Five types of –SH groups have been implicated, blocking of which either results in loss of catalytic activity or structural destabilization, but some are only manifest in the 'low molecular weight form', but not in the intact soluble or membraneous enzyme, others only in membraneous preparations [108,111]. Thus, their reactivity clearly depends on the conformation of the enzyme in different preparations and some become exposed only on dissociation of the subunits. As mentioned, mercurial-type –SH inhibitors nearly halve the amount of piericidin and rotenone bound at the specific site [22]. Interestingly, dithiothreitol reverses this effect of mercurial inhibitors and reopens the other combining site of piericidin and rotenone [112].

7. *Energy conservation site 1*

Although coupling site 1 had been recognized for many years to be in the NADH–ubiquinone reductase segment of the respiratory chain, its more precise localization came from the studies of Gutman et al. [113]. After early work demonstrated an oligomycin-sensitive effect of ATP [114] on the reappearance of a chromophore at 470–500 nm bleached by NADH [115], EPR studies at 13°C of phosphorylating inner membrane preparations, inhibited by piericidin but in the aerobic state, showed rapid reduction of all four then recognized Fe–S centers of the enzyme by NADH [113]. Because of the incomplete block by piericidin, centers 1, 3, and 4 were slowly reoxidized on exhaustion of the NADH but most of the signal due to reduced center 2 remained. Addition of ATP caused rapid reoxidation of center 2. These data were interpreted as indicating that ATP lowers the redox potential of this center, so that electrons can flow back to accumulated NAD^+, resulting in reoxidation of all of the Fe–S clusters of the enzyme. This places coupling site 1 near Fe–S center 2, either between center 2 and Q or, as earlier suggested [113], between center 2 and the low potential centers.

These observations were viewed by Mitchell [116] and Garland [117] as supporting the chemiosmotic hypothesis. The suggestion of Gutman et al. [113] that ATP lowered the potential of center 2 in the presence of NAD^+/NADH was later confirmed by Ohnishi [118] and by others, although Ohnishi also reported that ATP alters the redox potential of the postulated center 1a and, curiously, that if the experiments were repeated with the succinate/fumarate couple in lieu of the NAD^+/NADH pair, ATP seems to raise the potential of center 2. The effect of ATP on the putative center 1a is difficult to evaluate because no other investigator has seen this shift in potential. The effect of ATP in the presence of succinate/fumarate, on the other hand, may well reflect lack of equilibration of the system with the electrodes used.

On the basis of the observations of the effect of ATP on the redox state of cluster 2, chemiosmotic coupling schemes were proposed for coupling site 1 [116,117]. Many attempts have been made at measuring the proton to electron stoichiometry of the site using kinetic (e.g., refs. [119–122]) or thermodynamic (e.g., refs. [123,124]) techniques, each study having its own limitations or problems (for comments, see refs. [5] and [124]). However, the consensus from the variety of methods used both in intact mitochondria and in ETP_H is a ratio of $2H^+/e$. In other words, four protons must be pumped across the membrane for each NADH molecule oxidized, which means that a simple loop mechanism is impossible.

More elaborate schemes proposing mechanisms to account for the stoichiometry are theoretical exercises, because, at present, there is no solid evidence for any of them, and the exact pathway of electrons through the prosthetic groups is still unknown. Schemes [3,76,118] placing FMN after some of the Fe–S groups seem unlikely, because direct reduction of FMN by NADH occurs even in flavoprotein fragments which lack functional Fe–S clusters. A role of FMN in the transfer of electrons to a proton channel or pump (e.g., ref. [125]) is more likely and is consistent with the location of FMN on the matrix face of the membrane but in contact with both the hydrophobic shell and transmembraneous iron–sulfur proteins (e.g., refs. [37,57]).

Whereas evidence for a Q-cycle mechanism is very strong for coupling site 2 in complex III, it still remains a postulation for site 1. In the high molecular weight soluble enzyme containing no detectable Q, Fe–S clusters were rapidly reduced and exhibited normal line shape [29] but the Q reduction properties were different; so this does not exclude a Q-cycle mechanism. In contrast, complex I has 2–4 tightly bound Q per FMN [81]. On the positive side, the Q reduction by both the small and large forms of NADH-Q oxidoreductase in *N.crassa* was interpreted to suggest that Q was involved as a redox carrier on both sides of center 2 [5,54], which would indicate two pools of bound Q, as required by a Q-cycle mechanism. However the data do not exclude the possibility that the Q reductase of the small form is artefactual, similar to that found in the various NADH–cytochrome *c* reductase preparations. The best evidence for two Q pools in mammalian NADH-Q oxidoreductase is the observation of two Q-radical signals from tightly bound Q [81]. The major pool (75%) was anionic and its appearance was highly rotenone sensitive, requiring 1 mol rotenone bound/FMN for complete inhibition. The other radical was in the neutral form and much less sensitive to rotenone inhibition. More recently a Q-radical signal which appears only when oligomycin is added to coupled submitochondrial particles has been reported [125], reaching a maximum of 0.5 per center 2. The relaxation characteristics suggested interaction with center 2, lending further weight to a role for center 2 as the electron donor to Q, but probably via this tightly bound Q rather than in equilibrium with the Q pool [125]. Since the signal is $\Delta\mu_{H^+}$ dependent, the authors postulated a Q-cycle mechanism. However, in the final analysis, the coupling mechanism remains a 'black box'.

8. *Diseases related to complex I deficiency*

One of the most important developments in the field during the past few years has been the growing recognition that a number of congenital or acquired diseases involve deficiency of NADH-ubiquinone oxidoreductase. Most of the relevant reports in the literature loosely refer to their findings as 'complex I deficiencies'. The diseases include mitochondrial myopathies, encephalopathy, lactic acidosis, stroke-like episodes (for reviews see refs. [126–128]), and idiopathic Parkinson's disease [129,130]. Since some of the reports in the literature were based on small clinical samples, the evidence for the localization of the defect within discrete regions of the enzyme has not always been rigorous. Thus, enzyme assays were performed in some cases on frozen–thawed samples of tissues obtained considerable time after death, resulting in uncertain post-mortem changes, while in other studies biopsy samples were used and rapidly processed to minimize such changes.

The most reliable data for pinpointing the defect to given subunits of the enzyme came from immunoblotting assays. In other reports, however, the interpretations were extended [26,128] to localizing the defect in the hypothetical 'iron–sulfur protein' or 'hydrophobic protein' postulated by Hatefi [131], or even to specific Fe–S centers of the enzyme within given subunits, apparently unaware that the reports on their purported distribution within individual subunits [77] lacks solid experimental basis or even that, as discussed above, there is no sound evidence for the existence of the 'iron–sulfur' and 'hydrophobic' regions as discrete entities in the native enzyme.

It is nevertheless clear that in the heterogeneous group of the metabolic diseases, called mitochondrial myopathies, several instances have been documented of decreased activity of NADH–ubiquinone oxidoreductase, accompanied by the decline or absence of two or more of its subunits [128–130,132,133]. Interestingly, the defect has been reported to be tissue-specific in some cases, with lowered activity in skeletal muscle but normal values in liver biopsy samples [133].

Growing evidence that NADH-ubiquinone oxidoreductase is the ultimate target of MPP^+, the oxidation product of the neurotoxic amine MPTP, which causes Parkinsonian symptoms in primates, led to an intensive search for evidence for the deficiency of this enzyme in the nigrostriatum of Parkinsonian patients. These efforts have produced impressive immunochemical and biochemical evidence for a specific deficiency of the enzyme in the striatum of patients who died with Parkinson's disease [129,130].

Acknowledgements

Work in this laboratory is supported by grants from the National Institutes of Health (HL-16251), the National Science Foundation (DMV-9020115), and by the Department of Veterans Affairs. We thank our many colleagues for the discussions which helped us produce this manuscript.

References

1 Hatefi, Y., Haavik, A.G. and Griffiths, D.E. (1962) J. Biol. Chem. 237, 1676–1680.
2 Cremona, T. and Kearney, E.B. (1964) J. Biol. Chem. 239, 2328–2334.
3 Ragan, C.I. (1987) Curr. Top. Bioenerg. 15, 1–36.
4 Beinert, H. and Albracht, S.P.J. (1982) Biochim. Biophys. Acta 683, 245–277.
5 Weiss, H., Friedrich, T., Hofhaus, G. and Preis, D. (1991) Eur. J. Biochem. 197, 563–576.
6 Edelhoch, H., Hayaishi, O. and Teply, L.J. (1952) J. Biol. Chem. 197, 97–104.
7 Mahler, H.R., Sarkar, N.K., Vernon, L.P. and Alberty, R.A. (1952) J. Biol. Chem. 199, 585–597.
8 Watari, H., Kearney, E.B. and Singer, T.P. (1963) J. Biol. Chem. 238, 4063–4073.
9 Cremona, T., Kearney, E.B., Villavicencio, M. and Singer, T.P. (1963) Biochem. Z. 338, 407–442.
10 Mackler, B. (1961) Biochim. Biophys. Acta 50, 141–146.

11 DeBernard, B. (1957) Biochim. Biophys. Acta 23, 510–515.
12 Pharo, R.L., Sordahl, L.A., Vyas, S.R. and Sanadi, D.R. (1966) J. Biol. Chem. 241, 4771–4780.
13 King, T.E. and Howard, R.L. (1962) J. Biol. Chem. 237, 1686–1692.
14 Singer, T.P. and Gutman, M. (1971) Adv. Enzymol. 34, 79–153.
15 Hatefi, Y. and Stempel, K.E. (1969) J. Biol. Chem. 244, 2350–2357.
16 Hatefi, Y. (1978) Methods Enzymol. LIII, 11–14.
17 Machinist, J.M. and Singer, T.P. (1965) Proc. Natl. Acad. Sci. USA 53, 467–474.
18 Ackrell, B.A.C., Ball, M.B. and Kearney, E.B. (1980) J. Biol. Chem. 255, 2761–2769.
19 Paech, C., Friend, A. and Singer, T.P. (1982) Biochem. J. 203, 477–481.
20 Gutman, M., Singer, T.P. and Casida, J.E. (1970) J. Biol. Chem. 245, 1992–1997.
21 Leonard, K., Haiker, H. and Weiss, H. (1987) J. Mol. Biol. 194, 277–286.
22 Friedrich, T., Hofhaus, G., Ise, W., Nehls, U., Schmitz, B. and Weiss, H. (1989) Eur. J. Biochem. 180, 173–180.
23 Grossman, S., Cobley, J.G., Singer, T.P. and Beinert, H. (1974) J. Biol. Chem. 249, 3819–3826.
24 Coles, C.J., Gutman, M. and Singer, T.P. (1974) J. Biol. Chem. 249, 3814–3818.
25 Biggs, D.R., Nakamura, H., Kearney, E.B., Rocca, E. and Singer, T.P. (1970) Arch. Biochem. Biophys. 137, 12–29.
26 Yagi, T. (1991) J. Bioenerg. Biomembranes 23, 211–225.
27 Minakami, S., Cremona, T., Ringler, R.L. and Singer, T.P. (1963) J. Biol. Chem. 238, 1529–1537.
28 Rossi, C., Cremona, T., Machinist, J.M. and Singer, T.P. (1965) J. Biol. Chem. 240, 2634–2643.
29 Beinert, H., Palmer, G., Cremona, T. and Singer, T.P. (1965) J. Biol. Chem. 240, 475–480.
30 Hatefi, Y. (1973) Biochem. Biophys. Res. Commun. 50, 978–984.
31 Hatefi, Y. and Hanstein, W. G. (1973) Biochemistry 12, 3515–3522.
32 Van Belzen, R., van Gaalen, M.C.M., Cuypers, P.A. and Albracht, S.P.J. (1990) Biochim. Biophys. Acta 1017, 152–159.
33 Hatefi, Y. and Bearden, A.J. (1976) Biochem. Biophys. Res. Commun. 69, 1032–1038.
34 Ragan, C.I. and Racker, E. (1973) J. Biol. Chem. 248, 2563–2569.
35 Ragan, C.I. and Racker, E. (1973) J. Biol. Chem. 248, 6876–6882.
36 Fukushima, T., Decker, R.R. Anderson, W.M. and Spivey, H.O. (1989) J. Biol. Chem. 266, 16483–16488.
37 Gondal, J.A. and Anderson, W.M. (1985) J. Biol. Chem. 260, 12690–12694.
38 Runswick, M.J., Fearnley, I.M., Skehel, J.M. and Walker, J.E. (1991) FEBS Lett. 286, 121–124.
39 Galante, Y.M. and Hatefi, Y. (1979) Arch. Biochem. Biophys. 192, 559–568.
40 Cremona, T. and Kearney, E. (1963) Nature 600, 542–544.
41 Von Bahr-Lindström, H., Galante, Y.M., Persson, M. and Jornvall, H. (1983) Eur. J. Biochem. 134, 145–150.
42 Ragan, C.I., Galante, Y.M. Hatefi, Y. and Ohnishi, T. (1982) Biochemistry 21, 590–594.
43 Ohnishi, T., Ragan, C.I. and Hatefi, Y. (1985) J. Biol. Chem. 260, 2782–2788.
44 Orme–Johnson, N.R., Hansen, R.E. and Beinert, H. (1974) J. Biol. Chem. 249, 1922–1927.
45 Pilkington, S.J. and Walker, J.E. (1989) Biochemistry 28, 3257–3264.
46 Chen, S. and Guillory, R.J. (1981) J. Biol. Chem. 256, 8318–8323.
47 Deng, P.S.K., Hatefi, Y. and Chen, S. (1990) Biochemistry 29, 1094–1098.
48 Chen, S. and Guillory, R.J. (1985) Biochem. Biophys. Res. Commun. 129, 584–590.
49 Chen, S. and Guillory, R.J. (1984) J. Biol. Chem. 259, 5945–5953.
50 Runswick, M.J., Gennis, R.B., Fearnley, I.M. and Walker, J.E. (1989) Biochemistry 28, 9452–9459.
51 Ragan, C.I., Galante, Y.M. and Hatefi, Y. (1982) Biochemistry 21, 2518–2524.
52 Fearnley, I.M., Runswick, M.J. and Werner, S. (1990) EMBO J. 8, 665–672.
53 Attardi, G., Chomyn, A., Doolittle, R.F., Mariottini, P. and Ragan, C.I. (1986) Cold Spring Harbor Symp. Quant. Biol. LI, 103–114.
54 Wang, D.-C., Meinhardt, S.W. Sackmann, U., Weiss, H. and Ohnishi, T. (1991) Eur. J. Biochem., 197, 257–264.
55 Earley, F.G.P. and Ragan, C.I. (1984) Biochem. J. 224, 525–534.
56 Earley, F.G.P., Patel, S.D., Ragan, C.I. and Attardi, G. (1987) FEBS Lett. 219, 108–113.

57 Cleeter, M.W.J., Banister, S.H. and Ragan, C.I. (1985) Biochem. J. 227, 467–474.
58 Patel, S.D. and Ragan, C.I. (1988) Biochem. J. 256, 521–528.
59 Yagi, T. and Hatefi, Y. (1988) J. Biol. Chem. 263, 16150–16155.
60 Vuokila, P.T. and Hassinen, I.E. (1988) Biochem. J. 49, 339–344.
61 Tuschen, G., Sackmann, U., Nehls, U., Haiker, H., Buse, G. and Weiss, H. (1990) J. Mol. Biol. 213, 845–857.
62 Smith, S. and Ragan, C.I. (1980) Biochem. J. 185, 315–326.
63 Patel, S.D., Cleeter, M.W.J. and Ragan, C.I. (1988) Biochem. J. 256, 529–535.
64 Friedrich, T., Strohdeicher, M., Hofhaus, G., Preis, D., Sahm, H. and Weiss, H. (1990) FEBS Lett. 265, 37–40.
65 Hall, R.E. and Hare, J.F. (1990) J. Biol. Chem. 265, 16484–16490.
66 Boekema, E.J., van Breeman, J.F.L., Keegstra, W., Bruggen, E.F.J. and Albracht, S.P.J. (1982) Biochim. Biophys. Acta 679, 7–11.
67 Huennekens, F.M., Felton, S.P., Rao, N.A. and Mackler, B. (1961) J. Biol. Chem. 236, PC57.
68 Lusty, C.J., Machinist, J.M. and Singer, T.P. (1965) J. Biol. Chem. 240, 1804–1810.
69 Paech, C., Reynolds, J.G., Singer, T.P. and Holm, R.H. (1981) J. Biol. Chem. 256, 3167–3170.
70 Ingledew, W.J. and Ohnishi, T. (1980) Biochem. J. 186, 111–117.
71 Albracht, S.P.J., Dooijewaard, G., Leeuwerik, F.J. and Van Swol, B. (1977) Biochim. Biophys. Acta 459, 300–317.
72 Kowal, A.T., Morningstar, J.E., Johnson, M.K., Ramsay, R.R. and Singer, T.P. (1986) J. Biol. Chem. 261, 9239–9245.
73 Ohnishi, T. (1975) Biochim. Biophys. Acta 387, 475–490.
74 Ohnishi, T., Leigh, J.S., Ragan, C.I. and Racker, E. (1974) Biochem. Biophys. Res. Commun. 56, 775–782.
75 Salerno, J.C., Ohnishi, T., Lim, J., Widger, R. and King, T.E. (1977) Biochem. Biophys. Res. Commun. 75, 618–624.
76 Krishnamoorthy, G. and Hinkle, P.C. (1988) J. Biol. Chem. 263, 17566–17575.
77 Ohnishi, T., Blum, H., Galante, Y.M. and Hatefi, Y. (1981) J. Biol. Chem. 256, 9216–9220.
78 Albracht, S.P.J., Leeuwerik, F.J. and Van Swol, B. (1979) FEBS Lett. 104, 197–200.
79 Bakker, P.T.A. and Albracht, S.P.J. (1986) Biochim. Biophys. Acta 850, 413–422.
80 Albracht, S.P.J. and Bakker, P.T.A. (1986) Biochim. Biophys. Acta 850, 423–428.
81 Suzuki, H. and King, T.E. (1983) J. Biol.Chem. 258, 352–358.
82 Gutman, M., Singer, T.P. and Casida, J.E. (1970) J. Biol. Chem. 245, 1992–1997.
83 Takeshige, K. and Minakami, S. (1979) Biochem. J. 180, 129–135.
84 Kang, D., Narabayashi, H., Sata, T. and Takeshige, K. (1983) J. Biochem. 94, 1301–1306.
85 Kean, E.A., Gutman, M. and Singer, T.P. (1971) J. Biol. Chem. 246, 2346–2353.
86 Ernster, L., Jalling, O., Löw, H. and Lindberg, O. (1955) Exp. Cell Res. Suppl. 3, 124–132.
87 Chance, B. (1956) in: P.H. Gaehlber (Ed.) Enzymes: Units of Biological Structure and Function, Academic Press, New York, pp. 465–482.
88 Hatefi, Y. (1968) Proc. Natl. Acad. Sci. USA 60, 733–740.
89 Minakami, S., Ringler, R.L. and Singer, T.P. (1962) J. Biol. Chem. 237, 569–576.
90 Horgan, D.J., Singer, T.P. and Casida, J.E. (1968) J. Biol. Chem. 243, 834–843.
91 Palmer, G., Horgan, D.J., Tisdale, H., Singer, T.P. and Beinert, H. (1968) J. Biol. Chem. 243, 844–847.
92 Horgan, D.J., Ohno, H., Singer, T.P. and Casida, J.E. (1968) J. Biol. Chem. 243, 5967–5976.
93 Ragan, C.I. (1976) Biochim. Biophys. Acta. 456, 249–290.
94 Shimomura, Y., Kawada, T. and Suzuki, M. (1989) Arch. Biochem. Biophys. 270, 573–577.
95 Gerth, K., Jansen, R., Reifenstahl, G., Höfle, G., Irschik, H., Kunze, B., Reichenbach, H. and Thierbach, G. (1983) J. Antibiotics 36, 1150–1156.
96 Chung, K.H., Cho, K.Y., Asami, Y., Takahashi, N. and Yoshida, S. (1989) Z. Naturforsch. 44c, 609–616.
97 Nicklas, W.J., Vyas, I. and Heikkila, R.E. (1985) Life Sci. 36, 2503–2505.
98 Ramsay, R.R., Salach, J.I., Dadgar, J. and Singer, T.P. (1986) Biochem. Biophys. Res. Commun. 135, 269–276.

99 Ramsay, R.R., Dadgar, J., Trevor, A. and Singer, T.P. (1986) Life Sci. 39, 581–588.
100 Ramsay, R.R., Krueger, M.J., Youngster, S.K., Gluck, M.R., Casida, J.E. and Singer, T.P. (1991) J. Neurochem. 56, 1184–1190.
101 Gluck, M.R., Nicklas, W.J., Ramsay, R.R. and Singer, T.P., to be published.
102 Ramsay, R.R., Mehlhorn, R.J. and Singer, T.P. (1989) Biochem. Biophys. Res. Commun. 159, 983–990.
103 Sayre, L.M., Wang, F. and Hoppel, C.L. (1989) Biochem. Biophys. Res. Commun. 161, 809–818.
104 Ramsay, R.R., Youngster, S.K., Nicklas, W.J., McKeown, K.A., Jin, Y.-Z., Heikkila, R.E. and Singer, T.P. (1989) Proc. Natl. Acad. Sci. USA 86, 9168–9172.
105 Ramsay, R.R., Krueger, M.J., Youngster, S.K. and Singer, T.P. (1991) Biochem. J. 273, 481–484.
106 Anderson, W.M., Patheja, H.S., Delinck, D.L., Baldwin, W.A., Smiley, S.T. and Chen, L.B. (1989) Biochem. Intern. 19, 673–685.
107 Davies, K.J.A. and Doroshow, J.H. (1986) J. Biol. Chem. 261, 3060–3067.
108 Mahler, H.R. and Elowe, D.G. (1954) J. Biol. Chem. 210, 165–179.
109 Cremona, T. and Kearney, E.B. (1965) J. Biol. Chem. 240, 3645–3652.
110 Minakami, S., Schindler, F.J. and Estabrook, R.W. (1964) J. Biol. Chem. 239, 2042–2048.
111 Gutman, M., Mersmann, H., Luthy, J. and Singer, T.P. (1970) Biochemistry 9, 2678–2687.
112 Gutman, M., Kearney, E.B., Mayr, M. and Singer, T.P. (1971) Physiol. Chem. Phys. 3, 319–335.
113 Gutman, M., Singer, T.P. and Beinert, H. (1972) Biochemistry 11, 556–562.
114 Gutman, M. and Singer, T.P. (1970) Biochemistry 9, 4750–4758.
115 Bois, R. and Estabrook, R.W. (1969) Arch. Biochem. Biophys. 129, 362–369.
116 Mitchell, P. (1972) in: S.G. van der Bergh, P. Borst, L.L.M. van Deenen, J.C. Riemersma, E.C. Slater and J.M. Tager (Eds.) Mitochondria/Biomembranes, Vol. 28, Elsevier, Amsterdam, pp. 353–370.
117 Garland, P.B., Clegg, R.A., Downie, J.A., Gray, T.A., Lawford, H.G. and Skyrme, J. (1972) in: S.G. van der Bergh, P. Borst, L.L.M. van Deenen, J.C. Riemersma, E.C. Slater and J.M. Tager (Eds.) Mitochondria/Biomembranes, Vol. 28, Elsevier, Amsterdam, pp. 105–117.
118 Ohnishi, T. (1976) Eur. J. Biochem. 64, 91–103.
119 Pozzan, T., Miconi, V., Di Virgilio, F. and Azzone, G.F. (1979) J. Biol. Chem. 254, 10200–10205.
120 Vercesi, A., Reynafarje, B. and Lehninger, A.L. (1978) J. Biol. Chem. 253, 6379–6385.
121 Di Virgilio, F. and Azzone, G.F. (1982) J. Biol. Chem. 257, 4106–4113.
122 Beavis, A.D. (1987) J. Biol. Chem. 262, 6165–6173.
123 Lemasters, J.J., Grunwald, R. and Emaus, R.K. (1984) J. Biol. Chem. 259, 3058–3063.
124 Brown, G.C. and Brand, M.D. (1988) Biochem. J. 252, 473–479.
125 Kotlyar, A.B., Sled, V.D., Burbaev, D.Sh., Moroz, I.A. and Vinogradov, A.D. (1990) FEBS Lett. 264, 17–20.
126 Ichiki, T., Tanaka, M., Kobayashi, M., Sugiyama, N., Suzuki, H., Nishikimi, M., Ohnishi, T., Nonaka, I., Wada, Y. and Ozawa, T. (1989) Pediatr. Res. 25, 194–201.
127 Capaldi, R.A. (1988) Trends Biochem. Sci. 13, 144–148.
128 Morgan-Hughes, J.A., Schapira, A.H.V., Cooper, J.M. and Clark, J.B. (1988) J. Bioenerg. Biomembranes 20, 365–382.
129 Mizuno, Y., Ohta, S., Tanaka, M., Takamiya, S., Suzuki, K., Sato, T., Oya, H., Ozawa, T. and Kagawa, Y. (1989) Biochem. Biophys. Res. Commun. 163, 1450–1455.
130 Schapira, A.H.V., Cooper, J.M., Dexter, D., Clark, J.B., Jenner, P. and Marsden, C.D. (1990) J. Neurochem. 54, 823–827.
131 Hatefi, Y. (1985) Annu. Rev. Biochem. 54, 1015–1069.
132 Moreadith, R.W., Cleeter, M.W.J., Ragan, C.I., Batshaw, M.L. and Lehninger, A.L. (1987) J. Clin. Invest. 79, 463–467.
133 Watmough, N.J., Birch-Machin, M.A., Bindoff, L.A., Aynsley-Green, A., Simpson, K., Ragan, C.I., Sherratt, H.S.A. and Turnbull, D.M. (1989) Biochem. Biophys. Res. Commun. 160, 623–627.
134 Hatefi, Y., Galante, Y.M., Stiggall, D.L. and Ragan, C.I. (1979) Methods. Enzymol. 56, 577–602.
135 Anderson, S., de Bruijn, M.H.L., Coulson, A.R., Eperon, I.C., Sanger, F. and Young, I.G. (1982) J. Mol. Biol. 156, 683–706.

L. Ernster (Ed.) *Molecular Mechanisms in Bioenergetics*
© 1992 Elsevier Science Publishers B.V. All rights reserved

CHAPTER 7

Progress in succinate:quinone oxidoreductase research

LARS HEDERSTEDT[1] and TOMOKO OHNISHI[2]

[1]*Department of Microbiology, University of Lund, Sölvegatan 21, S-223 62 Lund, Sweden and,*
[2]*Department of Biochemistry and Biophysics, University of Pennsylvania, A606 Richard Building, Philadelphia, PA 19104, U.S.A.*

Contents

1. Introduction from a historical perspective 164
   1.1. Scope of review 164
   1.2. Definition of enzyme types 164
   1.3. Selected historical hallmarks 164
2. General composition of succinate:quinone oxidoreductases 166
3. Genetics, expression and biogenesis 167
   3.1. Gene organization 167
   3.2. Regulation of gene expression 168
   3.3. Enzyme biogenesis 172
4. Structure and function of succinate:quinone oxidoreductases 173
   4.1. Present stage of knowledge 173
   4.2. Dicarboxylate active site 173
      4.2.1. Substrate specificity 173
      4.2.2. Residues at the dicarboxylate binding site 174
   4.3. Intramolecular electron transfer 177
      4.3.1. Flavin 177
      4.3.2. Iron–sulfur clusters 179
      4.3.3. Cytochrome *b*/anchor polypeptides 185
   4.4. Quinone active site 188
5. Which are the directions of future research? 190
Acknowledgements 191
References 192

Abbreviations used

CD, circular dichroism; CRP, cAMP receptor protein; ENDOR, electron-nuclear double resonance; EPR, electron paramagnetic resonance; FRD, soluble subcomplex of QFR; LEFE, linear electric field effect; MCD, magnetic circular dichroism; NMR, nuclear magnetic resonance; QFR, quinol:fumarate reductase; $Q_s^{\cdot-}$, semiquinone bound to SQR, Q_2, 2,3-dimethoxy-5-

methyl-1,4-benzoquinone-6-di-(2-methyl-2-butene); SDH, soluble subcomplex of SQR; SQR, succinate:quinone reductase (complex II); TTFA, 2-thenoyltrifluoroacetone.

1. Introduction from a historical perspective

1.1. Scope of review

In this review we present progress made mainly within the last decade in our understanding of the genetics, biogenesis, structure and function of succinate:quinone oxidoreductases. Work on this class of enzymes has involved a vast amount of experimental efforts in many laboratories. An excellent, comprehensive, review on this topic has very recently been written by Ackrell et al. [1]. Other reviews on various aspects of succinate:quinone oxidoreductases are found in refs. [2–10].

As in many other fields of biological research rapid advances have resulted from the increased use of a molecular biologist's approach, i.e. a combination of molecular genetics, biochemistry and biophysical techniques. Among the latter are cryogenic MCD (Magnetic Circular Dichroism) [9] and cyclic voltammetry [11,12]. New aspects have been approached, some 'old' questions have been answered and new problems have appeared. Before we describe these developments and discoveries in detail, it is important to define the type of enzymes to be discussed and also to put the more recent findings into a historical perspective.

1.2. Definition of enzyme types

Succinate:quinone oxidoreductases (E.C.1.3.5.1) are membrane bound enzymes that can catalyze the oxidation of succinate to fumarate coupled to the reduction of a quinone and the reduction of fumarate to succinate coupled to the oxidation of quinol. Succinate:quinone reductase (named also complex II), which here is abbreviated as SQR, is present in strictly aerobic cells and in vivo predominantly catalyzes the oxidation of succinate. Quinol:fumarate reductase (QFR) is found in anaerobic cells respiring with fumarate as terminal electron acceptor. SQR and QFR are not only able to catalyze the same reaction, they also have a similar composition with generally 3 to 4 polypeptides and with several different redox prosthetic groups; a covalently bound FAD, 3 iron–sulfur clusters (designated center 1, center 2 and center 3)*, and in many cases a cytochrome *b*. Water soluble subcomplexes that catalyze succinate oxidation with artificial electron acceptors have been obtained from SQR and QFR complexes. These water soluble forms, which we designate SDH (Succinate DeHydrogenase) or FRD (Fumarate ReDuctase), depending if they originate from SQR or QFR, respectively, consist of only two polypeptides.

1.3. Selected historical hallmarks

Succinate oxidase activity was observed in extracts of frog muscle already in 1909 [13]. The enzyme has since been studied intensively in a variety of organisms and, for instance, the general concept of competetive inhibition of enzymes has emerged from studies on malonate

* These centers are often named S-1, S-2 and S-3 in SQR, whereas in QFR they are named FR-1, FR-2 and FR-3.

inhibition of succinate oxidation [14]. The first enzyme preparations and characterizations were described in the late fifties [15–17], and a preparation of SQR was reported in 1959 [18]. The enzyme preparations were found to contain non-heme iron and acid-labile sulfide [19,20]. The soluble enzyme form, SDH, containing two non-identical polypeptides was isolated pure in 1971 [21]. In 1970 FAD was discovered to be attached to SDH via histidyl N(3)-8α riboflavin linkage [22]. The g=1.94 EPR (Electron Paramagnetic Resonance) signal of reduced center 1, a [2Fe–2S] cluster, was detected in 1960 [23,24] and the g=2.01 EPR signal of iron–sulfur center 3 was observed [25] and identified as a component of SDH in 1975 by cryogenic EPR spectroscopy [26]. Iron–sulfur center 2 was observed in 1976 by EPR spectroscopy as a spin-coupled component to center 1 [27]. Center 2 was subsequently identified as a $[4Fe–4S]^{(1+,2+)}$ cluster by MCD [28] and cryogenic EPR [29], whereas center 3 was identified as a $[3Fe–4S]^{(0,1+)}$ cluster by LEFE (Linear Electric Field Effect) EPR [30] and MCD [28]. One or two small hydrophobic polypeptides present in succinate:quinone oxidoreductases have been shown by reconstitution experiments [31–39] and by the use of bacterial mutants [40–43] to anchor SDH or FRD to the membrane and also to be necessary for the reactivity with quinone/quinol. Protoheme IX, when present in the complex, is a prosthetic group of the anchor polypeptide(s).

Mutants of bacteria defective in SDH activity were first reported in the sixties [44,45]. Analysis of *Escherichia coli* mutants revealed the presence of two enzymes in this organism able to oxidize succinate and reduce fumarate [46,47]. During the subsequent decade, mutations resulting in defective SQR and QFR were mapped to specific loci, *sdh* and *frd*, respectively, [48–50] and mutations were mapped relative to each other within loci [51,52]. In 1979–1980 the *frd* region of the *E. coli* chromosome was cloned as a spin-off result from studies on ampicillin resistance [53] and by complementation of Frd mutants [54,55]. The nucleotide sequence of the *frd* region was published in 1982 [56–58] and the cloning of the *E. coli sdh* genes was reported the same year [59]. The global regulatory proteins Fnr and ArcA/ArcB in *E. coli* were discovered in 1976 [60] and in 1988 [61,62], respectively, partly through the use of *frd* and *sdh* as reporter systems. The nucleotide sequence of the *E. coli sdhCDAB* genes was reported in 1984 [63,64] and the first eukaryotic sequences of genes encoding the small subunit of SDH

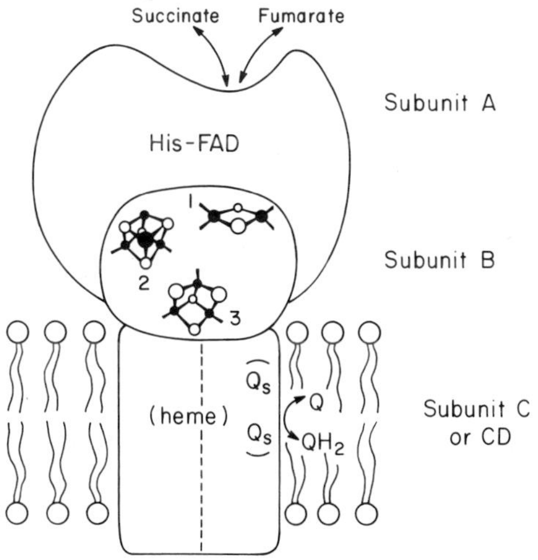

Fig. 1. General composition and topology in the membrane of succinate:quinone oxidoreductases.

were presented in 1989–1990 [65–67]. Overproduction of *E. coli* FRD was first reported in 1979 [68], *Bacillus subtilis* SQR subunits were expressed from cloned genes in *E. coli* in 1985 [69] and the *E. coli* SQR was overproduced and purified in 1989 [70]. Nucleotide sequences of mutant *sdh* and *frd* genes appeared in 1986 [71,72]. *E. coli* and *B. subtilis* strains with site-directed mutations in QFR and SQR, respectively, were introduced in 1989 [73,74].

2. General composition of succinate:quinone oxidoreductases

As a part of the respiratory system, succinate:quinone oxidoreductases are firmly bound to membranes, the cytoplasmic membrane in prokaryots and the mitochondrial inner membrane in eukaryots. Figure 1 presents the general composition and arrangement of these enzymes in the membrane. Two subunits, designated A and B (or Fp and Ip; FlavoProtein and Iron–sulfur Protein), are anchored to the membrane by one, subunit C, or two, subunits C and D, hydrophobic polypeptides. At least some succinate:quinone oxidoreductases can be split into an anchor polypeptide(s) fraction and a water-soluble heterodimer, SDH or FRD. Both SDH and FRD, if protected against strong oxidants, can reassociate (be reconstituted) with the anchor part with concomittant appearance of quinone oxidoreductase activity.

Succinate:quinone oxidoreductase has after solubilization from the membrane with deter-

TABLE I

Composition of succinate:quinone oxidoreductase in different organisms

Enzyme		No. of subunits (kDa)[a]	Heme per FAD (mol/mol)	Sequence data[b] available	Refs.
Organism	Function				
Prokaryots:					
Escherichia coli	SQR	4 (64,27,14,13)	1	ABCD	[63,64,70,75,76]
Escherichia coli	QFR	4 (69,27,15,13)	0	ABCD	[38,39,56–58]
Bacillus subtilis	SQR	3 (65,28,23)	2	ABC	[77–81][d]
Wolinella succinogenes	QFR	3 (79,31,25)	2	ABC	[82–84]
Micrococcus luteus	SQR[c]	4 (72,30,17,15)	≥1	----	[85]
Paracoccus denitrificans	SQR	4 (65,29,13,12)	1	----	[86]
Desulfobulbus elongatus	SQR	3 (69,28,22)	1	---	[87]
Sulfolobulos acidocaldarius	SQR	4 (66,31,28,13)	0	----	[88]
Eukaryots:					
Mammals	SQR	≥4 (70,27,15,13)	1	AB--	[1,33,66,67,89–95,100]
Ascaris suum	QFR	4 (66,27,12,11)	1	-B--	[96,97][e]
Paragonimus westermani	SQR	4 (69,27,14,12)	1	----	[97]
Neurospora crassa	SQR	3 (72,28,14)	1	---	[98]
Mung bean	SQR	4 (67,30,15,13)	present	----	[99]

[a] The relative mass of subunits are given except in the cases where the amino acid sequence or nucleotide sequence of the corresponding gene is known.
[b] Complete nucleotide or amino acid sequence data.
[c] Quinone reductase activity not demonstrated.
[d] See also C. Hägerhäll et al., Biochemistry (1992), in press.
[e] See also K. Kita et al., unpublished data.

gent been isolated from many different organisms as summarized in Table I. The dicarboxylate (succinate/fumarate) binding site is located on subunit A, which also carries the covalently bound FAD moiety. FAD serves as the primary electron acceptor in the enzyme at succinate oxidation and the terminal electron donor at fumarate reduction. Subunit B contains the [2Fe–2S] cluster and most likely also the [3Fe–4S] cluster and the [4Fe–4S] cluster (discussed in Section 4.3.2). The anchor polypeptide(s), possibly in conjunction with subunit B, are required for quinone reduction and quinol oxidation. Depending on the source of the enzyme (Table I) the anchor polypeptide(s) is a cytochrome *b* with one or two heme groups. Flavin, iron–sulfur clusters, and possibly heme and bound quinone are involved in electron transfer within the enzyme, i.e. between the succinate/fumarate and the quinone/quinol active sites.

3. Genetics, expression and biogenesis

3.1. Gene organization

Structural genes for bacterial SQR and QFR are designated *sdh* and *frd*, respectively. The gene organization and complete sequence are presently known for the *E. coli* [63,64] and *B. subtilis* [79,80] *sdh* genes, and for the *E. coli* [56–58], *Proteus vulgaris* [101] and *Wolinella (Vibrio) succinogenes* [83,84] *frd* genes. In all cases the 3 or 4 structural genes are clustered on the chromosome and co-transcribed into a single mRNA molecule of about 3.5 kb (Fig. 2). The *E. coli sdhCDAB* locus is located at 17 min on the chromosomal map in a gene cluster that encodes polypeptides of four different Krebs' cycle enzymes [102,103]. The *E. coli frdABCD* locus is located at 94 min [3,103] in a region that also harbours *fumB* and *aspA*, which encode a fumarase and an aspartase, respectively, that in anaerobic cells can serve to produce fumarate from malate or aspartate [104]. The gene downstream of *frdABCD* in the chromosome varies among closely related bacteria like those in the family Enterobacteriacae, i.e. the downstream gene is *ampC* in *E. coli* (Fig. 2) and *Shigella sonnei, ampR* in *Enterobacter cloacae* and *Citrobacter freundii* and genes of unidentified proteins in *P. vulgaris* [101,105]. Information on the gene organization of *frd* and *sdh* genes in strictly anaerobic and strictly aerobic bacteria is only available for *W. succinogenes* (anaerobe) and *B. subtilis* (aerobe) (Fig. 2).

Mutations in human chromosome 1 can result in SDH deficiency [106]. Subunit A antigen is lacking in mitochondria from such mutant cells, but it has not been shown that the mutations are located in a structural gene for SQR. Using synthetic oligonucleotides, designed on the basis of known amino acid sequences, the DNA polymerase chain reaction has been exploited to obtain genomic or cDNA clones encoding part (about 60%) of subunit B of SQR from a variety of eukaryotic cells; human (fibroblasts), mouse, rat (brain), *Drosophila melanogaster, Arabidopsis thaliana, Schizosaccharomyces pombe* and *Saccharomyces cerevisiae* [65]. A genomic DNA fragment containing the complete subunit B gene for *S. cerevisae* SQR was later isolated and sequenced [66]. cDNA clones encoding subunit B of human liver SQR [67] and adult *Ascaris suum* SQR (K. Kita, unpublished data) and a clone encoding the subunit A of bovine heart SQR [100] have been isolated using subunit specific antibodies, and sequenced. A cDNA clone from a human placenta DNA library encoding a subunit A of SQR with a sequence similar to that of *B. subtilis* SQR has been isolated [93]. A mutant gene encoding subunit B of SQR of the basidomycete *Ustilago maydis* was recently isolated and sequenced [107].

These different eukaryotic clones are very useful tools for many types of experiments, including the chromosomal mapping of the A and B genes. The gene for the subunit B of *S. cerevisiae*

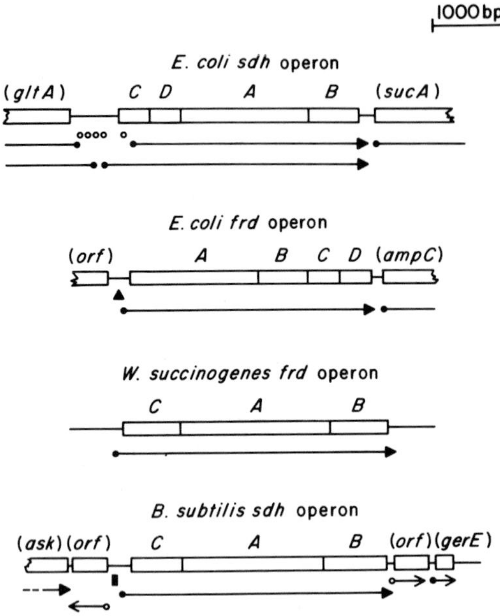

Fig. 2. Organization of *sdh* and *frd* genes in bacteria. The arrows indicate mRNAs; solid and open arrowheads denote confirmed and tentative transcripts, respectively. Open circles in the *E. coli sdh* promoter region indicate putative CRP binding sites. A filled triangle indicates an Fnr protein binding site in the promoter region of *E. coli frd.* A filled rectangle indicates a possible catabolite repressor site in the *B. subtilis sdh* promoter region. Sequence data are from refs. [56–58,63,64,79,80,83,84] and transcription data from refs. [83,84,109,112,122].

SQR hybridizes to chromosome VII [66], whereas that for subunit A maps to chromosome XI ([108], K.M. Robinson and B.D. Lemire, unpublished data). Thus, we may generally expect to find that the genes for polypeptides of SQR in eukaryotic cells are distributed on different chromosomes or are at least not clustered as in bacteria.

The deduced amino acid sequences of subunit A and subunit B, respectively, but generally not that of the anchor proteins, are conserved in SQR and QFR enzymes from different organisms. Figure 3 shows a dendrogram based on similarities in the primary sequence of subunit A. The *E. coli* SQR polypeptide shows the highest sequence similarity with that of bovine heart SQR and the lowest with that of *B. subtilis* SQR. Subunit A of *S. cerevisiae* SQR shows about 53 and 30% sequence identity to that of *E. coli* and *B. subtilis* SQR, respectively (K.M. Robinson and B.D. Lemire, unpublished data). Sequence comparisons using subunit B show roughly the same relative similarities as those obtained with subunit A.

3.2. Regulation of gene expression

The Krebs cycle has a dual role in aerobic respiring cells. It has an oxidative function in energy metabolism and it also provides intermediates for biosynthetic purposes, e.g. amino acid synthesis. To fulfil these functions under a variety of nutritional conditions, the individual enzymes of the cycle are carefully regulated. Expression of SQR in bacteria and yeasts is generally

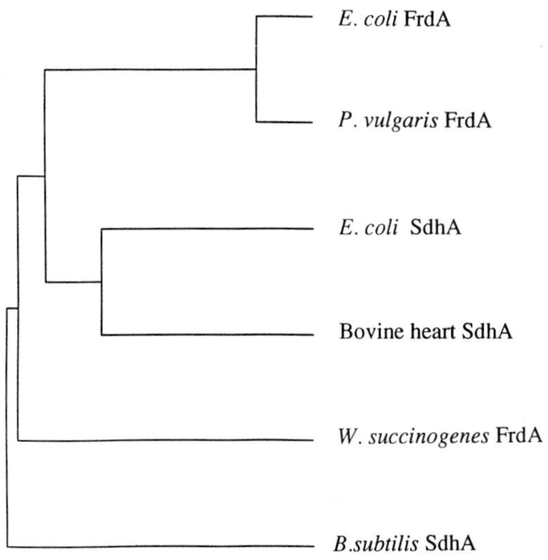

Fig. 3. Dendrogram for subunit A from different succinate:quinone oxidoreductases based on primary structure comparisons. The amino acid sequences as deduced from the DNA sequences were aligned using the program Pileup (UWGCG Version 7.0). References to sequence data are given in Fig. 4.

several-fold lower if the cells are grown in a broth medium supplemented with a high concentration of glucose compared to those grown without glucose. This is the phenomenon of catabolite repression which in *E. coli, B. subtilis* and *S. cerevisiae* is mainly an effect on transcription [66,109,110].

The *E. coli sdh* genes are transcribed into two polycistronic mRNAs, a major *sdhDAB* transcript of 3047 bases and a minor *sdhCDAB* transcript of 3493 bases (Fig. 2) [109]. The start site for the longer mRNA is at a position 219 bp upstream of the translational initiation codon for *sdhC*, and the shorter transcript starts at a position within *sdhC*. Both transcriptional start sites are preceded by an *E. coli* promoter-like sequence. Just downstream of *sdhB*, there are three inverted repeated palindromes but no typical sequence for a rho-independent terminator. Both *sdh* transcripts end at positions corresponding to sites between the first and the second palindrome. *sdh* mRNA was not detected in cells grown in the presence of glucose and from this it was concluded that transcription was repressed by glucose. Catabolite repression in Enterobacteria is mediated via cAMP and the cAMP receptor protein (CRP) [111]. Binding of the cAMP–CRP complex to CRP sites in the promoter region of catabolite repressed genes activates transcription of these genes probably by a direct interaction between the bound complex and the RNA polymerase. In glucose grown cells the cAMP concentration is low and consequently little cAMP–CRP complex can be formed and transcription is not activated. There are two potential CRP sites in the *E. coli sdh* promoter region; one at each promoter sequence (Fig. 2).

Only one *sdh* transcript has been detected in the Gram-positive bacterium *B. subtilis* [112]. Its start site is at a position 90 bp upstream from the *sdhC* translational start site. The transcription terminator is probably of the rho-independent type and located about 20 bp after the *sdhB* translational stop codon. The indicated start site for transcription is preceded by a promoter region, −35 (TTGACG) and −10 (TAAAAT), which is most likely recognized by RNA polymerase containing the *B. subtilis* major vegetative sigma-factor, SigA [113]. Single basepair changes in the −35 and −10 sequences, respectively, can essentially inactivate *sdh* promoter activity (Table II). Melin et al. have analyzed the effect of glucose on transcription and the stability of the *sdhCAB* mRNA in *B. subtilis* [110]. The steady-state concentration of the transcript is four times higher in cells grown in a broth medium compared to cells grown in the same medium supplemented with 1% glucose. This difference is due to different activity of the *sdh* promoter under different growth conditions. The mechanism for catabolite repression in Gram-positive bacteria is poorly understood but it is different from that in *E. coli* since cAMP is probably not involved [116]. A possible catabolite repressor operator sequence with twofold symmetry has been found in the promoter region of several *Bacillus* genes and in one case, (*amyE*), it has been shown by mutation analysis to be important for glucose repression [117]. Such a sequence is present upstream and partially overlapping the −35 sequence of the *sdh* promoter region. Very tentatively this sequence may bind a regulatory protein that modulates transcription initiation.

The cellular concentration of the *sdh* message in *B. subtilis* is strongly dependent on the growth stage of the cell and this has been found to be mainly the result of a variable stability of the transcript. The half-life ($t_{1/2}$) of *sdh* mRNA is 2 to 3 min in the exponential growth phase but only about 0.4 min in the early stationary growth phase [110]. This decrease in stability that occurs when the cells enter the stationary growth phase is not common for all transcripts in *B. subtilis*. For example, $t_{1/2}$ of the *odhAB* mRNA (encoding subenzymes of the 2-oxoglutarate dehydrogenase complex) seems unaffected by the growth stage [118] and $t_{1/2}$ of the *aprE* mRNA (encodes subtilisin) is more stable in early stationary growth phase cells than in exponential and late stationary growth phase cells [119]. The mechanism for degradation of the *sdh* mRNA is not known but the 5′-end of the transcript is important for stability [114]. The $t_{1/2}$ of the *sdh* mRNA is shorter than normal if the *sdhC* gene is less frequently translated, i.e. a mutation in

TABLE II

Mutations affecting expression of *Bacillus subtilis* succinate:menaquinone reductase

Mutated region	Mutation[a]	Membrane bound[b] SDH activity	Subunit A[c] Membrane	Total	*sdh* mRNA[d] detectable	Half-life of mRNA (min)[e] *sdhC*	*sdhA*	*sdhB*
–	wild type	100%	100%	100%	yes	2.6	1.9	3.0
Promoter (−35)	G^{142} to A	0.5%	<1%	<1%	no	–	–	–
Promoter (−10)	T^{163} to C	<1%	≈1%	≈1%	–	–	–	–
Shine–Dalgarno	G^{253} to A	8%	7–10%	90%	yes	0.5	0.7	1.3
sdhC coding region	C^{628} to T (Gln^{222} to stop; TAA)	n.d.[f]	n.d.	present	yes	1.8	1.5	1.8

Data from refs. [42,112,114,115].
[a] Nucleotides numbered as in ref. [79].
[b] Relative specific activity determined using phenazinemethosulfate and dichlorophenolindophenol.
[c] Subunit A was quantitated using subunit specific anti-serum. Total is in total cell extract.
[d] Transcripts detected by Northern blot or primer extension analysis.
[e] Half-lifes were determined on mRNA in exponentially growing cells.
[f] n.d.; not detectable.

the Shine–Dalgarno sequence, which decreases the translation of *sdhC* about tenfold and has only a minor polar effect on the translation of the following *sdhA* gene, destabilizes the transcript (Table II). By contrast, introduction of a stop codon in the middle of the *sdhC* gene does not cause a drastic decrease in stability of the *sdh* mRNA.

Information about transcription and regulation of eukaryotic SQR genes is emerging hand in hand with the increasing number of isolated genomic clones. The *S. cerevisiae* subunit B gene is transcribed as a 1.1 kb mono-cistronic mRNA apparently without introns [66]. Transcription starts about 50 bp from the translational start codon. A TATAA and a CCAAT promoter box are present at about positions −70 and −100, respectively, from the transcriptional start site. About 70 bp downstream of the open reading frame for subunit B there is a polyadenylation signal sequence TAATAAA. A mutant of *S. cerevisiae* with derepressed SDH activity has been described and may be mutated in a regulatory gene [120]. The promoter region of the *U. maydis* subunit B gene contains a CAAT box about 200 bp upstream from the protein coding sequence, but other promoter features are not apparent [107]. The sequenced *U. maydis* DNA fragment also lacks a polyadenylation signal.

Facultative bacteria, like *E. coli*, which contain both *frd* and *sdh* genes express these genes depending whether oxygen is available or not. The *frdABCD* operon is transcribed under anaerobic growth conditions in the presence of fumarate, but only if no terminal electron acceptor of higher redox potential, such as nitrate, is present [3,121,122]. The *sdhCDAB* operon is expressed only under aerobic conditions but QFR can functionally replace SQR to support growth of an *E. coli sdh* mutant [123,124]. A fascinating question is how the cell can sense the availability of oxygen in the environment and convey that information to gene regulatory systems. Two mini-reviews on this topic concerning *E. coli* have recently been published [125,126].

Anaerobic expression of the *frd* operon and many other genes encoding anaerobic respiratory enzymes in *E. coli* requires the *fnr* (Fumarate and Nitrate Reduction) gene product. Spiro and Guest have written a comprehensive review on Fnr [127] and here we will only mention some aspects relevant to *frd* regulation. Fnr is a 28 kDa DNA-binding protein which binds, probably as a dimer [134], to an about 20 bp long stretch 45 bp upstream from the *frdABCD* transcriptional start site (Fig. 2). The C-terminal portion of Fnr is homologous to the C-terminal part of CRP. The three-dimensional structure of CRP is known [128] and the sequence homologous with Fnr forms a helix–turn–helix DNA binding motif. In elegant site-specific mutagenesis experiments Guest et al. have demonstrated the close similarity in the DNA binding domain of Fnr and CRP [129]. The Fnr protein contains a cluster of four cysteine residues close to the N-terminus. Three of these cysteines are functionally important as shown by site-specific mutant analysis [130–132] and also by analysis of Fnr mutants [133]. These cysteines may bind an iron atom [134]. It has not been demonstrated how Fnr responds to anaerobiosis, but accumulated evidence [127,134,136] suggests that the redox state of the iron bound to Fnr determines the activity of the protein.

The NarL protein is part of a two-component regulatory system (together with NarX) and functions to repress expression of the *frd* genes when nitrate is present, i.e. it counteracts the effect of active Fnr (for reviews see refs. [125] and [126]). Another two-component regulatory system, ArcA/ArcB, regulates expression of the *sdh* genes [61,62]. Activated ArcA (Aerobic Respiratory Control) protein causes repression of the *sdhCDAB* operon in *E. coli*. The ArcA protein has in common with CRP and Fnr a helix–turn–helix DNA binding motif in the C-terminal part of the polypeptide. Recent data indicate that ArcA may have two separable functions, the function of respiratory control and a function in sex factor regulation (Sfr) [137]. Phosphorylation of the transmembrane ArcB protein is thought to signal to the soluble ArcA protein that oxygen is not available. Seemingly the ArcB protein ‘senses’ the degree of aerobiosis from components (oxidases) in the aerobic respiratory chain [138].

3.3. *Enzyme biogenesis*

The biogenesis of succinate:quinone oxidoreductases is rather complex because of the different prosthetic groups and of the fact that these enzymes are integral membrane proteins. In eukaryots there is one further complication in that all subunit polypeptides are synthesized in the cytoplasm and have to be translocated across the outer mitochondrial membrane and two subunits also have to pass the inner mitochondrial membrane. The polypeptides for both subunit A and subunit B of mitochondrial SQR are synthesized with a 10–50 residues long N-terminal signal sequence that is absent from the mature enzyme, as shown for *S. cerevisae* [66], *U. maydis* [107], sweet potato [139] and mammalian cells [67,140,141]. No data on eventual processing is reported for the anchor polypeptides of eukaryotic cells. In the bacterium *B. subtilis* the processing of subunit A and subunit B polypeptides is limited to removal of the initiating methionine [142]. The initiating methionine is retained in the *W. succinogenes* FrdA polypeptide [143]. Similarly, the transmembrane anchor polypeptides seem generally to be synthesized without a cleaved N-terminal signal sequence, i.e. only the methionine is removed from the *B. subtilis* SdhC polypeptide [142] and three residues are removed from the *E. coli* SdhC polypeptide [75].

Pulse-chase experiments with *B. subtilis* SQR [41] have shown that after translation has been completed subunits A and B first appear as soluble proteins whereas the hydrophobic anchor polypeptide has only been found membrane associated [115]. In the absence of a functional anchor part, subunits A and B accumulate in the cytoplasm as demonstrated by *B. subtilis* SQR and *E. coli* QFR mutants (cf. refs. [39,41,144]). In the case of *E. coli* QFR, the A and B subunits form an active FRD in the cytoplasm [68]. In *E. coli*, both FrdC and FrdD polypeptide segments are required for membrane binding of FRD [40,145]. Condon and Weiner have found the unexpected result that if the FrdC and FrdD polypeptides are expressed from separate transcripts, FRD is bound to the membrane but in a defective way such that quinol cannot be oxidized [40]. The reason for this phenomenon is obscure. Tubular intracellular membrane structures enriched in QFR (90% of the total protein is QFR) result when the *E. coli frdABCD* operon is overexpressed to high levels [146,147]. QFR enriched membranes are optically transparent and have been used for an analysis of the CD (circular dichroic) properties of the enzyme complex [148].

FAD is, or at least can be, covalently bound to the soluble A polypeptide before it associates with subunit B and becomes anchored to the membrane [149]. Also FAD-analogs can be incorporated, as demonstrated with a yeast riboflavin requiring mutant [150]. It is not known whether the FAD attachment mechanism is autocatalytic, as has been demonstrated for *Arthrobacter oxidans* 6-hydroxy-D-nicotine oxidase (this enzyme also contains FAD bound in 8α-N(3) histidyl linkage) [151], or requires protein factors not present in the final enzyme. The *B. subtilis* SdhA polypeptide synthesized in *E. coli* cells does not become flavinylated despite normal processing at the N-terminus [142]. This observation speaks for a requirement of specific factors for flavinylation to occur. Subunit B or the anchor polypeptide(s) are not necessary for flavinylation of subunit A [152]. Brandsch and Bichler have demonstrated that ^{14}C-FAD can be bound to subunit A of *E. coli* SQR and QFR in vitro in soluble extracts from cells starved for flavin if a Krebs cycle intermediate (citrate, isocitrate, fumarate or succinate) is present [153]. An interpretation of these results is that these intermediates serve as allosteric effectors to enhance covalent binding (by autocatalysis?) of FAD. Proteins containing covalently bound FAD have in eukaryots only been found in the mitochondrion, e.g. monoamine oxidase, L-gulonolactone oxidase, dimethylglycine dehydrogenase and SQR [154,155]. It is not known when during translocation into the mitochondrion flavin is covalently attached, but in analogy with the heme in cytochrome *c* it can be assumed that it occurs after membrane passage.

Very little is known about how iron–sulfur clusters are formed in vivo, both for complex and for simple (e.g. ferredoxins) iron–sulfur proteins (cf. ref. 156]). Rhodanese has been suggested to play a role [157,158], but convincing evidence is lacking [159]. Iron–sulfur clusters of ferredoxins can be formed by mixing apo-protein with relatively high concentrations of sulfide, a thiol and ferrous iron under anaerobic conditions. For a plant ferredoxin, ATP seems required for formation of the holo-protein [160]. Evidently much more has to be learned about iron and sulfide metabolism and storage before biogenesis of iron–sulfur clusters can be addressed in detail. Experiments in that direction have been performed with rats where iron-starvation resulted in a decrease in the number of SQR molecules formed rather than an accumulation of unassembled subunits or inactive enzyme [161]. The data may suggest that expression of SQR polypeptides is regulated directly or indirectly by iron availability, but could also be explained by rapid degradation of defective enzyme. Interestingly, only low levels of SQR and aconitase (also an iron–sulfur protein) were found in mitochondria of skeletal muscle from a young Swedish man with a muscle oxidative defect [162]. These mitochondria contained iron deposits which may indicate a defect in iron metabolism affecting only Krebs cycle iron–sulfur containing enzymes, i.e. the NADH dehydrogenase complex was found at normal levels. Effects of iron-limitation on *E. coli frd* gene transcription have been analyzed and found to occur only at extreme (non-physiological) iron-deficiency [136].

4. *Structure and function of succinate:quinone oxidoreductases*

4.1. *Present stage of knowledge*

Structural information on succinate:quinone oxidoreductases is essential for a detailed understanding of catalysis and intra-molecular electron transfer by these enzymes. Direct structural data by X-ray crystallography is lacking, primarily because of the general difficulty to obtain diffracting crystals of membrane proteins. The soluble AB subcomplex of these enzymes is a more promising candidate for direct structural analysis. Significant structural and functional information on the redox components and the substrate binding sites is however available to a large extent as the result of work involving molecular genetic and spectroscopic techniques. Next we discuss these advances, beginning with subunit A, continuing with subunit B and ending with the anchor/cytochrome *b* part.

4.2. *Dicarboxylate active site*

4.2.1. *Substrate specificity*

Succinate, both D- and L-malate, and L-chlorosuccinate, in the form of di-anions are substrates for SQR, whereas D-chlorosuccinate, malonate, and oxaloacetate are not substrates [163]. The two latter dicarboxylates are instead potent competitive enzyme inhibitors. Oxidation of succinate to yield fumarate is a *trans*-dehydrogenation which explains why only the L-steroisomer of chlorosuccinate functions as substrate. Malate is superficially a poor substrate because malate is oxidized to enol-oxaloacetate at the active site and this product causes (suicide) inhibition of enzyme activity [164]. Most preparations of SQR contain, when isolated, tightly bound oxaloacetate, which blocks enzyme activity and may complicate enzyme studies. The bound oxaloacetate can be removed by e.g. reduction of the enzyme or treatment with inorganic anions at elevated temperatures [1,2].

Detailed enzyme kinetic parameters have recently been determined for the mammalian SQR [165,166] (reviewed in ref. [2]) and *E. coli* QFR [167]. Interestingly, both these enzymes bind substrate with different affinity depending on whether the enzyme is oxidized or reduced. For example, K_d for succinate in oxidized beef heart SQR is about 10 μM whereas in the reduced enzyme it is 160 μM. In contrast, K_d for fumarate is higher (fourfold) in the oxidized enzyme compared to the reduced enzyme. These different affinities seem logical from the point of view of enzyme function and indirectly indicate that the structure of the dicarboxylate-binding site is different depending on the redox state of the enzyme (FAD, iron–sulfur clusters). Malonate bound to the active site is suggested to mimic the succinate/fumarate transition state; malonate binding is not significantly affected by the redox state of the enzyme [2]. The covalently bound FAD is functionally and presumably also structurally intimately connected to the dicarboxylate active site. Interestingly, cyclic voltammetry data with the mammalian SDH [11] shows that the enzyme is capable of fumarate reduction only in a narrow window of applied potential (maximal activity at −85 mV). An interpretation of this gating effect is that substrate entry or product release can only occur when the active site is oxidized (FAD in oxidized form). Mutants that lack flavin but contain iron–sulfur clusters with normal physico-chemical properties have been isolated and are enzymatically inactive [71,168]. It is not known whether this type of mutant has an intact dicarboxylate binding site.

4.2.2. Residues at the dicarboxylate substrate binding site

In Fig. 4A are shown, schematically, regions in the primary structure of subunit A of succinate:quinone oxidoreductases which appear important for dicarboxylate substrate binding or catalysis. The figure is based upon the combined results obtained so far by chemical modification experiments, amino acid sequence comparisons (Fig. 4B), and both 'random' and site-directed mutagenesis analysis (Table III).

Bovine heart SDH is known for many years to be very sensitive to substances that react with thiols, e.g. the enzyme is inactivated by *N*-ethylmaleimide and *p*-chloromercuribenzoate [1,171,172]. Substrate or substrate analogs, such as malonate, confer protection against inactivation by thiol modifying reagents. *E. coli* SQR [74], *E. coli* QFR [170,173] and *Wolinella succinogenes* QFR [174] share this property with the mammalian SDH. The results are explained by the presence of one cysteine residue localized at, or close to, the dicarboxylate active site. Analogous chemical modification studies have helped to identify substrate protectable histidine residues (diethylpyrocarbonate treatment and photooxidation of beef heart SDH [175,176]; photooxidation of *E. coli* QFR [170]) and arginine residues (diacetyl and phenylglyoxal treatment of beef heart SQR [177]; diacetyl treatment of *E. coli* QFR [170]; diacetyl treatment of *B. subtilis* SQR [74]).

A comparison of the deduced amino acid sequence for subunit A from six succinate:quinone oxidoreductases indicates ten apparently conserved segments (Fig. 4; segments a to j). The comparison reveals four positions with a conserved histidine residue (the His-FAD excluded) and twelve positions with a conserved arginine residue, but not any cysteine residue which is fully conserved. At one position (in segment g) there is a cysteine residue in five of the six sequences. The exception is the *B. subtilis* SdhA polypeptide which contains an alanine residue at this position (residue 252). *B. subtilis* SQR is not sensitive to thiol modifying reagents [74]. Site-directed mutant *B. subtilis* enzyme (SdhA/A252C) with a cysteine instead of the alanine

A

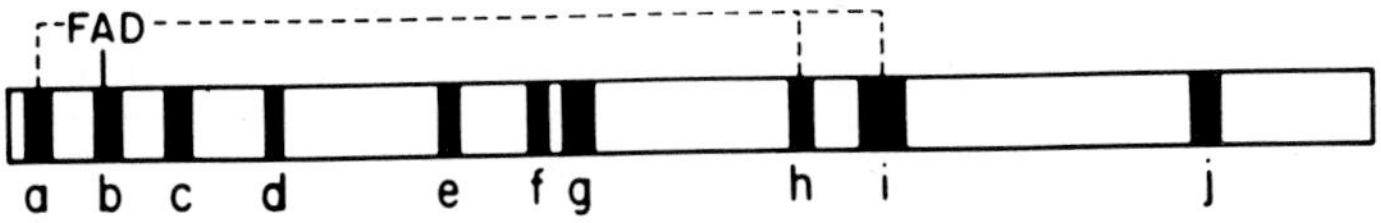

B

```
                              B.s.SdhA47
                    E.c.FrdA44
E.c.    FrdA  ALISKVYPMR SHTVAAEGG. .SAAVA...Q DHDSFEYHFH DTVAGGDWLC EQDVVDYFVH  87
P.v.    FrdA  ALISKVYPMR SHTVAAEGG. .SAAVT...Q AHDSYDFHFN DTVSGGDWLC EQDVVDYFVE  87
E.c.    SdhA  ALLSKVFPTR SHTVSAQGGI TVALGN...T HEDNWEWHMY DTVKGSDYIG DQDAIEYMCK  89
Bovine  SdhA  ACVTLLFPTR SHTVAAQGGI NAALGN...M EEDNWRWHFY DTVKGSDWLG DQDAIHYMTE 145
W.s.    FrdA  IVLSLIPVKR SHSAAAQGGM QASLGNSKMS DGDNEDLHFM DTVKGSDWGC DQKVARMFVN  91
B.s.    SdhA  KLFSIVPVKR SHSVCAQGGI NGAVNT..KG EGDSPWEHFD DTVYGGDFLA NQPPLKAMCE  86
A.s.    SdhA             SHTLAAQGGI NAA
   Consensus  ---------R SH---A-GG- ---------- --D----H-- DTV-G-D--- -Q--------
                            b                          c
```

```
                    B.s.SdhA116
E.c.    FrdA  ..GSVNVRRF GGMKIE.... ....RTWFAA DKTGFHMLHT LFQTSLQF.. .PQIQRFDEH 155
P.v.    FrdA  ..GSVNVRRF GGMKIE.... ....RTWFAA DKTGFHMLHT LFQTSLKY.. .PQIQRFDEH 155
E.c.    SdhA  ..GRIYQRPF GGQSKNFG.G EQAARTAAAA DRTGHALLHT LYQQNLKN.. ..HTTIFSEW 163
Bovine  SdhA  ..GKIYQRAF GGQSLKFGKG GQAHRCCCVA DRTGHSLLHT LYGRSLRY.. ..DTSYFVEY 220
W.s.    FrdA  RHGLIHSRDF GGTKK..... ...WRTCYTA DATGHTMLFA VANECLKL.. .GVS..IQDR 178
B.s.    SdhA  ..GLLDFRRF GGT....... .QHHRTAYAG ATTGQQLLYA LDEQVRRYEV AGLVTKYEGW 157
   Consensus  -G----R-F GG-------- ----R----- --TG---L-- ---------- ----------
                 d
```

```
                                E.c.FrdA247
                  E.c.FrdA232   B.s.SdhA252
E.c.    FrdA  RDMEFVQYHP TGLPGSGI.. LMTEGCRGEG G.ILVNKNGY RYLQDYGMGP ETPLGEPKNK YMELGPRDKV SQAFWHEWRK 300
P.v.    FrdA  RDMEFVQYHP TGLPGSGI.. LMTEGCRGEG G.ILVNKDGY RYLQDYGLGP ETPLGKPENK YMELGPRDKV SQAFWHEWRA 300
E.c.    SdhA  QDMEMWQFHP TGIAGAGV.. LVTEGCRGEG G.YLLNKHGE RFMERYAPNA K......... ..DLAGRDVV ARSIMIEIRE 298
Bovine  SdhA  QDLEFVQFHP TGITGAGC.. LITEGCRGEG G.ILINSQGE RFMERYAPVA K......... ..DLASRDVV SRSMTLEIRE 354
W.s.    FrdA  GNMEAVQFHP TPLFPSGI.. LLTEGCRGDG G.ILRRCGWT PIHADYEPEK K......... ..ELASRDVV SRRMIEHIRK 314
B.s.    SdhA  ANGEFIQIHP TAIPGDDKLR LMSESARGEG GRVWTYKDGK PWY....... ..FLEEKYPA YGNLVPRDIA TREIF.DVCV 296
   Consensus  ---E--Q-HP T--------- L--E--RGEG G--------- ---------- ---------- ---L--RD-- ----------
                     f             g
```

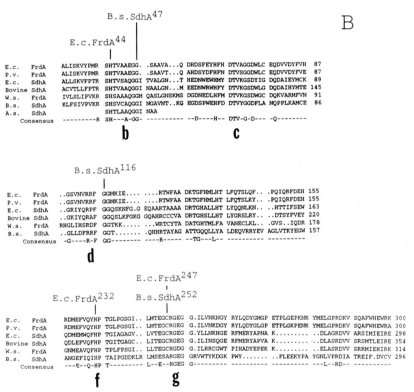

Fig. 4. (A) Generalized primary structure of subunit A of succinate:quinone oxidoreductases. Dotted lines to FAD indicate segments probably making contact with flavin in the folded protein. Segments with extensive sequence conservation are indicated a to j, and correspond in *E. coli* FrdA to segments around residues 15 (a), 45 (b), 70 (c), 115 (d), 195 (e), 230 (f), 250 (g), 355 (h), 380 (i) and 530 (j). FAD is covalently bound to the conserved histidine residue in segment b. (B) Amino acid sequence comparisons of segments in subunit A that have been studied by mutant analysis in *E. coli* QFR (*E.c.* FrdA) or *B. subtilis* SQR (*B.s.* SdhA) (see Table III); superscripts give amino acid residue number. Numbers at the end of the sequence refer to the number of the last residue shown. The sequence data are from: *E. coli* FrdA, ref. [56]; *P. vulgaris* FrdA, ref. [101]; *E. coli* SdhA, ref. [64]; bovine heart SdhA, ref. [100]; *W. succinogenes* FrdA, ref. [83]; *B. subtilis* SdhA, ref. [80]; adult *A. suum* muscle SdhA, ref. [182]. The alignments were created using the program Pileup (UWGCG Version 7.0) and visual inspection.

TABLE III

Properties of succinate:quinone oxidoreductases mutated in subunit A

Protein	Mutation	Enzyme properties	Ref.
E. coli FrdA	H44R	Inactive, contains non-covalently bound FAD	[73]
	H44S	Active in fumarate reduction but inactive in succinate oxidation. Contains non-covalently bound FAD.	[73]
	H44Y	Same as for H44S	[73]
	H44C	Same as for H44S.	[73]
B. subtilis SdhA	G47D	Inactive in succinate oxidation, lacks FAD but has all 3 Fe–S clusters with normal properties.	[71]
	G116E	Inactive in succinate oxidation, has covalently bound FAD and all 3 Fe–S clusters with normal properties.	[71,169]
E. coli FrdA	H232S	Catalytically defective; inactive in succinate oxidation and low fumarate reductase activity.	[170]
	C247S	Active. Not sensitive to thiol modifying reagent.	[170]
	C247A	Same as for C247S.	[170]
	R248H	Catalytically defective; <2% of wild-type succinate oxidation and fumarate reduction activities.	[170]
	R248L	Same as for R248H.	[170]
B. subtilis SdhA	S251G	Active in succinate oxidation, but altered apparent K_d for malonate.	[74][a]
	A252C	Same as S251G, but also sensitive to thiol modifying reagent.	[74]
	S251G and A252C	Same as for A252C.	[74]

[a] B.A.C. Ackrell and L. Hederstedt, unpublished data.

residue is enzymatically active, sensitive to thiol modifying reagents and can be protected by substrate against inactivation of enzyme activity by such reagents (Table III and ref. [74]). These experimental data show that the cysteine residue in segment g (when present) is the reactive thiol which when modified blocks enzyme function, probably by excluding dicarboxylate binding. This conclusion is corroborated by *E. coli* QFR mutant studies demonstrating that replacement of the conserved cysteine residue in FrdA by an alanine is accompanied with decreased sensitivity to thiols (Table III) [170].

Before any extensive subunit A amino acid sequence information was available Kotlyar and Vinogradov postulated that the high reactivity of the thiol in mammalian SDH is an effect of a vicinal arginine residue [177]. Next to the generally conserved cysteine in the sequence of subunit A we find a conserved arginine residue (Fig. 4B). Replacement of this arginine with histidine or leucine in *E. coli* QFR (Table III) results in a mutant enzyme with less than 2% of wild-type activity [170]. One interpretation is that the guanidium group of this arginine forms a bidentate ionic pair with one of the carboxyl groups of the dicarboxylate (succinate/fumarate) and thereby orients the substrate at the active site. Diacetyl modification data with wild-type and mutant *E. coli* QFR indicates that another arginine is also functionally important, maybe by forming an ionic pair with the other carboxyl of the substrate [170]. Among the other eleven conserved arginine residues in subunit A this other arginine may be that of segment d. Very speculatively, residues in this segment could be close to the active site, since *B. subtilis* SQR with a glutamate at position 116 instead of the conserved glycine residue in segment d (Fig. 4B) is enzymatically inactive but contains a full set of prosthetic groups with seemingly normal properties [71,169] (Table III). It should be noted that residues essential for enzyme activity can also be situated in subunit B.

Close to the discussed active-site region (segment g) in the primary sequence of subunit A there is a conserved triad His–Pro–Thr (segment f), that is present in other oxidoreductases [178]. Mutant *E. coli* QFR with a Ser instead of His in this triad cannot oxidize succinate but most interestingly it has close to normal (70%) fumarate reductase activity [170]. Enzyme with similar properties is obtained by photooxidation using Rose bengal, a procedure that generally causes oxidation of His [170]. The bovine heart mitochondrial SDH is also inactivated by photooxidation with Rose bengal, and protected against inhibition by substrate, but in those experiments fumarate reductase activity was not analyzed [176]. Mammalian SQR contains a catalytically important group with a pKa of around 7.0 [175]. This group is suggested to be an imidazole and possibly is the His in the conserved triad of segment f. The unprotonated imidazole of the histidine may have a proton receptor function at succinate oxidation and the protonated imidazole a proton donor function at fumarate reduction [175]. The drastic effect on succinate oxidation compared to fumarate reduction by replacing the His in the His–Thr–Pro sequence with Ser could indicate that the imidazole is very important for subtracting the proton from succinate, but that Ser or another residue can function as proton donor at fumarate reduction. In that context it would be interesting to know if the pH dependency curve for fumarate reductase activity of wild-type and this mutant QFR are different.

In summary, available data indicate that residues in the conserved segments f, g and possibly d in subunit A contribute to the dicarboxylate binding site. As more data on wild-type and mutant SdhA and FrdA accumulate we can refine and correct the one-dimensional model for subunit A. Hopefully, the accumulated data can ultimately be combined with three-dimensional data.

4.3. Intramolecular electron transfer

4.3.1. Flavin

FAD is covalently bound to a histidine residue in the A polypeptide via a N(3)-8α-riboflavin linkage as demonstrated experimentally for bovine heart SQR [22], *E. coli* QFR [179] and *W. succinogenes* QFR [180]. It is likely that the linkage between FAD and the polypeptide is the same in all succinate:quinone oxidoreductases. The amino acid sequence of a flavin-peptide is known from mammalian [181] and *A. suum* SQR [182] and with the help of these sequences the histidine residue with covalently bound FAD has been identified in the primary structure of subunit A from different organisms (Fig. 4). The histidine is located close to the N-terminal end of the polypeptide in a conserved segment (segment b). This sequence is not conserved in other types of enzymes also containing His-N(3)-8α-FAD [155], which suggests that the conserved sequence of segment b is more related to enzyme function than FAD binding. Mutant *B. subtilis* SQR, containing an aspartate residue instead of a conserved glycine in segment b (SdhA/G46D, see Fig. 4 and Table III) lacks flavin and, as expected, this mutant enzyme is enzymatically inactive. Possibly the aspartyl side-group sterically hinders FAD to be bound. Three other conserved segments in the subunit A polypeptide (Fig. 4A, segments a, h and i) probably contact the AMP part of the FAD molecule in the folded protein (see Fig. 13 of ref. [3]). From comparisons with other FAD containing proteins of known three-dimensional

[1,4,9,206]. Here we shall focus on the general thermodynamic profile of redox centers, and more recent progress on the ligand structure and spin–spin interactions of the active centers.

As can be seen in Table V, all SQRs and QFRs have so far been found to contain the same iron–sulfur cluster composition, but the clusters have different thermodynamic parameters. The SQRs have an apparent midpoint redox potential for center 1 and 3 high enough to be reducible by the substrate, succinate ($E_{m,7}$=+30 mV). The corresponding potential for center 2 is extremely low and this cluster is not reducible by succinate. In QFRs, both centers 1 and 3 have, compared to SQR, much lower midpoint potentials, which are almost equipotential to that of the menaquinone/menaquinol couple. It seems as if the E_m of center 3 in different succinate:quinone oxidoreductases is influenced by the quinone. For example, in bovine heart SQR, the more tightly the $Q_s^{\bullet-}$ pair is bound to the enzyme, the higher is the E_m of center 3. Similarly the E_m value of center 3 is higher in SQR systems in which ubiquinone (E_m=+112 mV [194,261]) rather than menaquinone (E_m=−74 mV [194,261]) functions as electron acceptor in the membrane.

The complete amino acid sequence of subunit B is known for succinate:quinone oxidoreductases of a handfull of organisms, including both eukaryots and prokaryots (Table I). In all cases, except for that of bovine heart SDH [94], the sequence of subunit B has been deduced from the DNA sequence. Comparison of the primary sequences disclose 10 or 11 conserved cysteine residues distributed in three locations in the polypeptide numbered I–III (Fig. 5). Assignment of iron–sulfur ligands based only on the primary sequence is often misleading. However, location I contains four (three in *E. coli* SdhB) conserved cysteine residues arranged as in typical plant-type $[2Fe–2S]^{(2+,1+)}$ ferredoxins, C XXXX C XX C...C, with uncharged small polar residues between the second and the third cysteine in the sequence. Subunit B of *E. coli* SQR has an aspartate (D) residue in place of the third cysteine (Fig. 5). The cysteine residues at locations II and III are arranged as C XX C XX C XXX CP and C XXXXX C XXX CP,

TABLE V

Midpoint potentials (mV) of the iron–sulfur clusters in succinate:quinone oxidoreductases

Enzyme	Quinone (acceptor/ donor)	Center 1	Center 2	Center 3	Ref.
SQR:					
Bovine heart	UQ	0	−260	+60 to +120	[26,27]
Escherichia coli	UQ	+10	−175	+65	[76]
Rhodobacter sphaeroides	UQ	+50	−250	+80	[207]
Micrococcus luteus	MQ	+70	−295	+10	[208]
Bacillus subtilis	MQ	+80	−240	−25	[71]
QFR:					
Escherichia coli	MQ	−20 to −79	−320	−50 to −70	[209,210]
Wolinella succinogenes	MQ	−59	<−250	−24	[188]

Midpoint potentials were determined by EPR-monitored mediator titrations (pH 7) and are expressed relative to the standard hydrogen electrode, except for the case of *W. succinogenes* QFR where the potentials for centers 1 and 3 were obtained by poising the potential using the fumarate/succinate couple. UQ = ubiquinone ($E_{m,7}$ = +112 mV); MQ = menaquinone ($E_{m,7}$ = −74 mV).

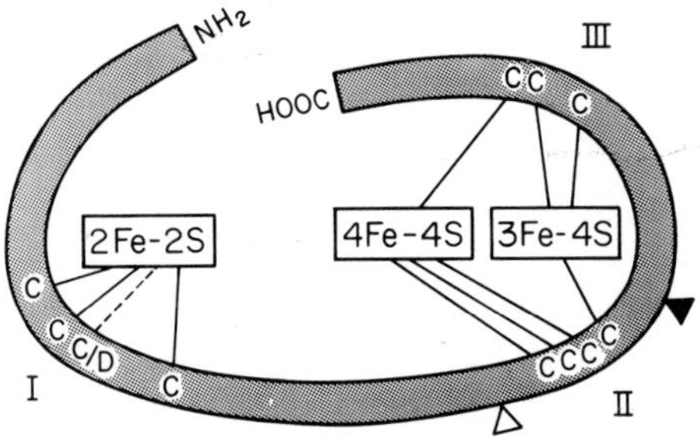

Fig. 5. Current model for ligation of the three iron–sulfur clusters in subunit B of succinate:quinone oxidoreductases. The empty and filled triangles indicate the position of the C-terminus of truncated subunit B of *B. subtilis* SQR [213] and *E. coli* QFR [212] mutants, respectively, discussed in Section 4.3.2.

respectively. These sequences are similar to those for the ligands of $[4Fe–4S]^{(2+,1+)}$ and $[3Fe–4S]^{(1+,0)}$ clusters in bacterial ferredoxins [211].

Subunit B contains all the ligands for the [2Fe–2S] cluster, as demonstrated by studies with *E. coli* FrdB [212], and the cysteine residues at locations II and III are not required for ligation of this cluster, as shown by the properties of a *B. subtilis* mutant with a truncated subunit B (Fig. 5 and ref. [213]). Experimental results with mutants having a truncated subunit B furthermore indicate that the N-terminal part of subunit B forms a separate [2Fe–2S] iron–sulfur cluster domain which can interact with subunit A [212,213]. In order to more directly identify ligand residues to the [2Fe–2S] cluster, Werth et al. [210] have individually replaced each of the cysteine residues in location I of *E. coli* FrdB by serine. Enzyme activities, spectroscopic properties and electrochemical properties of each mutant enzyme were analyzed. The four mutant enzymes were found to be active and to contain all iron–sulfur clusters. The mutations only affected the properties of center 1 which further supports the notion that the N-terminal arrangement of the four cysteine residues ligate the [2Fe–2S] cluster. As summarized in Fig. 6 the C57S, C62S and C77S mutations in FrdB caused a decrease in the E_m value of center 1 of about 100, 250 and 30 mV, respectively, and a decrease in QFR activity of about 90, 90 and 70%, respectively. The C65S mutation, by contrast, caused a 30 mV increase in the E_m of center 1 and a slightly increased QFR activity compared to wild type. The C57S and C62S mutations gave rise to considerable changes in the center 1 EPR line shape (to rhombic spectra) while the spectra of the remaining two mutant proteins showed axial spectra as in the wild-type enzyme. The C77S has only a minor effect on the spectrum of center 1 but conferred instability to the QFR. Werth et al. concluded from their study that the cysteine at position 65 is not essential for enzyme function even if it may ligate iron in the wild-type enzyme and proposed an alternative ligation scheme for the [2Fe–2S] cluster in subunit B; direct coordination of the cluster to three cysteines and perhaps one water molecule while the fourth ligand is hydrogen bonded to a cysteine or aspartate residue (in the case of *E. coli* SQR). Experiments involving the ENDOR (Electron-Nuclear Double Resonance) technique are in progress to test this proposed ligation scheme for center 1 (personal communication from G. Cecchini). This hypothetical ligand structure is very attractive since it can explain the much higher E_m value of center 1 in SQR and QFR relative to that of the same type of cluster in ferredoxins. The [2Fe–2S] iron–sulfur cluster in the cytochrome bc_1 complex, generally called the Rieske iron–sulfur cluster, has a midpoint redox potential which is higher than that of most ferredoxins. It also exhibits an extremely anisotropic EPR spectrum, which is characterized by a lower g_{av} (1.91) than that of the ferre-

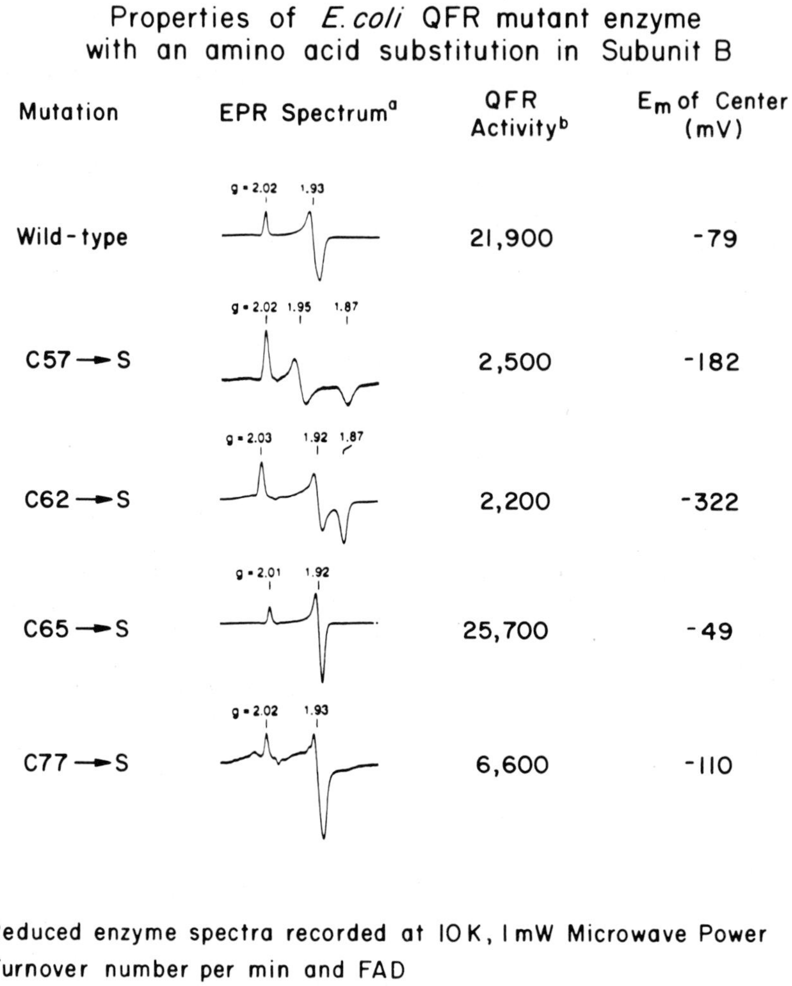
Properties of E. coli QFR mutant enzyme
with an amino acid substitution in Subunit B
Mutation
EPR Spectrum[a]
QFR Activity[b]
Em of Center (mV)
g = 2.02 1.93
Wild-type
21,900
-79
g = 2.02 1.95 1.87
C57 → S
2,500
-182
g = 2.03 1.92 1.87
C62 → S
2,200
-322
g = 2.01 1.92
C65 → S
25,700
-49
g = 2.02 1.93
C77 → S
6,600
-110
[a]Reduced enzyme spectra recorded at 10 K, 1 mW Microwave Power
[b]Turnover number per min and FAD

Fig. 6. Effects of mutagenesis of center 1 candidate ligands in subunit B of *E. coli* QFR. The X-band EPR spectra of dithionite reduced cytoplasmic membrane preparations with amplified amounts of QFR were obtained under the following conditions: temperature, 10 K; microwave power, 1 mW. Data are reproduced with permission from ref. [210].

doxin-type [2Fe–2S] clusters (g_{av} about 1.96). Spectroscopically similar clusters are also found in oxygenases of certain bacteria [214]. Based on detailed EPR spectral analysis, Blumberg and Peisach [215] suggested that this type of cluster may have ligands which are less electron donating than sulfur. Subsequently, various spectroscopic analyses indicated involvement of at least one nitrogen ligand [216]. Recently, analysis of ^{15}N- and ^{14}N-ENDOR spectra of the iron–sulfur cluster in the phthalate dioxygenase of *Pseudomonas cepacia* [217] and in the cytochrome bc_1 complex of *Rhodobacter capsulatus* [218] have established that both 'Rieske-type' and the Rieske iron–sulfur clusters are liganded by two nitrogen and two cysteine residues. From the

E. coli FrdB mutant analysis it was proposed [210] that the cysteines at position 57 and 62 in FrdB are ligands to the valence localized Fe(II), because substitutions at these positions induced the strongest spectral changes (Fig. 6). This proposal is consistent with recent ligand assignments by NMR (Nuclear Magnetic Resonance) analysis of plant-type [2Fe–2S] ferredoxins, where the two first cysteine residues in the sequence ligate the valence localized Fe(II) [219]. Regarding the FrdB center 1 study of Werth et al. [210] the following possibility can as yet not be completely excluded. In the mutants C57S, C62S or C77S, the non-ligand cysteine at position 65 may replace the substituted third ligand with resulting rearrangements of the protein backbone. Such a phenomenon has been observed with the [4Fe–4S] cluster of *Azotobacter vinelandii* ferredoxin I, where substitution of the liganding cysteine at position 20 to alanine did not eliminate the cluster. X-ray crystallographic analysis of the mutant revealed that a nearby cysteine (position 24) had been recruited as a substituting ligand [220].

Cecchini's and Johnson's groups in collaboration have performed interesting site-specific mutation studies also on the cysteine residues at location III [221], close to the C-terminal of *E. coli* FrdB (Fig. 5). The three conserved cysteines at this location (positions 204, 210 and 214) were individually changed to serine. All the resulting mutant enzymes turned out to be deficient in QFR activity and membrane binding and to have lost centers 2 and 3, but not center 1. Thus, assembly of the [3Fe–4S] cluster (and perhaps also the [4Fe–4S] cluster) seems to be necessary for membrane attachment of both subunit A and B, and for sustaining electron transfer between the enzyme and quinone. These results are consistent with the well-known importance of center 3 for functional membrane binding of mammalian SDH [1]. A replacement of the valine at position 207 in *E. coli* FrdB by cysteine turned out to provide more conclusive results. This V207C mutant enzyme was found to assemble in the membrane as a functional QFR and interestingly contains *two* [4Fe–4S] clusters, i.e. the trinuclear cluster was transformed into a tetranuclear cluster as a result of the mutation. This is not totally unexpected since a 'consensus' arrangement of cysteinyl residues for tetranuclear clusters was created in the mutant. A similar finding, but in the opposite direction, was recently made by Rothery and Weiner in experiments with *E. coli* dimethylsulfoxide reductase; a [4Fe–4S] cluster was converted to a [3Fe–4S] cluster by replacing cysteine residue 102 in DmsB to serine or an aromatic amino acid residue [222].

EPR spectra of dithionite reduced wild-type and FrdB/V207C mutant *E. coli* QFR are shown in Fig. 7. The wild-type spectrum (top) represents 1.8 spins per FAD while in the mutant spectrum (bottom) there is 2.7 per FAD. In the wild type, the reduced center 3 (S=2 state) is EPR silent while in the mutant, the tetranuclear cluster ($S=\frac{1}{2}$ state) converted from center 3 is EPR detectable in addition to center 1 and 2 spins. The two $[4Fe–4S]^{1+}$ clusters in the mutant QFR are also magnetically coupled and show EPR signals distinct from that of the spin-coupling between the $[4Fe–4S]^{1+}$ and the $[3Fe–4S]^{0}$ clusters in the wild-type enzyme [221].

Considerable information on the spatial organization of redox active centers in SQR and QFR systems has been obtained from the analysis of their magnetic interactions. The distance between the covalently bound flavin and center 1, and between centers 1 and 2 are estimated as 12–18 Å [185] and 9–12 Å [223], respectively. The latter distance was estimated prior to the final identification of center 2 as a tetranuclear cluster with $g_{x,y,z} \approx 1.85$, 1.93–1.94, 2.06 [29,221,224,225], but seems approximately correct [198]. Spin–spin interaction between centers 1 and 3 ($S=\frac{1}{2}$ state) can be observed in *Micrococcus luteus* [208] and *B. subtilis* [71] SQR, due to the more positive E_m of center 1 compared to center 3, indicating a 10–20 Å distance between the [2Fe–2S] and the [3Fe–4S] clusters. This is further confirmed by the slower spin relaxation of magnetically isolated center 1 in the *E. coli* QFR FrdB/V207C mutant [221]. The extremely broad wing EPR signals of dithionite reduced *E. coli* QFR at g=2.17 and 1.65 (Fig. 7, top

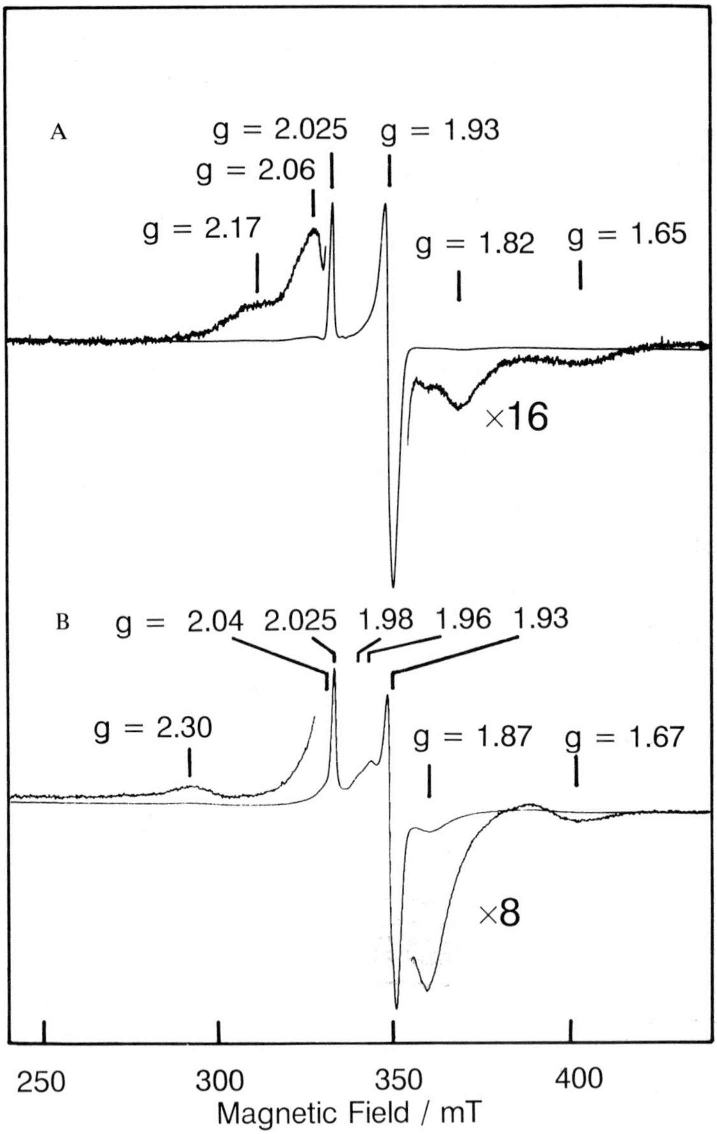

Fig. 7. EPR spectra of dithionite reduced cytoplasmic membrane preparations of *E. coli* containing amplified amounts of (A) wild type or (B) mutant (FrdB/V207C) QFR. EPR conditions were as in Fig. 6. Reproduced with permission from ref. [221].

spectrum) are not observable with the FRD lacking active center 3 [4,221]. This suggests strong spin-coupling between center 2 and 3, indicating an even closer mutual distance than between centers 1 and 3. The very rapid spin relaxation of the split signal of the $Q^{\bullet -}_s$ pair and their non-split g=2 EPR signal (see Section 4.4) provides evidence for the proximity of $Q^{\bullet -}_s$ and center

3 or/and cytochrome *b*. Thus, each redox active center in succinate:quinone oxidoreductase interacts with at least one neighboring redox component. In this connection, the recent studies with the *E. coli* FrdB V207C mutant (which contains two tetranuclear clusters) provided novel information for a clearer understanding of EPR spectra and inter-cluster spin–spin interactions in these rather complex systems [221].

4.3.3. *Cytochrome b/anchor polypeptide(s)*

Succinate:quinone oxidoreductase of most organisms contains an endogenous cytochrome *b* (Table I). This cytochrome has received relatively little attention in the past, as discussed before [226]. Molecular properties of heme(s) in five relatively well characterized enzymes are compared in Table VI. SQR of bovine heart mitochondria, *E. coli* and *Ascaris suum* adult worm muscle has one heme group, whereas *B. subtilis* SQR and *W. succinogenes* QFR contain two hemes. Each cytochrome *b* component in the di-heme cytochromes exhibits unique thermodynamic and spectroscopic properties. One has a high midpoint redox potential, is reducible by succinate and in the reduced state it shows (in most cases) a symmetrical α-band peak in the light-absorption spectrum. The second cytochrome *b* component has a low redox potential, it can be reduced only by strong reductants like dithionite, and it shows an asymmetrical (split) α-band absorption peak in the reduced state. The EPR spectra of the two low-spin heme-irons are also distinct, at least in the case of *B. subtilis* SQR (Table VI). The apparent midpoint redox potential of the single heme-iron in bovine heart and *E. coli* SQR is −180 mV [228,239] and +36 mV [70], respectively. These values explain why succinate can reduce cytochrome *b* only in the *E. coli* enzyme. From the data of Table VI it seems as if the properties of the cytochrome *b* in *E. coli* SQR and in bovine SQR are similar to those of the high-potential and low-potential cytochrome *b* component, respectively, in the di-heme cytochromes. Cytochrome *b* of *A. suum* QFR (and also of *Paragonimus westermani* SQR [97]) seems to have intermediate properties, i.e. it shows an apparent redox potential of −34 mV [232], that may account for the only partial reduction by succinate, and shows a split α-band absorption peak in the reduced state otherwise characteristic for the very low-potential cytochrome *b*s.

CD spectra in the Soret region of di-heme cytochrome *b* seemingly depend on exciton coupling between the hemes and interaction of heme with aromatic residues [240]. The di-heme cytochrome *b* of *W. succinogenes* QFR [241] and *B. subtilis* SQR (M. Degli Esposti and L. Hederstedt, unpublished experiments) both show CD spectra of a type closely resembling that of the di-heme cytochrome *b* of the cytochrome bc_1 complex [240]. Also the EPR spectrum of oxidized cytochrome *b* in *B. subtilis* SQR (g_{max} at 3.42 and 3.68) (C. Hägerhäll, R. Aasa, C. von Wachenfeldt and L. Hederstedt, unpublished data) is similar to that of bovine cytochrome bc_1 complex (g_{max} at 3.45 and 3.78) [242]. It is well documented (Chapter 9 of this volume) that the cytochrome *b* of bc_1 complexes has bis-histidine axial ligation to both hemes and spectroscopic data strongly indicate that the planes of the two ligating imidazole rings at the respective heme-irons are near perpendicular to each other [243]. The four histidine residues serving as axial heme-iron ligands in cytochrome *b* of bc_1 complexes are located pairwise in two consecutive transmembrane segments (for a recent review see ref. [244]). Bis-histidine ligation of heme-iron in *B. subtilis* SQR is strongly suggested from the EPR spectral data combined with near-infrared-MCD spectroscopic data. The NIR-MCD spectrum shows an intense peak (ΔE=380 $M^{-1}cm^{-1}$ at 4.2 K and 5 T) at about 1600 nm [245].

Amino acid sequence comparisons of cytochrome *b* of succinate:quinone oxidoreductases to the cytochrome *b* of bc_1 complexes show only insignificant similarities [79,84]. The anchor polypeptides, in contrast to the subunit A and B polypeptides of different SQR and QFR enzymes generally show little sequence conservation except for the pattern of hydrophobic

of SdhD. If the latter alternative proves correct, it would indicate that the analogous His residues in *B. subtilis* SQR, His^{70} and His^{155} (Fig. 8), ligate the high-potential heme whereas His^{28} and His^{113} ligate the low-potential heme. Such an arrangement is coherent with current structural models for cytochrome bc_1 complexes in which the high-potential heme is closest to the cytoplasmic/matrix side (negative side) of the membrane [246].

The single polypeptide cytochrome *b* of *W. succinogenes* QFR and *B. subtilis* SQR can be regarded as the result of a protein-fusion between the very C-terminal part of *E. coli* SdhC and the hydrophilic loop connecting transmembrane segments I and II in SdhD (Fig. 8). The function of the transmembrane segment I of the SdhD polypeptide may only be to act as signal peptide for insertion of the polypeptide with the correct orientation in the *E. coli* cytoplasmic membrane. This hypothesis can be tested experimentally by analyzing if the cytochrome *b* of *E. coli* SQR is still formed after *sdhC* and a partially deleted *sdhD* gene (lacking the sequence encoding the first 40–50 residues) have been fused in frame.

The structural interaction between the anchor polypeptide(s) and SDH/FRD in succinate:quinone oxidoreductases is poorly understood. Chemical modification studies have indicated histidine [247], amino groups [248,249] and carboxyl groups [250] as essential for the binding of mammalian SDH to its membrane anchor. To obtain a detailed picture of how SDH is bound to the anchor polypeptide(s) it is necessary to identify these essential residues in the primary structure of the polypeptides. Using a genetic approach with *B. subtilis* SQR, a region in the cytochrome *b* (SdhC) possibly making contact with the A or B subunit has been identified, i.e. replacement of Gly^{168}, which is located in the loop connecting transmembrane segments IV and V (Fig. 8), with Asp results in a cytochrome with normal spectroscopic properties in the membrane but unable to bind subunits A and B [42]. Recent *E. coli* QFR mutant studies have provided information on the role of different segments of the anchor polypeptides in the binding of FRD [43,145]. Severely C-terminally truncated FrdC (lacking about 60 of the total 130 residues) and FrdD (lacking about 80 of the total 118 residues) were found still to bind FrdA and FrdB, albeit less efficiently than wild-type polypeptides, but they were not functional in physiological electron transfer. An intact [3Fe–4S] cluster in SDH/FRD is required for functional binding to the membrane anchor [26]. In *B. subtilis* SQR heme is essential for assembly of the enzyme [41,115]. Evidence for close interactions between heme and SDH subunits is indicated by the observation that the cytochrome *b* of mammalian SQR shows different light-absorption and EPR spectra depending on whether SDH is bound or not [228]. Minor spectral differences between cytochrome *b* in SQR and isolated cytochrome have also been found for *B. subtilis* SQR (C. Hägerhäll, R. Aasa, C. von Wachenfeldt and L. Hederstedt, unpublished data).

4.4. Quinone active site

As shown in early EPR studies, a thermodynamically highly stable ubiquinone pair ($Q_s^{\bullet-}$ $Q_s^{\bullet-}$ can be detected in situ in the SQR segment of the mitochondrial respiratory chain [195,229]. This pair may function as the two-electron gating system between SDH iron–sulfur clusters and the quinone pool, analogous to the $Q_A^{\bullet-}$ $Q_B^{\bullet-}$ pair in the bacterial photosynthetic reaction center [196]. The $Q_s^{\bullet-}$ species was first detected as spin-coupled split EPR signals with peaks at g=2.05, 1.99 and 1.97 (at temperatures below 15 K) partially overlapping with the center 3 EPR signals around g=2.01 [229]. Ubisemiquinone was shown to be one of the interacting species by Q extraction and replenishment experiments. Ruzicka et al. conducted excellent computer simulation studies on the split signals obtained with isolated SQR trapped kinetically at an intermediate redox state [229]. The best fit between observed and simulated data was obtained with a semiquinone (free radical) interaction overlapping with center 3 signals. Subsequently, Salerno and coworkers performed a simulation analysis of the redox titrations on the $Q_s^{\bullet-}$ rapidly relaxing g=2.00 EPR signal and of the low-temperature $Q_s^{\bullet-}$ $Q_s^{\bullet-}$ interaction EPR signals, as well as studies on the spin relaxation properties and thenoyltrifluoroacetone* sensitivity of these signals [195,251]. Their results suggest that both EPR signals arise from the same pool of very stable ubisemiquinone which is magnetically coupled with the nearby paramagnetic center 3 (or/and cytochrome *b*). These EPR features constitute sensitive probes for the study of structure alterations of the polypeptide environment around the stable $Q_s^{\bullet-}$ binding site. Quantitative analysis of the spin-coupled signals (at pH 7.4) gave E_{m1} (oxidized to semiquinone state) and E_{m2} (semiquinone to fully reduced state) values of +140 and +80 mV, respectively, which corresponds to K=10 (see Table IV) and strongly indicates the existence of binding sites for $Q_s^{\bullet-}$ $Q_s^{\bullet-}$ within the SQR segment of the mitochondrial respiratory chain. The $Q_s^{\bullet-}$ pair is highly ordered in oriented multilayers of bovine heart mitochondria with the quinone–quinone vector perpendicular to the plane of the inner membrane [252]. The spin-coupling characteristics of the $Q_s^{\bullet-}$ pair are retained in isolated succinate:cytochrome *c* reductase complex [193]. In preparations of isolated SQR, however, split quinone EPR signals are obtainable only as kinetically transient species by rapid freezing in the reoxidation process of reduced SQR [229]. Very recently, Yu's group [253] has isolated a type of bovine heart SQR preparation in which the spin-coupled split EPR signals can be obtained by poising the redox state after the addition of Q_2 to the enzyme preparation (see Fig. 9). This system makes it possible to analyze the Q_s binding sites using biophysical techniques, such as ^{1}H-ENDOR. Since the split EPR signals are caused by the interaction of two paramagnetic centers they are much more sensitive to perturbations of the environment than signals arising from a single group. A small shift (e.g. 40 mV) of the E_m of one quinone relative to the other would cause almost complete loss of the split signals. In addition, a slight inhomogeneity in the positions of the dipolar coupled quinones would cause lineshape changes. So far, no clear four-line split EPR signals have been observed in the bacterial SQR or QFR systems.

The cytochrome *b* content of different types of mammalian SQR preparations varies (Table VI, ref. [1]) and the spin-coupled split quinone EPR signal is not detectable in the preparations with a low cytochrome *b* content [254]. The g=2.00 signal from the $Q_s^{\bullet-}$ species was shown to be further enhanced by the presence of cytochrome *b*, indicating that spin-coupled $Q_s^{\bullet-}$ is located close to both center 3 and cytochrome *b* [254]. Carboxins block electron transfer between iron–sulfur clusters and quinone in mitochondrial SQR (for review see ref. [1]). A mutant *U. maydis* SQR resistant to carboxin has been found mutated in subunit B [107].

The *E. coli* QFR, in contrast to the SQR, contains no heme (Table I). Similar to the SdhC and SdhD polypeptides the FrdC and FrdD polypeptides of *E. coli* QFR each has three transmembrane segments. The importance of Glu^{29} and His^{82} in FrdC (both residues are predicted to be located on the cytoplasmic side of the membrane) for quinol oxidation and quinone reduction are indicated from mutant studies [72,145]. These residues may be involved in protonation and deprotonation reactions as in the case of Glu^{212} at the Q_B site in the photosynthetic reaction center [255]. The positive charge on His^{82} in FrdC may be necessary for stabilization of the anionic $QH^{\bullet-}$ or $Q^{\bullet-}$ species which is expected to be produced during the quinone catalytic cycle. It is expected that bacterial SQR and QFR will be useful systems to investigate

* 2-Thenoyltrifluoroacetone (TTFA) and also carboxin inhibit electron transfer between center 3 and the Q pool in mammalian mitochondrial SQR (for review see ref. [1]).

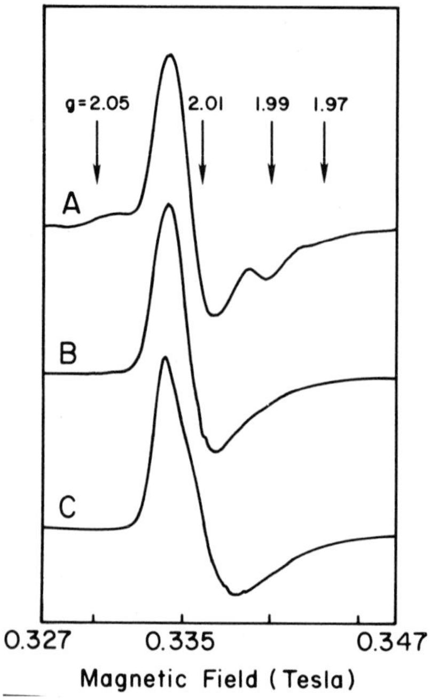

Fig. 9. EPR spectra of the spin-coupled split signals in bovine heart SQR poised with the fumarate/succinate couple (50/1). (A) Q_2 added. (B) without Q_2. (C) Q_2 and 2-thenoyltrifluoroacetone added. EPR conditions: temperature, 10 K; microwave power, 6.3 mW. Reproduced with permission from ref. [253].

the structure–function relationship in the molecular architecture of the '$Q_s^{\bullet-}$' site, which converts $n=1 \leftrightarrow n=2$ electron transfer steps.

5. *Which are the directions of future research?*

Many unexpected and exciting findings of general interest have emerged from studies on succinate:quinone oxidoreductase from different organisms. Looking back, it seems as whenever a problem concerning these enzymes was considered as solved a jack in the box has appeared with new questions concerning these rather complex enzymes catalyzing principally simple reactions. Continued investigations of the structure of SQR and QFR enzymes will hopefully provide detailed three-dimensional structural information, which is required for understanding the mechanisms of catalysis at the dicarboxylate and the quinone active sites as well as intra-molecular electron transfer (e.g. what is the role of cytochrome *b*, when present?). To date, two distinct models have been proposed in the literature to explain the presence in succinate:quinone oxidoreductases of an iron–sulfur cluster (center 2) with a midpoint potential approximately 250 mV lower than both the succinate/fumarate and the Q/QH_2 couples. As mentioned earlier in this chapter (Section 4.3.1) Cammack et al. have proposed a dual pathway model for electron transfer through cardiac SQR [199,200,239]. It was assumed that the low-potential center 2 and cytochrome *b* mediate electron transfer between $Q/QH^{\bullet}$ and $FADH^{\bullet}/FAD$ and that centers 1 and 3 provide a higher-potential pathway, linking $QH^{\bullet}/QH_2$ and

$FADH_2/FADH^{\bullet}$. In Section 4.3.1 we have discussed that the thermodynamically stable intermediate states observed for the $FAD/FADH_2$ and the Q/QH_2 couples can function as $n=2 \leftrightarrow n=1$ convertors by themselves and it is not necessary, as pointed out before [1], to have dual electron transfer pathways. A linear electron transfer pathway from $FAD/FADH_2$ via centers 1 and 3 to the Q/QH_2 couple has been suggested [1,27,198] and the possibility of anti-cooperative redox interactions between centers 1 and 2 and between centers 2 and 3 have been proposed [1,198]. The midpoint potential of center 2 determined by equilibrium redox titrations, is an apparent potential that may be lower than the intrinsic potential, depending on the interaction energy. Transient electron transfer through an intermediate iron–sulfur cluster, center 2, would facilitate electron transfer between centers 1 and 3. However, results of mutagenesis studies on centers 1 and 3 of *E. coli* QFR [210,221] strongly suggest that the midpoint potential of center 2 is intrinsically lower than −250 mV. Its potential, as measured in equilibrium redox titrations, is not dramatically altered when the potentials of center 1 and 3 are lowered by 250 mV and 280 mV, respectively. An alternative mechanism has therefore been proposed by Manodori et al. [221]; 'Center 2 does not function directly in the electron transfer between the covalently bound FAD and quinone. Center 2 is a vestige of the evolution of iron–sulfur subunits from an 8Fe ferredoxin, which is considered to be the most primitive form of iron–sulfur protein, while it is necessary for structural integrity and membrane binding of the enzyme'. Thus, these authors propose a linear electron transfer pathway through the iron– sulfur clusters of QFR and SQR involving only centers 1 and 3. Although this pathway is optimal, short-cut pathways that bypass either center 1 or center 3 become operative if the potential of either cluster is dramatically lowered.

Besides functional studies it can be anticipated that much more effort will be spent on the aspects of biogenesis (especially in eukaryotic cells), degradation, genetic regulation (differentiation) and clinical medicine. Regarding the last aspect, cases of SQR or SDH deficiency in human skeletal mitochondria have been identified [162,256,258], but the primary cellular defect in any of these individuals has not been elucidated. With genes for eukaryotic SQR subunits available it is possible to analyze whether the defect is a mutation in a structural gene for the enzyme or is a disorder on the level of gene expression or posttranslational processing (mitochondrial import and incorporation of prosthetic groups). The fact that mitochondrial defects can be found confined to specific types of tissues opens up the question of isoenzymes, as for example seemingly exists for the pyruvate dehydrogenase [259]. However, there is no strong biochemical [100,260] or genetic [100] evidence for the existence of different forms of SQR in humans, even though the gene for subunit A isolated from a human placenta cDNA library encodes a polypeptide which is more like the *B. subtilis* subunit A than that of bovine heart [100] or *E. coli* SQR [93]. We can look forward with excitement to the outcome of future research on succinate:quinone oxidoreductases.

Acknowledgements

We are grateful to Drs. B.A.C. Ackrell, F.A. Armstrong, G. Cecchini, R.B. Gennis, W.J. Ingledew, M.K. Johnson, K. Kita, B.D. Lemire, D.M. Turnbull, C.-A. Yu and their coworkers for generously providing unpublished data. Our own work presented in this chapter has been supported by grants from the Swedish Natural Science Research Council, the Swedish Medical Research Council and Emil and Wera Cornells Stiftelse (LH) and from the National Science Foundation, research grant DMB-8819305 (TO).

*References**

1 Ackrell, B.A.C., Johnson, M.K., Gunsalus, R.P. and Cecchini, G. (1992) in: F. Mueller (Ed.) Chemistry and Biochemistry of flavoenzymes, Vol. 3, CRC Press, Boca Racon, FL, pp. 229–297.
2 Vinogradov, A.D. (1986) Biokhimiya 51, 1944–1973.
3 Cole, S.T., Condon, C., Lemire, B.D. and Weiner, J.H. (1985) Biochim. Biophys. Acta 811, 381–403.
4 Ohnishi, T. (1987) Curr. Top. Bioenerg. 15, 37–65.
5 Hederstedt, L. and Rutberg, L. (1981) Microbiol. Rev. 45, 542–555.
6 Hatefi, Y. (1985) Annu. Rev. Biochem. 54, 1015–1069.
7 Jaramillo, L.R.D. and Escamilla, J.E. (1990) Rev. Latinoam. Microbiol. 32, 229–248.
8 Kita, K., Takamiya, S., Furushima, R., Villagra, E., Wang, H., Aoki, T. and Oya, H. (1991) in: T. Sato and S. DiMauro (Eds.), Mitochondrial Encephalomyopathies, Raven, New York.
9 Singer, T.P. and Johnson, M.K. (1985) FEBS Lett. 190, 189–198.
10 Kröger, A. (1980) in: C.J. Knowles (Ed.) Diversity of Bacterial Respiratory Systems, Vol. II, CRC Press, Boca Raton, FL, pp. 1–17.
11 Sucheta, A., Ackrell, B.A.C., Cochran, B. and Armstrong, F.A. (1992) Nature, 356, 361–362.
12 Armstrong, F.A. (1990) Struct. Bonding (Berlin) 72, 137–221.
13 Thunberg, T. (1909) Skand. Arch. Physiol. 22, 430–436.
14 Webb, J.L. (1966) Enzyme and Metabolic Inhibitors, Academic Press, New York, pp. 1–244.
15 Singer, T.P. and Kearney, E.B. (1955) Biochim. Biophys. Acta. 15, 151–153.
16 Wang, T.Y., Tsou, C.L. and Wang, Y.L. (1956) Sci. Sin. 5, 73–90.
17 Kelin, D. and King, T.E. (1958) Nature 181, 1520–1522.
18 Ziegler, D.M. and Doeg, K.A. (1959) Biochem. Biophys. Res. Commun. 1, 344–349.
19 Singer, T.P., Kearney, E.B. and Zastrow, N. (1955) Biochim. Biophys. Acta 17, 154–155.
20 Massey, V. (1957) J. Biol. Chem. 229, 763–770.
21 Davis, K.A. and Hatefi, Y. (1971) Biochemistry 10, 2509–2516.
22 Walker, W.H. and Singer, T.P. (1970) J. Biol. Chem. 245, 4224–4225.
23 Beinert, H. and Sands, R.H. (1960) Biochem. Biophys. Res. Commun. 3, 41–46.
24 Sands, R.H. and Beinert, H. (1960) Biochem. Biophys. Res. Commun. 3, 47–52.
25 Beinert, H, Ackrell, B.A.C., Kearney, E.B. and Singer, T.P. (1975) Eur. J. Biochem. 54, 185–194.
26 Ohnishi, T., Winter, D.B., Lim, J. and King, T.E. (1976) J. Biol. Chem. 251, 2105–2109.
27 Ohnishi, T., Salerno, J.C., Winter, D.B., Lim, J., Yu, C.A., Yu, L. and King, T.E. (1976) J. Biol. Chem. 251, 2094–2104.
28 Johnson, M.K., Morningstar, J.E., Bennett, D.E., Ackrell, B.A.C. and Kearney, E.B. (1985) J. Biol. Chem. 260, 7368–7378.
29 Maguire, J.J., Johnson, M.K., Morningstar, J.E., Ackrell, B.A.C. and Kearney, E.B. (1985) J. Biol. Chem. 260, 10909–10912.
30 Ackrell, B.A.C., Kearney, E.B., Mims, W.B., Peisach, J. and Beinert, H. (1984) J. Biol. Chem. 259, 4015–4018.
31 Baginsky, M.L. and Hatefi, Y. (1969) J. Biol. Chem. 244, 5313–5319.
32 Capaldi, R.A., Sweetland, J. and Merli, A. (1977) Biochemistry 16, 5707–5710.
33 Ackrell, B.A.C., Ball, M.B. and Kearney, E.B. (1980) J. Biol. Chem. 255, 2761–2769.
34 Hatefi, Y. and Galante, Y.M. (1980) J. Biol. Chem. 255, 5530–5537.
35 Vinogradov, A.D., Gavrikkov, V.G. and Gavrikova, E.V. (1980) Biochim. Biophys. Acta 592, 13–27.
36 Yu, C.-A. and Yu, L. (1980) Biochemistry 19, 3579–3585.
37 Unden, G. and Kröger, G. and Kröger, A.(1981) Eur. J. Biochem. 120, 577–584.
38 Lemire, B.D., Robinson, J.J. and Weiner, J.H. (1982) J. Bacteriol. 152, 1126–1131.
39 Cecchini, G., Ackrell, B.A.C., Deshler, J.O. and Gunsalus, R.P. (1986) J. Biol. Chem. 261, 1808–1814.
40 Condon, C. and Weiner, J.H. (1988) Mol. Microbiol. 2, 43–52.

* The reference list was completed in November 1991.

41 Hederstedt, L. and Rutberg, L. (1980) J. Bacteriol. 144, 941–951.
42 Fridén, H., Rutberg, L., Magnusson, K. and Hederstedt, L. (1987) Eur. J. Biochem. 168, 695–701.
43 Cecchini, G., Thompson, C.R., Ackrell, B.A.C., Westenberg, D.J., Dean, N. and Gunsalus, R.P. (1986) Proc. Natl. Acad. Sci. USA 83, 8898–8902.
44 Hirsch, C.A., Rasminsky, M., Davis, B.D. and Lin, E.C.C. (1963) J. Biol. Chem. 238, 3770–3774.
45 Fortnagel, P. and Freese, E. (1968) J. Bacteriol. 95, 1431–1438.
46 Spencer, M.E. and Guest, J.R. (1974) J. Bacteriol. 117, 947–953.
47 Spencer, M.E. and Guest, J.R. (1973) J. Bacteriol. 114, 563–570.
48 Rutberg, B. and Hoch, J.A. (1970) J. Bacteriol. 104, 826–833.
49 Creaghan, I.T. and Guest, J.R. (1972) J. Gen. Microbiol. 71, 207–220.
50 Cole, S.T. and Guest, J.R. (1979) FEMS Microbiol. Lett. 5, 65–67.
51 Ohné, M., Rutberg, B. and Hoch, J.A. (1973) J. Bacteriol. 115, 738–745.
52 Hederstedt, L., Magnusson, K. and Rutberg, L. (1982) J. Bacteriol. 152, 157–165.
53 Edlund, T., Grundström, T. and Normark, S. (1979) Mol. Gen. Genet. 173, 115–125.
54 Cole, S.T. and Guest, J.R. (1980) Mol. Gen. Genet. 178, 409–418.
55 Lohmeier, E., Hagen, D.S., Dickie, P. and Weiner, J.H. (1981) Can. J. Biochem. 59, 158–164.
56 Cole, S.T. (1982) Eur. J. Biochem. 122, 479–484.
57 Cole, S.T., Grundström, T., Jaurin, B., Robinson, J.J. and Weiner, J.H. (1982) Eur. J. Biochem. 126, 211–216.
58 Grundström, T. and Jaurin, B. (1982) Proc. Natl. Acad. Sci. USA 79, 1111–1115.
59 Spencer, M.E. and Guest, J.R. (1982) J. Bacteriol. 151, 542–552.
60 Lambden, P.R. and Guest, J.R. (1976) J. Gen. Microbiol. 97, 145–160.
61 Iuchi, S. and Lin, E.C.C. (1988) Proc. Natl. Acad. Sci. USA 85, 1888–1892.
62 Iuchi, S., Cameron, D.C. and Lin, E.C.C. (1989) J. Bacteriol. 171, 868–873.
63 Darlison, M.G. and Guest, J.R. (1984) Biochem. J. 223, 507–517.
64 Wood, D., Darlison, M.G., Wilde, R.J. and Guest, J.R. (1984) Biochem. J. 222, 519–534.
65 Gould, S.J., Subramani, S. and Scheffler, I.E. (1989) Proc. Natl. Acad. Sci. USA 86, 1934–1938.
66 Lombardo, A., Carine, K. and Scheffler, I.E. (1990) J. Biol. Chem. 265, 10419–10423.
67 Kita, K., Oya, H., Gennis, R.B., Ackrell, B.A.C. and Kasahara, M. (1990) Biochem. Biophys. Res. Commun. 166, 101–108.
68 Cole, S.T. and Guest, J.R. (1979) Eur. J. Biochem. 102, 65–71.
69 Magnusson, K., Hederstedt, L. and Rutberg, L. (1985) J. Bacteriol. 162, 1180–1185.
70 Kita, K., Vibat, C.R.T., Meinhardt, S., Guest, J.R. and Gennis, R.B. (1989) J. Biol. Chem. 264, 2672–2677.
71 Maguire, J.J., Magnusson, K. and Hederstedt, L. (1986) Biochemistry 25, 5202–5208.
72 Weiner, J.H., Cammack, R., Cole, S.T., Condon, C., Honore, N., Lemire, B.D. and Shaw, G. (1986) Proc. Natl. Acad. Sci. USA 83, 2056–2060.
73 Blaut, M., Whittaker, K., Valdovinos, A., Ackrell, B.A.C., Gunsalus, R.P. and Cecchini, G. (1989) J. Biol. Chem. 264, 13599–13604.
74 Hederstedt, L. and Hedén, L.-O. (1989) Biochem. J. 260, 491–497.
75 Murakami, H., Kita, K., Oya, H. and Anraku, Y. (1985) FEMS Microbiol. Lett. 30, 307–311.
76 Condon, C., Cammack, R., Patil, D.S. and Owen, P. (1985) J. Biol. Chem. 260, 9427–9434.
77 Hederstedt, L., Holmgren, E. and Rutberg, L. (1979) J. Bacteriol. 138, 370–376.
78 Hederstedt, L. (1980) J. Bacteriol. 144, 933–940.
79 Magnusson, K., Phillips, M.K., Guest, J.R. and Rutberg, L. (1986) J. Bacteriol. 166, 1067–1071.
80 Phillips, M.K., Hederstedt, L., Hasnain, S., Rutberg, L. and Guest, J.R. (1987) J. Bacteriol. 169, 864–873.
81 Lemma, E., Hägerhäll, C., Geisler, V., Brandt, U., von Jagow, G. and Kröger, A. (1991) Biochim. Biophys. Acta 1059, 281–285.
82 Unden, G., Hackenberg, H. and Kröger, A. (1980) Biochim. Biophys. Acta 591, 275–288.
83 Lauterbach, F., Körtner, C., Albracht, S.P.J., Unden, G. and Kröger, A. (1990) Arch. Microbiol. 154, 386–393.

169 Hederstedt, L. (1987) in: D.E. Edmondson and McCormick (Eds.) Flavins and Flavoproteins, Walter de Gruyter, Berlin, pp. 729–735.
170 Schröder, I., Gunsalus, R.P., Ackrell, B.A.C., Cochran, B. and Cecchini, G. (1991) J. Biol. Chem. 266, 13572–13579.
171 Kenney, W.C. (1975) J. Biol. Chem. 250, 3089–3094.
172 Vinogradov, A.D., Gavrikova, E.V. and Zuevsky, V.V. (1976) Eur. J. Biochem. 63, 365–371.
173 Robinson, J.J. and Weiner, J.H. (1982) Can. J. Biochem. 60, 811–816.
174 Unden, G. and Kröger, A. (1980) 117, 323–326.
175 Vik, S.B. and Hatefi, Y. (1981) Proc. Natl. Acad. Sci. USA 78, 6749–6753.
176 Hederstedt, L. and Hatefi, Y. (1986) Arch. Biochem. Biophys. 247, 346–354.
177 Kotlyar, A.B. and Vinogradov, A.D. (1984) Biochem. Int. 8, 545–552.
178 Guest, J.R. and Rice, D.W. (1984) in: R.C. Bray, P.C. Engel and S.G. Mayhew (Eds.) Flavins and Flavoproteins, Walter de Gruyter, Berlin, pp. 111–124.
179 Weiner, J.H. and Dickie, P. (1979) J. Biol. Chem. 254, 8590–8593.
180 Kenney, W.C. and Kröger, A. (1977) FEBS Lett. 73, 239–243.
181 Kenney, W.C., Walker, W.H. and Singer, T.P. (1972) J. Biol. Chem. 247, 4510–4513.
182 Furushima, R., Kita, K., Takamiya, S., Konishi, K., Aoki, T. and Oya, H. (1990) FEBS Lett. 263, 325–328.
183 Schulz, G.E., Schirmer, R.H. and Pai, E.F. (1982) J. Mol. Biol. 160, 287–308.
184 Barber, M.J., Eichler, D.C., Solomonson, L.P. and Ackrell, B.A.C. (1987) Biochem. J. 242, 89–95.
185 Ohnishi, T., King, T.E., Salerno, J.C., Blum, H., Bowyer, J.R. and Maida, T. (1981) J. Biol. Chem. 256, 5577–5582.
186 Clark, W.M. (1972) Oxidation–Reduction Potentials of Organic Systems, Krieger, New York.
187 Bonomi, F., Pagani, S., Cerletti, P. and Giori, C. (1983) Eur. J. Biochem. 134, 439–445.
188 Unden, G., Albracht, S.P.J. and Kröger, A. (1984) Biochim. Biophys. Acta 767, 460–469.
189 Ackrell, B.A.C., Kearney, E.B. and Edmonson, D. (1975) J. Biol. Chem. 250, 7114–7119.
190 Ackrell, B.A.C., Cochran, B. and Cecchini, G. (1989) Arch. Biochem. Biophys. 268, 26–34.
191 Blankenhorn, G. (1976) Eur. J. Biochem. 67, 67–80.
192 Mitchell, P. (1976) J. Theor. Biol. 62, 327–367.
193 Ohnishi, T. and Trumpower, B.L. (1980) J. Biol. Chem. 255, 3278–3284.
194 Sohnorf, U. (1966) Thesis no. 3871, Eidgenossische Technische Hochschule, Zürich, Switzerland.
195 Salerno, J.C. and Ohnishi, T. (1980) Biochem. J. 192, 769–781.
196 Wraight, C.A. (1979) Biochim. Biophys. Acta 548, 309–327.
197 Bowyer, J.R. and Ohnishi, T. (1975) in: J. Lenaz (Ed.) Coenzyme Q, J. Wiley, pp. 409–432.
198 Salerno, J.C. (1991) Biochem. Soc. Trans. 19, 599–605.
199 Cammack, R. (1987) in: H. Matsubara, Y. Katsube and K. Wada (Eds.) Iron–Sulfur Protein Research, Springer, Berlin, pp. 40–54.
200 Cammack, R., Crowe, B.A. and Cook, N.D. (1986) Biochem. Soc. Trans. 14, 1207–1208.
201 Walsh, C., Fisher, J., Spencer, R., Graham, D.W., Aston, W., Brown, J.E., Brown, R.D. and Rogers, E.F. (1978) Biochemistry 17, 1942–1951.
202 Edmondson, D.E. and Singer, T.P. (1973) J. Biol. Chem. 248, 8144–8149.
203 He, S.H., DerVartanian, D.V. and LeGall, J. (1986) Biochem. Biophys. Res. Commun. 135, 1000–1007.
204 Edmondson, D.E., Ackrell, B.A.C. and Kearney, E.B. (1981) Arch. Biochem. Biophys. 208, 69–74.
205 Williamson, G. and Edmondson, D.E. (1985) Biochemistry 24, 7790–7797.
206 Beinert, H. (1990) FASEB J. 4, 2483–2491.
207 Ingledew, W.J. and Prince, R.C. (1977) Arch. Biochem. Biophys. 178, 303–307.
208 Crowe, B.A., Owen, P., Patil, D.S. and Cammack, R. (1983) Eur. J. Biochem. 137, 191–196.
209 Cammack, R., Patil, D.S., Condon, C., Owen, P., Cole, S.T. and Weiner, J.H. (1985) in: R.C. Bray, P.C. Engel and S.G. Mayhew (Eds.) Flavin and Flavoproteins, Walter de Gruyter, Berlin, pp. 551–552.
210 Werth, M.T., Cecchini, G., Mandori, A., Ackrell, B.A.C., Schröder, I., Gunsalus, R.P. and Johnson, M.K. (1990) Proc. Natl. Acad. Sci. USA 87, 8965–8969.

211 Beinert, H. and Thomson, A.J. (1983) Arch. Biochem. Biophys. 222, 333–361.
212 Johnsson, M.K., Kowal, A.T., Morningstar, J.E., Oliver, M.E., Whittaker, K., Gunsalus, R.P., Ackrell, B.A.C. and Cecchini, G. (1988) J. Biol. Chem. 263, 14732–14738.
213 Aevarsson, L. and Hederstedt, L. (1988) FEBS Lett. 232, 298–302.
214 Fee, J.A., Kuila, D., Mather, M.W. and Yoshida, T. (1986) Biochem. Biophys. Acta 853, 153–185.
215 Blumberg, W.E. and Peisach, J. (1974) Arch. Biochem. Biophys. 162, 502–512.
216 Ohnishi, T. and Salerno, J.C. (1982) in: T. Spiro (Ed.) Iron–Sulfur Proteins, Metal Ions in Biology Vol. III, Wiley, New York, pp. 285–327.
217 Gurbiel, R.J., Batie, C.J., Sivaraja, M., True, A.E., Fee, J.A., Hoffman, B.M. and Ballou, D.P. (1989) Biochemistry 28, 4861–4871.
218 Gurbiel, R.J., Ohnishi, T., Robertson, D.E., Daldal, F. and Hoffman, B. (1991) Biochemistry 30, 11579–11584.
219 Dugad, L.B., LaMar, G.N., Banci, L. and Bertini, I. (1990) Biochemistry 29, 2263–2271.
220 Martin, A.E., Burgess, B.K., Stout, C.D., Cash, V.L., Dean, D.R., Jensen, G.M. and Stephens, P.J. (1990) Proc. Natl. Acad. Sci. USA 87, 598–602.
221 Manodori, A., Cecchini, G., Schröder, I., Gunsalus, R.P., Werth, M.T. and Johnson, M.K. (1992) Biochemistry, 31, 2703–2712.
222 Rothery, R.A. and Weiner, J.H. (1991) Biochemistry 30, 8296–8305.
223 Salerno, J.C., Lim, J., King, T.E., Blum, H. and Ohnishi, T. (1979) J. Biol. Chem. 254, 4828–4835.
224 Cammack, R., Patil, D.S. and Weiner, J.H. (1986) Biochim. Biophys. Acta 870, 545–551.
225 Johnson, M.K., Morningstar, J.E., Cecchini, G. and Ackrell, B.A.C. (1985) Biochem. Biophys. Res. Commun. 131, 756–762.
226 Fridén, H. and Hederstedt, L. (1987) in: S. Papa, B. Chance and L. Ernster (Eds.) Cytochrome Systems. Molecular Biology and Bioenergetics, pp. 641–647.
227 Hatefi, Y., Galante, Y.M., Stigall, D. and Ragan, C.I. (1979) Methods Enzymol. 55, 577–602.
228 Yu, L., Xu, J.-X., Haley, P.E. and Yu, C.-A. (1987) J. Biol. Chem. 262, 1137–1143.
229 Ruzicka, F., Beinert, H., Scheplev, K.L., Dunham, W.R. and Sands, R.H. (1975) Proc. Natl. Acad. Sci. USA 72, 2886–2890.
230 Orme-Johnson, N.R., Hansen, R.E. and Beinert, H. (1971) Biochem. Biophys. Res. Commun. 45, 871–878.
231 Davis, K.A., Hatefi, Y., Poff, K.L. and Butler, W.L. (1973) Biochim. Biophys. Acta 325, 341–356.
232 Takamiya, S., Kita, K., Matsuura, K., Furushima, R. and Oya, H. (1990) Biochem. Int. 21, 1073–1080.
233 Oya, H. and Kita, K. (1989) in: C. Bryant (Ed.) Comparative Biochemistry of Parasitic Helminths, Chapman and Hall, London, pp. 35–53.
234 Kita, K., Takamiya, S., Furushima, R., Ma, Y., Suzuki, H., Ozawa, T. and Oya, H. (1988) Biochim. Biophys. Acta 935, 130–140.
235 Hederstedt, L. and Andersson, K.K. (1986) J. Bacteriol. 167, 735–739.
236 Unden, G. and Kröger, A. (1986) Methods Enzymol. 126, 387–399.
237 Unden, G. and Kröger, A. (1981) Eur. J. Biochem. 120, 577–584.
238 Kröger, A. and Innerhofer, A. (1976) Eur. J. Biochem. 69, 497–506.
239 Cammack, R., Maguire, J.J. and Ackrell, B.A.C. (1987) in: S. Papa, B. Chance and L. Ernster (Eds.) Cytochrome Systems. Molecular Biology and Bioenergetics. Plenum Press, New York.
240 Degli Esposti, M., Palmer, G. and Lenaz, G. (1989) Eur. J. Biochem. 182, 27–36.
241 Degli Esposti, M., Crimi, M., Körtner, C., Kröger, A. and Link, T. (1991) Biochim. Biophys. Acta 1056, 243–249.
242 Salerno, J.C. (1984) J. Biol. Chem. 259, 2331–2336.
243 Walker, F.A., Huynh, B.H., Scheidt, W.R. and Osvath, S.R. (1986) J. Am. Chem. Soc. 108, 5288–5297.
244 Trumpower, B.L. (1990) Microbiol. Rev. 54, 101–129.
245 Fridén, H., Cheesman, M.R., Hederstedt, L., Andersson, K.K. and Thomson, A.J. (1990) Biochim. Biophys. Acta 1041, 207–215.
246 Yun, C.-H., Crofts, A. and Gennis, R.B. (1991) Biochemistry 30, 6747–6754.

247 Paudel, H.K., Yu, L. and Yu, C.-A. (1991) Biochim. Biophys. Acta 1056, 159–165.
248 Yu, L. and Yu, C.-A. (1981) Biochim. Biophys. Acta 637, 383–386.
249 Choudhry, Z.M., Gavrikova, E.V., Kotlyar, A.B., Tushurashvili, P.R. and Vinogradov, A.D. (1985) FEBS Lett. 182, 171–175.
250 Xu, J.-X., Yu, L. and Yu, C.-A. (1987) Biochemistry 26, 7674–7679.
251 Ingledew, W.J., Salerno, J.C. and Ohnishi, T. (1976) Arch. Biochem. Biophys. 177, 176–184.
252 Salerno, J.C., Blum, H. and Ohnishi, T. (1979) Biochim. Biophys. Acta 547, 270–281.
253 Miki, T., Yu, L. and Yu, C.-A. (1992) Biochem. Biophys. 293, 61–66.
254 Xu, Y., Salerno, J.C. and King, T.E. (1987) Biochem. Biophys. Res. Commun. 144, 315–322.
255 Paddock, M.L., Rongey, S.H., Fehr, G. and Okamura, M.Y. (1989) Proc. Natl. Acad. Sci. USA 86, 6602–6606.
256 Desnuelle, C., Birch-Machin, M., Pellissier, J.F., Bindoff, L.A., Ackrell, B.A.C. and Turnbull, D.M. (1989) Biochem. Biophys. Res. Commun. 163, 695–700.
257 Schapira, A.H.V., Cooper, J.M., Morgan-Hughes, J.A., Landon, D.N. and Clarke, J.B. (1990) New Engl. J. Med. 323, 37–42.
258 Rivner, M.H., Shamsnia, M., Swift, T.R., Trefz, J., Roesel, R.A., Carter, A.L., Yanamura, W. and Hommes, F.A. (1989) Neurology 39, 693–696.
259 Russell, G.C. and Guest, J.R. (1991) Biochim. Biophys. Acta 176, 225–232.
260 Shaw, M.-A., Edwards, Y. and Hopkinson, D.A. (1981) Biochem. Genet. 19, 741–756.
261 Thauer, R.K., Jungermann, K. and Decker, K. (1977) Bacteriol. Rev. 41, 100–180.

L. Ernster (Ed.) *Molecular Mechanisms in Bioenergetics*
© 1992 Elsevier Science Publishers B.V. All rights reserved

CHAPTER 8

Mitochondrial ubiquinol–cytochrome *c* oxidoreductase

GEORG BECHMANN, ULRICH SCHULTE and HANNS WEISS

Institut für Biochemie, Heinrich-Heine-Universität Düsseldorf, Universitätsstraße 1, W-4000 Düsseldorf 1, Germany

Contents

1. Introduction 199
2. Protein and redox components 200
 2.1. Cytochrome *b* 201
 2.2. Cytochrome c_1 202
 2.3. Iron–sulfur protein 203
 2.4. Subunits without prosthetic groups 203
3. Three-dimensional structure and topography of subunits 204
4. Biogenesis 205
 4.1. Regulation of gene expression 205
 4.2. Mitochondrial protein import 205
 4.3. Assembly 206
5. Electron and proton transport 206
 5.1. Proton-motive ubiquinone-cycle 206
 5.2. Ubiquinol-oxidation center o 208
 5.3. Ubiquinone-reduction center i 208
6. Role of subunit I in protein processing 209
7. Outlook 211
References 212

1. Introduction

Quinol–cytochrome *c* (plastocyanine) oxidoreductase, also called bc_1 (b_6f)-complex, is one of the most ancient and ubiquitous electron transfer complexes. It is found in mitochondria [1], chloroplasts and cyanobacteria [2,3], photosynthetic bacteria and many other prokaryotes [3,4]. It is minimally composed of three subunits, cytochrome $b(b_6)$ carrying a low- and a high potential heme group, cytochrome $c_1(f)$, and a high-potential iron–sulfur (FeS)-protein traditionally called Rieske-protein. The general function of the complex is electron transfer between two mobile redox carriers, ubiquinol and cytochrome $c(c_2)$ in mitochondria and some bacteria, or plastoquinol and plastocyanine in chloroplasts and cyanobacteria. This electron transfer is

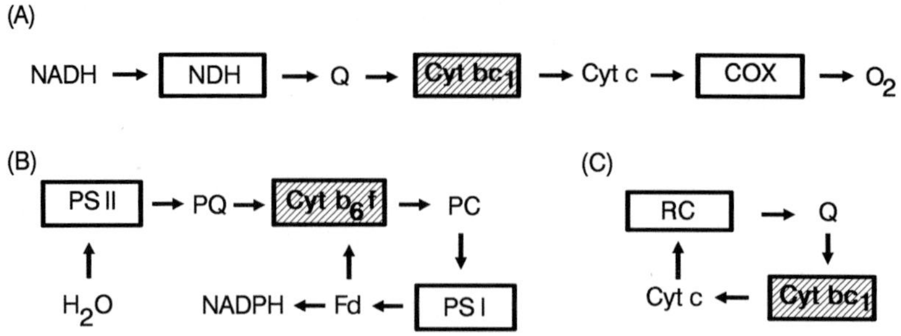

Fig. 1. Functional interaction of the $bc_1(b_6f)$-complex with other electron-transfer complexes in (A) mitochondria; (B) photosystems II and I in chloroplasts or cyanobacteria, and (C) photosynthetic reaction center in purple nonsulfur bacteria. The following abbreviations are used: NDH, NADH-ubiquinone oxidoreductase; Q, ubiquinone; Cytc, cytochrome *c*; COX, cytochrome *c* oxidase; PS, photosystem; PQ, plastoquinone; PC, plastocyanin; Fd, ferrodoxin.

coupled to proton translocation, thus generating proton-motive force in the form of an electrochemical proton potential which can drive ATP synthesis. Figure 1 shows schematically the reaction pathways in which the complex is involved. In mitochondria and some bacteria, the bc_1-complex connects NADH-ubiquinone oxidoreductase (and other dehydrogenases) with cytochrome *c* oxidase (Fig. 1A). In chloroplasts or cyanobacteria, the corresponding b_6f-complex is centered between the photosystems II and I (Fig. 1B). During cyclic electron transfer, electrons are cycled between the b_6f-complex and photosystem I. A light-driven cyclic electron flow is also found in photosynthetic purple nonsulfur bacteria, involving reaction center and the bc_1-complex (Fig. 1C).

This chapter will concentrate on the mitochondrial bc_1-complex, in the following referred to as cytochrome reductase. Data on the related chloroplast or bacterial complex will be included where helpful for the better understanding of the mitochondrial complex. Cytochrome reductase is the best understood among the proton translocating respiratory chain complexes of mitochondria, and a large amount of literature concerning structure, biogenesis and catalytic mechanism has accumulated. It is therefore not the aim of this chapter to give a comprehensive survey of this literature. We have rather attempted to focus on topics that we believe have had the greatest impact in recent years for the understanding of cytochrome reductase. The reader is referred to previous reviews or book articles focusing on the mechanism [5–7], structure [8], comparative aspects [2–4], and biogenesis [9,10] of the complex.

2. *Protein and redox components*

Mitochondrial cytochrome reductase has been isolated from different mammalian tissues, the yeast *Saccharomyces cerevisiae*, the filamentous fungus *Neurospora crassa* [1,8 and references therein] and higher plants [11,12]. The original isolation procedures used cholate and deoxycholate for protein extraction and different salts for fractional protein precipitation. This method yields preparations of high purity and enzymatic activity although in aggregated states. By the use of nonionic detergent (Triton X-100), in combination with chromatographic techniques, cytochrome reductase is isolated in monodisperse state. Using analytical ultracentrifugation, neutron scattering or laser-light scattering, and correcting the data for protein bound phospholipid and detergent, molecular weights of approximately 500 000 have been obtained [8 and references therein]. This value is about twice the sum of the molecular masses of the subunits (Table I), in agreement with the assumption of a dimeric state of the complex (see below). Subunit compositions differ slightly in various organisms. The subunits with redox centers, cytochrome *b*, cytochrome c_1 and the Rieske iron–sulfur protein, show the highest sequence similarity including the subunits of bacterial (Table I) and plastidal [2] complexes. The occurrence of several additional subunits, without redox groups, is a characteristic feature of mitochondrial cytochrome reductase. Eight such subunits are found in the mammalian complex, and six in the fungal complex [8] (Table I). These subunits are not required for electron transfer and proton translocation per se, since they are not found in the mechanistically very similar bacterial bc_1-complexes [33,34]. Poor sequence similarities of the non-redox subunits of different organisms (Table I) suggest a less well defined function.

2.1. Cytochrome b

Cytochrome *b* is the only subunit of cytochrome reductase coded by the mitochondrial genome. It has been sequenced from a large variety of different organisms (*A. transmontanus* [35], *A. nidulans* [36], *B. taurus* [16], *G. gallus* [37], *D. melanogaster* [38], *D. yakuba* [39], *H. sapiens* [40], *L. tarentolae* [41], *M. musculus* [42], *N. crassa* [43], *O. bertiana* [44], *O. sativa* [45], *P. lividus* [46], *P. tetraurelia* [47], *R. norvegicus* [48], *S. cerevisiae* [17], *S. pombe* [49], *S. purpuratus* [50], *T. brucei* [51], *T. aestivum* [52], *V. faba* [53], *X. laevis* [54], *Z. mays* [55]). Using the extensive sequence information, a folding model has been predicted, with eight helices spanning the membrane. Two heme moieties are attached to the apoprotein by four conserved histidines on opposite sides of the membrane, forming the low-potential cytochrome b_L (b_{566}, $E_{m,7} = -40$ mV) near the positive outer surface, and the high-potential cytochrome b_H (b_{562}, $E_{m,7} = +40$ mV) near the negative inner surface [56–58]. This model has now been confirmed by studies of mutants resistant to the inhibitors antimycin A, diuron, funiculosin and HQNO or

TABLE I

Subunit composition of ubiquinol–cytochrome *c* oxidoreductase. Molecular weights and pairwise degrees of sequence identity of subunits from *Bovine taurus* (B), *Saccharomyces cerevisiae* (S) and *Paracoccus denitrificans* (P) are shown. (Residues 1–200 of cytochrome c_1 from *P. denitrificans* were not considered for determination of sequence identity)

Subunit		Molecular mass (kDa)			Sequence identity (%)			Ref.
		B	S	P	B/S	B/P	S/P	
I	(Core protein 1)	35.8	48.1		24			[13,14]
II	(Core protein 2)	46.5	38.7		19			[13,15]
III	(Cytochrome *b*)	43.7	44.0	50.1	50	41	42	[16–18]
IV	(Cytochrome c_1)	27.9	27.4	44.7	57	28	35	[19,20,18]
V	(FeS-protein)	21.7	20.1	20.3	54	39	44	[21,22,18]
VI		13.4	14.4		32			[23,24]
VII		9.5	12.3		18			[25,26]
VIII		9.2	14.5		32			[27,28]
IX		8.0						[29]
X		7.2	7.3		27			[30,31]
XI		6.4						[32]

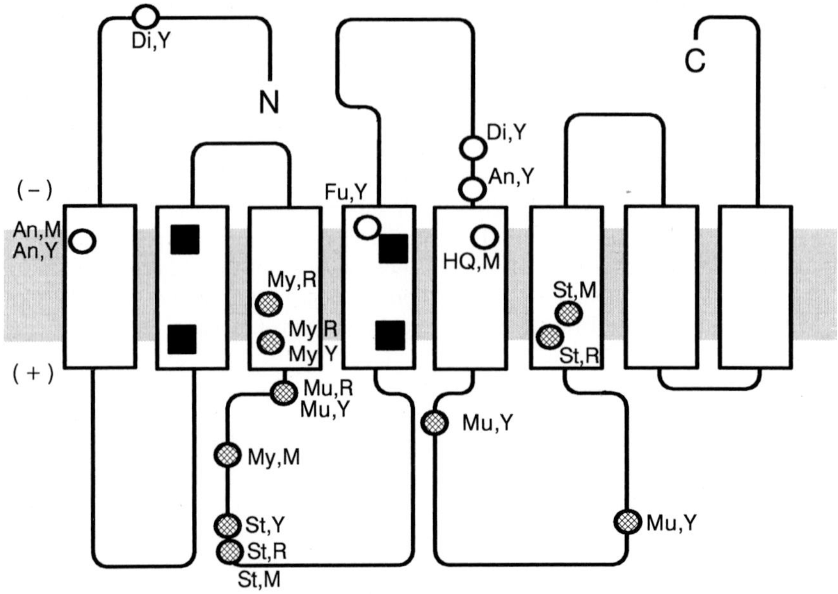

Fig. 2. Eight-helix model of cytochrome *b* with the positions of the mutations leading to inhibitor-resistance. The following abbreviations are used: An, antimycin A; Di, diuron; Fu, funiculosin; HQ, HQNO; My, myxothiazol; St, stigmatellin. Y, M and R refer to yeast, mouse and *Rhodobacter*. The black squares indicate histidines that ligate the two heme moieties. Mutations conferring resistance to inhibitors of electron transfer to cytochrome b_L or b_H are symbolized by open or shaded circles, respectively. The shaded section indicates the membrane. The model combines data reported by Di Rago et al. [60 and references therein], Daldal et al. [61] and Howell and Gilbert [62].

myxothiazol, stigmatellin and mucosin, respectively, which specifically interfere with either of the two sides of cytochrome *b*, blocking electron transfer to either cytochrome b_H or cytochrome b_L [59], (see below). Sequence positions of the mutations leading to inhibitor resistances are located in two loops on either of the two membrane sides [60–62 and refs. therein] (Fig. 2). This topology is further supported by analysis of gene fusions constructed with cytochrome *b* from *Rhodobacter* [63].

2.2. *Cytochrome c_1*

The primary structure of mitochondrial cytochrome c_1 is known for *B. taurus* [19], *E. gracilis* [64], *H. sapiens* [65], *N. crassa* [66], and *S. cerevisiae* [20]. The heme moiety is covalently bound to two cysteines near the N-terminus. A fifth ligand of the heme-iron is a conserved histidine next to one of the cysteines [19]. Cytochrome c_1 displays a pH independent $E_{m,7}$ of about +230 mV. The subunit is anchored in the membrane by one membrane-spanning helix near the C-terminus. A water-soluble cytochrome c_1 preparation can be obtained by removing this C-terminal sequence stretch. This was done by mild protease treatment of isolated, detergent-solubilized cytochrome c_1 [67] or by expression of a truncated cytochrome c_1 gene in yeast [68] and *Rhodobacter* [69]. The water-soluble cytochrome c_1 fragment could not be assembled into the cytochrome reductase complex, but displayed essentially the same spectroscopic properties as the unmodified protein, and is able to transfer electrons to cytochrome *c*.

2.3. Iron–sulfur protein

This cytochrome reductase subunit is traditionally called Rieske FeS-protein. It is distinguished from other binuclear FeS-proteins by its high midpoint redox potential, of +280 mV. Sequences of the mitochondrial protein are known for *B. taurus* [21], *N. crassa* [70], *R. norvegicus* [71], and *S. cerevisiae* [22]. The subunit is anchored in the membrane by a C-terminal hydrophobic helix, which can be removed by protease treatment releasing a water-soluble domain with an intact FeS-cluster [72]. The exact residues ligating this cluster have not yet been identified. In analogy with ferredoxin-type proteins, four cysteines arranged in two pairs have been proposed as ligands [70]. These cysteines have been found in all sequences determined so far, including the Rieske FeS-proteins of the bacterial bc_1-complexes [4]. Spectroscopic data [73], as well as association of the cluster with an ionizable group with a pKa of 8 [74], suggest involvement of at least one histidine in the binding of the FeS-cluster. Sequence studies of mutated Rieske proteins in *S. cerevisiae* lacking the FeS-cluster point to three cysteines and one histidine as ligands [75]. Temperature-sensitive mutants reveal, besides a more or less severe reduction in enzymatic activity an altered sensitivity to myxothiazol and UHDBT [76,77].

2.4. Subunits without prosthetic groups

The two large subunits I and II, which contribute about 40% of the total protein of mitochondrial cytochrome reductase, are traditionally called core proteins. They are hydrophilic proteins located peripherally, closely associated to each other at the matrix side of the complex (see below). The sequences of subunit I from *B. taurus* [13], *H. sapiens* [78], *N. crassa* [79], *R. norvegicus* [80], and subunit II from *S. cerevisiae* [14,15] have been determined. Although related bacterial and plastidal complexes contain no such subunits, the mitochondrial enzyme requires both subunits for activity. Reconstitution experiments with *N. crassa* cytochrome reductase indicate that the complex is inactive when it lacks subunits I and II [81]. Yeast deletion mutants lacking subunits I or II can form only trace amounts of cytochrome reductase, giving rise to speculations that these subunits are required for the assembly of the cytochrome reductase complex [14,15] (see below). Temperature-sensitive mutations in subunit I of *S. cerevisiae* destabilize cytochrome reductase at the nonpermissive temperature, leaving cytochrome *b* inaccessible for electrons [82]. Although the roles of subunits I and II concerning electron and proton transport remain enigmatic, the involvement of subunit I in processing of imported proteins has been shown [79] (see below).

All additional small subunits of bovine and yeast cytochrome reductase have been sequenced (Table I). However, the functions assigned to these subunits remain speculative. Subunits VIII and X of the bovine complex have been reported to be tightly associated to cytochrome c_1, and subunit VIII has therefore been termed 'hinge protein', suggesting its involvement in mediating interaction between cytochrome c_1 and cytochrome *c* [27]. But complex formation between these two cytochromes is also seen in the absence of subunit VIII [83]. A mutated yeast cytochrome reductase lacking the homologous subunit shows only minor deviations from the wild-type complex concerning interaction with cytochrome *c* [84]. Involvement of this subunit in the regulation of dimer activity has also been proposed [85]. It may be worth mentioning that subunit VIII contains an unusually high proportion of acidic residues, part of which are clustered as eight consecutive glutamate residues in the bovine subunit [27] and 25 glutamate and

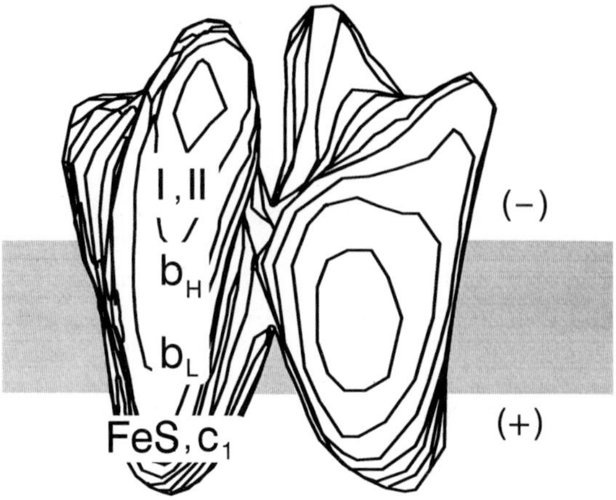

Fig. 3. Side-view of cytochrome reductase, parallel to the plane of the membrane. The density map was determined by electron microscopy of membrane crystals. The approximate topography of subunits I and II, cytochrome *b* with its two heme groups, b_H and b_L, cytochrome c_1 and the Rieske FeS-protein are shown. The shaded section indicates the membrane. (Modified from Weiss and Leonard [90].)

aspartate residues in the homologous yeast subunit [28]. Subunit VI of the bovine cytochrome reductase has been labeled with a ubiquinone analogue, and therefore postulated to act as a ubiquinone-binding protein [86]. However bacterial, bc_1-complexes do not require such a subunit for interaction with ubiquinone or ubiquinol [4].

3. Three-dimensional structure and topography of subunits

Very recently two groups have reported progress in preparing three-dimensional crystals of bovine heart cytochrome reductase suitable for X-ray diffraction analysis and the fine structure of the complex may soon be available [87,88]. So far, information about the three-dimensional structure has come only from electron microscopy of membrane crystals of the whole complex and of a subcomplex lacking subunits I, II and V, using in both cases preparations from *N. crassa* [89,90]. These structures were quantitatively compared with neutron scattering data obtained from preparations in hydrogenated and deuterated detergent [91]. The resolution was only 3 nm, but the correlation between the structures of the whole complex and the subcomplex, together with the predicted folding of subunits allowed the assignment of subunits to the different sections of the structure.

The dimeric complex consists of two elongated monomeric units, which are connected by a two-fold axis running perpendicular to the membrane (Fig. 3). About one-half of the protein is located in a large peripheral section protruding 7 nm into the mitochondrial matrix. This large peripheral section must be constituted essentially by the subunits I and II because the section is missing in the structure of the subcomplex that is devoid of both subunits. Furthermore, subunits I and II are water-soluble and contain no hydrophobic sequence stretch long enough to span the membrane. Approximately one-third of the total cytochrome reductase protein is located in the membrane, and contact between the monomeric units is mainly restricted to this membrane section. Cytochrome *b* makes up most of this membrane section, thus forming the center of the complex. Only one-fifth of the total protein protrudes into the intermembrane space of mitochondria, and this small peripheral section most likely includes the hydrophilic domains of cytochrome c_1 and the FeS-protein. The structural studies have given no information about the location of the smaller subunits lacking redox groups [90].

4. *Biogenesis*

Biogenesis of mitochondrial cytochrome reductase is initiated by translation of nuclear-coded subunits in the cytoplasm and of cytochrome *b* in the mitochondrion. The cytoplasmically synthesized subunits are then imported into the mitochondrion and directed to their final membrane location. Prosthetic groups are attached and the subunits are assembled to yield the functional complex.

4.1. Regulation of gene expression

Expression of nuclear-coded and mitochondrially coded subunits is most likely constitutive in obligate aerobic organisms. Significant regulation dependent on environmental conditions has not been observed. In contrast, the expression of cytochrome reductase in the facultatively anaerobic yeast *S. cerevisiae* is highly regulated. Catabolite repression keeps synthesis low as long as mitochondrial respiration is not needed, e.g. at low oxygen or high glucose concentrations [10,92]. An important mediator of catabolite repression is heme, which by itself is expressed in dependence of oxygen [93]. Regulation by heme has been shown for subunits I and II as for subunits of the other respiratory chain complexes, but not for ATPase subunits [94,95]. Binding sites for multiple factors including general transcription factors (GF1 and GF2) and heme activator proteins (HAP2 and HAP3) have been mapped upstream of the transcription initiation site of the structural genes of subunits II and VIII (12.3 kDa) [95]. There seems to be no tight coupling between the expression of the different nuclear-coded subunits, since overexpression of single subunits from multicopy plasmids is not restricted [96]. Very little is known about coordination of synthesis of nuclear-coded subunits and mitochondrially coded cytochrome *b* [9,10].

The biogenesis of cytochrome *b* has been studied in great detail in yeast. By characterizing mutants affected in cytochrome *b* formation, a number of proteins involved in RNA processing, translation and maturation of the protein have been identified [10,97]. The product of the nuclear gene *ABC1* (for activity of *bc1*) has been proposed to affect cytochrome *b* translation as well as assembly of the whole complex [98]. Similarly, yeast mutants have been described that are deficient in the assembly of cytochrome oxidase or ATPase, but which are not related to subunits of these complexes [99–101]. So far no relationship between all these mutants has been found.

4.2. Mitochondrial protein import

Mitochondrial import of cytochrome reductase subunits is believed to follow the general import pathway [102,103]. Most subunits carry N-terminal extensions which are cleaved off in the matrix. In *S. cerevisiae*, subunits VII to IX are synthesized without a presequence [104,31], while in *N. crassa* only subunit VI is not processed [105].

Cytochrome c_1 and the FeS-protein are processed in two steps [20,106]. The second step occurs at the outer surface of the inner membrane, after insertion of the intermediate into the membrane. The prosthetic groups of the two subunits seem to be attached to the proteins in the matrix before the second cleavage step takes place [107,108]

4.3. Assembly

The process of assembly of cytochrome reductase has been studied by characterizing yeast mutants lacking single subunits of the complex. Such mutants were isolated from *pet* strains or were constructed by gene disruption [109,110]. Only deletion of subunit VI (14.5 kDa) does not affect the enzymatic activity at normal ionic strength and cytochrome *c* concentration, and mutant cells show wild-type growth on nonfermentable carbon source [84,110]. Deletion of any other cytochrome reductase subunit leads to respiratory deficiency accompanied by more or less severe reduction of the steady-state levels of single subunits. mRNA levels of the subunits are not affected in the mutants, and expression of single subunits seen in pulse-label experiments is at wild-type levels [109,111,112]. It was therefore concluded that when single subunits are absent, the assembly of the remaining subunits is affected, and that the decrease of the steady-state levels of some subunits is caused by increased protease sensitivity of the unassembled proteins. The protease-sensitive subunits VII (14.4 kDa) and VIII (12.3 kDa) and cytochrome *b* are stabilized only if both are present, as well as cytochrome c_1 and subunits I and II. Protection of the FeS-protein requires expression of all other subunits except subunit VI. Absence of the FeS-protein itself has no considerable impact on the steady-state levels of the other subunits. Tentative models for the assembly process have been put forward [10,109]. In general, a subcomplex formed by the subunits I and II is joined by a subcomplex composed of the cytochromes *b* and c_1 and of subunits VII and VIII. This assembly intermediate is completed by the addition of subunit IX (7.3 kDa) and finally, of the FeS-protein and subunit VI. Preformation of a subcomplex containing cytochrome c_1 and subunits VI and IX has been suggested, although this seems to be no prerequisite for interaction of cytochrome c_1 with the other subunits.

The way of attachment of redox groups to the subunits has remained largely undefined. As mentioned earlier, cytochrome c_1 and the FeS-protein are believed to incorporate their respective redox group before integration into the complex. Maturation of cytochrome *b* requires the formation of a not yet defined subcomplex. In yeast mutants lacking one of the subunits I, II, VII or VIII, cytochrome *b* is present only as apoprotein. Formation of holocytochrome *b* does however not require cytochrome c_1 [109].

5. Electron and proton transport

5.1. Proton-motive ubiquinone-cycle

The overall reaction catalyzed by mitochondrial cytochrome reductase is

$$\text{ubiquinol} + 2\ \text{ferricytochrome}\ c + 2\ \text{H}^+_\text{N} \rightarrow \text{ubiquinone} + 2\ \text{ferrocytochrome}\ c + 4\ \text{H}^+_\text{P},$$

where the subscripts N and P designate the negative inner and positive outer side of the mitochondrial inner membrane. The ubiquinone-cycle (Q-cycle) originally proposed by Mitchell [113] is now widely accepted as model for this electron and proton transport reaction. According to the Q-cycle model, cytochrome reductase contains two ubiquinone-catalytic centers, the ubiquinol-oxidation center o and the ubiquinone-reduction center i. They are corre-

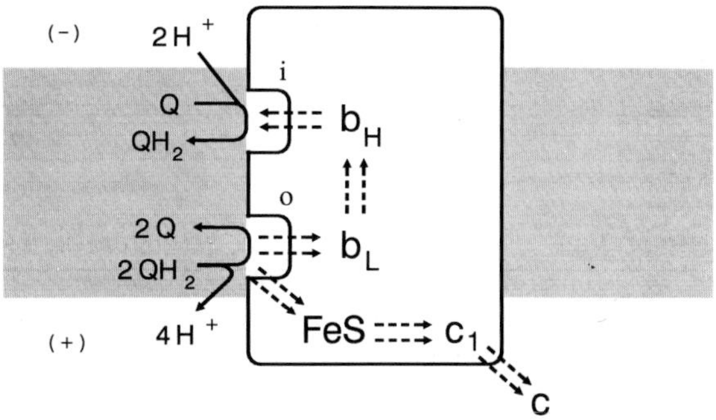

Fig. 4. 'Double turnover Q-cycle', modified from Crofts [6]. The full lines indicate the pathways of quinol, quinone and protons, the broken lines the pathways of the electrons. For further symbols see text. The shaded section indicates the membrane.

lated with the positive outer and the negative inner side, respectively, of the mitochondrial inner membrane. Oxidation of ubiquinol at centre o leads to the release of two protons into the intermembrane space and the transfer of two electrons, the first to cytochrome *c* via the FeS-protein and cytochrome c_1, and the second across the membrane via cytochromes b_L and b_H to ubiquinone at center i. The ubisemiquinone thus formed at center i remains protein-bound until a second ubiquinol is oxidized at center o and a second electron is transferred to center i to complete the reduction to ubiquinol. The two protons required for the formation of ubiquinol are taken up from the matrix space. One complete Q-cycle requires two redox turns at the reaction center o and two successive electron transfers to center i. For two ubiquinol molecules thus oxidized at center o, one ubiquinone is reduced at center i, four protons are released into the intermembrane space, and two protons are taken up from the matrix space (Fig. 4). The H^+/e^- ratio is thus two, in agreement with the stoichiometry measured in mitochondria [114–116] and bacteria [117]. However, this stoichiometry might decrease with an increasing electrical membrane potential as has been shown recently for the isolated cytochrome reductase of *N. crassa* [118]. This decline might be of physiological significance since an incomplete energy coupling has to exist to optimize the energy transforming process [119].

Some variants of the above 'double-turnover Q-cycle' [6,120] have been proposed. The 'double Q-cycle' proposes that the complex works as a dimer in which the two centers i of the dimer cooperate in the reduction of ubiquinone to ubiquinol [7,121,122]. Another model originally termed '*b*-cycle' has later been modified and called 'SQ(semiquinone)-cycle' [123]. It differs from the Q-cycle in that the semiquinone formed during the second turn at center o moves fast to center i and acts as the oxidant there.

The structural prerequisite for all these variants of the Q-cycle are the two ubiquinone reaction sites, centers o and i, located on opposite sides of the membrane. The most powerful evidence for the existence of these two sites comes from inhibitor studies showing that, first, the electron transfer reactions occurring at either of the two sites can be independently inhibited; and second, that mutations leading to inhibitor resistance are expressed on the two opposite membrane sides of cytochrome *b* [59] (Fig. 2).

5.2. *Ubiquinol-oxidation center o*

The reaction at the ubiquinol oxidation center o can be described by the equation

$$QH_2 + (FeS_{ox}/Cyt\ b_{L\ ox}) \rightarrow Q + (FeS_{red}/Cyt\ b_{L\ red}{-}nH^+) + (2{-}n)H^+_P.$$

The release coefficient n ($0 < n < 1$) of the protons depends on the ability of cytochrome b_L to serve as a redox-linked proton acceptor. In chromatophores one of the protons is retained by the reduced cytochrome b_L and released only after its reoxidation by cytochrome b_H. This view is supported by determination of different redox properties of the cytochromes b as a function of the pH-value [124–127].

Based on the finding that removal of the FeS-protein eliminates the rapid reduction of cytochrome c_1 as well as of cytochrome b_L by ubiquinol [127], it has been assumed that ubiquinol is the immediate reductant for the Rieske FeS-cluster, and that the transiently formed ubisemiquinone reduces cytochrome b_L. The QH_2/QH-couple has a redox potential of +280 mV and the QH/Q-couple has one of −160 mV, which is sufficiently low to reduce cytochrome b_L [127,128].

There are several ways to inhibit the electron transfer reactions at center o. If the FeS-cluster and cytochrome c_1 are kept reduced, and the reaction center i is inhibited by antimycin A (see below), cytochrome b_L cannot be reduced by ubiquinol. The block is removed when an oxidant of cytochrome c_1 is added. This phenomenon which has been called oxidant-induced reduction [127,129], is the consequence of the branched electron route from center o to the FeS-cluster as well as to cytochrome b_L (Fig. 4). Despite its negative redox potential, the QH/Q-couple is not able to reduce the FeS-cluster. To explain the thermodynamically anomalous bifurcation of the fixed-sequential electron flow, a catalytic switch has been proposed [130].

There is a multitude of natural and synthetic compounds available that specifically inhibit the reaction at center o. According to their different modes of action, they have been classified into two groups [59]. Inhibitors of group 1 are various β-methoxyacrylates (MOA), including the antibiotics myxothiazol and stigmatellin, which show the highest affinities of this group (K_d $<10^{-10}$ M and $<10^{-11}$ M, respectively) [131–133]. Myxothiazol induces a red shift of the cytochrome b_L α-absorbance band [132,134] consistent with the assumption that center o is structurally more closely related to cytochrome b_L than to cytochrome b_H. Group 2 inhibitors are quinone analogues. Typical examples are UHDBT [135], UHNQ [136], HMHQQ [137] and DBMIB [138,139]. They do not exert a red shift of cytochrome b_L and probably compete with ubiquinone for a binding site at the FeS-protein [140]. Group I inhibitors are able to displace group II inhibitors, an observation used as further evidence for center o being formed jointly by cytochrome b_L and the FeS-protein [141,142].

5.3. *Ubiquinone-reduction center i*

The redox reaction occurring at center i is

$$2\ Cyt\ b_{H\ red} + Q + 2\ H^+_N \rightarrow 2\ Cyt\ b_{H\ ox} + QH_2.$$

There are still some uncertainties about this reaction, arising from the fact that ubiquinone is a two-electron acceptor and cytochrome b_H a one-electron donor. The antibiotic antimycin A, the cytochrome reductase inhibitor with the highest affinity [117,143], prevents the formation of the transient ubisemiquinone as well as the final ubiquinol [144]. The shift of the α-absorbance band of reduced cytochrome b_H by antimycin A suggested that the inhibitor binds close to cytochrome b_H long before sequence data of antimycin A-resistant mutants confirmed this assumption (Fig. 2).

Detection by EPR spectroscopy of the formation of an antimycin A-sensitive ubisemiquinone indicated a sequential reaction at center i [145,146]. For a complete 'double-turnover Q-cycle', ubiquinone serves as oxidant after the first turn, and a bound ubisemiquinone as oxidant after the second turn. According to the 'SQ-cycle', a ubisemiquinone would be the only oxidant for cytochrome b_H at center i [123]. In the 'double-Q cycle', each oxidation turn takes place at one of the two monomeric units of the dimeric complex [7,121,122].

Further information about the reduction of ubiquinone by cytochrome b_H comes from studies of the steady-state electron transfer from duroquinol to 2,3-dimethoxy-5-decyl-6-methyl-benzoquinone [147,148]. Center i catalyzes this quinol/quinone transhydrogenation reaction according to a ping-pong mechanism. As a possible mechanism, two molecules duroquinol may each reduce one cytochrome b_H of the complex dimer, and the two molecules durosemiquinone formed may then react with each other in a disproportionation reaction. Subsequently two molecules of 2,3-dimethoxy-5-decyl-6-methyl-benzoquinone are reduced by the two cytochromes b_H of the dimer, and the semiquinones disproportionate in the same way as the durosemiquinones. This would imply that redox equilibration between the two cytochromes b_H within the dimeric complex occurs via two semiubiquinones, each bound at one center i of the dimer [148].

6. *Role of subunit I in protein processing*

Completely unexpected was the finding that subunits I and II of mitochondrial cytochrome reductase are members of a protein processing family [79].

During or after import of nuclear-coded proteins into the mitochondrion, N-terminal extensions directing the proteins into the organelle are cleaved off by an endopeptidase called matrix processing peptidase (MPP). The activity of MPP is strongly increased by a related protein called processing-enhancing protein (PEP) which by itself has no protease activity [102,103,149]. Sequence comparison and comparison of the processing enhancing activity showed that in *N. crassa* PEP and cytochrome reductase subunit I are one and the same protein. Cytochrome reductase subunit II yields no enhancement of processing [79].

In *S. cerevisiae*, the only other organism with known sequences of both PEP and subunit I, the two proteins are different with only moderate degree of sequence identity (Fig. 5, Table II). By aligning the sequences of the proteins involved in processing, MPP and PEP, and the subunits I and II of cytochrome reductase these proteins can be grouped as a family [13,78,79]. Very recently insulinase from *Drosophila* and human and protease III of *E. coli* have been suggested as further members of this family [151]. Figure 5 shows an alignment of the N-terminal sequences of putative members of this new protein family, displaying most of the conserved sequence positions.

The family is very heterogeneous, linking proteins with no apparent structural or functional relationship, like cytochrome reductase subunit II and insulin-degrading enzymes. A close connection between MPP, PEP and subunit I is apparent not only from the high degree of sequence similarity but also from the functional interaction seen in precursor processing. Sequence similarity of subunits I and II of cytochrome reductase is only moderate and although the two proteins are structurally associated [90,91], their functional interaction remains to be shown. The proteolytic enzymes of the family MPP, insulinase and protease III are all specific

[61,63,69,160,161]. Similarly, new information can be expected from studies of revertants of inhibitor resistant cytochrome *b* mutants and of proteins able to suppress nuclear *pet* mutations in yeast [98,162]. The question about the function of the non-redox subunits can only be answered by studies of the mitochondrial complex. The involvement of subunit I in protein processing has raised new questions on the link between mitochondrial bioenergetics and biogenesis. Understanding of the involvement of chaperone-like proteins in assembly and maturation of the complex is still largely fragmentary [98]. Yeast mutants are expected to be the most important tools in elucidating these processes. However, with the unique ability to repress the whole system of mitochondrial oxidative phosphorylation, yeast might not always give the same answers to questions of mitochondrial biogenesis as do mammals or obligate aerobic fungi.

References

1 Hatefi, Y. (1985) Annu. Rev. Biochem. 54, 1015–1065.
2 Hauska, G., Hurt, E., Gabellini, N. and Lockau, W. (1983) Biochim. Biophys. Acta 726, 97–133.
3 Scherer, S. (1990) TIBS 15, 458–462.
4 Trumpower, B.L. (1990) Microbiol. Rev. 54, 101–129.
5 Rich, P. (1984) Biochim. Biophys. Acta 768, 53–79.
6 Crofts, A.R. (1985) in: A.N. Matonosi (Ed.) The Enzymes of Biological Membranes 2nd edn, New York, Plenum Press, pp. 347–382.
7 De Vries, S. (1986) J. Bioenerg. Biomembranes 18, 195–224.
8 Weiss, H. (1987) Curr. Top. Bioenerg. 15, 67–90.
9 Tzagoloff, A. and Myers, A.M. (1986) Annu. Rev. Biochem. 55, 249–285.
10 Grivell, L.A. (1989) Eur. J. Biochem. 182, 477–493.
11 Nakajima, T., Maeshima, M. and Asahi, T. (1984) Agric. Biol. Chem. 48, 3019–3025.
12 Berry, A.B., Huang, L.-S. and DeRose, V.J. (1991) J. Biol. Chem. 266, 9064–9077.
13 Gencic, S., Schägger, H. and von Jagow, G. (1991) Eur. J. Biochem. 199, 123–131.
14 Tzagoloff, A., Wu, M. and Crivellone, M. (1986) J. Biol. Chem. 261, 17163–17169.
15 Oudshorn, P., van Steeg, H., Swinkels, B.W., Schoppink, P. and Grivell, L.A. (1987) Eur. J. Biochem. 163, 97–103.
16 Anderson, S., de Bruijn, M.H.L., Coulson, A.R., Eperon, J.C., Sanger, F. and Young, I.G. (1982) J. Mol. Biol. 156, 683–717.
17 Nobrega, F.G. and Tzagoloff, A. (1980) J. Biol. Chem. 255, 9828–9837.
18 Kurowski, B. and Ludwig, B. (1987) J. Biol. Chem. 262, 13805–13811.
19 Wakabayashi, S., Takeda, H., Matsubara, H., Kim, C.H. and King, T.E. (1982) J. Biochem. 91, 2077–2085.
20 Sadler, I., Suda, K., Schatz, G., Kaudewitz, F. and Haid, A. (1984) EMBO J. 3, 2137–2143.
21 Schägger, H., Borchart, U., Machleidt, W., Link, T.A. and von Jagow, G. (1987) FEBS Lett. 219, 161–168.
22 Beckmann, J.D., Ljungdahl, P.O., Lopez, J.L. and Trumpower, B.L. (1987) J. Biol. Chem. 262, 8901–8909.
23 Wakabayashi, S., Takao, T., Shinionishi, Y., Kuramitsu, S., Matsubara, H., Wang, T., Zhang, Z. and King, T.E. (1985) J. Biol. Chem. 260, 337–343.
24 De Haan, M., van Loon, A.P.G.M., Kreike, J., Vaessen, R.T.M.J. and Grivell, L.A. (1984) Eur. J. Biochem. 138, 169–177.
25 Borchart, U., Machleidt, W., Schägger, H., Link, T.A. and von Jagow, G. (1986) FEBS Lett. 200, 81–86.
26 Maarse, A.C. and Grivell, L.A. (1987) Eur. J. Biochem. 165, 419–425.
27 Wakabayashi, S., Matsubara, H., Kim, C.H. and King, T.E. (1982) J. Biol. Chem. 252, 9335–9344.

28 Van Loon, A.P.G.M., de Groot, R.J., de Haan, M., Dekker, A. and Grivell, L.A. (1984) EMBO J. 3, 1039–1043.
29 Borchart, U., Machleidt, W., Schägger, H., Link, T.A. and von Jagow, G. (1985) FEBS Lett. 191, 125–130.
30 Schägger H., von Jagow, G., Borchart, U. and Machleidt, W. (1983) Z. Physiol. Chem. 364, 307–311.
31 Phillips, J.D., Schmitt, M.E., Brown, T.A., Beckmann, J.D. and Trumpower, B.L. (1990) J. Biol. Chem. 265, 20813–20821.
32 Schägger, H., Borchart, U., Aquila, H., Link, T.A. and von Jagow, G. (1985) FEBS Lett. 190, 89–94.
33 Weiss, H., Leonard, K. and Neupert, W. (1990) TIBS 15, 178–180.
34 Yang, X. and Trumpower, B.L. (1988) J. Biol. Chem. 263, 11962–11970.
35 Brown, J.R., Gilbert, T.L., Kowbel, D.J., O'Hara, P.J., Buroker, N.E., Beckenbach, A.T., Smith, M.J. (1989) Nucleic Acids Res. 17, 4389.
36 Waring, R.B., Davies, R.M., Lee, S., Grisi, E., Berks, M.M. and Scazzochio, C. (1981) Cell 27, 4–11.
37 Desjardins, P. and Morais, R. (1990) J. Mol. Biol. 212, 599–634.
38 Garesse, R. (1988) Genetics 118, 649–663.
39 Clary, D.O., Wahleithner, J.A. and Wolstenholme, D.R. (1984) Nucleic Acids Res. 12, 3747–3762.
40 Anderson, S., Bankier, A.T., Barrell, B.G., de Bruin, M.H.L., Coulson, A.R., Drouin, J., Eperon, I.C., Nierlich, D.P., Roe, B.A., Sanger, F., Schreier, P.H., Smith, A.J.H., Staden, R. and Young, I.G. (1981) Nature 290, 457–465.
41 Feagin, J.E., Shaw, J.M., Simpson, L. and Stuart, K. (1988) Proc. Natl. Acad. Sci. USA 85, 539–543.
42 Bibb, M.J., van Etten, R.A., Wright, C.T., Walberg, M.W. and Clayton, D.A. (1981) Cell 26, 167–180.
43 Citterich, M.H., Morelli, G.F. and Macino, G. (1983) EMBO J. 2, 1235–1242.
44 Schuster, W., Unseld, M., Wissinger, B. and Brennicke, A. (1990) Nucleic Acids Res. 18, 229–233.
45 Kaleikau, E.K., Andre, C.P., Doshi, B. and Walbot, V. (1990) Nucleic Acids Res. 18, 372.
46 Cantatore, P., Roberti, M., Rainaldi, G., Gadaleta, M.N. and Saccone, C. (1989) J. Biol. Chem. 264, 10965–10975.
47 Pritchard, A.E., Seilhamer, J.J., Mahlingan, R., Sable, C.L., Ventusi, S.E. and Cummings, D.J. (1990) Nucleic Acids Res. 18, 173–180.
48 Koike, K., Kobayashi, M., Yaginuma, K., Taira, M., Yoshida, E. and Imai, M. (1982) Gene 20, 177–185.
49 Lang, B.F., Ahne, F. and Bonen, L. (1985) J. Mol. Biol. 184, 353–366.
50 Jacobs, H.T., Elliott, D.J., Math, V.B. and Farquharson, A. (1988) J. Mol. Biol. 202, 185–217.
51 Benne, R., de Vries, B.F., van den Burg, J. and Klaver, B. (1983) Nucl. Acids Res. 11, 6925–6941.
52 Gualberto, J.M., Lamattina, L., Bonnard, G., Weil, J.-H. and Grienenberger, J.-M. (1989) Nature 341, 660–662.
53 Wahleithner, J.A. and Wolstenholme, D.R. (1988) Nucleic Acids Res. 16, 6897–6913.
54 Dunon-Bluteau, D., Volowitch, M. and Brunn, G. (1985) Gene 36, 65–78.
55 Dawson, A.J., Jones, V.P. and Leaver, C.J. (1984) EMBO J. 3, 2107–2113.
56 Saraste, M. (1984) FEBS Lett. 166, 367–373.
57 Widger, W.R., Cramer, W.A., Herrmann, R.G. and Trebst, A. (1984) Proc. Natl. Acad. Sci. USA 81, 674–678.
58 Brasseur, R. (1988) J. Biol. Chem. 263, 12571–12575.
59 Von Jagow, G. and Link, T.A. (1986) Methods Enzymol. 126, 253– 271.
60 Di Rago, J.-P., Perea, J. and Colson, A.-M. (1990) FEBS Lett. 263, 93–98.
61 Daldal, F., Tokito, M.K., Davidson, E. and Faham, M. (1989) EMBO J. 8, 3951–3961.
62 Howell, N. and Gilbert, K. (1988) J. Mol. Biol. 203, 607–618.
63 Yun, C.-H., van Doren, S.R., Crofts, A. and Gennis, R.B. (1991) J. Biol. Chem. 266, 10967–10973.
64 Mukai, K., Toyosaki, H., Yoshida, M., Yao, Y., Wakabayashi, S. and Matsubara, H. (1989) Eur. J. Biochem. 178, 649–656.
65 Nishikimi, M., Ohta, S., Suzuki, H., Tanaka, T., Kikkawa, F., Tanaka, M., Kagawa, Y. and Ozawa, T. (1988) Nucleic Acids Res. 16, 3577.
66 Römisch, J., Tropschug, M., Sebald, W. and Weiss, H. (1987) Eur. J. Biochem. 164, 111–117.

151 Affholter, J.A., Fried, V.A. and Roth, R.A. (1988) Science 242, 1415–1418.
152 Kuo, W.L., Gehm, B.D. and Rosner, M.R. (1990) Mol. Endocrinol. 4, 1580–1591.
153 Claverie-Martin, F., Diaz-Torres, M.R. and Kushner, S.R. (1987) Gene 54, 185–195.
154 Kleiber, J., Kalousek, F., Swaroop, M. and Rosenberg, L.E. (1990) Proc. Natl. Acad. Sci. USA 87, 7978–7982.
155 Pollock, R.A., Hartl, F.-U., Cheng, M.Y., Ostermann, J., Horwich, A. and Neupert, W. (1988) EMBO J. 7, 3493–3500.
156 Schneider, H., Arretz, M., Wachter, E. and Neupert, W. (1990) J. Biol. Chem. 265, 9881–9887.
157 Witte, C., Jensen, R.E., Yaffee, M.P. and Schatz, G. (1988) EMBO J. 7, 1439–1447.
158 Yaffee, M.P., Ohta, S. and Schatz, G. (1985) EMBO J. 4, 2069–2074.
159 Ou, W.-J., Akio, J., Okazaki, H. and Omura, T. (1989) EMBO J. 8, 2605–2612.
160 Atta-Asafo-Adjei, E. and Daldal, F. (1991) Proc. Natl. Acad. Sci. USA 88, 492–496.
161 Yun, C.-H., Crofts, A. and Gennis, R.B. (1991) Biochemistry 30, 6747–6754.
162 Di Rago, J.-P., Netter, P. and Slonimski, P.P. (1990) J. Biol. Chem. 265, 3332–3339.

L. Ernster (Ed.) *Molecular Mechanisms in Bioenergetics*
© 1992 Elsevier Science Publishers B.V. All rights reserved

CHAPTER 9

Cytochrome oxidase: notes on structure and mechanism

TUOMAS HALTIA and MÅRTEN WIKSTRÖM

Helsinki Bioenergetics Group, Department of Medical Chemistry, University of Helsinki, Siltavuorenpenger 10, SF-00170 Helsinki, Finland

Contents

1. Introduction 217
2. Structure of the metal sites 218
   2.1. Cu_A 218
   2.2. The haem of cytochrome *a* 221
   2.3. The binuclear dioxygen reduction site 221
3. Relationships of the redox centres to the protein framework 224
   3.1. Subunit I 226
   3.2. Subunit II 226
   3.3. Subunit III 227
   3.4. Further notes on membrane topology 227
4. Electron transfer and the reduction of dioxygen 228
5. Proton translocation and the conservation of energy 231
   5.1. Some principles of the proton pump 231
   5.2. Catalytic steps linked to conservation of energy 231
   5.3. How might proton pumping occur? 233
Acknowledgements 234
References 234

1. Introduction

Cytochrome *c* oxidase is the terminal O_2-reducing respiratory enzyme in the inner mitochondrial membrane of all eukaryotic cells. It is also present in the cell membrane of many aerobic bacteria. Apart from its central function in respiration, its activity is coupled to proton translocation across the inner mitochondrial (or bacterial) membrane, by which respiratory energy is conserved as an electrochemical proton gradient [1,2].

A major development in cytochrome oxidase research in the 80s was the elucidation of the primary structures of the enzyme's subunits, mostly by cloning and sequencing of the genes coding for these subunits in several different species of eukaryotes and prokaryotes (see ref. [3]). Somewhat earlier, electron microscopy and image reconstruction work with orientated multilayers of the enzyme was initiated. This approach has now led to an improved three-dimensional picture of the enzyme in the membrane, though the resolution is still modest [4]. The primary structure information on the enzyme subunits now includes several species of bacteria. This work has led to a surprising finding: whilst cytochrome *c* oxidase is unique in the mitochondria of eukaryotic cells, it is but a member of a family of related enzymes of which several interesting variants are found in bacteria [191]. Studies on these variations, which not only include protein structures, but also variations in the prosthetic groups and the electron donor, have already yielded valuable structural and functional information [3,5–17].

A major controversy in the research on cytochrome oxidase in the period between 1977 and 1985 concerned whether or not the enzyme functions as a proton pump. Although this argument has been settled [18] there are still claims by Papa and coworkers [19] that proton-pumping would occur only in a narrow 'window' of electron transfer rate through the enzyme. For example, according to these workers the oxidase does not pump protons during uninhibited respiration with succinate in mitochondria (in contrast with, e.g., refs. [21,22]).

Two further lines of development have been significant in the last few years. First, high-resolution time-resolved spectroscopy, and studies of energy-dependent reversal of the oxidase reaction, have together significantly contributed to the elucidation of the mechanism of reduction of O_2 to water by the enzyme (see [23]). Second, the increasing use of recombinant DNA techniques has already yielded the first site-directed mutants of bacterial oxidases.

Here we will make a survey of some of the most recent developments in the field, which we will attempt to integrate into the pre-existing frame of knowledge. Due to the publication of some 300 articles per year on cytochrome oxidase it has not been possible to cite all the relevant data. Instead, we will rely heavily on comprehensive review articles published earlier.

2. Structure of the metal sites

Depending on the source, cytochrome *c* oxidase contains three to thirteen subunits, of which three are mitochondrially encoded in most eukaryotes. The genes for the rest are in the nucleus. Bacterial oxidases typically contain the counterparts of the mitochondrially encoded subunits plus, in some cases, one or possibly more polypeptides unique to bacteria.

The core structure of all cytochrome *c* oxidases is formed by two mitochondrially encoded subunits (I and II). It is virtually certain today that three of the four metal centres (haems *a*, a_3 and Cu_B) reside in subunit I, whereas the fourth, Cu_A, is in subunit II. The third subunit (III) is constantly found also in the bacterial enzymes purified using mild non-ionic detergents, such as dodecyl maltoside. It may thus be viewed as part of the core structure, although it does not contain any metal-binding sites. Below, we will discuss the structure and properties of the metal centres.

2.1. Cu_A

One of the coppers in the enzyme (Cu_A) has a characteristic X-band EPR signal with hardly any fine structure [24] and an unusual resistance to power saturation [25]. On this basis it was proposed that the unpaired spin of the oxidised form of Cu_A is delocalised onto one or two cysteine sulfurs, indicating a relatively covalent bonding between Cu and ligand cysteines [26,27]. Early protein sequencing of subunit II revealed a homology to a type 1 copper binding site (e.g. azurin and stellacyanin; ref. [28]). This sequence, which contains two cysteines and a histidine separated from one another by three residues, appears to be conserved among all known subunit II sequences of cytochrome *c* oxidases [29]. A conserved histidine, thought to be the fourth ligand, is located some 40 residues from this site towards the N-terminal [30].

ENDOR (Electron Nuclear Double Resonance) studies have been made using auxotrophic yeast mutants that are unable to synthesise histidine or cysteine [31]. Owing to this defect, an isotopically substituted amino acid present in the growth medium will be incorporated into the oxidase, and will bring about spectral fine structure if it is a ligand of the metal to be observed. The results showed that at least one cysteine and one histidine must be ligands of Cu_A [32]. Only one hyperfine line from ^{13}C was observed in the ENDOR spectrum of [β-^{13}C] cysteine-enriched oxidase. This means either single Cys ligation or ligation by two symmetrically positioned cysteines [27]. Such symmetric coordination would be expected to enhance the delocalisation of copper spin onto the sulfurs, explaining the unique properties of this metal site.

EXAFS (Extended X-ray Absorption Fine Structure) can identify the type of atoms within about 5 Å from the metal, provided that the neighbouring atoms are not too close in atomic number. It also yields estimates of distances between the atoms and thereby information about the metal coordination [33,34], although it does not directly indicate which atoms are true ligands. The EXAFS studies have been complicated by factors inherent to the technique, but also by the presence of two (or three, see below) coppers per enzyme molecule. Both coppers contribute to the EXAFS features, and ligand assignment requires other spectroscopic information, or deletion of one of the coppers. A total of 2 ± 1 sulfur (or chloride) and 6 ± 1 nitrogen (or oxygen) ligands to the two coppers have been identified by EXAFS [35]. Powers et al. [36] proposed that Cu_A is ligated by two sulfurs, one nitrogen (or oxygen) and one sulfur (or nitrogen, oxygen). Scott et al. [35] arrived at a similar model with two sulfurs and two nitrogens (or oxygen), which is in good agreement with the aforementioned ENDOR data. Specific removal of Cu_A allowed the determination of the contribution of Cu_B, and subsequent deconvolution to obtain the EXAFS of the Cu_A site [37]. This approach supported the above conclusion.

RR (Resonance Raman) bands attributable to Cu_A^{2+}-ligand stretches were recently described by Takahashi et al. [38]. MCD (Magnetic Circular Dichroism) and MCD-EPR double resonance spectroscopy have also been applied on Cu_A. The results suggest that visible and near-infrared absorption bands of Cu_A are due to ligand to metal charge transfer transitions, consistent with thiolate coordination, and have yielded the polarisation of the optical absorption bands with respect to one another [39].

EPR has been the classical method of studying Cu_A [24,40]. The EPR signal of Cu_A has an unusually low *g*-value (g_z, g_y, g_x are 2.18, 2.03, and 1.99, respectively, as measured at the Q-band; [40]), and has a broad featureless appearance that has made extraction of structural information difficult. In the past it appeared that no other copper centre had an environment similar to that of Cu_A [25,41] with the possible exception of copper-substituted alcohol dehydrogenase complexed with NADH [42]. It may be significant that in the latter, the metal binding site is formed by two cysteines and a histidine [43]. Only recently a truly homologous copper centre has been found in nitrous oxide reductase [44–48], which may have important implications for the structure of the Cu_A site (see below).

While the X-band EPR signal of Cu_A is almost devoid of fine structure (but see ref. [32]), a seven-line hyperfine pattern is observed using lower frequencies [44,45,49]). These spectra were interpreted in terms of two kinds of magnetic interactions, viz. a Cu–Cu interaction plus an interaction of copper with another paramagnet, possibly Fe of cytochrome *a* or the nuclear spin of a proton.

Two recent findings have revived the idea of a Cu–Cu interaction. First, the observation that cytochrome oxidase may contain three (and not only two) coppers per two haem irons [50–53], and second, the demonstration that the multicopper enzyme nitrous oxide reductase shows clear sequence homology to the putative Cu_A binding site and contains all the proposed ligands of the latter [47]. Comparison of EXAFS and MCD data on nitrous oxide reductase and cytochrome oxidase are also consistent with similar copper environments in both enzymes [46], but these data are averaged over all the coppers and may not be regarded as conclusive proof for the structure of a particular site. As it is very unlikely that Cu_A and Cu_B would interact magnetically, the finding of a 'third' copper prompted Kroneck et al. to suggest that the Cu_A site in subunit II is not mononuclear but composed of two interacting coppers with a formal valence of 1.5 and a total spin of 1/2 in the oxidised state. A binuclear site of this kind would behave as a one-electron acceptor and would become EPR-undetectable upon reduction. Interestingly, a form of nitrous oxide reductase produced by a mutant defective in biosynthesis of EPR-undetectable copper contains only 2–3 Cu/molecule, exhibits the multiline EPR signal of the normal enzyme (which has 8 Cu/molecule), and becomes EPR-undetectable by reduction with 0.5 e^-/copper [45,54].

The weak point of Kroneck's model is that it is not known whether the EPR-detectable copper centre of nitrous oxide reductase is really binuclear. The enzyme appears to be a homodimer with four coppers in each monomer, whereas the copper-deficient mutant enzyme might contain only one copper per monomer. If this is the case, then Cu–Cu interaction must occur between two Cu_A-like sites in different monomers (see, however, ref. [48]) and not between two coppers in one binuclear Cu_A-like site, as proposed for cytochrome oxidase. Moreover, Cu–Cu scattering is not observed in the EXAFS of either enzyme (see ref. [33]).

The copper EPR signals from the two enzymes are clearly not identical. Most notably, in cytochrome oxidase the hyperfine structure is observed only on the high-field side of the main signal, whereas in the reductase such signals are visible both on the high and low field sides. This and the lower *g*-value of the cytochrome oxidase Cu_A signal could be accounted for by assuming a weak magnetic interaction between Cu_A and Fe of cytochrome *a* [49,55].

However, some groups find only 2.5 coppers/2 irons in cytochrome oxidase [56–58], which would make a functional role of the 'third' copper unlikely. The latter authors claimed that all copper in excess of two/2Fe could be removed by EDTA after monomerisation of cytochrome oxidase while still retaining the characteristic EPR signal. Pan et al. [58] suggested that the 'third' copper is bound to subunit III and may reside at the interface of two monomers in the dimer. However, a 'third' copper is present in a bacterial oxidase from which subunit III has been removed [51,53].

Cytochrome *o* from *E. coli*, which exhibits strong primary structure homologies to the cytochrome *c* oxidases, lacks all four putative Cu_A ligands in subunit II [8], and also lacks the EPR signature of Cu_A [12]. The same is true for the recently discovered cytochrome aa_3-type quinol oxidases from *Bacillus subtilis* and *Sulfolobus acidocaldarius* [11,16,59]. This further strengthens the conclusion that the Cu_A site indeed resides in subunit II, and supports the idea, based on structural considerations, that the major function of Cu_A is to provide the enzyme with an acceptor of electrons from cytochrome *c* [30].

In summary, it has been firmly established that Cu_A is ligated by at least one cysteine and one histidine. It is noteworthy that the two cysteines in this subunit II domain are the only ones fully conserved in the enzyme. Especially the sequence homology, but also ENDOR and EXAFS data support the presence of a second cysteine ligand. The fourth ligand is likely to be a histidine, although a (distant) methionine ligand might fit the EXAFS data [29,33,34,36]. The possibility of the Cu_A site being a binuclear Cu–Cu centre must be left open at this time. One crucial test for this would be the crystallographic structure determination of nitrous oxide reductase, which is likely to contain a site most closely related to Cu_A.

2.2. *The haem of cytochrome a*

In contrast to Cu_A, which is not a constituent of the quinol oxidase subfamily of terminal oxidases, cytochrome *a*, or a corresponding low-spin haem, is a universal element of both the quinol and cytochrome *c* oxidases in the superfamily. EPR, ENDOR, EXAFS and resonance Raman experiments have all yielded consistent results indicating that the haem iron has a low-spin electronic configuration and has two histidine imidazoles as the axial ligands [33,36,60–62]. Resonance Raman studies also indicate the presence of water molecules in the vicinity of formyl and vinyl substituents of the haem [63]. NMR data have shown that haem *a* is exposed to solvent whereas haem a_3 is not [64].

The low-spin haem of cytochrome oxidase has a characteristic rhombic EPR signal with a component close to g=3. The three g-values of any low-spin haem can be related to the crystal field parameters designated tetragonal field strength and rhombicity. These describe the electronegativity of the ligands and their electronic geometry, respectively [65]. Several groups of low-spin haem proteins can be identified, each group having characteristic pairs of crystal field parameters. The haems belonging to different groups have different types of axial ligation of the haem iron [65]. Since a considerable number of new members of the terminal oxidase family have been found recently, we have calculated the crystal field parameters of their low-spin haem groups on the basis of their EPR spectra, and compared them to those determined earlier. In a Peisach–Blumberg plot (Fig. 1) the low-spin haems of all the oxidases fall within two groups, originally designated H and B. Both of these are thought to be examples of bis-imidazole ligation, their difference being due to different protonation states of the N(1) atom of the ligand histidines ([66]; H and B mean, respectively, that *both* histidines are deprotonated or protonated).

We note (Fig. 1) that in the oxidases from *P. denitrificans* and *R. sphaeroides* the low-spin haem appears to have *one* of the axial histidines in the deprotonated form, whereas this is not the case in the oxidases from other known species. In the EPR spectra this special property is seen as unusually low-lying g_z-value near 2.84 [67,68]. The deprotonation might be a consequence of strong hydrogen bonding. Ferrous cytochrome *a* has been reported to have two different conformers depending on the coordination state of haem a_3 [69,70]. It is tempting to speculate that these conformers might represent different protonation states of one of the ligand histidines. The H-bonding of a ligand histidine could modulate the relative orientations of the histidine and haem planes [71]. The possible functional role of such modulation is not understood at present, but it may be related to a critical 'electron-queueing' role of the low-spin haem during function [23].

The identity of the histidines that ligate the low-spin haem axially has recently been established by mutagenesis work [72–74], and we will return to the protein-structural aspects of this work below.

2.3. *The binuclear dioxygen reduction site*

In most, if not all, oxidases catalysing the reduction of O_2 to water, the O_2-reduction site comprises two or three closely apposed metal ions. In the terminal respiratory oxidase family haem iron (cytochrome a_3) and a nearby copper ion (Cu_B) form a binuclear centre, which is directly involved in the $4e^-$ reduction process where the O=O double bond is broken and two water molecules are generated. The two paramagnetic metals are located within 5Å from one another in the ferric/cupric form of the site, forming a spin-coupled, normally EPR-undetectable, bimetallic centre [75]. The spin-coupling is thought to be mediated by a shared bridging biochemical data (see Capaldi, [146]). However, it should be noted that in the gene fusion experiments it is assumed that the topology of the fusion protein corresponds to that of the native subunit complexed with other subunits in the native enzyme.

A general difficulty is encountered when trying to fit together the number of transmembranous helices inferred from data described above, with the dimensions of the enzyme deduced from electron microscopy of two-dimensional oxidase crystals. The former indicates 12, 2 and 7 transmembrane helices, respectively, in subunits I, II and III, plus several more in the additional nuclear-encoded subunits. The minimum number of membrane-spanning helices then becomes 21 [146]. This is far in excess of the 15–17 [146] or 12–16 [4] helices that the results from analysis of 2D crystals would accommodate. One explanation for this discrepancy is that helices predicted to be transmembranous by hydropathy plots are not always actually spanning the membrane. In fact, the gene fusion experiments do not necessarily support the transmembranous nature of the two N-terminal helices of subunit I in cytochrome *o* (see ref. [128]). Instead, this N-terminal segment of subunit I shows some homology to an extramembranous domain of thyroperoxidase [147]. Thus the number of membrane-spanning helices may be reduced, and the N-termini of subunits I of both cytochrome *o* and cytochrome aa_3 would be brought to the outside of the membrane. The topology of the central hydrophobic segments VI and VII of subunit I also remains somewhat ambiguous (cf. ref. [128]).

Two other segments for which there is no topological information are the two N-terminal helices of subunit III. These contain few conserved residues and are actually absent from cytochrome *o* oxidases of bacilli and cytochrome *o* from *E. coli* [3,9,16]). However, these oxidases appear to have two additional hydrophobic segments at the C-termini of their subunit I sequences, making definitive conclusions hard to draw.

Taken together, the picture emerges that three of the oxidase's four metal centres are located about 30–50% into the membrane dielectric, and within a structural domain composed of four or perhaps five well-conserved transmembranous segments of subunit I. The fourth centre, when present, resides in a separate periplasmic domain of subunit II, electrically very close to the aqueous phase outside the membrane. This structural arrangement suggests that Cu_A is the entryport of electrons from cytochrome *c*, and that the major electrogenicity of electron transfer occurs between Cu_A and the low-spin haem. It may be of interest to note that the four copper ions in ascorbate oxidase are organised in a similar fashion. In this enzyme, a type 1 copper centre bound to a separate domain feeds electrons to a trinuclear dioxygen-binding copper centre located 12-15 Å away [148].

4. *Electron transfer and the reduction of dioxygen*

During steady-state electron transfer in coupled mitochondria the redox state of Cu_A closely follows that of cytochrome *c* [149]. In kinetic experiments, oxidation of both Cu_A and cytochrome *c* proceed with similar rates [150], and Cu_A and haem *a* equilibrate electrons rapidly [151,152]. After pulse radiolysis in the absence of cytochrome *c*, Cu_A is the first centre to become reduced; its reoxidation exhibits the same time course as reduction of haem *a* [153]. These experiments support the notion based on structural analysis, that Cu_A is indeed the primary acceptor of electrons from cytochrome *c*. However, studies with Cu_A-depleted enzyme show that electrons from cytochrome *c* can still enter haem *a* [154,155]. Whether this indicates a physiologically significant branch in the electron transfer pathway is not known. In the past several groups have favoured such branched or cytochrome *a*-centered pathways [156,157], but this has been questioned [2,150]. Alternatively, the reduction of haem *a* in the Cu_A-depleted enzyme might reflect an electron transfer pathway that is artificially introduced by the procedure required to remove the copper. It should be noted here that the quinol oxidases operate without Cu_A, although the quinol-binding site of these enzymes might be formed by a domain intrinsic to subunit II (cf. ref. [158]). Conversely, cytochrome *c* oxidases do not normally oxidise ubiquinol. However, it is interesting that oxidation of tetrahalogenated *p*-benzoquinols can be induced by treatment of the enzyme with strongly positively charged compounds, such as polylysine [159].

Cytochrome *a* is the donor of electrons into the binuclear site [160,161]. There are reasons to believe that Cu_B is the primary electron acceptor and that direct electron transfer between the haems does not occur at physiological velocities ([23], but see below). Our model has incorporated this by assuming that the two haems are orientated with respect to one another in a way that maximally disfavours direct electron transfer between them. Interestingly, Cu_B and haem *a* may be ligated by two histidines (H-408 and H-410) predicted to reside on opposite sides of the same transmembranous segment (segment X, see Fig. 2). In the amino acid sequence these histidines are separated by a single fully conserved phenylalanine (F-409). In addition, position 404 (Fig. 2) is always occupied by either Phe or Tyr. It is thus conceivable that electron transfer from the low-spin haem to Cu_B could take place via these aromatic residues. If so, it is not difficult to imagine how this electron transfer could be structurally modulated by subtle conformational changes of helix X.

The mechanism of O_2 reduction at the binuclear site is now relatively well understood. Its elucidation has been based on two independent approaches, viz. (i) time-resolved low temperature and room-temperature spectroscopy (optical, EPR, resonance Raman), where the reaction of the reduced enzyme with O_2 is initiated by photolysis of the CO-derivative of haem a_3 in the presence of O_2, and (ii) partial reversal of the catalytic reaction cycle in mitochondria and mitoplasts, by means of a high electrochemical proton gradient, under which conditions catalytic intermediates accumulate. Together, these approaches (see ref. [23], for details and references) have yielded a model of the catalytic O_2 cycle, shown here in Fig. 3. Although Fig. 3 is nearly self-explanatory, some of its details require comment.

We start considering the O_2 reaction from the state **R**, where the binuclear centre has been reduced by the first two electrons out of the four in all required to reduce O_2 to water. In the first bimolecular step, an O_2 molecule binds to the ferrous haem a_3 iron, forming an 'oxy' intermediate (**A**) analogous to oxyhaemoglobin or oxymyoglobin, first detected at low temperatures by Chance et al. [162]. Next, the O_2 molecule is effectively 'trapped' at the site by rapid transfers of the first two electrons to it from iron and copper, yielding the 'peroxy' intermediate **P**. Whilst the binding per se of O_2 is not very tight ($K_d \sim 0.3$ mM), the trapping event raises the apparent affinity for dioxygen considerably ($K_m \sim 0.1$ μM). The peroxy intermediate has not been observed in the forward reaction of the fully reduced enzyme with O_2, in which O_2 binding is characteristically very quickly followed by electron transfer from haem *a* to the binuclear site. However, **P** accumulates in reversed electron transfer experiments as a two-electron oxidation product of the ferric/cupric site plus water [163,164], and also during forward electron transfer at high protonmotive force [165]. Moreover, **P** is observed subsequent to the oxy intermediate when the 'mixed valence' enzyme (binuclear site reduced; haem *a* and Cu_A oxidised) reacts with dioxygen (compound C [162]), and when the oxidised ferric/cupric binuclear site (**O**; see Fig. 3) reacts with H_2O_2 [166,167]. The very fast oxidation of haem *a* in the reaction of the fully reduced enzyme with O_2 may, in fact, be artifactual in the sense that fully reduced enzyme is very unlikely to occur significantly in the aerobic steady state [23].

The formation of **P** is thought to be followed by transfer of the third electron from haem *a* to Cu_B and protonation of the peroxide. Then the O–O bond is broken, triggered by electron energy (generation of membrane potential) although its contribution to overall energy conservation is small (~19%) in comparison with the 'power stroke' (cf. ref. [181]). This situation might be somewhat different in the proton-pumping quinol oxidases (e.g. cytochrome *o* of *E. coli*). Here, the electron donor, a hydrophobic quinol, may be located near the middle of the membrane dielectric so that electron transfer to the binuclear site is electroneutral per se. However, the protons released from quinol upon oxidation (1 H^+/e^-) appear in the outer positively charged aqueous domain, together with the truly translocated protons [2,182]. Hence, electron transfer from the donor to the binuclear site would also in this case be an electrogenic process, although its basis might differ from the cytochrome *c* oxidase case.

The third distinct phase comprises the combined O_2-binding and trapping events (**R** → **A** → **P**, Fig. 3). This phase, which includes electron transfers within the binuclear site, is not coupled to conservation of energy. The O_2-trapping event is highly exergonic ($\Delta G'_o$ ~−6 kcal/mol), providing the major driving force for the whole catalytic cycle in conditions where a high electrochemical proton gradient prevails [23,165].

Protonation of reduced dioxygen to water takes place from the electrically negatively charged side (inside) of the mitochondrial or bacterial membrane [183]. This occurs in the steps **P** → **F** and **F** → **O** ($2H^+$ each; see Fig. 3), as determined in reversed electron transfer experiments. However, the latter protonation may be incomplete, as it depends on the prevailing local pH [168], and could thus in some conditions of high local pH be completed later during the 'electron-filling' phase. It is noteworthy that the proton uptake also contributes to energy conservation, generating both membrane potential and ΔpH.

Babcock and Wikström [23] have recently argued that the proton translocation events are most likely to be coupled to the reactions of the binuclear site in states **P** and **F**, triggered by electron transfer through Cu_B, i.e. from the low-spin haem to the haem a_3 iron-oxygen site. Cu_B was proposed to be the so-called redox element of the proton pump. However, there is a dilemma associated with this proposal if it is further assumed (as above) that all electron transfer from haem *a* to haem a_3 occurs via Cu_B: why is electron transfer through Cu_B coupled to proton-pumping in the 'power stroke' but not in the 'electron-filling' phases of the catalytic cycle? Basically, there are two ways out of this dilemma, both of which are founded on the fact that the structure of the bimetallic site is quite different during these two phases of catalysis. During 'electron-filling' haem a_3 iron has high spin, but upon binding of O_2 to state **R** (Fig. 3) it becomes low spin, and ligated to relatively strong field oxygenous ligands. Furthermore, as indicated by the EXAFS studies (cf. above), the haem a_3 iron–Cu_B distance is lengthened as a result of reduction of the metals, also indicating a considerable conformational change in the site. Babcock and Wikström [23] suggested that electron transfer to the haem a_3 iron may always take place via Cu_B, but that the described structural modulation of the binuclear site allows this electron transfer to be either coupled to or decoupled from proton translocation. Alternatively, the structural modulation might allow direct electron transfer from haem *a* to haem a_3 (by-passing Cu_B) specifically during the 'electron-filling' phase of catalysis. As discussed above (and see Fig. 2), even a small structural perturbation could change the relative orientations of the two haems and thus enhance the probability of direct electron transfer between them [193].

Whereas the above arguments may locate the chemical reaction that drives proton pumping, and the switch controlling it, almost nothing is known about the proton transfers themselves and how they are coupled to the electron transfers. Below, we will discuss how far one might go in imagining possible mechanisms on the basis of present structural and functional knowledge.

enzyme might reflect an electron transfer pathway that is artificially introduced by the procedure required to remove the copper. It should be noted here that the quinol oxidases operate without Cu_A, although the quinol-binding site of these enzymes might be formed by a domain intrinsic to subunit II (cf. ref. [158]). Conversely, cytochrome *c* oxidases do not normally oxidise ubiquinol. However, it is interesting that oxidation of tetrahalogenated *p*-benzoquinols can be induced by treatment of the enzyme with strongly positively charged compounds, such as polylysine [159].

Cytochrome *a* is the donor of electrons into the binuclear site [160,161]. There are reasons to believe that Cu_B is the primary electron acceptor and that direct electron transfer between the haems does not occur at physiological velocities ([23], but see below). Our model has incorporated this by assuming that the two haems are orientated with respect to one another in a way that maximally disfavours direct electron transfer between them. Interestingly, Cu_B and haem *a* may be ligated by two histidines (H-408 and H-410) predicted to reside on opposite sides of the same transmembranous segment (segment X, see Fig. 2). In the amino acid sequence these histidines are separated by a single fully conserved phenylalanine (F-409). In addition, position 404 (Fig. 2) is always occupied by either Phe or Tyr. It is thus conceivable that electron transfer from the low-spin haem to Cu_B could take place via these aromatic residues. If so, it is not difficult to imagine how this electron transfer could be structurally modulated by subtle conformational changes of helix X.

The mechanism of O_2 reduction at the binuclear site is now relatively well understood. Its elucidation has been based on two independent approaches, viz. (i) time-resolved low temperature and room-temperature spectroscopy (optical, EPR, resonance Raman), where the reaction of the reduced enzyme with O_2 is initiated by photolysis of the CO-derivative of haem a_3 in the presence of O_2, and (ii) partial reversal of the catalytic reaction cycle in mitochondria and mitoplasts, by means of a high electrochemical proton gradient, under which conditions catalytic intermediates accumulate. Together, these approaches (see ref. [23], for details and references) have yielded a model of the catalytic O_2 cycle, shown here in Fig. 3. Although Fig. 3 is nearly self-explanatory, some of its details require comment.

We start considering the O_2 reaction from the state **R**, where the binuclear centre has been reduced by the first two electrons out of the four in all required to reduce O_2 to water. In the first bimolecular step, an O_2 molecule binds to the ferrous haem a_3 iron, forming an 'oxy' intermediate (**A**) analogous to oxyhaemoglobin or oxymyoglobin, first detected at low temperatures by Chance et al. [162]. Next, the O_2 molecule is effectively 'trapped' at the site by rapid transfers of the first two electrons to it from iron and copper, yielding the 'peroxy' intermediate **P**. Whilst the binding per se of O_2 is not very tight ($K_d \sim 0.3$ mM), the trapping event raises the apparent affinity for dioxygen considerably ($K_m \sim 0.1$ μM). The peroxy intermediate has not been observed in the forward reaction of the fully reduced enzyme with O_2, in which O_2 binding is characteristically very quickly followed by electron transfer from haem *a* to the binuclear site. However, **P** accumulates in reversed electron transfer experiments as a two-electron oxidation product of the ferric/cupric site plus water [163,164], and also during forward electron transfer at high protonmotive force [165]. Moreover, **P** is observed subsequent to the oxy intermediate when the 'mixed valence' enzyme (binuclear site reduced; haem *a* and Cu_A oxidised) reacts with dioxygen (compound C [162]), and when the oxidised ferric/cupric binuclear site (**O**; see Fig. 3) reacts with H_2O_2 [166,167]. The very fast oxidation of haem *a* in the reaction of the fully reduced enzyme with O_2 may, in fact, be artifactual in the sense that fully reduced enzyme is very unlikely to occur significantly in the aerobic steady state [23].

The formation of **P** is thought to be followed by transfer of the third electron from haem *a* to Cu_B and protonation of the peroxide. Then the O–O bond is broken, triggered by electron

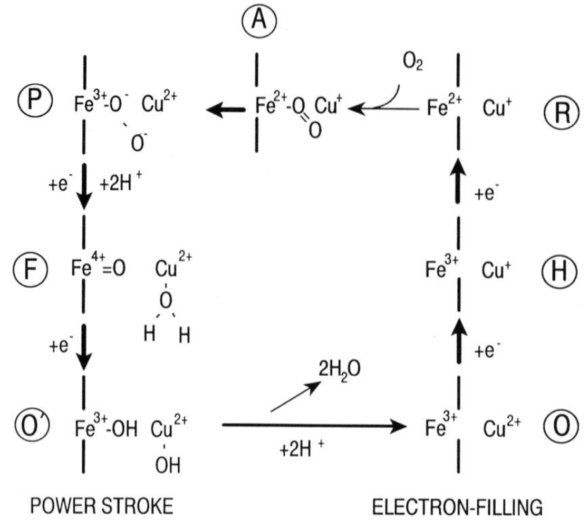

Fig. 3. The catalytic O_2 reduction cycle of the binuclear centre (adapted from refs. [23,189]. The binuclear site is depicted as haem Fe and Cu, corresponding to haem a_3 and Cu_B. The oxidised centre in the lower right corner (state **O**) is reduced stepwise by the first two electrons, derived from ferrocytochrome *c*, in the 'electron-filling' phase of the cycle. This yields the reduced centre (**R**). **R** binds O_2 to the haem iron forming the 'oxy' intermediate (**A**). This forces the haem iron into a low spin state from having been high spin in the intermediates **O**, **H** and **R**, as indicated schematically by the position of Fe relative to the haem plane. Highly exergonic and irreversible internal electron transfer traps dioxygen as the 'peroxy' intermediate (**P**, in which the O–O structure is probably bound 'end on' to the iron, without bridging to copper (see ref. [23]). Next, the third electron is transferred to haem a_3 iron via Cu_B, and one H^+ is taken up, producing a protonated ferrous-peroxy-cupric intermediate [157]; not shown). Then the O–O bond is broken, accompanied by uptake of another H^+, and the 'ferryl' species (**F**) is formed. The fourth electron from cytochrome *c* is transferred via Cu_B to the iron in **F**, forming a ferric/cupric hydroxide form of the oxidised centre (**O′**, [160,164,168]). This is finally protonated to produce state **O** with release of water. Proton pumping is not depicted in the figure, but occurs exclusively during the 'power stroke' phase (see ref. [180] and the text).

transfer from Cu_B to Fe and further protonation; O–O bond scission may be preceded by a ferrous-peroxy-cupric intermediate (not shown) having a distinctive copper EPR signal [157]. This produces a tetravalent iron-oxo (ferryl) intermediate (**F**), and a water molecule that remains coordinated to Cu_B [83,164,168]. The ferryl intermediate was first described in reversed electron transfer experiments, where it is formed by one-electron oxidation of the ferric/cupric state (**O**, Fig. 3; [163]). Subsequently, it has been observed also in time-resolved studies of the reaction of the fully reduced enzyme with O_2 [169], and the ferryl structure has been verified by resonance Raman spectroscopy [160,170–172].

The fourth electron is transferred from haem *a* to Cu_B, followed by reduction of Fe^{4+} to Fe^{3+} to produce the oxidised state **O** and water. An alkaline intermediate (**O′**) has been detected on the basis of the pH-dependence of this process [168], and the ferric hydroxide structure has been demonstrated by RR spectroscopy [160]. Finally, the cycle is completed by the transfers of the first two electrons to 'refill' the binuclear site, by which state **R** is regenerated and a new dioxygen cycle can occur.

With fully reduced soluble enzyme dispersed in detergent, one molecule of O_2 is converted to water in less than 100 µs (i.e. **R** + O_2 → **O** + 2 H_2O). In contrast, in the aerobic steady state the enzyme turns over at maximal velocities corresponding to a time scale of milliseconds. TN_{max} values around 500 s^{-1} are common for the isolated enzyme, or in mitochondria using artificial reductants such as tetramethyl-*p*-phenylenediamine (TMPD). Physiologically, however, the oxidase turnover is strongly limited by the input of electrons from the reducing foodstuffs, and turnovers faster than ~50 s^{-1} are hardly encountered. Moreover, spectroscopic features assignable to the oxygen intermediates are not normally seen during steady-state operation (but see ref. [165]), and the **O** state of the bimetallic site appears to be highly populated. These observations indicate that the 'electron-filling' reactions (Fig. 3) constitute the most effective rate-limiting steps of the catalytic cycle.

5. *Proton translocation and the conservation of energy*

5.1. *Some principles of the proton pump*

In conceptual models of proton translocation, which will not be reviewed here (but see, e.g. refs. [18,102,173–176] the proton–electron coupling may be classified as either direct or indirect. This refers to the type of physical association between the redox and proton transfer elements of the pump. Direct coupling implies that proton translocation occurs in close proximity of the redox element. This might take place, for example, by redox-dependent protonations/deprotonations of metal ligands, including water [21,177]. In indirect mechanisms, the proton transfer element is located at a distance from the redox element, and the coupling is mediated by long-range conformational changes (see ref. [178]). It is important to note, however, that independent of the degree of directness of coupling, a redox-linked proton pump mechanism must include conformational transitions that modulate input/output modes of both electron and proton transfer [173,176]. This is the crucial feature that distinguishes such mechanisms from the classical redox loop mechanism, concretised by the 'ubiquinone cycle' of the *bc*-type respiratory complexes [179]. The redox loop requires no such conformational modulations because the centres of 'output' and 'input' protonic and electronic function are separated in space and functionally connected to one another by specific electron and hydrogen transfer pathways. As pointed out by Wikström et al. [2], proton-pumping in cytochrome *c* oxidase would be very difficult to explain by the redox loop model because the enzyme does not contain redox carriers of hydrogen (such as ubiquinone). This limitation was suggested to be overcome by a model involving the O_2/H_2O_2 couple as the H-carrier [18], but this was subsequently abandoned due to insufficient experimental support [177].

5.2. *Catalytic steps linked to conservation of energy*

The catalytic O_2-cycle can be divided into three distinct sets of partial reactions that differ in their relative contributions to energy conservation. The 'electron-filling' phase (**O** → **R**, see Fig. 3) is not coupled to proton translocation. In contrast, the 'power stroke' phase (**P** → **O**) is linked to translocation of $4H^+$ in such a way that the **P** → **F** and **F** → **O** transitions are each coupled to translocation of $2H^+$ [180]. However, the electron transfers from cytochrome *c* to the binuclear site, which occur in both these reaction phases, are electrogenic since the binuclear site is located about halfway into the membrane dielectric, whereas cytochrome *c* is in the outer aqueous phase. Thus also the 'electron-filling' phase is associated with conservation of energy (generation of membrane potential) although its contribution to overall energy conservation is small (~19%) in comparison with the 'power stroke' (cf. ref. [181]). This situation might be somewhat different in the proton-pumping quinol oxidases (e.g. cytochrome *o* of *E. coli*). Here, the electron donor, a hydrophobic quinol, may be located near the middle of the membrane dielectric so that electron transfer to the binuclear site is electroneutral per se. However, the protons released from quinol upon oxidation (1 H^+/e^-) appear in the outer positively charged aqueous domain, together with the truly translocated protons [2,182]. Hence, electron transfer from the donor to the binuclear site would also in this case be an electrogenic process, although its basis might differ from the cytochrome *c* oxidase case.

The third distinct phase comprises the combined O_2-binding and trapping events (**R** → **A** → **P**, Fig. 3). This phase, which includes electron transfers within the binuclear site, is not coupled to conservation of energy. The O_2-trapping event is highly exergonic ($\Delta G'_o$ ~−6 kcal/mol), providing the major driving force for the whole catalytic cycle in conditions where a high electrochemical proton gradient prevails [23,165].

Protonation of reduced dioxygen to water takes place from the electrically negatively charged side (inside) of the mitochondrial or bacterial membrane [183]. This occurs in the steps **P** → **F** and **F** → **O** ($2H^+$ each; see Fig. 3), as determined in reversed electron transfer experiments. However, the latter protonation may be incomplete, as it depends on the prevailing local pH [168], and could thus in some conditions of high local pH be completed later during the 'electron-filling' phase. It is noteworthy that the proton uptake also contributes to energy conservation, generating both membrane potential and ΔpH.

Babcock and Wikström [23] have recently argued that the proton translocation events are most likely to be coupled to the reactions of the binuclear site in states **P** and **F**, triggered by electron transfer through Cu_B, i.e. from the low-spin haem to the haem a_3 iron-oxygen site. Cu_B was proposed to be the so-called redox element of the proton pump. However, there is a dilemma associated with this proposal if it is further assumed (as above) that all electron transfer from haem *a* to haem a_3 occurs via Cu_B: why is electron transfer through Cu_B coupled to proton-pumping in the 'power stroke' but not in the 'electron-filling' phases of the catalytic cycle? Basically, there are two ways out of this dilemma, both of which are founded on the fact that the structure of the bimetallic site is quite different during these two phases of catalysis. During 'electron-filling' haem a_3 iron has high spin, but upon binding of O_2 to state **R** (Fig. 3) it becomes low spin, and ligated to relatively strong field oxygenous ligands. Furthermore, as indicated by the EXAFS studies (cf. above), the haem a_3 iron–Cu_B distance is lengthened as a result of reduction of the metals, also indicating a considerable conformational change in the site. Babcock and Wikström [23] suggested that electron transfer to the haem a_3 iron may always take place via Cu_B, but that the described structural modulation of the binuclear site allows this electron transfer to be either coupled to or decoupled from proton translocation. Alternatively, the structural modulation might allow direct electron transfer from haem *a* to haem a_3 (by-passing Cu_B) specifically during the 'electron-filling' phase of catalysis. As discussed above (and see Fig. 2), even a small structural perturbation could change the relative orientations of the two haems and thus enhance the probability of direct electron transfer between them [193].

Whereas the above arguments may locate the chemical reaction that drives proton pumping, and the switch controlling it, almost nothing is known about the proton transfers themselves and how they are coupled to the electron transfers. Below, we will discuss how far one might go in imagining possible mechanisms on the basis of present structural and functional knowledge.

5.3. How might proton pumping occur?

By looking at the composition and sequences of various oxidase operons from organisms as distantly related as possible, a pattern of common structural themes has emerged. Since it seems reasonable to assume that proton translocation occurs by very similar, if not identical, mechanisms in any member of the large oxidase family of enzymes, these constant structures are likely to define the minimum requirements of the proton pump.

With the above in mind, it is clear that the proton-pumping machinery is to be found among the subunits I–III, which form the catalytic core of this class of enzymes. Of these, subunit III can be safely disregarded as directly involved in proton-pumping, as discussed above. Subunit II appears quite variable in amino acid sequence although the general secondary structure may be preserved (see ref. [3], and above). The Cu_A site in this subunit is not invariant either, being absent in the proton-pumping quinol oxidases. These simple considerations immediately lead to the conclusion that the most critical structures responsible for the proton pump mechanism are probably to be found in subunit I. This is consistent with Cu_B and the bimetallic site being part of this subunit.

A further comparison can be made to bacteriorhodopsin, the light-driven proton pump of *Halobacterium halobium*. Here, a Schiff base of the chromophore and several acidic amino acids are essential for the proton-pumping mechanism (see ref. [184] and refs. therein). Saraste [3] has pointed out the conserved acidic residues of subunit I, only two of which appear close to the binuclear site. Neither of these (E-275, D-401; see helices VI and X in Fig. 2) is conserved in the oxidase from *Sulfolobus* [59], and only E-275 is conserved in cytochrome *o* from *E. coli* (where it is E-286). It is not known yet whether the *Sulfolobus* enzyme pumps protons, though that would seem likely. However, after mutation of the E-286 of cytochrome *o* to glutamine in *E. coli*, the mutant spheroplasts were found to pump protons as efficiently as in the wild type (J. Thomas and A. Puustinen, unpublished). Therefore, this residue is also unlikely to be intimately involved in proton translocation.

We conclude that the proton-pumping mechanism of the oxidases does not reveal any obvious resemblance to the acidic residue-dependent mechanism in bacteriorhodopsin. Considering conservation of residues in subunits I from all known species, including the most recently described *Sulfolobus* oxidase, one arrives at the perhaps surprising conclusion that only some 35 amino acids are fully conserved, i.e. much less than 10%. Six of these are the metal-ligating histidines discussed above. Apart from these, the suggested catalytic core of subunit I (Fig. 2) includes the following as conserved potential proton-transfer elements: one tyrosine (Y-277 in helix VI), two tryptophans (W-269 in helix VI, W-320 in helix VII), three threonines/serines (T-341, T-348 and T-358, all in helix VIII; T-348 is replaced by a serine in the *Sulfolobus* enzyme), and one lysine (K-351 in helix VIII, which is replaced by a threonine in *Sulfolobus*).

Based on the discussion above, it may be surmised that proton/electron coupling is relatively direct, and might involve interactions between the ligands of Cu_B and conserved residues in helix VIII. We suggest that the input/output switching (see ref. [176]) is achieved by obligatory changes in Cu_B coordination and geometry upon oxidoreduction (cf. refs. [23,177,185]. Cu^{1+} complexes often assume tetrahedral geometry. However, several geometries are found for Cu^{2+} complexes, e.g. distorted octahedral with a planar array of four ligands and much longer bonds to axial ligands due to the Jahn–Teller effect [186]. This geometrical switching could, moreover, be thought to be coupled in a controlled fashion to protonation/deprotonation of histidine and possibly water ligands of Cu_B. These ligands may be envisioned to form transient hydrogen bonds with the abundant threonine (serine) OH groups in helix VIII (cf. ref. [56]), which could form parts of both input and output proton channels. Such hydrogen bonding, which could additionally involve structured water, must then be of a dynamic nature, breaking and reforming continuously in a controlled fashion during catalysis. Since it is chemically difficult to envisage how a single proton-translocating residue could effectuate the translocation of $2H^+$ per event (ref. [180], and cf. above), we are probably dealing with two groups that function synchronously. This requirement would easily be met by two Cu_B ligands, the steric properties of which must change in synchrony with the redox state of the metal.

Acknowledgements

This work has been supported by grants from the Academy of Finland and the Magnus Ehrnrooth Foundation (to T.H.), and the Sigrid Jusélius Foundation (to M.W.). We thank Hilkka Vuorenmaa and Lauri Ekman for help with the figures, and Y. Anraku, G.T. Babcock, R.B. Gennis, T. Kitagawa, T. Mogi, and M. Saraste for kindly providing us with manuscripts prior to publication.

References

1 Wikström, M. (1977) Nature 266, 271–273.
2 Wikström, M., Krab, K. and Saraste, M. (1981) Cytochrome Oxidase – A Synthesis, Academic Press, New York.
3 Saraste, M. (1990) Q. Rev. Biophys. 23, 331–366.
4 Valpuesta, J.M., Henderson, R. and Frey, T.G. (1990) J. Mol. Biol. 214, 237–251.
5 Raitio, M., Jalli, T. and Saraste, M. (1987) EMBO J. 6, 2825–2833.
6 Steinrücke, P., Steffens, G.C.M., Panskus, G., Buse, G. and Ludwig, B. (1987) Eur. J. Biochem. 167, 431–439.
7 Haltia, T., Puustinen, A. and Finel, M. (1988) Eur. J. Biochem. 172, 543–546.
8 Chepuri, V., Lemieux, L., Au, D.C.T. and Gennis, R.B. (1990) J. Biol. Chem. 265, 11185–11192.
9 Ishizuka, M., Machida, K., Shimada, S., Mogi, A., Tsuchiya, T., Ohmori, T., Souma, Y., Gonda, M. and Sone, N. (1990) J. Biochem. 108, 866–873.
10 Sone, N., Shimada, S., Okumori, T., Souma, Y., Gonda, M. and Ishizuka, M. (1990) FEBS Lett. 262, 249–252.
11 Lauraeus, M., Haltia, T., Saraste, M. and Wikström, M. (1991) Eur. J. Biochem. 197, 699–705.
12 Puustinen, A., Finel, M., Haltia, T., Gennis, R.B. and Wikström. M. (1991) Biochemistry 30, 3936–3942.
13 Puustinen, A. and Wikström, M. (1991) Proc. Natl. Acad. Sci. USA 88, 6122–6126.
14 Sone, N. and Fujiwara, Y. (1991) FEBS Lett. 288, 154–158.
15 Mather, M.W., Springer, P. and Fee, J.A. (1991) J. Biol. Chem. 266, 5025–5035.
16 Saraste, M., Metso, T., Nakari, T., Jalli, T., Lauraeus, M. and van der Oost, J. (1991) Eur J. Biochem. 195, 517–525.
17 Wu, W., Chang, C.K., Varotsis, C., Babcock, G.T., Puustinen, A. and Wikström, M. (1992) J. Am. Chem. Soc., in press.
18 Mitchell, P., Mitchell, R., Moody, A.J., West, I.C., Baum, H. and Wrigglesworth, J. (1985) FEBS Lett. 188, 1–7.
19 Capitanio, N., Capitanio, G., De Nitto, E., Villani, G., Papa, S. (1991) FEBS Lett. 288, 179–192.
20 Papa, S., Capitanio, N., Capitanio, G., De Nitto, E. and Minuto, M. (1991) FEBS Lett. 288, 183–186.
21 Wikström, M. and Krab, K. (1978) in: G. Schäfer and M. Klingenberg (Eds.) Energy Conservation in Biological Membranes, Springer, Berlin, pp. 128–139.
22 Wikström, M. and Saraste, M. (1984) in: L. Ernster (Ed.) Bioenergetics, Elsevier, Amsterdam, pp. 49–94.
23 Babcock, G.T. and Wikström, M. (1992), Nature 356, 301–309.

24 Beinert, H., Griffiths, D.E., Wharton, D.C. and Sands, R.H. (1962) J. Biol. Chem. 237, 2337–2346.
25 Vänngård, T. (1972) in: Swarz, H.M., Bolton, J.R. and Borg, D.C. (Eds.) Biological Applications of Electron Spin Resonance, Wiley, New York, pp. 411–447.
26 Chan, S.I., Bocian, D.F., Brudwig, G.W., Morse, R.H. and Stevens, T.H. (1979) in: King, T.E., Orii, Y., Chance, B. and Okunuki, K. (Eds.) Cytochrome Oxidase Elsevier, Amsterdam, pp. 177–188.
27 Martin, C.T., Scholes, C.P. and Chan, S.I. (1988) J. Biol. Chem. 263, 8420–8429.
28 Steffens, G.J. and Buse, G. (1979) Hoppe-Seyler's Z. Physiol. Chem. 360, 613–619.
29 Covello, P.S. and Gray, M.W. (1990) FEBS Lett. 268, 5–7.
30 Holm, L., Saraste, M. and Wikström, M. (1987) EMBO J. 6, 2819–2823.
31 Blair, D.F., Martin, C.T., Gelles, J., Wang, H., Brudwig, G.W., Stevens, T.H. and Chan, S.I. (1983) Chem. Scr. 21, 43–53.
32 Stevens, T., Martin, C.T., Wang, H., Brudwig, G., Scholes, C.P. and Chan, S.I. (1982) J. Biol. Chem. 257, 12106–12113.
33 Scott, R.A. (1989) Annu. Rev. Biophys. Biophys. Chem. 18, 137–158.
34 Powers, L. and Kincaid, B.M. (1989) Biochemistry 28, 4461–4468.
35 Scott, R.A., Schwarz, J.R. and Cramer, S.P. (1986) Biochemistry 25, 5546–5555.
36 Powers, L., Chance, B., Ching, Y. and Angiolillo, P. (1981) Biophys. J. 34, 465–498.
37 Li, P.M., Gelles, J., Chan, S.I., Sullivan, R.J. and Scott, R.A. (1987) Biochemistry 26, 2091–2095.
38 Takahashi, S., Ogura, T., Itoh-Shinzawa, K., Yoshikawa, S. and Kitagawa, T. (1991) J. Am. Chem. Soc. 113, 9400–9401.
39 Greenwood, C., Thomson, A.J., Barrett, C.P., Peterson, J., George, G.N., Fee, J.A. and Reichardt, J. (1988) Annu N.Y. Acad. Sci. 550, 47–52.
40 Aasa, R., Albracht, S.P.J., Falk, K.-E., Lanne, B. and Vänngård, T. (1976) Biochim. Biophys. Acta 422, 260–272.
41 Peisach, J. and Blumberg, W.E. (1974) Arch. Biochem. Biophys. 165, 691–708.
42 Maret, W., Zeppezauer, M., Desideri, A., Morpurgo, L. and Rotilio, G. (1981) FEBS Lett. 136, 72–74.
43 Eklund, H., Nordström, B., Zeppezauer, E., Söderlund, G., Ohlsson, I., Boiwe, T., Söderberg, B.-O., Tapia, O., Bränden, C.-I. and Åkeson, Å. (1976) J. Mol. Biol. 102, 27–59.
44 Kroneck, P., Antholine, W.A., Riester, J. and Zumft, W.G. (1988) FEBS Lett. 242, 70–74.
45 Kroneck, P., Antholine, W.A., Kastrau, D.H.W., Buse, G., Steffens, G.C.M. and Zumft, W. G. (1990) FEBS Lett. 268, 274–276.
46 Scott, R.A., Zumft, W.G., Coyle, C.L. and Dooley, D.M. (1989) Proc. Natl. Acad. Sci. USA 86, 4082–4086.
47 Viebrock, A. and Zumft, W.G. (1988) J. Bacteriol. 170, 4658–4668.
48 Zhang, C., Hollocher, T.C., Kolodziej, A.F. and Orme-Johnson, W.H. (1991) J. Biol. Chem. 266, 2199–2202.
49 Froncisz, W., Scholes, C.P., Hyde, J.S., Wei, Y.-H., King, T.E., Shaw, R.W. and Beinert, H. (1979) J. Biol. Chem. 254, 7482–7484.
50 Bombelka, E., Richter, F.-W., Stroh, A. and Kadenbach, B. (1986) Biochem. Biophys. Res. Commun. 140, 1007–1014.
51 Steffens, C.G.M., Biewald, R. and Buse, G. (1987) Eur. J. Biochem. 164, 295–300.
52 Öblad, M., Selin, E., Malmström, B., Strid, L., Aasa, R. and Malmström, B.G. (1989) Biochim. Biophys. Acta 975, 267–270.
53 Buse, G. and Steffens, C.G.M. (1991) J. Bioenerg. Biomembranes 23, 269–289.
54 Riester, J., Zumft, W.G. and Kroneck, P.M.H. (1989) Eur. J. Biochem. 178, 751–762.
55 Brudwig, G.W., Blair, D.F. and Chan, S.I. (1984) J. Biol. Chem. 259, 11001–11009.
56 Yoshikawa, S. and Caughey, W.S. (1990) J. Biol. Chem. 265, 7945–7958.
57 Yoshikawa, S., Tera, T., Takahashi, Y., Tsukihara, T. and Caughey, W.S. (1988) Proc. Natl. Acad. Sci. USA 85, 1354–1358.
58 Pan, L.-P., Li, Z., Larsen, R. and Chan, S.I. (1991) J. Biol. Chem. 266, 1367–1370.
59 Lübben, M., Kolmerer, B. and Saraste, M. (1992) EMBO J. 11, 805–812.
60 Peisach, J. (1978) in: Dutton, P.L., Leigh, J.S. and Scarpa, A. (Eds.) Frontiers of Biological Energetics

Vol. 2, Academic Press, New York, pp. 873–881.
61 Martin, C.T., Scholes, C.P. and Chan, S.I. (1985) J. Biol. Chem. 260, 2857–2861.
62 Babcock, G.T. (1988) in: Spiro, T.G. (Ed.) Biological Applications of Raman Spectroscopy, Wiley, New York, pp. 293–346.
63 Sassaroli, M., Ching, Y., Dasgupta, S. and Rousseau, D.L. (1989) Biochemistry 28, 3128–3132.
64 Viola, R.E., Shaw, R.W., Ransom, S.C. and Villafranca, J.J. (1983) Arch. Biochem. Biophys. 220, 106–115.
65 Blumberg, W.E. and Peisach, J. (1971) in: Probes of Structure and Function of Macromolecules and Membranes, Vol.II, Probes of Enzymes and Hemoproteins, Academic Press, New York, pp. 215–229.
66 Babcock, G.T., van Steelandt, J., Palmer, G., Vickery, L.E. and Salmeen, I. (1979) in: King, T.S., Orii, Y., Chance, B. and Okunuki, K. (Eds.) Cytochrome Oxidase, Elsevier, Amsterdam, pp. 105–115.
67 Albracht, S.P.J., van Verseveld, H.W., Hagen, W.R. and Kalkman, M.L. (1980) Biochim. Biophys. Acta 593, 173–186.
68 Hosler, J.P., Tecklenburg, M.J., Atamian, M., Revzin, A., Ferguson-Miller, S. and Babcock, G.T. (1991) Biophys. J. 59, 470a.
69 Sherman, D., Kotake, S., Ishibe, N. and Copeland, R.A. (1991) Proc. Natl. Acad. Sci. USA 88, 4265–4269.
70 Copeland, R.A. (1991) Proc. Natl. Acad. Sci. USA 88, 7281–7283.
71 Moore, G.R. and Pettigrew, G.W. (1990) Cytochromes *c*; Evolutionary, Structural and Physicochemical Aspects, Springer, Berlin.
72 Lemieux, L.L., Calhoun, M.W., Thomas, J.W., Ingledew, W.J. and Gennis, R.B. (1992) J. Biol. Chem. 267, 2105–2113.
73 Minagawa, J., Mogi, T., Gennis, R.B. and Anraku, Y. (1992) J. Biol. Chem. 267, 2096–2104.
74 Shapleigh, J.P., Hosler, J.P., Tecklenburg, M.J., Kim, Y., Babcock, G.T. Gennis, R.B. and Ferguson-Miller, S. (1992) Proc. Natl. Acad. Sci. USA, 89, 4786–4790.
75 Van Gelder, B.F. and Beinert, H. (1969) Biochim. Biophys. Acta 189, 1–24.
76 Thomson, A.J., Johnson, M.K., Greenwood, C. and Gooding, P.E. (1981) Biochem. J. 193, 687–697.
77 Kent, T.A., Münck, E., Dunham, W.R., Filter, W.F., Findling, K.L., Yoshida, T. and Fee, J. A. (1982) J. Biol. Chem. 257, 12489–12492.
78 Kent, T.A., Young. L.J., Palmer, G., Fee, J.A. and Münck, E. (1983) J. Biol. Chem. 258, 8543–8546.
79 Reinhammar, B., Malkin, R., Jensen, P., Karlsson, B., Andreasson, L.-E., Aasa, R., Vänngård, T. and Malmström, B.G. (1980) J. Biol. Chem. 255, 5000–5003.
80 Karlsson, B. and Andreasson, L.-E. (1981) Biochim. Biophys. Acta 635, 73–80.
81 Karlsson, B., Aasa, R., Vänngård, T. and B.G. Malmström (1981) FEBS Lett. 131, 186–188.
82 Cline, J., Reinhammar, B., Jensen, P., Venters, R. and Hoffman, B.M. (1983) J. Biol. Chem. 258, 5124–5128.
83 Hansson, Ö., Karlsson, B., Aasa, R., Vänngård, T. and Malmström, B.G. (1982) EMBO J. 1, 1295–1297.
84 Ingledew, W.J. and Bacon, M. (1991) Biochem. Soc. Trans. 19, 613–616.
85 Shaw, R.W., Hansen, R.E. and Beinert, H. (1978) J. Biol. Chem. 253, 6637–6640.
86 Armstrong, F., Shaw, R.W. and Beinert, H. (1983) Biochim. Biophys. Acta 722, 61–71.
87 Woodruff, W.H., Einarsdottir, O., Dyer, R.B., Bagley, K.A., Palmer, G., Atherton, S.J., Goldbeck, R.A., Dawes, T.D. and Kliger, D.S. (1991) Proc. Natl. Acad. Sci. USA 88, 2588–2592.
88 Blokzijl-Homan, M.F.J. and van Gelder, B.F. (1971) Biochim. Biophys. Acta 234, 493–498.
89 Stevens, T.H. and Chan, S.I. (1981) J. Biol. Chem. 256, 1069–1071.
90 Goodman, G. (1984) J. Biol. Chem. 259, 15094–15099.
91 Goodman, G. and Leigh, J.S., Jr. (1987) Biochim. Biophys. Acta 890, 360–367.
92 Einarsdottir, O., Choc, M.G., Weldon, S. and Caughey, W.S. (1988) J. Biol. Chem. 263, 13641–13654.
93 Rousseau, D.L., Singh, S., Ching, Y. and Sassaroli, M. (1988) J. Biol. Chem. 263, 5681–5685.
94 Salerno, J.C., Bolgiano, B., Poole, R.K., Gennis, R.B. and Ingledew, W.J. (1990) J. Biol. Chem. 265, 4364–4368.
95 Jones, M.G., Bickar, D., Wilson, M.T., Brunori, M., Colosimo, A. and Sarti, P. (1984) Biochem. J. 220, 57–66.

96 Ellis, W.J., Wang, H., Blair, D.F., Gray, H.B. and Chan, S.I. (1986) Biochemistry 25, 161–167.
97 Scholes, C.P. and Malmström, B.G. (1986) FEBS Lett. 198, 125–129.
98 Wang, H., Blair, D.F., Ellis, W.R., Jr., Gray, H.B. and Chan, S.I. (1986) Biochemistry 25, 167–171.
99 Alleyne, T.A. and Wilson, M.T. (1987) Biochem. J. 247, 475–484.
100 Fan, C., Bank, J.F., Dorr, R.G. and Scholes, C.P. (1988) J. Biol. Chem. 263, 3588–3591.
101 Kornblatt, J.A., Hoa, G.H.B. and Heremans, K. (1988) Biochemistry 27, 5122–5128.
102 Malmström, B.G. (1990) Arch. Biochem. Biophys. 280, 233–241.
103 Malmström, B.G. (1990) Chem. Rev. 90, 1247–1260.
104 Haltia, T. (1992) Biochim. Biophys. Acta, 1098, 343–350.
105 Li, P.M., Morgan, J.E., Nilsson, T., Ma, M. and Chan, S.I. (1988) Biochemistry 27, 7538–7546.
106 Wikström, M.K.F., Harmon, H.J., Ingledew, W.J. and Chance, B. (1976) FEBS Lett. 65, 259–277.
107 Blair, D.F., Ellis, W.R., Wang, H., Gray, H.B. and Chan, S.I. (1986) J. Biol. Chem. 261, 11524–11537.
108 Hendler, R.W. and Westerhoff, H.V. (1992), Biophys. J., in press.
109 Blair, D.F., Bocian, D.F., Babcock, G.T. and Chan, S.I. (1982) Biochemistry 21, 6928–6935.
110 Scott, R.A., Li, P.M. and Chan, S.I. (1988) Annu N. Y. Acad. Sci. 550, 53–58.
111 Powers, L., Chance, B., Ching, Y. and Lee, C.-P. (1987) J. Biol. Chem. 262, 3160–3164.
112 Powers, L., Chance, B., Ching, Y., Muhoberac, B., Weintraub, S. T. and Wharton, D.C. (1982) FEBS Lett. 138, 245–248.
113 Moody, A.J., Brandt, U. and Rich, P.R. (1991) FEBS Lett. 293, 101–105.
114 Moody, A.J., Cooper, C. and Rich, P.R. (1991) Biochim. Biophys. Acta 1059, 189–207.
115 Finel, M. (1989) Oligomeric Structure and Subunit Requirement of Proton-Translocating Cytochrome Oxidase, PhD Thesis, University of Helsinki.
116 Finel, M. and Wikström, M. (1986) Biochim. Biophys. Acta 851, 99–108.
117 Sone, N. and Kosako, T. (1986) EMBO J. 5, 1515–1519.
118 Moody, A.J. and Rich, P.R. (1989) Biochim. Biophys. Acta 973, 29–34.
119 Müller, M., Schläpfer, B. and Azzi, A. (1988) Biochemistry 27, 7546–7551.
120 Blum, H., Harmon, H.J., Leigh, J.S., Salerno, J.C. and Chance, B. (1978) Biochim. Biophys. Acta 502, 1–10.
121 Erecinska, M., Wilson, D.F. and Blasie, J.K. (1979) Biochim. Biophys. Acta 545, 352–364.
122 Salerno, J.C. and Ingledew, W.J. (1991) Eur. J. Biochem. 198, 789–792.
123 Wikström, M., Saraste, M. and Penttilä, T. (1985) in: A.N. Martonosi (Ed.) The Enzymes of Biological Membranes, Vol. 4, Plenum, New York pp. 111–148.
124 Malmström, B.G. (1989) FEBS Lett. 250, 9–21.
125 Goodman, G. and Leigh, J.S., Jr. (1985) Biochemistry 24, 2310–2317.
126 Hinkle, P. and Mitchell, P. (1970) J. Bioenerg. 1, 45–60.
127 Rich, P.R., West, I.C. and Mitchell, P. (1988) FEBS Lett. 233, 25–30.
128 Chepuri, V. and Gennis, R.B. (1990) J. Biol. Chem. 265, 12978–12986.
129 Capaldi, R.A., Malatesta, F. and Darley-Usmar, V. M. (1983) Biochim. Biophys. Acta 726, 135–148.
130 Imai, M., Shimada, H., Watanabe, Y., Matsushima-Hibiya, Y., Makino, R., Koga, H., Horiuchi, T. and Ishimura, Y. (1989) Proc. Natl. Acad. Sci. USA 86, 7823–7827.
131 Von Heijne, G. (1991) J. Mol. Biol. 218, 499–503.
132 Bisson, R., Jacobs, B. and Capaldi, R.A. (1980) Biochemistry 19, 4173–4178.
133 Millet, F., Darley-Usmar, V. and Capaldi, R.A. (1982) Biochemistry 21, 3857–3862.
134 Finel, M. (1988) FEBS Lett. 236, 415–419.
135 Casey, R.P., Thelen, M. and Azzi, A. (1980) J. Biol. Chem. 255, 3994–4000.
136 Prochaska, L.J., Bisson, R., Capaldi, R.A., Steffens, G.C.M. and Buse, G. (1981) Biochim. Biophys. Acta 637, 360–373.
137 Haltia, T., Saraste, M. and Wikström, M. (1991) EMBO J. 10, 2015–2021.
138 Haltia, T., Finel, M., Harms, N., Nakari, T., Raitio, M., Wikström, M. and Saraste, M. (1989) EMBO J. 8, 3571–3579.
139 Solioz, M., Carafoli, E. and Ludwig, B. (1982) J. Biol. Chem. 257, 1579–1582.
140 Hendler, R.W., Pardhasaradhi, K., Reynafarje, B. and Ludwig, B. (1991) Biophys. J. 60, 415–423.

141 Brunori, M., Antonini, G., Malatesta, F., Sarti, P. and Wilson, M.T. (1987) Eur. J. Biochem.169, 1–8.
142 Finel, M. and Wikström, M. (1988) Eur. J. Biochem. 176, 125–129.
143 Rigell, C.W., de Saussure, C. and Freire, E. (1985) Biochemistry 26, 4366–4371.
144 Rigell, C.W. and Freire, E. (1987) Biochemistry 26, 4366–4371.
145 Morin, P.E., Diggs, D. and Freire, E. (1990) Biochemistry 29, 781–788.
146 Capaldi, R.A. (1990) Arch. Biochem. Biophys. 280, 252–262.
147 Libert, F., Ruel, J., Ludgate, M., Swillens, S., Alexander, N., Vassart, G. and Dinsart, C. (1987) EMBO J. 6, 4193–4196.
148 Messerschmidt, A., Rossi, A., Ladenstein, R., Huber, R., Bolognesi, M., Gatti, G., Marchesini, A., Petruzzelli, R. and Finazzi-Agro, A. (1989) J. Mol. Biol. 206, 513–529.
149 Morgan, J.E. and Wikström, M. (1991) Biochemistry 30, 948–958.
150 Hill, B.C. (1991) J. Biol. Chem. 266, 2219–2226.
151 Antalis, T.M. and Palmer, G. (1982) J. Biol. Chem. 257, 6194–6206.
152 Morgan, J.E., Li, P.M., Jang, D.-J., El-Sayed, M.A. and Chan, S.I. (1989) Biochemistry 28, 6975–6983.
153 Kobayashi, K., Une, H. and Hayashi, K.J. (1989) J. Biol. Chem. 264, 7976–7980.
154 Gelles, J. and Chan, S.I. (1985) Biochemistry 24, 3963–3972.
155 Pan, L.-P., Hazzard, J.T., Lin, J., Tollin, G. and Chan, S.I. (1991) J. Am. Chem. Soc. 113, 5908– 5910.
156 Clore, G.M., Andreasson, L.-E., Karlsson, B. and Malmström, B.G. (1980) Biochem. J. 185, 139–154.
157 Blair, D.F., Witt, S.N. and Chan, S.I. (1985) J. Am. Chem. Soc. 107, 7389–7399.
158 Dueweke, T.J. and Gennis, R.G. (1991) Biochemistry 30, 3401–3406.
159 Jacobs, E.E., Andrews, E.C. and Crane, F.L. (1964) in: T.E. King, H.S. Mason and M. Morrison, (Eds.) Oxidases and Related Redox Systems, Vol. 2, Wiley, New York, pp. 784–803.
160 Han, S., Ching, Y.-C. and Rousseau, D.L. (1990) Biochemistry 29, 1380–1384.
161 Oliveberg, M. and Malmström, B.G. (1991) Biochemistry 30, 7053–7057.
162 Chance, B., Saronio, C. and Leigh, J.S., Jr. (1975) J. Biol. Chem. 250, 9226–9237.
163 Wikström, M. (1981) Proc. Natl. Acad. Sci. USA 78, 4051–4054.
164 Wikström, M. (1987) Chemica Scripta 27B, 53–58.
165 Wikström, M. and Morgan, J.E. (1992) J. Biol. Chem. 267, 10266–10273.
166 Wrigglesworth, J.M. (1984) Biochem. J. 217, 715–719.
167 Vygodina, T.V. and Konstantinov, A.A. (1989) Biochim. Biophys. Acta 973, 390–398.
168 Wikström, M. (1988) Chemica Scr. A28, 71–74.
169 Orii, Y. (1988) Annu N.Y. Acad. Sci. 550, 105–117.
170 Varotsis, C. and Babcock, G.T. (1990) Biochemistry 29, 7357–7362.
171 Ogura, T., Takahashi, S., Shinzawa-Itoh, K., Yoshikawa, S. and Kitagawa, T. (1990) J. Biol. Chem. 265, 14721–14723.
172 Ogura, T., Takahashi, S., Shinzawa-Itoh, K., Yoshikawa, S. and Kitagawa, T. (1991) Bull. Chem. Soc. Jpn. 64, 2901–2907.
173 Wikström, M. and Krab, K. (1979) Biochim. Biophys. Acta 549, 177–222.
174 Blair, D.F., Gelles, J. and Chan, S.I. (1986) Biophys. J. 50, 713–733.
175 Malmström, B.G. (1985) Biochim. Biophys. Acta 811, 1–12.
176 Krab, K. and Wikström, M. (1987) Biochim. Biophys. Acta 895, 25–39.
177 Mitchell, P. (1988) Annu N.Y. Acad. Sci. 550, 185–198.
178 Williams, R.J.P. (1987) FEBS Lett. 226, 1–7.
179 Mitchell, P. (1976) J. Theor. Biol. 62, 327–367.
180 Wikström, M. (1989) Nature 338, 776–778.
181 Slater, E.C. (1990) in: Kim., C.H. and Ozawa (Eds.) Bioenergetics. Molecular Biology, Biochemistry and Pathology, Plenum, New York, pp. 163–170.
182 Puustinen, A., Finel, M., Virkki, M. and Wikström, M. (1989) FEBS Lett. 249, 163–169.
183 Wikström, M. (1988) FEBS Lett. 231, 247–252.
184 Henderson, R., Baldwin, J.M., Ceska, F., Zemlin, F., Beckmann, E. and Downing, K.H. (1990) J. Mol. Biol. 213, 899–929.

185 Mitchell, P. (1987) FEBS Lett. 222, 235–245.
186 Cotton, F.A. and Wilkinson, G. (1988) Advanced Inorganic Chemistry, 5th edn., Wiley, New York.
187 Raitio, M., Pispa, J.M., Metso, T. and Saraste, M. (1990) FEBS Lett. 261, 431–435.
188 Falk, J.E. (1964) Porphyrins and Metalloporphyrins, Elsevier, Amsterdam, p. 6.
189 Wikström, M. and Babcock, G.T. (1990) Nature 348, 16–17.
190 Moody, A.J. and Rich, P.R. (1990) Biochim. Biophys. Acta 1015, 205–215.
191 Saraste, M., Holm, L., Lemieux, L., Lübben, M. and van der Oost, J. (1991) Biochem. Soc. Trans. 19, 608–612.
192 Buse, G., Steffens, C.G.M. Biewald, R., Bruch, B. and Hensel, S. (1987) in: S. Papa, B. Chance and L. Ernster (Eds.) Cytochrome Systems Molecular Biology and Bioenergetics, Plenum, New York, pp. 261–270.
193 Haltia, T. (1992) Cytochrome *c* Oxidase. Biochemical, Genetic and Spectroscopic Studies using the Enzyme from *Paracoccus denitrificans*, PhD Thesis, University of Helsinki.
194 Brown, S., Moody, A.J., Jeal, A.E., Bourne, R.M., Mitchell, J.R. and Rich, P.R. (1992) EBEC Short Rep. 7, 39.

L. Ernster (Ed.) *Molecular Mechanisms in Bioenergetics*
© 1992 Elsevier Science Publishers B.V. All rights reserved

CHAPTER 10

Cytochrome *c* oxidase: tissue-specific expression of isoforms and regulation of activity

BERNHARD KADENBACH and ACHIM REIMANN

Fb Chemie, Biochemie der Philipps-Universität, Hans-Meerwein-Straße, 3550 Marburg, Germany

Contents

1. Introduction 241
2. Comparison of cytochrome oxidases from prokaryotes and eukaryotes 242
 2.1. *Paracoccus denitrificans* 242
 2.2. *Escherichia coli* 243
 2.3. Other bacteria 243
3. Evolution of eukaryotic cytochrome *c* oxidase (COX) 244
 3.1. Subunit composition 244
 3.2. Isoforms of nuclear-coded subunits 245
 3.3. COX isozymes 246
4. Genes and protein structure of nuclear-coded subunits 246
 4.1. cDNAs, functional genes and pseudogenes 247
 4.2. Protein sequences of nuclear-coded subunits 247
 4.3. Import into mitochondria and assembly 250
5. Regulation of COX activity 252
 5.1. Isosteric effectors 252
 5.2. Allosteric effectors 253
 5.3. Regulation via expression of tissue-specific isoforms 254
 5.4. Electrochemical potential 256
6. Conclusion 257
Acknowledgements 257
References 258

1. Introduction

Cytochrome oxidase is the only enzyme that reduces dioxygen to water without the formation of poisonous intermediates like H_2O_2 or the superoxide radical. The reaction is accompanied by formation of a transmembrane electrochemical potential that can be utilized in the synthesis of ATP. Since the oxidation of glucose to CO_2 and H_2O yields almost 20 times more

ATP than fermentation to lactate, cell respiration, and thus cytochrome oxidase, the terminal electron acceptor of the respiratory chain, has been of central importance for the evolution of aerobic life.

Cytochrome oxidase evolved differently in prokaryotes and eukaryotes. In prokaryotes a diversification occurred into different forms, with a low but variable number of subunits, different cytochrome components, using different electron donors, and variable proton-pumping activities. By contrast, evolution of eukaryotic cytochrome oxidase was characterized by a strict conservation of the three catalytic subunits, encoded on mitochondrial DNA, but an extensive specification of the enzyme by increasing the number of nuclear-coded, regulatory subunits, which are partly expressed in developmental and tissue-specific forms. Since all mitochondrial cytochrome oxidases use cytochrome *c* as substrate, they are designated cytochrome *c* oxidase (COX). For the prokaryotic enzyme the term cytochrome oxidase will be used, because different electron donors are used as substrates.

Isozymes of cytochrome oxidase occur in prokaryotes as well as in eukaryotes, but the design is different. In prokaryotes, various genes code for the catalytic subunits of isozymes. In eukaryotes the same gene occurs for each catalytic subunit on mitochondrial DNA. The isozymes are assembled in mitochondria together with different isoforms of nuclear-coded subunits.

In this review, information on bacterial isozymes of cytochrome oxidase is summarized, and the structure, genes and regulatory function of nuclear-coded subunits of mitochondrial COX isozymes are discussed. Several reviews have appeared on the catalytic mechanism of COX (refs. [1–3], see also Haltia and Wikström, this volume) and on the structure and genes of mitochondrially coded subunits [4–8].

2. *Comparison of cytochrome oxidases from prokaryotes and eukaryotes*

The cytochrome oxidases of prokaryotes [9,10] are either closely related to the eukaryotic enzyme, e.g. cytochrome oxidase of *Paracoccus denitrificans*, which contains the same prosthetic groups and can accept electrons from the same donor [11], or differ in both the heme components and electron donor, e.g. cytochrome *o* ubiquinol oxidase from *Escherichia coli* [12]. Nevertheless, comparison of amino acid sequences has clearly established their evolutionary relationship [12]. The diversification of cytochrome oxidase in prokaryotes was apparently due to adaptation to different environmental conditions, e.g. supply of nutrients, temperature, pH, etc. Besides different types of cytochrome oxidases, isoforms of the same type were also found to occur in the same bacterium.

2.1. *Paracoccus denitrificans*

The catalytic properties of isolated cytochrome oxidase from aerobically grown *P. denitrificans* are almost identical to those of the mammalian enzyme [13,14]. This is due to its similar structure, consisting of three subunits which are homologous to subunits I–III of the mammalian enzyme [15], and containing the same prosthetic groups: two heme *a* molecules (cytochrome aa_3) and one Cu ion (Cu_B) in subunit I [16,17], and another Cu ion (Cu_A) in subunit II [11,18]. The four redox centers enable cytochrome oxidase to transfer simultaneously four electrons to dioxygen, which is bound at the binuclear cluster cytochrome a_3/Cu_B. After vectorial uptake of four protons from the inner phase, two molecules of water are formed [1,2,19].

In addition the enzyme pumps protons out of the cell with a stoichiometry of 1 H^+/e^- [20]. Subunit III, which carries no prosthetic group and can easily be separated from the functional enzyme [11], was suggested to participate in proton pumping, like subunit III of the mammalian enzyme [21,22], and to function in the assembly of the enzyme [23]. Recently, however, the proton-pumping function of subunit III could be excluded by site-directed mutagenesis studies [24]. A third copper atom was found per monomer in cytochrome oxidase from mammals [25–27], *Paracoccus* [26] and *Bacillus cereus* [28]. More recent data suggest a stoichiometry of 5Cu/4Fe/2Zn/2Mg per dimer of mitochondrial COX [29,30]. The extra Cu ion was suggested to play a structural role in enzyme dimerization [30].

Despite the large catalytic similarity between cytochrome oxidases from *Paracoccus* and mitochondria, they differ in the physiological electron donor. While the mitochondrial enzyme uses the soluble 12 kDa cytochrome c_{550}, the *Paracoccus* enzyme accepts in vivo only electrons from a membrane-bound 22 kDa cytochrome c_{552} [31], although the soluble cytochrome *c* does also occur. In fact, an enzymatically active 'supercomplex' has been isolated from *Paracoccus*, containing cytochrome bc_1, c_{552} and aa_3 [32].

Analysis of the gene structure of one cytochrome oxidase operon in *Paracoccus* has led to the identification of three additional open reading-frames between the genes for subunits II and III [15]. Two of these open reading-frames were strongly suggested to code for two independent assembly factors [31].

Recently another gene for subunit I was found in *Paracoccus*, which differs only slightly from the normal gene [33]. This would lead to an isozyme of cytochrome oxidase differing only in subunit I. The meaning of expressing two very similar isozymes in *Paracoccus* is unknown.

At least two other cytochrome oxidases are expressed in *Paracoccus*, depending on the growth conditions [34]: cytochrome *o*, which occurs also in *E. coli* as the main oxidase, and cytochrome *co*, containing hemes *c* and *b* as prostetic groups [35]. In contrast to cytochrome aa_3 and cytochrome *o*, cytochrome *co* does not pump protons [34].

2.2. *Escherichia coli*

The main cytochrome oxidase of *E. coli* is cytochrome *o* [10], containing two hemes *b* and only one Cu ion [36]. Under low oxygen conditions another oxidase, cytochrome *d*, is expressed, which contains three heme groups [37], but is unable to pump protons [36]. The two subunits of cytochrome *d* show no relationship with any known protein of respiratory electron transport complexes [38]. Cytochrome *o* transfers electrons from ubiquinol to dioxygen, omitting the mitochondrial intermediates cytochrome bc_1 and cytochrome *c*, which are absent in *E. coli* [39]. The isolated enzyme contains four subunits [40]. Cytochromes *o* of *Paracoccus* and *E. coli* pump protons with a similar stoichiometry as the mitochondrial cytochrome *c* oxidase [36,41]. Sequencing of the operon for cytochrome *o* [12] has led to the identification of five open reading-frames (cyo A–E), of which cyo B, A and C code for proteins homologous to subunits, I, II and III of the mitochondrial oxidase. Cyo E and D show homology to genes of the oxidase operon in *Paracoccus* [15], *Bacillus subtilis* [42] and the thermophilic bacterium PS3 [43].

2.3. Other bacteria

In *Thermus thermophilus* two cytochrome oxidases have been identified. Cytochrome ba_3 consists of a single 35 kDa polypeptide containing one heme *a* and one heme *b* molecule and two Cu ions [44]. The other oxidase, cytochrome c_1aa_3, consists of two protein subunits (55 and 33 kDa). The larger peptide corresponds to subunit I of the *Paracoccus* enzyme, while the smaller subunit represents a fused protein between subunit II and cytochrome *c* [45].

Two different cytochrome oxidases were also found in *Bacillus cereus.* A two-subunit (51 and 31 kDa) enzyme (cytochrome aa_3) contains two heme *a* molecules and three Cu ions. The other oxidase (cytochrome caa_3), which is only expressed in sporulating cells, contains in addition covalently bound heme *c* [28] similar to cytochrome c_1aa_3 from *T. thermophilus.*

Two isoforms of cytochrome oxidase were also described in *Bacillus subtilis.* The aa_3-600 oxidase [46] is different from another cytochrome *c* oxidase (caa_3-605), for which an operon with five structural genes has been sequenced [42]. Three genes code for subunits I–III, one may code for an assembly factor, and the fifth gene codes for a fourth subunit occurring in some bacterial cytochrome oxidases. The corresponding gene has also been found in the cytochrome oxidase operons of *E. coli* [12] and the thermophilic bacterium PS3 [43]. In fact, the corresponding protein has been identified in the isolated proton-pumping oxidase of PS3 as the fourth subunit [47].

A completely different cytochrome oxidase was isolated from the archaebacterium *Sulfolobus acidocaldarius*, which grows at 75–80°C and between pH 2–3. This enzyme consists of a single subunit (38 kDa), containing two hemes *a* and two Cu ions [48]. Since no cytochrome *c* occurs in this organism, the endogenous caldariella quinone is assumed to function as the physiological electron donor.

Finally, a cytochrome oxidase of the acidophilic iron-oxidizing *Thiobacillus ferrooxidans* should be mentioned. The isolated enzyme shows an unusual activity optimum at pH 3.5 [49].

3. *Evolution of eukaryotic cytochrome c oxidase (COX)*

According to the endosymbiotic theory [50] eukaryotic cells originate from the genetic symbiosis of a fermentative host cell with an oxygen-respiring bacterium. Most of the genes of the respiring bacterium, which have become the mitochondrial organelle, have then been transferred to the genome of the host cell.

During evolution mitochondria retained some genetic autonomy by their nucleus-independent, maternally inherited genome (mtDNA), coding in mammals for only thirteen proteins, which are all components of energy-transducing (proton-pumping) enzyme complexes, including seven subunits of NADH dehydrogenase, cytochrome *b* of the cytochrome bc_1 complex, subunits I–III of COX and two subunits of the ATP synthase [51,52]. This reduction of the mitochondrial genome to a minimal information content for essential components of oxidative phosphorylation is contrasted by a large amplification of the nuclear genome via gene duplication and independent evolution of multiple isozymes and pseudogenes. The 'catalytic' subunits I–III of COX are encoded on mtDNA and thus restricted in further evolution. In contrast, the 'regulatory' nuclear-coded subunits evolved in two respects: (i) increase of subunit number in the COX complex with increasing evolutionary stage of the organism. (ii) Doubling and independent evolution of tissue- and developmentally specific isogenes for some COX subunits. The genes of tissue-specific isoforms duplicated long before the radiation of mammalian lineages, i.e. about 240 as compared to 75 million years ago, respectively [53].

3.1. Subunit composition

The number of 1–4 subunits in bacterial cytochrome oxidases is contrasted by 7–13 subunits found in COX from eukaryotic organisms (see Table I). In COX from sweet potato more than

7 subunits may occur, because this number is only based on SDS-PAGE [60]. Similarly, the number of 9–10 subunits, identified in COX from the hearts of fish and birds by SDS-PAGE [61], very probably represents an underestimate. The increase in the number of nuclear-coded subunits via gene duplication and independent evolution was concluded for the subunit pairs VIa/VIc and VIIa/VIIb of mammalian COX from cross-reactivity of monoclonal antibodies and from hybridization at low stringecy of the corresponding cDNAs (VIa/VIc) [62]. A common evolutionary ancestor was also suggested for subunit VIIa of yeast and for VIc and VIII from mammals, which both revealed 15% sequence homology to yeast subunit VIIa [63]. In addition, yeast subunit VIII showed 27 and 37% homology to bovine subunits VIIa and VIIc, respectively [64]. Recently we have isolated a monoclonal antibody which reacts to the same extent with bovine COX subunits Va and Vb (Zimmermann and Kadenbach, unpublished), suggesting evolutionary relationship also between these two subunits.

The cDNAs for subunits IV [57], VI [56] and VIIe [58] of COX from *Dictyostelium discoideum* have been characterized and shown to be evolutionarily related to those of subunits Va, Vb and VIc [59] of the mammalian enzyme, respectively.

3.2. *Isoforms of nuclear-coded subunits*

The first indication for tissue-specific differences of nuclear-coded subunits came from different apparent molecular weights of isolated COX from liver and heart of the rat [66] and of beef, pig and chicken [67]. The tissue-specific structure of these subunits was supported by different immunological reactivity [68,69] and different reactivity of carboxylic groups [70] and SH-groups [71]. Differences in amino acid sequences were first described for subunits VIa [72] and VIIa and VIII [73] of COX from bovine heart and liver (see also ref. [74]). Meanwhile the corresponding cDNAs were isolated and characterized for the liver- and heart-type of subunit VIa from rat (see ref. [75]), corrected in (ref. [76]), subunit VIII from bovine [77] and subunit VIIa from bovine [78].

TABLE I

Subunit structure of COX from different species. The number of known subunit isoforms and the nomenclature of subunits which differs from the mammalian one are indicated. In the references the alignment with mammalian subunits is given

Subunit	Mammals	S. Potato	Yeast	*Dictyostelium*	*Paracoccus*
I	1	1	1	1	1
II	1	1	1	1	1
III	1	1	1	1	1
IV	1	1 IV	2 V [55]	1 VI [56]	–
Va	1	1 Va	1 VI [55]	1 IV [57]	–
Vb	1	1 Vb	1 IV [55]	1 V	–
VIa	2	–	–	–	–
VIb	1	–	–	–	–
VIc	1	1 Vc [54]	1 VIIa [55]	2 VII [59]	–
VIIa	2	–	1 VII [55]	–	–
VIIb	1	–	–	–	–
VIIc	1	–	1 VIII [55]	–	–
VIII	2	–	–	–	–

In human skeletal muscle, two different isoforms for subunit VIIa were identified by N-terminal amino acid sequencing [79], but in human heart only the liver-type of subunit VIII could be found [80,81]. Immunological data suggest the occurrence of two isoforms of subunit VIa also in human heart [82], as found in rat heart [81].

For COX subunits other than VIa, VIIa and VIII only indirect evidence exists for the occurrence of isoforms in mammals. With monospecific antibodies to all nuclear-coded subunits of rat liver COX, except Va, ELISA titrations were performed with isolated mitochondria from rat tissues, including fetal tissues from liver, heart and skeletal muscle. Large differences in immunological reactivities were also found between adult and fetal tissues for subunits IV, Va and VIIc [69]. For COX subunit VIc the occurrence of a fetal isoform was suggested from studies on the expression of the corresponding gene in primary embryonic rat myoblasts and myotubes in culture [83].

In the slime mold *Dictyostelium discoideum* the smallest of the seven subunits of COX was shown to occur in different isoforms (VIIe and VIIs), depending on the growth conditions [84]. Oxygen was identified as the only factor that switches the expression of one isoform into the other [85]. The two isoforms were shown to derive from a common ancestor related to mammalian subunit VIc [59].

In yeast two isoforms of subunit V (Va and Vb) were identified [65] and the two corresponding genes were characterized [86]. The expression of these subunits, which are homologous to mammalian subunit IV, is regulated differently by oxygen and heme [87,88]. The other five nuclear-coded subunits of yeast are encoded by single copy genes [64].

Isoforms of COX subunits were also shown to occur in plants. The enzyme from hypocotyls and roots of mung bean showed different immunological reactivity for one nuclear-coded subunit [89]. Similarly, the isolated COX from wheat germ and wheat seedling showed different subunit composition (7–8 subunits), suggesting development- or tissue-specific isozymes [90].

3.3. COX isozymes

The principle of COX isozyme formation in mammals always implies the assembly of 13 different polypeptides in stoichiometric amounts. For at least three subunits (VIa, VIIa and VIII), two isoforms can replace each other, resulting in eight different isozymes. In fact, several COX isozymes have been identified in addition to those found in bovine liver and heart [74]. A summary of presently known COX isozymes in mammals is given in Table II. The liver isoforms of subunits VIa, VIIa and VIII are also found in COX from bovine kidney and brain [82]. It appears that most tissues express the liver-type (constitutive) isoforms, while only in tissues with high oxidative activity, e.g. heart, muscle and brown fat, the heart isoforms are expressed.

In addition to tissue-specific isozymes, also species-specific forms occur. Thus, in COX of bovine heart only the heart isoforms were found for subunits VIa, VIIa and VIII. In rat heart, two isozymes containing either VIa-h or VIa-l were identified at a ratio 2:1 [81], and in human heart only subunit VIII-l was found [80,81].

4. Genes and protein structure of nuclear-coded subunits

The protein sequences of nuclear-coded subunits of COX from various species are mainly obtained from cDNAs, which are currently studied in several laboratories. Therefore the following lists of protein sequences (Fig. 1) and of publications on genes (Table III) will not be

TABLE II

Tissue- and species-specific expression of COX isozymes[a]

		Liver	Heart	Skeletal muscle	Smooth muscle	Brown fat[b]
Rat	VIa	l	l+h	h	–	l
	VIIa	l	(l)[c]	–	–	–
	VIII	l	h	–	–	h
Bovine	VIa	l	h	h	l	–
	VIIa	l	h	h	h	–
	VIII	l	h	h	(l)+h	–
Human	VIa	l	l+h	h	–	–
	VIIa	l	h	l+h	–	–
	VIII	l	l	–	–	–

[a] Most data are taken from ref. [91].
[b] Ref. [92].
[c] 2 N-terminal amino acids were found to be identical to the liver-form [82].

complete. Nuclear genes are actively studied in particular with the aim to obtain more insight into the tissue- and development-specific expression of isogenes.

4.1. cDNAs, functional genes and pseudogenes

In Table III are listed the publications in which sequences of cDNAs, functional genes and pseudogenes for nuclear-coded subunits of mammalian COX have been described. References for publications on nuclear-coded genes of other eukaryotic organisms are given in Sections 3.1 and 3.2.

For many genes of nuclear-coded COX subunits gene families have been described, due to the occurrence of multiple pseudogenes, which mostly represent retroposons, characterized by the lack of introns, a small poly-A tail, and direct repeats of DNA sequences flanking the gene [122,123,133]. These pseudogenes are suggested to represent a genetic pool for the evolution of new isogenes [133]. The genes for tissue-specific isoforms of subunits VIa, VIIa and VIII have only little homology, lacking cross-hybridization in Southern blots [75,133]. It has been suggested that the genes for liver isoforms may represent constitutive genes, expressed in all tissues, while the expression of the heart isoforms may be under the control of heart- or muscle-specific transcription factors [78]. Since Northern blots indicate the occurrence of mRNA for the liver isoform of subunit VIII also in bovine heart and muscle, but SDS-PAGE or N-terminal amino acid sequencing identified only the heart-type of subunit VIII in heart and muscle, the liver isoforms were suggested to be regulated posttranscriptionally [82].

4.2. Protein sequences of nuclear-coded subunits

The amino acid sequences of nuclear-coded subunits of mammalian COX are presented in Fig. 1. The sequences were obtained either from protein sequencing or deduced from cDNAs, which include the presequences of precursor proteins.

The comparison of the amino acid sequences of isoforms confirms the earlier notion, based

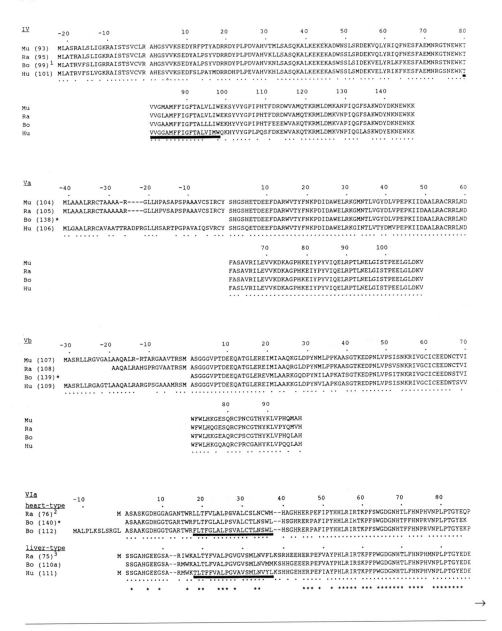

```
IV
            -20        -10                 10        20        30        40        50        60        70        80

Mu (93)     MLASRALSLIGKRAISTSVCLR AHGSVVKSEDYRFPTYADRRDYPLPDVAHVTMLSASQKALKEKEKADWNSLSRDEKVQLYRIQFNESFAEMNRGTNEWKT
Ra (95)     MLATRALSLIGKRAISTSVCLR AHGSVVKSEDYALPSYVDRRDYPLPDVAHVKLLSASQKALKEKEKADWSSLSRDEKVQLYRIQFNESFAEMNKGTNEWKT
Bo (99)1    MLATRVFSLIGRRAISTSVCVR AHGSVVKSEDYALPSYVDRRDYPLPDVAHVKNLSASQKALKEKEKASWSSLSIDEKVELYRLKFKESFAEMNRSTNEWKT
Hu (101)    MLATRVFSLVGKRAISTSVCVR AHESVVKSEDFSLPAYMDRRDHPLPEVAHVKHLSASQKALKEKEKASWSSLSMDEKVELYRIKFKESFAEMNRGSNEWKT

                                    90        100       110       120       130       140

Mu                                 VVGMAMFFIGFTALVLIWEKSYVYGPIPHTFDRDWVAMQTKRMLDMKANPIQGFSAKWDYDKNEWKK
Ra                                 VVGLAMFFIGFTALVLIWEKSYVYGPIPHTFDRDWVAMQTKRMLDMKVNPIQGFSAKWDYNKNEWKK
Bo                                 VVGAAMFFIGFTALLLIWEKHYVYGPIPHTFEEEWVAKQTKRMLDMKVAPIQGFSAKWDYDKNEWKK
Hu                                 VVGGAMFFIGFTALVIMWQKHYVYGPLPQSFDKEWVAKQTKRMLDMKVNPIQGLASKWDYEKNEWKK


Va
            -40       -30       -20       -10                 10        20        30        40        50        60

Mu (104)    MLAAALRRCTAAAA-R----GLLHPASAPSPAAAVCSIRCY SHGSHETDEEFDARWVTYFNKPDIDAWELRKGMNTLVGYDLVPEPKIIDAALRACRRLND
Ra (105)    MLAAALRRCTAAAAAR----GLLHPVSAPSPAAAVCSIRCY SHGSHETDEEFDARWVTYFNKPDIDAWELRKGMNTLVGYDLVPEPKIIDAALRACRRLND
Bo (138)*                                         SHGSHETDEEFDARWVTYFNKPDIDAWELRKGMNTLVGYDLVPEPKIIDAALRACRRLND
Hu (106)    MLGAALRRCAVAATTRADPRGLLHSARTPGPAVAIQSVRCY SHGSQETDEEFDARWVTYFNKPDIDAWELRKGINTLVTYDMVPEPKIIDAALRACRRLND

                                                      70        80        90        100

Mu                                                    FASAVRILEVVKDKAGPHKEIYPYVIQELRPTLNELGISTPEELGLDKV
Ra                                                    FASAVRILEVVKDKAGPHKEIYPYVIQELRPTLNELGISTPEELGLDKV
Bo                                                    FASAVRILEVVKDKAGPHKEIYPYVIQELRPTLNELGISTPEELGLDKV
Hu                                                    FASLVRILEVVKDKAGPHKEIYPYVIQELRPTLNELGISTPEELGLDKV


Vb
            -30       -20       -10                 10        20        30        40        50        60        70

Mu (107)    MASRLLRGVGALAAQALR-RTARGAAVTRSM ASGGGVPTDEEQATGLEREIMIAAQKGLDPYNMLPPKAASGTKEDPNLVPSISNKRIVGCICEEDNCTVI
Ra (108)                AAQALRAHGPRGVAATRSM ASGGGVPTDEEQATGLEREIMIAAQRGLDPYNMLPPKAASGTKEDPNLVPSVSNKRIVGCICEEDNCTVI
Bo (139)*                                   ASGGGVPTDEEQATGLEREVMLAARKGQDPYNILAPKATSGTKEDPNLVPSITNKRIVGCICEEDNSTVI
Hu (109)    MASRLLRGAGTLAAQALRARGPSGAAAMRSM ASGGGVPTDEEQATGLEREIMLAAKKGLDPYNVLAPKGASGTREDPNLVPSISNKRIVGCICEEDNTSVV

                                                80        90

Mu                                          WFWLHKGESQRCPNCGTHYKLVPHQMAH
Ra                                          WFWLHQGESQRCPNCGTHYKLVPYQMVH
Bo                                          WFWLHKGEAQRCPSCGTHYKLVPHQLAH
Hu                                          WFWLHKGQAQRCPRCGAHYKLVPQQLAH


VIa
                -10                 10        20        30        40        50        60        70        80
heart-type
Ra (76)2                  M ASASKGDHGGAGANTWRLLTFVLALPSVALCSLNCWM--HAGHHERPEFIPYHHLRIRTKPFSWGDGNHTLFHNPHVNPLPTGYEQP
Bo (140)*                   ASAAKGDHGGTGARTWRFLTFGLALPSVALCTLNSWL--HSGHRERPAFIPYHHLRIRTKPFSWGDGNHTFFHNPRVNPLPTGYEK
Bo (112)      MALPLKSLSRGL ASAAKGDHGGTGARTWRFLTFGLALPSVALCTLNSWL--HSGHRERPAFIPYHHLRIRTKPFSWGDGNHTFFHNPRVNPLPTGYEKP

liver-type
Ra (75)3                  M SSGAHGEEGSA--RIWKALTYFVALPGVGVSMLNVFLKSRHEEHERPEFVAYPHLRIRTKPFPWGDGNHTLFHNPHMNPLPTGYEDE
Bo (110a)                   SSGAHGEEGSA--RMWKALTLFVALPGVGVSMLNVMMKSHHGEEERPEFVAYPHLRIRSKPFPWGDGNHTLFHNPHVNPLPTGYEDE
Hu (111)                  M SSGAHGEEGSA--RMWKTLTFFVALPGVAVSMLNVYLKSHHGEHERPEFIAYPHLRIRTKPFPWGDGNHTLFHNPHVNPLPTGYEDE
```

→

on immunological data, that tissue-specificity overrides species-specificity [68]. This holds also for the presequences that are homologous in different species, but less homologous between isoforms in the same species.

```
VIb
                     10        20        30        40        50        60        70        80
Bo (115)     M AEDIQAKIKNYQTAPFDSRFPNQNQTRNCWQNYLDFHRCEKAMTAKGGDVSVCEWYRRVYKSLCPISWVSTWDDRRAEGTFPGKI
Hu (116)     M AEDMETKIKNYKTAPFDSRFPNQNQTRNCWQNYLDFHRCQKAMTAKGGDISVCEWYQRVYQSLCPTSWVTDWDEQRAEGTFPGKI

VIc
                     10        20        30        40        50        60        70
Ra (121)     M SSGALLPKPQMHDPLSKRLWVHIVGAFIVDLGVAAAHKFGAAKPRKKAYADFYRNHDPMKDFDEMRKAGVFRSVK
Bo (141)*       STAL-AKPQMRGLLARRLRFHIVGAFMVSLGFATFYKFAVAEKRKKAYADFYRNYDSMKDFEEMRKAGIFQSAK
Hu (124)     M -APEVLPKPRMRGLLARRLRNHMAVAFVLSLGVAALYKFRVADQRKKAYADSYRNYDVMKDFEEMRKAGIFQSVK

VIIa
                -20       -10                 10        20        30        40        50        60
heart-type
Ra (82)*                            FENKVPEKQKLFQEDNGMPVHLKGGT
Bo (142)*                           FENRVAEKQKLFQEDNGLPVHLKGGATDNILYRVTMTLCLGGTLYSLYCLGHAS---KK
Bo (78)        MRALRVSQALVRSFSSTARNR FENRVAEKQKLFQEDNGLPVHLKGGATDNILYRVTMTLCLGGTLYSLYCLGWASFPHKK
Hu (79)*                            FQNRVREKQKLFQEDNDIPLYLKGGIVDNILYRVTMTL

liver-type
Ra (125)     MLRNVLALRQIAQRTISTTSRRH FENKVPEKQKLFQEDNGMPVHLKGGTSDALLYRATMLLTVGGTAYAIYMLAMAAFPKKQN
Bo (126)     MLRNLLALRQIAKRTISTSSRRQ FENKVPEKQKLFQEDNGIPVHLKGGIADARLYRATLILTVGGTAYAMYELAVASFPKKQD
Hu (128)     MLRNLLALRQIGQRTISTASRRH FKNKVPEKQKLFQEDDEIPLYLKGGVADALLYRATMILTVGGTAYAIYELAVASFPKKQE

VIIb
               -30       -20       -10                 10        20        30        40        50
Bo (129)     MFNLRMFPLAKNALSRLRVQSIQQAVQAVARQ IHQKRAPDFHDKYGNAVLASGATFCVAVWVYMATQIGIEWNPSPVGRVTPKEWREQ

VIIc
                   -10                 10        20        30        40        50        60        70
Mu (130)     MLGQSIRRFTTSVVRR SHYEEGPGKNLPFSVENKWRLLAMMTVYFGSGFAAPFFIVRHQLLKK
Bo (131)     MLGQSIRRFTTSVVRR SHYEEGPGKNIPFSVENKWRLLAMMTLFFGSGFAAPFFIVRHQLLKKQSMRQIGRGAYQEVQSLKRINPLN
Hu (132)     MLGQSIRRFTTSVVRR SHYEEGPGKNLPFSVENKWSLLAKMCLYFGSAFATPFLVVRHQLLKT

VIII
                 -20       -10                 10        20        30        40
heart-type
Ra (81)*                            ISSKPAKSPTSAMDQAVGMSVIIAGFMVPAGWVLSHLE
Bo (134)     MLRLAPTV-RLLQAPLRGWVVPKAH ITAKPARTPTSPKEQAIGLSVTFLSFLLPAGWVLYHLDNYKKSSAA

liver-type
Ra (121)4    MSALTPLLLRSLTGPARRLMVPRAQ VHSKPPREQLGVLDITIGLTSCFVCCLLPAGWVLSHLESYKKRE
Bo (134)     MSVLTPLLLRGLTGPARRLPVPRAQ IHSKPPREQLGTMDIAIGLTSCFLCFLLPSGWVLSHMENYKKRE
Hu (135)     MSVLTPLLLRGLTGSARRLPVPRAK IHSLPPEGKLGIMELAVGLTSCFVTFLLPAGWILSHLETYRRPE
```

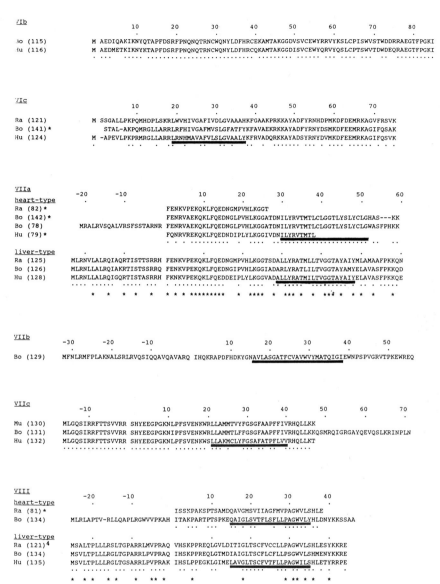

Fig. 1. Protein sequences of nuclear of nuclear coded subunits of mammalian COX. The protein sequences are either deduced from cDNAs (including presequences) or from protein sequencing, indicated by an asterix behind the reference. The beginning of the mature protein is indicated by a space. Identical amino acids are indicated by a point below the sequence. Identical amino acids between isoforms are indicated by an asterisk. The alignment of sequences was performed using the program Clustal, which is part of husar package on the genius net German Cancer Research Center Heidelberg. The husar package is an extended Unix version of the UWGCG (University of Wisconsin Genetic Computer Group) program package [113]. [1]Full length sequence from protein sequencing [137]. [2]Corrected cDNA from [75]. [3]The start methionine has been established by sequencing the 5' noncoding region (Mell, O., unpublished). [4]Full length cDNA in ref. [136].

4.3. *Import into mitochondria and assembly*

The structure and assembly of COX has been reviewed recently by Capaldi [6]. Understanding of targeting and sorting of cytoplasmically synthesized subunits and assembly together with the catalytic subunits I–III, which are synthesized in mitochondria, requires knowledge of their topology in the complex. In Fig. 1 are also indicated, by a horizontal bar, the membrane spanning domains of subunits IV, VIa,c, VIIa,b,c and VIII. Only subunits Va, Vb and VIb do not contain a hydrophobic region of about 20 amino acids [63]. The orientation of nuclear-coded subunits in the mitochondrial membrane, as determined by Capaldi and coworkers

TABLE III

Publications on genes of nuclear-coded subunits of mammalian COX

		cDNA	Functional gene	Pseudogene
IV	Mu	93	94	
	Ra	95,96	97,98	
	Bo	99[a]	100	100
	Hu	101		102
	Ch			103
Va	Mu	104		
	Ra	105		
	Hu	106		
Vb	Mu	107	114	
	Ra	108		
	Bo	109[a]		
	Hu	109	110	110
VIa-h	Ra	75,76		
	Bo	112		
VIa-l	Ra	75,76		
	Bo	110a		
	Hu	111		
VIb	Bo	115		
	Hu	116,117	118[a]	119
VIc	Ra	120[a],121	122	122,123
	Hu	124		
VIIa-h	Bo	78		
VIIa-l	Ra	125		
	Bo	126,78		
	Hu	128		
VIIb	Bo	129		
VIIc	Mu	130		
	Bo	131		
	Hu	132		
VIII-h	Bo	134		
VIII-l	Ra	121[a]		133
	Bo	134		
	Hu	135		

Numbers refer to references. Abbreviations: Mu, mouse, Ra, rat, Bo, bovine, Hu, human, Ch, chimpanzee.
[a] Incomplete gene (cDNA).

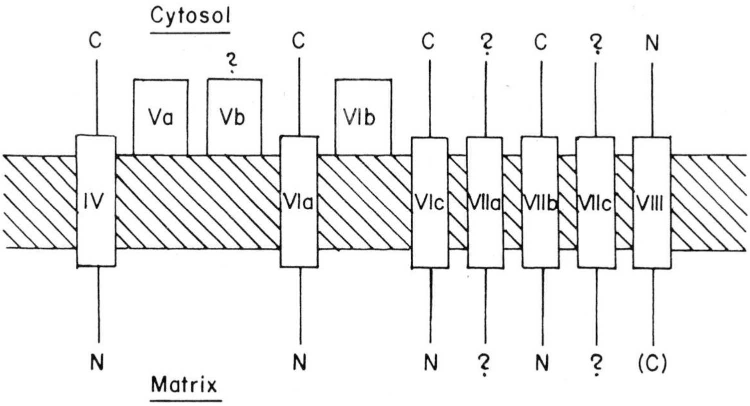

Fig. 2. Orientation of nuclear-coded subunits of mammalian COX in the mitochondrial membrane. The transmembraneous orientation of seven subunits has been deduced from hydropathy plots [63,145]. The arrangements of subunits IV, VIb and VIc [143] and subunits Va, VIa and VIII [144] has been determined by protease digestion and chemical labeling.

[143,144] is presented in Fig. 2. While most membrane-spanning subunits have their N-terminal parts oriented to the matrix, subunit VIII has an opposite orientation. At least two subunits (Va and VIb) are located at the intracristal side. Therefore, different targeting and sorting pathways must be assumed for nuclear-coded subunits. The targeting and sorting of proteins has been the subject of several reviews [146–149]. The precursors are kept unfolded in the cytosol by binding to chaperonins [150], which are related to heat-shock proteins [151]. Translocation into the matrix occurs via binding to a receptor [152,153] located in contact sites between the outer and inner mitochondrial membrane. The different locations of proteins in mitochondria require different sorting mechanisms (reviewed in ref. [149]). After translocation into the matrix, some proteins, like cytochrome c_1, located at the intermembrane space, are reexported by a conservative sorting pathway [154,155] which represents an ancestral sorting and assembly mechanism established in prokaryotes for periplasmic proteins.

As follows from Fig. 1, nuclear-coded subunits of mammalian COX have different targeting structures: subunit IV has a classical presequence of 22 amino acids, with positively charged and hydroxyl-containing amino acids and no negative charges [146]. Subunit Va has a 37 (rat) or 41 (human) amino acid presequence, containing basic and hydroxylated but also one acidic amino acid at position −24 in the human, not in the rat, presequence. In contrast to presequences of subunits IV and Vb, which have a 30 (mouse) or 31 amino acid presequence (human), it does not seem to have the potential to form an amphiphilic structure on a helical wheel, believed to be important for the import of cytosolic proteins into mitochondria [156,157].

Subunits VIa-l from rat and human and VIa-h from rat have no presequence. The mature protein begins after the start methionine, but different from subunit VIc it has an acidic amino acid close to the N-terminus. By contrast, a 12 amino acid presequence was found in VIa-h from bovine, which fulfills all criteria of a classical presequence [112]. Also subunits VIb from bovine and human have no presequence. The mature protein is the only subunit which is N-terminal acetylated. Subunit VIc also has no presequence, but the N-terminal region was suggested to function as a leader sequence, because it contains basic and hydroxylated, but no acidic residues [121]. Protein sequencing revealed in rat liver SGAL, and in rat heart SSGAL, as the mature N-terminus (Kadenbach and Lottspeich, unpublished). The lack of precursor sequences in only subunits VIa, VIb and VIc of most species suggest a common ancestral precursor of these subunits. Subunits VIIa, VIIb and VIIc have classical presequences with basic, hydroxylated and no acidic residues, but the length varies from 16 (VIIc) to 21 (VIIa-h), 23 (VIIa-l) and 32 amino acids (VIIb). The presequence of subunits VIIa and IV have features typical for a two-step processing with an Arg at position −10, a hydrophobic residue at −8, and a hydroxylated residue (Ser) at −5 [158]. Subunit VIII has a classical 25 amino acid presequence, but without characteristics of two-step processing, which should be expected, since the N-terminus is located on the cytosolic side (Fig. 2).

From these considerations it is clear that targeting and sorting of nuclear-coded COX subunits represent a complex and heterogeneous process. Very little is known about the assembly of these subunits with the three mitochondrially coded subunits, two heme *a* groups and 2.5 Cu, 1 Zn and 1 Mg ion per monomer, into the functional enzyme complex (see ref. [6]).

5. *Regulation of COX activity*

The reduction of oxygen to water by COX represents a strongly exergonic and practically irreversible process. The high turnover number of the isolated COX is contrasted by its low activity in vivo [159], indicating its pacemaker function in cellular energy metabolism [160]. Therefore, a regulation of activity according to the variable energy needs of cells and tissues must be expected. COX activity, i.e. electron transport from ferrocytochrome *c* to dioxygen, accompanied by the formation of an electrochemical (and/or protonchemical) potential across the membrane, could be regulated by (i) isosteric effectors (substrates), (ii) allosteric effectors, (iii) the membrane potential, and (iv) via expression of tissue-specific isoforms.

5.1. *Isosteric effectors*

The activity of COX is dependent on the concentration of its substrates, ferrocytochrome *c* and dioxygen, and the electrochemical potential across the membrane. Since cytochrome *c* is a protein, its electrostatic interaction with COX [161,162] will depend on the conformation of both the substrate and the enzyme. Moreover, specific binding sites for ATP, ADP and phosphate have been identified on cytochrome *c* [163,164]. Some effects of these anions on the kinetics of membrane-bound COX [165,166,167] could be explained by modification of the binding domain of cytochrome *c*.

The inhibition of COX activity by its reaction product ferricytochrome *c* has been known since 1956 [168] and shown to be of a competitive type [169,170]. The kinetics of COX, however, is complex and characterized by a non-Michaelis–Menten behavior, with increasing K_m values for ferrocytochrome *c* at increasing substrate concentration [171]. Recently we could demonstrate that the multiphasic kinetics of COX at low ionic strength disappears with increasing concentrations of ferricytochrome *c* [172]. It was suggested that ferricytochrome *c* interacts with both the catalytic and the regulatory binding site for ferrocytochrome *c*, thus inhibiting the total activity by competing with ferrocytochrome *c* at the catalytic site, and preventing the increase of K_m for ferrocytochrome *c* by displacing the substrate from the regulatory binding-site. The physiological meaning of these complex kinetics, however, remains to be established.

Little is known about the steady-state kinetics of COX at low oxygen concentrations. Because of the high affinity of COX to oxygen, only under very low oxygen concentrations an effect on the kinetics will be expected. Half-maximal reduction of cytochrome *c* in isolated pigeon heart mitochondria was found to depend on the metabolic activity and ranged from 0.27 tot 0.03 μM oxygen [173]. Recently complex kinetics was observed with isolated COX at low oxygen concentrations [174].

5.2. *Allosteric effectors*

An indication for a regulatory effect of 'cytosolic' ATP came from studies on the respiratory rate, the reduction level of COX and the proton-motive force of isolated mitochondria from yeast [127].

Early studies of the effect of ATP on the activity of mammalian COX [165,175] could not distinguish between an interaction with cytochrome *c* or with COX. The first proof for the influence of bound ATP on the kinetics of COX was presented by Hüther and Kadenbach [176], where ATP was covalently bound to COX by photoaffinity-labeling with 8-azido-ATP. An increase of K_m for cytochrome *c* of the reconstituted enzyme by bound ATP could be prevented when ATP but not ADP was present during the labeling of the solubilized enzyme before reconstitution.

An attempt to localize bound ATP at subunits IV and VIII (i.e. subunits VIIa,b,c + VIII) by Montecucco et al. [177] could not be reproduced in a later study [167]. Instead, an unspecific labeling of most subunits was obtained. We have found, that the specific labeling of a COX subunit with [β or γ-^{32}P]-labeled 8-azido- or 2-azido-ADP or -ATP depends on multiple, yet not well-defined parameters (see, e.g., ref. [167]). In some cases a specific labeling of subunit VIb or VIa from bovine heart was obtained when soluble COX or sonicated COX vesicles (with random orientation) were photoaffinity-labeled with 2-azido-[β-^{32}P]ADP [178] or 2-azido-[β, γ-^{32}P]ATP [179], respectively.

Further proof for an interaction of nucleotides and other ions with COX came from the demonstration of changes of the visible spectrum [180,181]. Since low concentrations of complexons like EDTA also gave spectral changes, it was assumed [182] that the chelator ATP might interact with the tightly bound 'non-redox' metal ions Mg, Zn and Cu [183,25,26]. ATP was also shown to induce a change in the water-exposed carboxylic groups of the enzyme, which are involved in binding of cytochrome *c*, as indicated by reaction with water-soluble carbodiimides [184].

An enhancement of activity with low, and an inhibition with high, anion concentrations was found with laurylmaltoside-solubilized [159], but not with membrane-bound, COX from bovine heart [185]. The maximal stimulation was obtained with different anions at about the same ionic strength [181]. Because this stimulation of activity was not found with COX from *Paracoccus* [186], we suggested that the activity of the mammalian enzyme is suppressed by nuclear-coded subunits [187], which are partly detached from the complex in the presence of high salt and detergent concentrations ([188,189], see also ref. [190]). In fact, an increased initial activity, but no further enhancement of activity at low KCl concentrations, was obtained after selective removal of the nuclear-coded subunit VIb [190a].

While the above experiments could not prove a physiological regulatory function of nucleotides on COX activity, a strong indication came from the effects of intraliposomal ATP and ADP on the kinetics of reconstituted COX from bovine heart. In spectrophotometric, but not in polarographic, assays the K_m for ferrocytochrome *c* was increased by 10 mM intraliposomal ATP and decreased by 10 mM intraliposomal ADP [191]. By contrast, no effect was obtained with 15 mM intraliposomal ATP or ADP on the kinetics of reconstituted COX from *Paracoc-*

cus [192], which lacks the nuclear-coded subunits. ATP and ADP apparently bind to nuclear-coded subunit(s) at the matrix side of COX from bovine heart and change the binding domain for cytochrome *c* at the cytosolic side via conformational changes through the lipid bilayer.

5.3. Regulation via expression of tissue-specific isoforms

A modulation of the activity of COX by isoforms of a nuclear-coded subunit was recently shown very clearly by Poyton and coworkers [193] with yeast isozymes. In yeast cells, expressing only subunit Vb isoform, COX has a higher turnover rate, and a higher rate of heme *a* oxidation than COX in yeast cells, expressing only subunit Va isoform.

Comparative studies on the catalytic activity of mammalian COX isozymes are scarce and controversial. Kinetic differences between isolated as well as membrane-bound COX from bovine heart and liver obtained in earlier studies [194,195] could not be confirmed in later investigations [196,197]. Recently, however, kinetic differences could be described for isolated COX from human heart, muscle, kidney and liver [198]. Also, the opposite effect of intraliposomal phosphate on V_{max} of reconstituted COX from bovine heart and liver suggested a tissue-specific response [195]. Now it became clear that fully activated COX isozymes, i.e., solubilized in laurylmaltoside, at high salt concentrations and in the absence of allosteric effectors, do not reveal kinetic differences with the polarographic method of assay. The tissue-specific response of COX isozymes becomes evident with the down-regulated, i.e. membrane-bound, enzymes by the effects of allosteric effectors, which interact with the nuclear-coded (tissue-specific) subunits, as previously suggested [159].

Recently we could show that intraliposomal ATP and ADP modify the kinetics of the bovine heart, but not of the bovine liver, enzyme [199]. In addition it was shown that ADP, in reconstituted COX from bovine heart, specifically abolished the effect of ferricytochrome *c*, which changes the multiphasic kinetics of both COX isozymes into monophasic ones [172]. The bovine heart and liver enzymes differ in subunits VIa, VIIa and VIII [74]. Recently Capaldi and coworkers could locate the N-terminal domain of subunit VIa on the matrix, and that of subunit VIII on the cytosolic side of COX [144]. Assuming a matrix orientation of the N-terminus also for subunit VIIa, as has been found for subunits IV, VIc and VIIb [143], the

TABLE IV

Charge distribution on the matrix and cytosolic side of subunits VIa, VIIa and VIII of COX from bovine heart and liver. The N-terminal domain of subunits VIa and VIIa was assumed to be located in the matrix, that of subunit VIII on the cytosolic side. +: number of positively charged amino acids (Arg + Lys); −: number of negatively charged amino acids (Asp + Glu)

Subunit	Cytosolic side			Matrix side		
	+	−	sum	+	−	sum
VIa-l [110a]	5	8	−3	2	2	0
VIa-h [140]	7	3	+4	3	1	+2
VIIa-l [78]	2	1	+1	5	5	0
VIIa-h [78]	2	0	+2	5	5	0
VIII-l [77]	2	2	0	3	2	+1
VIII-h [77]	3	2	+2	2	1	+1

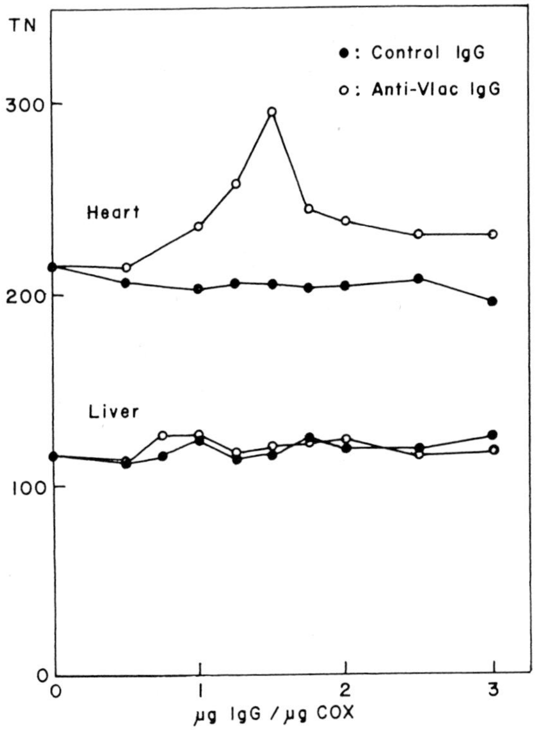

Fig. 3. Stimulation of the activity of COX from bovine heart, but not from bovine liver, with a monoclonal antibody (clone 86 from ref. [62]) which reacts with subunits VIa and VIc from heart, but only with subunit VIc from liver. The IgG of cell culture supernatants was purified by affinity chromatography with protein A-sepharose. The activity was measured polarographically in 10 mM K-Hepes, pH 7.4, 0.05% laurylmaltoside, 7 mM Tris-ascorbate, 1 mM TMPD, 10 µM cytochrome *c*, and 20 nM isolated COX from bovine heart or liver, as indicated.

charge distribution can be calculated for the three tissue-specific subunits, as shown in Table IV. Only for the matrix domain of subunit VIa is an excess of two positive charges obtained for the heart as compared to the liver isoform. Thus, subunit VIa could be responsible for the tissue-specific effect of negatively charged nucleotides on the kinetics of the bovine heart enzyme.

A direct indication of the tissue-specific involvement of bovine heart subunit VIa in regulating the activity of COX via binding of allosteric effectors, can be deduced from the experiments shown in Figs. 2 and 3, showing the effects of a monoclonal antiboy on subunit VIa-h, which cross-reacts with subunit VIc, but not with VIa-l [62]. About 30% stimulation of the activity of COX from bovine heart is obtained after addition of one mol antibody (~150 kDa) per mol COX (204 kDa). Further increase of the IgG/COX ratio leads to partial decrease of the stimulatory effect, which could be due to steric interaction of a second IgG molecule bound to subunit VIc. No effect of the monoclonal antibody is obtained with the bovine liver enzyme (Fig. 2). The IgG causes a conformational change of the catalytic center via binding to subunit VIa-h, as shown in Fig. 3. Addition of the antibody to bovine heart COX results in a red shift

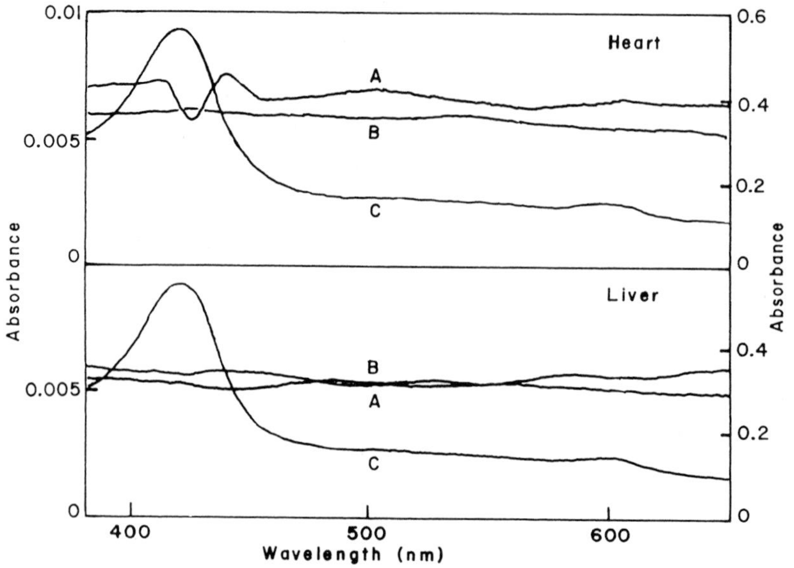

Fig. 4. Influence of a monoclonal antibody to COX subunits VIa+c of bovine heart on the visible spectrum of isolated COX from bovine heart and liver. (A) Difference spectrum 30 min after addition of 2 μM anti-subunit VIa+c IgG to the sample cuvette and 2 μM bovine standard IgG (Biorad) to the reference cuvette at room temperature. (B) Difference spectrum of 2 μM isolated COX in 10 mM K-Hepes, pH 7.4, 0.05% laurylmaltoside, present in both cuvettes. (C) Absolute spectrum of 1 μM isolated COX from bovine heart or liver, as indicated.

of the γ-band, indicating a change of the heme *a* liganding sphere (Fig. 4). No spectral change is obtained under identical conditions with the bovine liver enzyme (Anthony and Kadenbach, in preparation). The physiological meaning of the increase of the affinity of COX to its substrate at high ADP content in the matrix of mitochondria from heart (and muscle), but not from liver (and, e.g. kidney and brain), could be to stimulate oxidative phosphorylation under stress situations (high work load), which could be essential for survival of the organism.

Tissue-specific differences were also found with non-esterified fatty acids on the respiratory control ratios (RCR) of reconstituted COX from bovine heart and liver [200]. The partial decrease of RCR with the heart, but not with the liver, enzyme was taken to indicate the participation of the heart in nonshivering thermogenesis in mammals via allosteric interaction of non-esterified fatty acids with tissue-specific, nuclear-coded subunits.

5.4. Electrochemical potential

The reduction of dioxygen to water by COX involves the vectorial and stoichiometric uptake of four protons from the matrix, accompanied by the formation of an electrochemical potential. In addition protons are pumped from the matrix into the cytosol [1] creating a protonchemical potential. The stoichiometry of proton pumping, originally assumed to be 1 H^+/e^- [201] turned out to be variable, and to depend on the pH [202] and on the rate of respiration [203,204].

A variable H^+/e^- stoichiometry of COX is further supported by the demonstration of 'slippage' of proton translocation in COX of isolated mitochondria [205] and of reconstituted COX [206]. The local anesthetic bupivacaine was also shown to increase slippage in COX of isolated mitochondria [207].

The term 'respiratory control', originally introduced to describe the control of respiration in isolated mitochondria by the availability of the phosphate acceptor ADP [208], is generally used to describe the control of the rate of electron transport in proton pumps by $\Delta\tilde{\mu}_{H^+}$ across the membrane. The respiratory control ratio of reconstituted COX is estimated from the rate of respiration in the presence and absence of valinomycin and CCCP. The K^+-ionophores valinomycin [209] and nonactin, however, were shown to bind to COX in stoichiometric amounts, and to affect differently the respiratory control ratio [210], suggesting a modulation of the energy transduction in COX by specific binding of ionophores.

The control of electron flux through reconstituted COX by $\Delta\Psi$ across the membrane was recently studied in more detail [211,212]. It could be demonstrated that several internal electron transfer steps are controlled by $\Delta\Psi$. So far no studies have been performed on the contribution of nuclear-coded subunits to the control of internal electron transfer by $\Delta\Psi$. An insight into the function of nuclear-coded subunits in energy transduction could be obtained, e.g. by comparative studies of the effects of 'physiological' allosteric effectors on the relationship between respiratory rate and $\Delta\Psi$ [206] of reconstituted COX from *Paracoccus* and mammals. In particular, comparative studies of COX isozymes from the same species would be indicative.

6. Conclusion

By comparing cytochrome oxidases from bacteria, lower eukaryotes and mammals, a general principle of evolution becomes evident: the early established chemical principle, here the simultaneous transfer of four electrons from four redox centers to dioxygen, accompanied by vectorial uptake and translocation of protons across a membrane, was kept qualitatively unchanged, but quantitatively adapted to the different and variable needs of cells and organisms, depending on the environmental conditions. Thus, evolution of cytochrome oxidase was characterized by an increase in regulatory complexity. This suggestion [73] could now be verified by the demonstration of an increasing number of tissue-specific subunits, which, at least in part, regulate the rate of respiration and possibly the efficiency of energy transduction.

Many questions remain unanswered: how many isoforms of nuclear-coded subunits occur in mammals? Are there additional genes for developmentally specific subunits? What is the specific function of all ten nuclear-coded subunits, that are absent in the bacterial enzyme? Are there any allosteric effectors other than nucleotides? But it appears that we are now at the beginning to understand the multiple functions of the subunits of one of the most complex enzymes, which is essential for aerobic life.

Acknowledgements

We are thankful to Slike Grosch for technical assistance and to Roswitha Roller-Müller for excellent typing of the manuscript. This paper was supported by grants from the Deutsche Forschungsgemeinschaft (Ka 192/17-5), Wilhelm Sander Stiftung and Fonds der Chemischen Industrie.

References

1 Wikström, M., Krab, K. and Saraste, M. (1981) Cytochrome Oxidase: A Synthesis, Academic Press, New York.

2 Chan, I.S. and Li, P.M. (1990) Biochemistry 29, 1–12.

3 Malmström, B.G. (1990) Arch. Biochem. Biophys. 280, 233–241.

4 Wikström, M., Saraste, M. and Penttilä, T. (1985) in: A.N. Martonosi (Ed.) The Enzymes of Biological Membranes, Vol. 4, Plenum Press, New York, pp. 111–148.

5 Capaldi, R.A. (1990) Annu. Rev. Biochem. 59, 569–596.

6 Capaldi, R.A. (1990) Arch. Biochem. Biophys. 280, 252–262.

7 Bisson, R. (1990) in: G. Milazzo and M. Blank (Eds.) Bioelectrochemistry, Vol. III, Plenum Press., New York, pp. 3–53.

8 Saraste, M. (1990) Q. Rev. Biophys. 23, 331–366.

9 Poole, R.K. (1983) Biochim. Biophys. Acta 726, 205–243.

10 Anraku, Y. (1988) Annu. Rev. Biochem. 57, 101–132.

11 Ludwig, B. (1987) FEMS Microbiol. Rev. 46, 41–56.

12 Chepuri, V., Lemieux, L., An. D.C.-T. and Gennis, R.B. (1990) J. Biol. Chem. 265, 11185–11192.

13 Hendler, R.W., Pardhasaradhi, K., Reynafarje, B. and Ludwig, B. (1991) Biophys. J. 60, in press.

14 Pardhasaradhi, K., Ludwig, B. and Hendler, R.W. (1991) Biophys. J. 60, 408–414.

15 Raitio, M., Jalli, T. and Saraste, M. (1987) EMBO J. 6, 2825–2833.

16 Müller, M., Schläpfer, B. and Azzi, A. (1988) Biochemistry 27, 7546–7551.

17 Müller, M., Schläpfer, B. and Azzi, A. (1988) Proc. Natl. Acad. Sci. USA 85, 6647–6651.

18 Steffens, G.C.M., Buse, G., Oppliger, W. and Ludwig, B. (1983) Biochem. Biophys. Res. Commun. 116, 335–340.

19 Wikström, M. (1984) Nature 308, 558–560.

20 Van Verseveld, H.W., Krab, K. and Stouthamer, A.H. (1981) Biochim. Biophys. Acta 635, 525–534.

21 Prochaska, L.J. and Fink, P.S. (1987) J. Bioenerg. Biomemberanes 19, 143–166.

22 Brunori, M., Antonini, G., Malatesta, F., Sarti, P. and Wilson, M.T. (1987) Eur. J. Biochem. 169, 1–8.

23 Haltia, T., Finel, M., Harms, N., Nakari, T., Raitio, M., Wikström, M. and Saraste, M. (1989) EMBO J. 8, 3571–3579.

24 Haltia, T., Saraste, M. and Wikström, M. (1991) EMBO J. 10, 2015–2021.

25 Bombleka, E., Richter, F.-W., Stroh, A. and Kadenbach, B. (1986) Biochem. Biophys. Res. Commun. 140, 1007–1014.

26 Steffens, G.C.M., Biewald, R. and Buse, G. (1987) Eur. J. Biochem. 164, 295–300.

27 Yewey, G.L. and Caughey, W.S. (1987) Biochem. Biophys. Res. Commun. 148, 1520–1526.

28 Garcia-Horsman, J.A., Barquera, B. and Escamilla, J.E. (1991) Eur. J. Biochem. 199, 761–768.

29 Yoslikawa, S., Tera, T., Takahashi, Y., Tsukihara, T. and Caughey, W.S. (1988) Proc. Natl. Acad. Sci. USA 85, 1354–1358.

30 Pan, L.P., Li, Z., Larsen, R. and Chan, S.I. (1991) J. Biol. Chem. 266, 1367–1370.

31 Steinrücke, P., Gerhus, E. and Ludwig, B. (1991) J. Biol. Chem. 266, 7676–7681.

32 Berry, E. and Trumpower, B.L. (1985) J. Biol. Chem. 260, 2458–2467.

33 Raitio, M., Pispa, J.M., Metso, T. and Saraste, M. (1990) FEBS Lett. 261, 431–435.

34 Stouthamer, A.H. (1991) J. Bioenerg. Biomembranes 23, 163–185.

35 Bosma, G., Braster, M., Stouthamer, A.H. and van Verseveld, H.W. (1987) Eur. J. Biochem. 165, 657–663.

36 Puustinen, A., Finel, M., Haltia, T., Gennis, R.B. and Wikström, M. (1991) Biochemistry 30, 3936–3942.

37 Meinhardt, S.W., Gennis, R.B. and Ohnishi, T. (1989) Biochem. Biophys. Res. Commun. 975, 175–184.

38 Green, G.N., Fang, H., Lin, R.-J., Newton, G., Mather, M., Georgiou, C.D. and Gennis, R.B. (1988) J. Biol. Chem. 263, 13138–13143.

39 Anraku, Y. and Gennis, R.B. (1987) Trends Biochem. Sci. 12, 262–266.
40 Georgiou, C.D., Cokic, P., Carter, K., Webster, D.A. and Gennis, R.B. (1988) Biochim. Biophys. Acta 933, 179–183.
41 Puustinen, A., Finel, M., Virkki, M. and Wikström, M. (1989) FEBS Lett. 249, 163–167.
42 Saraste, M., Metso, T., Nakari, T., Jalli, T., Lauraeus, M. and van der Oost, J. (1991) Eur. J. Biochem. 195, 517–525.
43 Ishizuka, M., Machida, K., Shimada, S., Mogo, A., Tsuchiya, T., Ohmori, T., Souma, Y., Gonda, M. and Sone, N. (1990) J. Biochem. 108, 866–873.
44 Zimmermann, B.H., Nitsche, C.I., Fee, J.A., Rusnak, F. and Münck, E. (1988) Proc. Natl. Acad. Sci. USA 85, 5779–5783.
45 Buse, G., Hensel, S. and Fee, J.A. (1989) Eur. J. Biochem. 181, 261–268.
46 Lauraeus, M., Halüa, T., Saraste, M. and Wikström, M. (1991) Eur. J. Biochem. 197, 699–705.
47 Sone, N., Shimada, S., Ohmori, T., Souma, Y., Gonda, M. and Ishizuka, M. (1990) FEBS Lett. 262, 249–252.
48 Anemüller, S. and Schäfer, G. (1990) Eur. J. Biochem. 191, 297–305.
49 Kai, M., Yano, T., Fukumori, Y. and Yamanaka, T. (1989) Biochem. Biophys. Res. Commun. 160, 839–843.
50 Margulis, L. (1970) Origin of Eukaryotic cells, Yale University Press, New Haven, CT.
51 Anderson, S., Bankier, A.T., Barrell, B.G., de Bruijn, M.H.L., Coulson, A.R., Drouin, I., Eperon, I.C., Nierlich, D.P., Roe, B.A., Sanger, F., Schreier, P.H., Smith, A.J.H., Staden, R. and Young, I.G. (1981) Nature 290, 457–465.
52 Chomym, A., Mariotti, P., Cleeter, M.W.J., Ragan, C.I., Matsuno-Yai, A., Hatefi, Y., Doolittle, R.F. and Attardi, G. (1985) Nature 314, 592–597.
53 Saccone, C., Pesole, G. and Kadenbach, B. (1991) Eur. J. Biochem. 195, 151–156.
54 Nakagawa, T., Maeshima, M., Nakamura, K. and Asahi, T. (1990) Eur. J. Biochem. 191, 557–561.
55 Patterson, T.E., Trueblood, C.E., Wright, R.M. and Poyton, R.O. (1987) in: S. Papa, B. Chance and L. Ernster (Eds.) Cytochrome Systems: Molecular Biology and Bioenergetics, Plenum Press, New York, pp. 254–260.
56 Rizzuto, R., Sandona, D., Capaldi, R.A. and Bisson, R. (1991) Biochim. Biophys. Acta 1089, 386–388.
57 Rizzuto, R., Sandona, D., Capaldi, R.A. and Bisson, R. (1991) Biochim. Biophys. Acta 1090, 125–128.
58 Rizzuto, R., Sandona, D., Capaldi, R.A. and Bisson, R. (1990) Nucleic Acids Res. 18, 6711.
59 Capaldi, R.A., Zhang, Y.-Z., Rizzuto, R., Sandona, D., Schiavo, G. and Bisson, R. (1990) FEBS Lett. 261, 158–160.
60 Nakagawa, T., Maeshima, M., Muto, H., Kajiura, H., Hattori, H. and Asahi, T. (1987) Eur. J. Biochem. 165, 303–307.
61 Montecucco, C., Schiavo, G., Bacci, B. and Bisson, R. (1987) Comp. Biochem. Physiol. B 87, 851–856.
62 Schneyder, B., Mell, O., Anthony, G. and Kadenbach, B. (1991) Eur. J. Biochem. 198, 85–92.
63 Kadenbach, B., Kuhn-Nentwig, L. and Büge, U. (1987) Curr. Top. Bioenerg. 15, 113–161.
64 Wright, R.M., Trawick, J.D., Trueblood, C.E., Patterson, T.E. and Poyton, R.O. (1987) in: S. Papa, B. Chance and Ernster L. (Eds.) Cytochrome Systems: Molecular Biology and Bioenergetics, Plenum Press, New York, pp. 49–56.
65 Trueblood, C.E. and Poyton, R.O. (1987) Mol. Cell. Biol. 7, 3520–3526.
66 Merle, P. and Kadenbach, B. (1980) Hoppe-Seyler's Z. Physiol. Chem. 361, 1257–1259.
67 Kadenbach, B., Büge, U., Jarausch, J. and Merle, P. (1981) in: F. Palmieri, E. Quagliariello, N. Siliprandi and E.C. Slater (Eds.) Vectorial Reactions in Electron and Ion Transport in Mitochondria and Bacteria, North-Holland Biomedical Press, Amsterdam, pp. 11–23.
68 Jarausch, J. and Kadenbach, B. (1982) Hoppe-Seyler's Z. Physiol. Chem. 363, 1133–1140.
69 Kuhn-Nentwig, L. and Kadenbach, B. (1985) Eur. J. Biochem. 149, 147–158.
70 Kadenbach, B. and Stroh, A. (1984) FEBS Lett. 173, 374–380.
71 Stroh, A. and Kadenbach, B. (1986) Eur. J. Biochem. 156, 199–204.

72 Kadenbach, B., Hartmann, R., Glanville, R. and Buse, G. (1982) FEBS Lett. 138, 236–238.
73 Kadenbach, B. (1983) Angew. Chem. Int. Ed. Engl. 22, 275–282.
74 Yanamura, W., Zhang, Y.-Z., Takamiya, S. and Capaldi, R.A. (1988) Biochemistry 278, 4909–4914.
75 Schlerf, A., Droste, M., Winter, M. and Kadenbach, B. (1988) EMBO J. 7, 2387–2391.
76 Kadenbach, B., Hüther, F.-J., Büge, U., Schlerf, A. and Johnson, M.A. (1989) in: A. Azzi, Z. Drahota, J. Jaz and S. Papa (Eds.) Molecular Basis of Membrane Associated Diseases, Springer, Berlin, pp. 216–227.
77 Lightowlers, R., Ewart, G., Aggeler, R., Zhang, Y.-Z., Calavetta, L. and Capaldi, R.A. (1990) J. Biol. Chem. 265, 2677–2681.
78 Seelan, R.S. and Grossman, L.I. (1991) J. Biol. Chem. 266, 19752–19757.
79 Van Beeumen, J.J., van Kuilenbeurg, A.B.P., van Bun, S., van den Bogert, C., Tager, J.M. and Muijsers, A.O. (1990) FEBS Lett. 263, 213–216.
80 Van Kuilenburg, A.B.P., Muijsers, A.O., Demol, H., Dekker, H.L. and van Beeumen, J.J. (1988) FEBS Lett. 240, 127–132.
81 Kadenbach, B., Stroh, A., Becker, A., Eckerskorn, C. and Lottspeich, F. (1990) Biochim. Biophys. Acta 1015, 368–372.
82 Kennaway, N.G., Carrero-Valenzuela, R.D., Ewart, G., Balan, V.K., Lightowlers, R., Zhang, Y.-Z., Powell, B.R., Capaldi, R.A. and Buist, N.R.M. (1990) Pediatr. Res. 28, 529–535.
83 Lomax, M.I., Coucouvanis, E., Schon, E.A. and Barald, K.F. (1990) Muscle Nerve 13, 330–337.
84 Bisson, R. and Schiavo, G. (1986) J. Biol. Chem. 261, 4373–4376.
85 Schiavo, G. and Bisson, R. (1989) J. Biol. Chem. 264, 7129–7134.
86 Cumsky, M.G., Trueblood, C.E., Ko, C. and Poyton, R.O. (1987) Mol. Cell. Biol. 7, 3511–3519.
87 Poyton, R.O., Trueblood, C.E., Wright, R.M. and Farrell, L.E. (1988) Ann. N.Y. Acad. Sci. 550, 289–307.
88 Hodge, M.R., Kim, G., Singh, K. and Cumsky, M.G. (1989) Mol. Cell. Biol. 9, 1958–1964.
89 Nakagawa, T., Maeshima, M. and Asahi, T. (1989) Plant Cell Physiol. 29, 1297–1302.
90 Pfeiffer, W.E., Ingle, R.T. and Ferguson-Miller, S. (1990) Biochemistry 29, 8696–8701.
91 Anthony, G., Stroh, A., Lottspeich, F. and Kadenbach, B. (1990) FEBS Lett. 277, 97–100.
92 Kadenbach, B., Stroh, A., Becker, A., Eckerskorn, C. and Lottspeich, F. (1990) Biochim. Biophys. Acta 1015, 368–372.
93 Grossman, L.I. and Akamalsu, M. (1990) Nucleic Acids. Res. 18, 6454.
94 Carter, R.S. and Avadhani, N.G. (1991) Arch. Biochem. Biophys. 288, 97–106.
95 Goto, Y., Amuro, N. and Okazaki, T. (1989) Nucleic Acids. Res. 17, 2851.
96 Gopalan, G., Droste, M. and Kadenbach, B. (1989) Nucleic Acids Res. 17, 4376.
97 Yamada, M., Amuro, N., Goto, Y. and Okazaki, T. (1990) J. Biol. Chem. 265, 7687–7692.
98 Amuro, N., Yamada, M., Goto, Y. and Okazaki, T. (1990) Nucleic Acids Res. 18, 3992.
99 Lomax, M.J., Bachman, N.J., Nasoff, M.S., Caruthers, M.H. and Grossman, L.J. (1984) Proc. Natl. Acad. Sci. USA 81, 6295–6299.
100 Bachman, N.J., Lomax, M.I. and Grossman, L.I. (1987) Gene 55, 205–217.
101 Zeviani, M., Nakagawa, M., Herbert, I., Lomax, M.I., Grossman, L.J., Sherbany, A.A., Miranda, A.F., DiMauro, S. and Schon, E.A. (1987) Gene 55, 205–217.
102 Ewart, G., Lightowlers, R., Zhang, Y.Z., Balan, V.I., Kennaway, N. and Capaldi, R.A. (1990) Biochim. Biophys. Acta 1018, 223–224.
103 Lomax, M.I., Welch, M.D., Darras, B.T., Francke, U. and Grossman, L.I. (1990) Gene 86, 209–216.
104 Nielson, P.J., Ayane, M. and Kohler, C. (1989) Nucleic Acids. Res. 17, 6723.
105 Droste, M., Schon, E. and Kadenbach, B. (1989) Nucleic Acids Res. 17, 4375.
106 Rizzuto, R., Nakase, H., Zeviani, M., DiMauro, S. and Schon, E.A. (1988) Gene 69, 245–256.
107 Basu, A. and Avadhani, N.G. (1990) Biochim. Biophys. Acta 1087, 98–100.
108 Goto, Y., Amuro, N. and Okazaki, T. (1989) Nucleic Acids Res. 17, 6388.
109 Zeviani, M., Sakoda, S., Sherbany, A.A., Nakase, H., Rizzuto, R., Sammit, C.E., DiMauro, S. and Schon, E.A. (1988) Gene 65, 1–11.
110 Lomax, M.I., Hsieh, C.L., Darras, B.T. and Francke, U. (1991) Genomics 10, 1–9.
110a Ewart, G.D., Zhang, Y.-Z. and Capaldi, R.A. (1991) FEBS Lett. 292, 79–84.

111 Fabrizi, G.M., Rizzuto, R., Nakase, H., Mita, S., Kadenbach, B. and Schon, E.A. (1989) Nucleic Acids Res. 17, 5845.
112 Smith, E.O., Bellert, D.M., Grossman, L.I. and Lomax, M.I. (1991) Biochim. Biophys. Acta 1089, 266–268.
113 Devereux, J., Haeberli, P. and Smithies, O. (1984) Nucleic Acids Res. 12, 387–396.
114 Basu, A., Avadhani, N.G. (1991) J. Biol. Chem. 266, 15450–15456.
115 Lightowlers, R.N. and Capaldi, R.A. (1989) Nucleic Acids Res. 17, 5845.
116 Taanman, J.-W., Schrage, C., Ponne, N., Bolhuis, P., de Vries, H. and Agsteribbe, E. (1989) Nucleic Acids Res. 17, 1766.
117 Taanman, J.-W., Schrage, C., Ponne, N.J., Das, A.T., Bolhuis, P.A., DeVries, H. and Agsteribbe, E. (1990) Gene 93, 285–291.
118 Taanman, J.-W., Schrage, C., Bokma, E., Reuvekamap, P., Agsteribbe, E. and DeVries, H. (1991) Biochim. Biophys. Acta 1089, 283–285.
119 Taanman, J.-W., Schrage, C., Reuvekamp, P., Bijl, J., Hartog, M., DeVries, H. and Agsteribbe, E. (1991) Gene, in press.
120 Parimoo, S., Seelan, R.S., Desai, S., Buse, G. and Padmanaban, G. (1984) Biochem. Biophys. Res. Commun. 118, 902–909.
121 Suske, G., Mengel, T., Cordingley, M. and Kadenbach, B. (1987) Eur. J. Biochem. 168, 233–237.
122 Suske, G., Enders, C., Schlerf, A. and Kadenbach, B. (1988) DNA 7, 163–171.
123 Seelan, R.S. and Padmanaban, G. (1988) Gene 67, 125–130.
124 Otsuka, M., Mizuno, Y., Yoshida, M., Kagawa, Y. and Ohta, S. (1988) Nucleic Acids Res. 16, 10916.
125 Enders, C., Schlerf, A., Mell, O., Grossman, L.I., Kadenbach, B. (1990) Nucleic Acids Res. 18, 7143.
126 Seelan, R.S., Scheuner, D., Lomax, M.I. and Grossman, L.I. (1989) Nucleic Acids Res. 17, 6410.
127 Rigoulet, M., Guerin, B. and Denis, M. (1987) Eur. J. Biochem. 168, 275–279.
128 Fabrizi, G.M., Rizzuto, R., Nakase, H., Mita, S., Lomax, M.I., Grossman, L.I. and Schon, E.A. (1989) Nucleic Acids Res. 17, 7107.
129 Lightowlers, R., Takamiya, S., Wessling, R., Lindorfer, M. and Capaldi, R.A. (1989) J. Biol. Chem. 264, 16858–16861.
130 Akamatsu, M. and Grossman, L.I. (1990) Nucleic Acids Res. 18, 3645.
131 Aqua, M.S., Lomax, M.I., Schon, E.A. and Grossman, L.I. (1989) Nucleic Acids Res. 17, 8376.
132 Koga, Y., Fabrizi, G.M., Mita, S., Arnaudo, E., Lomax, M.I., Agua, M.S., Grossman, L.I. and Schon, E.A. (1990) Nucleic Acids Res. 18, 3992.
133 Cao, X., Hengst, L., Schlerf, A., Droste, M., Mengel, T. and Kadenbach, B. (1988) Ann. N.Y. Acad. Sci. 550, 337–347.
134 Lightowlers, L., Ewart, G., Aggeler, R., Zhang, Y.Z., Calavetta, L. and Capaldi, R.A. (1990) J. Biol. Chem. 365, 2677–2681.
135 Rizzuto, R., Nakase, H., Darras, B., Francke, U., Fabrizi, G.M., Mengel, T., Walsh, F., Kadenbach, B., DiMauro, S. and Schon, E.A. (1989) J. Biol. Chem. 264, 10595–10600.
136 Winter, M. (1991) Dissertation, Fachbereich Chemie, Philipps-Universität, Marburg.
137 Sacher, R., Steffens, G.J. and Buse, G. (1979) Hoppe-Seyler's Z. Physiol. Chem. 360, 1385–1392.
138 Tanaka, M., Hanin, M., Yasunobu, K.T., Yu, C.-A., Yu, L., Wei, Y.-H. and King, T.E. (1979) J. Biol. Chem. 254, 3879–3885.
139 Biewald, R. and Buse, G. (1982) Hoppe-Seyler's Z. Physiol. Chem. 363, 1141–1153.
140 Meinecke, L. and Buse, G. (1985) Biol. Chem. Hoppe-Seyler 366, 687–694.
141 Erdweg, M. and Buse, G. (1985) Hoppe-Seyler's Z. Physiol. Chem. 366, 257–263.
142 Meinecke, L. and Buse, G. (1986) Biol. Chem. Hoppe-Seyler 367, 67–73.
143 Zhang, Y.-Z., Lindorfer, M.A. and Capalid, R.-A. (1988) Biochemistry 27, 1389–1394.
144 Zhang, Y.-Z., Ewart, G. and Capaldi, R.A. (1991) Biochemistry 30, 3674–3681.
145 Lightowlers, R., Takamiya, S., Wessling, R., Lindorfer, M. and Capaldi, R.A. (1989) J. Biol. Chem. 264, 16858–16861.
146 Schatz, G. and Butow, R.A. (1983) Cell 32, 316–318.
147 Hay, R., Böhni, P. and Gasser, S. (1984) Biochim. Biophys. Acta 779, 65–87.
148 Hurt, E.C. and van Loon, A.P.G.M. (1986) Trends Biochem. Sci. 11, 204–207.

L. Ernster (Ed.) *Molecular Mechanisms in Bioenergetics*
© 1992 Elsevier Science Publishers B.V. All rights reserved

CHAPTER 11

The energy-transducing nicotinamide nucleotide transhydrogenase

YOUSSEF HATEFI and MUTSUO YAMAGUCHI

Division of Biochemistry, Department of Molecular and Experimental Medicine, The Scripps Research Institute, La Jolla, CA 92037, U.S.A.

Contents

1. Introduction 265
2. Structure 267
   2.1. General features 267
   2.2. Nucleotide binding sites 268
   2.3. Membrane topography 269
3. Catalytic properties 270
4. Mechanism of energy transduction 272
5. The proton channel of the transhydrogenase 278
6. The physiological role of the transhydrogenase 279
Acknowledgements 280
References 280

1. Introduction

The energy-transducing nicotinamide nucleotide transhydrogenases of mitochondria and bacteria are membrane-bound enzymes that catalyze the direct and stereospecific transfer of a hydride ion between the 4A position of NAD(H) and the 4B position of NADP(H). In mitochondria, the enzyme is embedded in the inner membrane with its nucleotide binding sites protruding into the matrix, and the transhydrogenation reaction is coupled to transmembrane proton translocation with a H^+/H^- stoichiometry of unity (see Eqn. 1 where H_c^+ and H_m^+ are cytosolic and matrix protons, respectively).

$$NADH + NADP + H_c^+ \leftrightarrows NAD + NADPH + H_m^+. \quad (1)$$

Because of the stereospecificity of hydride ion transfer, this type of transhydrogenase is referred to as AB-transhydrogenase. The known AB-transhydrogenases are integral membrane proteins, their scalar transhydrogenation reaction is coupled to proton translocation, they contain separate binding sites for NAD(H) and NADP(H), and have no prosthetic groups. In bacteria, another type of transhydrogenase is found, which is known as BB-transhydrogenase. The BB-type transhydrogenases are water-soluble, do not pump protons, are flavoproteins containing FAD, have a single substrate binding site, and are thought to be concerned with equilibrating the cellular NAD(H) and NADP(H) pools. They may also have another, as yet unknown, redox function. Space limitation does not allow further foray into the field of BB-transhydrogenases. Nor does it permit a survey of the early work on the AB-

```
Bovis     CSAPVKPGIPYKQLTVGVPKEIFQNEKRVALSPAGVQALVKQGFNVVVESGAGEASKFSDDHYRAAGAQIQGAKEVLASD    80
                          * * *   ** ***  *  *  * * ** * ******    * *     *** *     *  *
E. coli                MRIGIPRERLTNETRVAATPKTVEQLLKLGFTVAVESGAGQLASFDDKAFVQAGAEIVEGNSVWQSE    67

Bovis     LVVKVRAPMLNPTLGVHEADLLKTSGTLISFIYPAQNPDLLNKLSKRKTTVLAMDQVPRVTIAQGYDALSSMANIAGYKA   160
             ** **         *  **    ** *** ***** *  **  *  ** *** ***   **  ************ *
E. coli   IILKVNAPL......DDEIALLNPGTTLVSFIWPAQNPELMQKLAERNVTVMAMDSVPRISRAQSLDALSSMANIAGYRA   141

Bovis     VVLAANHFGRFFTGQITAAGKVPPAKILIVGGGVAGLASAGAAKSMGAIVRGFDTRAAALEQFKSLGAEPLEVDLKESGE   240
           * **  *******************    * ******  *** * ***** ****    **  * *** ** * **
E. coli   IVEAAHEFGRFFTGQITAAGKVPPAKVMVIGAGVAGLAAIGAANSLGAIVRAFDTRPEVKEQVQSMGAEFLELDFKEEAG   221
          FSBA↓  ↓TRP     ↓DCCD
Bovis     GQGGYAKEMSKEFIEAEMKLFAQQCKEVDILISTALIPGKKAPILFNKEMIESMKEGSVVVDLAAEAGGNFETTKPGELY   320
             **** **  ** *** *** * *****   ******* ** *   **  *** *** *****  *** * * ***
E. coli   SGDGYAKVMSDAFIKAEMELFAAQAKEVDIIVTTALIPGKPAPKLITREMVDSMKAGSVIVDLAAQNGGNCEYTVPGEIF   301

Bovis     .VHKGITHIGYTDLPSRMATQASTLYSNNITKLLKAISPDKDNFYFEVKDDFDFGTMGHVIRGTVVMKDGQVIFPAPTPK   399
              *   ******* *  ** * **  *   ***     **        ***      ****  *   *    *** *
E. coli   TTENGVKVIGYTDLPGRLPTQSSQLYGTNLVNLLKLLCKEKDG...NITVDFD....DVVIRGVTVIRAGEITWPAP.PI   373
                  TRP↓                 ↓TRP
Bovis     NIPQGAPVKQKTVAELEAEKAATITPFRKTMTSASVYTAGLTGILGLGIAAPNLAFSQMVTTFGLAGIVGYHTVWGVTPA   479
                   **   *   *   *  * **         *      *         *    * * **  ***  ** *  *
E. coli   QVSAQPQAAQKAAPEVKTEEKCTCSPWRKY.....ALMALAIILFGWMASVAPKEFLGHFTVFALACVVGYYVVWNVSHA   448

Bovis     LHSPLMSVTNAISGLTAVGGLVLMGGHLYPSTTSQGLAALATFISSVNIAGGFLVTQRMLDMFKRPTDPPEYNYLYLLPA   559
          ** ***********   ** *   *     *     *   *  * * ** *** ****** **
E. coli   LHTPLMSVTNAISGIIVVGALLQIGQGGWVSF....LSFIAVLIASINIFGGFTVTQRMLKMFRKN                 510
                                                  TRP↓                      ↓NEM
Bovis     GTFVGGYLASLYSGYNIEQIMYLGSGLCCVGFLAGLSTQGTARLGNALGMIGVAGGLAATLGGLKPCPELLAQMSGAMAL   639
                        *    *    *       *****   * * **  *  * *  * **     *          **
E. coli        MSGGLVTAAYIVAAILFIFS.......LAGLSKHETSRQGNNFGIAGMAIALIATI..FGPDTGNVGWILLAMVI    66
                                                            ↓Prot.K
Bovis     GGTIGLTIAKRIQISDLPQLVAAFHSLVGLAAVLTCIAEYIIEYPHFATDAAANLTKI...VAYLGTYIGGVTFSGSLVA   716
          ** **   **       * ***  ** *******   . *     *     *  *  *      **  ** *** ** **
E. coli   GGAIGIRLAKKVEMTEMPELVAILHSFVGLAAVLVGFNSYL....HHDAGMAPILVNIHLTEVFLGIFIGAVTFTGSVVA   142

Bovis     YGKLQGILKSAPLLLPGRHLLNAGLLAGSVGGIIPFMMDPSFTTGITCLGSVSALSAVMGETLTARIGGADMPVVITVLN   796
           *** *   * ** ** **  *   *  *    * *    *       *    *   * *  * * *********   **
E. coli   FGKLCGKISSKPLMLPNRHKMNLAALVVSFLLLIVFVRTDSVGLQVLALLIMTAIALVFGWHLVASIGGADMPVVVSMLN   222
                                                                     ↓Papain
Bovis     SYSGWALCAEGFLLNNNLLTIVGALIGSSGAILSYIMCVAMNRSLANVILGGYGTTSTAGGKPMEISGTHTEINLDNAID   876
          ******  * ** * * **   *** ************ *****   ** ** **     *   *  * * **
E. coli   SYSGWAAAAAGFMLSNDLLIVTGALVGSSGAILSYIMCKAMNRSFISVIAGGFGTDGSSTGDDQEV.GEHREITAEETAE   301
                          ↓NEM
Bovis     MIREANSIIITPGYGLCAAKAQYPIADLVKMLSEQGKKVRFGIHPVAGRMPGQLNVLLAEAGVPYDIVLEMDEINHDFPD   956
                * *******   * **** *     *   *  *********** **  ******* ************* ** *
E. coli   LLKNSHSVIITPGYGMAVAQAQYPVAEITEKLRARGINVRFGIHPVAGRLPGHMNVLLAEAKVPYDIVLEMDEINDDFAD   381
                                                           ↓FSBA
Bovis     TDLVLVIGANDTVNSAAQEDPNSIIAGMPVLEVWKSKQVIVMKRSLGVGYAAVDNPIFYKPNTAMLLGDAKKTCDALQAK  1036
          ** *********** *** ** * ***********   *** ***   *** * ** * * ** ** ****   **
E. coli   TDTVLVIGANDTVNPAAQDDPKSPIAGMPVLEVWKAQNVIVFKRSMNTGYAGVQNPLFFKENTHMLFGDAKASVDAILKA   461

Bovis     VRESYQK                                                                           1043
E. coli   L                                                                                  462
```

Fig. 1. Amino acid sequences of bovine and *E. coli* nicotinamide nucleotide transhydrogenases. The *E. coli* enzyme is composed of two subunits, α and β (see text). The sequences of the two enzymes have been aligned to give the minimum number of vacant spaces (dots). Asterisks show residue identities. On the bovine sequence, the residues modified by FSBA, DCCD, and NEM, and the bonds cleaved by trypsin (TRP), proteinase K (prot. K) and papain are marked by arrows. The *E. coli* sequence shown is that which includes Dr. Bragg's recent corrections (P.D. Bragg, private communication).

transhydrogenases. These areas have been covered in several excellent and extensive reviews [1–6], among them a chapter in volume 9 of the previous edition of this series [6]. The present article will concentrate on the more recent information regarding the structure and mechanism of action of the proton-translocating AB-transhydrogenases of mitochondria and bacteria.

2. Structure

2.1. General features

The transhydrogenases of bovine heart mitochondria [7–9], *Escherichia coli* cytoplasmic membrane [10], and *Rhodobacter capsulatus* chromatophores [11] have been purified, and the amino acid sequences of the bovine and *E. coli* enzymes have been deduced from the complementary DNA and the gene sequences, respectively (Fig. 1 and refs. [12,13]). The mitochondrial transhydrogenase is nuclearly encoded, extramitochondrially synthesized, and imported [14–16]. The mature bovine enzyme is a homodimer of monomer M_r=109 228, and is composed of 1043 amino acid residues [12]. The sequence of its signal peptide has also been deduced from the messenger RNA sequence [17]. It contains 43 residues and, like the presequences of other nuclearly encoded mitochondrial proteins, it is rich in basic (4 lysines, 2 arginines) and hydroxylated (4 threonines, 3 serines) amino acids (Fig. 2). The *E. coli* transhydrogenase is composed of two unlike subunits, α with 510 residues and M_r=54 000, and β with 462 residues and M_r=48 700 [13]. The *E. coli* enzyme has also been shown to be dimeric, consisting of 2 α and 2 β chains [18]. There is considerable sequence identity between the bovine and the *E. coli* transhydrogenases, especially in the nucleotide binding domains (65–70%) which have been determined for the bovine enzyme (see below). The bovine transhydrogenase has 71 residues more than the *E. coli* enzyme. Among these are 13 N-terminal and 6 C-terminal residues plus a stretch of 19 amino acids where the α-subunit of the *E. coli* transhydrogenase ends and its β-subunit begins. The published amino acid sequence of the α- and β-subunits of the *E. coli* enzyme contained a few inaccuracies, which have been recently corrected (P.D. Bragg, private communication). The sequences shown in Fig. 1 are the most recent. The transhydrogenase from *Rb. capsulatus* is also composed of two subunits with reported M_r values of 53 000 and 48 000 [11].

Hydropathy analysis of the amino acid sequence of the bovine transhydrogenase has indi-

M A N L L K T V V T G C S C P F L S N L G S C K V 25

L P G K K N F L R T F H T H R I L W C S A P V K P 50

G I P Y K Q L T V G V P K E I F Q N E K R V A L S 75

Fig. 2. Amino acid sequence of the signal peptide (first 43 residues) and the N-terminal region of the bovine transhydrogenase as derived from the messenger RNA sequence. The vertical arrow shows where the signal peptide ends and the mature transhydrogenase begins. From ref. [17] with permission.

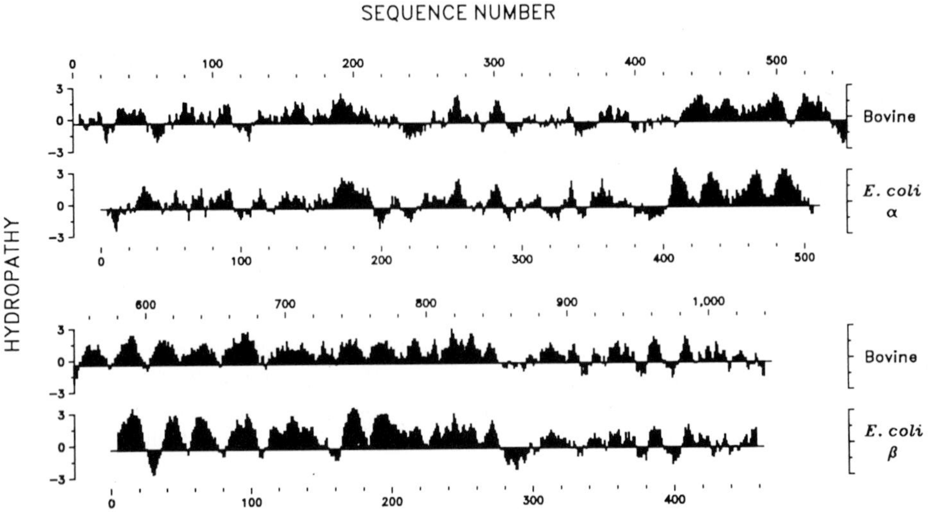

Fig. 3. Hydropathy profiles of the bovine and *E. coli* transhydrogenases. Hydropathy scores were calculated by the method of Kyte and Doolittle [27], using a setting of nine residues. Areas above and below the average hydropathy line indicate relative hydrophobic and hydrophilic regions, respectively. The amino acid sequences used are those given in Fig. 1.

cated that the enzyme monomer is composed of three domains: a 430-residue-long hydrophilic N-terminal domain, a 400-residue-long hydrophobic central domain, and a 200-residue-long hydrophilic C-terminal domain. Figure 3 compares the hydropathy profiles of the bovine and the *E. coli* transhydrogenases when the subunits of the latter are aligned with the bovine enzyme on the basis of sequence identity. It is seen that together the two subunits of the *E. coli* transhydrogenase present the same hydropathy profile as that of the bovine enzyme. The α subunit of the *E. coli* transhydrogenase has a 400-residue-long hydrophilic N-terminal domain and a 100-residue-long hydrophobic C-terminal domain, while its β-subunit starts with a 280-residue-long hydrophobic N-terminal domain and ends with a 180-residue-long hydrophilic C-terminal domain.

2.2. Nucleotide binding sites

It has been shown by Colman and coworkers [19,20] that *p*-fluorosulfonylbenzoyl-5'-adenosine (FSBA)* modifies proteins at their binding sites for ADP, ATP, NAD or NADH. When incubated with [^{3}H]FSBA, purified bovine transhydrogenase was also modified and inhibited

* Abbreviations used: FSBA, *p*-fluorosulfonylbenzoyl-5'-adenosine; DCCD, *N,N'*-dicyclohexylcarbodiimide; EEDQ, *N*-(ethoxycarbonyl)-2-ethoxy-1,2-dihydroquinoline; AcPyAD, 3-acetylpyridine adenine dinucleotide; AcPyADP, 3-acetylpyridine adenine dinucleotide phosphate; thio-NADP, thionicotinamide adenine dinucleotide phosphate; NEM, *N*-ethylmaleimide; SDS, sodium dodecyl sulfate; GSH, reduced glutathione; S-13, 5-chloro-3-*t*-butyl-2'-chloro-4'-nitrosalicylanilide.

[21]. Presence of NADH in the incubation medium greatly decreased the inhibition rate, while presence of NADPH slightly increased it. The tryptic peptide whose modification by [^{3}H]FSBA was greatly diminished in the presence of NADH was isolated and sequenced, and the modified residue was determined to be Tyr^{245} [22]. This residue is located within the N-terminal hydrophilic domain 24 (or 31) to 59 residues downstream of a $\beta\alpha\beta$ fold, which is considered to be the hallmark of binding sites for ADP, ATP, NAD or FAD [23]. When in the above experiments the incubation of the transhydrogenase with [^{3}H]FSBA was performed in the presence of NADH to protect the NAD binding site, then the reagent modified a single other site, albeit at a relatively slow rate. This second modified residue was determined to be Tyr^{1006} near the C-terminus of the protein. In agreement with conclusions from kinetic experiments [1,24,25], the above results suggested, therefore, that the bovine transhydrogenase possesses separate binding sites for NAD(H) and NADP(H), the former being in the N-terminal hydrophilic domain, and the latter in the C-terminal hydrophilic domain. Subsequent findings, which will be discussed below, were fully consistent with these assignments.

2.3. Membrane topography

Figure 4 depicts the membrane topography of the bovine mitochondrial transhydrogenase monomer as suggested by the hydropathy profile of the protein, immunochemical experiments, and use of proteolytic enzymes [26]. Antibodies raised to the N- and C-terminal hydrophilic domains, isolated after controlled proteolysis from purified transhydrogenase, indicated that epitopes from these domains were exposed on the matrix side of submitochondrial particles (inside-out inner membrane vesicles) but not on the cytosolic side of mitoplasts (mitochondria denuded of outer membrane). Furthermore, treatment of submitochondrial particles with tryp-

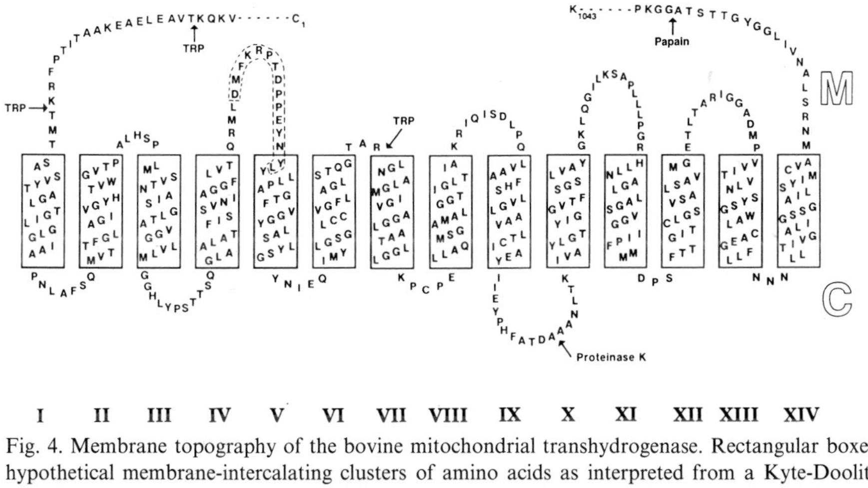

Fig. 4. Membrane topography of the bovine mitochondrial transhydrogenase. Rectangular boxes show hypothetical membrane-intercalating clusters of amino acids as interpreted from a Kyte-Doolittle hydropathy analysis of the amino acid sequence of the transhydrogenase. The abbreviated 430-residue-long N-terminal and the 200-residue-long C-terminal hydrophilic domains are shown protruding from the membrane on the matrix (M) side. Also largely exposed on the M side is the segment Asp^{540}–Leu^{554} which is outlined by a dashed line. The exposed protease-sensitive loop is shown on the cytosolic (C) side connecting the presumed membrane-intercalating clusters IX and X. The bonds cleaved by trypsin (TRP), papain and proteinase K are marked. For other details, see text and ref. [26].

sin or papain resulted in the release of soluble N- and C-terminal segments comparable in size to nearly the entire N- and C-terminal hydrophilic domains, which indicated that these domains are extramembranous and protrude from the inner membrane into the mitochondrial matrix. The cleavage sites of trypsin and papain are marked in Fig. 4. The finding that the N-and C-terminal hydrophilic domains protrude into the mitochondrial matrix agreed with the results described above regarding the binding sites of NAD(H) and NADP(H), and suggested that the nucleotide binding regions of the N- and C-terminal hydrophilic domains come together in the matrix to form the enzyme's catalytic site. In addition, it was shown that antibody to a synthetic pentadecapeptide corresponding to position Asp^{540}–Leu^{554} within the central hydrophobic domain of the bovine transhydrogenase reacted with submitochondrial particles, but not with mitoplasts, thus placing reactive epitopes from this segment outside of the membrane on the matrix side. When mitoplasts were treated with various proteolytic enzymes, the transhydrogenase molecule was either unaffected (by trypsin, α-chymotrypsin or papain) or bisected into a 72 kDa N-terminal and a 37 kDa C-terminal fragment (by proteinase K, subtilisin, thermolysin or pronase E). These results suggested that large, protease-sensitive masses of the transhydrogenase molecule are not exposed on the cytosolic side of the inner membrane. The proteinase K cleavage site was determined to be Ala^{690}–Ala^{691}, which, as shown in Fig. 4, is located in a cytosolic-side exposed loop between the hypothetical membrane-spanning clusters IX and X.

Although in Fig. 4 the arrangement of amino acids as extramembranous domains or loops and as membrane-intercalating clusters is derived from a Kyte–Doolittle [27] hydropathy analysis, it is clear from the foregoing that this arrangement agrees with the experimental results. It is, of course, possible that not all the 14 hydrophobic clusters shown in Fig. 4 are membrane-intercalated, as has been demonstrated by Gennis and coworkers for cluster IV of *Rhodobacter sphaeroides* cytochrome *b* [28]. However, the established positions of the N- and C-terminal hydrophilic domains and the Asp^{540}–Leu^{554} loop on the matrix side, and Ala^{690}–Ala^{691} on the cytosolic side, place certain restrictions on the position of the hydrophobic clusters between these extramembranous markers. Thus, in order to keep the sidedness of these markers intact, one would have to place out of the membrane a stretch of amino acids corresponding to two clusters between each pair of adjacent markers. This possibility is not in accord, however, with the extreme hydrophobicity of the clusters shown in Fig. 4. Therefore, it seems reasonable to assume that the arrangement depicted in Fig. 4 is a close approximation of the membrane topography of the bovine transhydrogenase. Essentially a similar picture has been proposed for the membrance topography of the *E. coli* enzyme (P.D. Bragg, private communication). The N-terminal hydrophilic domain of the α-subunit and the C-terminal hydrophilic domain of the β-subunit are considered to protrude into the cytoplasmic space, with the hydrophobic tail of α forming four membrane-spanning clusters and the hydrophobic N-terminal domain of β forming eight such clusters.

3. *Catalytic properties*

The catalytic properties of the transhydrogenase have been reviewed in extenso [1–6]. Following are some salient features. Hydride ion transfer between NAD(H) and NADP(H) is direct, and the kinetics is consistent with a mechanism involving the formation of a transient ternary complex [1,2,6]. Substrate addition, originally thought to be ordered [1,2], was later shown to be random [29–31]. In the absence of energy, the reaction shown in Eqn. 1 proceeds to an equilibrium point (K_{eq} = 0.79) dictated by the difference in the reduction potentials of NAD/

NADH and NADP/NADPH ($\Delta E'_0 \approxeq 5$ mV) [32,33]. Mitochondrial membrane energization increases the rate of NADPH production 10–12-fold, and shifts the equilibrium of the reaction toward product formation (apparent $K_{eq} \approxeq 500$ as determined from the initial rates of the forward and the reverse reactions) [2]. The pH optimum of the reaction is < 6.0 in the absence of energy, and 7.5 in the presence of energy [34]. The bovine transhydrogenase catalyzes NADPH to NADP transhydrogenation at a very slow rate, but not NADH to NAD transhydrogenation [35]. The NADPH to NADP reaction (the 3-acetylpyridine and the thionicotinamide analogues of NADP were used to allow spectrophotometric monitoring of hydride ion transfer) is accelerated by membrane energization as well as at acid pH values where the 2'-phosphate of NADP ($pK_a = 6.1$) would be expected to be protonated [35]. The purified bovine enzyme has been coincorporated in liposomes together with a preparation of complex V (ATP synthase) or bacteriorhodopsin, and high rates of ATP or light-driven transhydrogenation have been demonstrated [36–39]. The stoichiometry of mol NADPH formed/mol ATP utilized has been calculated to be three [36], which, on the basis of H^+/ATP = 3, agrees with the H^+/H^- stoichiometry of near unity that had been earlier estimated in direct measurements [3].

The bovine transhydrogenase is inhibited by a variety of protein-modifying reagents, including thiol modifiers [40–44], DCCD [9,45–49], EEDQ [45,48] FSBA [21,22], ethoxyformic anhydride, dansyl chloride, pyridoxal phosphate [50], tetranitromethane [51], 4-chloro-7-nitrobenzofurazan (Nbf-C1) [52], butanedione and phenylglyoxal [53]. It is also highly sensitive to trypsin [53–56]. Substrates show different effects on inhibition of transhydrogenase activity by these reagents, from strong protection against inhibition [9,21,45,50] to considerable stimulation of the inhibition rate [41,44,50,55,56]. As was discussed earlier, [^{3}H]FSBA was used in the absence and presence of NADH to affinity-label, respectively, the NAD(H) and the NADP(H) binding sites of the bovine transhydrogenase, and identify the modified residue at each site after isolation and sequencing of the labeled tryptic peptides.

Fisher, Rydström and their respective colleagues have published data suggesting that DCCD inhibits proton translocation somewhat more than hydride ion transfer from NADPH to NAD [47,57]. The former group has concluded that DCCD binds to the transhydrogenase outside the active site in the enzyme's proton binding domain [47], and the latter that H^- transfer and H^+ translocation by the transhydrogenase are not obligatorily linked [57]. By contrast, our results showed that the enzyme in mitochondrial membranes and in the purified state is completeley inhibited by DCCD with pseudo-first-order kinetics, and strongly protected against this inhibition by NAD(H), NAD analogs, 5'-AMP and 5'ADP [9,46]. Also, the DCCD-treated transhydrogenase did not bind to NAD-agarose [9]. More significantly it was shown that in submitochondrial particles the transhydrogenase-catalyzed hydride ion transfer from NADPH to acetylpyridine-NAD (AcPyAD) and the membrane potential induced by this reaction were inhibited in parallel by either DCCD or EEDQ, again with pseudo-first-order kinetics with respect to the duration of incubation of the particles with each inhitibor [45]. Consistent with these results, the site of DCCD binding was shown to be at the NAD binding domain of the enzyme at Glu^{257}, 12 residues downstream of the FSBA modification site [22] (Fig. 1). Clarke and Bragg [10] have also shown that DCCD inhibits the *E. coli* transhydrogenase by binding to the α-subunit, and that NADH protects the enzyme against this inhibition. Other studies suggested that the two carboxyl modifying reagents, DCCD and EEDQ, react at different sites within the NAD binding domain of the bovine transhydrogenase [48]. Thus, it was shown that 5'-AMP and 5'-ADP protected the enzyme strongly against inhibition by DCCD, and only weakly against inhibition by EEDQ. By contrast, NMNH offered no protection against DCCD, but strong protection against EEDQ [48]. These results suggested that DCCD modifies the transhydrogenase where the AMP moiety of NAD binds, whereas EEDQ reacts where the NMN portion of NAD resides. Another finding of interest was that inhibition of the bovine enzyme by [^{3}H]FSBA or [^{14}C]DCCD at 100% inhibition was accompanied by 0.5 mol label incorporation per transhydrogenase monomer (Fig. 5). These results suggested half-of-the- sites reactivity, which agreed with the dimeric nature of the enzyme [9,21].

It might also be added here that, based on the effect of substrates on the reactivity of the transhydrogenase with various thiol modifiers, Fisher and coworkers [3,40,41] have suggested that the bovine enzyme contains two types of sulfhydryl groups, one that is located near the NADP(H)-biding site, and another that is located peripherally and distant from this site. Furthermore, Robillard and Konings ([58], see also ref. [42]) have proposed a dithiol–disulfide interchange mechanism for proton translocation by the transhydrogenase, which involves such difficult chemistry as abstraction of two hydrogen atoms from a dithiol to form a disulfide, reduction of a second disulfide by these hydrogen atoms to form a dithiol, homolytic scission of one S–H bond to form a hydrogen atom and a sulfur free radical, and loss of an electron and a proton separately from the second S–H to form another sulfur-free radical which reacts with the first to reform a disulfide.

4. *Mechanism of energy transduction*

As indicated by the reverse of Eqn. (1), transhydrogenation from NADPH to NAD results in proton translocation from the matrix to the cytosolic side of the membrane. Indeed, it has been shown that the membrane potential so created can drive ATP synthesis [59]. However, as a proton pump, the transhydrogenase has certain unique features, which have provided an insight into its mechanism of energy transduction as well as the means for exploring this mechanism. These unique features are as follows:

(a) To translocate an ion and establish an electrochemical gradient, ion pumps utilize energy derived from a redox reaction, light, ATP hydrolysis, or the transmembrane gradient of a

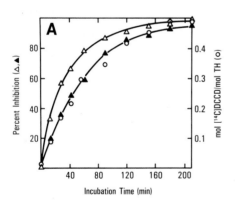

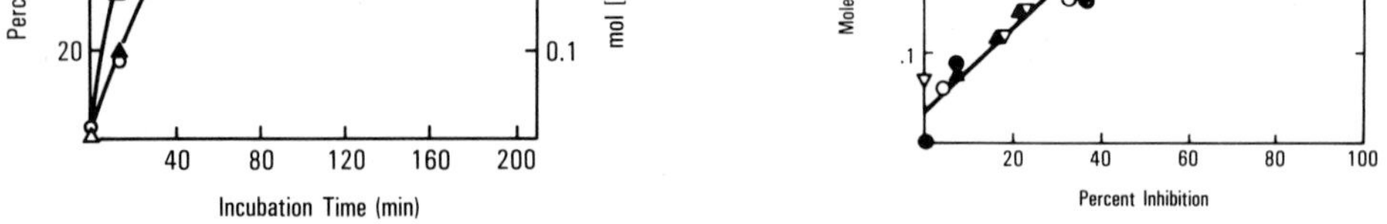

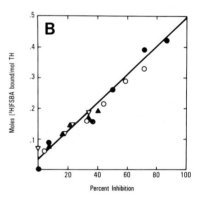

Fig. 5. Correlation of the degrees of inhibition and labeling of purified bovine transhydrogenase with [^{14}C]DCCD (A) and [^{3}H]FSBA (B). In (A), filled and open triangles show the extent of inhibition in the absence and presence, respectively, of NADPH, and open circles the extent of label incorporation from [^{14}C]DCCD. In (B), the degree of inhibition on the abscissa is correlated with the extent of radioactivity from [^{3}H]FSBA incorporated into the enzyme on the ordinate. Symbols indicate the absence (●) or presence of NADP (○), 5'-AMP (▲) or NADP + 5'-AMP (▽) in the incubation mixture. From refs. [9] and [21] with permission.

transported solute. The transhydrogenase does not fit into any one of these categories. Its nucleotide substrates and products are all on the same side of the membrane, and there is essentially no difference, in the midpoint potentials of the NADH/NAD and the NADPH/NADP couples ($\Delta E'_0 \simeq 5$ mV). This unique feature draws attention to the difference in the concentrations (i.e., the binding energies) of the reactants (NADPH + NAD) and products (NADP + NADH) as the only possible source of energy for uphill proton translocation.

(b) Unlike uphill electron transfer, net ATP synthesis, light emission via reverse photophosphorylation, etc., which do not proceed to any measurable extent in the absence of an energy supply, transhydrogenation from NADH to NADP (the forward reaction in Eqn. 1) occurs at a slow rate in the absence of a protonmotive force. This feature has made it possible to measure the effect of protonic energy on the kinetic and the thermodynamic parameters of the reaction, such as substrate K_m's, V_{max}, and the equilibrium constant (see Section 3 above).

(c) Again, unlike the scalar reactions catalyzed by the respiratory chain and the ATP synthase complexes, the scalar reaction catalyzed by the transhydrogenase does not involve uptake or release of protons. Hydride ion transfer between NAD and NADP is direct, the enzyme does not have a prosthetic group, and no protein residue participates as an intermediate in H^- transfer. Hence, it must be concluded that it is the protein itself that takes up protons on one side of the membrane and releases protons on the other side.

These unique characteristics of the transhydrogenase suggested the following mechanistic expectations. (i) In reverse transhydrogenation (see Eqn. 1), transduction of substrate binding energy to a proton electrochemical potential should involve protein conformation change. (ii) The conformation change of the protein should lead to pK_a changes of appropriate amino acid residues, which would result in proton uptake and release on opposite sides of the membrane. (iii) In $\Delta\mu_{H^+}$-promoted transhydrogenation from NADH to NADP (Eqn. 1 in the forward direction), the protonmotive force should alter enzyme conformation (presumably via protonation) and thereby increase the affinity of the enzyme for one or both substrates (i.e., NADH and NADP). Experimental data in agreement with these expectations are provided below.

(i) *Substrate-promoted conformation change of the transhydrogenase.* As was stated in Section 3, inhibition of the transhydrogenase by several modifiers is accelerated by substrates. These modifiers are NEM, DCCD, FSBA, ethoxyformic anhydride, dansyl chloride, and trypsin [9,21,41,44,50,55,56]. In all cases, the promoting substrate was NADPH. NADP also had some positive effect, except when the modifier was NEM, while NAD and NADH either had no effect or protected against inhibition (e.g., in the case of FSBA and DCCD, see above). These results suggested that NADP(H) binding results in conformation change of the transhydrogenase, thus making the targets more amenable to modification by the reagents mentioned. Close examination of the effect of substrates on inactivation of the transhydrogenase by trypsin showed the following. As seen in the pseudo-first-order plot in Fig. 6, NAD had no effect on the inactivation rate, NADH retarded it by 50% (because it protected a cleavage site, see Fig. 1), NADP nearly doubled it, while NADPH increased it sixfold [55]. The effect of NADPH was concentration dependent, with half-maximal stimulation at pH 7.5 occurring at 35 μM NADPH, which is close to the apparent K_m of the bovine transhydrogenase for NADPH (~20 μM) at the same pH. Figure 7 depicts on SDS-polyacrylamide gel the substrate effects on degradation of purified bovine transhydrogenase by trypsin. The devastating effect of NADPH is clear in lane 5. N-terminal sequencing of the obtained tryptic fragments showed two significant points regrading the effect of NADPH. The presence of NADPH greatly accelerated the cleavage rate of the Lys^{410}–Thr^{411} bond (see Figs. 1 and 4), which is located in the NAD-binding extramembranous arm of the enzyme, several hundred residues distant from the site of

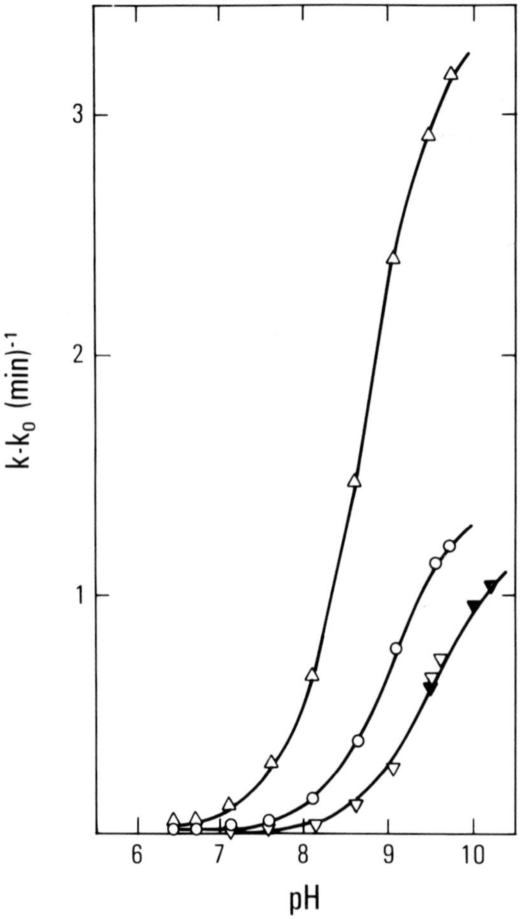

Fig. 8. pH dependence of the inactivation rate constant (k) of purified bovine transhydrogenase by N-ethylmaleimide in the absence of substrates (○) and presence of NADPH (△) or NADP (▽,▼). The values of k and k_o, pseudo-first-order rate constants, respectively, in the presence and absence of the inhibitor, were determined from semilogarithmic plots of the inactivation time course. From ref. [44] with permission.

absence of a protonmotive force. In its presence, the rate of transhydrogenation increases 10–12-fold, and the equilibrium of the reaction is shifted to the right from $K_{eq} = 0.79$ to an apparent $K_{eq} \simeq 500$ [2]. As discussed above, one would expect that the protonmotive force would change the conformation of the enzyme and increase its affinity for NADH, NADP or both. While physicochemical data of the sort discussed in part (i) are not yet available for the effect of protonic energy on transhydrogenase conformation, there is suggestive evidence that the enzyme does undergo energy-dependent conformation change.

As mentioned earlier, we had shown that the bovine transhydrogenase can catalyze at a slow rate transhydrogenation from NADPH to AcPyADP or thio-NADP, and that these reactions were energy promoted [35]. We, therefore, wondered whether the reaction shown in Eqn. 2 was also accelerated by a protonmotive force.

$$\text{NADPH} + [^{14}\text{C}]\text{NADP} \leftrightarrows \text{NADP} + [^{14}\text{C}]\text{NADPH}. \qquad (2)$$

The progress of the reaction was followed by measuring the increase of radioactivity in the column-separated NADPH fraction as a function of time, and it was found that energization of submitochondrial particles increased the initial rate of this reaction about sixfold [60]. This result is of interest, because, except for the radioactivity in NADP and NADPH, there is no difference at any point in the course of this reaction in the nature and concentration of reactants and products. Hence, no fraction of the consumed energy can be conserved. This means that, from a strictly thermodynamic viewpoint, the unrecoverable energy utilized in this reaction could be ascribed to an entropic component, most likely the cyclic conformation change of the enzyme itself.

Consistent with this possibility, it was shown that in the respiration-driven reaction NADH → AcPyADP progressive uncoupling resulted in an increase of the apparent K_m for AcPyADP and a decrease of V_{max}, with $\ln(V_{max}/K_m)$ decreasing linearly with increasing uncoupler concentration (Fig. 9) [61]. The results suggested that at 90% uncoupling with S-13 there was a decrease in the binding energy of AcPyADP by about 10 kJ/mol. Similarly, it was shown that in transhydrogenation from NADH to AcPyADP or thio-NADP the apparent K_m for all substrates decreased as V_{max} was increased upon membrane energization [34]. Under the same conditions of temperature and pH as in the experiments of Fig. 9, ΔG values calculated from $\ln(V_{max}/K_m)$ in going from nonenergized to respiration-energized conditions were again in the range 9–10 kJ/mol for all the substrates indicated. A comment regarding the V_{max}/K_m change with NADH as the variable substrate is in order. It was discussed above that NAD and

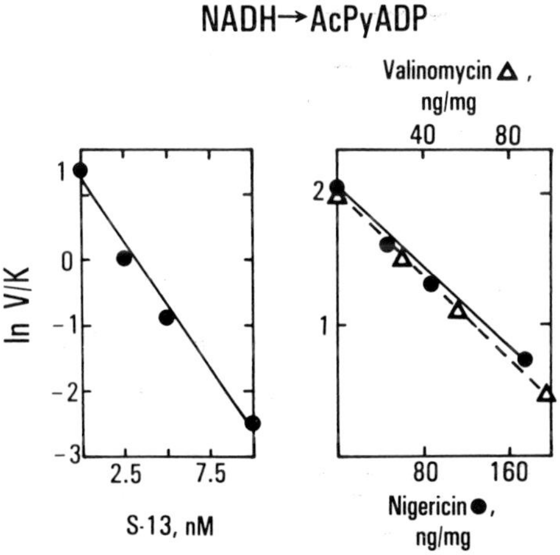

Fig. 9. Effect of partial uncoupling on $\ln(V_{max}/K_m)$ for NADH to AcPyADP transhydrogenation. Where indicated, 0.5 µg nigericin/mg of submitochondrial particle protein plus variable amounts of valinomycin (top abscissa, △) or 0.5 µg valinomycin plus variable amounts of nigericin (bottom abscissa, ●) were added. The variable substrate was AcPyADP. Adapted from ref. [61] with permission.

components. The F_1 sector catalyses the ATP hydrolysis and is water soluble and the F_0 sector is located in the membrane and is able to passively conduct protons. While this separation has been very useful experimentally it represents a problem when considering assembly. Presumably all cells have evolved a mechanism for assembling the ATP synthase without, at any stage, forming an open proton pore.

Due to space limitations, we have decided to review the structural and assembly aspects of the ATP synthase from particular sources. This chapter therefore does not represent a comprehensive review of the literature and for that we apologise.

2. *Subunit composition and organisation of ATP synthase*

The multiple functional aspects of ATP synthases are shared amongst many subunits (Table I). In general, it is possible to discriminate two major types of enzyme complex organisation. The first is the more simple bacterial and chloroplast type, which has a clearly demarcated F_1 sector carrying nucleotide binding and catalytic properties, and a membrane-embedded F_0 sector which contains the proton channel and, in conjunction with F_1, the energy-transducing or coupling functions of the complex. In such complexes the five different subunits comprising F_1 function together with three different F_0 subunits in bacterial ATP synthase (F_0 in chloroplasts has four subunits since there are two distinct *b*-subunits). Each F_0 subunit has at least one well-defined transmembrane helix. Second, mitochondrial ATP synthase (mtATPase) complexes have a similar F_1 organisation to that of bacteria, with five different subunits. Mitochondrial F_1 functions in the context of an F_0 sector comprising three hydrophobic proteins (proteolipids), which are presumably integral membrane proteins; there is a further panel of subunits whose interaction with the membrane is less well defined. According to the classification of Nagley [1], utilised here, these subunits are denoted F_A (associated components), being part of neither the clearly defined soluble F_1 sector nor the membrane integral F_0 sector within which the proton channel lies. It is likely that F_A proteins make physical and functional links between F_0 and F_1; their roles therefore encompass assembly, energy transduction or coupling functions, as well as modulating activity of the enzyme complex as a whole. Presumably, mitochondria are capable of enzyme control appropriate to the particular demands of the energy metabolism of that organelle as opposed to the cellular milieu of bacteria or the photosynthetic environment of the chloroplast.

Table I presents a compilation of information concerning the composition, genetic specification and stoichiometry of subunits of ATP synthase complexes, selected as representative of major biological and organisational subdivisions. The examples chosen are those dealt with in more detail below in this article. The two 'prokaryote' types, *E. coli* and chloroplast, represent the simpler F_0F_1-ATPase organisation in very different molecular cell biological contexts. The mitochondrial types are represented first by the yeast system, where molecular genetic analysis coupled with studies on import and assembly in isolated organelles have made very substantial contributions to our understanding of the biogenesis of the complex. Second, the bovine heart mtATPase has, of course, historically been the favourite for detailed biochemical dissection of subunit composition and function. This, together with reconstitution work, has been aimed largely at an understanding of the mechanistic bioenergetics of the complex.

As summarised in Table I it is evident that some subunit types are clearly homologous between bacteria, chloroplasts and mitochondria. In other instances there are apparent structural relationships between subunits assigned to different sectors of the complexes from the various biological subdivisions. Thus, F_1 sectors have the general organisation $\alpha_3\beta_3\gamma\delta\varepsilon$. The

TABLE I

Subunit composition, genetic specification and stoichiometry of F_0F_1-ATP synthase from different sources. Subunits are aligned horizontally according to perceived relationships based on sequence homology or other structural predictions (see text for details)

Sector	Bacteria (*E. coli*)			Chloroplast			Mitochondria				Approx. subunit mass (kDa)[i]
	Subunit	Gene[a]	Stoichio-metry[b]	Subunit	Gene[c]	Stoichio-metry[d]	Yeast		Bovine subunit[g]	Stoichio-metry[h]	
							Subunit	Gene[f]			
F_1	α	*uncA*	3	α	*atpA*$_c$	3	α	*ATP1*	α	3	55
	β	*uncB*	3	β	*atpB*$_c$	3	β	*ATP2*	β	3	52
	γ	*uncG*	1	γ	*atpC*	1	γ		γ	1	33
	δ	*uncH*	1	δ	*atpD*	1	(see OSCP)		(see OSCP)		20
	ε	*uncC*	1	ε	*atpE*$_c$	1	δ		δ	1	15
							ε		ε	1	6
F_0	*a*	*uncB*	1	IV(a)[e]	*atpI*$_c$	1	Su6	*oli2*$_M$	Su6$_M$	?	28
	b	*uncF*	2	I(b)	*atpFc*	1	(see P25)		(see *b*)		17
				II(b')	*atpG*	1					14
	c	*uncE*	6–12	III(c)	*atpH*$_c$	6–12	Su9	*oli1*$_M$	Su9	?	8
							Su8	*aap1*$_M$	A6L$_M$	1	7
F_A							P25	*ATP4*	*b*	2	22
							OSCP	*ATP5*	OSCP	1	21
							P18	*ATP7*	*d*	1	18
									e	?	8
									F_6	2	9
							INH	*INH1*	INH	1	9
							9KD	*STF1*		1	7
							15KD	*STF2*		1	10

[a]Walker et al. [3]; a recent genetic map uses the designation *atp* for the ATP synthase genes in *E. coli* [199]. [b]Reviewed in Senior [148]. [c]Reviewed in Glaser and Norling [102]; genes are in nuclear DNA of plants except those marked $_c$ which are in chloroplast DNA. [d]Reviewed in Glaser and Norling [102]. [e]Subunit designation in parentheses as used for photosynthetic bacteria. [f]Genes are in nuclear DNA of *Saccharomyces cerevisiae*, except those marked $_M$ which are in mitochondrial DNA. References for individual subunits are as follows: *ATP1*, Takeda et al. [200]; *ATP2*, Takeda et al. [77]; *oli2*, Macino and Tzagoloff [201]; *oli1*, Linnane and Nagley [202]; *aap1*, Macreadie et al. [87]; *ATP4*, Velours et al. [5]; *ATP5*, Uh et al. [90]; *ATP7*, Norais et al. [93]; *INH1*, Ichikawa et al. [203]; *STF1*, Akashi et al. [204]; *STF2*, Yoshida et al. [205]. [g]α, β, γ, δ, ε, Walker et al. [3]; Su6, A6L, Fearnley and Walker [206]; Su9, Sebald and Hoppe [56]; *b*, *d*, Walker et al. [4]; OSCP, F_6, INH, Walker et al. [2]; *e*, Walker et al. [6]. [h]Compilation of data for both yeast and bovine systems presented in Gregory and Hess [207], Hekman et al. [62], Muraguchi et al. [64], Okada et al. [15], Stutterheim et al. [208], Todd et al. [209] and Walker et al. [3]. A question mark indicates subunits for which no reliable stoichiometric data are available. [i]Subunit masses are shown as a broad generality across a set of homologous subunits.

counterpart of the bacterial δ-subunit appears in fact to be OSCP (an F_A subunit) in mitochondria [2]. Additionally, the mitochondrial δ-subunit shares structural features with bacterial and chloroplast ε [3]. The mitochondrial ε-subunit has no evident counterpart in bacterial or chloroplast systems. Bacterial and mitochondrial F_0 has three subunits, but not all are parallel in identity [1]. Subunits *a* and *c* clearly have structural and functional counterparts in the mitochondrial F_0 subunits 6 and 9, respectively. By contrast, bacterial *b* is not found as such in mitochondrial F_0; the third such subunit in mitochondria is the much shorter and hydrophobic subunit 8 or A6L in fungal (yeast) or mammalian (bovine) systems, respectively. Suggestions have been made that mitochondria contain a subunit with loose structural similarity to bacterial subunit *b*, based on weak sequence homology and comparative hydropathy plots [4,5]. As there is no evidence that the suggested bacterial *b* homologue, denoted P25 in yeast and *b* in the bovine system, is explicitly integral to the membrane, these mitochondrial subunits are classified into F_A. The divergence in the genetics and molecular biology of subunit *b* can be seen in an indirect way comparing *E. coli* on the one hand and chloroplasts (and photosynthetic bacteria) on the other. The former has two identical *b*-subunits while the latter have two related but distinct polypeptides. In the case of chloroplasts, each *b* subtype is encoded in a different cellular compartment. This may provide a clue to the divergent evolutionary path taken by the bacterial *b*-subunit, as opposed to its now distantly related mitochondrial counterpart which is nuclearly encoded.

Amongst the F_A components of yeast and bovine mtATPase can be recognised some cognate proteins, whilst others may be characteristic of fungal or mammalian mtATPase complexes. Common subunits are bovine *b*/yeast P25, OSCP, bovine *d*/yeast P18, and the inhibitor protein. Components characteristic of the beef heart complex include subunits F_6 and *e*; the latter subunit may require further investigation to establish its status as an authentic component of the mtATPase complex [6]. A concentrated effort has been made by Walker et al. [6] to detect a number of subunits in preparations of beef heart mtATPase reported by other investigators. None of the following were detected: factor B [7], the ADP/ATP translocase [8–10], the uncoupler-binding protein [11], a protein of size 30 kDa [12] and the phosphate carrier [13]. Moreover, homologues of the two protein factors characterised in the yeast system [14,15] that stabilise and facilitate the binding of the ATPase inhibitor were not detected in the bovine complex isolated by Walker et al. [6] or by others.

3. Mitochondrial ATP synthase

3.1. Biogenesis and assembly of mitochondrial ATP synthase subunits

The biogenesis of mitochondrial ATP synthase requires contributions from both the mitochondrial and nuclear genomes. At least two subunits are encoded in mitochondrial DNA of all organisms so far examined. These are subunit 6, equivalent to bacterial subunit *a*, and subunit 8 (or A6L in the bovine system), which has no apparent bacterial homologue. In some fungi, including the yeast *Saccharomyces cerevisiae*, subunit 9 (bacterial homologue subunit *c*) is encoded in mtDNA. By contrast, genes encoding subunit 9 in bovine cells and all other metazoa as well as filamentous fungi, such as *Neurospora crassa*, are located in the nucleus. Interestingly, the mRNAs expressed by two bovine subunit 9 genes are expressed in a tissue-specific manner [16]. The encoded gene products consist of identical subunit 9 moieties each with a different N-terminal leader [16].

All other gene products comprising subunits of mtATPase are encoded in the nuclear

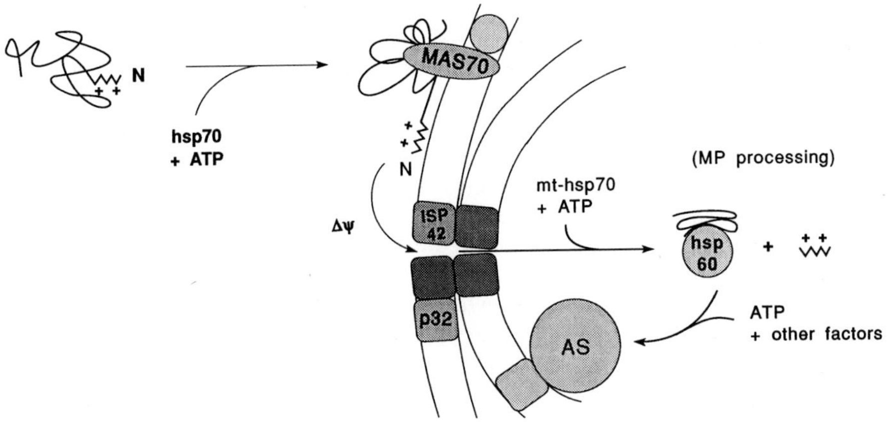

Fig. 1. Import and assembly of nucleus-encoded subunits in yeast. Details of the import of precursor proteins, their interactions with components of the import apparatus and hsp proteins, processing by matrix protease (MP) and subsequent assembly into mtATPase (AS) are described in the text. In the precursor, the cleavable N-terminal leader peptide is indicated by a zig-zag line (with associated positive charges); the imported subunit, as such, is indicated by a smooth line.

genome except for the α-subunit of F_1 of plant mitochondrial mtATPase which is synthesised within mitochondria [17,18]. Interestingly, the gene encoding subunit α has been found to be duplicated in mtDNA in some plant sources: fertile maize, sorghum and *Oenothera* [19,20]. In mammals there are at least two nuclear genes for subunit α [21,22].

With constituent subunits encoded in two different subcellular compartments there would seem to exist an obvious need for coordination of synthesis to achieve efficient assembly [23]. The limited coding capacity of mitochondrial genes means that this coordination must be under nuclear gene control. In yeast, the study of this nuclear involvement in mitochondrial gene expression has been greatly advanced by the analysis of *pet* mutants, defined by nuclear mutations causing the loss of respiratory function (reviewed by Tzagaloff and Dieckmann [24]). Recently, several nuclear genes have been identified in yeast whose products are implicated in regulation of expression of individual mitochondrially synthesised subunits by acting at different levels of control including subunit translation and assembly (see below).

The majority of mitochondrial proteins are imported into the organelle from the cytosol. In Fig. 1 is summarised the pathway for import and assembly of such proteins. (More detailed descriptions of the individual steps involved in this pathway can be found in several excellent reviews [25–28].) The study of imported mtATPase subunits such as β, and to a lesser extent subunit 9, has played key roles in the dissection of these pathways. Most of the nuclearly encoded mtATPase subunits are synthesised as precursors that carry an N-terminal targeting sequence (presequence) that directs the protein into mitochondria. In the case of a hydrophobic integral membrane protein such as nuclearly encoded subunit 9 [29], the presequence plays a role in maintaining the solubility of the hydrophobic passenger protein in the aqueous environment of the cytosol [30]. Also important in this respect is the molecular chaperone cytosolic hsp70 that maintains precursors in a translocation competent state [31] by acting as an unfoldase, interacting with precursors via repeated cycles of binding and release requiring the hydrolysis of ATP [32]. ATP itself is not required for the subsequent precursor–receptor interaction and does not appear to be directly involved in the translocation process [33].

The precursor protein in a translocation-competent state then binds to the mitochondrial surface by interaction with a receptor protein, designated MAS70 in yeast [34], although other, as yet unidentified, proteins are implicated in the import process. The MAS70 protein is enriched at sites of close contact between the two mitochondrial membranes at which translocation occurs. Following the initial interaction with surface receptors, precursors are then inserted into the import channel complex that includes the import site protein 42 (ISP42). An energised inner membrane is critical for precursor translocation. The electrochemical potential, $\Delta\Psi$, provides the energy for the translocation of the presequence [27] across the membrane. The $\Delta\Psi$-driven movement of the leader sequence through the import channel causes the polypeptide chain traversing the membranes to become extended [35,36]. Translocation of the precursor through the import channel also involves interaction with the mitochondrial hsp70 [37,38]. Mitochondrial hsp70 binds to the polypeptide chain emerging on the matrix side of the membrane [38], thereby promoting the $\Delta\Psi$-independent translocation of the major part of the polypeptide [26]. The $\Delta\Psi$-mediated transfer of the leader sequence is viewed as a 'trigger' for translocation followed by binding to, and ATP-dependent release from, intramitochondrial hsp70 which drives import of the precursor.

Once the protein has reached the mitochondrial matrix its N-terminal targeting sequence is cleaved. The matrix protease is comprised of two non-identical subunits, one is the peptidase and the other a processing-enhancing protein [28,39]. The strict temporal sequence, if any, of intramitochondrial processing events in relation to hsp70 and hsp60 interactions has not yet been fully elucidated.

Interaction with mitochondrial hsp60 is required for the mature polypeptide to achieve its native conformation and for the assembly of oligomeric proteins. The disengagement of the non-folded imported polypeptides from hsp60 also involves ATP-dependent catalysis that appears to be responsible for the subsequent folding process and final release from hsp60 itself [40]. The individual mtATPase components must then be assembled into the requisite oligomeric arrays. F_1 subunits probably assemble within the matrix, eventually to be associated with the mitochondrial inner membrane (see below). F_0 and F_A subunits interact with the membrane in a more direct manner.

An unresolved question concerns the possible interaction of mitochondrially encoded polypeptides with mitochondrial hsp60. Since the mitochondrially encoded subunits of mtATPase are integral membrane components of the F_0 sector and are hydrophobic in nature, the efficient assembly of the complex may be facilitated by their direct insertion into the inner membrane from mitochondrial ribosomes associated with the matrix face of the membrane. This site of synthesis could conceivably avoid the need for such hydrophobic subunits to undergo interactions with other imported subunits and the hsp60 complex in the more hydrophilic milieu of the mitochondrial matrix. Indeed it may be asked whether mitochondrially encoded subunit 9 in *S. cerevisiae* follows the same intramitochondrial trafficking route of imported subunit 9 as in *N. crassa*, the latter presumably involving hsp60. Moreover, the recoding of mitochondrial mtATPase subunits for expression by the nucleo-cytosolic system of yeast [41] (see below) raises a parallel question for subunit 8 expressed naturally or via allotopic expression.

It is noteworthy that some mtATPase subunits lack a cleavable leader sequence (subunit *d* of the beef heart complex and P18 and the 15 kDa stabilising factor of the yeast complex) and thus are analogous to the more well characterised examples of proteins in this class such as the ADP/ATP carrier [42], which are processed to the extent of removal of the initiator methionine residue. Although the ADP/ATP carrier precursor is imported without cleavage of a leader sequence many facets of its import are the same as for cleavable precursors. However, a different import receptor is used on the mitochondrial surface and there is independence from

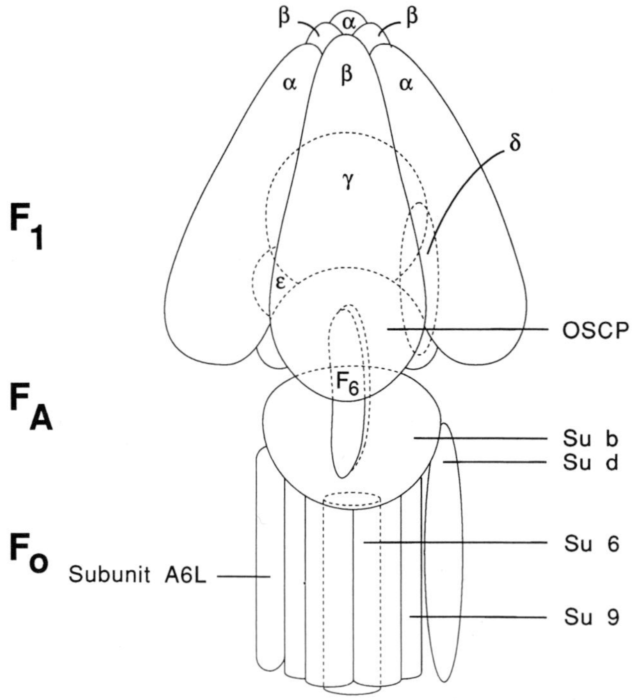

Fig. 2. Subunit arrangement of beef heart mtATPase (after Joshi and Burrows [52], with subunit nomenclature amended to be consistent with Table I). Thus, subunits *b* and *d* are referred to as 24 kDa and 20 kDa proteins, respectively, by Joshi and Burrows [52]. Subunit 8 is termed A6L in the figure. The model is itself based on a previous schematic [195], following cross-linking and nearest-neighbour studies [52]; see text for further details.

hsp60 in the matrix. The import of mtATPase non-cleavable precursors has not been studied in sufficient detail to establish whether these latter features of import are shared with or are unique to the ADP/ATP carrier.

3.2. Structure and assembly of the beef heart mtATPase complex

The beef heart complex is the mitochondrial enzyme complex for which the most complete picture of its structural organisation exists (see reviews by Hatefi [43], Godinot and Di Pietro [44], Senior [45], Penefsky and Cross [46]). In this section we highlight some recent contributions to an understanding of the structure and assembly of mammalian mtATPase, focussing on the bovine heart system, where most of the relevant work has been carried out. Studies on other mammalian systems, such as rat [47,48], porcine [49] or human [50,51] enzymes are also in progress. A scheme of the structural organisation of bovine mtATPase is shown in Fig. 2.

The nearest-neighbour relationships of bovine mtATPase subunits have been investigated by Joshi and Burrows [52] using the chemical cross-linking approach. In addition, these authors investigated the accessibility of F_0 subunits at the membrane surface by determining their susceptibility to degradation by trypsin. The inferences made from these studies, combined with data derived from earlier studies in the same laboratory on the topology of the OSCP and F_6 subunits [53], have been used to generate a model for mtATPase subunit arrangement [52]. In this model (cf. Fig. 2) the stoichiometry of F_0 was assumed to be $OSCP_1$, $(F_B)_2$, $(F_6)_2$, su *b* $(24\ kDa)_1$, su *d* $(20\ kDa)_1$, su 6_1, su 9_6. The various subunits were organised into three different regions referred to as the 'head', the 'stalk' and the 'base' piece. In the head piece the α- and β-subunits were depicted as extending to the surface of the membrane; this accounts for their interactions with the F_6 and OSCP subunits, and also the previously published observations of shielding by F_1 of trypsin-sensitive domains of the OSCP, F_6 and *b* subunits [53]. The stalk region was envisaged to comprise single copies of the F_1 subunits γ, δ and ε, the F_A subunits OSCP, F_B, as well as two copies of the F_6 subunit. The placement of subunits is in accord with the observed ability to make cross-links (Fig. 2). Subunit *b* was positioned directly above the proton channel because of its reported requirement for passive proton conductance [54]. Subunit F_B was not explicitly included in the depicted model of the complex [52] because of a lack of experimental evidence on its precise location. It was, however, suggested that it might be located near subunits 9 and *b*, based on a proposal that F_B is required for passive proton conductance through F_1-depleted mitochondrial membranes [55]. However, the status of F_B as an authentic component of the bovine complex has been questioned [6].

In Fig. 2 the membrane-located base piece is depicted as containing one copy each of subunits 6, 8 and *d*, and six copies of subunit 9 arranged to form a ring, as suggested by Sebald and Hoppe [56]. By analogy with fungal subunit 9, it is likely that the N- and C-termini of subunit 9 face the intermembrane space [57] with the hydrophilic loop facing F_1 towards the matrix [58]. Subunit 6 is placed within this ring following the earlier proposals for the arrangement of its *E. coli* equivalent subunit *a* [59]. Subunit A6L is arbitrarily placed outside the ring and shown traversing the membrane once. Subunit *d* is placed next to the subunit 9 ring and positioned such that it would be partially exposed on the inner mitochondrial membrane surface. This membrane-interactive location is based on the reported hydrophobic amino acid stretch in the primary structure reported by Walker et al. [4] and the observed resistance to solubilisation by aqueous and chaotropic reagents [52]. The exposure of part of this subunit on the matrix side of the inner membrane surface is in accord with observed cross-links between subunit *d* and F_6 [52]. The representation of subunit *d* and A6L as containing definitive transmembrane helices must, however, remain tentative as this is primarily based on hydropathy plots [4,60]. Recent findings have questioned this assumption in the case of yeast subunit 8. Site-directed mutagenesis studies of yeast subunit 8, the inferred homologue of mammalian A6L, have shown subunit 8 to be fully functional when carrying within the putative transmembrane stem additional positive charged residues which would prevent membrane integration of this subunit, at the same time destroying its proteolipid character (T. Papakonstantinou, R.J. Devenish and P. Nagley, unpublished). Nevertheless, the C-terminus of subunit 8 [61] and A6L [62] appears to face the matrix component.

A different approach to determining the mtATPase complex structure has recently been reported by Hatefi and colleagues [62,63]. The topography of a number of subunits was studied with subunit-specific antibodies raised to subunits *b*, *d*, 6, F_6, A6L, OSCP, 9 and the inhibitor protein (IF_1). Exposure of these subunits in both inverted and right-side-out inner mitochondrial membrane preparations was investigated both by direct antibody binding and by determining the susceptibility of the subunits to degradation by various proteases. The latter technique involved SDS-PAGE of membrane digests followed by immunoblotting with the subunit-specific antibodies [62]. Subunits *b*, *d*, F_6, A6L and OSCP were shown to be exposed on the matrix side of the inner mitochondrial membrane. Antibodies were unable to recognise any exposed surfaces on the cytosolic side. Subunits 6 and 9 were not detectable on either side of the inner membrane. The stoichiometry of several subunits, including the inhibitor protein, relative to F_1 was determined by quantitative immunoblotting. It was concluded [62] that in bovine heart mtATPase per mol of F_1 there are 1 mol each of subunit *d*, OSCP and the inhibitor protein and 2 mol each of subunits *b* and F_6. Determination of stoichiometries for subunits 6 and 9 was not reported. Note that these stoichiometries are largely in agreement with those assumed by Joshi and Burrows [52] except that subunit *b* is suggested to be present in two copies by Hekman et al. [62]. A6L is suggested to represent 1 mol per mol mtATPase [64].

Papa and colleagues have concentrated on detailing the structural and functional characteristics of subunits contributing to the stalk sector of the bovine mtATPase complex, using dissociation and reconstitution approaches. Information on the *b*-subunit (designated PVP protein by Papa and colleagues) was obtained by selective enzymatic digestion of membrane proteins and reconstitution with the isolated native components [54,65–67]. OSCP and subunit 6 in submitochondrial particles can be digested with trypsin only after removal of F_1. Subunit *b* undergoes selective cleavage of the C-terminal region of the protein. Progressive digestion of subunit *b* protein by trypsin is linearly associated to depression of proton conduction [54]. Reconstitution of trypsin-digested particles, or liposomes containing membrane-associated subunits isolated from such particles, with isolated subunit *b* resulted in the recovery of proton conduction and in a re-acquired sensitivity to oligomycin and DCCD. This finding would implicate subunit *b* directly in an energy transduction or coupling role. By contrast, the C-terminally truncated subunit *b* was not able to reconstitute the same activities.

Using this approach, OSCP, F_6 and the C-terminal region of subunit *b* were shown to be collectively required for functionally correct binding of F_1 to the membrane and coupling ATPase activity to transmembrane proton conduction. The data (summarised by Papa et al. [68]) indicate that subunit *b* is essential for correct functional organisation of the membrane proton channel, because DCCD binding, which results in inhibition of proton conduction, does so only in the presence of subunit *b*. Thus, subunit *b* may be thought to play a central role in organising the proton channel and controlling proton conduction to the catalytic and/or allosteric sites in F_1. In this view, dynamic interactions between subunit 9 (the DCCD-binding protein), subunit *b* and subunit 6 could be envisaged as being mechanistically involved in proton conduction [65] (cf. the description of Cox et al. [59] of the proton conduction mechanism through bacterial F_0). Futher, it has been proposed that subunit *b* has a role in the function of the proton gate of the complex with subunit γ of F_1. The interaction between these two subunits is proposed to take place through disulphide bridging [69]. The relevance of these observations to other systems, such as yeast, warrants closer investigation, particularly since the single cysteine residue of the beef heart protein implicated in the gating process is absent in the yeast P25 protein [5], the inferred homologue of bovine subunit *b* (Table I).

3.3. Assembly of the yeast mtATPase complex

The application of yeast, *S. cerevisiae*, to study the assembly of mtATPase has proved very fruitful. The utility of this system for molecular genetic approaches has been combined with studies of mtATPase assembly both in vivo and in isolated mitochondria to develop new paradigms for investigating the assembly process, thus generating the emerging picture of the sequential and deterministic interactions of subunits. A particular focal point has been the assembly of those three F_0 subunits encoded by mtDNA; more limited data relating to F_1 and F_A subunit assembly are also considered.

An important tool in this assembly work has been the use of antibodies against F_1 components. Earlier work entailed use of polyclonal sera raised against the F_1 sector as a whole or individual subunits such as β. The more recent use of monoclonal antibodies against subunits β [70] or α [71], which immunoadsorb the assembled holoenzyme without apparent dissociation of constituent subunits, avoids some of the pitfalls encountered when polyclonal antisera are employed, since the latter may contain antibodies which actually dissociate the assembled complex [70].

In cytoplasmic petite (*rho*^{0}) mutants of yeast which completely lack mtDNA and, therefore, produce no F_0 subunits, the mitochondria contain assembled F_1 which has a cold-labile ATP-ase activity [72,73]. On the basis of the temperature-dependence of enzyme kinetics it was concluded that this assembled F_1 has a loose association with the mitochondrial membrane in spite of the lack of an assembled F_0 sector [73].

An assembly pathway for the individual F_0 sector subunits of the yeast complex has been deduced by analysis of a series of *mit*$^-$ mutants bearing early frameshift mutations in each of the mitochondrial genes encoding subunits 9, 8 and 6 [74]. The paradigm introduced for this study was to radiolabel the proteins of mutant cells, grown in a chemostat to minimise glucose repression effects, then to analyse the nature of mtATPase subunits in mitochondrial lysates that were immunoadsorbed to an immobilised anti-β monoclonal antibody [70,75]. In *oli2 mit*$^-$ mutants, effectively devoid of subunit 6, the assembly of subunits 8 and 9 was apparently unaffected. In the absence of subunit 8 (*aap1* mutants), only subunit 9 was observed to be assembled, although synthesis of subunit 6 still continued [76]. Analysis of those *oli1 mit*$^-$ mutants in which subunit 9 is not assembled, revealed that subunits 8 and 6 cannot be assembled even though they are still synthesised. Hence the assembly pathway [75] for the F_0 sector subunits was proposed to involve the sequential addition of subunits 9, 8 and 6 to an independently formed F_1 sector assembled from imported subunits.

The suggestion has been made that there is dependence on F_1 for proper F_0 assembly. Analysis of respiration rates in an F_1 β-subunit null mutant indicated that a functional proton channel formed by F_0 proteins is not detected in the mitochondrial inner membrane [77]. It remains to be established whether a functional proton transporting channel is not assembled as such, or is present but controlled by another protein prior to the assembly step involving F_1 β-subunit.

Other factors, not themselves part of the final assembled complex, may act to coordinate the assembly of F_0 and F_1. Thus, mutations in the yeast nuclear *ATP10* gene [78] confer a phenotype suggestive of a defect in the coupling of the F_1 sector to the F_0 sector. The biochemical lesion affects the F_0 sector itself. Although an effect on translation of a mitochondrially encoded F_0 sector subunit cannot be completely excluded, the available evidence suggests that the *ATP10* gene product functions in the assembly of the F_0 sector. The product of the *ATP10* gene has been found to be associated with the inner mitochondrial membrane but does not co-fractionate with the F_1 sector or the F_0F_1 complex [78].

At the level of expression of individual genes, control of F_0 assembly may be exerted through nuclear regulation of synthesis of the mitochondrially encoded subunits. Recent reports have indicated that the biosynthesis of subunit 9 in mitochondria is controlled by the products of the nuclear genes *AEP1*, *AEP2* [79,80] and *ATP13* (apparently the same locus as *AEP2*; [81]), which are required for stabilisation of the transcript (*AEP2*) or its translation into subunit 9 (*AEP1*). Additional nucleus-encoded proteins are implicated as effectors of biosynthesis in mitochondria of subunits 8 and 6 [82,83], in this case involving concerted action of two nuclear gene products acting at the level of maturation or translation of the dicistronic transcript [1] encoding both subunits 8 and 6.

In order to investigate the assembly properties of F_0 sector components more fully, use is now being made of an alternative strategy for manipulation of these mitochondrial genes and

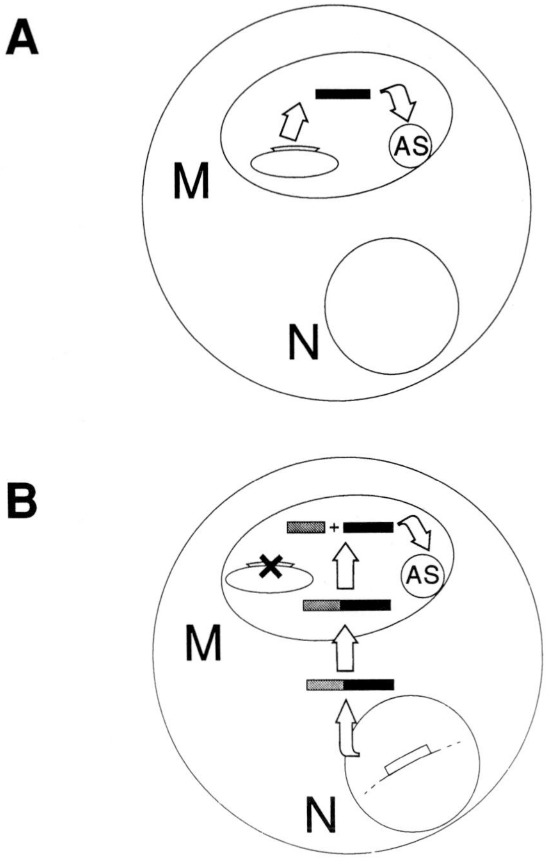

Fig. 3. Transfer to nucleus of mitochondrial gene encoding subunit 8 of ATP synthase. (A) In normal yeast cells, subunit 8 (solid block) is biosynthesised and delivered to ATP synthase (AS) entirely within mitochondria (M) [1]. Other genes in both mitochondria and the nucleus (N) specify various components of this multi-subunit enzyme. (B) In the yeast cell with subunit 8 transplanted to the nucleus, the mitochondrial gene specifying subunit 8 is inactive (X). Subunit 8 is now specified by an artificial nuclear gene, in which codons were corrected for optimum compatibility with the nucleo-cytosolic gene expression system [41]. Subunit 8, bearing a targeting signal (hatched block) for sorting to mitochondria, is biosynthesised in the cytosol, then imported into mitochondria. Following removal of the targeting signal [196], subunit 8 is delivered to ATP synthase inside the mitochondria [197]. This system, called allotopic expression [41], allows systematic site-directed mutagenesis studies on subunit 8 (see text).

their products, denoted allotopic expression [41]. The strategy involves relocation to the nucleus (Fig. 3) of the relevant mitochondrial gene specifically redesigned for nucleo-cytosolic expression. By inclusion of an appropriate N-terminal leader sequence, the now nuclearly encoded mitochondrial protein, in the form of a chimaeric precursor, could be targeted back to the organelle both in vitro and in vivo. The functional relocation of the *aap1* gene providing allotopically expressed subunit 8 [1] has permitted a program of site-directed mutagenesis to be implemented [60,71,84–86].

The assays of such structural variants of subunit 8 encompass in vivo complementation tests of *aap1 mit⁻* mutants lacking mitochondrially encoded subunit 8 [87] and in vitro tests using isolated mitochondria in which tests are done of both the import and assembly of the radiolabelled subunit 8 variant. The development of a successful assembly test in vitro [88] required the use of mitochondria from cells which had been partially depleted of subunit 8 by deliberate transcriptional regulation of allotopic expression [88]. Similarly, depletion of endogenous OSCP is required for assembly of imported radiolabelled OSCP into mtATPase of isolated mitochondria [86].

Using these paradigms it has been established that the evolutionarily conserved positively charged C-terminal region of subunit 8 is required for assembly of this subunit into mtATPase [71,86], whilst its N-terminus may have some functional role [86]. Surprisingly, the hydrophobic medial region of subunit 8 can tolerate additional charged residues that do not prevent its function [86] but do destroy the proteolipid character of the protein, which questions the assumption that this protein in fact traverses the inner membrane [60,61,89].

The assembly of F_A components is dependent upon F_0 assembly. In the absence of a structurally intact F_0 sector the proteins encoded by the nuclear genes *ATP7* (P18) and *ATP4* (P25) were not observed to be associated with immunoabsorbed mtATPase complexes [75]. The absence of subunit 6 prevents P25 and P18 assembly. In these experiments, which involved radiolabelling proteins with ^{35}S-sulphate, the *ATP5* gene product (OSCP) could not be monitored because of the paucity of sulphur-containing amino acids in this protein [90].

Complementary experiments on the influence of F_A subunits on F_0 assembly utilise null mutants in the relevant yeast nuclear genes. Disruption of the *ATP4* gene results in a mtATPase complex lacking not only P25 but also subunit 6 [91,92], although subunits 8 and 9 were apparently assembled. This indicates a mutual interdependence of subunits 6 and P25 for assembly. The assembly status of P18 and OSCP in the *ATP4* null mutant has not yet been reported. Preliminary experiments have shown that disruption of the *ATP5* gene to generate an OSCP null mutant leads to non-assembly of P25 and P18 [86]. An *ATP7* null mutant lacking P18 has been generated [93], but little assembly data are available. There are also effects on the biosynthesis of F_0 subunit 6 resulting from the lack of P18 in mitochondria [93]. The availability of cloned genes for each of the F_A subunits P25, OSCP and P18, and corresponding null mutants, will facilitate detailed examination of the role of these subunits in mtATPase assembly and function.

Surprisingly, there is little information on F_1 assembly as such. The role of hsp60 and its general requirement for assembly of imported proteins such as subunit β has been mentioned above (Fig. 1). Indeed, hsp60 can often be found associated with the isolated mtATPase complex [94]. Other, more specific, assembly factors may exist. Thus the assembly of the F_1 α- and β-subunits (and presumably the other F_1 subunits) is dependent upon the products of two yeast nuclear genes denoted *ATP11* and *ATP12* [95]. In the mutant strains, subunits α and β are imported into mitochondria but do not separate from the membrane to assemble into a soluble F_1 sector. The function of the *ATP11* and *ATP12* gene products may be to catalyse a necessary chemical modification required for assembly or to confer the proper conformational structure thus mediating folding of the α- and β-subunits [95].

Burns and Lewin [96] studied the rate of import of subunits and assembly of F_1 in vitro, employing a polyclonal anti-α antibody. They showed that newly imported subunits associate with each other almost immediately post-import and that intramitochondrial pools of unassembled subunits must be very small. It appeared that F_1 β-subunits were imported and assembled at a lower rate than the two other subunits α and γ, suggesting that F_1 assembly may be limited by the availability of β-subunits [96]. The additional role proposed for subunit α as a molecular chaperone in its own right, akin to hsp60 [97], has the potential to be systematically investigated in the yeast system.

4. Chloroplast ATP synthase

4.1. Structure

The F_0F_1 from chloroplasts and photosynthetic bacteria is thought to be structurally similar to the *E. coli* enzyme. The CF_1 portion has a subunit composition and stoichiometry identical to the *E. coli* enzyme, consisting of 3α, 3β, 1δ, 1ε and 1γ. Electron microscopy of CF_1 also shows similarities to the *E. coli* and mitochondrial ATPases, with a hexagonal arrangement of the α- and β-subunits and an asymmetric mass in the centre formed by the minor subunits [98]. The CF_0 portion of the enzyme consists of four subunits rather than the three found in *E. coli.* CF_0IV and CF_0III correspond to the *E. coli a-* and *c*-subunits, respectively. One copy of CF_0IV is present but the number of copies of CF_0III is unresolved. CF_0I and CF_0II both have a hydropathy profile similar to the *E. coli b*-subunit so it has been suggested that the *b*-subunit dimer in *E. coli* is replaced by a heterodimer of CF_0I and CF_0II in chloroplasts and photosynthetic bacteria [99–101]. The structure of CF_0F_1 has been recently reviewed [102].

The α- and β-subunits show the highest level of conservation between different species. The β-subunit contains the catalytic site for ATP synthesis and has the greatest homology to other species. The higher plant chloroplast β-subunit shows 60–70% identity to mitochondrial and bacterial β-subunits and about 80% homology to the cyanobacterial β-subunit [99,103]. Furthermore, isolated β-subunit from spinach [104] and *E. coli* [105] is able to reconstitute an active F_0F_1 with *Rhodospirillum rubrum* chromatophores from which the β-subunit has been removed.

The α-subunit is weakly homologous to the β-subunit in all species. It is thought that CF_1, like the bacterial and mitochondrial enzyme, has six nucleotide binding sites. Three of these are located on the β-subunits and it seems likely that the other three may be found at the interface between α and β (see refs. [102,106]).

In contrast to the α- and β-subunits, the sequences and roles of the minor subunits are not well conserved [107]. A complex of equal amounts of spinach α and β has a low but significant ATPase activity [104]. This is in contrast to *E. coli* where the smallest catalytic unit consists of 3α, 3β and γ [108]. Furthermore, δ and ε are not required for binding of CF_1 to CF_0, as they are in *E. coli* [109–112], although the complex is not active unless both δ and ε are present [113].

In CF_0F_1 the γ-subunit is required for light-dependent activation of the enzyme [114]. Activation is thought to be due to an extra domain present in γ from chloroplasts and photosynthetic bacteria which contains several cysteine residues. Reduction of a disulphide bond between two conserved cysteine residues appears to be involved. This may involve a conformational change affecting the interaction between γ and ε [110,115,116]. Besides a possible role in light-dependent activation, little is known about the function of the ε-subunit. It is able to act as an inhibitor of soluble CF_1 in a manner similar to *E. coli* ε [117].

The δ-subunit has been suggested to have a role in coupling proton translocation to ATP synthesis [118]. Although δ is not required for the binding of CF_1 to CF_0, it can bind to CF_0 and block proton flow through the pore [118]. Cross-reconstitution experiments have demonstrated that proton conduction through CF_0 can be blocked by either CF_1 (-δ) + *E. coli* δ or *E. coli* F_1 (-δ) + chloroplast δ, although the hybrid ATPases are not able to synthesise ATP [118]. In contrast, *E. coli* δ is not able to block the *E. coli* F_0, although it may have a role in the assembly of a functional F_0 [119]. The δ-subunit also has some homology to OSCP from the mitochondrial ATPase.

CF_0I and CF_0II share similar hydropathy profiles with each other and with the *b*-subunit of *E. coli*, although the sequence homology between the three subunits is low. The predicted structure is a single transmembrane helix at the N-terminus and an extended, largely helical C-terminus protruding into the stroma [120]. This topology has been confirmed by studies with proteases and antibody mapping [121]. CF_0I and CF_0II may be involved in connecting CF_0 to CF_1, with the *b*-subunit dimer found in *E. coli* being replaced by one copy each of CF_0I and CF_0II in chloroplasts and photosynthetic bacteria.

CF_0III is equivalent to the *c*-subunit from *E. coli*. Again, the sequence homology between the two subunits is low, but both are thought to form a helical hairpin structure which is largely embedded in the membrane [120]. An acidic residue is found in the centre of helix II in all species and in *E. coli* this residue is required for proton translocation [122,123]. CF_0III is present in multiple copies in CF_0F_1. An oligomer of this subunit can be isolated from CF_0F_1 but there is uncertainty as to the number of copies of CF_0III in each oligomer [124,125]. As well as functioning in the proton pore, CF_0III may also be involved in the binding of CF_1 to CF_0. Feng and McCarty [126] have shown that CF_0III remains attached to CF_1, even when no other CF_0 subunits are present. In mutants of *Clamydomonas reinhardtii* which lack CF_0III, α and β fail to bind to the membrane [127]. This is consistent with work on *E. coli* in which mutations in the bend between the two helices result in reduced F_1 binding [128].

CF_0IV is equivalent to the *a*-subunit in *E. coli*. There is some conservation of sequence, particularly in the C-terminal third of the molecule, and the hydropathy profiles are also similar [120]. Amino acid residues in the conserved region of this subunit have been implicated in proton translocation [129–133] and some of these are conserved in CF_0IV [120,134,135]. This suggests that CF_0IV may have a similar role. However, Feng and McCarty [136] have proposed that the main function of CF_0IV is structural.

4.2. Assembly

Assembly of a CF_0F_1 in the chloroplast results from coordinated expression from the nuclear and chloroplast genomes. Subunits CF_0II, γ and δ are encoded in the nucleus and are synthesised as higher-molecular-weight precursors which are imported in the chloroplast and processed. The remaining subunits are encoded by the chloroplast genome and in higher plants exist in two operons, one including CF_0IV, CF_0III, CF_0I and α and the other containing β and ε (see ref. [102]). In *Chlamydomonas reinhardtii*, the chloroplast encoded genes are separated from each other but their precise locations have not yet been determined [137]. Two models of assembly of CF_0F_1 have been proposed in which the nuclear subunits regulate the assembly process [138,139]. The chloroplast-encoded subunits would either associate individually with the thylakoid membrane [138] or form partial, inactive assemblies [139] with the final assembly of the CF_0F_1 complex occurring in response to, and being limited by, import of the nuclear encoded subunits. While little direct evidence exists to support these models, it is clear from a number of approaches that excess unassembled subunits do not accumulate in the chloroplast [140–143].

A number of studies have found that mutants unable to synthesise only one or two subunits were affected in the assembly of the other subunits [127,144–146]. For example, in mutants of *Chlamydomonas reinhardtii* which failed to synthesise either CF_0I, CF_0IV or ε, the α- and β-subunits were synthesised but were not found associated with the membrane in significant amounts [127]. The membranes of these mutants, as well as of other mutants which lacked α

and β, were not proton leaky, indicating that a functional CF_0 was not formed. In one mutant which overproduced α and β, excess α was degraded unless complexed with β [127]. Biekmann and Feierabend [146] demonstrated that proteolysis of γ and δ occurred in 70S ribosome deficient rye. In these plants, no chloroplast-encoded subunits could be synthesised. Since only the processed forms of γ and δ were observed, proteolysis must have occurred in the chloroplast after import and processing of the precursors.

The studies with *Chlamydomonas reinhardtii* mutants discussed above [127] also suggested that levels of some subunits in the chloroplast could affect synthesis of others. For example, in one mutant the synthesis of β appeared to be controlled by the level of α. Control of chloroplast gene expression is also exerted by the nucleus, as some *Chlamydomonas reinhardtii* mutants, affected in nuclear genes, were unable to synthesise one or more chloroplast-encoded CF_0F_1 subunits [127]. Other chloroplast genes were unaffected.

A recent finding which may have some bearing on assembly of CF_0F_1, is that the α-subunit may have chaperonin-like activity [121]. Experiments involving F_0F_1-ATPase from *Rhodospirillum rubrum* from which β had been removed showed that small amounts of α (<5% of the level of β) were required for reconstitution of an active hybrid using tobacco or lettuce β-subunit. The α-subunit from several sources has recently been shown to contain two short conserved sequences characteristic of chaperonins [97].

The CF_0F_1 from photosynthetic bacteria shows considerable homology with that from higher plants but does not present the same assembly problem of coordination of expression between two genomes. The genes are organised in a similar way to the chloroplast, with β and ε being found in a separate operon from the other genes, which form one large operon [99,135]. Little is known about the assembly of CF_0F_1 from photosynthetic bacteria.

5. *Bacterial ATP synthases*

5.1. *Gross structure of Escherichia coli F_0F_1-ATPase*

The bacterial F_0F_1-ATPase structure that is best understood is that from *Escherichia coli*. This section will therefore concentrate on results obtained with the *E. coli* enzyme and attempt to refer only to publications that give definitive information concerning the F_0F_1-ATPase structure. Readers are referred to the excellent reviews of Walker et al. [147], Senior [148] and Fillingame [149] for a wider ranging discussion of *E. coli* F_0F_1-ATPase structure/function relationships. The genes encoding the eight subunits of the F_0F_1-ATPase of *E. coli* form the *unc* (or *atp*) operon. The genes are in the order *BEFHAGDC*, encoding respectively the *a*-, *c*- and *b*-subunits of F_0 and the δ-, α-, γ-, β- and ε-subunits of F_1. The complete amino acid sequence of each of the subunits was deduced from the nucleotide sequence (see ref. [147]). The stoichiometry of the subunits is $a(1)$, $b(2)$, $c(9)$, $\alpha(3)$, $\beta(3)$, $\gamma(1)$, $\delta(1)$ and $\varepsilon(1)$, although there remains some doubt about the number of *c*-subunits (see ref. [148]).

An understanding of the gross structure of the *E. coli* F_0F_1-ATPase can best be gained from a consideration of a series of papers from Capaldi's laboratory, in which the technique of cryoelectron microscopy was used. The F_1-ATPase portion (ECF_1) was found to be a roughly spherical molecule about 95 Å in diameter. Six elongated protein densities (3α, 3β) about 90 × 30 Å form an hexagonal barrel, the central cavity of which is partially occluded by a compact protein density [150]. A particular monoclonal antibody against the α-subunit, when mixed with ECF_1, resulted in a uniform orientation of the ECF_1-mAB complexes in the amorphous ice layer enabling the averaging of images obtained from the electron microscope. The hexagonal structure of the ECF_1 was confirmed with the pattern of anti-α mAB binding, indicating an alternate arrangement of the 3α- and 3β-subunits [151]. The protein density partially occluding the central cavity was asymmetrically located and was shown to be associated with a β- rather than an α-subunit. The size of this central density could be reduced by trypsin treatment, which has been shown to cleave the γ-subunit and release the δ- end ε-subunits. Antibodies against the γ-, δ- or ε-subunits were able to bind to the ECF_1, in addition to the anti-α mAB. The antibodies against the smaller subunits appeared to be superimposable on one of the β-subunits. However binding of the α-, γ- and δ-antibodies simultaneously indicated that the γ- and δ-subunits were associated with different β-subunits [151]. A further set of experiments in which ECF_1 interacted with mABs against α and ε enabled the position of the ε-subunit to be related to the position of the asymmetrically located protein density in the central cavity. The relative positions of the ε-subunit and the central mass (thought to be the N-terminal domain of the γ-subunit) were randomised with respect to each other and to the β-subunits with which they are associated [152]. However, particular relative positions are favoured depending on the ligands bounds at the catalytic sites. In the presence of ADP, Mg^{2+}, and P_i, for example, the central density and the ε-subunit are predominantly associated with the same β-subunit. The environment of the ε-subunit was also shown to be ligand dependent by determining rates of trypsin cleavage and degree of cross-linking to the β-subunit [153]. Experiments with engineered ε in which two serine residues were replaced by cysteine and then coupled to neighbouring subunits by a cross-linker also indicated that the ε-subunit environment was ligand dependent (R.A. Capaldi, personal communication). It would appear, then, that at least one of the smaller subunits (and maybe more than one) are mobile elements with respect to the α- and β-subunits in ECF_1 preparations.

The F_0F_1-ATPase reconstituted into membrane structures has also been examined in a layer of amorphous ice by cryoelectron microscopy [154]. The elongated α- and β-subunits of the ECF_1 formed a barrel-like structure and were perpendicular to the membrane bilayer. The ECF_1 appears to be connected to the membrane by a narrow stalk about 40 Å long and 25–30 Å thick [154]. Structure predictions for the b-subunit [155] suggest that there are two helices per subunit extending some 90 Å from the membrane surface with a third helix extending about 35 Å. With two b-subunits per assembly a cluster of six helices would be sufficient to account for the dimensions of the stalk. There is a suggestion in one of the averaged images (see Fig. 3c in ref. [154]) of an extension of either the α- or β-subunit towards the membrane surface. The method would not resolve single helices. The F_0 part of the reconstituted complex was mostly within the bilayer and was poorly defined.

5.2. *Structure of the E. coli F_0F_1-ATPase subunits*

There is not yet available a high-resolution structure, from X-ray or electron diffraction of 3D or 2D crystals, of the F_0F_1-ATPase or any part thereof. Any understanding of the structure of the subunits, other than the primary structure, has been gained indirectly from the effects of the various modifications being related principally to its function. Structure prediction methods are not sufficiently reliable to be of significant value, at least for the subunits of the F_1-ATPase portion. Duncan et al. [156] have argued that the adenylate kinase nucleotide binding region, as determined by X-ray crystallography and solution NMR, is sufficiently similar to the nucleotide-binding region of the β-subunit for it to be used in modelling. The resultant model (depicted in Fig. 2 of Duncan et al. [156]) has been modified to include a sixth β-strand, extending from about residue 325 to residue 335, adjacent to β-strand 5 in the parallel β-sheet (A. Senior, personal communication). This model gains support from the fact that it essentially

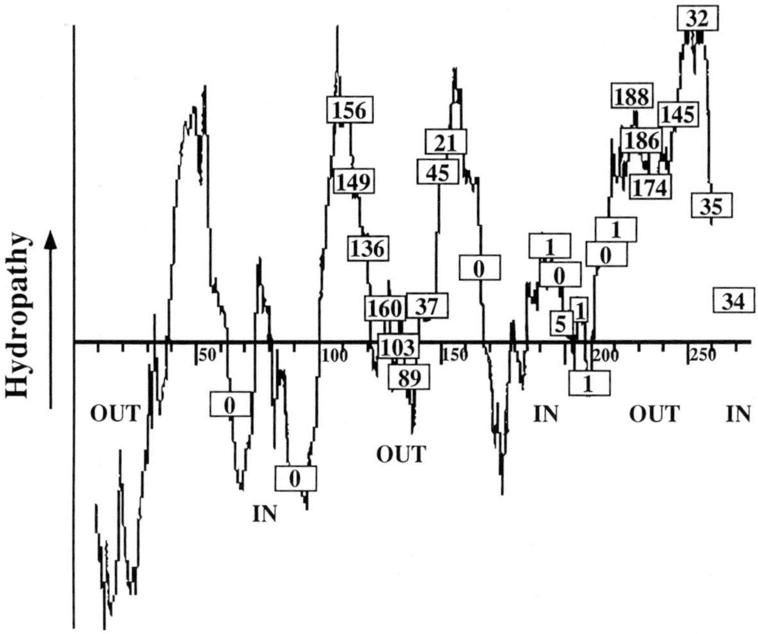

Fig. 4. Hydropathy profile of the *a*-subunit of the *E. coli* F_0F_1-ATPase using the GES scale [198] with a window of 19 amino acids. The amino acid residue number is indicated on the horizontal axis. The numbers in squares are alkaline phosphatase activities of fusion proteins taken from Lewis et al. [158]. The numbers are positioned at the point where the *a*-subunit is fused to the alkaline phosphatase. 'IN' and 'OUT' refer to the cytoplasmic and periplasmic sides of the membranes, respectively.

accommodates all the relevant genetic and chemical modification studies (see Fillingame [149] and Senior [148]).

The modelling of integral membrane proteins from amino acid sequences seems to be more reliable than that for soluble proteins. This may be due to the apparent accuracy of hydropathy profiles and to the validity of the assumption that portions of the protein traversing the membrane are α-helical. Assitance in modelling is also gained from the observation that the distribution of charged residues on either side of the membrane appears to be biased in response to the membrane potential [157]. Topology studies using fusion proteins [158,159] or labelling with the greasy compound 3-(trifluoromethyl)-3-(*m*-[^{125}I]iodophenyl) diazirine ([^{125}I]TID) [160,161] have also provided relevant information, although there are some difficulties with both these techniques. A mutational approach has also yielded valuable information. The rationalisation, in structural terms, of the effects of specific amino acid substitutions and particularly of second-site revertants has given information as to the relative positions of residues within and between subunits.

The F_0 or membrane portion of the F_0F_1-ATPase comprises three subunits of stoichiometry *a*:*b*:*c*=1:2:9 (see above). The *a*-subunit has a molecular weight of about 30 kDa and a number of structural models have been proposed (see Fillingame [149]). The hydropathy profile clearly indicates four transmembrane helices (Fig. 4) centred at about residues 50, 110, 155 and 250, with the suggestion of a fifth centred at about 220–230. A method for analysis of the topology

TABLE II

Effects of mutations in the *a*-subunit of *E. coli* F_0F_1-ATPase

Authors and reference	Uncoupled	Partial effects	No effect
Cain and Simoni [131]	$S^{206}L$		
	$H^{245}Y$		
Lightowlers et al. [129]	$R^{210}Q$		$K^{167}Q+K^{169}Q$
	$H^{245}L$		
Lightowlers et al. [130]	$E^{219}Q$		$E^{196}A$
			$K^{203}I$
Cain and Simoni [132]	$E^{219}Q,L$	$E^{219}H$	$E^{219}D$
		$H^{245}E$	
Eya et al. [210]	$P^{143}S$		
Vik et al. [211]	$P^{190}Q,R$	$E^{196}K,P$	$E^{196}D,H,Q,N,S,A$
		$G^{197}R$	$P^{190}N$
			$S^{199}A,T$
Howitt et al. [174]			$S^{199}A$
			$S^{202}A$
			$S^{206}A$
			$R^{61}Q$
Cain and Simoni [133]	$R^{210}K,I,V,E$	$N^{214}L$	$N^{214}V,Q,E$
	$N^{214}H$	$A^{217}H,L$	$L^{207}C,Y$
	$A^{217}R$	$G^{218}V,D$	$L^{211}Y,F$
			$G^{218}A$
Paule and Fillingame [212]		$D^{119}H$	
		$S^{152}F$	
		$G^{197}R$	
Howitt et al. [165]	$D^{44}N+D^{124}N$	$E^{219}H$	$D^{119}A$
	$D^{44}N+R^{140}Q$		$D^{146}N$
			$K^{65}Q+K^{66}Q$
			$K^{97}Q+K^{99}Q$
			$E^{219}H+R^{140}H$
			$D^{44}N$
			$D^{124}N$
			$R^{140}Q$
			$D^{124}N+R^{140}Q$
Vik et al. [213]			$Q^{181}H$
			$N^{184}D,Y,H$
			$H^{185}Y,Q$
			$N^{192}L,P,R.S,T,V$
Vik et al. [214]			$Y^{263}F,A,E,G,K,Q$
Eya et al. [164]	$E^{219}Q$	$Q^{252}L$	$S^{265}A$
	$R^{210}K,Q$	$H^{245}E$	$Q^{252}E$
		$P^{230}L$	
		$E^{219}H$	

of membrane proteins has been developed [162] in which the alkaline phosphatase protein is fused at various points along the membrane protein sequence. If the fusion occurs at a periplasmic loop then high alkaline phosphatase activity is obtained, if fused at a cytoplasmic loop then low alkaline phosphatase activity is obtained. A number of difficulties have arisen with this

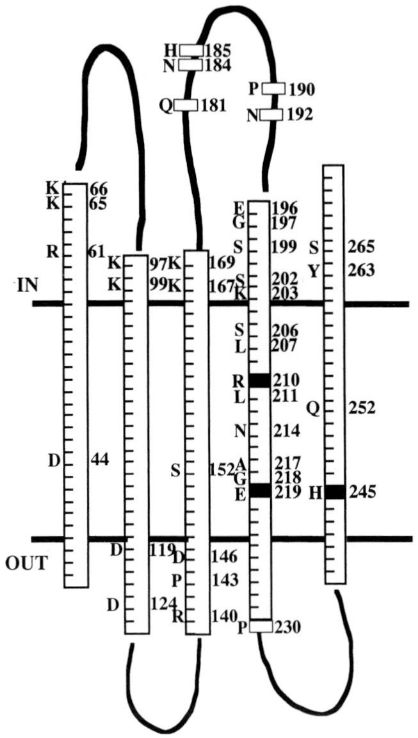

Fig. 5. A five-helical model of the *a*-subunit of the *E. coli* F_0F_1-ATPase. The amino acid residues indicated are those that have been altered by random or site-directed mutagenesis. The filled-in residues are those that, when 'conservatively' mutated, cause complete loss of activity (see text). 'IN' and 'OUT' refer to the cytoplasmic and periplasmic sides of the membranes, respectively. There is a preponderance of positively charged residues on the cytoplasmic side compared with the periplasmic side (11–6) and a preponderance of negatively charged residues on the periplasmic side compared with the cytoplasmic side (8–2).

technique (see ref. [163]), an example of which is the export of free alkaline phosphatase to the periplasm although originally fused to a cytoplasmic loop. Lewis et al. [158] encountered this problem when fusing alkaline phosphatase at various positions along the *a*-subunit. However, they found that the alkaline phosphatase remained predominantly fused if glucose was added to the growth medium. In Fig. 4 the alkaline phosphatase values obtained by Lewis et al. [158] have been superimposed on the hydropathy profile and clearly indicate the presence of a fifth helix centred at about residue 220.

There has been extensive mutational analysis of the function of the *a*-subunit with a number of laboratories contributing (see Table II). Structural information can also be obtained from many of these studies. Figure 5 indicates, with a five helical model, residues of the *a*-subunit that have been mutated either by random mutagenesis or site-directed mutagenesis. There have been two philosophies that have operated in the design of these experiments. One involves making a minimal change to a residue in order to test for the requirement of the particular property of that residue and the other involves substituting a residue by a range of amino acids in order to determine which amino acids can or cannot be tolerated in a particular position. For

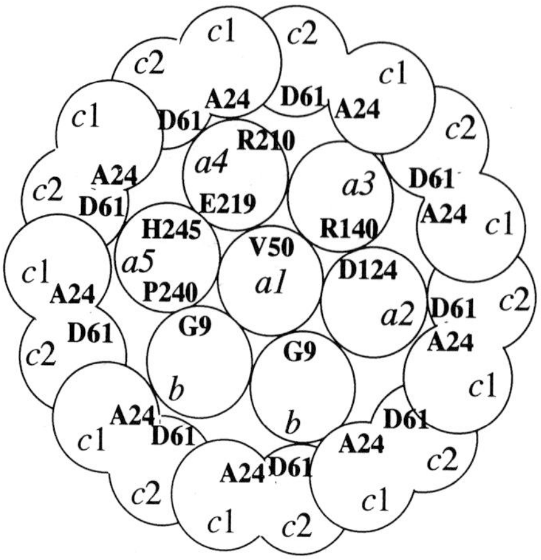

Fig. 6. An arrangement of the five putative transmembrane helices of the *a*-subunit and the single transmembrane helix from each of the *b*-subunits. The *a*- and *b*-subunits are placed within a ring of *c*-subunits. The interacting amino acid residues described in the text are indicated.

example if E^{219} of the *a*-subunit is replaced by glutamine then the F_0F_1-ATPase is nonfunctional [130,132,164]. The two amino acids are closely similar in size and the clear indication is that an acidic residue is required at this position. The observations that aspartate can replace glutamate and leucine cannot [132] are consistent with this conclusion. However, histidine allows a low level of activity [132,165] and it would appear that there is a minimum requirement for a charged amino acid at position 219. Given the placement of this charged residue, as part of helix 4, within the hydrophobic portion of the bilayer, a vital role is required for the proposed structure to be acceptable [129]. The same situation applies for the residue at position 245, with histidine being clearly the preferred residue, although some activity is retained if histidine is replaced by glutamate. An interaction between residues at positions 219 and 245 was inferred when it was shown that the low activity of the $E^{219}H$ mutant could be raised somewhat by the presence of the $H^{245}E$ mutation [132].

There is a cluster of charged residues on the periplasmic side of putative helices, 1, 2 and 3 (D^{44}, D^{124} and R^{140}) that appear to be involved in stabilising the *a*-subunit structure [165]. The replacement of glutamate by histidine at position 219 allows a low level of activity (see above); however this level returns to normal if R^{140} is replaced by histidine. It was concluded that the change in geometry of a putative D^{124}–R^{140} salt bridge allowed histidine at position 219 in the neighbouring helix to become a fully functional component of the proton pore [165]. With the exception of R^{210} in helix 4 and the charged residues referred to above, all other charged residues that have been substituted conservatively have been shown not to be required. All such residues are placed outside the hydrophobic portion of the bilayer in the five helix model (Fig. 5).

The replacement of R^{210} by lysine results in complete loss of activity. This mutation is dominant when used in complementation tests with mutations affecting the *c*-subunit. Without determining the basis for this dominance, revertants were sought which overcame the dominant effect of the lysine residue at position 210 (S.M. Howitt, unpublished). One such revertant involved replacement of G^{53} (helix 1) by aspartate, with the obvious interpretation that the effect of the lysine residue at position 210 was being neutralised by the formation of a salt bridge with the aspartate at position 53. If this interaction occurred in the fully folded subunit then the positioning of an aspartate in helix 1, that would enable neutralisation of the lysine, should be somewhat precise. Aspartate introduced by site-directed mutagenesis at positions 46, 50 or 53 completely neutralised the lysine effect, whereas aspartate on the other faces of putative helix 1 had a reduced or no effect (S.M. Howitt, unpublished). The position of helix 1 in the membrane is clearly indicated by the hydropathy profile (see above) whereas the positioning of helix 4 was dependent on the alkaline phosphatase fusion data from Lewis et al. [158]. The observations connecting helices 1 and 4 confirm the position of the latter in the membrane. It is possible to arrange the putative five transmembrane helices of the *a*-subunit to take into account all the interactions described above (see Fig. 6).

The *b*-subunit has a molecular weight of about 17 kDa and its amino acid sequence is characterised by a short hydrophobic sequence at the N-terminus with the remainder of the molecule being hydrophilic (see ref. [147]). The hydrophilic portion is predicted to be largely α-helical [147,166] with seventeen potential salt bridges stabilising an α-helical hairpin structure. The hydrophilic portion (residues 25 to 146) has been produced fused at the N-terminus with an octapeptide (S. Dunn, personal communication). Measurements made indicated that the protein existed in solution as a highly elongated dimer and was largely α-helical. The fusion protein also bound to F_1-ATPase and this complex was unable to bind to F_1-depleted membrane vesicles (S. Dunn, personal communication). While the *b*-subunit is required, together with the *a*- and *c*-subunits, for proton translocation through the F_0, (see ref. [148]) the only random *b*-subunit mutants described (G^9D, $G^{131}D$) affect assembly [167–169]. If the mutation resulting in the G^9D substitution is chromosomally located, then the mutant *b*-subunit is not assembled into the membrane. If all the F_0 subunits are plasmid encoded (including the G^9D mutation) then an active F_0F_1-ATPase is assembled. If, however, only the mutant *b*-subunit is plasmid encoded, then although assembled, the F_0F_1-ATPase is nonfunctional [169]. Functional second-site revertants were obtained which were located in the *a*-subunit ($P^{240}A$ or L) or *c*-subunit ($A^{62}S$) [170,171]. This data would suggest that the transmembrane N-terminal region of the *b*-subunit is adjacent to both helix 5 of the *a*-subunit and to helices 2 of the *c*-subunits (see Fig. 6).

An interesting mutational analysis of the putative β-turn region (N^{80}, K^{81}, R^{82}, R^{83}) of the *b*-subunit has been carried out by McCormick and Cain [171a]. The mutant screening did not distinguish between mutations that affect assembly and those that allow assembly but affect mechanism. The most striking feature however was that the substitution of any of three residues (A^{79}, R^{82}, Q^{85}) by proline caused loss of ATP-synthase activity. Chou-Fasman secondary structure prediction indicates a plausible β-turn at any of three overlapping positions (78–81, 80–83, 82–85). The $A^{79}P$ mutation greatly increases the probability of a β-turn at the 78–81 position, thus making it unlikely that the β-turn in the normal *b*-subunit is located at this position. The $R^{82}P$ mutation results in the prediction of a new β-turn at the 81–84 position and $Q^{85}P$ results in the loss of the predicted β-turn at the 82–85 position. While the single mutation $R^{83}P$ was not reported the results may be rationalised if the β-turn in the wild-type *b*-subunit occurs at the 82–85 position.

The *c*-subunit structure is thought to be in the form of a hairpin with two hydrophobic transmembrane helices separated on the cytoplasmic side of the membrane by a polar loop (see Senior [148], Fillingame [149]). There are two sets of mutants that give a clear indication of the positioning of the N-terminal helix with respect to the C-terminal helix. The first involved the primary mutation $P^{64}L$, which caused complete loss of function but not assembly and this effect could be partially overcome by a second mutation $A^{20}P$ [172]. The second, which is consistent

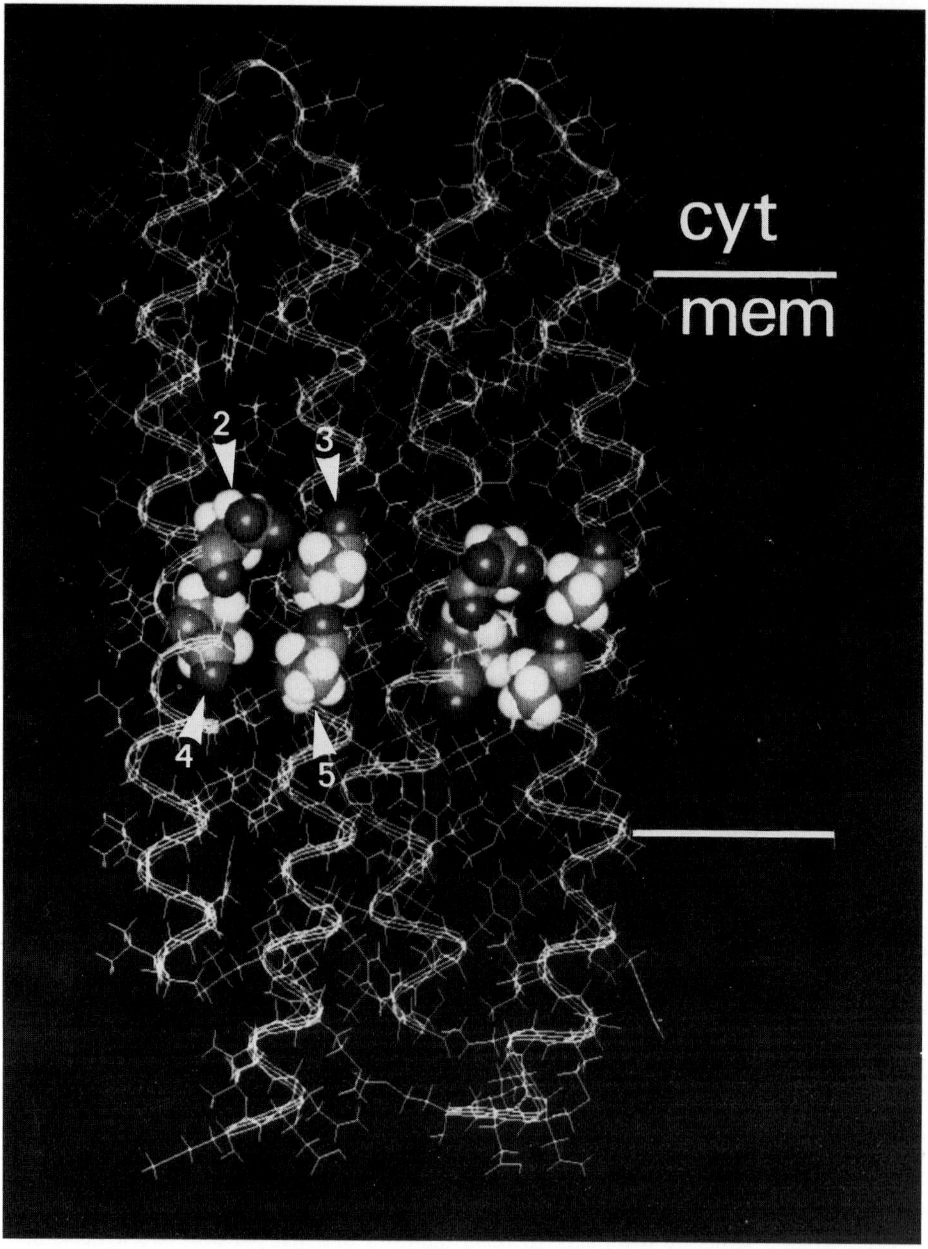

Fig. 7. A computer graphics model of a pair of *c*-subunits generated and energy minimised using BIOSYM software. The residues D^{61} (2), A^{24} (3), P^{64} (4) and A^{20} (5) are shown as CPK models.

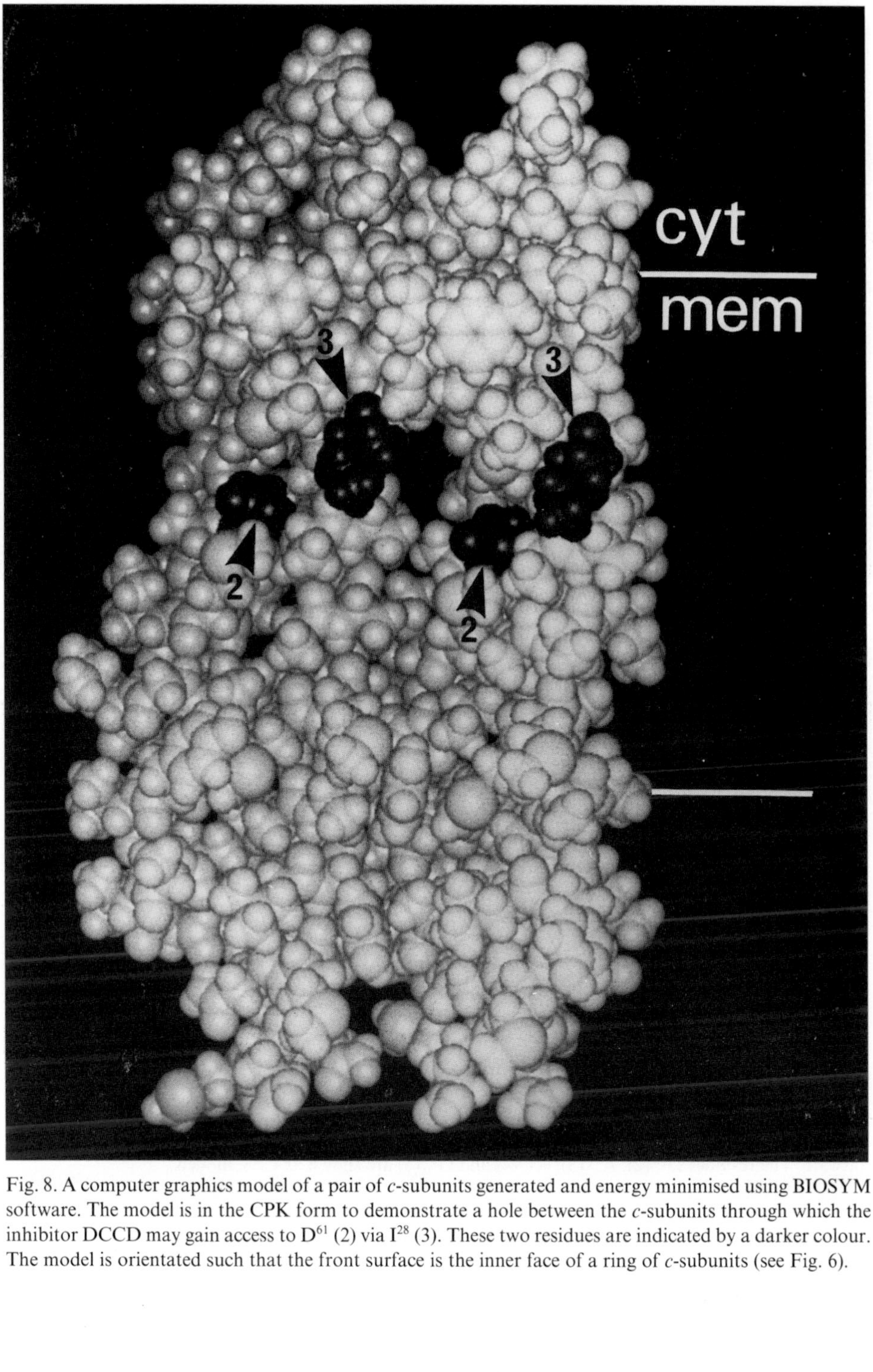

Fig. 8. A computer graphics model of a pair of *c*-subunits generated and energy minimised using BIOSYM software. The model is in the CPK form to demonstrate a hole between the *c*-subunits through which the inhibitor DCCD may gain access to D^{61} (2) via I^{28} (3). These two residues are indicated by a darker colour. The model is orientated such that the front surface is the inner face of a ring of *c*-subunits (see Fig. 6).

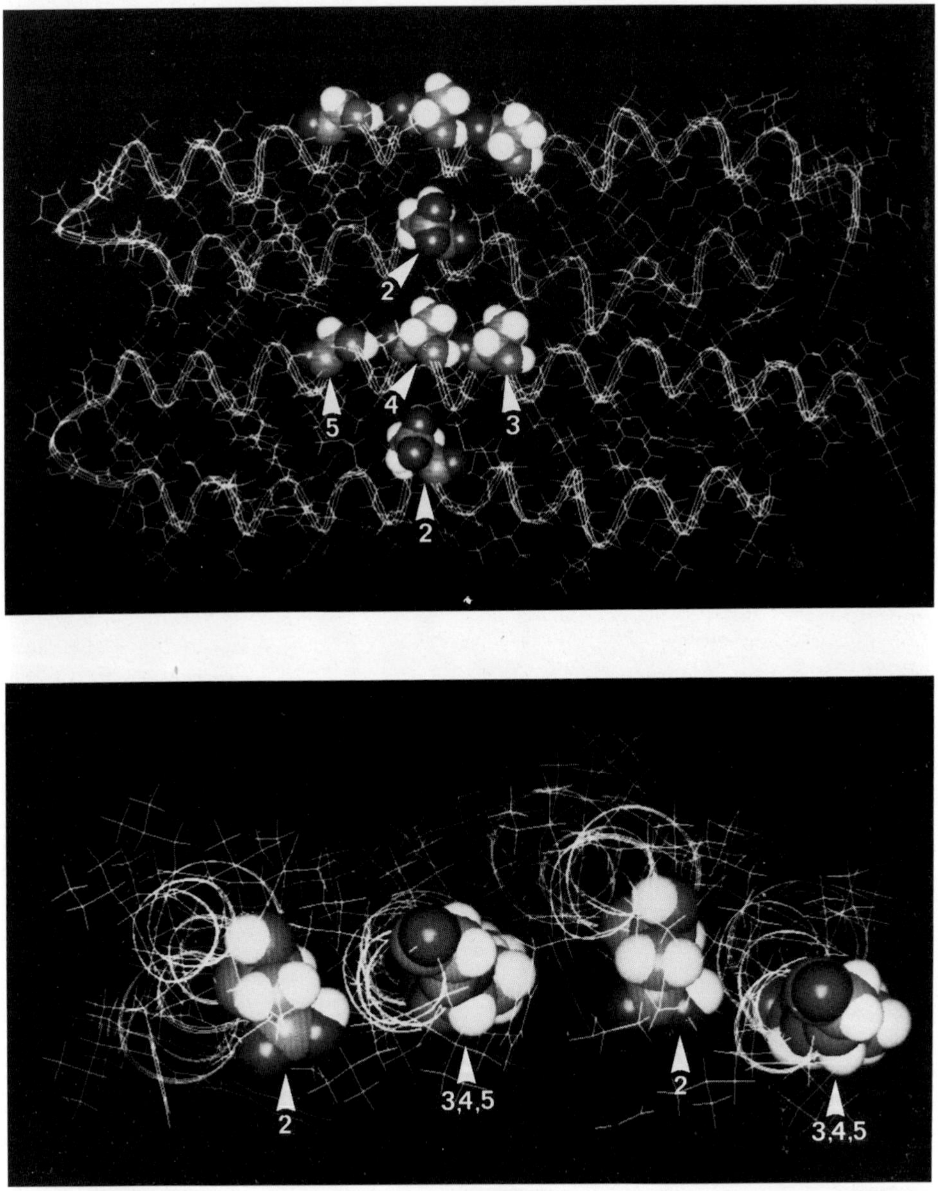

Fig. 9. A computer graphics model of a pair of *c*-subunits generated and energy minimised using BIOSYM software. The residues D^{61} (2), A^{21} (3), A^{25} (4) and G^{29} (5) are shown as CPK models.

with the first, involved the mutation $D^{61}G$, which could be partially reversed by a second mutation $A^{24}D$ [173]. This latter pair of mutations is made all the more significant by the observation that moving the aspartate from position 61 to positions 58, 60 or 62 caused loss of the ability to grow via oxidative phosphorylation on succinate media [173]. Aspartate^{61} is though to be a component of the proton pore, interacting with residues on helices 4 and 5 of the *a*-subunit [174] and clearly must be very close to A^{24}. A computer-generated model of a pair of *c*-subunits shows the relative positions of D^{61}, A^{24}, P^{64} and A^{20} (Fig. 7).

D^{61} reacts covalently with the inhibitor DCCD [175]. However, the reactivity of DCCD with D^{61} in normal ATP synthase is somewhat restricted since, in the P^{64}L mutant, the reactivity of D^{61} with DCCD increases 6–7-fold (G, Ash, 1981, PhD thesis, ANU). The *c*-subunits must therefore be arranged such that D^{61} has only limited access to the lipid of the membrane bilayer (Fig. 6). Hoppe et al. [176] identified the amino acid substitutions in six DCCD-resistant mutants of *E. coli*. Isoleucine at position 28 had been replaced by either valine or threonine with the latter replacement having the greater effect. The CPK computer-generated model of the pair of *c*-subunits (Fig. 8) indicates that, while there is close-packing between the *c*-subunits for most of their length, a hole occurs close to position 28 of one *c*-subunit and position 61 of the adjacent *c*-subunit. It may be through this hole that DCCD gains access from the bilayer to the reactive D^{61}.

Fimmel and coworkers [177–180] have mutated the small residues A^{21}, A^{25}, G^{29}, G^{32} and G^{27} to larger residues resulting in loss of function of the F_0F_1-ATPase. The first four of these residues lie on the helical face between *c*-subunits (Fig. 9) and increasing the size would interfere with the tight packing of adjacent *c*-subunits. The residues A^{21}, A^{25} and G^{29} are located in the vicinity of D^{61}. The residue G^{27} is very close to D^{61} in the same *c*-subunit and the effect of increasing the size of this amino acid is likely to be due to a direct effect on D^{61}.

Hoppe et al. [160] used the photoactivatable carbene precursor TID labelled with [^{125}I] to obtain information on membrane-embedded subunits of the *E. coli* F_0F_1-ATPase. However, labelling of the F_1-ATPase β-subunit was consistently observed in addition to labelling of the F_0 subunits *a*, *b* and *c*. It seems likely therefore that this reagent labels residues in hydrophobic pockets within proteins in addition to those that may be exposed to lipid. It would also seem possible that TID labelling may be prevented by lipid tightly bound to the membrane protein. Steffens et al. [161] have also used [^{125}I]TID to label proteoliposomes containing F_0 subunit *c*, an *ac* complex or an *abc* complex. The presence of the *a*-subunit prevented the labelling of the *c*-subunit methionine residues 16 and 17 and partially prevented the labelling of methionine 57. There is only one *a*-subunit and probably 9 *c*-subunits and this result provides strong support for the model in Fig. 6, which the *a*-subunit together with the *b*-subunits is located in the centre of a ring of *c*-subunits.

5.3. Coordination of protein subunit expression in E. coli

The subunits of the F_0F_1-ATPase of *E. coli* are synthesised at levels consistent with the different requirements for the different subunits. Since all the genes for the ATPase subunits are present in one operon, this implies some control over the level of translation of individual genes. Experiments either measuring the levels of individual subunits produced, or using *unc::lacZ* fusions to different genes, have demonstrated that the degree of translation of each gene is indeed related to the requirements of the final complex [181,182]. Control appears to occur through differences in the efficiency of translational initiation [183]. In particular, the *uncE* gene is translated at very high rates, consistent with the need for 9 copies of the *c*-subunit per F_0F_1-ATPase. Full efficiency of translation of this gene is dependent on the presence of about 30 bases upstream of the start codon and this region has also been shown to promote efficient translation of other genes (see ref. [183]).

Translational coupling is also evident in the *unc* operon. Some individually cloned genes

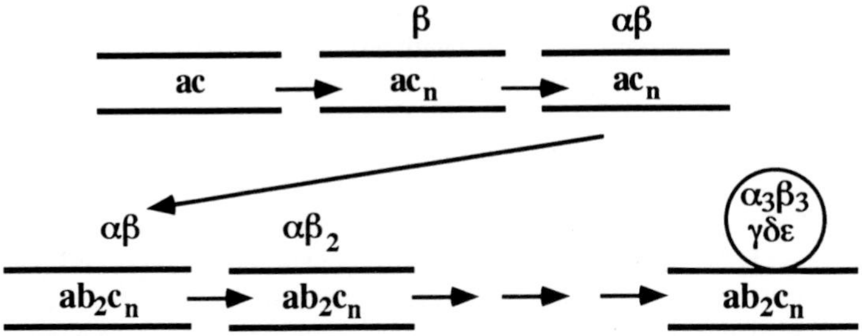

Fig. 10. The proposed pathway of assembly of F_0F_1-ATPase in *E. coli*. The evidence for this 'integrated membrane assembly pathway' is detailed by Cox and Gibson [155].

show significantly reduced rates of expression compared to the rate obtained when each gene is cloned with the preceding gene. Translational coupling occurs mainly in the middle part of the *unc* operon [184] and is due to *cis* rather than *trans* effects. Coupling has been shown to occur between the gene pairs *uncEF*, *uncFH*, *uncHA* and *uncAG* [184]. Stable hairpin structures which encompass the Shine–Dalgarno sequences and the start codons of coupled genes, are predicted to occur between each of the pairs of genes. This contrasts with the *uncB–E* intergenic region which is relatively free from secondary structure and thus is easily accessible to ribosomes [183].

The coupling between *uncH* and *uncA* has been studied in some detail. High levels of translation of *uncA* are dependent on the presence of *uncH* upstream. Two Shine–Dalgarno sequences are found between *uncH* and *uncA* and both are involved in base pairing which extends into *uncH*, forming stable hairpin structures. Site-directed mutagenesis of a construct containing *uncA* only has freed either one or both Shine–Dalgarno sequences from base pairing [184]. Releasing the more 5' sequence from base pairing was found to have the greatest effect on the level of translation of *uncA*. However, a further increase was observed when both changes were present, indicating that freeing up this region from secondary structure does enhance ribosome binding. In vivo, therefore, translation of *uncH* may interfere with secondary structure formation in the intergenic region, allowing greater binding of ribosomes to the *uncA* Shine–Dalgarno sequence and hence increased translation.

Mutagenesis studies on the *uncF* translation initiation region also show that changes which reduce the stability of predicted secondary structures increase translational efficiency [185]. Furthermore, translation of *uncH* also improved in such mutants, consistent with coupling between these two genes. The *uncF* gene starts with GUG rather than AUG; a mutant in which AUG was substituted for GUG showed increased translation, pointing to another mechanism for controlling the level of translation [185].

5.4. *Assembly of F_0F_1-ATPase in E. coli*

Sternweis and Smith (1977) proposed a model for the assembly of the *E. coli* F_0F_1-ATPase in which the F_0 and F_1 sectors assembled separately prior to the formation of the ATPase complex. A number of observations have been made which support this hypothesis (see ref. [155]). These observations include the ability to assemble an active F_1-ATPase and an active F_0 from purified subunits in vitro [108,186] and a relatively inefficient assembly from high copy number plasmids of the F_1-ATPase or F_0 in vivo in strains lacking chromosomal *unc* genes [187, 188].

The first indication that the F_1 and F_0 sectors of the F_0F_1-ATPase might not be assembled separately in vivo came from the properties of mutant strains which were unable to form α- and/or β-subunits of the F_0F_1-ATPase [189]. Such strains appeared unable to integrate the *b*-subunit into a functional F_0. These observations, among others, led to the proposal of the 'integrated membrane assembly pathway' (see Cox and Gibson [155]) which is depicted in Fig. 10. This assembly pathway has the attractive feature of not exposing the cell, at any stage, to open proton channels. The requirement for the presence of F_1 subunits in addition to the three F_0 subunits in order to form a functional F_0 has received support from a series of papers from Brusilow's laboratory [119,190–193].

Brusilow [190] reported the harmful effect on growth of *E. coli* of a high copy number plasmid carrying, in addition to the *uncB*, *-E*, *-F*, *-H*, *-D* and *-C* genes, an *uncA–uncG* fusion (α–γ1). The plasmid could only be isolated in a strain lacking the chromosomal *unc* genes and growth, even then, was poor. Evidence was obtained that the growth inhibition was due to a proton leaky F_0 channel and that the deleterious effect of the presence of the chromosomal *unc* genes was specifically due to the chromosomal *uncA* gene. It would seem likely therefore that efficient assembly of the F_0 from the genes on the plasmid required a normal α-subunit, from the chromosome, in addition to the β-subunit, from the plasmid, and the α–γ fusion prevented the normal F_1 assembly. The requirement for the β-subunit from the plasmid was not directly tested. The data obtained with chromosomal mutants affecting the α-, γ- and β-subunits could not be assessed as there was insufficient information about these mutations. The effect of the *uncA–uncG* fusion could be eliminated if a deletion was introduced into the plasmid *uncB* gene (encoding the *a*-subunit of the F_0) [194].

Pati and Brusilow [191] prepared strains in which the α–γ fusion plasmid (carrying the remaining genes of the *unc* operon) was accompanied by a second plasmid from which temperature-sensitive expression of either α or $\alpha + \gamma$ could be obtained. The host strains lacked the chromosomal *unc* operon. The induction of α in the presence of the α–γ fusion caused cessation of growth, a result consistent with the previous work (see above). The induction of α and γ was not lethal, presumably because a complete F_0F_1-ATPase was able to be formed rather than having any special role in regulating proton flow. The inducible lethal system allowed the isolation of membranes from the sick cells which were indeed proton permeable. A model was proposed [191] in which the F_0 subunits are assembled in a non-leaky form, and one or more α-subunits then bind to this complex and unblock the channel. This model is essentially very similar to the 'integrated membrane assembly pathway' in that it is very difficult to distinguish between an assembled non-leaky F_0 and a non-assembled F_0 and while the α-subunit was clearly shown to be required, a requirement for the β-subunit was not tested. The reconstitution experiments using crude F_1-ATPase preparations [192] would suggest that the F_0F_1-ATPase itself (or subunits therefrom) is able to catalyse the formation of an active F_0 from subunits already in the membrane. Pati et al. [193] made F_0 preparations from cells in which F_0 subunits were formed but F_1 subunits were not. These preparations, compared with positive controls, gave about 20% proton conductance although binding a similar amount of F_1-ATPase. The low proton conductance may have been due to a low percentage (20%) of normal F_0 or a uniform preparation of subunit aggregates with low proton permeability.

In summary, if high copy number plasmids carrying only the genes encoding the F_0 subunits *a*, *b* and *c*, are used then sufficient of these subunits are assembled in the membrane to give detectable proton translocating activity. It is also clear that the presence of the F_1-subunits α and δ (and probably also β) greatly facilitate the formation of proton permeable F_0 sectors from the high copy number plasmids. However in *E. coli* cells, in which the genes encoding the F_0 subunits are on the chromosome, an F_0 sector is not formed, due to the lack of incorporation

60 Nagley, P., Devenish, R.J., Law, R.H., Maxwell, R.J., Nero, D. and Linnane, A.W. (1990) in: C.H. Kim and T. Ozawa (Eds.) Bioenergetics, Molecular Biology, Biochemistry and Pathology, Plenum Press, New York, pp. 305–325.
61 Velours, J. and Guerin, B. (1986) Biochem. Biophys. Res. Commun. 138, 78–86.
62 Hekman, C., Tomich, J.M. and Hatefi, Y. (1991) J. Biol. Chem. 266, 13564–13571.
63 Hekman, C. and Hatefi, Y. (1991) Arch. Biochem. Biophys. 284, 90–97.
64 Muraguchi, M., Yoshihara, Y., Tunemitu, T., Tani, I. and Higuti, T. (1990) Biochem. Biophys. Res. Commun. 168, 226–231.
65 Guerrieri, F., Capozza, G., Houstek, J., Zanotti, F., Colaianni, G., Jirillo, E. and Papa, S. (1989) FEBS Lett. 250, 60–66.
66 Papa, S., Guerrieri, F., Zanotti, F., Houstek, J., Capozza, G. and Ronchi, S. (1989) FEBS Lett. 249, 62–66.
67 Houstek, J., Kopecky, J., Zanotti, F., Guerrieri, F., Jirillo, E., Capozza, G. and Papa, S. (1988) Eur. J. Biochem. 173, 1–8.
68 Papa, S., Guerrieri, F. and Zanotti, F. (1991) in: J.W. Gorrod, O. Albano, E. Ferrari and S. Papa (Eds.) Molecular Basis of Neurological Disorders and their Treatment, Chapman and Hall, London, pp. 15–30.
69 Papa, S., Guerrieri, F., Zanotti, F., Fiermonte, M., Capozza, G. and Jirillo, E. (1990) FEBS Lett. 272, 117–120.
70 Hadikusumo, R.G., Hertzog, P.J. and Marzuki, S. (1984) Biochim. Biophys. Acta 765, 258–267.
71 Grasso, D.G., Nero, D., Law, R.H.P., Devenish, R.J. and Nagley, P. (1991) Eur. J. Biochem. 199, 203–209.
72 Schatz, G. (1968) J. Biol. Chem. 243, 2192–2199.
73 Orian, J.M., Hadikusumo, R.G., Marzuki, S. and Linnane, A.W. (1984) J. Bioenerg. Biomembranes 16, 561–581.
74 Linnane, A.W., Lukins, H.B., Nagley, P., Marzuki, S., Hadikusumo, R.G., Jean-François, M.J.B., John, U.P., Ooi, B.G., Watkins, L., Willson, T.A., Wright, J. and Meltzer, S. (1985) in: E. Quagliariello, E.C. Slater, F. Palmieri, C. Saccone and A.M. Kroon (Eds.) Achievements and Perspectives of Mitochondrial Research, Elsevier, Amsterdam, pp. 211–222.
75 Hadikusumo, R.G., Meltzer, S., Choo, W.M., Jean-François, M.J., Linnane, A.W. and Marzuki, S. (1988) Biochim. Biophys. Acta 933, 212–222.
76 Marzuki, S., Watkins, L.C. and Choo, W.M. (1989) Biochim. Biophys. Acta 975, 222–230.
77 Takeda, M., Vassarotti, A. and Douglas, M.G. (1985) J. Biol. Chem. 260, 15458–15465.
78 Ackerman, S.H. and Tzagoloff, A. (1990) Proc. Natl. Acad. Sci. USA 87, 4986–4990.
79 Payne, M.J., Schweizer, E. and Lukins, H.B. (1991) Curr. Genet. 19, 343–351.
80 Finnegan, P.M., Payne, N.J., Keramidaris, E. and Lukins, H.B. (1991) Curr. Genet. 20, 53–61.
81 Ackerman, S.H., Gatti, D.L., Gellefors, P., Douglas, M.G. and Tzagoloff, A. (1991) FEBS Lett. 278, 234–238.
82 Manon, S. and Guerin, M. (1989) Biochim. Biophys. Acta 985, 127–132.
83 Pelissier, P.P., Camougrand, N.M., Manon, S.T., Velours, G.M. and Guerin, M. (1991) J. Biol. Chem., 267, 2467–2473.
84 Law, R.H.P., Farrell, L.B., Nero, D., Devenish, R.J. and Nagley, P. (1988) FEBS Lett 236, 501–505.
85 Nero, D., Ekkel, S.M., Wang, L., Grasso, D.G. and Nagley, P. (1990) FEBS Lett. 270, 62–66.
86 Devenish, R.J., Galanis, G., Papakonstantinou, T., Law, R.H.P., Grasso, D.G., Helfenbaum, L. and Nagley, P. (1992) in: S. Papa, A. Azzi and J.M. Tager (Eds.) Adenine Nucleotides in Cellular Energy Transfer and Signal Transduction, Birkhauser, Basel, pp. 1–12.
87 Macreadie, I.G., Novitski, C.E., Maxwell, R.J., John, U.P., Ooi, B.G., McMullen, G.L., Lukins, H.B., Linnane, A.W. and Nagley, P. (1983) Nucl. Acids Res. 11, 4435–4451.
88 Law, R.H.P., Devenish, R.J. and Nagley, P. (1990) Eur. J. Biochem. 188, 421–429.
89 Velours, J., Ezparza, M., Hoppe, J., Sebald, W. and Guerin, B. (1984) EMBO J. 3, 207–212.
90 Uh, M., Jones, D. and Mueller, D.M. (1990) J. Biol. Chem. 265, 19047–19052.
91 Velours, J., Arselin, G., Paul, M.F., Galante, M., Durrens, P., Aigle, M. and Guerin, B. (1989) Biochimie 71, 903–915.

92 Paul, M.F., Velours, J., Arselin de Chateaubodeau, G., Aigle, M. and Guerin, B. (1989) Eur. J. Biochem. 185, 163–171.
93 Norais, N., Prome, D. and Velours, J. (1991) J. Biol. Chem. 266, 16541–16549.
94 Gray, R.E., Grasso, D.G., Maxwell, R.J., Finnegan, P.M., Nagley, P. and Devenish, R.J. (1990) FEBS Lett. 268, 265–268.
95 Ackerman, S.H. and Tzagoloff, A. (1990) J. Biol. Chem. 265, 9952–9959.
96 Burns, D.J. and Lewin, A.S. (1986) J. Biol. Chem. 261, 12066–12073.
97 Luis, A.M., Alconada, A. and Cuezva, J.M. (1990) J. Biol. Chem. 265, 7713–7716.
98 Boekema, E.J., Xiao, J. and McCarty, R.E. (1990) Biochim. Biophys. Acta Bio-Energetics 1020, 49–56.
99 Cozens, A.L. and Walker, J.E. (1987) J. Mol. Biol. 194, 359–383.
100 Falk, G. and Walker, J.E. (1988) Biochem. J. 254, 109–122.
101 Gabellini, N., Gao, Z., Eckerskorn, C., Lottspeich, F. and Oesterhelt, D. (1988) Biochim. Biophys. Acta 934, 227–234.
102 Glaser, E. and Norling, B. (1991) Curr. Top. Bioenerg. 16, 223–263.
103 Curtis, S.E. (1987) J. Bacteriol. 169, 80–86.
104 Avital, S. and Gromet-Elhanan, Z. (1991) J. Biol. Chem. 266, 7067–7072.
105 Richter, M.L., Gromet-Elhanan, Z. and McCarty, R.E. (1986) J. Biol. Chem. 261, 12109–12113.
106 Xue, Z., Zhou, J.-M., Melese, T., Cross, R.L. and Boyer, P.D. (1987) Biochemistry 26, 3749–3753.
107 Berkich, D.A., Williams, G.D., Masiakos, P.T., Smith, M.B., Boyer, P.D. and LaNoue, K.F. (1991) J. Biol. Chem. 266, 123–129.
108 Dunn, S.D. and Futai, M. (1980) J. Biol. Chem. 255, 113–118.
109 Andreo, J.L., Patrie, W.J. and McCarty, R.E. (1982) J. Biol. Chem. 257, 9968–9975.
110 Patrie, W.J. and McCarty, R.E. (1984) J. Biol. Chem. 259, 11121–11128.
111 Xiao, J.P. and McCarty, R.E. (1989) Biochim. Biophys. Acta 967, 203–209.
112 Feng, Y. and McCarty, R.E. (1990) J. Biol. Chem. 265, 5104–5109.
113 Richter, M.L., Snyder, B., McCarty, R.E. and Hammes, G.G. (1985) Biochemistry 24, 5755–5763.
114 Ketcham, S.R., Davenport, J.W., Warncke, K. and McCarty, R.E. (1984) J. Biol. Chem. 259, 7286–7293.
115 Richter, M.L., Patrie, W.J. and McCarty, R.E. (1984) J. Biol. Chem. 259, 7371–7373.
116 Feng, Y. and McCarty, R.E. (1985) Arch. Biochem. Biophys. 238, 61–68.
117 Nelson, N. and Racker, E. (1972) J. Biol. Chem. 247, 7657–7662.
118 Engelbrecht, S. and Junge, W. (1990) Biochim. Biophys. Acta 1015, 379–390.
119 Angov, E., Ng, T.C.N. and Brusilow, W.S.A. (1991) J. Bacteriol. 173, 407–411.
120 Hudson, G.S., Mason, J.G., Holton, T.A., Koller, B., Cox, G.B., Whitfeld, P.R. and Bottomley, W. (1987) J. Mol. Biol. 196, 283–298.
121 Otto, J. and Berzborn, R. (1989) FEBS Lett. 250, 625–628.
122 Hoppe, J., Schairer, H.U. and Sebald, W. (1980) FEBS Lett. 109, 107–111.
123 Hoppe, J., Schairer, H.U., Friedl, P. and Sebald, W. (1982) FEBS Lett. 145, 21–24.
124 Fromme, P., Boekema, E.J. and Graber, P. (1987) Z. Naturforsch. 42c, 1239–1245.
125 Lill, H. and Junge, W. (1989) FEBS Lett. 244, 15–20.
126 Feng, Y. and McCarty, R.E. (1990) J. Biol. Chem. 265, 12481–12485.
127 Lemaire, C. and Wollman, F.A. (1989) J. Biol. Chem. 264, 10235–10242.
128 Fraga, D. and Fillingame, R.H. (1991) J. Bacteriol. 173, 2639–2643.
129 Lightowlers, R.N., Howitt, S.M., Hatch, L., Gibson, F. and Cox, G.B. (1987) Biochim. Biophys. Acta 894, 399–406.
130 Lightowlers, R.N., Howitt, S.M., Hatch, L., Gibson, F. and Cox, G.B. (1988) Biochim. Biophys. Acta 933, 241–248.
131 Cain, B.D. and Simoni, R.D. (1986) J. Biol. Chem. 261, 10043–10050.
132 Cain, B.D. and Simoni, R.D. (1988) J. Biol. Chem. 263, 6606–6612.
133 Cain, B.D. and Simoni, R.D. (1989) J. Biol. Chem. 264, 3292–3300.
134 Cozens, A.L., Walker, J.E., Phillips, A.L., Huttly, A.K. and Gray, J.C. (1986) EMBO J. 5, 217–222.
135 Hennig, J. and Herrmann, R.G. (1986) Mol. Gen. Genet. 203, 117–128.

136 Feng, Y. and McCarty, R.E. (1990) J. Biol. Chem. 265, 12474–12480.
137 Lemaire, C. and Wollman, F.A. (1989) J. Biol. Chem. 264, 10228–10234.
138 Nelson, N., Nelson, H. and Schatz, G. (1980) Proc. Natl. Acad. Sci. USA 77, 1361–1364.
139 Herrin, D. and Michaels, A. (1985) Arch. Biochem. Biophys. 237, 224–236.
140 De Heij, H.T., Jochemsen, A.-G., Willemsen, P.T.J. and Groot, G.S.P. (1984) Eur. J. Biochem. 138, 161–168.
141 Merchant, S. and Selman, B.R. (1984) Plant Physiol. 75, 781–787.
142 Woessner, J.P., Masson, A., Harris, E.H., Bennoun, P., Gillham, N.W. and Boynton, J.E. (1984) Plant Mol. Biol. 3, 177–190.
143 Lemaire, C., Wollman, F.A. and Bennoun, P. (1988) Proc. Natl. Acad. Sci. USA 85, 1344–1348.
144 Robertson, D., Woessner, J.P., Gillham, N.W. and Boynton, J.E. (1989) J. Biol. Chem. 264, 2331–2337.
145 Sears, B.B. and Herrmann, R.G. (1985) Curr. Genet, 9, 521–528.
146 Biekmann, S. and Feierabend, J. (1985) Eur. J. Biochem. 152, 529–535.
147 Walker, J.E., Saraste, M. and Gay, N.J. (1984) Biochim. Biophys. Acta 768, 164–200.
148 Senior, A.E. (1990) Annu. Rev. Biophys. Biophys. Chem. 19, 7–41.
149 Fillingame, R.H. (1990) in: T.A. Krulwich (Ed.) The Bacteria: a Treatise on Structure and Function, Academic Press, New York, pp. 345–391.
150 Gogol, E.P., Lücken, U., Bork, T. and Capaldi, R.A. (1989) Biochemistry 28, 4709–4716.
151 Gogol, E.P., Aggeler, R., Sagermann, M. and Capaldi, R.A. (1989) Biochemistry 28, 4717–4724.
152 Gogol, E.P., Johnston, E., Aggeler, R. and Capaldi, R.A. (1990) Proc. Natl. Acad. Sci. USA 87, 9585–9589.
153 Mendel-Hartvig, J. and Capaldi, R.A. (1991) Biochemistry 30, 1278–1284.
154 Lucken, U., Gogol, E.P. and Capaldi, R.A. (1990) Biochemistry 29, 5339–5342.
155 Cox. G.B. and Gibson, F. (1987) Curr. Top. Bioenerg. 15, 163–175.
156 Duncan, T.M., Parsonage, D. and Senior, A.E. (1986) FEBS Lett. 208, 1–6.
157 Diesenhofer, J. and Michel, H. (1989) Science 1463–1473.
158 Lewis, M.J., Chang, J.A. and Simoni, R.D. (1990) J. Biol. Chem. 265, 10541–10550.
159 Bjorbæk, C., Foërsom, V. and Michelsen, O. (1990) FEBS Lett. 260, 31–34.
160 Hoppe, J., Brunner, J. and Jorgenson, B.B. (1984) Biochemistry 23, 5610–5616.
161 Steffens, K., Hoppe, J. and Altendorf, K. (1988) Eur. J. Biochem. 170, 627–630.
162 Manoil, C. and Beckwith, J. (1986) Science 233, 1403–1408.
163 Erhmann, M., Boyd, D. and Beckwith, J. (1990) Proc. Natl. Acad. Sci. USA 87, 7574–7578.
164 Eya, S., Maeda, M. and Futai, M. (1991) Arch. Biochem. Biophys. 284, 71–77.
165 Howitt, S.M., Lightowlers, R.N., Gibson, F. and Cox. G.B. (1990) Biochim. Biophys. Acta 1015, 264–268.
166 Senior, A.E. (1983) Biochim. Biophys. Acta 726, 81–95.
167 Jans, D.A., Fimmel, A.L., Hatch, L., Gibson, F. and Cox, G.B. (1984) Biochem. J. 221, 43–51.
168 Jans, D.A., Hatch, L., Fimmel, A.L., Gibson, F. and Cox, G.B. (1985) J. Bacteriol. 162, 420–426.
169 Porter, A.C., Kumamoto, C., Aldape, K. and Simoni, R.D. (1985) J. Biol. Chem. 260, 8182–8187.
170 Kumamoto, C.A. and Simoni, R.D. (1986) J. Biol. Chem. 261, 10037–10042.
171 Kumamoto, C.A. and Simoni, R.D. (1987) J. Biol. Chem. 262, 3060–3064.
171a McCormick, K.A. and Cain, B.D. (1991) J. Bacteriol. 173, 7240–7248.
172 Fimmel, A.L., Jans, D.A., Langman, L., James, L.B., Ash, G.R., Downie, J.A., Gibson, F. and Cox. G.B. (1983) Biochem. J. 213, 451–458.
173 Miller, M.J., Oldenburg, M. and Fillingame, R.H. (1990) Proc. Natl. Acad. Sci. USA 87, 4900–4904.
174 Howitt, S.M., Gibson, F. and Cox, G.B. (1988) Biochim. Biophys. Acta 936, 74–80.
175 Sebald, W., Hoppe, J. and Wachter, E. (1979) in: E.M. Klingenberg, F. Palmieri and E. Quagliariello (Eds.) Function and Molecular Aspects of Biomembrane Transport. Elsevier, Amsterdam, pp. 63–74.
176 Hoppe, J., Schairer, H.U. and Sebald, W. (1980) Eur. J. Biochem. 112, 17–24.
177 Fimmel, A.L., Jans, D.A., Hatch, L., James, L.B., Gibson, F. and Cox, G.B. (1985) Biochim. Biophys. Acta 808, 252–258.
178 Fimmel, A.L. and Fordham, S.A. (1989) Biochim. Biophys. Acta 978, 299–304.

179 Fimmel, A.L. and Norris, U. (1989) Biochim. Biophys. Acta 986, 257–262.
180 Fimmel, A.L., Karp, P.E. and Norris, U. (1990) Biochem. J. 269, 303–308.
181 Brusilow, W.S., Klionsky, D.J. and Simoni, R.D. (1982) J. Bacteriol. 151, 1363–1371.
182 McCarthy, J.E.G. (1990) Mol. Microbiol. 4, 1233–1240.
183 McCarthy, J.E.G. (1988) J. Bioenerg. Biomembranes 20, 19–39.
184 Hellmuth, K., Rex, G., Surin, B., Zinck, R. and McCarthy, J.E.G. (1991) Mol. Microbiol. 5, 813–824.
185 Klionsky, D.J., Skalnik, D.G. and Simoni, R.D. (1986) J. Biol. Chem. 261, 8096–8099.
186 Schneider, E. and Altendorf, K. (1985) EMBO J. 4, 515–518.
187 Aris, J.P., Klionsky, D.J. and Simoni, R.D. (1985) J. Biol. Chem. 260, 11207–11215.
188 Klionsky, D.J. and Simoni, R.D. (1985) J. Biol. Chem. 260, 11200–11206.
189 Cox, G.B., Downie, J.A., Langman, L., Senior, A.E., Ash, G., Fayle, D.R.H. and Gibson, F. (1981) J. Bacteriol. 148, 30–42.
190 Brusilow, W.S.A. (1987) J. Bacteriol. 169, 4984–4990.
191 Pati, S. and Brusilow, W.S. (1989) J. Biol. Chem. 264, 2640–2644.
192 Solomon, K.A. and Brusilow, W.S. (1988) J. Biol. Chem. 263, 5402–5407.
193 Pati, S., Brusilow, W.S.A., Deckers-Hebestreit, G. and Altendorf, K. (1991) Biochemistry 30, 4710–4714.
194 Humbert, R., Brusilow, W.S., Gunsalus, R.P., Klionsky, D.J. and Simoni, R.D. (1983) J. Bacteriol. 153, 416–422.
195 Hamamoto, T. and Kagawa, Y. (1985) in: A.N. Martonosi (Ed.) The Enzymes of Biological Membranes, Plenum Press, New York, pp. 149–176.
196 Gearing, D.P. and Nagley, P. (1986) EMBO J. 5, 3651–3655.
197 Nagley, P., Farrell, L.B., Gearing, D.P., Nero, D., Meltzer, S. and Devenish, R.J. (1988) Proc. Natl. Acad. Sci. USA 85, 2091–2095.
198 Engelman, D.M., Steitz, T.A. and Goldman, A. (1986) Annu. Rev. Biophys. Biophys. Chem. 15, 321–353.
199 Bachmann, B.J. (1990) Microbiol. Rev. 54, 130–197.
200 Takeda, M., Chen, W.J., Saltzgaber, J. and Douglas, M.G. (1986) J. Biol. Chem. 261, 15126–15133.
201 Macino, G. and Tzagoloff, A. (1980) Cell 20, 507–517.
202 Nagley, P. and Linnane, A.W. (1987) in: T. Ozawa and S. Papa (Eds.) Bioenergetics: Structure and Function of Energy Transducing Systems, Japan Sci. Soc. Press, Tokyo, pp. 191–204.
203 Ichikawa, N., Yoshida, Y., Hashimoto, T., Ogasawara, N., Yoshikawa, H., Imamoto, F. and Tagawa, K. (1990) J. Biol. Chem. 265, 6274–6278.
204 Akashi, A., Yoshida, Y., Nakagoshi, H., Kuroki, K., Hashimoto, T., Tagawa, K. and Imamoto, F. (1988) J. Biochem. Tokyo 104, 526–530.
205 Yoshida, Y., Sato, T., Hashimoto, T., Ichikawa, N., Nakai, S., Yoshikawa, H., Imamoto, F. and Tagawa, K. (1990) Eur. J. Biochem. 192, 49–53.
206 Fearnley, I.M. and Walker, J.E. (1986) EMBO J. 5, 2003–2008.
207 Gregory, R. and Hess, B. (1981) FEBS Lett. 129, 210–214.
208 Stutterheim, E., Henneke, M. and Berden, J. (1981) Biochim. Biophys. Acta 634, 271–278.
209 Todd, R.D., Griesenbeck, T.A. and Douglas, M.G. (1980) J. Biol. Chem. 255, 5461–5467.
210 Eya, S., Noumi, T., Maeda, M. and Futai, M. (1988) J. Biol. Chem. 263, 10056–10062.
211 Vik, S.B., Cain, B.D., Chun, K.T. and Simoni, R.D. (1988) J. Biol. Chem. 263, 6599–6605.
212 Paule, C.R. and Fillingame, R.H. (1989) Arch. Biochem. Biophys. 274, 270–284.
213 Vik, S.B., Lee, D., Curtis, C.E. and Nguyen, L.T. (1990) Arch. Biochem. Biophys. 282, 125–131.
214 Vik, S.B., Lee, D. and Marshall, P.A. (1991) J. Bacteriol. 173, 4544–4548.

L. Ernster (Ed.) *Molecular Mechanisms in Bioenergetics*
© 1992 Elsevier Science Publishers B.V. All rights reserved

CHAPTER 13

The reaction mechanism of F_0F_1-ATP synthases

RICHARD L. CROSS

Department of Biochemistry and Molecular Biology, State University of New York, Health Science Center at Syracuse, Syracuse, New York 13210, U.S.A.

Contents

1. Introduction 317
2. Structure 318
 2.1. Subunit composition 318
 2.2. Structure of F_0 318
 2.3. Strucutre of F_1 319
 2.4. Structural model for F_0F_1 320
3. Mechanism 320
 3.1. The binding change mechanism 320
 3.2. The number of functional catalytic sites 321
  3.2.1. Introduction 321
  3.2.2. Does uni-site catalysis occur at a normal catalytic site? 322
  3.2.3. Do the two K_m's reflect catalysis at a second and third site? 323
  3.2.4. The stoichiometry for inhibition of F_1 by affinity probes is unlikely to reveal the number of functional sites 323
  3.2.5. Attempts to demonstrate the equivalence of catalytic sites on F_1 324
 3.3. The rotary mechanism 325
  3.3.1. Introduction 325
  3.3.2. Experiments with soluble F_1 325
  3.3.3. Experiments with membrane-bound F_0F_1 326
Acknowledgements 327
References 327

1. Introduction

ATP synthesis by oxidative phosphorylation and photophosphorylation is catalyzed by a complex of proteins, F_0F_1, [1–6] found embedded in the membranes of eubacteria (BF_0F_1), mitochondria (MF_0F_1), and chloroplasts (CF_0F_1). An electrochemical gradient appears to serve as an obligatory intermediate in coupling the exergonic oxidation/reduction reactions of respiratory and photosynthetic electron-transport chains to the endergonic synthesis of ATP by the synthases (see the Skulachev chapter, this volume). Close cousins of the traditional F_0F_1 complexes are found as ATP synthases in archaebacteria (AF_0F_1) and as H^+-translocating ATPases in the vacuolar membranes of eukaryotic cells (VF_0F_1). Because of the close evolutionary relationship between vacuolar and archaebacterial F_0F_1, it has been suggested that eukaryotic cells evolved following invasion by eubacteria of archaebacterial host cells. The invaders eventually became mitochondria and chloroplasts, whereas the AF_0F_1 of the host cells internalized, becoming the vacuolar ATPases of endomembrane systems [7–9].

The main focus of this review will be recent advances in our understanding of the mechanism of ATP synthesis by F_0F_1. Since the development of mechanistic models has been strongly influenced by advances in our knowledge of structure, a brief review of structural features of the complex precedes discussion of mechanism. For additional information about the structure and assembly of ATP synthases, refer to the chapter by Cox et al. (this volume).

2. Structure

2.1. Subunit composition

Using two independent radioisotope labeling methods, Foster and Fillingame [10] showed the subunit composition of *E. coli* BF_0F_1 to be $\alpha_3\beta_3\gamma\delta\varepsilon ab_2c_{9-12}$. The uncertainty in the stoichiometry of subunit *c* may have important consequences regarding the energetics of ATP synthesis. It has been suggested that during one complete turnover of the F_0F_1 complex, each catalytic site, located on β, makes one ATP and each subunit *c*, containing a carboxyl group essential for proton conduction, transports one proton. Hence, the coupling ratio H^+/ATP may be equal to the ratio of stoichiometries for subunits *c* and β [11]. Initially, the H^+/ATP ratio was thought to be two for ATP synthases; however, a majority of laboratories that do such measurements now favor a value of three [12,13]. This would be consistent with nine *c*-subunits, i.e., $9c/3\beta = 3$. However, a value of four for the H^+/ATP ratio has also been reported [14] suggesting twelve *c*-subunits. Twelve copies of *c* would be consistent with the presence of six double-sized *c*-subunits in VF_0F_1 [11,15,16].

In order to simplify the study of individual reaction steps in the synthase's mechanism, F_0F_1 has been separated into a soluble hydrophilic portion, F_1, containing the α, β, γ, δ, and ε subunits, and a membrane-embedded hydrophobic part, F_0, containing the *a*, *b*, and *c* subunits. F_0 functions as a proton channel whereas F_1 contains the catalytic sites for ATP synthesis. Uncoupled from its energy source, soluble F_1 is only capable of net ATP hydrolysis; hence, it is often referred to as an ATPase [17].

2.2. Structure of F_0

Although *E. coli* F_0 is the best characterized, and having only three different subunits, appears to be the simplest, progress is also being made in defining the more complex F_0 found in mitochondria [18–20] and chloroplasts [21,22].

Soon after the amino-acid sequences of *E. coli* F_0 subunits had been deduced [23–25], attempts were made to predict their secondary and tertiary structures [26]. Subunit *a* is the largest, having 5 to 8 transmembrane α-helical segments [27,28]. The *b*-subunit is predicted to have a single hydrophobic transmembrane segment at its N-terminus which anchors it to the membrane, whereas the bulk of the subunit protrudes through the center of F_1 as an extended, doubled-over, α-helical chain [29–31]. Subunit *c* is predicted to have two transmembrane α-helices [26] with the short connecting segment facing F_1 [32].

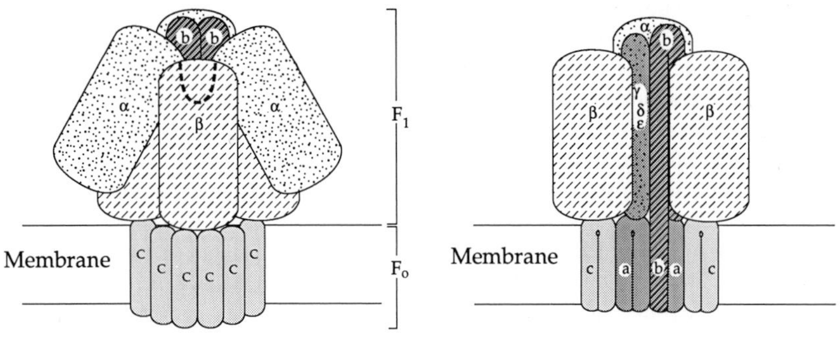

Fig. 1. A speculative model for the structure of *E. coli* F_0F_1. (Left) The complete complex. The dotted line represents one potential nucleotide binding site. (Right) The complex cut in half.

Isolated chloroplast *c*-subunits appear in electron micrographs to form tube-like aggregates composed of stacked rings, each containing about twelve subunits [33]. The diameter of the ring (62 Å) is the same as that of F_0. The central cavity would appear to provide just enough space to accommodate one *a*-subunit and the tails of two *b*-subunits. In the same study, CF_0F_1 was run on an SDS gel without prior heating in the presence of SDS. No 8 kDa band corresponding to monomeric *c*-subunits was observed, but there was a band at 100 kDa. When this band was extracted, heated with SDS, and again applied to an SDS gel, a single band was seen at 8 kDa. These results also support a complex of twelve *c*-subunits (i.e., 8 kDa × 12 = 96 kDa) [33].

Cross-linking of F_1-stripped *E. coli* membranes results in the formation of *b–b*, *a–b*, and *a–b–b* products [31]. This confirms a stoichiometry of 1:2 for *a:b* and indicates that these subunits are in close contact. Genetic evidence also supports an interaction between subunits *a* and *b*. Mutants that partially suppress the effects of a mutation in subunit *b* were found to be due to amino-acid substitutions in subunit *a* [34].

Taken together, the results suggest that F_0 consists of a complex of one *a*-subunit and two *b*-subunits enclosed by a ring of up to twelve *c*-subunits (Fig. 1). Evidence in apparent conflict with this model comes from studies in which the quaternary structure of F_0 was probed using lipophilic group-specific reagents and photolabels [35–37]. An unproven assumption in the interpretation of these results was that only protein surfaces in contact with lipids would be labeled. Since all three subunits were labeled, it was concluded that F_0 must be highly asymmetric. However, it seems possible that the hydrophobic probes may also have been able to intercalate between hydrophobic transmembrane α-helical segments of F_0 subunits. Indeed, some of them labeled hydrophobic domains on F_1 [35,37].

2.3. *Structure of* F_1

Until recently, the most detailed information regarding the tertiary and quaternary structure of F_1 was provided by electron microscopy. Images of monodispersed F_1 show six globular masses in a hexagonal array [38–41]. Immunoelectron microscopy identified the six masses as alternating α- and β-subunits [42]. A seventh mass, smaller than the others, is sometimes seen partially filling the central cavity. This has been shown to be due primarily to the γ-subunit [43–46]. Using monoclonal antibodies, the central mass was found to be associated with one of the β-subunits [47]. The remaining free space seen in the central cavity of F_1 may be filled in native F_0F_1 by the two *b*-subunits that protrude from the membrane surface.

One area of disagreement in interpreting the results of these studies has been the nature of the association of α- and β-subunits and their shapes. Originally, it was suggested that the large subunits of F_1 were globular and that the hexamer of alternating α and β consisted of a stacked dimer of trimers offset by 60° and slightly interdigitated [38,46]. However, cryoelectron microscopy [41] and small-angle neutron scattering [48] suggest that α and β are elongated ellipsoids. In the hexameric array, α and β would be found in equivalent positions forming a hollow cylinder. The latter view has several advantages. A symmetrical arrangement of α and β relative to F_0 would be easier to understand in terms of mechanism if, as some believe, progenitor F_0F_1 had six catalytically active β-like subunits [11,49]. Also, the catalytic site on β is further from the surface of the membrane [50] than is possible for a globular β that makes contact with F_0 [51,52].

The recently published 3.6 Å X-ray crystallographic structure for rat-liver MF_1 [53] suggests that the actual structure is something of a compromise between the two views summarized above. The α- and β-subunits do appear to be offset as originally proposed but only by about 15 Å, and their shapes are somewhat elongated. In addition, the β-subunits do not appear to make contact with each other but the α-subunits are tilted outward at an angle of about 30° (Fig. 1).

2.4. *Structural model of F_0F_1*

In constructing the model shown in Fig. 1, F_1 was oriented in such a way as to place F_0 in contact with that side of the complex having β-subunits protruding beyond the α-subunits. A close proximity of β to F_0 is suggested by results from Simoni's laboratory showing that β can be cross-linked to subunit *a* [51] and that isolated β binds to F_0 [52]. On the side of F_1 positioned furthest from the membrane in Fig. 1, three water-accessible pockets are seen in the X-ray crystallographic structure [53] at the point where α-subunits interface with β (dotted line). These are likely candidates for the nucleotide binding sites and their positioning is consistent with fluorescence transfer measurements showing the sites to be far from the surface of the membrane [50].

One structural feature detected in electron micrographs not shown in Fig. 1 is a stalk that appears to connect F_1 to F_0 [33,54,55]. The stalk is about 30 Å wide and would accommodate about seven helices, four of which would be due to the two *b*-subunits. However, in view of evidence that β-subunits are in direct contact with F_0 [51,52] and *c*-subunits with F_1 [56], it seems possible that the staining or freezing of membranes in preparation for electron microscopy may cause F_1 to partially slip off the rod-like aggregate of proteins that fills the central cavities of the α/β hexamer and the ring of *c*-subunits.

3. *Mechanism*

3.1. *The binding change mechanism*

Over the years, the binding change mechanism has become the most widely used model for describing ATP synthesis by F_0F_1-ATP synthases. As illustrated in Fig. 2, the five main features of the model are as follows:

(a) Energization from proton transport is not required to make ATP at the catalytic site but

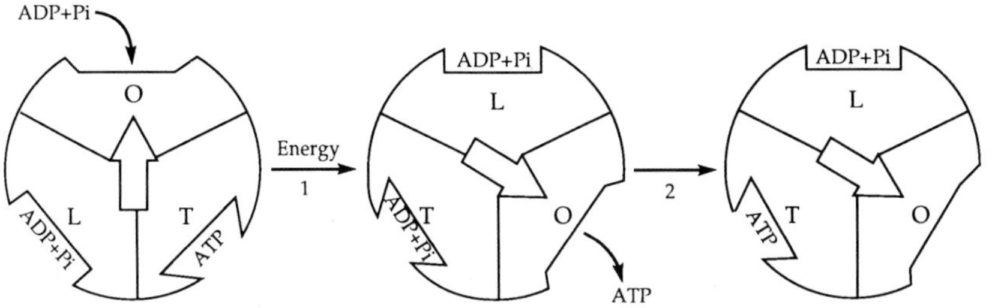

Fig. 2. The binding change mechanism for ATP synthesis by F_0F_1. The central arrow, representing $\gamma\delta\varepsilon b_2$, rotates in step 1 relative to three $\alpha\beta$ pairs, which, in this illustration, remain stationary. The rotation of the asymmetric central mass forces the three catalytic sites to undergo the conformational changes associated with substrate binding and product release. T, L, and O stand for tight, loose, and open conformations and refer to the affinities of the catalytic sites for ligand. In step 2, ATP forms spontaneously from tightly bound ADP and P_i.

rather to promote its release from the site [57]. This reflects the need to disrupt the many favorable interactions between protein and ligand that contribute to the very tight binding of ATP [58].

(b) Substrate binding is also associated with the energization step [59].

(c) Energy-linked substrate binding and product release occur simultaneously at separate but interacting catalytic sites [60].

(d) There are three interacting catalytic sites [61–63].

(e) The binding changes required for catalysis occur as a result of the rotation of an asymmetric aggregate of subunits [64–67], perhaps $\gamma\delta\varepsilon ab_2$, relative to a complex of the remaining subunits, $\alpha_3\beta_3c_{12}$. Like the bacterial flagellar motor [68], this rotation is driven by protons moving down an electrochemical gradient.

At the current time, there seems to be fair agreement regarding features (a) through (c) listed above. However, the question of the number of functional catalytic sites (d) is under active investigation, and there is little direct evidence for a rotary mechanism (e). The remainder of this chapter will focus on recent results that pertain to these two controversial issues. For a discussion of earlier relevant studies see ref. [69].

3.2. *The number of functional catalytic sites*

3.2.1. *Introduction*

The presence of three functional catalytic sites might normally be assumed for an enzyme having three copies of the catalytic subunit. However, it has been suggested that the interaction of single-copy F_1 subunits with one $\alpha\beta$ pair renders one catalytic site permanently non-equivalent or nonfunctional [70]. Hence several laboratories favor models having only two functional sites [70–73].

Based on our finding that three out of a total of six nucleotide binding sites on MF_1 exchange rapidly with medium nucleotide during catalysis [62], we proposed a model having three functional, interacting catalytic sites [61]. In support of this model, evidence was presented that MF_1 can hydrolyze ATP in three different modes where one, two or three sites may turn over simultaneously [63]. Uni-site catalysis occurs when substrate is present in amounts substoi-

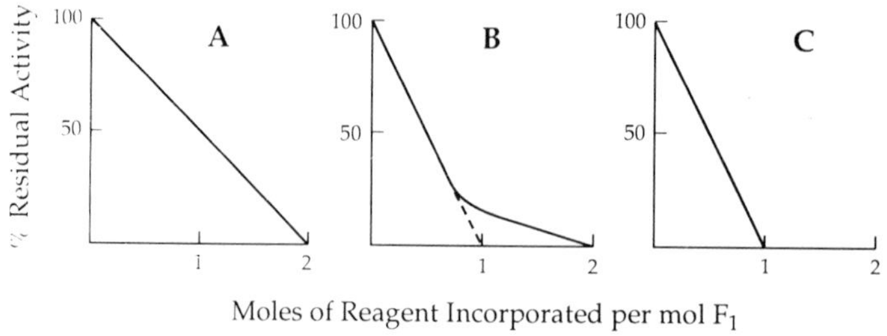

Fig. 3. Predicted inhibition curves for a three-site binding change model. Results anticipated for modification by a probe that, when incorporated into a single catalytic site, (A) has no effect on residual promoted catalysis at the remaining two sites, (B) partially impairs promoted catalysis at the remaining sites, or (C) fully inhibits promoted catalysis at the remaining sites.

catalysis would be 2 mol/mol (Fig. 3A). Examples of such reagents may include azido-naphthoyl-ADP [101], fluoroaluminum- and fluoroberyllium-nucleotide diphosphate complexes [102] and diadenosine oligophosphate compounds [103]. In addition, reconstituted hybrid enzymes having one defective β [104] or one defective α [105] showed considerable residual multi-site catalysis.

(b) Another type of probe may exist that partially impairs cooperative interactions at unmodified sites when incorporated into a single site. This would result in a nonlinear inhibition curve, where most of the activity would be lost on a line extrapolating to 1 mol/mol with loss of the remaining activity following a line extrapolating to 2 mol/mol. (Fig. 3B). Examples of such probes may include BzATP 1[06] and 2-azido-ATP [107].

(c) A third class of affinity probe may exist which, when incorporated into one site, freezes the enzyme's conformation in such a way that promoted catalysis at the remaining two sites is fully blocked. This would result in a linear inhibition curve extrapolating to 1 mol/mol. (Fig. 3C) [108,109].

Differentiating probes in class (c) from those in class (b) may not be possible using a coupled-enzyme assay. Since multi-site catalysis is so much faster than uni-site catalysis, a very sensitive hydrolysis assay would be required to rule out any residual promoted catalysis after incorporation of one mol of reagent per mol F_1. A greater source of error would be heterogeneity in the labeling of F_1. Except perhaps by using a method developed by Miwa et al. [104], it might not be possible to obtain a pure sample of singly labeled F_1 without some residual unmodified F_1 or doubly labeled F_1.

Another potential complication could arise if the modification reaction showed positive cooperativity. If the second or third catalytic site were modified at a rate much faster than the first, the stoichiometry for inhibition would appear to be 2 or 3 mol/mol even though all capacity for multi-site catalysis might be lost upon modification of the first site. This consideration is not likely to apply to the modification of F_1 catalytic sites by nucleotide affinity probes since the sites show strong negative cooperativity in binding nucleotides.

3.2.5. Attempts to demonstrate the equivalence of catalytic sites on F_1

When a single β-subunit of CF_1 is labeled by 2-azido-ATP bound at a catalytic site, subsequent modification by DCCD results in the exclusive labeling of the other two β-subunits. However, when CF_1 is first labeled by DCCD, allowed to hydrolyze 2-azido-ATP in the dark, and then photolyzed, both the DCCD-labeled and unlabeled β are modified by the azido nucleotide. The results indicate that the β-subunit tagged by DCCD becomes randomized during catalytic turnover [110]. In this same study, evidence was presented which supports the participation of three catalytic sites. First, the site modified by 2-azido-ATP cleaved the probe; hence it was a catalytic site. In addition, hydrolysis of medium ATP by the modified enzyme showed positive catalytic cooperativity, indicating promoted catalysis by the two remaining unmodified sites [110]. In an extension of this work [107], a single CF_1 β-subunit was modified by 2-azido-ATP and its effect on the intermediate $P_i \leftrightarrow$ HOH oxygen exchange was measured. This exchange occurs during ATP hydrolysis as a result of reversible cleavage of ATP at the catalytic site. With unmodified CF_1, a single reaction pathway is seen for the exchange [111], indicating the presence of equivalent interacting catalytic sites. However, with the modification of a single catalytic site, two new pathways are observed, indicating that the two unmodified catalytic sites retain activity but see the modified site differently. Substrate modulation of the exchange reaction remains, demonstrating residual bi-site catalysis [107].

A third approach applied to the chloroplast enzyme was to use fluorescence transfer to measure the distance between a reference point – covalently incorporated lucifer yellow – and two different catalytic sites, occupied by fluorescent nucleotide analogs [112–114]. The results showed that during ATP hydrolysis, the properties of the two distinct catalytic sites randomize. Interestingly, this randomization decayed with a half-time of 2 h, suggesting that either the sites are not equivalent in the resting state or modification by lucifer yellow induces asymmetry.

3.3. The rotary mechanism

3.3.1. Introduction

F_0 subunits *a* and *c* contain essential residues that are thought to participate in the translocation of protons [27,115–118]. How then is it possible that a single copy of subunit *a* is able to coordinate with 9 to 12 copies of subunit *c* in performing this function? Similarly, F_1 has three copies of the catalytic subunit but only single copies of γ, δ, and ε. How is it possible that despite this structural asymmetry, the enzyme shows a single mechanistic pathway in the oxygen exchange reactions [111,119,120]? One potential answer to the latter question is that there is just a single functional catalytic site [99,121]. However, compelling evidence has been presented for the presence of at least two interacting catalytic sites [60,63,122,123]. If there were only two functional sites, they would still see the asymmetric core of the protein differently and might be expected to display multiple catalytic pathways.

A plausible answer to both questions comes from the suggestion that the asymmetric core of the complex, perhaps $\gamma\delta\varepsilon ab_2$, rotates relative to a symmetric outer cylinder, $\alpha_3\beta_3c_{12}$ [64,66,67,124–126]. Hence, all three catalytic sites would pass through equivalent states during turnover, except that they would do so 120° out of phase. In addition, the *a*-subunit would transport a proton in cooperation with one *c*-subunit at a time during its movement around the inside surface of the ring of *c*-subunits. The binding changes necessary for substrate binding and product release would result from conformational changes required to accommodate movement of the asymmetric core (Fig. 2). Rotation would be driven by the movement of protons down an electrochemical gradient where a gating process would obligatorily couple rotation to the completion of the proton channel.

3.3.2. Experiments with soluble F_1

One approach that has been employed to test the possibility of rotation has been to cross-link the single-copy subunits of F_1 to α or β and measure the effect on ATP hydrolysis. In one such study, it was found that when amino groups on α of *E. coli* BF_1 are cross-linked to γ or δ, partial inhibition occurs which can be reversed by cleaving the disulfide bond present in the middle of the cross-linking reagent [127]. Although these results are consistent with a required rotation of the small subunits of F_1 with regard to the large subunits, they could also be the result of an inhibition of required conformation changes. In this type of experiment, a negative result would be more definitive than a positive one. Such negative results have in fact been obtained. When *E. coli* BF_1 is passed through a centrifuge column without DTT or EDTA, the δ-subunit cross-links to α by an intersubunit disulfide bond [128,129]. Cross-linked BF_1 shows full activity. With CF_1, the cross-linking of γ to α and β gave considerably less than proportional inhibition of activity [130]. Although these results provide evidence that rotation of $\gamma\delta\varepsilon$ relative to $\alpha_3\beta_3$ is not required for activity by the soluble enzyme, there are results which suggest that rotation can in fact occur. Using antibodies to identify various subunits of *E. coli* BF_1 in electron micrographs, Capaldi and colleagues demonstrated movement of the central mass with regard to specific reference points following ATP hydrolysis [131].

In summary, it would appear that rotation is not required for ATP hydrolysis by the soluble enzyme. However, this may be the result of dissociating F_1 from interacting F_0 subunits. It is possible that two *b*-subunits must fill the central cavity of F_1 in order for hydrolysis to be obligatorily coupled to rotation. The fact that rotation can occur at all with soluble enzyme [131] suggests that it may be required for catalysis by the intact coupled F_0F_1-ATP synthase.

3.3.3. Experiments with membrane-bound F_0F_1

CF_0F_1 was covalently labeled with a fluorescent probe and reconstituted into membranes. Rotation of the molecule was measured by phosphorescence emission anisotropy and found to be independent of catalytic turnover. The authors concluded that rotation has no role in catalysis [132]. However, the time resolution of their measurements was not adequate to detect a possible effect of catalysis on rotation. Any motion caused by catalysis, which occurs on a time scale of milliseconds, would have been much too slow to have affected the decay of phosphorescence anisotropy, which occurs on a time scale of microseconds.

In another study, MF_0F_1 having IgG attached to the α-subunit was exposed to anti IgG's. No change in the ATP synthesis rate was observed even though the increase in mass totaled about 600 kDa [133]. The authors concluded that rotation was unlikely to be required for catalysis. However, again the time resolution appears not to have been adequate to detect such rotation. The molecular motion of the attached antibody would be rapid compared to the millisecond time scale required for catalysis. In addition, the center of the F_0F_1 complex may rotate relative to $\alpha_3\beta_3$; thus, the attachment of additional mass to α might not be expected to impede that rotation.

In summary, the idea that F_0F_1 functions as a truncated bacterial flagellar motor where rotation drives the endergonic binding changes remains attractive. However, the concept needs to be tested on the intact ATP synthase. It is possible that the asymmetric properties observed for soluble F_1 will not be observed with F_0F_1 [134,135]. Also a rotary mechanism would likely require sequential firing of the catalytic sites, and unlike soluble F_1, no residual promoted catalysis would be expected for membrane-bound enzyme having one modified site. Finally, cross-linking, which has little effect in some cases on soluble F_1, would be predicted to strongly inhibit the intact synthase.

Acknowledgements

Work cited from the author's laboratory was supported by Research Grant GM23152 from the National Institutes of Health, United States Public Health Service.

References

1 Senior, A.E. (1988) Physiol. Rev. 68, 177–231.
2 Futai, M., Noumi, T. and Maeda, M. (1989) Annu. Rev. Biochem. 58, 111–136.
3 Boyer, P.D. (1989) FASEB J. 3, 2164–2178.
4 Tiedge, H. and Schafer, G. (1989) Biochim. Biophys. Acta 977, 1–9.
5 Fillingame, R.H. (1990) in: T.A. Krulwich (Ed.) the Bacteria 12, Academic Press, New York, pp. 345–391.
6 Penefsky, H.S. and Cross, R.L. (1991) Adv. Enzymol. 64, 173–214.
7 Nelson, N. and Taiz, L. (1989) Trends Biochem. Sci. 14, 113–116.
8 Gogarten, J.P., Kibak, H., Dittrich, P., Taiz, L., Bowman, E.J., Bowman, B.J., Manolson, M.F., Poole, R.J., Date, T., Oshima, T., Konishi, J., Denda, K. and Yoshida, M. (1989) Proc. Natl. Acad. Sci. USA 86, 6661–6665.
9 Forgac, M. (1989) Physiol. Rev. 69, 765–796.
10 Foster, D.L. and Fillingame, R.H. (1982) J. Biol. Chem. 257, 2009–2015.
11 Cross, R.L. and Taiz, L. (1990) FEBS Lett. 259, 227–229.
12 Ferguson, S.J. and Sorgato, M.C. (1982) Annu. Rev. Biochem. 51, 185–217.
13 Berry, E.A. and Hinkle, P.C. (1983) J. Biol. Chem. 258, 1474–1486.
14 Rumberg, B., Schubert, K., Strelow, F. and Tran-Anh, T. (1990) in: M. Baltscheffsky (Ed.) Curr. Res. Photosynth. III, Kluwer Academic Publishers, Dordrecht, pp. 125–128.
15 Arai, H., Terres, G., Pink, S. and Forgac, M. (1988) J. Biol. Chem. 263, 8796–8802.
16 Mandel, M., Moriyama, Y., Hulmes, J.D., Pan, Y.-C.E., Nelson, H. and Nelson, N. (1988) Proc. Natl. Acad. Sci. USA 85, 5521–5524.
17 Penefsky, H.S., Pullman, M.E., Datta, A. and Racker, E. (1960) J. Biol. Chem. 235, 3330–3336.
18 Joshi, S. and Burrows, R. (1990) J. Biol. Chem. 265, 14518–14525.
19 Walker, J.E., Lutter, R., Dupuis, A. and Runswick, M.J. (1991) Biochemistry 30, 5369–5378.
20 Hekman, C., Tomich, J.M. and Hatefi, Y. (1991) J. Biol. Chem. 266, 13564–13571.
21 Feng, Y. and McCarty, R.E. (1990) J. Biol. Chem. 265, 5104–5109.
22 Mitra, B. and Hammes, G.G. (1990) Biochemistry 29, 9879–9884.
23 Gay, N.J. and Walker, J.E. (1981) Nucl. Acids Res. 9, 3919–3926.
24 Kanazawa, H., Mabuchi, K., Kayano, T., Noumi, T., Sekiya, T. and Futai, M. (1981) Biochem. Biophys. Res. Commun. 103, 613–620.
25 Nielsen, J. Hansen, F.G., Hoppe, J. Friedl, P. and von Meyenburg, K. (1981) Mol. Gen. Genet. 184, 33–39.
26 Senior, A.E. (1983) Biochim. Biophys. Acta 726, 81–95.
27 Cox, G.B. Fimmel, A.L., Gibson, F. and Hatch, L. (1986) Biochim. Biophys. Acta 849, 62–69.
28 Lewis, M.J., Chang, J.A. and Simoni, R.D. (1990) J. Biol. Chem. 265, 10541–10550.
29 Walker, J.E., Saraste, M. and Gay, N.J. (1982) Nature 298, 867–869.
30 Perlin, D.S., Cox, D.N. and Senior, A.E. (1983) J. Biol. Chem. 258, 9793–9800.
31 Hermolin, J., Gallant, J. and Fillingame, R.H. (1983) J. Biol. Chem. 258, 14550–14555.
32 Girvin, M.E., Hermolin, J., Pottorf, R. and Fillingame, R.H. (1989) Biochemistry 28, 4340–4343.
33 Boekema, E.J., Fromme, P. and Graber, P. (1988) Ber. Bunsenges. Phys. Chem. 92, 1031–1036.
34 Kumamoto, C.A. and Simoni, R.D. (1986) J. Biol. Chem. 261, 10037–10042.
35 Hoppe, J., Brunner, J. and Jorgensen, B.B. (1984) Biochemistry 23, 5610–5616.
36 Hoppe, J., Gatti, D., Weber, H. and Sebald, W. (1986) Eur. J. Biochem. 155, 259–264.
37 Aggeler, R., Zhang, Y.-Z. and Capaldi, R.A. (1987) Biochemistry 26, 7107–7113.

38 Tiedge, H., Schafer, G. and Mayer, F. (1983) Eur. J. Biochem. 132, 37–45.
39 Akey, C.W., Crepeau, R.H., Dunn, S.D., McCarty, R.E. and Edelstein, S.J. (1983) EMBO J. 2, 1409–1415.
40 Boekema, E.J., Berden, J.A. and van Heel, M.G. (1986) Biochim. Biophys. Acta 851, 353–360.
41 Gogol, E.P., Lucken, U., Bork, T. and Capaldi, R.A. (1989) Biochemistry 28, 4709–4716.
42 Tiedge, H., Lunsdorf, H., Schafer, G. and Schairer, H.U. (1985) Proc. Natl. Acad. Sci. USA 82, 7874–7878.
43 Boekema, E.J., van Heel, M. and Graber, P. (1988) Biochim. Biophys. Acta 933, 365–371.
44 Yoshimura, H., Endo, S., Matsumoto, M., Nagayama, K. and Kagawa, Y. (1989) J. Biochem. 106, 958–960.
45 Fujiyama, Y., Yokoyama, K., Yoshida, M. and Wakabayashi, T. (1990) FEBS Lett. 271, 111–115.
46 Boekema, E.J., Xiao, J. and McCarty, R.E. (1990) Biochim. Biophys. Acta 1020, 49–56.
47 Gogol, E.P., Aggeler, R., Sagermann, M. and Capaldi, R.A. (1989) Biochemistry 28, 4717–4724.
48 Ito, Y., Harada, M., Ohta, S., Kagawa, Y., Aono, O., Schefer, J. and Schoenborn, B.P. (1990) J. Mol. Biol. 213, 289–302.
49 Saraste, M., Gay, N.J., Eberle, A., Runswick, M.J. and Walker, J.E. (1981) Nucleic Acids Res. 9, 5287–5296.
50 Snyder, B. and Hammes, G.G. (1984) Biochemistry 23, 5787–5795.
51 Aris, J.P. and Simoni, R.D. (1983) J. Biol. Chem. 258, 14599–14609.
52 Aris, J.P. and Simoni, R.D. (1985) Biochem. Biophys. Res. Commun. 128, 155–162.
53 Bianchet, M., Ysern, X., Hullihen, J., Pedersen, P.L. and Amzel, L.M. (1991) J. Biol. Chem. 266, 21197–21201.
54 Tsuprun, V.L., Orlova, E.V. and Mesyanzhinova, I.V. (1989) FEBS Lett. 244, 279–282.
55 Lucken, U., Gogol, E.P. and Capaldi, R.A. (1990) Biochemistry 29, 5339–5343.
56 Feng, Y. and McCarty, R.E. (1990) J. Biol. Chem. 265, 12481–12485.
57 Boyer, P.D., Cross, R.L. and Momsen, W. (1973) Proc. Natl. Acad. Sci. USA 70, 2837–2839.
58 Grubmeyer, C., Cross, R.L. and Penefsky, H.S. (1982) J. Biol. Chem. 257, 12092–12100.
59 Rosing, J., Kayalar, C. and Boyer, P.D. (1977) J. Biol. Chem. 252, 2478–2485.
60 Kayalar, C., Rosing, J. and Boyer, P.D. (1977) J. Biol. Chem. 252, 2486–2491.
61 Cross, R.L. (1981) Annu. Rev. Biochem. 50, 681–714.
62 Cross, R.L. and Nalin, C.M. (1982) J. Biol. Chem. 257, 2874–2881.
63 Cross, R.L., Grubmeyer, C. and Penefsky, H.S. (1982) J. Biol. Chem. 257, 12101–12105.
64 Boyer, P.D. and Kohlbrenner, W.E. (1981) in: R. Selmon and S. Selmon-Reimer (Eds.) Energy Coupling in Photosynthesis, Elsevier, Amsterdam, pp. 230–240.
65 Gresser, M.J., Meyers, J.A. and Boyer, P.D. (1982) J. Biol. Chem. 257, 12030–12038.
66 Boyer, P.D. (1983) in: D.L.F. Lennon, F.W. Stratman and R.N. Zahlten (Eds.) Biochemistry of Metabolic Processes, Elsevier, Amsterdam, pp. 465–477.
67 Cox, G.B., Jans, D.A., Fimmel, A.L., Gibson, F. and Hatch, L. (1984) Biochim. Biophys. Acta 768, 201–208.
68 Berg, H.C. (1975) Nature 254, 389–392.
69 Cross, R.L. (1988) J. Bioenerg. Biomembranes 20, 395–405.
70 Amzel, L.M., McKinney, M., Narayanan, P. and Pedersen, P.L. (1982) Proc. Natl. Acad. Sci. USA 79, 5852–5856.
71 Bullough, D.A., Verburg, J.G., Yoshida, M. and Allison, W.S. (1987) J. Biol. Chem. 262, 11675–11683.
72 Leckband, D. and Hammes, G.G. (1987) Biochemistry 26, 2306–2312.
73 Berden, J.A., Hartog, A.F. and Edel, C.M. (1991) Biochim. Biophys. Acta 1057, 151–156.
74 Cunningham, D. and Cross, R.L. (1988) J. Biol. Chem. 263, 18850–18856.
75 Ebel, R.E. and Lardy, H.A. (1975) J. Biol. Chem. 250, 191–196.
76 Cerdan, E., Campo, M.L., Lopez-Moratalla, N. and Santiago, E. (1983) FEBS Lett. 158, 151–153.
77 Wong, S.-Y., Matsuno-Yagi, A. and Hatefi, Y. (1984) Biochemistry 23, 5004–5009.
78 Roveri, O.A. and Calcaterra, N.B. (1985) FEBS Lett. 192, 123–127.
79 Dunn, S.D., Zadorozny, V.D., Tozer, R.G. and Orr, L.E. (1987) Biochemistry 26, 4488–4493.

80 Al-Shawi, M.K. and Senior, A.E. (1988) J. Biol. Chem. 263, 19640–19648.
81 Milgrom, Ya.M. and Murataliev, M.B. (1987) FEBS Lett. 222, 32–36.
82 Fromme, P. and Graber, P. (1989) FEBS Lett. 259, 33–36.
83 Penefsky, H.S. (1988) J. Biol. Chem. 263, 6020–6022.
84 Milgrom, Ya.M. and Boyer, P.D. (1990) Biochim. Biophys. Acta 1020, 43–48.
85 Lotscher, H.-R., DeJong, C. and Capaldi, R.A. (1984) Biochemistry 23, 4140–4143.
86 Wu, D. and Boyer, P.D. (1986) Biochemistry 25, 3390–3396.
87 Ysern, X., Amzel, L.M. and Pedersen, P.L. (1988) J. Bioenerg. Biomembranes 20, 423–450.
88 Miwa, K. and Yoshida, M. (1989) Proc. Natl. Acad. Sci. USA 86, 6484–6487.
89 Yoshida, M. and Allison, W.S. (1990) J. Biol. Chem. 265, 2483–2487.
90 Xue, Z. and Boyer, P.D. (1989) Eur. J. Biochem. 179, 677–681.
91 Milgrom, Ya.M., Ehler, L.L. and Boyer, P.D. (1990) J. Biol. Chem. 265, 18725–18728.
92 Senior, A.E. (1981) J. Biol. Chem. 256, 4763–4767.
93 Guerrero, K.J., Xue, Z. and Boyer, P.D. (1990) J. Biol. Chem. 265, 16280–16287.
94 Moyle, J. and Mitchell, P. (1975) FEBS Lett. 56, 55–61.
95 Kasho, V.N. and Boyer, P.D. (1984) J. Bioenerg. Biomembranes 16, 407–419.
96 Penefsky, H.S. and Grubmeyer, C. (1984) in: S. Papa et al. (Eds.) Proton-ATPase (ATP Synthase): Structure, Function, Biogenesis. The F_0F_1 Complex of Coupling Membranes, ICSU Press, pp. 195–204.
97 Vignais, P.V. and Lunardi, J. (1985) Annu. Rev. Biochem. 54, 977–1014.
98 Ferguson, S.J., Lloyd, W.J., Lyons, M.H. and Radda, G.K. (1975) Eur. J. Biochem. 54, 117–126.
99 Wang, J.H. (1988) J. Bioenerg. Biomembranes 20, 407–422.
100 Sloothaak, J.B., Berden, J.A., Herweijer, M.A. and Kemp, A. (1985) Biochim. Biophys. Acta 809, 27–38.
101 Lubben, M., Lucken, U., Weber, J. and Schafer, G. (1984) Eur. J. Biochem. 143, 483–490.
102 Issartel, J.P., Dupuis, A., Lunardi, J. and Vignais, P.V. (1991) Biochemistry 30, 4726–4733.
103 Vogel, P.D. and Cross, R.L. (1991) J. Biol. Chem. 266, 6101–6105.
104 Miwa, K., Ohtsubo, M., Denda, K., Hisabori, T., Date, T. and Yoshida, M. (1989) J. Biochem. 106, 679–683.
105 Rao, R. and Senior, A.E. (1987) J. Biol. Chem. 262, 17450–17454.
106 Aloise, P., Kagawa, Y. and Coleman, P.S. (1991) J. Biol. Chem. 266, 10368–10376.
107 Xue, Z., Melese, T., Stempel, K.E., Reedy, T.J. and Boyer, P.D. (1988) J. Biol. Chem. 263, 16880–16885.
108 Noumi, T., Tagaya, M., Miki-Takeda, K., Maeda, M., Fukui, T. and Futai, M. (1987) J. Biol. Chem. 262, 7686–7692.
109 Noumi, T., Taniai, M., Kanazawa, H. and Futai, M. (1986) J. Biol. Chem. 261, 9196–9201.
110 Melese, T. and Boyer, P.D. (1985) J. Biol. Chem. 260, 15398–15401.
111 Kohlbrenner, W.E. and Boyer, P.D. (1983) J. Biol. Chem. 258, 10881–10886.
112 Nalin, C.M., Snyder, B. and McCarty, R.E. (1985) Biochemistry 24, 2318–2324.
113 Shapiro, A.B. and McCarty, R.E. (1988) J. Biol. Chem. 263, 14160–14165.
114 Shapiro, A.B. and McCarty, R.E. (1990) J. Biol. Chem. 265, 4340–4347.
115 Hoppe, J. and Sebald, W. (1984) Biochim. Biophys. Acta 768, 1–27.
116 Cain, B.D. and Simoni, R.D. (1986) J. Biol. Chem. 261, 10043–10050.
117 Lightowlers, R.N., Howitt, S.M., Hatch, L., Gibson, F. and Cox, G. (1988) Biochim. Biophys. Acta 933, 241–248.
118 Paule, C.R. and Fillingame, R.H. (1989) Arch. Biochem. Biophys. 274, 270–284.
119 Hutton, R.L. and Boyer, P.D. (1979) J. Biol. Chem. 254, 9990–9993.
120 Wood, J.M., Wise, J.G., Senior, A.E., Futai, M. and Boyer, P.D. (1987) J. Biol. Chem. 262, 2180–2186.
121 Hammes, G.G. (1983) Trends Biochem. Sci. 8, 131–134.
122 Choate, G.L., Hutton, R.L. and Boyer, P.D. (1979) J. Biol. Chem. 254, 286–290.
123 Grubmeyer, C. and Penefsky, H.S. (1981) J. Biol. Chem. 256, 3718–3727.
124 Oosawa, F. and Hayashi, S. (1984) J. Phys. Soc. Jpn. 53, 1575–1579.

125 Mitchell, P. (1985) FEBS Lett. 182, 1–7.
126 Cross, R.L., Cunningham, D. and Tamura, J.K. (1984) Curr. Top. Cell. Regul. 24, 335–344.
127 Kandpal, R.P. and Boyer, P.D. (1987) Biochim. Biophys. Acta 890, 97–105.
128 Tozer, R.G. and Dunn, S.D. (1986) Eur. J. Biochem. 161, 513–518.
129 Bragg, P.D. and Hou, C. (1986) Biochim. Biophys. Acta 851, 385–394.
130 Musier, K.M. and Hammes, G.G. (1987) Biochemistry 26, 5982–5988.
131 Gogol, E.P., Johnston, E., Aggeler, R. and Capaldi, R.A. (1990) Proc. Natl. Acad. Sci. USA 87, 9585–9589.
132 Musier-Forsyth, K.M. and Hammes, G.G. (1990) Biochemistry 29, 3236–3241.
133 Moradi-Ameli, M. and Godinot, C. (1988) Biochim. Biophys. Acta 934, 269–273.
134 Kironde, F.A.S. and Cross, R.L. (1987) J. Biol. Chem. 262, 3488–3495.
135 Bragg, P.D. and Hou, C. (1990) Biochim. Biophys. Acta 1015, 216–222.

L. Ernster (Ed.) *Molecular Mechanisms in Bioenergetics*
© 1992 Elsevier Science Publishers B.V. All rights reserved

CHAPTER 14

Inorganic pyrophosphate and inorganic pyrophosphatases

MARGARETA BALTSCHEFFSKY and HERRICK BALTSCHEFFSKY

Department of Biochemistry, Arrhenius Laboratories, Stockholm University, S-106 91 Stockholm, Sweden

Contents

1. Introduction 331
2. PPases from prokaryotes 333
 2.1. Membrane bound PPases 333
  2.1.1. The H^+-PPase from *Rhodospirillum rubrum* 333
  2.1.2. From other sources 335
3. PPases from eukaryotes 335
 3.1. Mitochondrial PPases 335
  3.1.1. Animal mitochondrial PPases 336
  3.1.2. Yeast mitochondrial PPases 336
 3.2. Vacuolar (tonoplast), Golgi and plasma membrane PPases 336
  3.2.1. Vacuolar PPases 336
  3.2.2. Golgi and plasma membrane PPases 338
4. Structures and mechanisms 338
 4.1. Primary structures 338
 4.2. Three-dimensional structures. Mechanisms 339
5. PP_i as energy and phosphate donor 341
6. Some evolutionary aspects 341
7. Summary and outlook 343
Acknowledgements 345
References 345

1. Introduction

In 1984, the metabolism of inorganic pyrophosphate (PP_i), with particular emphasis on the bioenergetic aspects of PP_i reactions, was covered in an earlier volume in this series [1]. Since then a number of remarkable developments have expanded the field of PP_i bioenergetics in several new and also old directions. Some of the progress has been in areas primarily concerned with other aspects of PP_i metabolism, but has, nevertheless, been of apparent bioenergetic significance.

This presentation will be rather sharply focussed on those sectors where much has happened since 1984 and which have not been extensively covered in earlier reviews. These reviews have dealt with: PP_i and inorganic polyphosphates as sources of energy and donors of phosphate in bacteria [2], H^+-PPases bound to the 'non-energy-coupling' tonoplast membranes isolated from plant cell vacuoles [3], PP_i in mitochondrial metabolism [4,5], and the possible roles of PP_i in the molecular evolution of biological energy conversion [6].

We would like to try to elucidate for the more general reader some basic questions concerning PP_i and its functions in living cells. One early question has been whether PP_i can be formed in ways analogous to the formation of adenosine triphosphate (ATP). Reactions 1 and 2:

$$2P_i + \text{energy} \rightleftharpoons PP_i + H_2O, \tag{1}$$

$$P_i + ADP + \text{energy} \rightleftharpoons ATP + H_2O, \tag{2}$$

illustrate the biological fact that two orthophosphate (P_i) molecules can conserve energy from a suitable energy source by giving PP_i in a similar way as one P_i and one adenosine diphosphate (ADP) molecule can conserve energy by giving ATP, the well-known primary molecular 'energy currency' in living cells. This has been known since 1966, when it was shown that light in the presence of P_i can drive reaction 1 to the right in photophosphorylation of isolated bacterial chromatophores [7] in a similar way [8] as it can drive reaction 2 to the right in the presence of both P_i and ADP.

Similarly, again, as is well-known with ATP, also PP_i can function as an energy donor in various cellular energy requiring reactions [9–14], and also, as might have been expected, as a donor of phosphate [1,9], also in protein phosphorylation [15–17]. Thus, another early question, whether the energy liberated when reaction 1 goes to the left is always 'lost' as heat or whether it can be energetically utilised, has been answered in favour of the utilitarians, as has the question of whether PP_i can phosphorylate. What PP_i, on the other hand, of course, cannot do, in contrast to the more complex and thus more versatile ATP, is to adenylate or pyrophosphorylate. These additional capabilities of ATP, especially the adenylation reactions involved in various biosynthetic pathways, provide an immediate explanation for its most central metabolic position. It has long been common knowledge that PP_i, the by-product of the adenylation reactions:

$$ATP + X \rightleftharpoons AMP\text{–}X + PP_i, \tag{3}$$

by being hydrolysed to $2P_i$,

$$PP_i + H_2O \rightarrow 2P_i, \tag{4}$$

fulfills an important metabolic role, even when the energy liberated is 'lost' as heat, facilitating the biosynthesis by pulling the different reactions of type 3 in the direction of adenylation.

One more question should be asked in this connection: are the energy (Eqn. 1, going to the left) and phosphate donor reaction

$$PP_i + X \rightleftharpoons P\text{–}X + P_i \tag{5}$$

characteristics of PP_i of major significance in connection with its central role in cellular metabolism and if so, in prokaryotes or eukaryotes or both, or are these functions only a remnant, a kind of leftover appendix as it were, from a possibly much more pronounced position during earlier stages of life and its evolution, perhaps only in connection with its distant origin from prebiotic systems, about 4×10^9 years ago? As will be evident below, we are of the opinion, based on the current evidence, that the roles of PP_i in these respects lie somewhere between the two extreme suggestions, both in prokarotes and in eukaryotes.

Let us briefly mention here some of the more recent progress. The integrally membrane-bound and proton pumping inorganic pyrophosphatase (PPase; H^+-PPase; PP_i synthase; H^+-PP_i synthase), which in chromatophores from *Rhodospirillum rubrum* catalyses both light-induced formation of PP_i [7,8] and its utilisation in several energy requiring reactions [1,10–14], has recently been further purified and characterised [18]. Two major new developments, discussed below in detail, are the demonstration of the presence of and their description in eukaryotic plant cells of vacuolar membrane-bound H^+-PPases [3] and the cloning of genes for cytoplasmic PPases from both prokaryotes [19,20] and eukaryotes [21–23], for a mitochondrial PPase [24] and now also for a vacuolar PPase [25]. Of particular importance, also as related to the primary structures of the PPases thus obtained, is the X-ray crystallographic work on the 3D structure of cytoplasmic PPase from *Saccharomyces cerevisiae* [26–28]. The recent finding of a similar antibody response in the membrane-bound H^+-PPases from chromatophores of *Rsp. rubrum* and from vacuoles of mung bean [29] is of interest both from a structure determination and an evolutionary point of view.

2. *PPases from prokaryotes*

2.1. *Membrane bound PPases*

2.1.1. *The H^+-PPase from Rhodospirillum rubrum*

After early evidence, proving that PP_i may be involved in the bioenergetic processes of the photosynthetic bacterium *Rsp. rubrum* [7,12], it was shown that there exists a separate enzyme in this organism, capable of electrogenic charge transfer in the form of H^+, through the plasma membrane [30–33]. Also several other photosynthetic bacteria appear to have functional membrane bound, H^+ translocating PPases [34–37]. So far, of these enzymes, the *Rsp. rubrum* PPase has been solubilised, extensively purified and characterised [18], and the enzyme from *Rhodopseudomonas palustris* has been solubilised, partially purified and characterised [38]. Very recently, a membrane bound PPase has been solubilised and isolated also from membranes of dark grown, respiratory cells of *Rsp. rubrum* [39]. This enzyme indeed appears to be very similar to, or identical with, the chromatophore-derived PPase. Much of the early work leading to the characterisation of the *Rsp. rubrum* enzyme, and the various energy requiring reactions which may be driven by PP_i as an alternative to ATP was extensively reviewed in a previous volume [1].

The method for purifying the PPase has recently been considerably improved [18], yielding a preparation which reproducibly retains the *N,N'*-dicyclohexylcarbodiimide (DCCD) sensitivity of the original, membrane bound enzyme. The protein, as isolated, consists of a single polypeptide with an apparent molecular weight of 56 kDa. This single polypeptide contains both the catalytic activity and the H^+ pumping activity, as was demonstrated by reconstitution experiments in liposomes [18]. The molecular weight may well be an underestimate since the protein is extremely hydrophobic, and it is thus possible that the true molecular weight is closer to about 75 kDa. Some indications suggest that the enzyme has a dimeric structure. Its hydrophobic nature is also shown by the requirement for phospholipid or detergent for obtaining significant hydrolysing activity. Further investigation of the activating function of several different phospholipids has shown that cardiolipin is the most potent activator, giving an almost tenfold increase in activity [18]. The amount of PPase is only about 1% of the total protein in the chromatophore membrane [18].

In the previous review the main emphasis was on the hydrolytic reaction of the enzyme. Up to then very few reports had dealt with the *Rsp. rubrum* PPase as a synthase, with as notable exceptions refs. [7,8,31]. This was mainly due to the fact that accurate measurement of PP_i synthesis was technically difficult, especially in the presence of ATP. The study of PP_i-synthesis became greatly facilitated with the new method [40] for its continuous monitoring via the extremely sensitive, stabilised firefly luciferase system for measuring ATP. This system was coupled with the ATP:sulphate adenylyltransferase reaction, which with adenosine-5′-phosphosulphate (APS) catalyses a virtually quantitative conversion of PP_i to ATP:

$$PP_i + APS \rightarrow ATP + SO_4. \tag{6}$$

The ATP thus formed may then be measured with the firefly luciferase technique. What is measured is the total concentration in the medium, but this would include also the ATP formed concomitantly in photophosphorylation. By omitting ADP from the reaction medium, ATP synthesis is eliminated, except for that obtained in the transferase reaction, and thus the method allows exact measurement of PP_i synthesis. With ADP present it is possible to evaluate the total phosphorylating capacity of the system. With this method it has been possible to study in great detail the synthase function of the membrane bound PPase from *Rsp. rubrum* as well as from some other photosynthetic bacteria [41,42].

One main characteristic, which has emerged from these studies is that the PPase is completely competent as a PP_i synthase under a variety of conditions, but always operating at high capacity on a lower energy level than ATP.

It has earlier been shown [31] that the PP_i synthesis is maximal at lower light intensities than ATP synthesis. In fact, at very low light intensities the rate of PP_i synthesis is up to twice that of ATP synthesis. This is very different from the ratio between the two reactions at high light intensities, where the PP_i synthesis only contributes 10–15% of the total photophosphorylation [41].

It was also shown [41] that electron transport in the cytochrome b/c_1 complex, when inhibited by antimycin, suffices to maintain a low rate of PP_i synthesis, about 30% of the maximal, whereas it is well known that ATP synthesis is almost completely inhibited. This supports the concept that PP_i, in agreement with its thermodynamic properties, is working at a lower energy level than ATP.

A finding apparently reflecting the physiological situation was the competition for the available $\Delta\mu_{H^+}$, between the PP_i synthase and the ATP synthase activities [41]. Inhibiting specifically one invariably enhanced the other, indicating that in vivo both enzymes are working simultaneously.

With an artificial pH gradient as the driving force for synthesis of PP_i and ATP, several main results have been obtained. One is that the yield of PP_i is about tenfold that of ATP. The second is that in contrast to the case with ATPase, there is no requirement for an activation by a $\Delta\Psi$ of the PPase. A third finding from this work is that the actual rate with which PP_i is released into the medium is much lower than the rate of appearance in the medium of ATP, even if the total amount of PP_i synthesised is much larger [43].

In flash-induced photophosphorylation [44] the yield of PP_i after a single 1 ms flash, is 150–200% of that of ATP under the same conditions. Also in this case the actual rate of PP_i synthesis was considerably lower than the rate of ATP synthesis, 0.23 μmol PP_i/min per μmol bacteriochlorophyll (Bchl) as compared with >1 μmol ATP/min per μmol Bchl. In contrast to PP_i synthesis under continuous illumination, flash induced PP_i synthesis is very sensitive to inhibition by valinomycin and K^+. This is probably due to the fact that the main component of $\Delta\mu_{H^+}$ after a single flash is the membrane potential. Under continuous illumination it has been shown that nigericin, which specifically causes the H^+ gradient to collapse, has a stronger effect on PP_i synthesis than on ATP synthesis [41]. This has been interpreted as indicating that ΔpH is the predominant driving force for PP_i synthesis, which is in line with the findings with an artificial pH gradient as the sole driving force for PP_i synthesis.

The interconvertibility of $\Delta\mu_{H^+}$ between a number of chromatophore-bound reactions and the synthesis/hydrolysis of PP_i (and ATP) has been further manifested by the demonstration of the synthesis of PP_i and ATP by reversal of the energy-linked transhydrogenase reaction [45], albeit at a low rate. As was the case at very low light intensities, the ATP synthesis proceeds at a lower rate than PP_i synthesis, about 70%, which shows that the available energy is very low.

The rates of PP_i synthesis and hydrolysis, respectively, have been studied as a function of $\Delta\mu_{H^+}$. There is a drastic increase in the synthesis rate, > 45-fold, when a $\Delta\mu_{H^+}$ is built up under illumination as compared to unenergised conditions, while the rate of hydrolysis decreases by a factor of 4–8 [46]. As has earlier been shown for ATPase [47], the forward and reverse rate constants are not altered by the same factor with a change in the electrochemical potential. This indicates an asymmetric architecture of the enzyme in the membrane, with the catalytic site localised on the side of the membrane that becomes negative during energisation.

A recent hypothesis about the physiological role of *Rsp. rubrum* H^+-PPase suggests that the main role of the enzyme is to maintain a proton motive force in darkness through the H^+ translocating hydrolysis of PP_i [48].

2.1.2. From other sources

Through the years there have been various indications for the presence of H^+-PPases in several species of photosynthetic bacteria [34–38]. These came from experiments using PP_i as an energy source for NAD^+ reduction in *Rhodopseudomonas viridis* [34], or for carotenoid band shift in *Rhodobacter sphaeroides* [35] and *Chromatium vinosum* [37]. Uncoupler stimulated PPase activity has been found in *Chr. vinosum* [37] and in *Rps. palustris* [38]. PP_i has been shown to drive ATP synthesis in chromatophores from *Rps. palustris* [38]. The PPase from *Rps. palustris* seems to be rather similar to the *Rsp. rubrum* enzyme, with one notable exception. It is not inhibited by fluoride to any significant extent. Light-induced PP_i synthesis has recently been found in *Rps. viridis* [42] both under continuous light and after single light flashes. The rate of PP_i synthesis is, however, considerably lower than in *Rsp. rubrum* chromatophores.

3. PPases from eukaryotes

3.1. Mitochondrial PPases

With reference to already cited review articles [1,4,5], our treatment of mitochondrial PPases will essentially be limited to certain progress which has not been covered in these articles. In general, new information on metal and substrate binding and effects of inhibitors and uncouplers, has been obtained. Cloning and sequence determination has been possible so far in only one case, with the mitochondrial PPase from *S. cerevisiae* [24].

3.1.1. *Animal mitochondrial PPases*

In recent studies aimed at further elucidating the properties of the PPases of animal mitochondria, two forms of PPases were described, a multisubunit membrane and a matrix form, with the matrix PPase representing the catalytic part of the membrane PPase [49]. Ca^{2+} was found to inhibit the PPases of bovine heart and rat liver mitochondria uncompetitively, with respect to substrate $MgPP_i$ at pH 8.5 [49]. At pH 7.2 the bovine heart PPase was noncompetitively inhibited. Only about 10 μM Ca^{2+} was required to decrease the maximal velocities at pH 8.5 and 0.4 mM Mg^{2+}. The possible physiological significance of this inhibition of mitochondrial PPases by comparatively low concentrations of Ca^{2+}, for example with respect to hormonal stimulation of the oxidative metabolism [50], is still an open question.

The catalytic part of the membrane PPase in rat liver mitochondria had earlier been reported to consist of two subunits with molecular weights 28 and 35 kDa [51]. Detailed kinetic models for the activities of PPases from rat liver mitochondria and cytosol are valuable but outside the scope of this review, and are thus only referred to [52].

3.1.2. *Yeast mitochondrial PPase*

Early studies have shown that PP_i in yeast mitochondria is capable of driving energy requiring reversed electron transport from *c*-type to *b*-type cytochrome [53]. Later, yeast mitochondria were reported to make PP_i in oxidative phosphorylation [54]. Recently a yeast mitochondrial PPase was characterised [55] and cloned [24]. Some of its functional properties will be discussed below, its structural characteristics will be presented in Section 4.1.

The yeast mitochondrial PPase activity was strongly stimulated by uncouplers and ionophores. The stimulation was much higher in isotonic than in hypotonic suspension, indicating the requirement of intact membranes for maximal effects. The 20–50-fold higher PPase activities from a PPase overproducing strain were also strongly stimulated by uncoupler. These results clearly indicate that the yeast mitochondrial PPase is energy linked. As a hydropathy plot did not show any extended hydrophobic stretches, it is reasonable to assume that this mitochondrial PPase also performs its energy linked functions connected to integrally membrane bound polypeptides, situated in the inner mitochondrial membrane [24,55].

3.2. *Vacuolar (tonoplast), Golgi and plasma membrane PPases*

3.2.1. *Vacuolar PPases*

Tonoplast vesicles obtained from the vacuolar membrane of higher plant cells have been shown to contain two distinct proton pumps, an H^+-ATPase and an H^+-PPase. Their H^+-translocating and other properties were well reviewed [3] in 1987. Among the steps on the road to current basic knowledge about the H^+-PPase of vacuolar membranes, which can incisively be described as *known sites for energy transduction but not for energy coupling*, were the early demonstration of a microsomal membrane-bound K^+-stimulated PPase from sugar beet [56], and the chromatographic resolution of two enzymes, the H^+-PPase and the H^+ ATPase, from tonoplast membranes of red beet, confirming that they are separate entities [57].

In the last few years our knowledge about the vacuolar H^+-PPases, which are ubiquitously found in vacuolar membranes of higher plants and also of yeast [58] and primitive unicellular algae [59], has developed rapidly. Yeast vacuoles have recently been shown to contain two distinct PPases: a homotrimeric membrane bound H^+-PPase glycoprotein with a subunit M = 41 kDa and a homotrimeric soluble PPase glycoprotein with a subunit M = 28 kDa [59a].

The active subunits of vacuolar H^+-PPases from mung bean [60] and red beet [61] were identified, partially purified and characterised in 1989. Their molecular masses were given as 73 kDa and 64 kDa, respectively. Earlier 64 kDa had been given as the molecular weight of a microsomal H^+ PPase from *Triticum* [62].

The mung bean PPase was inhibited by DCCD. It was shown to bind to the enzyme, which thus was suggested to contain a region forming a proton channel in itself [60]. All but one of the first 30 amino-terminal amino acids were determined by protein sequencing [60]. There is homology to the *A. thaliana* H^+-PPase but not to known cytoplasmic or mitochondral PPases.

N-ethylmaleimide (NEM) was bound to, and found to inhibit, the PPase activity of red beet (storage root) with pseudo-first-order kinetics. Whereas $MgPP_i$ almost fully protected the PPase from inhibition of NEM, free PP_i increased the inhibitory potency of NEM nearly twofold (Mg alone gave 20% protection) [61]. These results bear some similarity to early results with the H^+-PPase from *Rsp. rubrum* [63].

The K^+-dependence of the vacuolar H^+-PPase [64] indicated that a function of the enzyme was to translocate K^+ across the tonoplast. This possibility was recently tested by the patch clamp method using whole vacuoles, which allowed the monitoring of PP_i-dependent currents. A selective, vectorial activation of the H^+-PPase by cytosolic K^+ was demonstrated [65], which is in line with the in vivo direction of the electrochemical gradient for K^+ over the tonoplast. This strongly supports the assumption that the H^+-PPase can drive K^+ transport.

The vacuolar H^+-PPase from both mung bean [66] and pumpkin seedlings [67] appears to exist in dimeric form in the membrane.

In a recent comparison [68] of the effects of Ca^{2+} and Mg^{2+} ions on solubilised and purified H^+-PPase from mung bean vacuoles with those on other PPases, the expected importance of Mg^{2+} ions for both enzyme and substrate and the inhibition of the H^+-PPase activity by Ca^{2+} ions were established. The enzyme was stabilised and activated by Mg^{2+} ions, with the apparent K_m for $MgPP_i$ being about 130 μM. In view of the inhibitory effect of Ca^{2+} on the vacuolar H^+-PPase, the recent report that a PP_i-driven Ca^{2+}/H^+ antiport exists in tonoplasts from maize primary roots deserves particular attention [69]. Another interesting report has been about a PP_i activated ATP-dependent H^+ transport in the tonoplast of mesophyll cells of the CAM plant *Kalancoë daigremontiana* [70]. Preenergisation with PP_i, only when the PPase was active, strongly stimulated the initial rates of ATP-dependent H^+ transport, whereas PP_i-dependent H^+ transport was not stimulated with ATP. The PP_i activated H^+ transport shows properties of both PP_i- and ATP-induced H^+ transport activities, for example the chloride and malate stimulation of ATP-dependent and the nitrate stimulation and the KF inhibition of PP_i-dependent H^+ transport activities. Clearly, these recent results suggest that PP_i in membranes containing both H^+-PPase and H^+-ATPase may be of hitherto unknown importance in the kinetic regulation of ATPase activity.

The antibody to the vacuolar H^+-PPase did not react with rat liver mitochondrial or yeast cytosolic PPases, and the antibody to the yeast cytosolic PPase did not react with the vacuolar enzyme [68]. In this light, the recent demonstration that there is immunological cross-reactivity between the mung bean vacuolar H^+-PPase and the chromatophore membrane bound H^+-PPase, but not the soluble PPase, from *Rsp. rubrum* [29] is of special significance. This cross-reactivity between an energy transducing H^+-PPase from a plant vacuolar membrane and an energy transducing and energy coupling H^+-PPase from the chromatophore membrane of a photosynthetic bacterium indicates some similarity between the structural requirements for the respective energy linked functions of these prokaryotic and eukaryotic membrane bound H^+-PPases.

The very recent cloning and sequencing of cDNAs encoding the H^+-PPase from the vacuolar membrane of *A. thaliana* is a most significant step forward in our knowledge about proton pumping PPases [25]. The predicted protein consists of 770 amino acids (80 kDa) and appears to contain at least 13 transmembrane spans. It lacks sequence homology with other known H^+ pumps or PP_i metabolising enzymes. Due to the demonstrated immunological cross-reactivity between the mung bean vacuolar H^+-PPase and the H^+-PPase from *Rsp. rubrum* chromatophores [29], and the reported cross-reactivity between the *A. thaliana* H^+-PPase and that from chromatophores [25], one may speculate that these prokaryotic and eukaryotic H^+-PPases constitute a distinct family of H^+-PPases.

3.2.2. Golgi and plasma membrane PPases

Three Mg^{2+}-PPases were recently shown to exist in three different membrane fractions obtained from microsomes of Jerusalem artichoke (*Helianthus tuberosus* L.) tubers by separation on a linear sucrose gradient: tonoplast (d = 1.11–1.12), Golgi (d = 1.14–1.15) and plasma membrane (d = 1.17–1.18) Mg^{2+}-PPases [71]. The activities overlapped with corresponding ATPase activities and, as expected, were not affected by vanadate. In agreement with this result are other indications that Golgi [72] and plasma [73] membranes are sites for PPase activity. The degree of structural and functional similarity between Golgi and plasma membrane PPases and vacuolar and bacterial chromatophore H^+-PPases is an important research problem for the near future.

4. Structures and mechanisms

4.1. Primary structures

In the last few years cloning and DNA sequencing have led to clarification of the complete primary structures of several PPases. Since 1988, when the amino acid sequence of the soluble, cytoplasmic PPase from *Escherichia coli* was determined [19], the sequences of the cytoplasmic PPases from *Saccharomyces cerevisiae* [21], *Kluyveromyces lactis* [22], *Shizosaccharomyces pombe* [74], the thermophilic bacterium PS-3 [20] and *Arabidopsis thaliana* [23], the mitochondrial PPase from *Saccharomyces cerevisiae* [24] and now also the vacuolar H^+-PPase from *Arabidopsis thaliana* [25], have become known. Whereas the mitochondrial PPase belongs to the same protein family as all the known soluble prokaryotic and eukaryotic PPases [24], the known vacuolar H^+-PPase is very different [25].

An archaebacterial, soluble PPase has very recently been isolated, purified to homogeneity and sequenced [75]. This enzyme, from *Thermoplasma acidophilum*, is extremely specific for Mg^{2+}-PP_i, the pH optimum is 6.2–6.5 and the temperature optimum 85°C. The native enzyme is a homo-hexamer of a 21.451 kDa monomer. It shows striking similarity to the PPase from the eubacterium PS-3 [75].

A soluble, cytoplasmic PPase from the phototrophic bacterium *Rps. palustris* has been purified to apparent homogeneity and kinetically characterised. It is dimeric under non-dissociating conditions and consists of two 32 kDa subunits. The enzyme is specific for PP_i and is activated by divalent cations. Interestingly, Ca^{2+} is a more potent activator than Mn^{2+} and Mg^{2+} [76].

A comparison between the primary structures of the yeast mitochondrial PPase and four cytoplasmic PPases, three from different yeasts and the fourth from *E. coli*, showed that the mitochondrial PPase is rather closely related to the three very closely related soluble PPases from yeasts, whereas all the four yeast PPases were much more distantly related to the *E. coli* PPase [24]. A detailed comparison of the amino acid sequences of the cytoplasmic PPases from *E. coli* and *S. cerevisiae* had already shown that the smaller, eubacterial enzyme, when aligned

```
-HPETKAV----GDNDPLDVLEIG-        K. lactis
-HPETKAV----GDNDPIDVLQIG-        S. cerevisiae (cytoplasmic)
-HPETKA----KGDSDPLDVCEIG-        S. pombe
-HKLGKCDVALKGDNDPLDCCEIG-        S. cerevisiae (mitochondrial)
-IPRTICE-----DSDPMDVLVLM-        A. thaliana
-HTLSL-------DGDPVDVLVPT-        E. coli
-NTLAL-------DGDPLDILVIT-        PS-3
```

Fig. 1. Aligned segments of one mitchondrial and six cytoplasmic PPases. The aligned Asp residues appear to be functional in the active site. The first and the third of the three Cys residues of the mitchondrial PPase have their counterparts in the segments *A. thaliana* and *S. pombe*, respectively.

with residues 28–225 of the yeast enzyme, was only moderately identical (22–27%, depending on the alignment parameter choice) with them, but that 14 to 16 of some 17 putative active site residues were identical [77].

As was also pointed out, the PPases from these widely divergent species, one prokaryotic and one eukaryotic, thus seem to have conserved functional amino acids within the context of substantial overall sequence variation [77]. One region with particularly high similarity score and high density of putative active site residues is that of the yeast PPase residues 115–122 [77]. Part of this segment has earlier been observed to be similar to conserved sequences in H^+-ATPases [78], especially with respect to the amino acid sequence

–Asp(Glu)–Pro–hydrophobic–Asp(Glu)–.

Both in adenine nucleotide and in PP_i binding enzymes the latter dicarboxylic acid of this sequence has been suggested to bind to a metal involved in the reversible phosphorylation reaction [26,79].

The sequence discussed above is of particular interest when a larger segment of the yeast mitochondrial PPase is compared with the corresponding parts of the known soluble PPases. As is shown in Fig. 1, the PPase from yeast mitochondria uniquely has this segment, nearly adjacent to some additional amino acid residues only found in the mitochondrial enzyme, surrounded by three Cys residues. The first of them is also found in the *A. thaliana* sequence and the third in the *S. pombe* sequence, which are adjacent to the yeast mitochondrial sequence in Fig. 1. As is shown in Fig. 2, a 15 amino-acid loop may possibly be found in a presumed disulphide bonding between the first and third of the Cys residues of the mitochondrial PPase [24], with some pronounced similarity to a corresponding loop in several ligand gated ion channels (LGICs) [80]. Reading clockwise in the potential loop one finds the requirement for an h (hydrophobic residue) and L (Leu), as well as a –P–h–D– sequence of the LGICs, satisfied in the mitochondrial PPase, however with the –P–h–D– sequence occurring two sites later in the mitochondrial PPase than in the LGICs. The possible significance of these similarities is still unknown, and computer modelling of the mitochondrial PPase based on the known 3D structure of the homologous yeast cytoplasmic PPase indicates that the distance between the two Cys residues, which would form the disulphide and the loop, is too large for it to be formed [81].

4.2. Three-dimensional structures. Mechanisms

A fuller understanding of the mechanisms involved in PPase reactions will have to await the finding of high resolution 3D structures of these enzymes in a further extension of the work

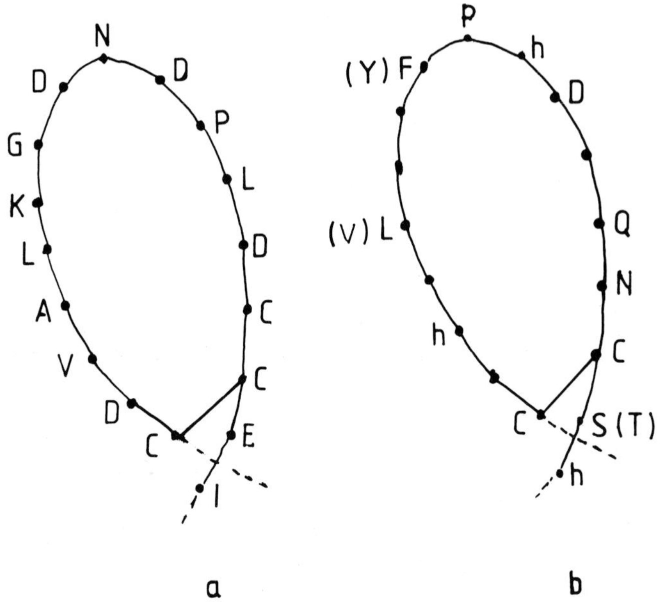

Fig. 2. Potential Cys–Cys loop similarities and differences between the yeast mitochondrial PPase (A) and ligand gated ion channels (B).

with the yeast cytoplasmic PPase [26–28]. An early model for the catalytic mechanism of the PPase [82] included leaving group activation and charge shielding of the electrophilic phosphoryl group allowing attack by an anionic nucleophile (an Asp or Glu residue). It has been complemented with detailed comparisons of active site residues [77]. Chemical modification [83] and site-specific mutation [84,85] have provided important new information about the functional and structural roles of the selected amino acid residues. For example, the strict requirement of Asp^{102} (the last amino acid in *E. coli* PPase of the four-amino acid-sequence above) for catalytic activity was demonstrated by substituting this Asp with Val or Glu, which resulted in, respectively, complete or nearly complete elimination of activity [84].

Recent studies of the binding of phosphate and Mg^{2+} to yeast cytoplasmic PPase suggest that the active roles of the enzyme- and substrate-bound Mg^{2+} are different for the hydrolysis and synthesis of PP_i [86]. The discovery and investigation of tightly bound PP_i in hexameric *E. coli* PPase indicated that this soluble prokaryotic enzyme has two types of PP_i-binding sites (one high-affinity and one low-affinity site, only the latter being located at the active site region) [87]. The possible relevance of this finding in connection with the corresponding situation in H^+-ATPases remains to be elucidated.

The early claim that the Arg^{77} residue of the yeast cytoplasmic PPase strongly binds a phosphate residue [82] was supported by the 3D studies [26,28]. These have provided the first rather detailed picture of the active site, which is characterised by containing several polar groups, with acidic groups in contact with bound Mg^{2+} ions occupying central positions for the PPase activity. More detailed descriptions of the active site and the reaction mechanism will need the availability of a higher resolution 3D structure. Very desirable would also be 3D structures of other PPases, for example from prokaryotes and mitochondria.

5. PP_i as energy and phosphate donor

It has long been known, and has been discussed above and also in earlier reviews [1–6], that PP_i is capable of functioning as a biological donor of useful chemical energy and of a phosphate residue in various phosphorylation reactions. As compared to ATP, PP_i releases less free energy during its hydrolysis, which has been the main reason for sometimes naming it 'poor man's ATP'. Another factor determining the actual physiological significance of these capabilities of PP_i, is the 'steady-state' concentration of PP_i in living cells.

There is considerable variation in estimates of cellular PP_i concentrations. Under saturating energy supply, exponentially growing bacterial (*E. coli*) cells may keep the intracellular PP_i concentration at about 500 μM [88]. In animal and plant cells one finds significantly lower values. Rat tissues were found to contain from 2.2 ± 0.1 nmol/g wet weight (skeletal muscle) up to 11.1 ± 0.3 and 13.1 ± 0.7 nmol/g wet weight heart and liver, respectively [89]. For higher plant cells one finds PP_i levels in pea and corn tissues from 5–39 nmol/g fresh tissue weight, depending on the tissue [90]. Considering the various constants given for the actual reactions one is led to the conclusion that cellular energetics, and phosphorylations, at least in bacteria and plants, may be assumed to have appreciable contributions from PP_i in those cases where the appropriate enzymes are active (see, for example, ref. [3]).

The situation has been well elucidated in a detailed study of the subcellular localisation of PP_i and PPase in spinach leaves and wheat mesophyll protoplasts [91]. PP_i was found to be located predominantly in the cytosol, where its concentration was 200–300 μM. The PPase was, however, largely, if not exclusively, located in the chloroplast. By taking into account earlier data on the cytosolic levels of phosphate and metabolic intermediates it was shown that the PP_i:fructose-6-phosphate transferase and the UDP-glucose pyrophosphorylase are close to the thermodynamic equilibrium in vivo. By comparison of the PP_i levels with the reported electrical and pH gradients across the tonoplast membrane it could be shown that the free energy released by the hydrolysis of PP_i was sufficient to allow the suggestion that it could function as a secondary energy donor in the cytosol of plant cells. In addition to establishing this fact, the investigation demonstrated the importance of being able to determine actual PP_i concentrations in different cellular compartments.

The comparatively high cytosolic concentration of PP_i in spinach leaves is of interest also in relation to its recently observed capability to function as a phosphate donor, an alternative for ATP, in protein kinase reactions of the photosynthetic apparatus [16]. The similarities found between PP_i induced phosphorylation of spinach thylakoid proteins and of yeast mitochondrial protein [17] would seem to indicate similar physiological roles for PP_i in kinase reactions of these organelles.

6. Some evolutionary aspects

In 1965 it was pointed out that 'generation of the phosphate group potential might have originated with inorganic pyrophosphate as the primitive group carrier' [92].

Soon after our demonstration of light-induced phosphorylation of P_i with P_i to PP_i in bacterial chromatophores [7] and of the capacity of PP_i to be an energy donor in this system for several energy requiring reactions [12], we discussed in some detail [93] the possible role of PP_i as an early energy carrier, perhaps preceding ATP. In this connection also more high-molecular-weight inorganic polyphosphates have been discussed.

More recent considerations have included a link in early energy conversion reactions between PP_i and a 'thioester world' [94], as exemplified by an acetylthioester, and with acetylphosphate as the link:

$$PP_i \rightleftharpoons \text{Acetylphosphate} \rightleftharpoons \text{Acetyl-S-R}. \tag{7}$$

(The three configurations $-\overset{\|}{\underset{|}{P}}-O-\overset{\|}{\underset{|}{P}}-$, $-\overset{\|}{C}-O-\overset{\|}{\underset{|}{P}}-$ and $-\overset{\|}{C}-S-R$ are well known to be more or less equivalent energetically).

Independent support for the possibility of such early energy transfer at the pre-nucleotide level has recently appeared from two sources, of more geological than biochemical nature. They both concern the availability of condensed inorganic phosphates in the early, prebiological world, where even the availability of scarcely occurring phosphate compounds as such is considered to be a problem.

The first of these two findings shows the relative stability of PP_i under natural conditions, and highlights the well-known fact that a thermodynamically feasible hydrolytic reaction does not have to occur at an easily measurable rate, due to the barrier posed by a high energy of activation. It is the demonstration in 1988 of a PP_i-containing mineral, canaphite, with the structure $CaNa_2P_2O_7 \cdot 4H_2O$ [95].

The second is very recently found evidence for the production of PP_i and higher polyphosphates, both in hot volcanic gas (from several fumaroles at Mount Usu, Hokkaido, Japan) collected at temperatures of 540–690°C, and in model experiments simulating magmatic conditions [96]. In a typical sample obtained by column chromatography of a volcanic gas condensate, there was 1 μM P_i before and 6–9 μM P_i after hydrolysis. After neutralisation, filtration and ion exchange chromatography of a sample, the quantitative analysis showed the ratio P_i:PP_i:PPP_i to be 3:1.2:1 (1.13, 0.45 and 0.37 μM), respectively. The production of water-soluble pyrophosphate and polyphosphate occurs through partial hydrolysis of P_4O_{10}. This appears to be the only so far identified geological process, which could lead to the above-mentioned condensed phosphate compounds on the primitive earth.

With sources for the continuous production and recycling of PP_i and higher inorganic polyphosphates by volcanic processes thus identified [96], these compounds emerge even more strongly than before as potential candidates for being pivotal sources for reactions requiring energy and/or phosphate in connection with the origin and early evolution of life on earth. Table I shows a comparison of PP_i, inorganic polyphosphates and ATP, as potential candidates for prebiological and early biological carriers and donors of energy and phosphate. On

TABLE I

Evaluation, with four parameters, of PP_i, inorganic polyphosphate and ATP as potential candidates for being prebiological and/or early biological carriers and donors of energy and phosphate

Parameter	PP_i	Inorganic polyphosphate	ATP
Known biological energy donor	+	–	+
Known biological phosphate donor	+	+	+
Known occurrence as mineral	+	–	–
Indicated continuous formation on the early earth	+	+	–
Sum of +	4	2	2

the basis of the four selected parameters, PP_i clearly emerges as the most likely candidate among these compounds, and other possible candidates appear to be difficult to find.

What further links to this new and seemingly plausible background for prebiological and early biological energy conversion and phosphorylation would appear to be of particular importance to seek in currently living organisms? It may well be possible to find additional links in investigations of enzymes and pathways of energy conversion and phosphate metabolism in various archaebacteria as well as in photo- and chemolithotrophic bacteria.

7. *Summary and outlook*

A picture is currently emerging, not in the least thanks to the recent clonings of the mitochondrial [24] and vacuolar [25] PPases, of two fundamentally different types, and evolutionary pathways, of PPases. The sequence homologies between cytoplasmic PPases from archaebacteria, eubacteria and eukaryotes, and the yeast mitochondrial PPase, which appears to be energy linked, indicate conservation of early basic PPase characteristics. On the other hand, the vacuolar PPase was found to be very different [27].

The immunological cross-reactivity between photobacterial and plant vacuolar H^+-PPases would appear to allow the guess that they belong to a new, distinct family of energy linked PPases. It remains to be seen how closely related these prokaryotic and eukaryotic enzymes are when the *Rsp. rubrum* primary structure becomes known. This should also allow scrutiny of the possibility that the bacterial PPase contains regions uniquely necessary for its energy coupling function.

Figure 3 is an attempt to summarise the discussion by presenting in schematic form a very tentative picture of two classes of PPases, which are of significance in connection with energy conversion. Although some evolutionary relationships exist between PPases [24,77], the scheme should not be read in an evolutionary sense.

Assuming a relationship between the yeast mitochondrial PPase (M = 32 kDa) and the catalytic subunit(s) of the animal mitochondrial PPases (28 and/or 35 kDa), recalling that the active forms of several cytoplasmic prokaryotic PPases are known to have six identical catalytic subunits and some eukaryotic two, and recalling also that the α- and β-subunits of F_1-ATPases seem to have a common, catalytic ancestor, one may ask some quite unconventional questions. Have there been, are there still existing, cells with six identical catalytical subunits in membrane bound, energy transferring PPases and/or ATPases? Has there been a gene duplication, increasing subunit size from about 28–32 kDa to about 56–64 kDa, the former found in present-day cytoplasmic and mitochondrial PPases, the latter in α- and β-subunits of F_1-ATPases? As long as great gaps in our knowledge remain about details of the active sites and other structural properties of PPases and F_1-ATPases, the answers will be lacking.

Two observations could, however, be of relevance in this context. The first is that the well conserved, catalytically important stretch –Asp(Glu)–Pro–hydrophobic–Asp(Glu)– of all known PPases except the vacuolar one, occurs twice, once per half-molecule, in the β-subunits of many F_1-ATPases. The second is the recent evidence indicating a high degree of conservation of elements of the active site in ATP- and PP_i-dependent 6-phosphofructo-1- kinases (PFKs), suggesting that PP_i binds to the same site in PP_i-dependent PFK of *Propionibacterium freudenreichii* as the ATP site in ATP-dependent PFK of *E. coli* [97]. So, one may with continued fascination keep looking for clear indications of evolutionary links between energy coupling PPases and ATPases.

As was pointed out in the introduction, certain criteria have steered our decisions about

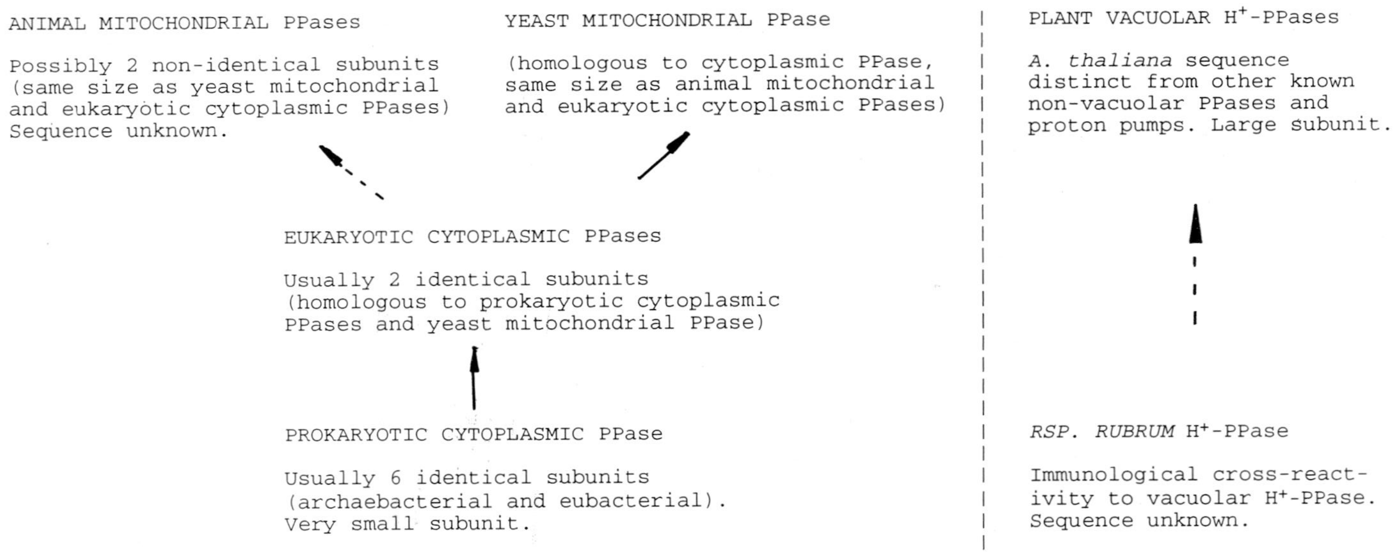
ANIMAL MITOCHONDRIAL PPases
Possibly 2 non-identical subunits (same size as yeast mitochondrial and eukaryotic cytoplasmic PPases) Sequence unknown.
YEAST MITOCHONDRIAL PPase
(homologous to cytoplasmic PPase, same size as animal mitochondrial and eukaryotic cytoplasmic PPases)
EUKARYOTIC CYTOPLASMIC PPases
Usually 2 identical subunits (homologous to prokaryotic cytoplasmic PPases and yeast mitochondrial PPase)
PROKARYOTIC CYTOPLASMIC PPase
Usually 6 identical subunits (archaebacterial and eubacterial). Very small subunit.
PLANT VACUOLAR H+-PPases
A. thaliana sequence distinct from other known non-vacuolar PPases and proton pumps. Large subunit.
RSP. RUBRUM H+-PPase
Immunological cross-reactivity to vacuolar H+-PPase. Sequence unknown.

Fig. 3. Very tentative scheme for two apparent classes of PPases, each of significance in connection with energy conversion. The solid arrows indicate existence of known homologies, the dashed arrows possible homologies.

what to include and what not, in this review. Thus, some sectors of PP_i metabolism have been excluded, such as PP_i in human disease (chondrocalcinosis [98], osteoporosis [99]) and its therapy (of urolithiasis [100]), due to insufficient bioenergetic connection. Also, higher plant chloroplast and mitochondrial PPases have been left out, due to comparatively limited fundamental progress, nothwithstanding, for example, new information about uncoupler stimulation [101] and energy-dependent formation of PP_i [102], respectively.

In general, a reasonable working hypothesis appears to be that, where PP_i-specific reactions are found, they may be of at least potential physiological significance. The so far discovered roles for PP_i in energy transduction and energy coupling, as well as the regulatory effect seen by PP_i on ATP-induced energy transduction, clearly indicate that, in bioenergetic systems, PP_i is involved in other ways than as participant in a 'futile cycle'.

Summarising, as has been discussed above, active and productive research on PP_i and PPases has in recent years led to a wealth of new information of bioenergetic interest. With the PPases involved either in both energy transduction and energy coupling or solely in energy transduction, the new knowledge may be claimed to be more bioenergetically significant than with soluble PPases. However, the rapidly advancing research on molecular structures and reaction mechanisms of soluble PPases may well, in our opinion, become increasingly valuable also from a bioenergetic point of view – for example, as comparatively uncomplicated model systems for basic parameters of biological energy coupling and phosphorylation, involving PP_i but with possible repercussions also for the systems involving ATP.

Finally, whereas about a decade ago only a few laboratories were pursuing investigations on PP_i and PPases, today research groups in more than ten countries are contributing to the current advances. The increasingly active international network which at present is developing between these laboratories may well be making further acceleration possible of the emergence of answers to the numerous questions still remaining, about how cellular PPases contribute to the formation and utilisation of PP_i.

Acknowledgements

Very important contributions to the more recent work presented from our research groups have been made by the former graduate students, Drs. Jeff Boork, Maria Lundin, Beston F. Nore, Pål Nyrén and Åke Strid. We would like also to thank Profs. A.P. Halestrap (Bristol), R. Lahti (Turku), M. Maeshima (Sapporo), P.A. Rea (Philadelphia), G. Schäfer (Lübeck) and J.E. Seegmiller (La Jolla) for valuable discussions and for kindly providing us with material prior to its publication.

This review was written during our 'sabbatical' at the Department of Biochemistry, University of Oxford. We wish to thank Professor George K. Radda and his staff for their active interest and support.

References

1 Baltscheffsky, M. and Nyrén, P. (1984) in: L. Ernster (Ed.) Bioenergetics, Ch. 6, Elsevier, Amsterdam, pp. 187–206.
2 Wood, H.G. (1985) Curr. Top. Cell. Regul. 26, 355–369.
3 Rea, P.A. and Sanders, D. (1987) Physiol. Plant. 71, 131–141.
4 Halestrap, A.P. (1989) Biochim. Biophys. Acta 973, 355–382.

5 Mansurova, S.E. (1989) Biochim. Biophys. Acta 977, 237–247.
6 Baltscheffsky, H., Lundin, M., Luxemburg, C., Nyrén, P. and Baltscheffsky, M. (1986) Chem. Scr. B 26, 259–262.
7 Baltscheffsky, H., von Stedingk, L.-V., Heldt, H.W. and Klingenberg, M. (1966) Science 153, 1120–1123.
8 Baltscheffsky, H. and von Stedingk, L.-V. (1966) Biochem. Biophys. Res. Commun. 22, 722–728.
9 Siu, P.M.L. and Wood, H.G. (1962) J. Biol. Chem. 237, 3044–3051.
10 Baltscheffsky, M., Baltscheffsky, H. and von Stedingk, L.-V. (1966) Brookhaven Symp. Biol. 19, 246–257.
11 Keister, D.L. and Yike, N.J. (1967) Biochem. Biophys. Res. Commun. 24, 519–525.
12 Baltscheffsky, M. (1967) Nature 216, 241–243.
13 Keister, D.L. and Yike, N.J. (1967) Arch. Biochem. Biophys. 121, 415–422.
14 Keister, D.L. and Minton, N.J. (1971) Arch. Biochem. Biophys. 147, 330–338.
15 Lam, K.S. and Kasper, C.B. (1980) Proc. Natl. Acad. Sci. USA 77, 1927–1931.
16 Pramanik, A.M., Bingsmark, S., Lindahl, M., Baltscheffsky, H., Baltscheffsky, M. and Andersson, B. (1991) Eur. J. Biochem. 198, 183–186.
17 Pereira da Silva, L., Lindahl, M., Lundin, M. and Baltscheffsky, H. (1991) Biochem. Biophys Res. Commun. 178, 1359–1364.
18 Nyrén, P., Nore, B.F. and Strid, Å. (1991) Biochemistry 30, 2883–2887.
19 Lahti, R., Pitkäranta, T., Valve, E., Ilta, I., Kukko-Kalske, E. and Heinonen, J. (1988) J. Bacteriol. 170, 5901–5907.
20 Ichiba, T., Takenaka, O., Samejima, T. and Hachimori, A. (1990) J. Biochem. 108, 572–578.
21 Kolakowski, L.F., Schloesser, M. and Cooperman, B.S. (1988) Nucl. Acids Res. 22, 10441–10452.
22 Stark, M.J.R. and Milner, J.S. (1989) Yeast 5, 35–50.
23 Kieber, J.J. and Singer, E.R. (1991) Plant Mol. Biol. 16, 277–284.
24 Lundin, M., Baltscheffsky, H. and Ronne, H. (1991) J. Biol. Chem. 266, 12168–12172.
25 Sarafian, V., Kim, Y., Poole, R.J. and Rea, P.A. (1992) Proc. Natl. Acad. Sci. USA 89, 1775–1779.
26 Kuranova, I.P., Terzyan, S.S., Voronova, A.A., Smirnova, E.A., Vainshtein, B.K., Höhne, W.E. and Hansen, G. (1983) Bioorg. Khim. 9, 1611–1619.
27 Terzyan, S.S., Voronova, A.A., Smirnova, E., Kuranova, I.P., Nekrasov, Y.V., Arutyunyun, E.G., Vainshtein, B.K., Höhne, W. and Hansen, G. (1984) Bioorg. Khim. 10, 1469–1482.
28 Kuranova, I.P. (1988) Biokhimiya 53, 1821–1827.
29 Nore, B.F., Sakai-Nore, Y., Baltscheffsky, M., Maeshima, M. and Nyrén, P. (1991) Biochem. Biophys. Res. Commun. 181, 962–967 .
30 Johansson, B.C., Baltscheffsky, M. and Baltscheffsky, H. (1971) in: G. Forti, M. Avron and B.A. Melandri (Eds.) Proc. 2nd Int. Photosynthesis Res. Vol. 2, Junk, The Hague, pp. 1203–1209.
31 Guillory, R.J. and Fisher, R.R. (1972) Biochem. J. 129, 471–481.
32 Barsky, E.L., Bonch-Osmolovskaya, E.A., Ostroamov, S.A., Samuilov, V.D. and Skulachev, V.P. (1975) Biochim. Biophys. Acta 387, 388–395.
33 Johansson, B.C. (1975) Doctoral Thesis, University of Stockholm.
34 Jones, O.T.G. and Saunders, V.A. (1972) Biochim. Biophys. Acta 275, 427–436.
35 Sherman, L.A. and Clayton, R.K. (1972) FEBS Lett. 22, 127–132.
36 Knobloch, K. (1975) Z. Naturforsch. 30c, 771–776.
37 Knaff, D.B. and Carr, J.W. (1979) Arch. Biochem. Biophys. 173, 379–384.
38 Schwarm, H.-M., Vigenschow, H. and Knobloch, K. (1986) Biol. Chem. Hoppe-Seyler 367, 127–133.
39 Romero, I., Gomez-Priego, A. and Celis H. (1991) J. Gen. Microbiol. 137, 2611–2616.
40 Nyrén, P. and Lundin, A. (1985) Anal. Biochem. 151, 504–509.
41 Nyrén, P., Nore, B.F. and Baltscheffsky, M. (1986) Biochim. Biophys. Acta 851, 276–282.
42 Nyrén, P., Nore, B.F. Salih, G.F. and Strid, Å. (1990) in: M. Baltscheffsky, (Ed.) Current Research in Photosynthesis, Vol.III, Kluwer Academic Publishers, Dordrecht, pp. 23–28.
43 Strid, Å., Karlsson, I.-M. and Baltscheffsky, M. (1987) FEBS Lett. 224, 348–352.
44 Nyrén, P., Nore, B.F. and Baltscheffsky, M. (1986) Photobiochem. Photobiophys. 11, 189–196.
45 Nore, B.F., Husain, I., Nyrén, P. and Baltscheffsky, M. (1986) FEBS Lett. 200, 133–138.

46 Strid, Å., Nyrén, P., Boork, J. and Baltscheffsky, M. (1986) FEBS Lett. 196, 337–340.
47 Boork, J., Strid, Å. and Baltscheffsky, M. (1985) FEBS Lett. 180, 314–316.
48 Nyrén, P. and Strid, Å. (1991) FEMS Microbiol. Lett. 77, 265–270.
49 Baykov, A.A., Volk, S.E. and Unguryte, A. (1989) Arch. Biochem. Biophys. 273, 287–291.
50 Davidson, A.M. and Halestrap, A.P. (1987) Biochem. J. 246, 715–723.
51 Volk, S.E. and Baykov, A.A. (1984) Biochim. Biophys. Acta 791, 198–204.
52 Unguryte, A., Smirnova, I.N. and Baykov, A.A. (1989) Arch. Biochem. Biophys. 273, 292–300.
53 Baltscheffsky, M. (1968) in: J. Järnefelt (Ed.) Regulatory Functions of Biological Membranes, Biochim. Biophys. Acta Library 11, 277–286.
54 Mansurova, S.E., Ermakova, S.A., Zvyagilskaya, R.A. and Kulaev, I.S. (1975) Mikrobiologiya 44, 874–879.
55 Lundin, M., Deopujari, S.W., Lichko, L., Pereira da Silva, L. and Baltscheffsky, H. (1992) Biochim. Biophys. Acta 1098, 217–223.
56 Karlsson, J. (1975) Biochim. Biophys. Acta, 356–363.
57 Rea, P.A. and Poole, R.J. (1986) Plant Physiol. 81, 126–129.
58 Kulakovskaya, T.V., Lichko, L.P. and Okorokov, L.A. (1989) Sov. Plant Physiol. 36, 1–5.
59 Ikeda, M., Satoh, S., Maeshima, M., Mukohata, Y. and Moritani, C. (1991)
59a Lichko, L. and Okorokov, L. (1991) Yeast 7, 805–812.
60 Maeshima, M. and Yoshida, S. (1989) J. Biol. Chem. 264, 20068–20073.
61 Britten, C.J., Turner, J.L. and Rea, P.A. (1989) FEBS Lett. 256, 200–206.
62 Malsowski, P. and Maslowska, H. (1987) Biochem. Physiol. Pflanz. 182, 73–84.
63 Randahl, H. (1979) Eur. J. Biochem. 102, 251–256.
64 Walker, R.R. and Leigh, R.A. (1981) Planta 153, 150–155.
65 Davies, J., Rea, P.A. and Sanders, D. (1991) FEBS Lett. 278, 66–68.
66 Maeshima, M. (1990) Biochem. Biophys. Res. Commun. 168, 1157–1162.
67 Sato, M.H., Maeshima, M., Ohsumi,Y. and Yoshida, M. (1991) FEBS Lett. 290, 177–180.
68 Maeshima, M. (1991) Eur. J. Biochem. 196, 11–17.
69 Chanson, A. (1991) J. Plant Physiol. 137, 471–476.
70 Marquardt-Jarzyk, G. and Lüttge, U. (1990) Bot. Acta 103, 203–213.
71 Petel, G. and Genraud, M. (1989) J. Plant Physiol. 134, 466–470.
72 Chanson, A., Fichmann, J., Spear, D. and Taiz, L. (1985) Plant Physiol. 79, 159–164.
73 Williams, L.E., Nelson, S.J. and Hall, J.L. (1990) Planta 182, 532–539.
74 Kawasaki, I., Adachi, N. and Ikeda, H. (1990) Nucl. Acids Res. 18, 5888.
75 Richter, O. and Schäfer, G. (1992) Eur. J. Biochem., in press..
76 Schwarm, H.-M., Vigenschow, H. and Knobloch, K. (1986) Biol. Chem. Hoppe-Seyler 367, 119–126.
77 Lahti, R., Kolakowski, L.F., Heinonen, J., Vihinen, M., Pohjanoksa, K. and Cooperman, B.S. (1990) Biochim. Biophys. Acta 1038, 338–345.
78 Baltscheffsky, H., Alauddin, Md., Falk, G. and Lundin, M. (1987) Acta Chem. Scand. Ser. B 41, 106–107.
79 Fry, D.C., Kuby, S.A. and Mildvan, A.S. (1986) Proc. Natl. Acad. Sci. USA 83, 907–911.
80 Schofield, P.R., Darlison, M.G., Fujita, N., Burt, D.R., Stephenson, F.A., Rodriguez, H., Rhee, L.M., Ramachandran, J., Reale, V., Glencorse, T.A., Seeburg, P.H. and Barnard, E.A. (1987) Nature 328, 221–227.
81 Vihinen, M., Lundin, M. and Baltscheffsky, H. (1992) Biochem. Biophys. Res. Commun. 186, 122–128.
82 Cooperman, B.S. (1982) Methods Enzymol. 87, 526–548.
83 Samejima, T., Tamagawa, Y., Kondo, Y., Hachimori, A., Kaji, H., Takeda, A. and Shiroya, Y. (1988) J. Biochem. 103, 766–772.
84 Lahti, R., Pohjanoksa, K., Pitkäranta, T., Heikinheimo, P., Salminen, T., Meyer, P. and Heinonen, J. (1990) Biochemistry 29, 5761–5766.
85 Lahti, R., Salminen, T., Latonen, S., Heikinheimo, P., Pohjanoksa, K. and Heinonen, J. (1991) Eur. J. Biochem. 198, 293–297.

86 Smirnova, I.N., Shestakov, A.S., Dubnova, E.B. and Baykov, A.A. (1989) Eur. J. Biochem. 182, 451–456.
87 Shestakov A.A., Baykov, A.A. and Avaeva, S.M. (1990) FEBS Lett. 262, 194–196.
88 Kukko-Kalske, E., Lintunen, M., Karjalainen, M., Lahti, R. and Heinonen, J. (1989) J. Bacteriol. 171, 4498–4500.
89 Cook, G.A., O'Brien, W.E., Wood, H.G., King, M.T. and Veech, R.L. (1978) Anal. Biochem. 91, 557–565.
90 Smyth, D.A. and Black C.L. (1984) Plant Physiol. 75, 862–864.
91 Weiner, H., Stitt, M. and Heldt, H.W. (1987) Biochim. Biophys. Acta 893, 13–21.
92 Lipmann, F. (1965) in: S.W. Fox (Ed.) The Origins of Prebiological Systems, Academic, New York, pp. 259–280,
93 Baltscheffsky, H. (1971) in: R. Buvet and C. Ponnamperuma (Eds.) Molecular Evolution, Vol. 1, North–Holland, Amsterdam, pp. 466–477.
94 De Duve, C. (1991) Blueprint for a Cell: The Nature and Origin of Life, Neil Patterson, Burlington, pp. 150–166.
95 Rouse, R.C., Peacor, D.R. and Freed, R.L. (1988) Am. Mineral. 73, 168–171.
96 Yamagata, Y., Watanabe, H., Saitoh, M. and Namba, T. (1991) Nature 352, 516–519.
97 Ladror, U.S., Gollapudi, L., Tripathi, R.L., Latshaw, S. and Kemp, R.G. (1991) J. Biol. Chem. 266, 16550–16555.
98 Lust, G., Faure, G., Netter, P., Gaucher, A. and Seegmiller, J.E. (1981) Arthritis Rheum. 24, 1517–1521.
99 Ryan, L.M., Kurup, I., Rosenthal, A.K. and McCarty, D.J. (1989) Arch. Biochem. Biophys. 272, 393–399.
100 Ryall, R.L., Harnett, R.M. and Marshall, V.R. (1981) Clin. Chim. Acta 112, 349–356.
101 Baltscheffsky, M., Pramanik, A., Lundin, M., Nyrén, P. and Baltscheffsky, H. (1991) in: M. Baltscheffsky (Ed.) Current Research in Photosynthesis, Vol. III, Kluwer Academic Publishers, Dordrecht. pp.197–200.
102 Kowalczyk, S. and Maslowski, P. (1984) Biochim. Biophys. Acta 766, 570–575.

L. Ernster (Ed.) *Molecular Mechanisms in Bioenergetics*
© 1992 Elsevier Science Publishers B.V. All rights reserved

CHAPTER 15

Mitochondrial calcium transport

CHRISTOPH RICHTER

Laboratory of Biochemistry I, Swiss Federal Institute of Technology, Universitätstr. 16, CH-8092 Zürich, Switzerland

Contents

1. Cellular Ca^{2+} homeostasis 349
2. Mitochondrial Ca^{2+} transport 350
2.1. Ca^{2+} uptake 350
2.2. Ca^{2+} release 351
2.2.1. Na^{+}-dependent release 351
2.2.2. Na^{+}-independent release 352
2.2.2.1. Unstimulated release 352
2.2.2.2. Stimulated release 352
2.2.2.3. Pyridine nucleotide-linked release 352
2.2.2.4. Protein ADP-ribosylation as trigger of the Na^{+}-independent release 353
2.2.2.5. Ca^{2+} release occurs from intact mitochondria 353
2.2.2.6. 'Pore' formation is not required for Ca^{2+} release 354
2.2.3. Inhibitors of the Na^{+}-independent Ca^{2+} release 355
3. Summary 356
Acknowledgements 356
References 356

Abbreviations

EGTA: ethylenebis(oxyethylenenitrilo) tetraacetic acid; ER: endoplasmic reticulum; $\Delta\Psi$: mitochondrial transmembrane potential; MIBG: *m*-iodobenzylguanidine; P_i: inorganic phosphate; SDS-PAGE: sodium dodecylsulfate-polyacrylamide gel electrophoresis; SMP: submitochondrial particles; SR: sarcoplasmic reticulum; tbh: *tert*-butylhydroperoxide.

1. Cellular Ca^{2+} homeostasis

Free calcium ions play a prominent role in the regulation of many enzyme systems, including those responsible for muscle contraction, nerve impulse transmission, neuronal activity, blood clotting, and modulation of hormone action. Specifically the intracellular Ca^{2+} has the fundamental function of an intracellular messenger, and, therefore, its fine-tuning is required. The concentration of ionized cytoplasmic calcium, which in resting cells is kept between 0.1 and 0.2 μM, can rapidly increase by up to one order of magnitude when transducing hormonal messages.

The intracellular Ca^{2+} concentration is regulated by binding to nonmembraneous proteins, and by membrane-bound transport systems. Besides buffering the intracellular Ca^{2+} concentration, the soluble proteins also participate in signal processing and are therefore called 'Ca^{2+}-modulated proteins'. Examples are troponin C, parvalbumin, and calmodulin. The main burden of regulation is, however, carried by Ca^{2+}-specific proteins which transport the ion across membranes. Transport proteins were found in the plasma membrane, the endoplasmic (sarcoplasmic) reticulum (ER, SR), in mitochondria, and very recently also in the nucleus. The transport systems differ in affinity and capacity for Ca^{2+} transport, and probably also in their importance for the maintenance of physiological as well as for the development of pathological states of the cell.

The importance of mitochondria as buffers of cytosolic Ca^{2+} under physiological conditions is probably minor since the mitochondrial Ca^{2+} affinity and uptake rate are much smaller than those of the ER or SR. This is corroborated by electron probe X-ray microanalysis experiments, which show that mitochondria in situ contain much less Ca^{2+} than previously deduced from measurements with isolated mitochondria. Indeed, it is now generally accepted that mitochondria normally do not contain more than about 1 to 2 nmol of Ca^{2+} per mg of protein. On the other hand, isolated mitochondria can accumulate enormous amounts of Ca^{2+} in the presence of P_i (inorganic phosphate) due to the formation of insoluble calcium phosphate complexes in the matrix. It is important to note that a massive short-term loading with Ca^{2+} and P_i disturbs the mitochondrial functions very little. Deposits of Ca^{2+} complexes in mitochondria in situ were also observed in electron microscopy and electron probe X-ray microanalyses when the plasma membrane had been permeabilized. It therefore appears that mitochondria take up and buffer the cytosolic Ca^{2+} when its concentration increases to levels which allow the operation of the mitochondrial low-affinity uptake system. Mitochondria can thereby act as safety devices against toxic increases of cytosolic Ca^{2+}, as for example, when the plasma membrane Ca^{2+} pump is compromised due to, e.g., lack of ATP. This is important, since net influx of Ca^{2+} across the plasma membrane appears to be a frequent and early common mechanism by which some cell types are killed.

2. *Mitochondrial Ca^{2+} transport*

2.1. *Ca^{2+} uptake*

Ca^{2+} uptake by mitochondria, discovered more than 30 years ago [1,2] has been reviewed comprehensively [3–13]. For about two decades the regulation of cytosolic Ca^{2+} levels was thought to be its main role. Electron probe X-ray microanalysis then showed [14,15] that mitochondria in situ contain much less Ca^{2+} than deduced from early in vitro studies. Also, the kinetic properties of the mitochondrial Ca^{2+} uptake (see below) rule out these organelles as important short-term regulators of cytosolic Ca^{2+}. It is now believed that the main task of mitochondrial Ca^{2+} is the regulation of intramitochondrial Ca^{2+}-dependent pyruvate dehydrogenase, NAD-dependent isocitrate dehydrogenase, and α-ketoglutarate dehydrogenase [16–18].

In the presence of ATP and P_i mitochondria can take up more than 3 μmol Ca^{2+}/mg of protein due to the formation of insoluble calcium phosphate deposits in the matrix ('matrix loading'). In the absence of P_i the rate and extent of uptake are much lower ('limited loading'). Ca^{2+} uptake is generally believed to be an active process, driven by the transmembrane potential ($\Delta\Psi$) (but see below). Several other divalent cations such as Sr^{2+}, Mn^{2+}, Ba^{2+}, Fe^{2+}, Pb^{2+}, and lanthanides are sequestered by the same mechanism [19,20]. Mg^{2+} uptake is tissue specific, and independent of the Ca^{2+} uptake pathway in heart but Ca^{2+}-dependent in liver and brain [21].

Ca^{2+} uptake occurs electrogenically via the 'uniporter', i.e., without charge compensation. Despite much effort, no protein engaged in Ca^{2+} uptake has been identified with certainty. Best studied [22–26] are glycoproteins with M_r of 33 000–42 000, which bind Ca^{2+} in a ruthenium red- and La^{3+}-sensitive fashion. Ca^{2+} uptake is modulated by removal from or addition of glycoprotein to mitochondria, and anti-glycoprotein antibodies inhibit Ca^{2+} uptake.

The polycations ruthenium red ($[Ru_3O_2(NH_3)_{14}]Cl_6 \cdot 4H_2O$) [27] and hexamine cobalt [28] inhibit rather specifically and potently Ca^{2+} uptake by isolated mitochondria, as do Sr^{2+}, Mn^{2+}, Ba^{2+}, Fe^{2+}, and lanthanides. Finally, β-blockers [29], guanidines [30], and diuretics [31] inhibit Ca^{2+} uptake, particularly in mitochondria of heart [29] and kidney [31]. The most commonly used inhibitor in isolated mitochondria, but also in permeabilized cells, is the polysaccharide stain ruthenium red, which inhibits the uniporter noncompetitively with a K_i of about 30 nM. At submicromolar cytosolic Ca^{2+} concentrations, Mg^{2+} is the most important physiological modulator of mitochondrial Ca^{2+} uptake. Mg^{2+} lowers the affinity of the uniporter for Ca^{2+} several-fold to a K_m of about 30 μM [32,33], which is much too high to play a role in cytosolic Ca^{2+} regulation under physiological conditions.

The ability of mitochondria to take up and store transiently large amounts of Ca^{2+} without impunity, together with the relatively low Ca^{2+} affinity and sluggish Ca^{2+} uptake suggest that mitochondria act as safety devices in cases of cellular emergency, such as chemical intoxication or hypoxia/reperfusion injury.

An issue again brought to the forefront recently is the nature of the driving force of Ca^{2+} entry. Mitochondria have long been thought to accumulate Ca^{2+} actively [34–37] in response to the $\Delta\Psi$ established by the activity of the respiratory chain, ATP hydrolysis, or by coupling to K^+ efflux. This traditional view is now challenged by Gunter and coworkers (for review, see ref. [37]). They conclude that Ca^{2+} enters mitochondria by facilitated diffusion down the electrochemical gradient ('passive uniporter').

2.2. *Ca^{2+} release*

Two separate Ca^{2+} release pathways exist in mitochondria, one Na^+-dependent, the other Na^+-independent. Selective inhibitors were important in the identification of the two release pathways. These pathways coexist in the same mitochondrion, but their relative importance depends on the tissue origin of the organelle. Why Nature developed the two separate pathways is presently not clear.

2.2.1. *Na^+-dependent release*

In mitochondria of heart, brain, skeletal muscle, adrenal cortex, brown fat, and most tumor tissue the Na^+-dependent Ca^{2+} efflux mechanism predominates [38–40]. It operates as a Ca^{2+}/nNa^+ exchanger. Heart mitochondria also catalyze a Ca^{2+}/Sr^{2+} and a Ca^{2+}/Ca^{2+} exchange [38,41,42]. V_{max} of the release in heart mitochondria is 18 nmol Ca^{2+}/mg protein and min at 25°C [39]. Half-maximal velocity requires about 10 mM Na^+ and 10 nmol Ca^{2+}/mg protein [39,43]. Ca^{2+} efflux is sigmoidally dependent on the external Na^+ concentration [43]. Numerous inhibitors of the Na^+-dependent Ca^{2+} release have been identified: Mg^{2+} [44,45], Mn^{2+} [46], trifluoperazine [47], lipophilic cations [48], Ca^{2+} channel blockers (e.g., refs. [49,50]), and external Ca^{2+} [51]. Cyclosporine A, a potent inhibitor of the Na^{+}-independent release, leaves the Na^{+}-dependent mechanism unaffected [52].

2.2.2. Na^{+}-independent release

2.2.2.1. Unstimulated release. The predominant Ca^{2+} release mechanism in liver, kidney, lung, and smooth muscle mitochondria is Na^{+}-independent [38,53]. It operates by Ca^{2+}/nH^{+} exchange [54–57], but the exact stoichiometry of Ca^{2+} and H^{+} is presently not worked out. The release velocity is sigmoidally dependent on the intramitochondrial Ca^{2+} load [58]. Without stimulators of this release pathway, V_{max} is about 1.2 nmol Ca^{2+}/min and mg protein, and K_m for Ca^{2+} is about 8.4 nmol/mg protein [58]. It should be remembered that the intramitochondrial free Ca^{2+} concentration is proportional to a Ca^{2+} load of up to about 55 nmol Ca^{2+}/mg protein [59]. The Na^{+}-independent release can be inhibited by Sr^{2+} [60,61], Mn^{2+} [46], uncouplers [62–64], lipophilic cations [58], and lasalocid A [65]. Other inhibitors will be discussed below.

The Ca^{2+}/nH^{+} exchange was initially considered to operate passively, i.e., energy-independently. This view has been challenged. According to Gunter's group [53,66], the currently available evidence strongly argues for an active Na^{+}-independent Ca^{2+} release from mitochondria.

2.2.2.2. Stimulated release. A large number of agents (compiled in ref. [37]) stimulate (a Na^{+}-independent) Ca^{2+} release from mitochondria. The agents comprise such diverse compounds as heavy metal ions, redox cyclers, prooxidants, sulfhydryl reagents, uncouplers, and P_i. Of these compounds only those linked to the oxidation (and hydrolysis) of mitochondrial pyridine nucleotides are established as true release agents because only they promote release from intact mitochondria (discussed below). In the author's view valid criteria for intactness of mitochondria during Ca^{2+} release are the maintenance of inner membrane integrity as determined by $\Delta\Psi$ measurements, and the specificity of Ca^{2+} movement across the inner membrane, i.e., it must not be accompanied by movement of other solutes such as K^{+}, sucrose, or proteins. The prooxidant-dependent Ca^{2+} release from mitochondria meets these criteria, and is accompanied by protein ADP-ribosylation. This release has recently been reviewed [67,68]. Therefore, the following sections (2.2.2.3 to 2.2.2.5) are limited to its most salient features. For most of the original references the reader is referred to the aforementioned reviews.

2.2.2.3. Pyridine nucleotide-linked Ca^{2+} release. The first evidence that Ca^{2+} release from mitochondria is linked to the oxidation of mitochondrial pyridine nucleotides came from Lehninger and coworkers who showed that enzymatic oxidation of NAD(P)H by acetoacetate or oxaloacetate promotes release whereas reduction of $NAD(P)^{+}$ by β-hydroxybutyrate prevents it. Shortly thereafter Lötscher et al. [69] showed that hydroperoxides promote Ca^{2+} release, and that the release requires an active glutathione peroxidase, giving the first clear evidence that the release is not simply due to mitochondrial damage. Since then, a variety of other prooxidants has been identified whose ability to stimulate Ca^{2+} release is secondary to pyridine nucleotide oxidation and hydrolysis.

Small amounts (a few nmol per mg protein) of *tert*-butylhydroperoxide (tbh) or hydrogen peroxide induce oxidation of mitochondrial pyridine nucleotides by an enzyme cascade consisting of glutathione peroxidase EC 1.11.1.9), glutathione reductase (NAD(P)H: oxidized glutathione oxidoreductase; EC 1.6.4.2), and the energy-linked transhydrogenase ($NAD(P)^{+}$ transhydrogenase; EC 1.6.1.1). Concomitantly with the oxidation of pyridine nucleotides, hydroperoxides induce Ca^{2+} release from mitochondria, as shown for the organelles isolated from liver, heart, brain, and kidney.

Bellomo et al. [70] have reported that menadione (2-methyl-1,4-naphthoquinone) impairs the ability of mitochondria to take up and retain Ca^{2+}. Work by Richter and coworkers has subsequently demonstrated [71] that menadione does not impair the integrity of mitochondria but stimulates Ca^{2+} release by a mechanism involving pyridine nucleotide oxidation and hydrolysis. The link between menadione and pyridine nucleotides is enzymatic and engages predominantly three diaphorases [71].

Upon exposure to alloxan (2,4,5,6-tetraoxypyrimidine), mouse and rat liver [72] mitochondria release Ca^{2+}. As with the hydroperoxides and menadione, the release of Ca^{2+} induced by alloxan is accompanied by pyridine nucleotide oxidation and hydrolysis, and occurs from intact mitochondria. Furthermore, both alloxan-induced pyridine nucleotide hydrolysis and Ca^{2+} release are strongly inhibited by ATP, cyclosporine A, and 4-hydroxynonenal (see below). However, the mechanism by which alloxan induces pyridine nucleotide oxidation in mitochondria is different from that of hydroperoxides or menadione, since oxidation by alloxan does not involve enzymes to any significant extent [72].

Pyridine nucleotide-linked Ca^{2+} release from intact mitochondria is also induced, e.g., by divicine, *N*-methyl-4-phenylpyridine, hydroperoxyeicosatetraenoic acids, or iron ions.

Oxidation alone of pyridine nucleotides is not sufficient to induce Ca^{2+} release. This is shown with ATP and three other recently discovered inhibitors of Ca^{2+} release (discussed below). In the presence of ATP the hydroperoxide-induced pyridine nucleotide oxidation is even accelerated due to the stimulation of the energy-linked transhydrogenase, yet pyridine nucleotide hydrolysis and Ca^{2+} release are inhibited. Similar observations were made for the menadione- and alloxan-induced Ca^{2+} release. Importantly, ATP also inhibits the NAD^+ glycohydrolase activity and protein ADP-ribosylation in inner mitochondrial membranes (submitochondrial particles, SMP). Neither a decrease of reduced, nor an increase of oxidized, mitochondrial glutathione favor Ca^{2+} release. Thus, pyridine nucleotide hydrolysis is an obligatory and sufficient prerequisite of Ca^{2+} release from intact liver mitochondria. Finally, both pyridine nucleotide hydrolysis and Ca^{2+} release show the same sigmoidal dependence on the mitochondrial Ca^{2+} load. The functional link between pyridine nucleotide hydrolysis and Ca^{2+} release may be protein ADP-ribosylation in the inner mitochondrial membrane (see below).

2.2.2.4. Protein ADP-ribosylation as trigger of the Na^+-independent Ca^{2+} release. The intramitochondrial pyridine nucleotide hydrolysis and the release of nicotinamide from mitochondria exposed to prooxidants suggested the existence of a NAD^+ glycohydrolase (EC 3.2.2.5) in mitochondria. Indeed, an enzyme with this activity was localized on the inner side of the inner mitochondrial membrane from where it was subsequently isolated and purified to apparent homogeneity on SDS-PAGE. The NAD^+ glycohydrolase has an apparent M_r of about 64 000. It is inhibited by ATP.

Incubation of SMP with NAD^+ leads to monoADP-ribosylation of a protein with an M_r of about 32 000. The modifying ADP-ribose residue turns over rapidly.

Although protein ADP-ribosylation receives growing attention, its significance for the physiological functioning of cells is not clear. Since oxidation and subsequent hydrolysis of pyridine nucleotides leads to release of Ca^{2+} from liver mitochondria, and since NAD^+ hydrolysis is accompanied by protein ADP-ribosylation in the isolated inner mitochondrial membrane, Richter and coworkers proposed that the regulation of the Na^+-independent Ca^{2+} release pathway occurs by protein mono(ADP-ribosylation) [73] (see Fig. 1), and provided direct evidence for this [74].

2.2.2.5. Ca^{2+} release occurs from intact mitochondria. In principle, Ca^{2+} can leave mitochondria in three ways: by nonspecific leakage through the inner membrane, by reversal of the uniport influx carrier, and by the Na^+-dependent or -independent release pathway.

The specifity of the $NAD(P)^+$-linked Ca^{2+} efflux was initially questioned by some groups,

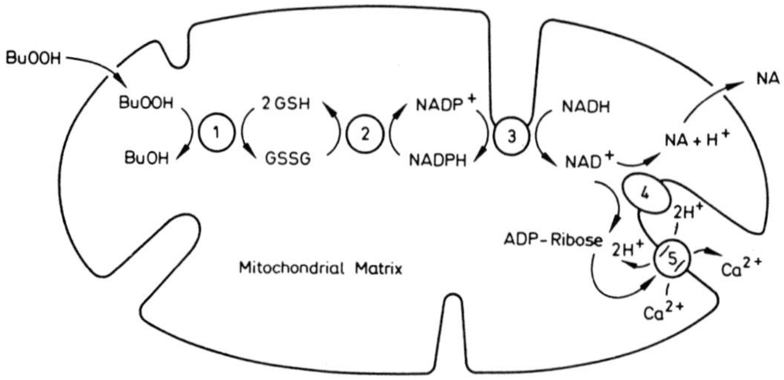

Fig. 1. Prooxidant-induced ADP-ribosylation and Ca^{2+} release from mitochondria. As a representative prooxidant, *tert*-butylhydroperoxide (BuOOH) is shown (cf. ref. [69]). Menadione is coupled to the mitochondrial pyridine nucleotides via three diaphrases (cf. ref. [71]), whereas alloxan oxidizes the nucleotides predominantly non-enzymatically (cf. ref. [72]). The Ca^{2+} release inhibitors ATP, 4-hydroxynonenal, and cyclosporine prevent ADP-ribosylation by inhibiting NAD^+ hydrolysis; *m*-iodobenzylguanidine intercepts ADP-ribose (see Section 2.2.3). 1, glutathione peroxidase; 2, glutathione reductase; 3, energy-linked pyridine nucleotide transhydrogenase; 4, NAD^+ glycohydrolase; 5, Ca^{2+}/H^+ antiporter, regulated by ADP-ribosylation. NA, nicotinamide.

who argued that Ca^{2+} release stimulated by tbh, oxaloacetate, or acetoacetate is preceded by mitochondrial large amplitude swelling and collapse of $\Delta\Psi$ due to a nonspecific increase in membrane permeability. However, several lines of evidence clearly show that nonspecific permeability changes can be dissociated from pyridine nucleotide redox changes and Ca^{2+} fluxes (summarized in ref. [68]; see also the following sections).

2.2.2.6. 'Pore' formation is not required for Ca^{2+} release. It has recently been suggested that Ca^{2+} release induced by hydroperoxides and/or P_i is accompanied by opening of a 'pore' [75–77]. Operation of the putative 'pore', termed by others 'permeability transition' [78], is characterized by general leakiness of the inner mitochondrial membrane as measured by K^+ and protein release from and sucrose entry into mitochondria. Another characteristic important feature of the postulated 'pore' is that it is closed by ethylenebis(oxyethylenenitrilo) tetraacetic acid (EGTA) [75].

Neither the unstimulated nor the hydroperoxide-dependent Ca^{2+} release requires 'pore' formation [79] (Fig. 2). Hydroperoxides induce Ca^{2+} release from mitochondria in the presence of EGTA, and Ca^{2+} release is neither accompanied by sucrose entry into nor by K^+ release from or swelling of mitochondria. The Ca^{2+}-uptake inhibitor ruthenium red, when added to Ca^{2+}-loaded mitochondria, also prevents the unspecific solute fluxes, yet hydroperoxide-dependent Ca^{2+} release occurs [79]. Cyclosporine A, the high-affinity inhibitor of the Na^+-independent Ca^{2+} release, also inhibits Ca^{2+} release in the presence of EGTA (Fig. 2).

In the absence of EGTA or ruthenium red, the hydroperoxide-dependent Ca^{2+} release is accompanied by sucrose entry, K^+ release and swelling of mitochondria [75–79]. Whether this reflects formation of a 'pore' or the initial, reversible phase of the notorious mitochondrial damage due to continuous Ca^{2+} release and re-uptake (Ca^{2+} 'cycling') remains to be established.

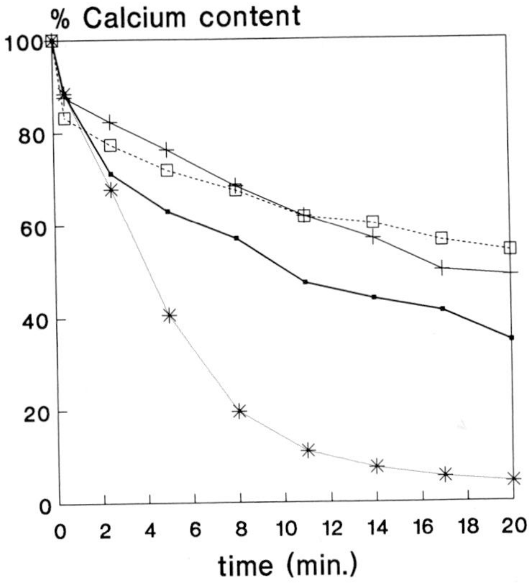

Fig. 2. Spontaneous and *tert*-butylhydroperoxide-induced Ca^{2+} release from mitochondria in the presence of EGTA and its inhibition by cyclosporine. Mitochondria were loaded with 40 nmol (=100%) of Ca^{2+}/mg protein. At time 0 min, EGTA alone or in combination with *tert*-butylhydroperoxide (tbh) was added, and the mitochondrial Ca^{2+} content was measured by Millipore filtration. (●), EGTA alone; (+), EGTA plus 1 μM cyclosporine A; (*), EGTA plus tbh; (□), EGTA plus tbh plus 1 μM cyclosporine A.

2.2.3. Inhibitors of the Na^+-independent Ca^{2+} release

As mentioned, some cations, uncouplers, and lasalocid A can inhibit the Na^+-independent Ca^{2+} release. Several other inhibitors, all of which interfere directly or indirectly with ADP-ribosylation in mitochondria, have now been identified.

Early studies established that the oxidation of mitochondrial pyridine nucleotides is necessary but not sufficient for Ca^{2+} release. For example, ATP stimulates the prooxidant-induced pyridine nucleotide oxidation, yet it inhibits their hydrolysis, protein ADP-ribosylation, and Ca^{2+} release [80,81]. Three newly discovered inhibitors corroborate the proposal that ADP-ribosylation is required for Ca^{2+} release. 4-Hydroxynonenal [82] and cyclosporine A [52,83] do not prevent prooxidant-mediated pyridine nucleotide oxidation, but inhibit both the hydrolysis of oxidized intramitochondrial pyridine nucleotides and Ca^{2+} release. Similarly, *m*-iodobenzylguanidine (MIBG), a substrate for mono(ADP-ribosylation) reactions and an inhibitor of ADP-ribosyltransferase, affects neither pyridine nucleotide oxidation nor its hydrolysis, but prevents Ca^{2+} release [84,85] presumably by competing for ADP-ribose with the mitochondrial acceptor proteins.

The mode of action of the three novel inhibitors is not only important for the understanding of the molecular mechanism of Ca^{2+} release from mitochondria in vitro, but may also explain some of their in vivo effects. The ADP-ribose acceptor MIBG is used in radio-iodinated form for the detection and therapy of human tumors. It will be important to determine whether the killing of the tumor cells is caused by MIBG's interference with cellular ADP-ribosylation, e.g., in mitochondria, or by radiation. 4-Hydroxynonenal, a product generated during lipid peroxidation, is a compound with cytotoxic, hepatotoxic, mutagenic, genotoxic, and other, (e.g., chemotactic) properties. Its in vivo concentrations (0.5–5 μM) are not drastically lower than those required (10 μM) to inhibit Ca^{2+} release from isolated mitochondria. Finally, cyclosporine A, a cyclic unadecapeptide, has several pharmacological properties, such as antiparasitic and antimalarial activities, and also the capacity to reverse multidrug resistance in tumors. Most important, and clinically most relevant, is its unique immunosuppressive effect on certain immunocompetent cells, and the nephrotoxicity and hepatotoxicity encountered upon cyclosporine A treatment.

Cyclosporine A and its derivatives are presently receiving great attention. Out of the five derivatives (cyclosporine A, B, C, dihydro D, and H) tested so far as inhibitors of NAD^+ hydrolysis and Ca^{2+} release, only cyclosporine H is inactive (Schweizer, Schlegel, and Richter, unpublished results). Interestingly, the only difference between cyclosporine A and H is the presence of a D-valine residue (cyclosporine H) instead of a L-valine residue (cyclosporine A) in position 11 of the peptides. When the derivatives were tested for inhibition of the matrix-located peptidyl-prolyl *cis-trans* isomerase [86], it was found (Schweizer, Schlegel, and Richter, unpublished results) that again only cyclosporine H does not inhibit the enzyme. This strongly suggests that the Na^+-independent Ca^{2+} release engages this enzyme and further documents the specificity of this pathway.

3. *Summary*

Mitochondrial Ca^{2+} transport has been studied for more than 30 years. The initial view that mitochondria act as buffers of cytosolic Ca^{2+} under physiological conditions gave way to the present opinion that mitochondrial Ca^{2+} transport participates in the regulation of intramitochondrial processes, and that it enables the organelles to act as safety devices against toxic increases of cytosolic Ca^{2+}.

Ca^{2+} enters and leaves mitochondria via different pathways. Whereas the understanding of the uptake pathway has not improved much since the seventies, there has been considerable progress in the study of the release pathways, in particular the Na^+-independent one. Cyclosporine A and some of its derivatives are high-affinity inhibitors of this pathway, and promise to be tools in the search for the components involved in the Na^+-independent release.

Acknowledgements

The author's work cited here was generously supported by the Schweizerisch Nationalfonds, the Schweizerische Krebsliga, and by donations from Sandoz AG and Hoffmann-La Roche, Basel, Switzerland. I thank Drs. Schlegel, Schweizer, and Walter for critically reading the manuscript, Dr. Suter for editorial work, and Dr. Winterhalter for his interest and support.

References

1 Vasington, F.D. and Murphy, P.A. (1961) Fed. Proc. Fed. Am. Soc. Exp. Biol. 20, 146.
2 DeLuca, H.F. and Engström, G.W. (1961) Proc. Natl. Acad. Sci. USA 47, 1744–1750.
3 Lehninger, A.L., Carafoli, E. and Rossi, C.S. (1967) Adv. Enzymol. 29, 259–320.
4 Bygrave, F.L. (1977) Curr. Top. Bioenerg. 6, 259–318.
5 Carafoli, E. and Crompton, M. (1977) Curr. Top. Membr. Transp. 10, 151–216.
6 Nicholls, D.G. (1978) Biochem. J. 176, 463–474.

7 Saris, N.E.L. and Akerman, K.E.O. (1980) Curr. Top. Bioenerg. 10, 104–179.
8 Nicholls, D.G. and Akerman, K.E.O. (1982) Biochim. Biophys. Acta 683, 57–88.
9 Hansford, R.G. (1985) Rev. Physiol. Biochem. Pharmacol. 102, 1–72.
10 Crompton, M. (1985) Curr. Top. Membr. Transp. 25, 231–276.
11 Denton, R.M. and McCormack, J.G. (1985) Am. J. Physiol. 249 (Endocrinol. Metab. 12) E543–E554.
12 Fiskum, G. (1985) Cell Calcium 6, 25–37.
13 Carafoli, E. (1987) Annu. Rev. Biochem. 56, 395–433.
14 Somlyo, A.P., Somlyo, A.V. and Shuman, H. (1979) J. Cell Biol. 81, 316–335.
15 Somlyo, A.P., Bond, M. and Somlyo, A.V. (1985) Nature 314, 622–625.
16 Denton, R.M., Randle, P.J. and Martin, B.R. (1972) Biochem. J. 128, 161–163.
17 Denton, R.M., Richards, D.A. and Chin, J.G. (1978) Biochem. J. 176, 899–906.
18 McCormack, J.G. and Denton, R.M. (1979) Biochem. J. 180, 533–544.
19 Drahota, Z., Gazzotti, P., Carafoli, E. and Rossi, C.S. (1969) Arch. Biochem. Biophys. 130, 267–273.
20 Vainio, H., Mela, L. and Chance, B. (1970) Eur. J. Biochem. 12, 387–391.
21 Diwan, J.J. (1987) Biochim. Biophys. Acta 895, 155–156.
22 Sottocasa, G.L., Sandri, G., Panfili, E. and de Bernard, B. (1971) FEBS Lett. 17, 100–105.
23 Sottocasa, G.L., Sandri, G., Panfili, E., de Bernard, B., Gazzotti, P., Vasington, F.D. and Carafoli, E. (1972) Biochem. Biophys. Res. Commun. 47, 808–813.
24 Carafoli, E. and Sottocasa, G.L. (1974) in: L. Ernster, R.W. Estabrock and E.C. Slater (Eds.) Dynamics of Energy-Transducing Membranes, Elsevier, Amsterdam, pp. 455–469.
25 Prestipino, G., Ceccarelli, D., Conti, F. and Carafoli, E. (1974) FEBS Lett. 45, 99–103.
26 Sandri, G., Panfili, E. and Sottocasa, G.L. (1976) Biochem. Biophys. Res. Commun. 68, 1272–1279.
27 Moore, C.L. (1971) Biochem. Biophys. Res. Commun. 42, 405–418.
28 Tashmukhamedov, B.A., Gagelgans, A.I., Mamatkulov, K. and Makhmuddova, E.M. (1972) FEBS Lett. 28, 239–242.
29 Noack, E. and Greef, K. (1971) Experientia 27, 810–811.
30 Davidoff, F. (1974) J. Biol. Chem. 249, 6406–6415.
31 Gemba, M. (1974) Jpn. J. Pharmacol. 24, 271–277.
32 Scarpa, A. and Graziotti, P. (1973) J. Gen. Physiol. 62, 756–772.
33 Crompton, M., Sigel, E., Salzmann, M. and Carafoli, E. (1976) Eur. J. Biochem. 69, 429–434.
34 Chance, B. (1965) J. Biol. Chem. 240, 2729–2748.
35 Scarpa, A. and Azzone, G.F. (1970) Eur. J. Biochem. 12, 328–335.
36 Rottenberg, H. and Scarpa, A. (1974) Biochemistry 13, 4811–4819.
37 Gunter, T.E. and Pfeiffer, D.R. (1990) Am. J. Physiol. 258 (Cell Physiol. 27), C755–C786.
38 Crompton, M., Künzi, M. and Carafoli, E. (1977) Eur. J. Biochem. 79, 549–558.
39 Crompton, M., Moser, R., Lüdi, H. and Carafoli, E. (1978) Eur. J. Biochem. 82, 25–31.
40 Murphy, A.N. and Fiskum, G. (1988) in: D.R. Pfeiffer, J.B. McMillin and S. Little (Eds.) Cellular Ca^{2+} Regulation. Advances in Experimental Medicine and Biology, Vol. 232, Plenum Press, New York, pp. 139–150.
41 Crompton, M. (1980) Biochem. Soc. Trans. 8, 261–262.
42 Crompton, M., Heid, I. and Carafoli, E. (1980) FEBS Lett. 115, 257–259.
43 Crompton, M., Capano, M. and Carafoli, E. (1976) Eur. J. Biochem. 69, 453–462.
44 Clark, A.F. and Roman, I.J. (1980) J. Biol. Chem. 255, 6556–6558.
45 Rizzuto, R., Bernardi, P., Favaron, M. and Azzone, G.F. (1987) Biochem. J. 246, 271–277.
46 Gunter, K.K., Gavin, C.E. and Gunter, T.E. (1989) Biophys. J. 55, 571a.
47 Hayat, L.H. and Crompton, M. (1985) FEBS Lett. 182, 281–286.
48 Wingrove, D.E. and Gunter, T.E. (1986) J. Biol. Chem. 261, 15166–15171.
49 Wolkowicz, P.E., Michael, L.H., Lewis, R.M. and McMillin-Wood, J. (1983) Am. J. Physiol. 244, H644–H651.
50 Chiesi, M., Rogg, H., Eichenberger, P., Gazzotti, P. and Carafoli, E. (1987) Biochem. Pharmacol. 36, 2735–2740.
51 Hayat, L.A. and Crompton, M. (1982) Biochem. J. 202, 509–518.
52 Schlegel, J., Meier, P., Kass, G.E.N. and Richter, C. (1991) Biochem. Pharmacol. 42, 2193–2197.

53 Rosier, R.N., Tucker, D.A., Meerdink, S., Jain, I. and Gunter, T.E. (1981) Arch. Biochem. Biophys. 210, 549–564.
54 Akerman, K.E.O. (1978) Arch. Biochem. Biophys. 189, 256–262.
55 Brand, M.D. (1985) Biochem. J. 225, 413–419.
56 Fiskum, G. and Cockrell, R.S. (1978) FEBS Lett. 92, 125–128.
57 Fiskum, G. and Lehninger, A.L. (1979) J. Biol. Chem. 254, 6236–6239.
58 Wingrove, D.E. and Gunter, T.E. (1986) J. Biol. Chem. 261, 15159–15165.
59 Coll, K.E., Joseph, S.K., Corkey, B.E. and Williamson, J.R. (1982) J. Biol. Chem. 257, 8696–8704.
60 Saris, N.-E. and Bernardi, P. (1983) Biochim. Biophys. Acta 725, 19–24.
61 Gunter, T.E., Wingrove, D.E., Banerjee, S. and Gunter, K.K. (1988) in: D. Pfeiffer, J.B. McMillin and S. Little (Eds.) Cellular Ca^{2+} Regulation. Advances in Experimental Medicine and Biology, Vol. 232, Plenum Press, New York, pp. 1–14.
62 Gunter, T.E., Gunter, K.K., Puskin, J.S. and Russell, P.R. (1978) Biochemistry 17, 339–345.
63 Bernardi, P. and Azzone, G.F. (1982) FEBS Lett. 139, 13–16.
64 Ligeti, E. and Lukacs, G.L. (1984). J. Bioenerg. Biomembranes 16, 101–113.
65 Da Silva, L.P., Bernardes, C.F. and Vercesi, A.E. (1984) Biochem. Biophys. Res. Commun. 124, 80–86.
66 Gunter, K.K., Zuscik, M.J. and Gunter, T.E. (1991) J. Biol. Chem. 266, 21640–21648.
67 Richter, C. and Frei, B. (1988) Free Radical Biol. Med. 4, 365–375.
68 Richter, C. and Kass, G.E.N. (1991) Chem. Biol. Interact. 77, 1–23.
69 Lötscher, H.R., Winterhalter, K.H., Carafoli, E. and Richter, C. (1979) Proc. Natl. Acad. Sci. USA 76, 4340–4344.
70 Bellomo, G., Jewell, S.A. and Orrenius, S. (1982) J. Biol. Chem. 257, 11558–11562.
71 Frei, B., Winterhalter, K.H. and Richter, C. (1986) Biochemistry 25, 4438–4443.
72 Frei, B., Winterhalter, K.H. and Richter, C. (1985) J. Biol. Chem. 260, 7394–7401.
73 Richter, C., Frei, B. and Schlegel, J. (1985) in: F.R. Althaus, H. Hilz and S. Shall (Eds.) ADP-ribosylation of Proteins, Springer, Berlin, pp. 530–535.
74 Frei, B. and Richter, C. (1988) Biochemistry 27, 529–535.
75 Al-Nasser, I. and Crompton, M. (1986) Biochem. J. 239, 19–29.
76 Crompton, M., Ellinger, H. and Costi, A. (1988) Biochem. J. 255, 357–360.
77 Crompton, M. and Costi, A. (1990) Biochem. J. 266, 33–39.
78 Broekemeier, K.M., Dempsey, M.E. and Pfeiffer, D.R. (1989) J. Biol. Chem. 264, 7826–7830.
79 Schlegel, J., Schweizer, M. and Richter, C. (1992) Biochem. J., 285, 65–69.
80 Hofstetter, W., Mühlebach, T., Lötscher, H.R., Winterhalter, K.H. and Richter, C. (1981) Eur. J. Biochem. 117, 361–367.
81 Richter, C., Winterhalter, K.H., Baumhüter, S., Lötscher, H.R. and Moser, B. (1983) Proc. Natl. Acad. Sci. USA 80, 3188–3192.
82 Richter, C. and Meier, P. (1990) Biochem. J. 269, 735–737.
83 Richter, C., Theus, M. and Schlegel, J. (1990) Biochem. Pharmacol. 40, 779–782.
84 Richter, C. (1990) Free Radical Res. Commun. 8, 329–334.
85 Weiss, M., Kass, G.E.N., Orrenius, S. and Moldeus, P. (1992) J. Biol. Chem. 267, 804–809.
86 Halestrap, A.P. and Davidson, A.M. (1990) Biochem. J. 268, 153–160.

L. Ernster (Ed.) *Molecular Mechanisms in Bioenergetics*
© 1992 Elsevier Science Publishers B.V. All rights reserved

CHAPTER 16

Metabolite carriers in mitochondria

REINHARD KRÄMER[1] and FERDINANDO PALMIERI[2]

[1]*Institut für Biotechnologie I, Forschungszentrum Jülich, Jülich, Germany and*
[2]*Department of Pharmaco-Biology, Laboratory of Biochemistry and Molecular Biology, University of Bari, Bari, Italy*

Contents

1. Introduction and overview 359
2. Metabolic significance of mitochondrial metabolite carriers 362
3. State of identification by purification and reconstitution 363
4. Structural studies at the protein and DNA level 365
4.1. Primary structure 365
4.2. Secondary structure and membrane topology 367
4.3. Organ specificity 369
5. Functional studies 369
5.1. Kinetic mechanism: functional schemes 371
5.2. Functional family 373
5.3. Possible channel structure 375
6. Correlation of structure and function 376
7. Biogenesis and import of mitochondrial carriers 377
8. Conclusion and perspectives 378
Acknowledgements 379
References 379

1. Introduction and overview

For many metabolic processes in the cell a concerted function of both the cytosolic and the mitochondrial compartment is necessary. Due to the presence of nonspecific pore proteins, the outer mitochondrial membrane is permeable to small molecules, such as the metabolites discussed in this review. The inner mitochondrial membrane, on the other hand, is impermeable to most solutes. Thus, the presence of specific transporters catalyzing the import into and the export out of the matrix space is essential for mitochondrial function. However, it would be too simple to explain the purpose of these carriers solely as the facilitation of the flux through the permeability barrier of the inner membrane. Although it is true for several carriers, e.g. those involved in the import of substrates like pyruvate, malate or fatty acids, particular transporters have important functions in regulating the balance between cytosol and mitochondrial matrix with respect to the phosphorylation potential (ADP/ATP carrier) or the redox potential (aspartate/glutamate carrier). Carrier proteins represent an important share of the protein content of mitochondria. The most abundant carrier in heart mitochondria, the ADP/ATP carrier, alone comprises more than 10% of the total inner mitochondrial membrane protein [1].

Up to now, the presence of at least twelve transporters for different solutes has been demonstrated in mammalian mitochondria [2–6], catalyzing transport of nucleotides, phosphate, amino acids, carboxylic acids and acylcarnitines (see Table I). Although discussed in more detail in a later chapter in this book, the uncoupling protein from brown fat mitochondria will also be mentioned here because of its obvious similarity to the family of metabolite carriers with respect to both structure and function. It is furthermore obvious that besides these 'major' carriers, a large variety of other transporters in the inner mitochondrial membrane must exist, e.g. for the import of various nucleotides and coenzymes.

The pioneering studies starting about 25 years ago were concerned with the identification of the different carriers on the basis of studies in intact mitochondria, using the methods of mitochondrial swelling [7], of direct transport measurement by silicone centrifugation [8,9], or filtration [10,11]. Detailed investigations of transport kinetics in mitochondria, as well as those

TABLE I

Metabolic significance of major mitochondrial carriers

Carrier species	Important substrates	Main metabolic significance	Organ distribution
1. ADP/ATP	ADP, ATP	energy transfer, phosphate transfer, oxidative phosphorylation	ubiquitous
2. phosphate	phosphate	phosphate transfer, oxidative phosphorylation	ubiquitous
3. aspartate/glutamate	aspartate, glutamate, cysteinesulfinate	malate/asp shuttle, urea synthesis, gluconeogenesis, cysteine degradation	most cells
4. oxoglutarate	oxoglutarate, malate, succinate	malate/asp shuttle, gluconeogenesis from lactate, isocitrate/oxoglutarate shuttle	most cells
5. pyruvate	monocarboxylates, ketone bodies	citric acid cycle, gluconeogenesis	ubiquitous
6. citrate	citrate, isocitrate, phosphoenolpyruvate, malate, succinate	lipogenesis, gluconeogenesis, isocitrate/oxoglutarate shuttle	mainly liver
7. dicarboxylate	phosphate, malate, succinate, sulfate, sulfite, thiosulfate	gluconeogenesis, urea synthesis, sulfur metabolism	mainly liver
8. carnitine	acylcarnitine, carnitine	fatty acid oxidation	ubiquitous
9. glutamate	glutamate	urea synthesis	mainly liver
10. ornithine	ornithine, citrulline	urea synthesis	mainly liver and kidney
11. glutamine	glutamine	glutamine degradation	liver and kidney
12. branched keto acids	branched chain keto acids	amino acid degradation	high in muscle/heart

carried out later using isolated and reconstituted proteins, were only possible by applying inhibitor stop methods [9].

Around 1970, several laboratories started to focus on the purification of mitochondrial carrier proteins. The first carrier to be purified by the use of adsorption chromatography on hydroxyapatite was the ADP/ATP carrier [12]. This procedure has proven to be suitable also for purification of most of the other mitochondrial carriers [6]. Although the ADP/ATP carrier was identified during purification by tight binding to its specific inhibitor, carboxyatractylate, virtually all the other carriers have been isolated by monitoring their activity after functional reconstitution. Since the reconstitution of the ADP/ATP carrier, the first carrier that was reconstituted from the purified protein [13], this method has been used for isolation and purification of all the other carriers too (with one exception [14]). In the course of the following years, nearly all 'major' carriers from mammalian mitochondria have been isolated and reconstituted.

Once mitochondrial carriers were available both as purified proteins and in a functionally active state, further investigations were directed mainly in two areas of research, which are still in focus today. On the one hand, the important field of structural studies was opened by the elucidation of the primary structure of the ADP/ATP carrier by protein sequencing [15]. Later, DNA sequencing was also successfully employed [16]. The combined data on the structure of several mitochondrial carriers revealed the presence of a common structural 'family' of mitochondrial carrier proteins [17–20], which on the basis of topological considerations is in fact thought to represent a 'sub-family' within the larger kingdom of carrier proteins in general [21–25]. On the other hand, concomitantly with the elucidation of a close relation of mitochondrial carrier proteins to each other on a structural basis, efforts were undertaken to elucidate the transport mechanism of mitochondrial carriers in more detail. In intact mitochondria, elaborate kinetic analyses, exceeding the determination of basic kinetic parameters, were carried out for several carriers, i.e. the ADP/ATP carrier [26–32], the oxoglutarate carrier [33,34], the aspartate/glutamate carrier [35–37], the phosphate carrier [38] and the dicarboxylate carrier [39,40]. In addition to these studies, in later years the transport mechanism of several mitochondrial carriers was examined in detail by using functionally reconstituted transporters [6]. Also in this case, recent data suggest a carrier 'family', based on common kinetic function.

In view of the broad field of research on mitochondrial carriers we cannot review all the important topics, as discussed above. This of course holds true also with respect to the overwhelming amount of literature. After a basic introduction, we thus mainly intend to describe the progress in defining and characterizing these transporters at the molecular level. We will focus on the important developments with respect to the elucidation of both the structural and the functional properties of mitochondrial metabolite transporters, i.e. topics which in the end will hopefully lead to finding the desired structure–function correlation. Most data mentioned will describe the carriers from mammalian mitochondria. However, especially in the section concerning the structure–function relationship, new developments with respect to yeast mitochondria will be included. For a detailed overview of the earlier literature, especially with respect to the influence on particular metabolic fluxes in the cell, the reader is referred to reviews on mitochondrial carriers in general [2,3,6,41–46], on transport in plant mitochondria [47,48], and on particular mitochondrial carriers, i.e. the ADP/ATP carrier [1,49–53], the phosphate carrier [54–57], the pyruvate carrier [58] and the uncoupling protein [59,60].

2. *Metabolic significance of mitochondrial metabolite carriers*

The metabolite carriers from mitochondria which are well known until now are listed in Table I. There are numerous possibilities for their classification, which have been used in other reviews. In Section 5, the classification according to their type of mechanism and energetic regulation will be discussed. In this section, their metabolic significance will be briefly mentioned; for an extensive description, the reader is referred to the reviews cited above [2,5,41,46].

The metabolic significance of these carriers is reflected in their organ distribution in mammals. The central function of mitochondria is production of ATP, thus the first two carriers involved in energy transfer and oxidative phosphorylation, the ADP/ATP carrier and the phosphate carrier, are present in all mitochondria. Also, the main carriers for import of reducing equivalents or substrates for oxidative phosphorylation into the mitochondrial matrix are widely distributed. This includes the aspartate/glutamate carrier, the oxoglutarate carrier, the pyruvate carrier and the carnitine carrier. Several carriers, on the other hand, have a limited distribution. This reflects their main importance in special functions, e.g. gluconeogenesis (dicarboxylate carrier), fatty acid synthesis (citrate carrier), or urea synthesis (glutamate carrier, glutamine carrier, ornithine carrier).

Another important differentiation of mitochondrial carriers is based on their inhibition by different reagents. These have been the main tools for identification of these carriers. For this wide field of investigations, the reader is also referred to earlier excellent reviews [2,5,46]. Inhibitors have been very useful for distinguishing these carriers in situ in mitochondria and for studying their kinetics by using the inhibitor stop method [9] both in intact mitochondria and in reconstituted systems [61]. Most successful in this respect have been the cases where specific inhibitors were found for particular carrier proteins, e.g. the atractylosides and bongkrekate for the ADP/ATP carrier (in this case even side specific) [1,31], benzenetricarboxylate for the citrate carrier [62], phtalonate for the oxoglutarate carrier [63], and α-cyano-3-hydroxy-cinnamate for the pyruvate carrier [64].

The overall activities of these carriers are of course very different because of differences in their importance in different metabolic fluxes [3]. The highest activity can be attributed to the phosphate carrier, followed by the other carriers with relatively high activity, like the ADP/ATP carrier, the aspartate/glutamate carrier, the carnitine carrier, the pyruvate carrier (and the glutamine carrier in liver and kidney). Because of the different molecular activities, the transport activity as determined in the intact organelle, however, does not of course provide a correct basis for estimating the number of copies of these carriers. This becomes especially obvious when comparing the two carriers involved in phosphate transfer, the ADP/ATP carrier and the phosphate carrier. The high activity of the phosphate carrier in mitochondria in vivo is mainly due to the high turnover, about 70 mmol min^{-1} g^{-1} protein at 25°C [65], whereas the high total activity of the mitochondrial ADP/ATP carrier is based on the fact that up to 13% of the inner mitochondrial membrane (in heart) is composed of this protein [1].

The carriers mentioned here have, as already stated above, a specific function in particular metabolic pathways. This is reflected by the fact that, if a particular metabolic flux is considered, one or a few carriers are specifically important or in some cases even controlling the fluxes under consideration. This has been elegantly shown for example for the possibility of controlling function of the ADP/ATP carrier in respiration [46,66,67], of gluconeogenesis in liver by the pyruvate carrier [68,69], and of urea synthesis in liver and kidney by glutamine uptake [46,70,71].

Finally, we want to emphasize that the list of carriers presented here is by no means complete. There are definitely many carriers within the mitochondrial membrane in addition to those shown in Table I, the presence of which is perfectly clear. This can be concluded in part from theoretical (metabolic) considerations and in part from experimental evidence in intact mitochondria. In this review we focus on molecular data, which are not available in those cases. Therefore, some of these in fact essential carriers will only briefly be mentioned here. Not only have nucleotides to be exchanged between the mitochondrial matrix and the cytosol in energy metabolism, but also net uptake of various kinds of nucleotides has to occur. At least one carrier of this type, a Mg-dependent ATP uptake carrier, has been identified [72,73]. It is also obvious that a wide variety of cofactors, coenzymes and additional compounds which are not synthesized in mitochondria, also have to be taken up. Experimental evidence has already been presented for uptake systems for *N*-acetylglutamate [74], for coenzyme A [75], for thiamine [76], for glutathione [77], and for spermine [78].

3. State of identification by purification and reconstitution

Even in view of the fact that some mitochondrial carriers have been successfully analyzed using the methods of molecular genetics (see Section 4), the identification of carrier proteins by isolation, purification and reconstitution still remains the main tool for functional characterization and also for important structural studies. The different kinds of methods used for successful solubilization of mitochondrial carriers as well as those used for purification and finally for functional reconstitution have been extensively reviewed recently [6]. In the last three years, however, significant progress has been achieved in this field of research.

For solubilization of mitochondrial carrier proteins, mainly nonionic detergents with long polyoxyethylene tails, characterized by low critical micellar concentrations, have been used [6]. Ionic detergents, and those with high critical micellar concentration in particular, lead to inactivation of this type of carrier. In spite of numerous efforts, however, the suitability of a particular detergent under specific solubilization conditions cannot be explained by a convincing rationale, but is still mainly based on empirical observations [6]. Frequently, lipids have to be added during solubilization in order to improve the stability in the presence of detergents.

A striking feature with respect to purification is the particular usefulness of hydroxyapatite in chromatographic procedures. With very few exceptions all mitochondrial carrier proteins have been purified using hydroxyapatite at least in the major step in the different procedures. In spite of this general concept, it has to be pointed out that successful purification in many cases needed the application of additional important 'tricks' for modifying the hydroxyapatite chromatography, i.e. use of special lipids, particular ionic conditions and appropriate design of elution buffers. Additionally, in most cases further chromatographic steps had to be applied (see below and ref. [6]).

So far ten mitochondrial carriers have been successfully purified to the point where they show a single homogeneous band in SDS-PAGE. Table II gives an overview of the purification of these carriers. In this table only the first publication reporting successful purification and/or functional reconstitution of a particular carrier is mentioned.

In addition to the overview in Table II, several remarks have to be made. A main breakthrough in completing the list of purified metabolite carriers from mitochondria has actually been made in the last few years. The first successful isolation and reconstitution of the ADP/ATP carrier [12,13], the uncoupling protein [14,91,92], the phosphate carrier [79,80], and the oxoglutarate carrier [82,83] were achieved in the years 1975 to 1985. From 1988 to date the purification in functionally active state of the dicarboxylate carrier [84], the citrate carrier [85,86], the pyruvate carrier [87], the carnitine carrier [88], the aspartate/glutamate carrier [89]

TABLE II

Purification and reconstitution of mitochondrial carriers from mammalian tissues

Carrier species	Source[a]	Purified by[b]	Mol. wt. (SPAGE)[b]	Identification by
1. ADP/ATP	BHM [12]	HA-chrom.	30 kDa	inhib. binding [12] reconstitution [13]
2. uncoupling protein	BFM [14]	HA-chrom./ultracentr.	32 kDa	GTP-binding
3. phosphate	PHM [79]	HA/affin. chrom.	33 kDa	reconstitution NEM-binding
	BHM [80]	HA-chrom.	34 kDa	SPAGE[b]
	RLM [81]	HA and other chrom.	33 kDa	reconstitution NEM-binding[b]
4. oxoglutarate	PHM [82]	HA/celite chrom.	31.5 kDa	reconstitution
	BHM [83]	HA/celite chrom.	31.5 kDa	reconstitution
	RLM [84]	HA-chrom.	32.5 kDa	reconstitution
5. dicarboxylate	RLM [84]	HA/celite chrom.	28 kDa	reconstitution
6. citrate	RLM [85]	HA/celite chrom.	30 kDa	reconstitution
	BLM [86]	HA/silica chrom.	37 kDa	reconstitution
7. pyruvate	BHM [87]	HA/affin. chrom.	34 kDa	reconstitution
8. carnitine	RLM [88]	HA/celite chrom.	32.5 kDa	reconstitution
9. aspartate/glutamate	BHM [89]	HA/celite chrom.	31.5 kDa	reconstitution
10. ornithine	RLM [90]	HA-chrom. and others	33.5 kDa	reconstitution

In Table II only the first publication reporting complete purification and or functional reconstitution is mentioned. For an overview describing different stages of purification, different methods of enrichment from other sources as well as data about partially purified carrier proteins see ref. [6] and text.

[a] Abbreviations used for the different sources of mitochondria: BHM, beef heart; BLM, beef liver; RLM, rat liver; PHM, pig heart; BFM, brown fat.

[b] HA: hydroxyapatite, SPAGE: SDS-polyacrylamide gel electrophoresis; NEM: *N*-ethylmaleimide.

and the ornithine carrier [90] have been reported. The arduous efforts in isolating functionally active carriers have been especially successful recently when specific modifications of the general isolation procedure for mitochondrial carriers were applied.

This general purification scheme involves (i) solubilization with a nonionic detergent, (ii) chromatography on hydroxyapatite, recovering the respective carrier in the eluate, and (iii) final purification using further chromatographic procedures [6]. Varying this general scheme, the following modifications have been applied to achieve complete purification of individual carriers. Procedures applying to step (i) are: preextraction of mitochondria, variation in the type of nonionic detergent, particular pH and ionic conditions during solubilization, and addition of specific lipids to the solubilization buffer. Step (ii) has been varied by applying hydroxyapatite under different conditions (for example dry columns) and significant variation in the protein/hydroxyapatite ratio, and by partial removal of the solubilizing detergent before chromatography on hydroxyapatite. The modifications of step (iii) include a large variety of different column materials, i.e. celite, silica gel, various Matrex affinity gels, SH-affinity chromatography and binding to affinity columns with the ligand 2-cyano-4-hydroxycinnamic acid. In addition, in some cases a specific design of buffers, including presaturation of the respective columns was necessary.

Besides further completion of the list shown in Table II, the successful purification of the pyruvate carrier and the aspartate/glutamate carrier solved another problem. Based on labeling with phenylmaleimide, a molecular mass of 15 kDa was reported for the mitochondrial pyruvate carrier [69]. The correct molecular mass of the functionally active protein now turned out to be 34 kDa [87]. The partial purification of the aspartate/glutamate carrier by a rather complicated HPLC method led to definite enrichment of a protein with an apparent molecular mass of about 68 kDa [93]. After complete purification, controlled by functional reconstitution of the aspartate/glutamate carrier, it turned out that the previous observation was probably due to dimerization under these conditions, and the aspartate/glutamate carrier has in fact a molecular mass of about 31.5 kDa [89]. Thus all mitochondrial metabolite carriers isolated until now, also including the pyruvate carrier and the aspartate/glutamate carrier, as well as the recently isolated ornithine carrier [90], fall into a very narrow range of apparent molecular masses between 28 and 34 kDa.

In addition to the now nearly complete list of successfully purified 'major' mitochondrial carrier proteins, there seems to be at least one other candidate which is expected to be added to the list in the near future. Solubilization and purification of the carrier catalyzing uptake of branched chain keto acids led to a preparation consisting of only two bands with molecular masses of 39 and 41 kDa. However, both still show activity of the aminotransferase and the corresponding carrier protein [94,95].

The pressure on obtaining molecular data about the carrier function has directed the investigation of mitochondrial carriers to switch to yeast mitochondria, which are more easily accessible to the approach by molecular genetics (see Sections 4 and 5). This has obviously prompted isolation and characterization of metabolite carriers from yeast mitochondria. This task, however, is definitely more difficult as compared to the sources of mitochondria mentioned in Table II. This is, on the one hand, due to difficulties in isolation of intact mitochondria and, on the other hand, due to the presence of large amounts of mitochondrial outer membrane porin in the detergent extracts. Successful purification of the ADP/ATP carrier [96] (see Sections 4 and 5), the phosphate carrier [97] and partial purification of the pyruvate carrier [98] have been reported so far.

4. Structural studies at the protein and DNA level

4.1. Primary structure

It is only ten years ago since structural studies on membrane carrier proteins started with elucidation of the sequence of the ADP/ATP carrier from bovine heart, which in fact was the first metabolite carrier of which the amino acid sequence was determined [15]. Since then, the primary structure of ADP/ATP carriers from other sources, namely from mitochondria of *Neurospora crassa* [99], of *Saccharomyces cerevisiae* [100], of *Zea mays* [101] and recently also of human tissues (see below), has been established by DNA sequencing. The amino acid sequence of the uncoupling protein from brown fat and the phosphate carrier from bovine heart were determined both by amino acid analysis [17,19] and DNA sequencing [18,102]. Also the phosphate carriers from rat liver [103], human heart [104] and yeast [105] have been sequenced at the DNA level. Recently the oxoglutarate carrier from cow [20] and man [106] joined this family as the fourth carrier.

From analysis of the respective primary structures and from comparison of these different carriers several important aspects became clear (Fig. 1).

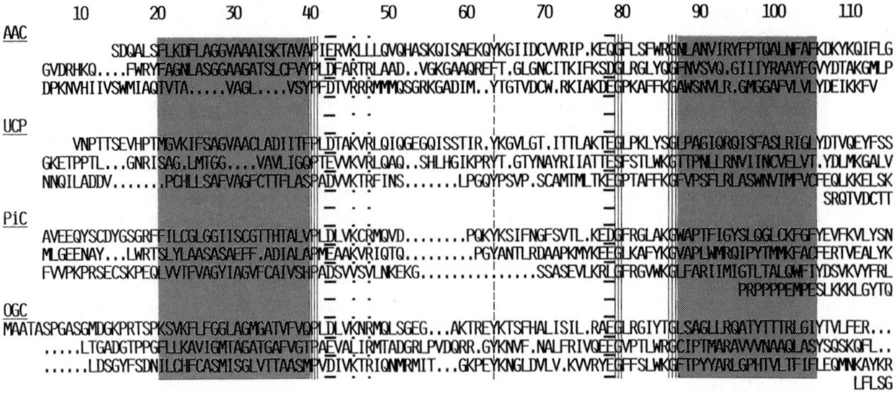

Fig. 1. Alignment of the amino acid sequences of four mitochondrial carriers. The hydrophobic regions are shaded. Abbreviations: AAC, ADP/ATP carrier; UCP, uncoupling protein; PIC, phosphate carrier; OGC, oxoglutarate carrier.

(i) The hydropathy plots derived from the sequence data are now generally accepted to be in agreement with the presence of six transmembrane α-helices when taking into consideration possible amphipathic helices [17,18,20,107]. It has to be mentioned, however, that the correct number of membrane-spanning regions is not yet absolutely clear, and is mainly based on homology considerations. In the case of AAC, for example, although the N-terminus was shown to be exposed to the cytosolic side [108], the orientation of the C-terminus has still to be determined.

(ii) A striking feature of mitochondrial carriers in general is their tripartite structure, i.e. an apparent triplication within the primary structure [107]. This can be realized by 'diagon plots', in this case used to identify internal sequence identity. Obvious similarities exist between three domains of about 100 amino acid residues each (Fig. 1) within the respective carrier proteins. Furthermore, the analysis of this triplicate structure is one of the most convincing arguments so far for the presence of in total six transmembrane helices in the carrier proteins of the 'mitochondrial family', in spite of some doubts on the interpretation of the hydropathy plots.

(iii) Based on this analysis, the four carrier proteins of which the structure is given in Fig. 1, show clear sequence similarity. This becomes even more evident when comparing the predicted to the determined data on the respective secondary structures (see below). This putative homology is the main basis for the now generally accepted hypothesis of a mitochondrial carrier family, arguing for a common ancestor gene with a size of about 100 amino acids [17,107]. All these carriers, which are encoded by nuclear genes, may have evolved first by gene triplication and subsequent diversification. The structural similarity shown in Fig. 1 is another reason for the uncoupling protein being included in the mitochondrial metabolite carrier family, even though it is a proton transporter.

The common motif of a tripartite structure composed of three segments of about 100 residues together with conserved sequence motifs, mainly located in the large matrix loops, has even been used to identify genes and proteins, respectively, as belonging to the mitochondrial carrier family. This was done even when their function and significance was not known, and held true for two proteins from yeast mitochondria [109] and a gene from man [110].

The mitochondrial carrier family can be integrated into the still increasing number of carrier families in general. It becomes obvious that both prokaryotic and eukaryotic membrane carriers, with few exceptions, show in common a number of 12 transmembrane helices, as shown by numerous studies [21–25]. It is therefore quite obvious that the mitochondrial carriers can be included into this scheme by assuming that they function as dimers. Thus the functional entity also accounts for 12 transmembrane helices. It should, however, be emphasized that the hypothesis of mitochondrial carriers being functional dimers was not in any sense basically the result of homology considerations, since dimerization of mitochondrial carrier proteins had been deduced many years before. There are in fact several good experimental arguments for mitochondrial carriers being functional dimers. These include the binding stoichiometry of inhibitors and of substrates [1] to the ADP/ATP carrier. It should be mentioned, however, that binding experiments using different adenine nucleotide analogues [111,112] were interpreted in favor of a functional tetramer. In addition to the above-mentioned argument, cross-linking studies with the ADP/ATP carrier [113], and especially with the uncoupling protein [114] and the oxoglutarate carrier [106] are in favor of functional dimers. Finally, mitochondrial carrier proteins were shown to be dimers when isolated in the solubilized state [115,116]. Although these arguments are of course not unequivocal proofs, when taken together, the functional dimeric state of mitochondrial carriers seems to be relatively well established.

4.2. Secondary structure and membrane topology

Based on the location of transmembrane helices as predicted from hydropathy plots, as well as supported from studies using chemical modification, antibody probes and protease cleavage, a certain secondary structure of mitochondrial carriers can be envisaged. As shown in Fig. 2, for the phosphate carrier both the N-terminal and the C-terminal regions are facing the external, cytoplasmic surface. These data have been mainly provided by the use of peptide-specific antibodies and by specific protease cleavage [106,117,118]. The interconnecting hydrophilic loops are of considerable length. Those protruding into the matrix volume are significantly larger than the cytosolic ones.

The hydropathy plots of all mitochondrial carriers of which the sequence is known so far, seem to be essentially very similar [20], yet in the case of the ADP/ATP carrier there still exist controversial data, or interpretations, on the topology, especially of the matrix loops. Nevertheless, there is agreement that the ADP/ATP carrier also contains six transmembrane helices and that both the N- and the C-terminus (not proven in this case) are facing the cytosol, similar to that described above for the PIC. The topological studies of the ADP/ATP carrier have been based on experiments with specific amino acid reagents [119], on labeling with covalently binding substrate and inhibitor analogues [120,121], on specific peptide bond cleavage [108], and on the use of anti-N-terminal antibodies [108]. Due to conflicting results of amino acid modifications, especially of lysine residues by pyridoxalphosphate, there exist differing interpretations of the location of particular residues generally supposed to be situated within matrix loops. Attempts have been made to accomodate this situation in a positive way, by assuming that parts of the loops facing the matrix side protrude into the membrane part, thus forming a hydrophilic region [53]. This pathway was then supposed to form part of the transport channel. The location of the respective amino acids is more or less in agreement with the proposal of ligand-binding sites, a result obtained independently by chemical probing [120,121].

If we compare these results with the level of knowledge on the topology of several other membrane carrier proteins, both in bacteria and in eukaryotes, we realize that this level is still somewhat lower for mitochondrial carriers. The membrane topology of bacterial metabolite carriers, e.g. the lactose carrier (LacY) [122], the membrane subunit of the maltose transport

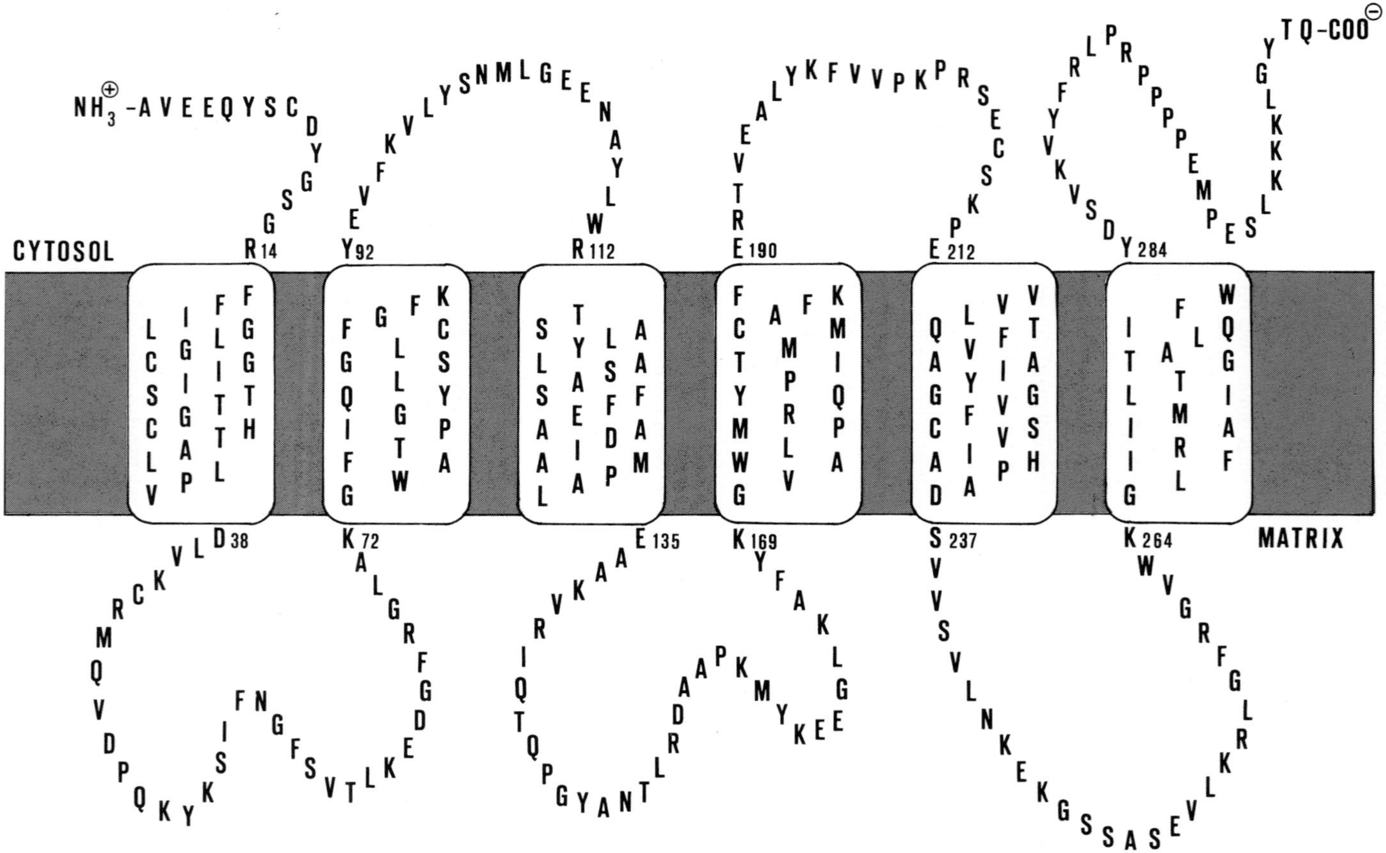

Fig. 2. Model for the transmembrane arrangement of the phosphate carrier from bovine heart mitochondria, as suggested in refs. [117] and [106]. The N- and C-terminal regions of the polypeptide chain are located on the external cytoplasmic surface. Six α-helices cross the membrane. The loop containing the cleavage site of Arg-endoprotease (Arg^{140} and/or Arg^{152}) protrudes into the matrix. The loops containing Lys^{96} and Lys^{198}, the peptide bonds of which are susceptible to Lys-endoprotease, are exposed towards the cytosol.

system (MalF) [123] or the family of sugar carriers in bacteria [24], was elucidated by using mainly the advantage of phoA- and lacZ-fusions. This made possible, for example, an exact location of particular amino acids at the membrane surface region or in adjacent helices. A comparable state of analysis was achieved also for mammalian glucose carriers [124].

4.3. Organ specificity

As in the case of many other proteins some mitochondrial carriers also exist in isoforms encoded by different isogenes. Early evidence for organ-specific isoforms came from immunological studies of the ADP/ATP carrier [125,126]. Later the occurrence was shown of at least three different genes in man, coding for organ specific ADP/ATP carrier species [127]. Also in cow [128], yeast [129,130], and corn [131] different genes for the ADP/ATP carrier have been found. The three human genes were predominantly found in heart, liver and fibroblasts, respectively [127,132–134]. The expression of one bovine ADP/ATP carrier gene predominates in heart, another in intestine [128]. It is furthermore now clear that the expression of these genes is sensitive to particular metabolic conditions of the cell. This has been observed both in man [135] and in yeast [130,136]. The most interesting question of course concerns a possible functional difference of these various isoforms. Except for yeast, this question has not been solved so far (see Section 6).

With respect to the other members of the mitochondrial carrier family, limited data are available. Only one gene has been found for the phosphate carrier in cow [18] and in yeast [105], for the uncoupling protein in mouse [137] and man [138], and for the oxoglutarate carrier in cow and man [106]. The single isoform that has been detected for the oxoglutarate carrier is expressed at least in heart, liver and brain [20].

5. Functional studies

An impressive number of publications on the function of carrier proteins in intact mitochondria has by now accumulated. Most of these have already been excellently reviewed [2,5]. Essentially not much new information has been obtained from intact mitochondria. Thus, with respect to these results, the reader is referred to the above-mentioned reviews. Some basic data are also included in Table III, where the 'major' mitochondrial carriers are listed according to the respective category of mechanism, in similar order as introduced by LaNoue and Schoolwerth [2].

In the last few years, however, new data on mitochondrial carriers in reconstituted systems have become available, due to the fact that many of these carriers have been isolated only recently. In addition to the overview on transport mechanism (Table III), we would like to emphasize some interesting developments mainly based on functional studies with isolated and reconstituted proteins. The use of proteoliposomes for the analysis of the carrier protein function has some obvious advantages [6]. Interfering activities of other proteins (enzymes) and carriers are in general avoided when using purified or at least partially purified carrier proteins. It is very helpful for kinetic studies that the concentration of substrates can be varied freely on both sides of the membrane and particular problems with binding or possibly microcompartmentation [144,145] can be avoided. On the other hand, there are of course also drawbacks. In this respect, the somewhat undefined state of activity after isolation and purification, the a priori unknown orientation of the inserted proteins, and the possibility of missing a regulatory compound should be mentioned.

TABLE III

Energetic regulation and kinetic mechanism of mitochondrial carrier proteins

Carrier species	Transport mode	Driving force[a]	Kinetic mechanism
A. Electrophoretic carriers			
1. adenine nucleotide	antiport	$\delta\Psi$	controversial[b] data
2. aspartate/glutamate	antiport	$\delta\Psi + \delta pH$	sequential
3. uncoupling protein	uniport[c]	$\delta\Psi$ (+ δpH)[c]	–
B. Electroneutral, proton compensated carriers			
4. phosphate	symport (antiport)[e]	δpH[g]	not known[d]
5. pyruvate	symport/antiport	δpH[g]	not known[d]
6. glutamate	symport/(antiport)[e]	δpH[g]	not known[d]
7. branched keto acids	symport/antiport	δpH[g]	not studied
C. Electroneutral exchange carriers			
8. oxoglutarate/malate	antiport	–	sequential
9. dicarboxylate/phosphate	antiport	–	sequential
10. citrate/malate	antiport	–	sequential
11. ornithine	antiport[f]	not conclusive[f]	not studied
D. Neutral carriers			
12. carnitine	antiport/uniport	–	not studied
13. glutamine	uniport[h]	–	not studied

[a] Only driving forces besides the chemical gradient of substrates itself are mentioned.
[b] For discussion of the ADP/ATP carrier mechanism see text.
[c] Only transport of protons is referred to here, thus the δpH is a substrate gradient in this case.
[d] For discussion of the molecular mechanism of electroneutral, proton compensated carriers see ref. [5]. The studies are not detailed enough to define the kinetic mechanism.
[e] In these cases antiport means homologous exchange of the same substrate from both sides, which in general takes place in this type of carrier ('exchange diffusion').
[f] It is not yet clearly elucidated whether antiport of ornithine against citrulline is the physiological function of this carrier. Both electrogenic transport in an uniport mode [139] or as ornithine/phosphate exchange [140] and electroneutral ornithine/citrulline [141] or ornithine/proton exchange [142] have been suggested. In the case of the purified and reconstituted carrier, ornithine/citrulline exchange was found to be the predominant function [90].
[g] The pH-gradient is a driving force only when uniport is considered.
[h] Although being electroneutral, uptake of glutamine was shown to depend linearly on the proton gradient across the inner mitochondrial membrane [143].

Table III lists the mitochondrial carriers according to their basic mechanism category. For an elaborate definition of the transport mode and the experimental differentiation of the respective driving forces previous reviews should be consulted [2–5,44]. These basic data have already been well established by studies in intact mitochondria and do not need any extension or correction. In addition, some new developments mainly in the exact kinetic definition and in the interpretation of these mechanisms have been described in the recent literature. This holds true especially for the oxoglutarate carrier [83,146], the dicarboxylate carrier [147,148], the citrate carrier [149,150] and the carnitine carrier [151], where new data on the optimization of the reconstitution procedure, on the exact determination of basic parameters like V_{max}, K_m

and activation energy [65], and on particular interactions with specific phospholipids have been described.

5.1. Kinetic mechanism: functional schemes

In view of the very similar primary and secondary structures of mitochondrial carriers elucidated so far (see Section 4), it is somewhat surprising to find such large differences in the type of substrates, including one of the largest substrates transported (ATP) and the most simple one (proton). Furthermore, extensive variations in mechanisms are also observed, i.e. unidirectional transport, symport and antiport. It has to be emphasized here that it is absolutely justified to include the uncoupling protein into the family of metabolite carriers, not only due to its similar structure but also because of its functional properties concerning molecular activity and kinetics. Additional functions of this protein with respect to anion transport are obviously important for describing the correct mechanism and its physiological significance [59,60,152,153]. They will be treated in detail elsewhere (see Chapter 20, this book). The data concerning different mitochondrial carriers have been combined in a hypothesis on the development of functional schemes [60,154]. The basic kinetic scheme of the majority of mitochondrial carrier proteins, exemplified here by exchange of ADP and ATP by the adenine nucleotide carrier, is shown in Fig. 3A. This translocation cycle describing coupled antiport, can be easily modified to a scheme representing substrate/H^+-symport or substrate/OH^--antiport, respectively, representing, for example, the mechanism of the phosphate carrier (Fig. 3B). The scheme may ultimately be even reduced to a simple uniporter (Fig. 3C), i.e. the uncoupling protein, catalyzing unidirectional transport of protons [60]. Although these models in fact seem to be very simple, the functional consequences of these schemes and of the underlying mechanisms are by no means trivial. The observed variations in the mechanisms by which the widely different substrates of these carriers are transported, provided the basis for developing a fascinating new concept of the mechanism and energy involved in carrier catalysis [154,155], based on a fundamental comparison with the principles of the mechanism of enzyme catalysis [156,157]. An essential feature of this hypothesis concerns the extent of conformational changes of the carrier protein necessary to translocate the substrate. The large positive free energy involved in this process, i.e. the change in the conformation of the carrier protein, is compensated by the intrinsic binding energy of the substrate/carrier interaction. The resulting difference in the energy profile when comparing the liganded and the unliganded carrier thus limits the probability of the carrier without bound substrate undergoing an appropriate conformational change, i.e. the probability for a unidirectional transport. It therefore seems only plausible that transport of the extremely large substrate ATP, or of carboxylates in the case of other mitochondrial carriers, is catalyzed by antiport mechanisms. One can rationalize that small substrates like protons or possibly phosphate, on the other hand, which do not exert an extensive constraint on the carrier protein, could be transported by uniport mechanisms [60,155].

An exception to this qualitative rule, when applied to mitochondrial carrier proteins, seems to be the carnitine carrier, which, in spite of the large substrate transported (acylcarnitine), can catalyze both uniport and antiport [158,159]. The upper part of Fig. 3D can be used for explaining unidirectional uptake of carnitine by the carnitine carrier. This variation of the general mechanism would in principle also be a model for the pyruvate carrier and the branched chain keto acid carrier, which catalyze both uniport and antiport as well. However, since the primary structure of these carriers and thus their putative membership of the mitochondrial carrier family is not yet known, this idea is purely hypothetical.

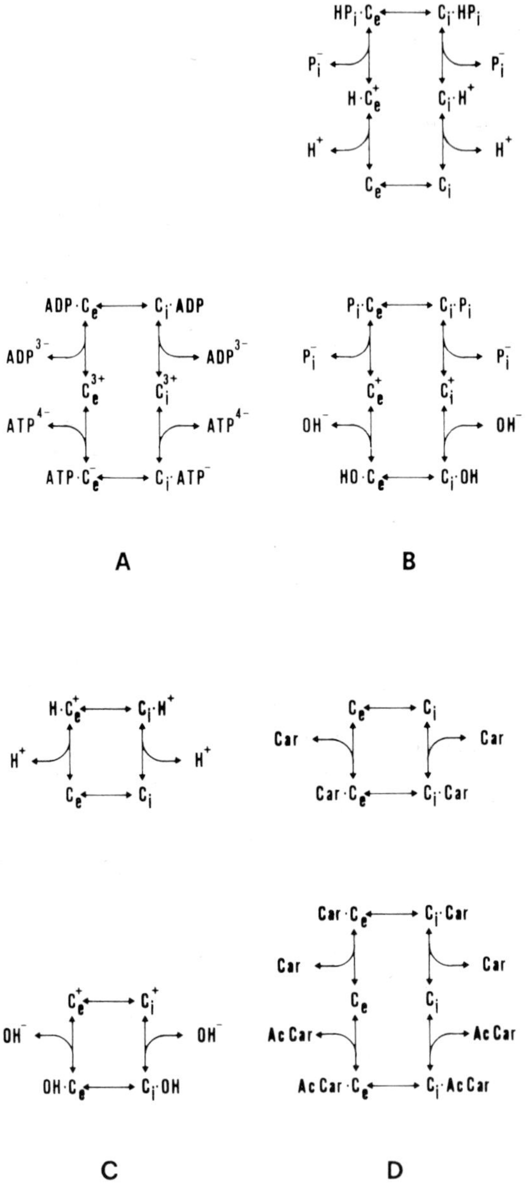

Fig. 3. Kinetic schemes of transport cycles catalyzed by various mitochondrial carriers. The figure is basically derived from a functional description presented for the uncoupling protein [60]. The examples of kinetic schemes for the different type of carriers are: (A) adenine nucleotide carrier; (B) phosphate carrier, assuming either H^+-cotransport (upper part) or OH^- antiport (lower part); (C) uncoupling protein, modifications as explained for (B); (D) carnitine carrier, kinetic scheme for uniport function (upper part) and antiport function (lower part). The indices mean C_e, carrier in a conformational state with the substrate binding site facing the cytosolic side; C_i, conformational state with the substrate binding site facing the matrix side.

If the scheme described in Fig. 3, in fact, correctly describes a correlation between the mechanisms of various mitochondrial carriers, it becomes clear that substrate binding sites (e.g. the nucleotide binding site of the ADP/ATP carrier) must be assumed to have changed into regulatory sites (GTP binding site in the uncoupling protein). Furthermore, protons (or OH^- ions) are treated as normal substrates. The models as discussed above have still to be proved in many aspects by detailed kinetic studies and by appropriate correlation to structural properties of the respective proteins (see Section 6). Finally, it should be mentioned that these schemes, when taken as simple as drawn in Fig. 3, pose some severe problems in connection with the data on a functional family, based on a particular kinetic mechanism. This will be discussed extensively in Section 5.2 (see below).

5.2. *Functional family*

Mainly due to recent data obtained with mitochondrial carriers in reconstituted systems, another interesting simplification evolves with respect to the function of this carrier family. As clearly pointed out in a previous review [5], basically a ping-pong type of mechanism for these kinds of carriers could be expected, what was in fact concluded based on data for the ADP/ATP carrier [31] and the aspartate/glutamate carrier [36]. However, studies on the oxoglutarate carrier [34] and the ADP/ATP carrier [29] already gave arguments in favor of a sequential type of mechanism for binding and transport. In the case of an antiport carrier, in ping-pong kinetics, the first substrate (transport substrate from side a) dissociates from the carrier before the second substrate (transport substrate from the side b) binds to the carrier. A well-known example for ping-pong kinetics is the transaminase reaction [160]. In principle, Fig. 3A is in agreement with a typical ping-pong type of mechanism. On the other hand, a so-called sequential mechanism would imply binding of the second substrate (transport substrate on side b) before the first substrate (substrate from side a) is released, thus a ternary complex between the carrier and two substrates would have to be involved in the transport cycle.

For reconstituted carrier proteins from mitochondria, until now sequential binding and transport kinetics have been shown conclusively for the aspartate/glutamate carrier [161], the oxoglutarate carrier [146], the dicarboxylate carrier [162] and the citrate carrier [163]. These data have been corroborated for the aspartate/glutamate carrier also in intact mitochondria [164]. Although the kinetic mechanism of all carriers mentioned above, when reconstituted into liposomes, proved to be basically of sequential type, some complications were recognized. In the case of the dicarboxylate carrier a complex binding behavior involving two types of binding sites for phosphate and dicarboxylates, respectively, was observed both in mitochondria [39] and in the reconstituted system [148]. In spite of this situation, a sequential mechanism could be shown when using antiport substrates of one class of ligands [162]. In the case of the carnitine carrier, in addition to antiport activity, uniport activity can be observed under physiological conditions both in intact mitochondria [158] and in the reconstituted system [159].

Since for the most extensively studied carrier in mitochondria, the ADP/ATP carrier, kinetic data in favour of a sequential mechanism have also been reported [29,165], it is tempting to assume a homogeneous functional family, based on a common kinetic mechanism. However, we have to take into account that the interpretation of kinetic data has obvious limitations and may only show one aspect of the functional properties of these carrier proteins. It has to be considered that the behavior of the ADP/ATP carrier with respect to substrate and inhibitor binding clearly indicates the presence of two different states of the carrier protein, depending on the side at which the ligands are bound [31]. These results have therefore been interpreted in terms of reorientation of the ligand binding site. Also for the phosphate carrier two significantly different states have been suggested, based on the observation of different accessibility of SH-groups to different reagents [166] or to the same reagent under different conditions [167]. To summarize, these results were mainly based on defined functional states of the ADP/ATP carrier and on the exclusiveness of accessibility of binding sites for particular ligands, and were interpreted as a 'single binding center gated pore' mechanism [53]. This functional model basically does not fit into a sequential type of mechanism [31]. In fact, this holds equally true for the two alternatives of both binding to two independent sites at the two opposite sides of the membrane, as discussed for the aspartate/glutamate carrier [161], and formation of a ternary complex with two substrates bound at one side, as suggested for the ADP/ATP carrier [29] and the aspartate/glutamate carrier [35]. In the case of the electroneutral, proton compensated carriers (see Table III), the kinetic studies reported were not detailed enough to provide a basis for the decision between a ping-pong or a sequential type of mechanism (see also ref. [5]).

In spite of these considerations, when summarizing the data obtained from the various studies in reconstituted systems, as discussed above, a pattern seems to evolve where the mitochondrial carriers, besides forming a structural family (cf. Section 4), also constitute a homogeneous group characterized by an identical mechanism, i.e. a sequential mode of transport [146,164]. We would like to emphasize that the distinction between ping-pong type and sequential type kinetics is not trivial or merely fun for the kineticist; in fact it has important structural consequences. If a carrier protein accepting two substrates, e.g. a classical antiport

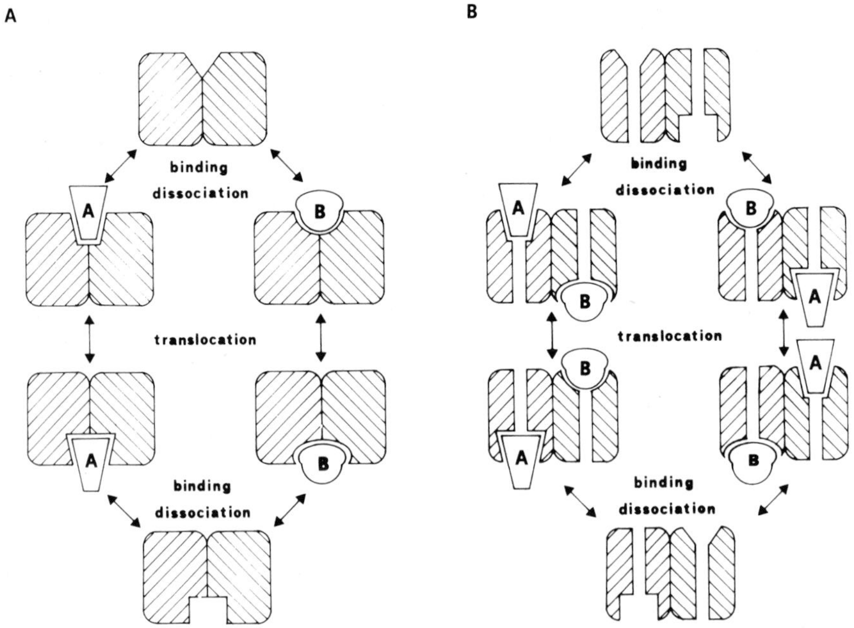

Fig. 4. Functional models for mitochondrial exchange carriers. The different substrates A and B are bound to the active sites at different sides of the membrane. (A) 'Single site gated pore mechanism' as suggested for the adenine nucleotide carrier [53]. The single binding site is exposed either at the inside or at the outside of the carrier protein. The transport pathway is formed between the two monomers. The kinetic mechanism is of the ping-pong type. (B) Transport mechanism as suggested for the aspartate/glutamate carrier [161]. The dimer formed contains two binding sites, exposed at one side of the membrane each. A separate, preformed channel in each monomer is assumed. The kinetic mechanism is sequential.

carrier, functions according to a ping-pong mechanism, basically only one binding site is necessary, alternatively exposed to the two different sides of the membrane [1]. On the other hand, if a sequential type of kinetics is detected including binding at opposite sides at a given time within the transport cycle, two binding sites must be exposed and occupied (Fig. 4). It has to be taken into account that the problem could in principle be solved by simply assuming a functional dimer composed of two monomers with one substrate binding site and one transport channel each. This, however, would be in contrast with the postulate assuming a transport channel between the two putative subunits [53,113], a hypothesis not proved so far.

5.3. Possible channel structure

Coupled antiport always implies the possibility of uncoupling or 'slippage' of transport. This may not only be an unphysiological accident, but can also be of important physiological significance. It should be clear that we do not consider here the observation of simultaneous uptake and exchange, as for example in the case of the phosphate carrier. That kind of activity is observed in general for that type of carriers, e.g. also for the pyruvate carrier and the glutamate carrier, and may be due to 'exchange diffusion' catalyzed by uniport carriers. However, at least in the case of the carnitine carrier an interesting switching between a heterologous antiport mode (carnitine/acylcarnitine) and a uniport mode (uptake) occurs [158,159]. Surprisingly, a dramatic functional switch upon modification of specific cysteine residues has been observed also in the case of classically coupled antiport carriers, namely the aspartate/glutamate carrier, and to a lesser extent, also the ADP/ATP carrier [168], as well as for the carnitine carrier [169]. The results clearly indicate that the respective carrier proteins can change their transport mode from coupled antiport to unspecific uniport, although clearly carrier mediated, after chemical modification of specific amino acid residues within the active site [170]. This functional switching was fully reversible. The results have been interpreted in terms of an indication for the presence of an intrinsic channel within the antiport carrier protein. It has to be pointed out that this structural model can in principle also be reconciled with a gated pore mechanism [171]. This interpretation is only possible, however, when assuming two binding sites, either at the entrance/exit of one pore, or located at two parallel pores within the functional unit of the carrier.

It is interesting to relate these observations to the evolving general concept of carrier proteins. Mainly based on results from bacterial carriers, it has been suggested that evolution possibly started with a simple pore carrier, evolving to more and more complicated mechanisms [22]. One can thus assume that a fixed channel within a carrier protein would be a common motif for all carrier molecules, even for complicated ones like strictly coupled antiport carriers. These ideas are further supported by recent findings in the case of bacterial phosphotransferase systems, catalyzing so-called 'group translocation', which originally was thought to function by mechanisms completely different from general substrate carriers. Also these systems now, surprisingly, were shown to contain an intrinsic pore, basically catalyzing facilitated diffusion of the substrate [23,172,173]. The speculations on the presence of a preformed channel pathway within mitochondrial carriers have of course to be substantiated by further functional and especially structural studies leading to a molecular description of this putative channel.

6. Correlation of structure and function

In order to understand correctly how a particular solute is transported from one side of the membrane to the other, which in fact is the ultimate aim in carrier research, it is indispensable to gain insight at the molecular level into the correlation between elucidation of the structural properties of carrier proteins and the results from functional studies using kinetic approaches. This direction of work is currently the most fascinating one, particularly in the field of research on carrier proteins. Although still far from a true understanding, significant progress has been achieved, for example, in the detailed studies on bacterial secondary carriers like lactose permease [174,175,176], binding protein systems [177], and phosphotransferase systems [25,172,178], using mainly the molecular genetic approach, sometimes not equally well supplemented by kinetic data. In general, one should realize that the level of analysis in the mitochondrial field with respect to structure–function correlations is definitely not as advanced as in the examples mentioned above, mainly owing to obvious difficulties in applying molecular genetics properly. Another reason, however, is the fact that only few laboratories are continuing functional studies on mitochondrial carrier proteins by elaborate kinetic approaches.

The first studies in this direction were the experiments for defining specific binding sites of the ADP/ATP carrier, based on topological arrangement of the carrier in combination with chemical modification. Lysine-specific reagents [119], as well as covalently binding analogues of the substrates [112,121,179,180], and the inhibitor atractylate [120] were used in these studies. At least one region within a central hydrophilic loop was recognized to be involved in ligand binding [121,180] and possibly a second one in the third triplicate. On the basis of lysine labeling, amino acids involved in the 'translocation path' were tentatively identified [53]. Particular regions of the protein, i.e. the N-terminus [108,120] and specific residues in several loop regions [108,119], were found to be differentially exposed to chemical modification, antibody binding, or protease cleavage, depending on different conformational states. In summary, these studies are not fully conclusive, since the exact topology of the ADP/ATP carrier is still controversial. They are of course additionally complicated owing to the somewhat uncertain situation with respect to the type of multimeric state of these carrier proteins.

There is very little information available with respect to structure–function correlation in the case of the other mitochondrial carriers. Different chemical probes have been used to differentiate binding sites at the uncoupling protein for substrate (H^+) and inhibitor (nucleotides) [181–183]. Molecular interpretation of the results on modification of the carrier function due to specific modification of particular residues in the case of the aspartate/glutamate carrier [170] and of the carnitine carrier [169] are hampered by the fact that the primary structures of these carriers are still not resolved.

The most promising development so far again combines the genetic and the kinetic approach. Since expression of mitochondrial carriers in bacteria, particularly in *E. coli*, in the functionally active state has failed so far, the obvious strategy to overcome this problem was to use yeast mitochondria. The application of elaborate genetic methods was thereby made possible. This route again was started with work on the ADP/ATP carrier [184,185]. Yeast mitochondria express three different isoforms of the ADP/ATP carrier [129,130,186]. Under aerobic conditions mainly the *AAC2* gene is expressed [129]. Different deletions of the two main *AAC* genes in aerobic yeast, site-directed mutagenesis of these genes [184] and kinetic analysis using a reconstituted system [185] revealed that only the *AAC2* gene product can account for the physiological function of ADP/ATP-transport, due to both low abundance and reduced activity of the *AAC1* gene product. Recent experiments using site-directed mutations in the

AAC2 gene have provided a first insight into the role of some possibly important amino acid residues. Especially substution of arginines were found to be functionally important [187]. Essential arginines located within the transmembrane helices were identified, possibly forming part of the translocation path. Further arginines are clustered in a triplet motif at the third hydrophilic loop, which is exposed to the matrix side, and which is assumed to be involved in substrate binding and recognition, respectively.

In contrast to these results obtained in yeast, it has not been possible so far to elucidate functional differences of the isoforms expressed in different mammalian tissues (see Section 4), nor to isolate these isoforms in a functionally active state. It is known that certain energetic parameters, e.g. the membrane potential and the ADP/ATP ratio, are different in heart and liver, respectively [188,189]. Thus, it would be very interesting to know whether these structurally different ADP/ATP carrier isoforms, expressed in these organs, in fact show differences in the transport function.

7. Biogenesis and import of mitochondrial carriers

Since protein import is described in detail in a separate chapter in this book, we will briefly review here only the data that are of particular importance for import of carrier proteins and which may differ to some extent from the data on the general import mechanism.

Mitochondrial metabolite carriers are products of nuclear genes and have thus to be imported into the mitochondrial inner membrane. The ADP/ATP carrier, the uncoupling protein, the oxoglutarate carrier and the yeast phosphate carrier are synthesized without a cleavable presequence [20,105], whereas the phosphate carrier in cow, human and rat was found to carry a presequence of 49 [18,104] or 44 [103] amino acids. The proteins without a cleavable presequence carry the targeting information within the mature part.

The import of the ADP/ATP carrier precursor form, which is present as a high-molecular-weight complex in the cytosol [190], needs a special import receptor in the outer membrane [191] and uses the general mitochondrial import pathway, as described in detail elsewhere (see ref. [192] and Chapter 20 of this book). Effective import depends on particular regions within the mature part of the protein. These regions can be identified by causing them to act as targeting signals when fused to other proteins [100]. Additionally, the precursor and the mature form of the ADP/ATP carrier obviously differ in their particular conformational state [193]. The uncoupling protein was also expressed in vitro and imported into isolated mitochondria. Thereby at least two targeting regions within the uncoupling protein have been identified with possibly different functional importance during import [194,195].

The phosphate carrier from bovine heart and rat liver, which becomes proteolytically processed during import [196,197], also uses the general import pathway. A truncated form of the rat liver phosphate carrier that lacked 35 amino acid residues of the presequence, but retained net positive charge of the authentic precursor showed only very little capacity for import into mitochondria [197]. However, mature bovine phosphate carrier of which the presequence was completely deleted, was imported into mitochondria with about 40–70% of the original rate and was correctly inserted into the inner mitochondrial membrane [198]. Since the authentic presequence, when fused to other proteins, causes very low import efficiency, the major import information must be postulated to reside in the mature part also in the case of the PIC. Thus the correct function of the phosphate carrier presequence remains somewhat unclear, possibly exerting only a stimulatory role in import [198].

8. *Conclusion and perspectives*

It is obviously tempting to estimate the progress in the field reviewed by comparison with the last extensive review on the same topic, incidentally also published in this series [5]. Undoubtedly definite progress has been made in several respects, (i) in elucidating an increasing number of primary structures, (ii) in isolating, purifying and reconstituting more or less all major carrier proteins, and (iii) in elucidating the kinetic mechanism for a considerable number of the mitochondrial metabolite carriers. However, not only do we have to admit that we still have no idea how a particulate solute is translocated from one side of the membrane to the other, but also that we do not see a clear strategy for reaching this aim in the near future, unless the 3D-structure of a carrier protein becomes accessible. But this somewhat pessimistic view also holds true for any other carrier from any other source.

The elucidation of the primary structure of up to now in total four mitochondrial carrier proteins without any doubt was the most important breakthrough, thereby establishing the mitochondrial carrier family with respect to a common structure. The detailed elucidation of functional properties, mainly by analysis in reconstituted systems, provided important contributions to completing a correct view of mitochondrial metabolite carriers. Studies using chemical modification also significantly supplemented our understanding of functionally important regions or of basic functional principles like binding mechanisms, conformational changes, and transport pathway structures. The ultimate aim of all this work, i.e. to correlate specific functions with defined structural properties of particular carrier proteins, can only be achieved by creative combination of these methods, as long as a the 3D-structure of carriers is missing. There is no doubt, that the genetic approach, using specifically constructed mutants, is indispensable for this purpose. Unfortunately, the most obvious strategy, i.e. functional expression of these proteins in *E. coli*, the best known system and relatively easy to handle, has failed so far. Thus, the first successful results in this direction use the most appropriate alternative means, i.e. studying the function of these carriers in yeast, where the methods of molecular genetics are available. Further progress may be possible by expressing metabolite carriers from mammalian mitochondria in yeast. This approach seems to have been successfully applied recently in the case of the uncoupling protein [59,199], where functional expression in *Xenopus oocytes* and in chinese hamster ovary cells has also been reported [200,201]. We would like to emphasize, however, that in our opinion the usefulness of this genetic work essentially depends on a close connection with detailed functional studies with the obtained mutants and constructions.

Furthermore, we should bear in mind that, when studying the structure and function of a particular mitochondrial carrier in detail, we may miss important points of its significance and regulation in the physiological environment. That may be exemplified by the definite differences in complexity when studying simple kinetic properties of a particular carrier in intact mitochondria and in proteoliposomes, e.g. in the case of the oxoglutarate carrier [33,34,146]. Some of the differences not understood so far may be caused by experimental problems when using intact organelles, but there are additional conditions which may complicate a correct correlation to the data obtained in simplified systems. In this respect we would like to mention the problems of a possible microcompartmentation of substrates in the matrix space [11,144,202], which was repeatedly observed phenomenologically, at least in kinetic terms, and which is still not understood. Other examples are the functional significance of areas of contact between the outer and the inner membrane of mitochondria (contact sites) which may lead to compartmentalization of substrates within the intermembrane space [203], as well as interesting observations of regulation of carrier proteins by mitochondrial osmolarity [204] or hormones (see Chapter 20 in this book). Several of the above-mentioned complications may, furthermore, be due to direct interaction of carriers with particular metabolizing enzymes. This kind of interaction has been suggested for the ADP/ATP carrier [203,205–207], the aspartate/glutamate carrier [144,145], the glutamine carrier [208] and the pyruvate carrier [209].

Acknowledgements

The research in the authors' laboratories was supported by grants from the Deutsche Forschungsgemeinschaft (SFB 189), the Fonds der Chemischen Industrie, the C.N.R. Target Project 'Ingegneria genetica' and the Ministero dell'Università e della Ricerca Scientifica e Tecnologica (MURST). We are indebted to Dr. George Wolf for critical reading of the manuscript.

References

1 Klingenberg, M. (1976) in: A.N. Martonosi (Ed.) The Enzymes of Biological Membranes, Vol. 3, Plenum Press, New York, pp. 383–438.
2 LaNoue, K.F. and Schoolwerth, A.C. (1979) Annu. Rev. Biochem. 48, 871–922.
3 Meijer, A.J. and van Dam, K. (1981) in: S.L. Bonting and J.J.H.H.M. de Pont (Eds.) Membrane Transport, Elsevier, Amsterdam, pp. 235–255.
4 Palmieri, F. and Stipani, I. (1981) Biol. Zbl. 100, 515–525.
5 LaNoue, K.F. and Schoolwerth, A.C. (1984) in: L. Ernster (Ed.) Bioenergetics, Elsevier, Amsterdam, pp. 221–268.
6 Krämer, R. and Palmieri, F. (1989) Biochim. Biophys. Acta 974, 1–23.
7 Chappell, J.B. (1968) Br. Med. Bull. 25, 150–157.
8 Pfaff, E., Klingenberg, M., Ritt, E. and Vogell, W. (1968) Eur. J. Biochem. 5, 222–232.
9 Palmieri, F. and Klingenberg, M. (1979) Methods Enzymol. 56, 279–301.
10 LaNoue, K.F., Bryla, J. and Williamson, J.R. (1972) J. Biol. Chem. 247, 667–679.
11 Brandolin, G., Marty, I. and Vignais, P.V. (1990) Biochemistry 29, 9720–9727.
12 Riccio, P., Aquila, H. and Klingenberg, M. (1975) FEBS Lett. 56, 133–138.
13 Krämer, R. and Klingenberg, M. (1977) FEBS Lett. 82, 363–367.
14 Lin, C.S. and Klingenberg, M. (1980) FEBS Lett. 113, 299–303.
15 Aquila, H., Misra, D., Eulitz, M. and Klingenberg, M. (1982) Hoppe-Seyler's Z. Physiol. Chem. 363, 345–349.
16 Walker, J.E., Cozens, A.L., Dyer, M.R., Fearnley, I.M., Powell, S.J. and Runswick, M.J. (1987) Chem. Scr. B 27, 97–105.
17 Aquila, H., Link, T.A. and Klingenberg, M. (1985) EMBO J. 4, 2369–2376.
18 Runswick, M.J., Powell, S.J., Nyren, P. and Walker, J.E. (1987) EMBO J. 6, 1367–1373.
19 Aquila, H., Link, T.A. and Klingenberg, M. (1987) FEBS Lett. 212, 1–9.
20 Runswick, M.J., Walker, J.E., Bisacca, F., Iacobazzi, V. and Palmieri, F. (1990) Biochemistry 29, 11033–11040.
21 Henderson, P.J.F. (1987) Nature 325, 641–643.
22 Maloney, P.C. (1989) Philos. Trans. R. Soc. London Ser. B 326, 437–454.
23 Wu, L.-F. and Saier, M.H. Jr. (1990) Mol. Microbiol. 4, 1219–1222.
24 Henderson, P.J.F. (1991) Curr. Opinion Struct. Biol. 1, 590–601.
25 Saier, M.H. Jr. and Reizer, J. (1991) Curr. Opinion Struct. Biol. 1, 362–368.
26 Pfaff, E. and Klingenberg, M. (1968) Eur. J. Biochem. 6, 66–79.
27 Pfaff, E., Heldt, H.W. and Klingenberg, M. (1969) Eur. J. Biochem. 10, 484–493.

28 Duée, E.D. and Vignais, P.V. (1969) J. Biol. Chem. 244, 3932–3940.
29 Duyckaerts, C., Sluse-Goffart, C., Fux, J-P., Sluse, F.E. and Liebecq, C. (1980) Eur. J. Biochem. 106, 1–6.
30 Klingenberg, M., Grebe, K. and Appel, M. (1982) Eur. J. Biochem. 126, 263–269.
31 Klingenberg, M. (1985) in: A.N. Martonosi (Ed.) The Enzymes of Biological Membranes, Vol. 4, Plenum Press, New York, pp. 511–553.
32 Vignais, P.V., Block, M.R., Boulay, F., Brandolin, G. and Lauquin, G.J.M. (1985) in: G. Bengha (Ed.) Structure and Properties of Cell Membranes, Vol. II, CRC Press, Boca Raton, pp. 139–179.
33 Sluse, F.E., Ranson, M. and Liebecq, C. (1972) Eur. J. Biochem. 25, 207–217.
34 Sluse, F.E., Duyckaerts, C., Liebecq, S. and Sluse-Goffart, M. (1979) Eur. J. Biochem. 100, 3–17.
35 Murphy, E., Coll, K.E., Viale, R.O., Tischler, M.E. and Williamson, J.R. (1979) J. Biol. Chem. 254, 8369–8376.
36 LaNoue, K.F., Duszynski, J., Watts, J.A. and McKee, E. (1979) Arch. Biochem. Biophys. 195, 578–590.
37 Schoolwerth, A.C. and LaNoue, K.F. (1980) J. Biol. Chem. 255, 3403–3411.
38 Coty, W.A. and Pedersen, P.L. (1974) J. Biol. Chem. 249, 2593–2598.
39 Palmieri, F., Prezioso, G., Quagliariello, E. and Klingenberg, M. (1971) Eur. J. Biochem. 22, 66–74.
40 Crompton, M., Palmieri, F., Capano, M. and Quagliariello, E. (1975) Biochem. J. 146, 667–673.
41 Meijer, A.J. and van Dam, K. (1974) Biochim. Biophys. Acta 346, 213–244.
42 Fonyo, A., Palmieri, F. and Quagliariello, E. (1976) in: E. Quagliariello, F. Palmieri and T.P. Singer (Eds.) Horizons in Biochemistry and Biophysics, Vol. II, Addison-Wesley, Reading, MA, pp. 60–105.
43 Williamson, J.R. (1976) in: R.W. Hanson and M.A. Mehlman (Eds.) Gluconeogenesis, Wiley-Interscience, New York, pp. 165–220.
44 Palmieri, F. and Quagliariello, E. (1978) in: L. Wojtczak, E. Lenartowicz and J. Zborowski (Eds.) Bioenergetics at Mitochondrial and Cellular Levels, Nencki Institute of Experimental Biology, Warsaw, pp. 5–38.
45 Scarpa, A. (1979) in: G. Giebisch, D.C. Tosteson, H.H. Ussing and M.T. Tosteson (Eds.) Membrane Transport in Biology, Vol. II, Springer, Berlin, pp. 263–355.
46 Schoolwerth, A.C. and LaNoue, K.F. (1985) Annu. Rev. Physiol. 47, 143–171.
47 Wiskich, J.T. (1977) Annu. Rev. Plant Physiol. 28, 45–69.
48 Heldt, H.W. and Flügge, U.I. (1992) in: A.K. Tobin (Ed.) Metabolic Interaction of Organelles in Photosynthetic Tissues, Cambridge University Press, Cambridge, in press.
49 Klingenberg, M. (1979) Trends Biochem. Sci. 4, 249–252.
50 Vignais, P.V. (1976) Biochim. Biophys. Acta 456, 1–38.
51 Vignais, P.V. and Lauquin, J.M. (1979) Trends Biochem. Sci. 4, 90–92.
52 Vignais, P.V., Block, M.R., Boulay, F., Brandolin, G. and Lauquin, G.J.M. (1982) in: A.N. Martonosi (Ed.) Membranes and Transport, Vol. 1, Plenum Press, New York, pp. 405–413.
53 Klingenberg, M. (1989) Arch. Biochem. Biophys. 270, 1–14.
54 Fonyo, A., Ligeti, E., Palmieri, F. and Quagliariello, E. (1975) in: G. Gardos and G. Szasz (Eds.) Biomembranes, Structure and Function, Akademiai Kiado, Budapest and North-Holland, Amsterdam, pp. 287–306.
55 Durand, R., Briand, Y., Touraille, S. and Alziari, S. (1981) Trends Biochem. Sci. 6, 211–213.
56 Pedersen, P.L. and Wehrle, J.P. (1982) in: A.N. Martonosi (Ed.) Membranes and Transport, Vol. 1, Plenum Press, New York, pp. 645–663.
57 Wohlrab, H. (1986) Biochim. Biophys. Acta 853, 115–134.
58 Halestrap, A.P., Scott, R.D. and Thomas, A.P. (1980) Int. J. Biochem. 11, 97–105.
59 Klaus, S., Casteilla, L., Bouillaud, F. and Ricquier, D. (1991) Int. J. Biochem. 23, 791–801.
60 Klingenberg, M. (1990) Trends Biochem. Sci. 15, 108–112.
61 Krämer, R. and Klingenberg, M. (1979) Biochemistry 18, 4209–4215.
62 Robinson, B.R., Williams, G.R., Halpering, M.L. and Leznoff, C.C. (1971) Eur. J. Biochem. 20, 65–71.
63 Meijer, A.J., van Woerkom, G.M. and Eggelte, T.A. (1976) Biochim. Biophys. Acta 430, 53–61.
64 Halestrap, A.P. and Denton, R.M. (1974) Biochem. J. 138, 313–316.

65 Palmieri, F., Bisaccia, F., Capobianco, L. and Iacobazzi, V. (1990) Biochim. Biophys. Acta 1018, 147–150.
66 Gellerich, F.N., Bohnensack, R. and Kunz, W. (1983) Biochim. Biophys. Acta 722, 381–391.
67 Westerhoff, H.W., Plomp, P.J.A.M., Groen, A.K., Wanders, R.J.A., Bode, J.A. and van Dam, K. (1987) Arch. Biochem. Biophys. 257, 154–169.
68 Halestrap, A.P. (1978) Biochem. J. 172, 389–398.
69 Thomas, A.P. and Halestrap, A.P. (1981) Biochem. J. 196, 471–479.
70 Lenzen, C., Soboll, S., Sies, H. and Häussinger, D. (1987) Eur. J. Biochem. 106, 483–488.
71 Simpson, D.P. and Adam, W. (1975) Med. Clin. North Am. 59, 555–567.
72 Aprille, J.R. and Austin, J. (1981) Arch. Biochem. Biophys. 212, 689–699.
73 Nosek, M.T., Dransfield, D.T. and Aprille, J.R. (1990) J. Biol. Chem. 265, 8444–8450.
74 Meijer, A.J., van Woerkom, G.M., Wanders, R.J.A. and Lof, C. (1982) Eur. J. Biochem. 124, 325–330.
75 Tahiliani, A.G. and Neely, J.R. (1987) J. Biol. Chem. 262, 11607–11610.
76 Barile, M., Passarella, S. and Quagliariello, E. (1990) Arch. Biochem. Biophys. 280, 352–357.
77 Kurosawa, K., Hayashi, N., Sato, N., Kamada, T. and Tagawa, K. (1990) Biochem. Biophys. Res. Commun. 167, 367–372.
78 Toninello, A., Miotto, G., Siliprandi, D., Siliprandi, N. and Garlid, K. (1988) J. Biol. Chem. 263, 19407–19411.
79 De Pinto, V., Tommasino, M., Palmieri, F. and Kadenbach, B. (1982) FEBS Lett. 148, 103–106.
80 Kolbe, H.V.J., Costello, D., Wong, A., Lu, R.C. and Wohlrab, H. (1984) J. Biol. Chem. 259, 9115–9120.
81 Kaplan, R.S., Pratt, R.D. and Pedersen, P.L. (1986) J. Biol. Chem. 261, 12767–12773.
82 Bisaccia, F., Indiveri, C. and Palmieri, F. (1985) Biochim. Biophys. Acta 810, 362–369.
83 Indiveri, C., Palmieri, F., Bisaccia, F. and Krämer, R. (1987) Biochim. Biophys. Acta 890, 310–318.
84 Bisaccia, F., Indiveri, C. and Palmieri, F. (1988) Biochim. Biophys. Acta 933, 229–240.
85 Bisaccia, F., De Palma, A. and Palmieri, F. (1989) Biochim. Biophys. Acta 977, 171–176.
86 Claeys, D. and Azzi, A. (1989) J. Biol. Chem. 264, 14627–14630.
87 Bolli, R., Nalecz, K.A. and Azzi, A. (1989) J. Biol. Chem. 264, 18024–18030.
88 Indiveri, C., Tonazzi, A. and Palmieri, F. (1990) Biochim. Biophys. Acta 1020, 81–86.
89 Bisaccia, F., De Palma, A. and Palmieri, F. (1992) Biochim. Biophys. Acta 1106, 291–296.
90 Indiveri, C., Tonazzi, A. and Palmieri, F. (1992) Eur. J. Biochem. 207, 449–454.
91 Strieleman, P.J., Schalinske, K.L. and Shrago, E. (1985) Biochem. Biophys. Res. Commun. 127, 509–516.
92 Klingenberg, M. and Winkler, E. (1985) EMBO J. 4, 3087–3092.
93 Krämer, R., Kürzinger, G. and Heberger, C. (1986) Arch. Biochem. Biophys. 251, 166–174.
94 Hutson, S.M., Roten, S. and Kaplan, R.S. (1990) Proc. Natl. Acad. Sci. USA 87, 1028–1031.
95 Hutson, S.M., Wallin, R. and Roten, S. (1992) in: E. Quagliariello and F. Palmieri (Eds.) Molecular Mechanims of Transport, Abstracts book, Bari meeting, September 29 – October 1, 1991, p. 69.
96 Knirsch, M., Gawaz, M.P. and Klingenberg, M. (1989) FEBS Lett. 244, 427–432.
97 Guerin, B., Bukusoglu, C., Rakotomanana, F. and Wohlrab, H. (1990) J. Biol. Chem. 265, 19736–19741.
98 Nalecz, M.J., Nalecz, K.A. and Azzi, A. (1991) Biochim. Biophys. Acta 1079, 87–95.
99 Arends, H. and Sebald, W. (1984) EMBO J. 3, 377–382.
100 Adrian, G., McCammon, M.T., Montgomery, D.L. and Douglas, M.G. (1986) Mol. Cell. Biol. 6, 626–634.
101 Baker, A. and Leaver, C.J. (1985) Nucleic Acids Res. 13, 5857–5867.
102 Bouillaud, F., Weissenbach, J. and Riquier, D. (1986) J. Biol. Chem. 261, 1487–1490.
103 Ferreira, G.C., Pratt, R.D. and Pedersen, P.L. (1989) J. Biol. Chem. 264, 15628–15633.
104 Dolce, V., Fiermonte, G., Messina, A. and Palmieri, F. (1991) DNA Sequence 2, 131–134.
105 Phelps, A., Schobert, C.T. and Wohlrab, H. (1991) Biochemistry 30, 248–252.

106 Palmieri, F., Bisaccia, F., Capobianco, L., Dolce, V., Iacobazzi, V., Indiveri, C. and Zara, V. (1992) in: E. Quagliariello and F. Palmieri (Eds.) Molecular Mechanisms of Transport, Elsevier, Amsterdam, pp. 151–158.
107 Saraste, M. and Walker, J.E. (1982) FEBS Lett. 144, 250–254.
108 Brandolin, G., Boulay, F., Dalbon, P. and Vignais, P.V. (1989) Biochemistry 28, 1093–1100.
109 Wiesenberger, G., Link, T.A., von Ahsen, U. and Schweyen, R.J. (1991) J. Mol. Biol. 217, 23–38.
110 Zarilli, R., Oates, E.L., McBride, W., Lerman, M.I., Chan, J.Y., Santisteban, P., Ursini, M.V., Notkins, A.L. and Kohn, L.D. (1989) Mol. Endocrin. 3, 1498–1508.
111 Block, M.R. and Vignais, P.V. (1984) Biochim. Biophys. Acta 767, 369–376.
112 Vignais, P.V. and Lunardi, J. (1985) Annu. Rev. Biochem. 54, 977–1014.
113 Klingenberg, M. (1981) Nature 290, 449–454.
114 Klingenberg, M. and Appel, M. (1989) Eur. J. Biochem. 180, 123–131.
115 Hackenberg, H. and Klingenberg, M. (1980) Biochemistry 19, 548–555.
116 Lin, C.S., Hackenberg, H. and Klingenberg, M. (1980) FEBS Lett. 113, 304–306.
117 Capobianco, L., Brandolin, G. and Palmieri, F. (1991) Biochemistry 30, 4963–4969.
118 Ferreira, G.C., Pratt, C. and Pedersen, P.L. (1990) J. Biol. Chem. 265, 21202–21206.
119 Bogner, W., Aquila, H. and Klingenberg, M. (1986) Eur. J. Biochem. 161, 611–620.
120 Boulay, F., Lauquin, G.J.M., Tsugita, A. and Vignais, P.V. (1983) Biochemistry 22, 477–484.
121 Dalbon, P., Brandolin, G., Boulay, F., Hoppe, J. and Vignais, P.V. (1988) Biochemistry 27, 5141–5149.
122 Calamia, J. and Manoil, C. (1990) Proc. Natl. Acad. Sci. USA 87, 4937–4941.
123 Froshauer, S., Green, G.N., Boyd, D., McGovern, K. and Beckwith, J. (1988) J. Mol. Biol. 200, 501–511.
124 Carruthers, A. (1990) Phys. Rev. 70, 1135–1176.
125 Eiermann, W., Aquila, H. and Klingenberg, M. (1977) FEBS Lett. 74, 209–214.
126 Schultheiss, H.P. and Klingenberg, M. (1984) Eur. J. Biochem. 143, 599–605.
127 Houldsworth, J. and Attardi, G. (1988) Proc. Natl. Acad. Sci. USA 84, 377–381.
128 Powell, S.J., Medd, S.M., Runswick, M.J. and Walker, J.E. (1989) Biochemistry 28, 866–873.
129 Lawson, J.E. and Douglas, M.G. (1988) J. Biol. Chem. 263, 14812–14818.
130 Kolarov, J., Kolarova, N. and Nelson, N. (1990) J. Biol. Chem. 265, 12711–12716.
131 Bathgate, B., Baker, A. and Leaver, C.J. (1989) Eur. J. Biochem. 183, 303–310.
132 Neckelmann, N., Li, K., Wade, R.P., Shuster, R. and Wallace, D.C. (1987) Proc. Natl. Acad. Sci. USA 84, 7580–7584.
133 Battini, R., Ferrari, S., Kaczmarek, L., Calbretta, B., Chen, S. and Baserga, R. (1987) J. Biol. Chem. 262, 4355–4359.
134 Cozens, A.L., Runswick, M.J. and Walker, J.E. (1989) J. Mol. Biol. 206, 261–280.
135 Lunardi, J. and Attardi, G. (1991) J. Biol. Chem. 266, 16534–16540.
136 Drgon, T., Sabova, L., Nelson, N. and Kolarov, J. (1991) FEBS Lett. 289, 159–162.
137 Kozak, L.P., Britton, J.H., Kozak, U.C. and Wells, J.M. (1988) J. Biol. Chem. 263, 12274–12277.
138 Cassard, A.M., Bouillaud, F., Mattei, M.G., Hentz, E., Raimbault, S., Thomas, M. and Ricquier, D. (1990) J. Cell Biochem. 43, 255–264.
139 Gamble, J.G. and Lehninger, A.L. (1973) J. Biol. Chem. 248, 610–618.
140 Passarella, S., Atlante, A. and Quagliariello, E. (1990) Eur. J. Biochem. 193, 221–227.
141 Bradford, N.M. and McGivan, J.D. (1980) FEBS Lett. 113, 294–298.
142 McGivan, J.D., Bradford, N.M. and Beavis, A.D. (1977) Biochem. J. 162, 147–156.
143 Soboll, S., Lenzen, C., Rettich, D., Grundel, S. and Ziegler, B. (1991) Eur. J. Biochem. 197, 113–117.
144 Duszynski, J., Mueller, G. and LaNoue, K. (1978) J. Biol. Chem. 253, 6149–6157.
145 Schoolwerth, A.C. and LaNoue, K.F. (1980) J. Biol. Chem. 255, 3403–3411.
146 Indiveri, C., Dierks, T., Krämer, R. and Palmieri, F. (1991) Eur. J. Biochem. 198, 339–347.
147 Indiveri, C., Capobianco, L., Krämer, R. and Palmieri, F. (1989) Biochim. Biophys. Acta 977, 187–193.
148 Indiveri, C., Dierks, T., Krämer, R. and Palmieri, F. (1989) Biochim. Biophys. Acta 977, 194–199.

149 Bisaccia, F., De Palma, A., Prezioso, G. and Palmieri, F. (1990) Biochim. Biophys. Acta 1019, 250–256.
150 Glerum, D.M., Claeys, D., Mertens, W. and Azzi, A. (1990) Eur. J. Biochem. 194, 681–684.
151 Indiveri, C., Tonazzi, A., Prezioso, G. and Palmieri, F. (1991) Biochim. Biophys. Acta 1065, 231–238.
152 Nicholls, D.G. and Lindberg, O. (1973) Biochim. Biophys. Acta 549, 1–29.
153 Jezek, P., Orosz, D.E. and Garlid, K.D. (1990) J. Biol. Chem. 265, 19296–19302.
154 Klingenberg, M. (1991) in: S.A. Kuby (Ed.) A Study of Enzymes, Vol. 2, Mechanism of Enzyme Action, pp. 367–389.
155 Klingenberg, M. (1987) in: C.H. Kim, H. Tedeschi, J.J. Diwan and J.C. Salerno (Eds.) Advances in Membrane Biochemistry and Bioenergetics, Plenum Press, New York, pp. 389–399.
156 Jencks, W.P. (1975) Adv. Enzymol. 43, 319–410.
157 Jencks, W.P. (1980) Adv. Enzymol. 51, 75–106.
158 Pande, S.V. and Parvin, R. (1980) J. Biol. Chem. 255, 2994–3001.
159 Indiveri, C., Tonazzi, A. and Palmieri, F. (1991) Biochim. Biophys. Acta 1069, 110–116.
160 Cleland, W.W. (1963) Biochim. Biophys. Acta 67, 104–137.
161 Dierks, T., Riemer, E. and Krämer, R. (1988) Biochim. Biophys. Acta 943, 231–244.
162 Indiveri, C., Prezioso, G., Dierks, T., Krämer, R. and Palmieri, F. (1992), in preparation.
163 Bisaccia, F., De Palma, A., Dierks, T., Krämer, R. and Palmieri, F. (1992), Biochim. Biophys. Acta, submitted.
164 Sluse, F.E., Evens, A., Dierks, T., Duyckaerts, C., Sluse-Goffart, C.M. and Krämer, R. (1991) Biochim. Biophys. Acta 1058, 329–338.
165 Sluse, F.E., Sluse-Goffart, C.M. and Duyckaerts, C. (1989) in: A. Azzi, K.A. Nalecz, M.J. Nalecz and L. Wojtczak (Eds.) Anion Carriers of Mitochondrial Membranes, Springer, Berlin, pp. 183–195.
166 Klingenberg, M., Durand, R. and Guerin, B. (1974) Eur. J. Biochem. 42, 135–150.
167 Ligeti, E. and Fonyo, A. (1984) Eur. J. Biochem. 139, 279–285.
168 Dierks, T., Salentin, A., Heberger, C. and Krämer, R. (1990) Biochim. Biophys. Acta 1028, 268–280.
169 Indiveri, C., Tonazzi, A., Dierks, T., Krämer, R. and Palmieri, F. (1992) Biochim. Biophys. Acta, in press.
170 Dierks, T., Stappen, R., Salentin, A. and Krämer, R. (1992) Biochim. Biophys. Acta, 1103, 13–24.
171 Dierks, T., Salentin, A. and Krämer, R. (1990) Biochim. Biophys. Acta 1028, 281–288.
172 Lengeler, J. (1990) Biochim. Biophys. Acta 1018, 155–159.
173 Lolkema, J.S., Tenhoeveduurkens, R.H., Dijkstra, D.S. and Robillard, G.T. (1991) Biochemistry 30, 6716–6721.
174 Roepe, P.D., Consler, T.G., Menezes, M.E. and Kaback, H.R. (1990) Res. Microbiol. 141, 290–309.
175 Kaback, H.R., Bibi, E. and Roepe, P.D. (1990) Trends Biochem. Sci. 15, 309–314.
176 Brooker, R.J. (1990) Res. Microbiol. 141, 309–316.
177 Ames, G.F.-L. and Joshi, A.K. (1990) J. Bacteriol. 172, 4133–4137.
178 Meadow, N.D., Fox, D.K. and Roseman, S. (1990) Annu. Rev. Biochem. 59, 497–542.
179 Boulay, F., Brandolin, G. and Vignais, P.V. (1986) Biochem. Biophys. Res. Commun. 134, 266–271.
180 Mayinger, P., Winkler, E. and Klingenberg, M. (1989) FEBS Lett. 244, 421–426.
181 Rial, E., Arechaga, I., Sainz-de-la-Maza, E. and Nicholls, D.G. (1989) Eur. J. Biochem. 182, 187–193.
182 Jezek, P. and Drahota, Z. (1989) Eur. J. Biochem. 183, 89–95.
183 Katiyar, S.S. and Shrago, E. (1989) Proc. Natl. Acad. Sci. USA 86, 2559–2562.
184 Lawson, J.E., Gawaz, M., Klingenberg, M. and Douglas, M.G. (1990) J. Biol. Chem. 265, 14195–14202.
185 Gawaz, M., Douglas, M.G. and Klingenberg, M. (1990) J. Biol. Chem. 265, 14202–14208.
186 Boulay, F., Brandolin, G., Lauquin, G.J.M., Jolles, J., Jolles, P. and Vignais, P.V. (1979) FEBS Lett. 98, 161–164.
187 Klingenberg, M., Gawaz, M., Douglas, M.G. and Lawson, J.E. (1992) in: E. Quagliariello and F. Palmieri (Eds.) Molecular Mechanisms of Transport, Elsevier, Amsterdam, pp. 187–195.
188 Soboll, S., Scholz, H. and Heldt, H.W. (1978) Eur. J. Biochem. 87, 377–390.
189 Soboll, S. and Bünger, R. (1981) Hoppe-Seyler's Z. Physiol. Chem. 362, 125–132.
190 Zimmermann, R. and Neupert, W. (1980) Eur. J. Biochem. 109, 217–229.

191 Söllner, T., Pfaller, R., Griffiths, G., Pfanner, N. and Neupert, W. (1990) Cell 62, 107–115.

192 Pfanner, N. and Neupert, W. (1990) Annu. Rev. Biochem. 59, 331–353.

193 Schleyer, M. and Neupert, W. (1984) J. Biol. Chem. 259, 3487–3491.

194 Liu, X., Freeman, K.B. and Shore, G.C. (1988) J. Cell Biol. 107, 503–509.

195 Liu, X., Freeman, K.B. and Shore, G.C. (1990) J. Biol. Chem. 265, 9–12.

196 Zara, V., Rassow, J., Wachter, E., Tropschug, M., Palmieri, F., Neupert, W. and Pfanner, F. (1991) Eur. J. Biochem. 198, 405–410.

197 Pratt, R.D., Ferreira, G.C. and Pedersen, P.L. (1991) J. Biol. Chem. 266, 1276–1280.

198 Zara, V., Palmieri, F., Mahlke, K. and Pfanner, N. (1992) J. Biol. Chem. 267, 12077–12081.

199 Murdza-Inglis, D.L., Patel, H.V., Freeman, K.B., Jezek, P., Orosz, D.E. and Garlid, K.D. (1991) J. Biol. Chem. 266, 11871–11875.

200 Klaus, S., Casteilla, L., Bouillaud, F., Raimbault, S. and Ricquier, D. (1990) Biochem. Biophys. Res. Commun. 167, 784–789.

201 Casteilla, L., Blondel, O., Klaus, S., Raimbault, S., Diolez, P., Moreau, F., Boillaud, F. and Ricquier, D. (1990) Proc. Natl. Acad. Sci. USA 87, 5124–5128.

202 Murthy, M.S.R. and Pande, S.V. (1985) Biochem. J. 230, 657–663.

203 Brdiczka, D. (1991) Biochim. Biophys. Acta 1071, 291–312.

204 Halestrap, A.P. (1989) Biochim. Biophys. Acta 973, 355–382.

205 Adams, V., Bosch, W., Schlegel, J., Wallimann, T. and Brdiczka, D. (1989) Biochim. Biophys. Acta 981, 213–225.

206 Jacobus, W.E. (1985) Annu. Rev. Physiol. 47, 707–725.

207 Brooks, S.P.J. and Suelter, C.H. (1987) Arch. Biochem. Biophys. 257, 144–153.

208 Simpson, D.P. (1983) Kidney Int. 23, 785–793.

209 Pande, S.V. and Parvin, R. (1978) J. Biol. Chem. 253, 1565–1573.

L. Ernster (Ed.) *Molecular Mechanisms in Bioenergetics*
© 1992 Elsevier Science Publishers B.V. All rights reserved

CHAPTER 17

The uncoupling protein thermogenin and mitochondrial thermogenesis

JAN NEDERGAARD and BARBARA CANNON

The Wenner-Gren Institute, The Arrhenius Laboratories F3, Stockholm University, S-106 91 Stockholm, Sweden

Contents

1. Overview 386
2. Isolation of thermogenin and reconstitution of thermogenin activity 388
  2.1. The isolation of thermogenin 389
  2.2. Reconstitution of thermogenin activity 389
    2.2.1. First generation reconstitutions 389
    2.2.2. Second generation reconstitutions 389
3. The size of the functional unit 392
  3.1. Monomer or dimer? 392
  3.2. The unmasking phenomenon 392
4. What is transported by thermogenin? 393
  4.1. A carrier or a channel? 393
  4.2. Cl^- transport 395
  4.3. H^+ transport 397
    4.3.1. H^+ permeability and free fatty acids 397
  4.4. H^+ and Cl^- permeabilities: the same or not? 400
5. Suggested thermogenin structure(s) 400
  5.1. General structure 400
  5.2. The nucleotide binding site 400
    5.2.1. Agonists, partial agonists and antagonists 403
    5.2.2. Coupling of nucleotide binding to permeability inhibition 404
6. Control of thermogenin activity 404
  6.1. Suggested nonlipolytic control mechanisms 405
    6.1.1. Cytosolic alkalinization 405
    6.1.2. A decrease in total cytosolic nucleotide levels 406
  6.2. Suggested lipolytic control mechanisms 406
    6.2.1. The fatty acid hypothesis 406
    6.2.2. The acyl-CoA hypothesis 410
  6.3. Conclusion 410
7. A member of a family: the mitochondrial membrane carriers 411
8. The thermogenin gene 411
9. Regulation of the expression of thermogenin 414
  9.1. Tissue-specific thermogenin gene expression 414
  9.2. Physiological control of thermogenin gene expression 415

9.3. The synthesis of thermogenin 415
10. Perspectives 416
Acknowledgements 416
References 416

1. Overview

Brown adipose tissue is the only dedicated heat-producing tissue in the mammalian body. It is the activity of a single, unique protein – the uncoupling protein thermogenin – that is the molecular background for the heat-producing ability of the tissue (Fig. 1).

The activity of thermogenin has traditionally been studied in isolated brown-fat mitochondria, but it has become possible to isolate thermogenin itself, and this has opened new avenues for thermogenesis research. Thermogenin has successfully been reconstituted into lipid vesicles, and in these it displays most of the characteristics expected from earlier studies. It is likely that thermogenin is functional in the form of a homodimer. Thermogenin functions as a transporter (a carrier or a channel) over the mitochondrial membrane. The transported species

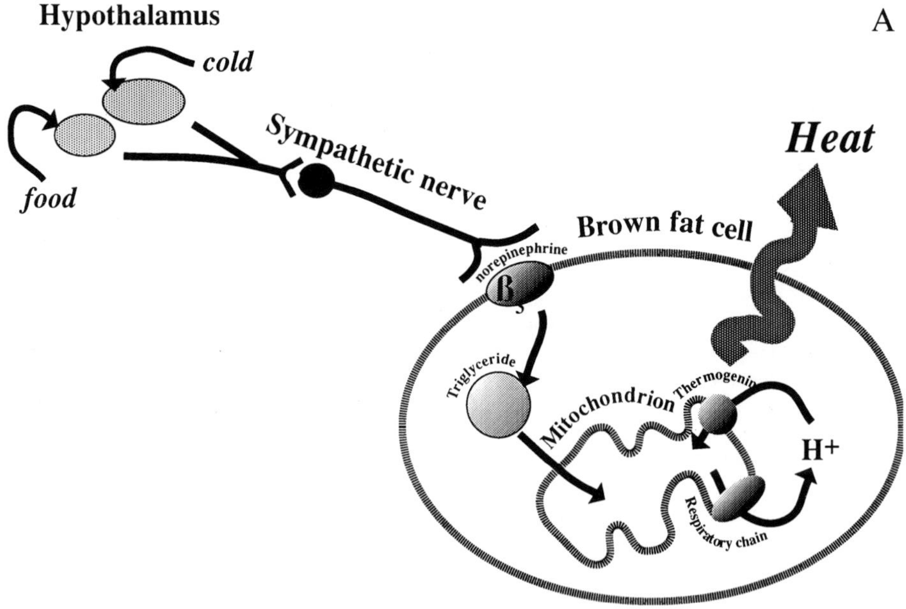

Fig. 1. Brown adipose tissue and thermogenin function. (A) Overview of physiological function (for details see ref. [10]). Brown adipose tissue is only found in mammals and is especially prominent in mammalian newborns [113], including human babies [133]. The activity of brown adipose tissue is controlled from the hypothalamic area. Adequate physiological stimuli include cold (especially the cold experienced immediately after birth), but also certain diets [134]. The tissue is activated via the sympathetic nervous system,

→

B

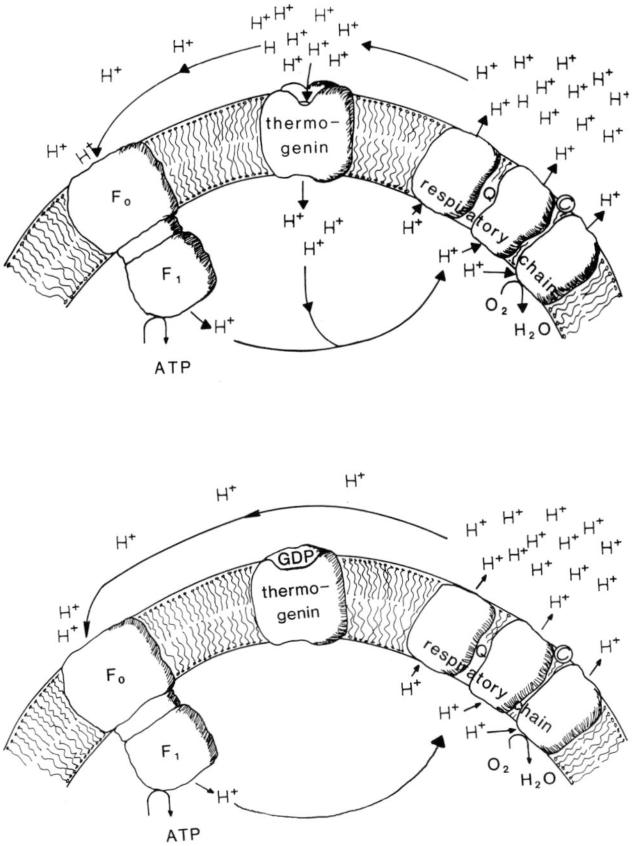

and norepinephrine is released at the cell membrane. The norepinephrine binds to adrenergic receptors of different kinds, but the β_3-receptor is the most relevant for acute thermogenic stimulation [83,135]. Receptor stimulation leads to an increase in cAMP, activation of protein kinase and stimulation of hormone-sensitive lipase, which in its turn leads to the release of free fatty acids from the triglyceride droplets in the cells [7] – all processes with equivalents in e.g. white adipose tissue. The fatty acids are transported to the mitochondria and oxidised by β-oxidation, as in other mitochondria. What makes brown-fat mitochondria unique is the presence of the uncoupling protein thermogenin in the mitochondrial membrane. (B) The inner membrane of brown-fat mitochondria. During oxidation of fatty acids, protons are pumped out by the respiratory chain. In brown-fat mitochondria, they may either re-enter through the ATP-synthetase (as in normal mitochondria), or through the uncoupling protein thermogenin. In the latter case, all chemical energy released by oxidation and temporarily stored in the mitochondrial proton electrochemical gradient is released in the form of heat. In many species (but not e.g. in lamb brown adipose tissue [136]), there is a low content of ATP-synthetase, relative to the amount of respiratory chain components (about 1/10 of the concentration in 'normal' mitochondria) [137,138]. This low content seems to be an effect of a posttranscriptional regulation: the amount of ATP-synthetase mRNA is normal or even high, but the proteins are either not synthesized or not incorporated [139]. This review concentrates on thermogenin itself; the basic properties of thermogenin are as summarized here in the figure: it is able to uncouple the brown-fat mitochondria by a transmembral transport which is at least functionally equivalent to H^+ transport, and the presence on thermogenin of a regulatory binding site, often referred to as the GDP-binding site but which is a purine nucleotide di/triphosphate binding site; thus, ATP is probably the dominant native ligand.

include halide anions as well as bulky, hydrophobic anions. The physiological function of thermogenin is probably to transport H^+; this may perhaps be in the form of the reverse transport of OH^-. All these transport activities may be inhibited by the binding of purine nucleotides to a site on thermogenin itself. This leads to conformational changes in thermogenin structure, inducing inhibition of transport. Physiologically, it is likely that a postlipolytic event activates thermogenin acutely, but the molecular nature of this process is still not understood. Thermogenin is a member of a growing family of mitochondrial carriers which all share a tripartite structure. The thermogenin gene, which is located in the nucleus, is also tripartite, and upstream regulatory elements have been identified. Thermogenin is only found in brown adipose tissue, and even there, the expression of the gene is not constitutive but is under physiological control. When thermogenin is synthesized, it is transported to and inserted into the mitochondrial membrane, without the participation of a cleavable signal sequence. Significant progress in the molecular understanding of thermogenin function can be expected from the newly acquired possibilities to modify the amino acid residues and study the function of the modified protein in reconstituted systems.

In the present review, we shall concentrate on the molecular bioenergetics of thermogenin and on developments in understanding after the previous review in this series [1]. A more historical perspective can be found elsewhere [2]. Other reviews on thermogenin [3,4,5], on brown-fat mitochondria [6], on brown-fat cells [7] and on brown adipose tissue in general [8,9,10] can be found elsewhere.

The start of a new era in thermogenesis research can be dated to the report by Klingenberg's group that it was possible to isolate thermogenin by a comparatively simple procedure [11]. The access to pure thermogenin gave rise to an entirely new series of studies. It had now become possible to examine the properties of thermogenin without interference from other mitochondrial proteins, especially in reconstitution studies. It also became possible to obtain antibodies against thermogenin, and through this, to isolate cDNA clones corresponding to thermogenin mRNA. This has allowed direct studies of the control of thermogenin gene expression and analysis of the gene itself.

Thus, today, in many respects amazing progress has been made. However, it was believed some fifteen years ago that if the amino acid sequence of thermogenin and some hints of the secondary structure would be available, the answers to many – by now 'classical' – questions in thermogenic bioenergetics would immediately be settled. Alas, this has not been the case, and it will be clear from this review that many of the classical questions remain as yet unsettled, despite the efforts of more than a decade in molecular bioenergetics.

In the present review, as in most textbooks in biochemistry and cell biology, we shall refer to the unique protein which confers to brown-fat mitochondria their unique property of being thermogenic, with the name *thermogenin*. As this protein works as an uncoupling protein, it is often referred to as *the* uncoupling protein (implicitly 'from brown adipose tissue'). However, the possibility that several other proteins, especially within the same family, may experimentally or physiologically exhibit 'uncoupling' properties, may make this term less appropriate. The thermogenin gene is known as *ucp*; it should not be confused with the gene name *unc* for 'uncoupling' which has been used in several unrelated systems.

2. *Isolation of thermogenin and reconstitution of thermogenin activity*

In several respects, present investigations of thermogenin function can be described as a dialogue between experiments performed with isolated/reconstituted thermogenin and experiments performed with isolated brown-fat mitochondria (and occasionally brown fat cells), in which the purity of the molecular experiments is balanced by the inherent problems of denaturation and loss of native environment during isolation.

Descriptions of isolation and investigation of mitochondria can be found elsewhere [12]. Basically, brown-fat mitochondria are prepared as other mitochondria, except that an initial centrifugation step (which sediments both nuclei and mitochondria) is often included. This step allows the lipid of the tissue to be discarded. Often albumin is present in different steps of the isolation procedure; due to the central role ascribed to free fatty acids in the regulation of thermogenin activity this scavenger of fatty acids may be of considerable significance in brown-fat research.

2.1. The isolation of thermogenin

The isolation procedure has been described in detail by Klingenberg and Lin [13,156] and is outlined in Fig. 2. It is in principle a very simple procedure, which is nearly identical to that earlier developed for the ATP/ADP carrier. Thus, it could be assumed that the ATP/ADP carrier would be co-isolated with thermogenin. However, Klingenberg and Lin [13] pointed out that in the thermogenin procedure, the ATP/APD carrier is left unprotected (carboxyatractylate is included in the isolation procedure for the ATP/ADP carrier), and that the use of room temperature helps to denature the ATP/ADP carrier. In practice, the thermogenin preparation is devoid of the ATP/ADP carrier.

2.2. Reconstitution of thermogenin activity

An expected advantage of the availability of isolated thermogenin would be the possibility to study thermogenin behaviour in reconstituted systems. However, it has not been easy to accomplish reconstitutions which fulfil the minimal requirement for being successful: the presence of GDP-inhibited H^+ permeability.

2.2.1. First generation reconstitutions

In the first reports of reconstitution, thermogenin solubilized with Triton-X100 was used. However, Triton-X100 makes the vesicles permeable and only small or insignificant effects of nucleotides could be seen [13,14]. Commonly used alternatives such as cholate or octylglucoside have been considered to denature the protein [13], but Klingenberg and Lin found that octyl-polyethylene oxide functioned better during reconstitution. The procedure is detailed in refs. [13,156]. In reconstitution experiments, this detergent is used by Klingenberg and Winkler also for the initial isolation of thermogenin; this preparation is reported to be 70% pure [15]. The preparation is mixed with egg yolk phospholipid, and the detergent is removed by Amberlite (which is then filtered away). It has been calculated that 2–5 thermogenin molecules may be incorporated per vesicle; Klingenberg and Lin report most of these to be right-side-out [13], but Jezek et al. [16] obtain a 50:50 distribution.

In this type of reconstituted system, the basic criterion for minimal successful reconstitution is met: the presence of GDP-inhibited H^+ permeability [13,15–17] (Fig. 3).

2.2.2. Second generation reconstitutions

In this type of reconstitution studies, thermogenin is not isolated from ‘real’ brown adipose tissue. Rather, thermogenin is expressed in a non-brown-fat system by the transfer to such a system of a cDNA sequence corresponding to thermogenin, inserted in a vector. The mito-

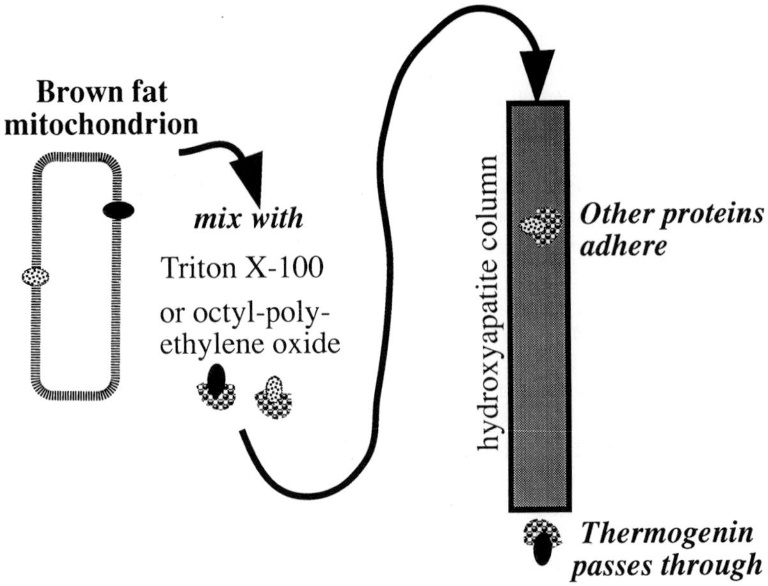

Fig. 2. The isolation of thermogenin. Thermogenin is isolated from brown-fat mitochondria, normally prepared from cold-acclimated animals. The mitochondria may be pretreated with a nonsolubilizing nonionic detergent (Brij or Lubrol) to remove loosely associated proteins [13]. In a comparison between different detergents, Triton-X100 was found to be the best [21]. When, instead, a more 'mild' procedure with octylglucoside was used, the resulting preparation when used in a reconstituted system demonstrated more of the properties expected of thermogenin, but the preparation was not pure, and the results therefore difficult to interpret [140,141]. Although it would seem in retrospect that most of the properties described even in these preparations were thermogenin-related, we shall not discuss these results here in detail, since methods have been developed which seem to combine the advantages of the purity of the Triton-X100 preparation method with the better reconstitution ability of the octylglucoside method [17,142]. Octyl-polythylene oxide may alternatively be used, especially for reconstitution experiments [143] (this gives a preparation which is about 70% pure). The detergent-solubilized protein extract is passed through a hydroxyapatite column at room temperature; apparently most other solubilized proteins remain on the hydroxyapatite, but the pass-through is 70% pure thermogenin [13]. The pass-through also contains a high amount of Triton-X100 and phospholipid, and these constituents may be removed by gel chromatography (Sephadex G-150) or sucrose gradient centrifugation. To follow thermogenin during the isolation procedure, GDP binding is determined [13]. The methods used for determination of GDP-binding capacity in intact mitochondria (filtration or centrifugation) cannot easily be used for the solubilized thermogenin, but other methods, especially equilibrium dialysis (which takes about 5 h) and anion exchange chromatography (which takes about 1 min) have been described [13].

chondria of the host are then analyzed. The main conclusion of these experiments is that the presence of thermogenin in the mitochondria of any cell is sufficient to transfer the basic features of brown-fat mitochondria to these mitochondria [18,19].

However, for further elucidation of thermogenin function, these systems are not optimal, and a powerful technique where thermogenin, expressed in a yeast cell, is reconstituted into

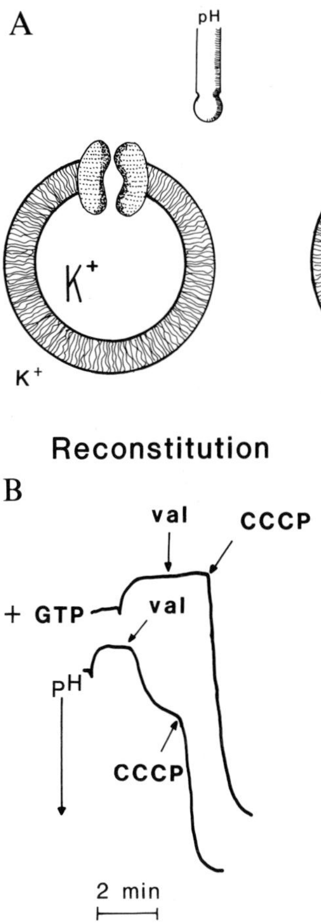

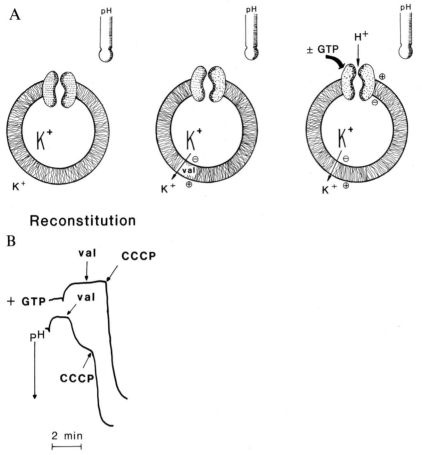

Fig. 3. Reconstitution of thermogenin activity in liposomes. A method [15] for examination of thermogenin-mediated H^+ permeability is shown in (A): a K^+ gradient is applied over the liposomal membrane, and the addition of the K^+ ionophore valinomycin induces a negative transmembrane potential; this potential will drive H^+ through thermogenin, and an increase in medium pH can be measured. If the reconstitution is successful, GDP addition leads to inhibition. In (B), traces from such a successful reconstitution are seen. In the lower trace, it is seen that no H^+ movement takes place before valinomycin is added; if subsequently an artificial uncoupler is added, further permeability is induced (also from vesicles without incorporated thermogenin). In the upper trace, it is seen that if GTP is present, valinomycin does not induce transmembrane H^+ transport (i.e. thermogenin activity is inhibited), but the artificial uncoupler still has full effect. Adapted from Klingenberg and Winkler [15].

liposomes has recently been developed [19]. This technique provides the experimental basis for the – as yet unexplored – potential of protein engineering: to modify the primary structure of thermogenin and investigate the effects of these modifications on thermogenin functions.

3. The size of the functional unit

3.1. Monomer or dimer?

Thermogenin in itself has a molecular weight of approximately 32 000, but it was observed already when thermogenin was first isolated that the solubilized protein migrates in ultracentrifugation as a homodimer [20]. Due to the hydrophobic nature of part of the protein, this association in pairs of the isolated protein is not unexpected. It is, however, also likely that the functional unit is the dimer. One argument for this is that the number of GDP-binding sites per solubilized molecule has been reported to be only about 0.5:1 [21] (although French et al. [22] found binding values which were close to 1:1). However, the reduced number of binding sites could also be secondary to a hydrophobic interaction occurring during the isolation, shielding one binding site. With thermogenin in its native state, in the mitochondria, some studies have found relations between amount of thermogenin (determined immunologically) and [^{3}H]GDP-binding capacity which were close to one binding site per thermogenin dimer [23], but other relationships have also been reported.

Another argument for the dimeric function is that a disulfide cross-link between two thermogenin units can be formed by treatment with a disulfide reagent [24]. With solubilized thermogenin such a result could still be explained based on hydrophobic pairing during isolation, and a better indication is, therefore, that it is also possible to form such cross-linked dimers in intact mitochondria [24]. This would indicate that the two thermogenin monomers are at least physically close. There are no indications that such cross-linked dimers form spontaneously. It is difficult to envisage how the molecule is arranged in space so that only one purine nucleotide binding site is exposed, but it may be pointed out that the functional unit of the ATP/ADP carrier is also considered to be a dimer.

3.2. The unmasking phenomenon

Related to the question of the number of binding sites per thermogenin molecule is the so-called 'unmasking' phenomenon. As first described by Desautels et al. [25], the phenomenon relates to an increase in the number of [^{3}H]GDP-binding sites in brown-fat mitochondria, occurring under circumstances when no new thermogenin can be synthesized (Fig. 4). This phenomenon also relates to thermogenin activity, as the observed Cl^- permeability is lower in mitochondria with masked [^{3}H]GDP-binding sites than in mitochondria with unmasked GDP-binding sites [26,27] (but contrast ref. [28]).

The cause of this phenomenon is still unknown, and basically two viewpoints may be found: the masking/unmasking could be due to a change in thermogenin itself, or it could be secondary to other alterations in the mitochondria – it could, e.g. be related to the physiologically induced mitochondrial swelling observed under some circumstances [29]. Indeed, a simple swelling treatment has by some [30], but not by others [31], been reported to unmask GDP binding.

This latter viewpoint should be contrasted to the idea that a permanent change in thermogenin itself is involved in the masking process. Swick and coworkers have repeatedly [32,33] presented evidence that the [^{3}H]GDP-binding capacity of thermogenin isolated from mitochondria obtained from animals with apparently masked or unmasked [^{3}H]GDP-binding sites is altered, in parallel with what is observed in isolated mitochondria. In extension of this, pretreatment of brown-fat mitochondria with Mg^{2+} before isolation of thermogenin doubled

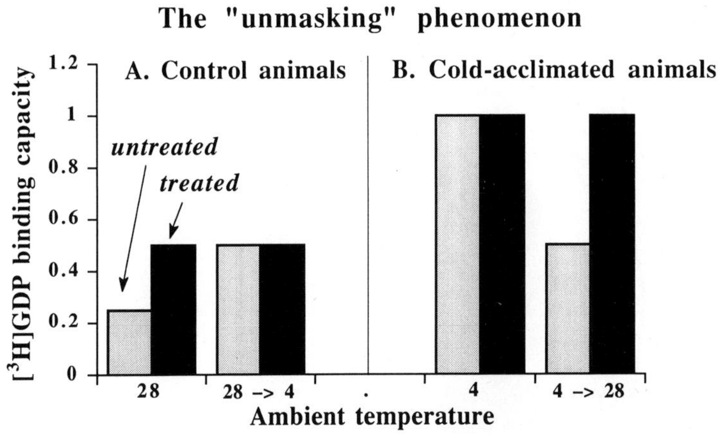

Fig. 4. The masking/unmasking phenomenon. This sketch is compiled from several investigations of the phenomenon. As seen, animals living at thermoneutral temperatures (A) (here indicated as 28°C) have a relatively low capacity for [^{3}H]GDP-binding, but if they are exposed for a short time to cold (here indicated as 4°C) (or e.g. injected with norepinephrine) they show a marked increase in [^{3}H]GDP-binding capacity; this increase has characteristics, including very rapid kinetics, which eliminates an increase in the amount of thermogenin as an explanation. In animals living permanently in 'cold' conditions (B), and this includes any temperature below thermoneutrality, including 'normal animal house conditions' (18–22°C), thermogenin is apparently not masked, but if these animals are transferred to 28°C, [^{3}H]GDP-binding sites disappear (become masked) very rapidly. Several treatments have been discussed to 'unmask' [^{3}H]GDP-binding sites in vivo; it should be noted that such treatments, to be relevant, should show the indicated pattern: clear effect in physiologically masked states and no effect in unmasked states.

the [^{3}H]GDP-binding capacity of the subsequently isolated protein [32,33]. Functionally, this corresponds to an unmasking phenomenon in that this was seen when mitochondria were isolated from animals living at thermoneutral temperatures (the only condition where unmasking can be seen), but generally (exception in ref. [34]) not when the mitochondria were isolated from cold-acclimated animals (where also all [^{3}H]GDP-binding sites on thermogenin seem to be unmasked in mitochondrial preparations). The authors suggest that these changes could be due to an enzyme-mediated change in thermogenin itself.

4. *What is transported by thermogenin?*

4.1. A carrier or a channel?

Thermogenin allows for a transmembrane flow of certain ions. The mechanism could either be described as a carrier or a channel; the channel would allow bulk flow whereas the carrier would move one molecule of the transported species per turnover. Both models have been suggested; due, e.g. to the similarity of thermogenin to the ATP/ADP carrier, a carrier model has often been promoted [4,156], but a channel has also been suggested [35,36], e.g. due to an observed linear effect of membrane potential on transport rate [36], and the high transport number at high potential (about 100 H^+ per second per molecule) [4].

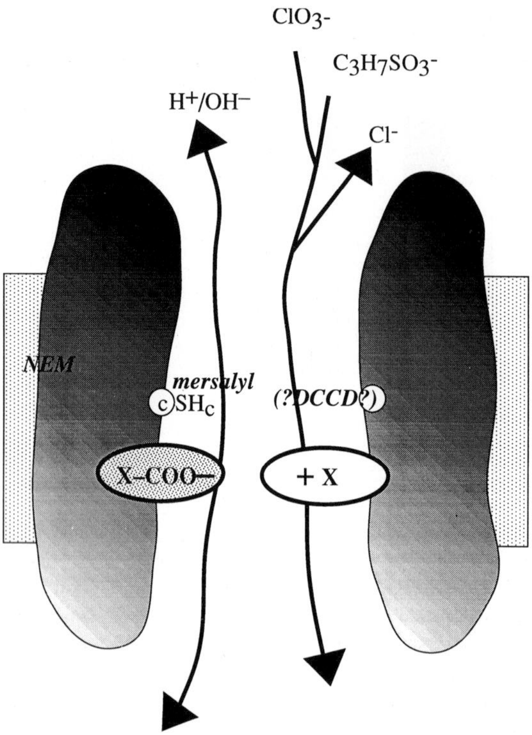

Fig. 5. Transport through thermogenin. The transport mechanism is here symbolized in the shape of a channel. Some properties of the carrier/channel have been characterized, especially effects of chemical modifiers have been examined. The results are not fully identical for Cl^- transport and for H^+ transport. The indication 'Cl^-' represents the family of classical halide compounds which show high permeability (Cl^-, Br^- and I^-, and also NO_3), also in the reconstituted system [16]. Surprisingly, the rate of halide permeation increases with increasing size (i.e. $Cl^- < Br^- < I^-$) [16]. In an extension of this, Jezek et al. [35] have found that a large family of fairly bulky anions may be transported by thermogenin in the reconstituted system. The apparently transported species includes alkylsulfonates (of type $C_3H_7SO_3^-$) with a chain length from two up to at least nine carbons, pyruvate (assessed in the presence of NEM to inhibit the classical pyruvate translocator), phosphate analogues (but apparently not phosphate itself), metavanadate and also the series chlorite, chlorate and perchlorate. All of these bulky anions are very rapidly transported in comparison with the classical permeant anion Cl^- [35]. These substances are not 'weak hydrophobic acids' and it is therefore unlikely that the permeation seen is that of the protonated form (as is the case for acetate), but they have a structure very similar to fatty acids (which have not been directly tested in this connection). The transport of these substances (Cl^-) is not pH dependent [16] and it therefore probably does not involve a positively charged residue in the pathway; in agreement with this, histidine reagents do not affect Cl^- permeability [4]. At low concentrations, *DCCD*, also in brown-fat mitochondria, binds to the F_0 part of the mitochondrial ATP synthetase [138], but at high concentrations the dominant part of DCCD binding is to thermogenin [144]. This was originally reported to lead to inhibition of Cl^- permeability (but not H^+ permeability) in isolated mitochondria [66,144], and the binding site was therefore supposed to be in the Cl^- channel. However, these observations have been critisized [39,145], and it would seem that DCCD induces a high initial unspecific permeability. The mitochondria are thus pre-swollen and exhibit

→

According to the carrier model, thermogenin could exist in two forms, an outer U_c and an inner U_m. In this model, thermogenin would be locked in the U_c position by GDP [37]. This kind of kinetic scheme resembles the kinetic schemes discussed for thermogenin's sister protein, the ATP/ADP carrier; so far few experiments have been interpreted in these terms.

In Fig. 5 we have drawn the transporter in the form of a channel, with two fluxes through it. This drawing is thus also related to the question as to whether there are independent pathways for H^+ and Cl^- (i.e. whether it is H^+ or OH^- which is the transported 'uncoupling' species) (Fig. 6).

4.2. Cl^- transport

Isolated brown-fat mitochondria show a high Cl^- permeability [38]. It has been established that the high Cl^- permeability observed in brown-fat mitochondria does not reflect the activity of the ubiquitous inner membrane anion carrier (IMAC) [39]; rather, both of these Cl^- transport mechanisms can be found in brown-fat mitochondria. It has, however, been pointed out that the Cl^- fluxes seen under certain conditions (e.g. in energized mitochondria) may involve also a contribution from the IMAC [39]. The high Cl^- permeability is also different from the anion channel observed by the patch-clamp technique in brown-fat mitochondria [40]; also this channel type is ubiquitous.

The Cl^- permeability may be estimated using mitochondrial swelling in KCl in the presence of the K^+ ionophore valinomycin [38]. It is advisable to start the swelling with valinomycin addition, as this allows for a routine check that nonspecific permeability has not been induced. The Cl^- permeability may also be followed using entry of labelled $^{36}Cl^-$ [38], or, in a reconstituted system, with a Cl^--sensitive dye (SPQ) trapped within a liposome [16].

It was earlier doubted that thermogenin in itself conferred Cl^- permeability to the mitochondria, as the first reconstituted systems did not transport Cl^- [15]. However, this tenet has not been upheld since Cl^- permeability has been observed in later reconstituted systems [16,19].

---

an apparent low permeability. In agreement with this, DCCD is without effect in a reconstituted system [16]. The binding site for DCCD on thermogenin has thus not been located; it is not in the purine nucleotide site, as it does not influence GDP binding. Whereas no agent has thus been identified to inhibit Cl^- permeability, there are apparently inhibitors of the H^+ permeability: certain sulfhydryl agents. Hydrophobic SH-reagents such as NEM (*N*-ethylmaleimide) do not inhibit H^+ permeability [146], but *mersalyl* inhibits H^+ permeability (but apparently not Cl^- permeability [146]), also in a reconstitued system [16]. Also DTNB inhibits H^+ permeability, both in intact brown-fat mitochondria and in a reconstituted system [147], apparently without affecting Cl^- permeability. However, the lack of effect on Cl^- permeability has been suggested to be due to an introduction of a non-thermogenin-related Cl^- permeability in the mitochondria by these agents; this new permeability would overshadow any inhibitory effect of sulfhydryl agents on Cl^- permeability [148]. The earlier implied difference in sulfhydryl reagent sensitivity between the Cl^- and the H^+ permeabilities thus remains unestablished. The cystein carrying the –SH group involved (here denoted $-SH_c$, following Jezek and Drahota's [147] terminology, c for channel) is not identified, but it is probably not cystein^{304} which can be used for cross-linking without effect on H^+ permeability. The channel does not demonstrate any rectifier effect [15] (H^+ transport is equally effective in both directions). Klingenberg has suggested the functional part of the channel to consist of a single centre [4], perhaps involved in a 'salt bridge' of the type shown here. In e.g. bacteriorhodopsin, the negatively charged part of the bridge is an aspartyl residue, but an alternative discussed by Klingenberg is that the site is not a part of an amino acid. Rather, the site may basically be unoccupied and a free fatty acid may be placed there, giving rise to the apparent fatty acid dependence of H^+ permeability [4].

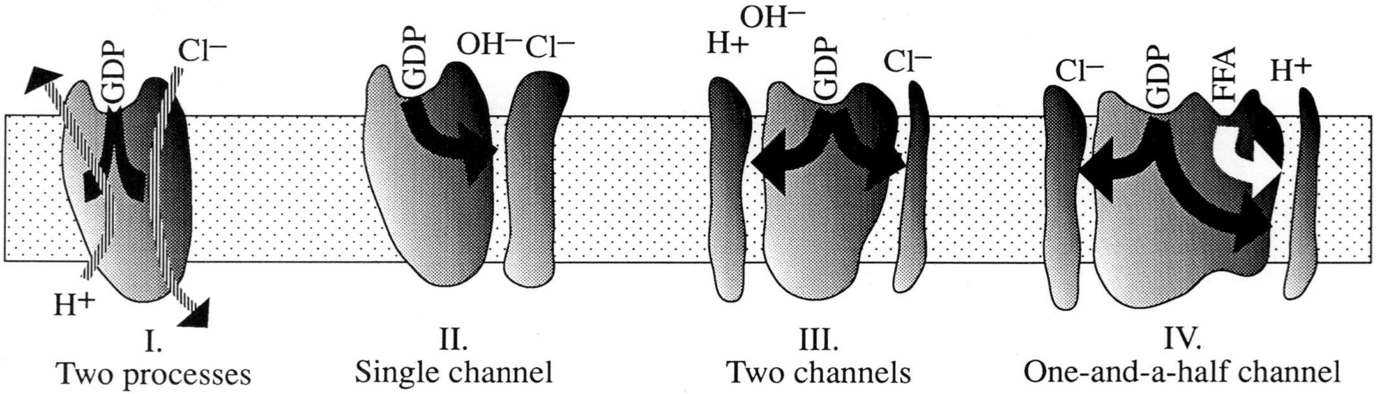

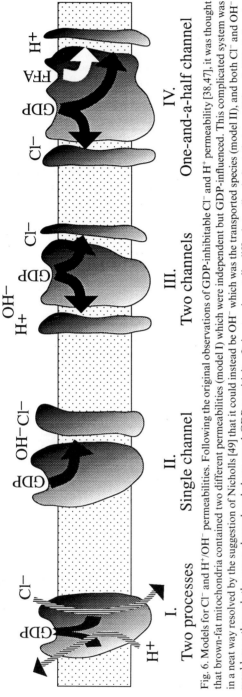

Fig. 6. Models for Cl^- and H^+/OH^- permeabilities. Following the original observations of GDP-inhibitable Cl^- and H^+ permeability [38,47], it was thought that brown-fat mitochondria contained two different permeabilities (model I) which were independent but GDP-influenced. This complicated system was in a neat way resolved by the suggestion of Nicholls [49] that it could instead be OH^- which was the transported species (model II), and both Cl^- and OH^- would pass through the same channel and show equal GDP sensitivity. It is experimentally difficult to discriminate between H^+ and OH^- being the transported species, although possibilities have been discussed [4]. It has been claimed that model II cannot explain all experimental results. It has especially been claimed that the GDP sensitivity is different for the two functions, i.e. Cl^- permeability is more easily inhibited than H^+ permeability [75,96,149]; in these experiments, the inhibition of H^+ permeability by GDP was not linearly related to GDP binding. Thus, model III (which is very close to model I) must be suggested (the two permeabilities could be on the same molecule or on slightly different molecules). The experiments behind this suggestion were performed in parallel in the same mitochondrial suspension, but different methods were used for the two parameters (swelling for Cl^- and pH electrode for H^+). When similar experiments were performed by Rial et al. [148], in parallel experiments but with both permeabilities investigated using swelling, the opposite relationship was observed: H^+ permeability was slightly more easily inhibited than Cl^- permeability (the small difference was suggested to be due to the effect of Cl^- on GDP binding), and a high correlation was found between GDP binding and H^+ permeability inhibition [15,150]. Another argument for model III has been that certain reagents have been claimed to inhibit only one of the permeabilities; as discussed in Fig. 5 these claims have not been substantiated. Thus, few arguments remain for the complicated model III, and for the simplistic model II also speak experiments which have been interpreted to demonstrate competition between OH^- and Cl^- permeation [38,151]. Further, both Cl^- and H^+ permeabilities are introduced by the reconstitution of thermogenin (expressed in yeast and isolated from there) into liposomes, and these two permeabilities have a similar GDP sensitivity in the reconstituted system [19]. However, as it has been claimed that H^+ but not Cl^- permeability can be induced by addition of free fatty acids to mitochondria [86], it may be necessary to introduce model IV, implying that free fatty acids induce an exclusively H^+ translocating pathway through thermogenin. It may be noted that the difference between this model and the catalytic effect of fatty acids discussed in Fig. 5 is not large.

The physiological significance of the halide/Cl^- permeability has in general been doubted, and it is generally thought that it is only an experimental reflection of the true 'uncoupling' permeability of thermogenin. It may, however, be relevant for the in situ mitochondrial swelling observed in certain physiological situations [29].

4.3. H^+ transport

Isolated brown-fat mitochondria show a high H^+ permeability, i.e. they are uncoupled. That this was the basis for the thermogenic property of these mitochondria was realized from the first isolations of brown-fat mitochondria [41–43]. The successfully conducted reconstitution experiments confirm that the GDP-sensitive H^+ permeability is a property of thermogenin which does not require the mediation by any other membrane component [15,16].

H^+ permeability may be followed either as changes in the external pH or as mitochondrial swelling, most often as valinomycin-induced swelling in a KAc medium, in which the H^+ outflux is made rate-limiting by the addition of the K^+ ionophore valinomycin [44]. As increased H^+ permeability may be induced in many unspecific ways, the precaution to start the swelling with valinomycin addition is always well taken. Indirectly, H^+ permeability may be followed as mitochondrial respiration (in many ways a physiologically very relevant estimation), and the H^+ conductance may be calculated if the proton motive force (or its approximation in the form of the mitochondrial membrane potential, a reasonable approximation when phosphate is present in the medium) is measured in parallel with respiration, and the H^+ current estimated from the respiration [44].

4.3.1. H^+ permeability and free fatty acids

That some relationship between H^+ permeability (uncoupling) and the presence of free fatty acids may exist has also been discussed since the first brown-fat mitochondria preparations [38,45–47]. The situation is still far from clear. Some of the issues under discussion may be understood as the choice between a formulation stating that fatty acids regulate thermogenin activity and another stating that the fatty acids are necessary for thermogenin to exhibit H^+ permeability. The navigation between these formulations is not uncomplicated, and the formulations are, of course, interrelated. In this section we shall concentrate on the possible necessity of fatty acids for H^+ transport (for the regulation formulation, see later).

The reconstituted systems have so far not clarified the situation. Originally, fatty acids were reported not to affect H^+ transport in the reconstituted system [15], and indeed GDP-sensitive H^+ permeability can be observed in these systems without addition of fatty acids [15,16,19]. It has, however, also recently been stated that H^+ translocation is dependent upon (re)addition of fatty acids to the reconstituted system [36] (provided that the fatty acids are first carefully removed). There is, thus, a possibility that the fatty acids are directly involved in the H^+ transport process.

The observation of Jezek and Garlid [35] that thermogenin may be able to transport large bulky, hydrophobic anions has enabled a new formulation of the H^+ transport process. Such anions have characteristics which are very close to those of free fatty acids, and indeed Skulachev [48] offered the hypothesis that the fatty acids are the true physiological substrates being transported via this mechanism over the mitochondrial membrane. In Skulachev's proposal they would then function as uncouplers, because they would enter the mitochondrion by membrane diffusion in their protonated state, but would be transported back in their negatively charged form through thermogenin; they would thus function as proton shuttles, with the net effect being the transfer of a proton across the mitochondrial membrane (i.e. uncoupling). Undoubtedly this hypothesis links the observations of Jezek and Garlid to the physiological role, but the scheme proposed remains a hypothesis.

In a different formulation by Klingenberg, the free fatty acids would function as 'localized H^+-supply buffers' [36]. An illustration of this hypothesis is included in Fig. 5. In this formulation, the free fatty acids would bind to a site within thermogenin and there form a part of a 'salt bridge'. In this position they would function as mediators of H^+ transport, in being a stepping stone for the H^+ during its membrane forming.

Thus, the difference in these two formulations may not be so large. Skulachev sees the fatty acids as themselves being transported and acting as shuttles whereas Klingenberg sees the fatty acid fixed and only the H^+ moving. In both cases, the fatty acids have been ascribed a catalytic role and they perform this by changing between the protonated and the unprotonated state. It

---

→

Fig. 7. Thermogenin structure. The figure is based on the compilation of conserved residues presented by Klaus et al. [58], but using the numbering system for the hamster/rat protein. Shaded circles indicate conserved hydrophobic residues, and the two heavy lines indicate the ends of the second and fourth exons. Some additional features may be pointed out.

The initial amino acids are – as seen – poorly conserved, in agreement with the fact that they are not necessary for thermogenin insertion into the mitochondrial membrane [130]. There is no experimental evidence for their localization to the cytosolic side.

In contrast, the last 11 amino acids are well conserved. There is strong evidence that the terminal end is on the cytosolic site of the structure. It is a strong antigen [152] and – being a conserved sequence – a good candidate for a general antigen for thermogenin for many species. Its accessibility is also seen from the fact that it is easily split by trypsin at lysine^{292} [65]; this tryptic cleavage site is of special interest as it becomes shielded from tryptic attack when an agonist is bound to the purine nucleotide binding site [65]; this conformational change is thus involved in the intraproteinous transmission of the signal. The accessibility of the C-terminal is also in agreement with the fact that the conserved cystein^{304} with disulfide-forming agents can be induced to form a thermogenin dimer [21,24]; this has no effect on the properties of the purine nucleotide binding site or on the H^+ permeability.

The α-helixes were originally defined from sided hydropathy plots (Fig. 8), and the end-points of the helixes were originally suggested to be defined with prolines [51]. However, not all the prolines originally used to define the end-points are conserved, and the exact start and end of most helixes are therefore uncertain. Several suggestions for the orientation of the α-helixes have been given. In an original, nonsymmetrical model, proposed by Klingenberg and coworkers [51], an extra transmembrane sequence (based on the amphiphilic character of a β-sheet) was suggested. However, the rabbit sequence makes the amphiphilic β-sheet less likely [55], and the structure with this extra transmembrane sequence is no longer advocated by Klingenberg [4]. A further modification of the nonsymmetrical model has been proposed by Klaus et al. [58]. We have here, however, drawn a structure which emphasizes the tripartite nature of the sequence. For this purpose we have defined the inward helices as the 21 residues preceding the circled prolines (32, 132, 231) which are fully conserved within the three repeats in all members of the mitochondrial carrier family. We have similarly defined the start of the outward helices based on the occurrence of a sequence which is reasonably conserved both within the repeats and within the family, i.e. the sequence Glu-Gly-X-X-X-X-X-Lys-Gly (first glycines 69, 168 and 262). In both cases, the resulting α-helices show an amphiphilic nature. There are, however, several complications with this structure: arginine^{238} is supposedly a second tryptic cleavage point [65], but this does not fit with this model, and the location of the purine nucleotide binding site at threonine^{259} is also on a nonexpected side of the membrane. Of the conserved seven cysteins, up to three can be reacted with sulfhydryl agents, leaving four cysteines which cannot be observed, even with SDS/urea-denaturated thermogenin [148]; as discussed by Rial et al. [148], these four missing cysteines may represent two putative disulphide bonds. However, no thermogenin structure which includes intramolecular disulfide bonds has so far been suggested.

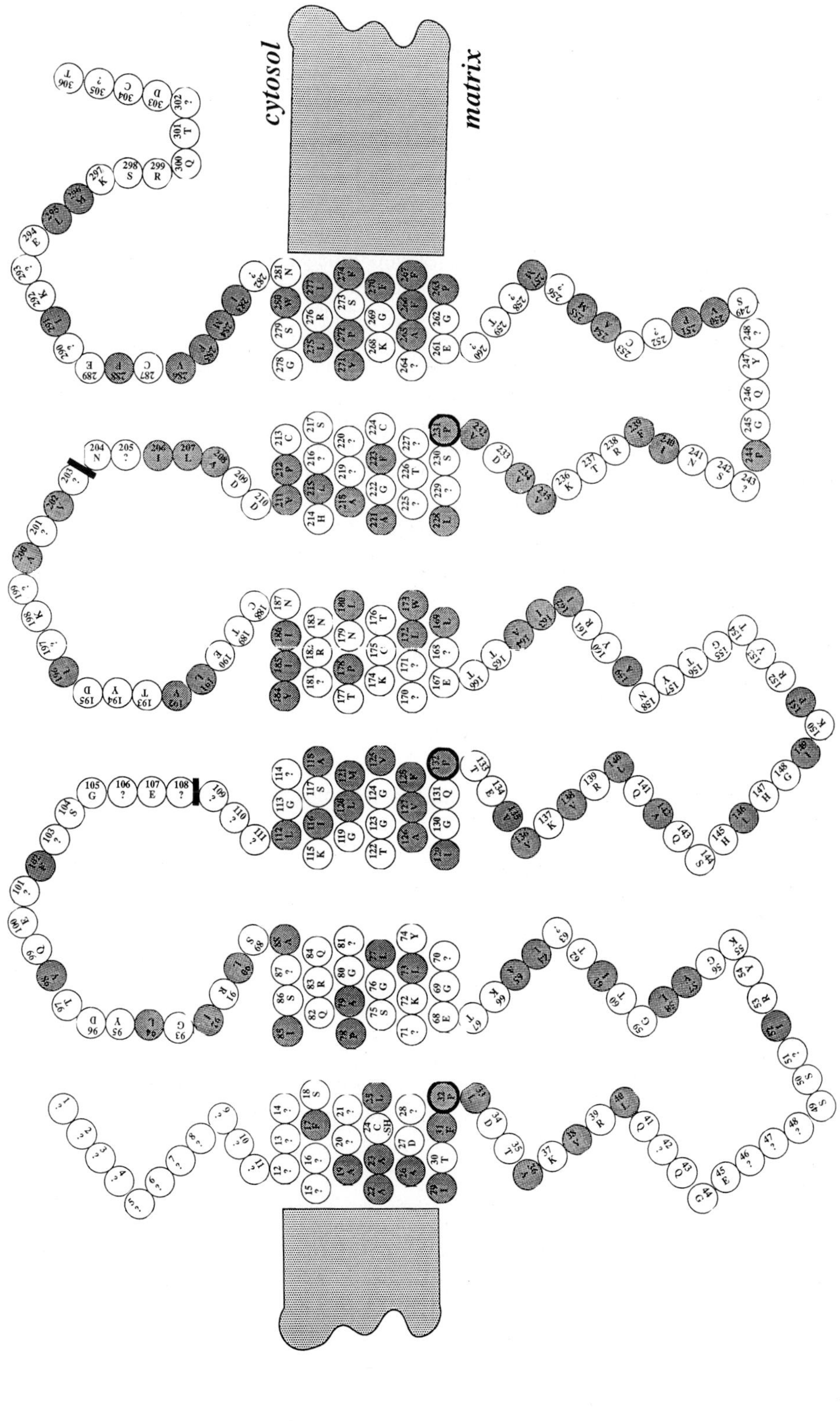
cytosol
matrix

may be added that Skulachev has extended the hypothesis to other mitochondrial membrane proteins and suggests that the presence of fatty acids may convert also the ATP/ADP carrier to an 'uncoupling protein' [48].

4.4. H^+ and Cl^- permeabilities: the same or not?

The association of both H^+ and Cl^- permeability with brown-fat mitochondria and with thermogenin raises the question whether these two permeabilities are different reflections of the same transmembrane pathway or two really different phenomena. This discussion is summarized in Fig. 6. Presently, most evidence would favour model II, i.e. there is only one channel which can transport anions, including OH^-; an outward OH^- transport is energetically identical to an inward H^+ transport and leads to uncoupling [49]. For the identity of the Cl^- and H^+/OH^- pathway also speaks the lack of pH sensitivity of both transports (between pH 6 and 8) [15,16,36,38], and that chymotrypsin treatment does not dissociate these two permeabilities [50]. Several other attempts have been made to modify the mitochondria in order to separate the two permeabilities, but none of those published so far have been able to withstand further scrutiny (Figs. 8 and 9). However, if free fatty acids activate thermogenin (see Section 6.2.1), the complicated model IV must be accepted, with an extra permeability exclusively for H^+ being introduced by the fatty acids; in another formulation, the presence of fatty acids may alter the properties of the existing channel to make it protonophoric, and the channel would thus be able to switch between the characteristics of the two permeabilities indicated in model IV.

5. Suggested thermogenin structure(s)

5.1. General structure

The amino acid sequence of thermogenin is now known from many species: hamster [51], rat [52,53], mouse [54], rabbit [55], ox [56] and man [57]. A compilation of the conserved sequence [58] can be seen in Fig. 7.

From the amino acid sequence, thermogenin 'structures' have been suggested, but these structures – as so many 'structures' presently drawn – are only based on computer analysis of amino acid sequences and on only little experimental data. No crystalline thermogenin has as yet been reported, and a structure from X-ray diffraction analysis is thus not within reach.

Aquila et al. [51] first pointed out the tripartite structure, which could be deduced from the amino acid sequence. They also pointed out that the hydrophobic domains which would be necessary for membrane spanning [51] could not be found. Only if the sequence is analyzed as an amphiphilic helix (Fig. 8) can transmembrane domains be suggested. The presence of α-helixes is confirmed by analysis of the infrared spectrum; the structure consists of about 50% α-helix, 30% β-structure, 15% β-turns and 7% unordered [59].

The structure shown in Fig. 7 is drawn on these premises.

5.2. The nucleotide binding site

The purine nucleotide binding site is the most well characterized topological unit on thermogenin. It was by the binding of photoaffinity labelled ATP to isolated brown-fat mitochondria that the existence of thermogenin was first indicated [60]. It has now been possible to localize

A

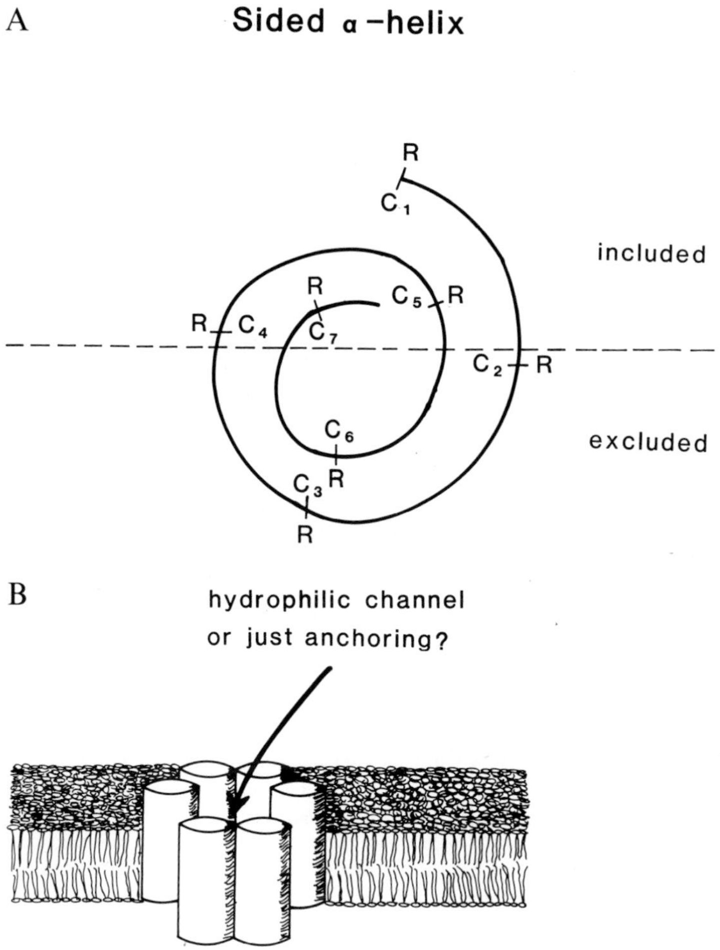

B

Fig. 8. The amphiphilic nature of thermogenin α-helices. (A) Only when analyzed as a sided amphiphilic helix can transmembrane α-helices be recognized in the primary structure of thermogenin. These helixes are thus characterized as having a hydrophobic and a hydrophilic part. For these calculations, only the sidegroups (R) on the indicated residues (C) are considered. (B) In the hydrophobic environment of the mitochondrial inner membrane, it is likely that the hydrophilic parts of the α-helices associate. It is not known whether this only has the effect of keeping the protein structurally together and anchoring it in the lipid layer, or whether the hydrophilic channel which would be expected to form through the membrane is directly involved in transport.

the nucleotide binding site on the thermogenin molecule [61]; it is threonine^{259}. Surprisingly, this site is located in a sequence, which, with the presently discussed thermogenin structure (Fig. 7), is on the matrix side of thermogenin. The site can apparently and quite unexpectedly also be labelled from the matrix side. This threonine^{259} site is close to a sequence from glutamic

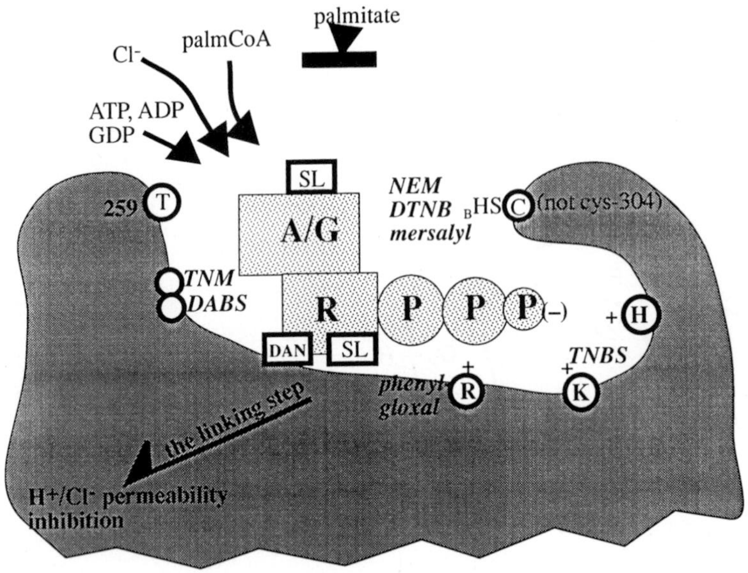

Fig. 9. The nature of the nucleotide binding site. The drawing summarizes several investigations concerning the purine nucleotide binding site. Mainly, the information is compiled from experiments in which the effect of reagents to modify [^{3}H]GDP-binding characteristics has been studied. Concerning the *binding of nucleotides* as such, the affinity of GDP in itself is in the order of 1 μM on the mitochondria. Other nucleotides with high affinity include the purine nucleotides ADP, GTP (not directly investigated), IDP, XDP and ATP, and the derivatives dGDP and 8-Br-ATP. Low-affinity binders include pyrimidine nucleotides, like UTP and UDP, as well as GMP-NH-P. All of these compete with bound [^{3}H]GDP for the site. Nonbinding analogues are the nucleotides AMP, cAMP, cGMP, the analogues AMP-SO_4, NAD(P)(H), AMP-CH_2-P, GDP-glucose, and PP_i, PPP_i, and cyclic triphosphate [60]. The ADP-analogue palmitoyl-CoA can bind (competitively with [^{3}H]GDP) to the nucleotide-binding site, both in intact mitochondria [66,87], with an affinity (K_i) of about 3 μM [66], and in a reconstituted system [17]. When ATP is modified by a spin-label (SL) at position 2',3' in the ribose moiety, it does not bind. However, when the spin label is attached to position 8 in the adenine moiety, the derivative both binds and inhibits Cl^- permeability, and the spin-labelled derivative shows a spectrum indicating that it has become immobilized [158]. When ATP is modified with the fluorescent compound DAN at the 3'-position in the ribose moiety, ATP still binds *but* it no longer inhibits H^+ permeability [64]. The *binding conditions* may modify purine nucleotide binding. Mg^{2+} leads to an apparently lower affinity (for GTP), probably due to complexing. Negative ions (Cl^-) [21] at high concentrations (in the 100 mM range) decrease the apparent affinity, probably due to anionic interaction. The bulky-anion family of apparent substrates for thermogenin transport observed by Jezek et al. [35] may also – for the same reason – affect GDP binding to the purine nucleotide site although this has not been demonstrated. Fatty acids are practically without effect on the purine nucleotide binding site [17,86]. The *protonation* state of both the binding nucleotide and of the charges within the site also affects the binding site. As indicated, occurrence of a terminal negative charge (−) at the tri/dinucleotides at high pH may be responsible for the decrease in affinity that occurs at pH above about 6.5 [63]. The presence of a histidine group in the binding site has been discussed from the effect of photobleaching [21] and on the proposal that this group which has a comparably high pI, is responsible for the somewhat steeper decrease in affinity for the purine triphosphates observed at pH higher than 7.2

→

acid[261] to glycine[269] which has been pointed out in several comparisons to be very similar to a sequence in the ATP/ADP carrier which is supposedly involved in nucleotide binding.

As any treatment which alters the GDP-binding characteristics may be suggested to affect the purine nucleotide binding site, a long list of properties implied from such experiments can be compiled. This is detailed in Fig. 9.

From the detailed studies, some general conclusions can be made. The interaction between thermogenin and the binding nucleotide involves interactions with all three parts of the nucleotide, although it is likely that the 'upper' part of the purine is the part which still protrudes.

5.2.1. Agonists, partial agonists and antagonists

By expaning the models for hormone receptors, it is possible that certain parts of the interphase between the purine nucleotides and thermogenin represent 'binding' or affinity sites only (with which both agonists and antagonists interact) and other parts represent effector sites, with which only agonists interact (but which, of course, are shielded if the binding site is occupied by an antagonist). Jezek et al. [62] have particularly pointed out that these interphases would involve specifiable amino acid residues, but no such amino acids have as yet been identified.

Most identified binding agents are full agonists, i.e. their binding affinity is parallel to their effectiveness as inhibitors of H^+ permeability. True antagonists are defined as those which bind well but which do not inhibit the associated permeabilities at all and partial antagonists do bind and inhibit permeabilities, but not to the extent expected from binding data. The analogue DAN-ATP (derivatized on the ribose moiety) apparently binds with an affinity close to that of

---

[63]. Albumin treatment can increase the affinity and capacity of brown-fat mitochondria for [^{3}H]GDP binding [153,154]. This may indicate that bound acyl-CoA derivatives can be removed (the removed species cannot be free fatty acids as these do not influence [^{3}H]GDP binding). Most further knowledge about the purine nucleotide binding site has been derived from experiments with *chemical modifiers*. In general, these modifiers also influence the ability of GDP to inhibit H^+ permeability. The only residue confirmed to be in the site is *threonine*[259], because this is the residue to which azido-ATP binds [61]. A photoaffinity-labelled coenzyme A derivative also binds to thermogenin [155]. The presence of an arginine at the site is demonstrated by inhibition of [^{3}H]GDP binding with either *phenylglyoxal* or 2,3-butanedione [142]; there is then no effect of GDP on H^+ permeability [17]. [^{35}S]DABS binds predominantly to thermogenin in brown-fat mitochondria [149]. Subsequent isolation of thermogenin from the labelled mitochondria indicated about two [^{35}S]DABS bound per thermogenin; the amino acid residues to which DABS binds covalently have not been identified. The binding of DABS does not in itself affect thermogenin H^+ or Cl^- permeability, but it leads to a diminished affinity of GDP for the site and – largely but perhaps not fully – in parallel with this, to a decreased ability of GDP to inhibit either permeability [149]. The presence of a sulfhydryl group is indicated by effects of *sulfhydryl reagents* on GDP binding and GDP effects. The –SH group is here denoted $-SH_B$, following the classification of ref. [147] (B for binding site). The binding of NEM, mersalyl or DTNB [147] abolished the inhibitory effect of GDP on the H^+ and Cl^- permeabilities and reduces the affinity for GDP [148]. The difference between the different sulfhydryl reagents is not fully clarified. The amino acid residue corresponding to the (or these) cysteine(s) has not as yet been identified, but the cysteine should *not* be cysteine[304], because this residue may be involved in thermogenin dimer formation without affecting GDP-binding properties [24]. The effect of the lysin-specific reagent TNBS on GDP-binding indicates the presence of a lysine residue [21]. Binding of TNM (tetranitromethane) also decreases the affinity of the affected sites for GDP (to about 10 μM) [70]; original claims that this also decouples [70] have not been sustained. Chymotrypsin lowers the affinity of the binding site by introducing a new affinity of about 20 μM (in comparison with the native 1 μM) for GDP [50].

ATP [63] but it does not recouple brown-fat mitochondria [64] and it does not induce the conformational change (hiding the tryptic cleavage site) which is seen upon agonist binding [65]. It must therefore be considered a full antagonist (the only one known). Palmitoyl-CoA binds well but only has a partial effect as an inhibitor of the permeabilities [66,67]; it is therefore a partial agonist. It is indeed remarkable that both these compounds are ATP derivatives with a substitution at the 3' position in the ribose unit (but CoA itself does not have these effects).

5.2.2. Coupling of nucleotide binding to permeability inhibition

As an effect of agonist binding to the purine nucleotide binding site, some kind of 'conformational change' would be expected to take place in the thermogenin structure. This conformational change would be involved in the intraproteinous pathway leading from nucleotide binding to the inhibition of H^+ and Cl^- permeability. There are indeed observations indicating that such conformational changes take place.

When the association kinetics of the agonist ATP and the antagonist DAN-ATP are compared, it is found that DAN-ATP binding is much more rapid than ATP binding, implying that the ATP binding is slowed down because of the conformational change induced [63].

Solubilized thermogenin in itself fluoresces, suggestedly due to the presence of tryptophan residues. The addition of GDP leads to a dose-dependent decrease in fluorescence, which may be indicative of a conformational change in thermogenin [22]. There is also a change in the infrared spectrum of solubilized thermogenin upon binding of GDP; the change may be due to a change in the β-turns [59].

Incubation of mitochondria (mitoplasts) with trypsin and chymotrypsin [68] leads to distinct cleavage products with a molecular weight of about 22 000 and 10 000, as observed with antibodies to thermogenin [69] or after post-digestion isolation [65]. The presence of GDP leads to a slower degradation [50,65,69] and perhaps to the occurrence of other degradation products [69]. The lower-molecular-weight product apparently contains part of the purine nucleotide binding site, as azido-ATP label remains with this fragment [69]. The protection by GDP against tryptic treatment reflects a conformational change induced by agonist binding: the shielding of a tryptic cleavage point. It is likely that it is lysine^{292} which physically changes position [65]. The antagonist DAN-ATP (which apparently binds to the purine nucleotide site) cannot protect against tryptic cleavage. During tryptic treatments, the mitochondria lose GDP-binding capacity and GDP-effect on H^+ and Cl^- permeabilities in parallel [50].

Several attempts have been made to dissociate the coupling between purine nucleotide binding and permeability inhibition. Original claims that TNM decouples [70] have not been sustained, and no decoupling agents are presently known.

6. *Control of thermogenin activity*

A thermogenin molecule which is always active or always inactive would not make much physiological sense. Thus, thermogenin activity is under physiological control. Despite much effort and despite the unravelling of the primary structure, a molecular understanding of this control has not been achieved. Here we shall summarize some of the discussions on this question. Cellularly, the activation hypotheses fall into two categories: those in which the adrenergically induced lipolysis is a mediating step between adrenergic receptor activation and thermogenin activation (Fig. 10A) and those in which this is not the case (Fig. 10B). Molecularly, the activation hypothesis also falls into two categories: those in which the activator functions by displacing the purine nucleotide bound (here ATP) (Fig. 10C), and those in which

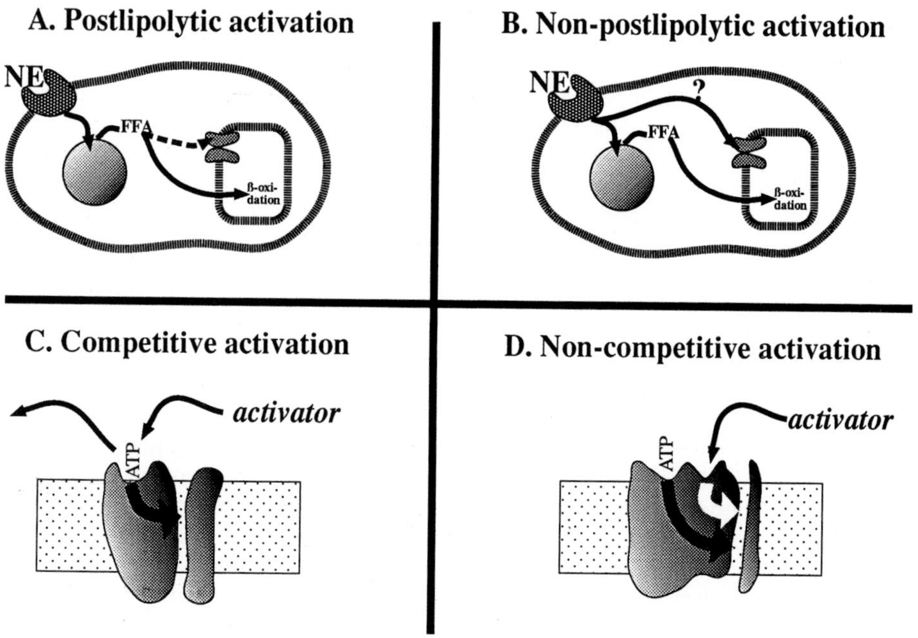

Fig. 10. Activation of thermogenin: two principles. Cellularly, the activator may be generated as a direct effect of lipolysis, thus giving rise to postlipolytic activation (A), or it may be generated independently of lipolysis (B). Molecularly, the activator may act by displacing the inhibitory ATP bound to the purine nucleotide binding site (C), or it may interact with another site on thermogenin, overriding the inhibition of thermogenin activity caused by ATP (D).

the purine nucleotide remains bound and the activator interacts with another site and overrides the inhibitory effect of the purine nucleotide (Fig. 10D).

6.1. Suggested nonlipolytic control mechanisms

6.1.1. Cytosolic alkalinization

This hypothesis was first suggested based on observations of the thermogenic potential of brown adipose tissue at alkaline pH [71] but it attained renewed interest when the effect of changes in pH on *purine nucleotide binding* of solubilized thermogenin was investigated in detail by Klingenberg [63]. A clear effect of pH on purine nucleotide binding to brown-fat mitochondria was observed already by Nicholls [72]. In solubilized thermogenin, the binding constant for ATP is 0.5 μM at pH 6.0 and increases to 32 μM at pH 7.5; at a pH of 7.7 or above the affinity decreases to about 100 μM [63,73] (in contrast, only a small change in affinity with pH was observed by French et al. [22]). The binding of ATP and probably GTP has been suggested to involve three pH- sensitive groups [63].

At higher pH, the ability of purine nucleotides to inhibit *Cl^- permeability* is diminished in parallel with the decrease in their binding affinity [74,75]. However, the effect is not fully that which would be expected for the physiological control mechanism; for GDP, e.g., even at pH 8.2 (!), the H^+ conducting activity of thermogenin is still only 50% activated in the presence of only 0.6 mM GDP. Considering the high (5–10 mM) concentrations of ATP in the cytosol, it would be implicated that the thermogenic potential of thermogenin under physiological conditions only would be exploited to a limited degree.

At higher pH, the ability of purine nucleotides to inhibit *H^+ permeability* is also diminished in parallel with the decrease in their binding affinity [74]. Thus, if a state of high pH is reached, the mitochondria would tend to become more uncoupled, and thermogenesis would proceed.

Whereas there is thus no doubt that in in vitro experiments, thermogenesis can be elicited through pH-dependent activation of thermogenin (via a decreased effectiveness of the inhibition by purine nucleotides), the physiological feasibility of this mechanism is less easy to evaluate. A physiological activation of thermogenin via this mechanism has as a prerequisite that a physiological mechanism for cellular alkalinization exists and can be evoked upon adrenergic stimulation of the cells. Experimental evidence for such a process is still lacking. Quite large changes in cytosolic pH would be necessary, judging from the studies referred to above. The cytosolic pH has not been studied much in brown-fat cells; the small alkalinization which has been observed has been ascribed to α_2-adrenergic stimulation [76], i.e. a pharmacological characteristic diametrically opposite to that needed for a physiological mediator of thermogenesis (β_3 and increase in cAMP levels). Thus, presently, cytosolic alkalinization cannot be considered as a major candidate for being the intracellular activator of thermogenin.

6.1.2. A decrease in total cytosolic nucleotide levels

A related hypothesis has been forwarded by LaNoue et al. [77]: a decrease in cytosolic purine nucleotide levels would activate thermogenin. The mechanism for such a decrease is not identified, and the total cellular level of ATP does not change much when thermogenesis is induced [78]; it may, however in some way be redistributed into the mitochondria [77]. However, as pointed out by Rial and Nicholls [3], a major problem with this hypothesis is the question how the ATP levels would be restored in order to inhibit thermogenesis when the adrenergic stimulation ceases: no known processes would then favour an increase in ATP level.

6.2. Suggested lipolytic control mechanisms

These suggestions for activation of thermogenin originate from the premolecular era, basically from very early experiments with isolated brown-fat cells. In these cells, thermogenesis (or – its equivalent [79] – increased oxygen consumption) can be induced by addition of norepinephrine [80–83]. The observation that the mere addition of fatty acids to isolated brown-fat cells can induce a thermogenic process which quantitatively and qualitatively seems identical to that induced by norepinephrine [80] was the origin of the lipolytic theory for thermogenesis regulation: it is the released fatty acids themselves or a fatty acid metabolite which activates thermogenin and thus lead to thermogenesis. That the thermogenic effect of fatty-acid addition to brown-fat cells is mediated by thermogenin is supported by the observation that the extent of thermogenesis observed is proportional to the thermogenin content of the cells [84].

6.2.1. The fatty acid hypothesis

The general concept in this hypothesis is simply that the free fatty acids themselves, liberated from the triglycerides within the tissue, interact with thermogenin to activate it. The hypothesis is summarised, e.g. in ref. [85].

Free fatty acids do not influence *purine nucleotide binding* [86] and they can thus not regulate H^+ permeability directly via interaction with the binding site and its control of permeability.

Similarly, free fatty acids do not (re)induce *Cl^- permeability* [86,87]; this indicates that any

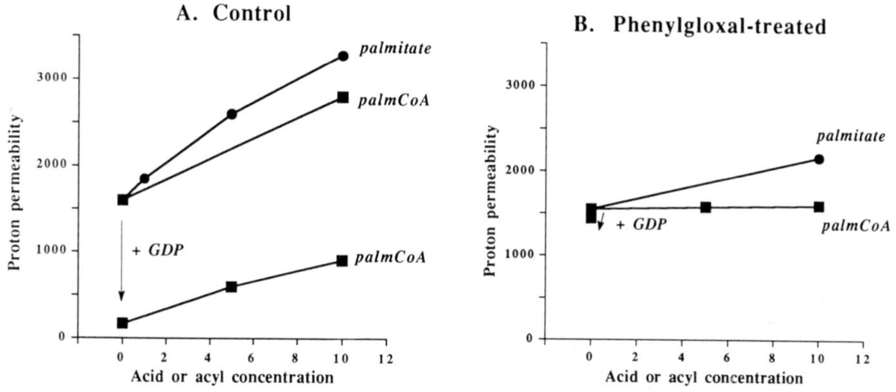

Fig. 11. Control of H^+ permeability in reconstituted thermogenin vesicles. The data are those of Katiyar and Shrago [17]. (A) It is seen how the initially high H^+ permeability is inhibited by the addition of GDP. It would seem that further addition of palmitoyl-CoA can overcome this inhibition, but the interpretation is complicated by the fact that palmitoyl-CoA has a stimulatory effect on H^+ permeability even in the absence of GDP (inhibition expected from mitochondrial results). Also palmitate can increase H^+ permeability. Results of experiments where palmitate is used in combination with GDP were not reported. (B) When the mitochondria have been pretreated with phenylgloxal which – probably by binding to an arginine residue – eliminates [^{3}H]GDP binding [142] and thus the GDP inhibition of H^+ permeability, the effect of palmitoyl-CoA is abolished as well, indicating the specificity of the palmitoyl-CoA interaction. The palmitate effect remains.

uncoupling effects observed are not due to the activation of this 'general' thermogenin channel.

Free fatty acids do, however, increase *H^+ permeability*, at least in isolated mitochondria [84,86,88]. This is true for fatty acids with 14 carbons and longer [86]. It has been shown that the increase in H^+ permeability induced by the addition of fatty acids to brown-fat mitochondria could be fully overcome by the addition of GDP; if instead an arteficial uncoupler (FCCP) (which does not work via thermogenin) was added, the increased permeability could not be overcome by GDP [86]. This experiment thus speaks for a specific interaction between free fatty acids and thermogenin, but as the fatty acids cannot compete with the bound nucleotides the effect must be understood as not being on the purine nucleotide binding site, but being on another, positive regulatory site. For a specific interaction of free fatty acids also speaks the observation that the H^+ permeability enhanced by fatty-acid addition, is diminished by degradation with chymotrypsin [50].

Free fatty acids do not alter the confirmation of solubilized thermogenin (as observed in the infrared spectrum) and would thus not – in contrast to GDP binding – seem to significantly alter thermogenin structure [59]; however, the lack of effect may be explained by the putative fatty acid binding site being hidden by the detergent in this preparation [59].

In reconstituted systems, it was initially stated that free fatty acids did not activate thermogenin [15]. However, more recently, effects of fatty acids (palmitate) have been observed [17,36] and stated to be dependent upon the incorporation of thermogenin into the liposomes [17] (see also Fig. 11).

The molecular mechanism by which free fatty acids act is not clarified. In bioenergetic terms, the effect has been formulated as a modulation of the break-point potential for the non-Ohmic conductance [86] but what this would mean in molecular terms has so far not been elucidated.

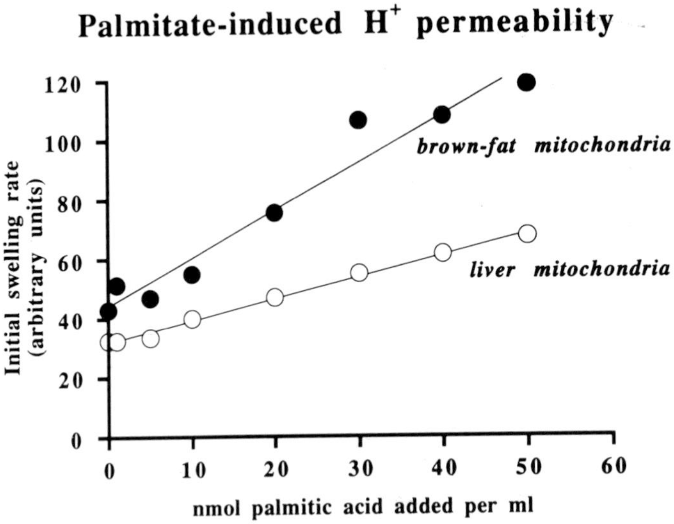

Fig. 12. A comparison of the effect of fatty acid (palmitate) on H^+ permeability in brown-fat and liver mitochondria. The mitochondria were prepared and tested fully in parallel, from cold-acclimated rats, following routine brown-fat mitochondrial procedures, including an albumin wash. The H^+ permeability was followed as swelling in a KAc medium, in the presence of valinomycin and 1 mM GDP (to couple the brown-fat mitochondria). Adapted from ref. [92].

Concerning the *physiological feasibility* of the fatty acid hypothesis, it is clear that fatty acids are released from the cellular triglycerides when thermogenesis is adrenergically stimulated [7]. A main consideration concerning the fatty acid hypothesis is instead the question whether the H^+ permeability induced by fatty acid addition to recoupled mitochondria has qualitative or quantitative characteristics which are sufficiently different from those observed in e.g. liver mitochondria, to justify the suggestion that an increased sensitivity to fatty acids is the result of the presence of thermogenin in brown-fat mitochondria.

It is often stated that brown-fat mitochondria are extremely sensitive to fatty acids and that a high H^+ conductance is induced by fatty acid addition [84,88,89], when compared to mitochondria from other sources. However, the interpretation of several of these observations [84,88] is complicated by the fact that there are also differences in the capacity of dehydrogenases or components of the respiratory chain between brown-fat mitochondria isolated from animals in different physiological states and between mitochondria from different sources; thus, the mere fact that a higher rate of respiration (and thus a higher H^+ conductance) is induced in brown-fat mitochondria by fatty acids may only mean that the brown-fat mitochondria have a higher respiratory capacity than other mitochondria (cf. observations in other systems [90]). Further, if the effect on the membrane potential is considered, there is no dramatic difference between brown-fat mitochondria and liver mitochondria during fatty acid infusion [91], and if a direct comparison is made between the ability of fatty acid to induce H^+ permeability in liver and brown-fat mitochondria, it turns out that only a marginal difference can be observed [92] (Fig. 12).

Indeed, that fatty acid can induce uncoupling of mitochondria in tissues other than brown adipose tissue is well known [93], and certain investigators even found that fatty acids added to

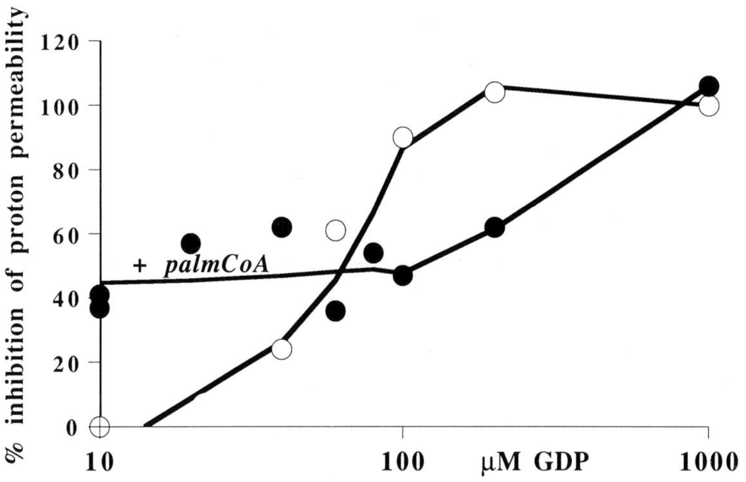

Fig. 13. Interaction between GDP and palmitoyl-CoA as regulators of H^+ permeability in brown-fat mitochondria. H^+ permeability was followed as swelling in a KAc buffer. Addition of increasing amounts of GDP (O) leads to H^+ permeability inhibition. The addition of 4 nmol/ml palmitoyl-CoA has in itself an inhibitory effect on H^+ permeability (●), but the presence of palmitoyl-CoA impedes the effect of GDP, i.e. palmitoyl-CoA functions as a partial agonist. Adapted from ref. [67].

or endogenously present in artificial phospholipid bilayer membranes provide H^+/OH^- permeabilities with some quantitative and qualitative characteristics similar to those discussed for so-called reintroduction of H^+ permeability by fatty acids in GDP-coupled brown-fat mitochondria, including increased permeability with increased pH, and reversibility with albumin [94]; these observations are important to consider in reconstitution experiments with thermogenin incorporated in artificial phospholipid bilayer membranes. As fatty acids may also increase H^+ permeability due to specific [48,95] or general interaction with incorporated proteins, 'the control experiment' for showing a 'specific' interaction of fatty acids with thermogenin in reconstituted systems is not easily defined.

Implicit in the fatty acid hypothetis is that a specific binding site on thermogenin for fatty acids must be postulated (fatty acids do not bind to the purine nucleotide binding site). Such a binding site has as yet not been identified but this may be a technical problem.

Finally, it may be considered intellectually unsatisfactory to accept that thermogenin is endowed with a purine nucleotide binding site with potential regulatory function, but that this site should always be saturated with purines and thus be *without physiological regulatory function*. The realization that thermogenin is evolutionarily not so distant from the ATP/ADP carrier may, however, facilitate the acceptance of such a situation. The – now somewhat degenerate – purine nucleotide binding site on thermogenin could be but a vestige of the active transport site on the ATP/ADP carrier, and could now have lost its functional role; it is, however, not certain that the nucleotide binding sites on thermogenin and on the ATP/ADP carrier are identical.

6.2.2. *The acyl-CoA hypothesis*

The hypothesis that the activator of thermogenin was the 'activated' fatty acids, i.e. the acyl-CoAs, was suggested based on the chemical similarity of CoA with the purine nucleotides [87]. Thus, it could be imagined that the acyl-CoAs formed after stimulation of lipolysis would interact with the nucleotides bound to the purine nucleotide binding site, and, if the acyl-CoA had antagonistic properties, could overcome ATP/GDP inhibition of the H^+ permeability. The hypothesis has mainly been tested with palmitoyl-CoA.

As expected, palmitoyl-CoA interacts with *purine nucleotide binding*. The initial observation, which was mainly of a qualitative character [87], has been substantiated, and the K_i for palmitoyl-CoA as a competitor for [^{3}H]GDP has been given to be 2–3 μM [66].

Palmitoyl-CoA has also effects on *Cl^- permeability*. In itself, it only partly inhibits Cl^- permeability; it thus has not full efficacy on the site [66,67]. However, when added in the presence of GDP it can reintroduce Cl^- permeability [66,67,87]. It therefore functions as a partial agonist, in practice overcoming GDP inhibition to about 60%.

Palmitoyl-CoA has also effects on *H^+ permeability*. It is seen in Fig. 13 that GDP can fully inhibit H^+ translocation in isolated mitochondria. However, in the presence of palmitoyl-CoA, GDP partly loses this ability [67] (thus, earlier statements that there is a difference between Cl^- and H^+ permeability in that palmitoyl-CoA has no effect in overcoming GDP-inhibition of H^+ permeability [96] could not be confirmed). In a reconstituted system, the data available also seem to indicate an activating effect of palmitoyl-CoA, although the expected partial agonist properties were not evident (Fig. 11) [17].

Concerning the *physiological feasibility*, it is clear that there is an increase in the level of acyl-CoA esters in the tissue when it is thermogenically active; thus, a fourfold increase in oleoyl-CoA was observed at the first point in time investigated, 3 h after initiation of a cold stress [97]. The kinetics have not been investigated at a shorter time interval, nor have they been investigated in e.g. the isolated brown-fat cell system, and the intracellular localization of the acyl-CoAs has not been determined.

The problem, similar to the case of the fatty acid hypothesis, again lies in the specificity. For example, during palmitoyl-CoA infusion there is not much difference in the effect on the membrane potential between brown-fat mitochondria and liver mitochondria [91]. The main question concerning observed palmitoyl-CoA effects in brown-fat mitochondria is the possibility that palmitoyl-CoA may induce nonspecific swelling. Undoubtedly, if sufficient palmitoyl-CoA is added, an apperently nonspecific swelling occurs, indicating that also K^+ permeability is induced [75,91], but it is possible that this represents a parallel effect of palmitoyl-CoA specifically on a K^+ permeability system (probably the K^+/H^+ exchanger) [67]. Further studies in the reconstituted system should clarify this question, and the recent development of a photoaffinity-labelled palmitoyl-CoA analogue and indications that it binds to thermogenin [98] may be of further help. A more functional argument against the palmitoyl-CoA hypothesis would be that due to the partial agonist nature of palmitoyl-CoA (Fig. 13), only half the potential thermogenin activity could ever be exploited and utilized for heat production through this mechanism. Again, this may simply be an evolutionary limitation, the consequence of which may have been overcome by an increase in the total number of thermogenin molecules per mitochondrion.

6.3. *Conclusion*

Unfortunately, despite all the advances in the understanding of the thermogenin structure, the mechanism through which it carries H^+ over the membrane, and the mechanism through which the rate of this transport is regulated, are still not understood. The most probable scheme today is still a postlipolytic activation of thermogenin and a transport of OH^- through thermogenin.

7. A member of a family: the mitochondrial membrane carriers

As pointed out in several recent reviews [4,58,99], thermogenin is a member of a family of mitochondrial membrane carriers, and it is possible to suggest an evolutionary pathway leading to the occurrence of thermogenin in the early mammalian ancestor. However, as the membrane carrier family is growing, the developmental outline (which, when based on the first three identified family members could simply be stated to be: phosphate carrier → ATP/ADP carrier → thermogenin) may have to be modified. Besides the three members mentioned, the family now includes also the oxoglutarate/malate carrier [100] and three proteins of which the functions are basically unknown (Fig. 14). One is the protein yielding specific antibodies in humans with Graves disease (hML-7) (hGT) and its rat equivalent (rF5-7) [101]. The two others are two yeast mitochondrial proteins MRS3 and MRS4 (Mitochondrial RNA Splicing) [102], which have been isolated from mutant analysis of non-RNA-splicing cells. All the mitochondrial membrane carrier family members with a known function are exchangers (not channels); the functions of the other members are unknown but it has been speculated that the MRS proteins are essential to transport into the yeast mitochondria substances (ions) which are necessary for the splicing process. We refer to all of these proteins as 'mitochondrial' but it is more likely that they originated at a time when the mitochondrial (and chloroplast) progenitors were still free-living unicellular organisms, with the need to take up nutrients from the surroundings.

The similarities between most of these proteins have been comprehensively compiled by Klaus et al. [58]; this compilation is extended in Fig. 14. Although it was thought until recently that the ATP/ADP carrier was the closest relative to thermogenin, a comparison between thermogenin and all other members of the family would indicate that the oxoglutarate/malate carrier is the closest now known relative to thermogenin (Fig. 14) and it is today even doubted that the nucleotide binding sites on thermogenin and the ATP/ADP carrier are homologues [61].

No low-stringency analysis of the mammalian genome for common sequences within the membrane carrier family has yet been published, but it is very likely that this family includes several other members, especially within the functional group of mitochondrial substrate carriers (see also the chapter on these in the present volume).

Discussions on protein families tend to concentrate upon similarities rather than differences. We would like to point out that for an understanding of the thermogenin function, not only the similarities between these family members but also their individualities may be of great importance. The presence of sequences in thermogenin conserved within different species but without similarity to sequences found in other family members may yield significant clues for the identification of important functional groups within the thermogenin molecule. We would here point to the final 11 amino acids in thermogenin, which are very well conserved between different species (Fig. 7), but which do not have a counterpart at all in any other identified member of the mitochondrial carrier family.

8. The thermogenin gene

The thermogenin gene *ucp* has been characterized in the mouse [54], the rat [103] and the human [57] genome. There is only one gene for thermogenin in the genome (as is the case for

A. Similarity dendogram of the mitochondrial carrier family

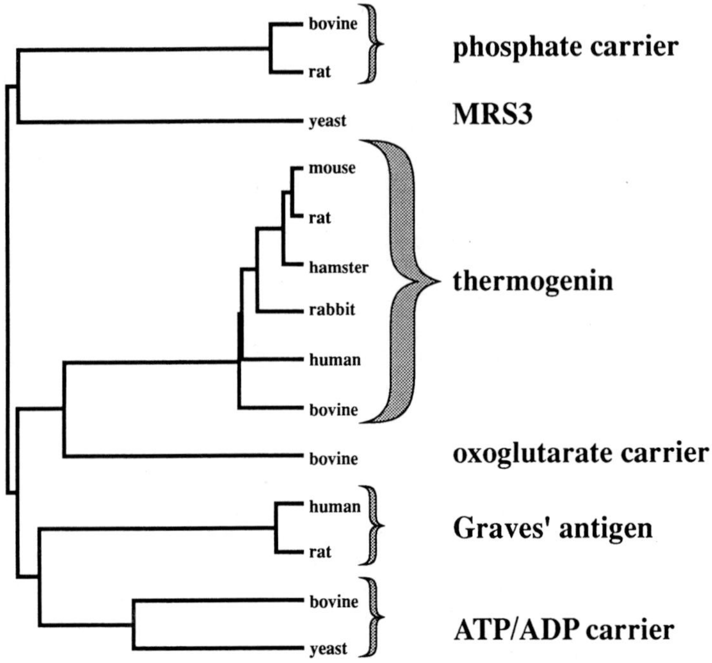

B. Similarity plot: thermogenin *versus* thermogenin

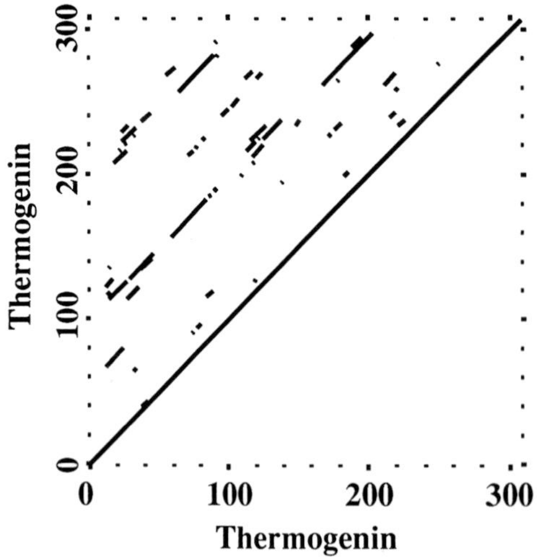

→

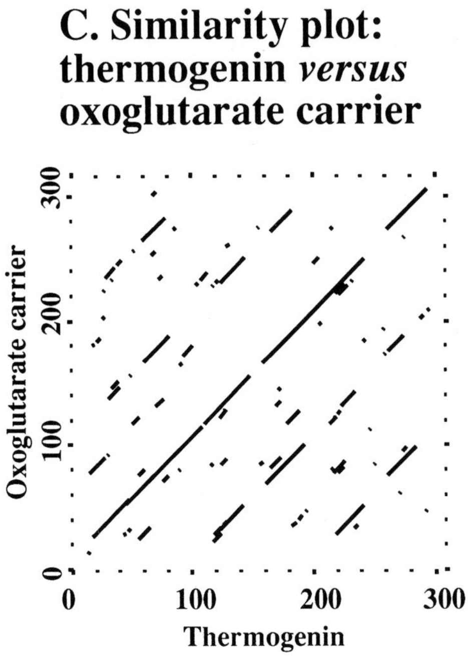

Fig. 14. Similarity analysis of the mitochondrial carrier family. (A) Similarity dendogram made by the 'pileup' alignment procedure of the University of Wisconsin, based on all relevant sequences present in the 'swissprotein' data base in Spring 1992. All recorded species variations are included, except for the ATP/ADP carrier where 11 species and isozyme variations are documented but only 2 are included here. Note that the carrier with the closest similarity to thermogenin is the oxoglutarate/malate carrier. (B) A 'dotplot' comparison of mouse thermogenin with itself showing the typical tripartite structure of the family. (C) A 'dotplot' comparison of thermogenin with its closest known relative, the oxoglutarate/malate carrier. Note the high degree of homology along the entire protein, and the evidence for the tripartite structure. The 'stringency' and 'window' parameters were identically defined for (B) and (C).

the oxoglutarate/malate transporter, but in contrast to the presence of several genes for the ATP/ADP carrier).

The mouse *ucp* gene is on chromosome 8 [104]. The human *ucp* gene has been ascribed to chromosome 4 [57]. However, Ceci et al. [105] have pointed out that a part of mouse chromosome 8 shows synteni with human chromosome 8 and another part with human chromosome 16q. The gene *ucp* is found just in the border zone between these two pieces of mouse chromosome 8. Although it is entirely possible that *ucp* and a short stretch around it could show synteni with human chromosome 4, it may be considered more likely that it should have followed some of the other large pieces of chromosome 8. It may be concluded that the localization of *ucp* on the human chromosome needs reconfirmation.

In Fig. 15, some details concerning the thermogenin gene organization are summarized. It may especially be noted that the gene is tripartite, just as the protein, as it consists of three times two exons [54,103,106].

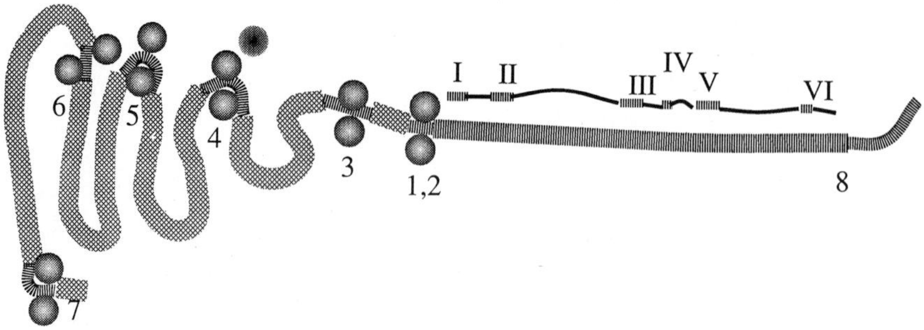

Fig. 15. The thermogenin gene. The figure, which is not fully to scale, is based on data from different species presented by Kozak et al. [54], Bouillaud et al. [103], Cassard et al. [57] and Boyer and Kozak [106]. The gene is organized in six exons (I–VI), and there is an interesting correlation between the suggested thermogenin structure and the gene arrangement, i.e. the exons I,II and III,IV and V,VI each correspond to one part of the tripartite structure. Within the gene, the homology is largest in the arrangement around exons III and V [103]. Boyer and Kozak could observe eight DNase-hypersensitive sites associated with the gene (1–8); of these, seven were upstream. Sites 1 and 2 probably correspond to general initiation sites. Based on results obtained with transgenic mice, site 4 is probably involved in tissue-specific expression. The DNase-hypersensitivity as such was not altered in different physiological states or tissues, but it is possible that positive and negative transcription factors (here illustrated by different globular symbols) may compete for such sites and thus determine gene expression.

9. Regulation of the expression of thermogenin

As compared to the other members of the mitochondrial carrier family, thermogenin possesses some interesting additional features: its extreme tissue-specific expression, and the ability to acutely regulate thermogenin expression in accordance with the physiological needs of the animal.

9.1. Tissue-specific thermogenin gene expression

Thermogenin itself [107,108] and thermogenin mRNA are exclusively located in brown adipose tissue. Even in cold-adapted mice, thermogenin mRNA is not found in any tissue except brown adipose tissue [104]. One report did state that a small but visible amount of thermogenin mRNA could be observed in livers of newborn or cold-exposed rats [109] but this observation has later been refuted [157].

In rats and mice, but not in most other species investigated, thermogenin mRNA is found in two lengths. This is due to the existence of two polyadenylation sites, but even though these are apparently also present, e.g. in rabbit, only the latter is used [55]. It is unlikely that the existence of the two different thermogenin sizes has any functional significance.

The determination of the cell clones to become brown-fat cells probably occurs rather early in development; even white adipose tissue in vivo [110] or in vitro [111,112], cannot express the thermogenin gene to any appreciable degree. Although determination thus occurs early, thermogenin gene expression comes late during gestation in altricial [113] species [114,115] and not until after a lag-phase after birth in 'immature' newborns such as hamsters [116,117].

9.2. Physiological control of thermogenin gene expression

In physiological situations in which more heat is needed, thermogenin gene expression is increased in the tissue, and an increase in thermogenin mRNA levels is rapidly seen [104,118–120]. An understanding of the cellular background for this phenomenon and of the whole *recruitment* process (the concerted increase in thermogenin amount, mitochondrial content and number of brown-fat cells occurring when an animal is adapting to a situation with increased demands on heat production) has been attempted in in vivo studies [10,121], but such studies have clear limitations for investigations of cellular processes.

For a more detailed study of the processes leading to increased thermogenin gene expression, investigations of brown-fat cells proliferating and differentiating in culture have therefore been pivotal. In such cultures, the addition of norepinephrine can lead to thermogenin gene expression [111,122,123] and to the synthesis of thermogenin itself [124,125]. Thus, norepinephrine has a dual function in brown-fat cells, in both promoting differentiation (i.e. thermogenin gene expression) and thermogenesis itself (thermogenin activation); surprisingly, norepinephrine may even stimulate cell division [126].

9.3. The synthesis of thermogenin

When synthesized from its mRNA, thermogenin has to be transported to the mitochondria and inserted into the inner mitochondrial membrane. Thermogenin has no cleavable targeting sequence (except for the N-terminal methionine) [53], a property it shares with some (the ATP/ADP carrier) but not all (the phosphate carrier) members of the mitochondrial membrane carrier family. Similar to what has been observed for the ATP/ADP carrier, somewhere in the first loop (the first 100 residues), information is found for targeting and insertion, and somewhere in the second two loops some weaker targeting information is found which does not lead to insertion [127]. The targeting must use general mechanisms, since in vitro synthesized thermogenin is taken up by heterologous (CHO-cell) mitochondria [128], and since thermogenin expressed even in very unrelated cell types is directed to the mitochondrial inner membrane [18,19,127,129].

If thermogenin cDNA is placed behind a matrix-targeting signal obtained from a matrix-bound enzyme, the synthesized thermogenin is imported into isolated (heart) mitochondria and the signal peptide cleaved off [127,130], but thermogenin is not inserted into the membrane. It is therefore most likely that the pathway for thermogenin insertion into the mitochondrial inner membrane does not go via the matrix compartment.

Once inserted into mitochondria, it would seem that thermogenin turns over as the entire mitochondrial membrane turns over, with a halflife in the order of about a week [131,132]. There is, however, some kinetic evidence for a pool of thermogenin with a more rapid turnover; this pool may represent thermogenin which has as yet not been fully inserted into the mitochondria [132].

How the synthesis of thermogenin is synchronized with total mitochondriogenesis during the recruitment process – the process which endows an animal with a higher capacity for heat production when the physiological need arises – is still unknown.

10. Perspectives

As understood from the above, there have been very significant developments in thermogenesis research since the isolation and purification of thermogenin [11] opened the way for molecular approaches. Presently, more effort is being devoted to the elucidation of the genetic control mechanisms than to an understanding of thermogenin activity as such. However, molecular techniques have now been developed [19] to allow for a molecular approach to thermogenesis research, since the thermogenic effects of modifications in thermogenin primary structure can now be examined.

Acknowledgements

Our own research on brown adipose tissue is supported by the Swedish Natural Science Research Council. We would like to thank M. Klingenberg and K.D. Garlid for access to manuscripts in press and for discussions.

References

1 Nedergaard, J. and Cannon, B. (1984) in: L. Ernster (Ed.) New Comprehensive Biochemistry (Bioenergetics), Vol. 9, Elsevier, Amsterdam, pp. 291–314.
2 Lindberg, O., Cannon, B. and Nedergaard, J. (1981) in: C.P. Lee, G. Schatz and G. Dallner (Eds.) Mitochondria and Microsomes, Addison-Wesley, Reading, MA, pp. 93–119.
3 Rial, E. and Nicholls, D.G. (1987) Revis. Biol. Cellular 11, 75–104.
4 Klingenberg, M. (1990) Trends Biochem. Sci. 15, 108–112.
5 Ricquier, D., Casteilla, L. and Bouillaud, F. (1991) FASEB J. 5, 2237–2242.
6 Nicholls, D.G. (1979) Biochim. Biophys. Acta 549, 1–29.
7 Nedergaard, J. and Lindberg, O. (1982) Int. Rev. Cytol. 74, 187–286.
8 Nicholls, D.G. and Locke, R.M. (1984) Physiol. Rev. 64, 1–64.
9 Cannon, B. and Nedergaard, J. (1985) Essays Biochem. 20, 110–164.
10 Trayhurn, P. and Nicholls, D.G. (1986) Brown Adipose Tissue, Edward Arnold, Paris.
11 Lin, C.S. and Klingenberg, M. (1980) FEBS Lett. 113, 299–303.
12 Cannon, B. and Lindberg, O. (1979) Methods Enzymol. F 55, 65–78.
13 Klingenberg, M. and Lin, C.S. (1986) Methods Enzymol. 126, 490–498.
14 Bouillaud, F., Ricquier, D., Gulik-Krzywicki, T. and Gary-Bobo, C.M. (1983) FEBS Lett. 164, 272–276.
15 Klingenberg, M. and Winkler, E. (1985) EMBO J. 4, 3087–3092.
16 Jezek, P., Orosz, D.E. and Garlid, K.D. (1990) J. Biol. Chem. 265, 19296–19302.
17 Katiyar, S.S. and Shrago, E. (1991) Biochem. Biophys. Res. Commun. 175, 1104–1111.
18 Casteilla, L., Blondel, O., Klaus, S., Raimbault, S., Diolez, P., Moreau, F., Bouillaud, F. and Ricquier, D. (1990) Proc. Natl. Acad. Sci. USA 87, 5124–5128.
19 Murdza-Inglis, D.L., Patel, H.V., Freeman, K.B., Jezek, P., Orosz, D.E. and Garlid, K.D. (1991) J. Biol. Chem. 260, 11871–11875.
20 Lin, C.S., Hackenberg, H. and Klingenberg, E.M. (1980) FEBS Lett. 113, 304–306.
21 Lin, C.S. and Klingenberg, M. (1982) Biochemistry 21, 2950–2956.
22 French, R.R., Gore, M.G. and York, D.A. (1988) Biochem. J. 251, 385–389.
23 Nedergaard, J. and Cannon, B. (1985) Am. J. Physiol. 248, C365–C371.
24 Klingenberg, M. and Appel, M. (1989) Eur. J. Biochem. 180, 123–131.
25 Desautels, M., Zaror-Behrens, G. and Himms-Hagen, J. (1978) Can. J. Biochem. 56, 378–383.

26 Trayhurn, P., Ashwell, M., Jennings, G., Richard, D. and Stirling, D. (1987) Am. J. Physiol. 252, E237–E243.
27 Nedergaard, J. and Cannon, B. (1992), submitted.
28 Swick, A. and Swick, R. (1986) Am. J. Physiol. 251, E192–E195.
29 Desautels, M. and Himms-Hagen, J. (1980) Can. J. Biochem. 58, 1057–1068.
30 Nedergaard, J. and Cannon, B. (1987) Eur. J. Biochem. 164, 681–686.
31 Milner, R.E. and Trayhurn, P. (1988) Biochem. Cell Biol. 66, 1226–1230.
32 Swick, A. and Swick, R. (1988) Am. J. Physiol. 255, E865–E870.
33 Henningfield, M.F. and Swick, R.W. (1991) Comp. Biochem. Physiol B 99, 821–825.
34 Henningfield, M.F. and Swick, R.W. (1989) Biochem. Cell Biol. 67, 108–112.
35 Jezek, P. and Garlid, K.D. (1990) J. Biol. Chem. 265, 19303–19311.
36 Klingenberg, M. (1991) Biophys. J. 59, 395a–395a.
37 Rial, E. and Nicholls, D.G. (1989) in: A. Azzi, K.A. Nalecz, M.J. Nalecz and L. Wojtczak (Eds.) Anion Carriers of Mitochondrial Membranes, Springer, Berlin, pp. 261–268.
38 Nicholls, D.G. and Lindberg, O. (1973) Eur. J. Biochem. 37, 523–530.
39 Jezek, P., Beavis, A.D., DiResta, D.J., Cousina, R.N. and Garlid, K.D. (1989) Am. J. Physiol. 257, C1142–C1148.
40 Klitsch, T. and Siemen, D. (1991) J. Membr. Biol. 122, 69–75.
41 Lindberg, O., de Pierre, J., Rylander, E. and Sydbom, R. (Eds.) (1966) Third FEBS Meeting, M 48, pp. 139.
42 Smith, R.E., Roberts, J.C. and Hittelman, K.J. (1966) Science 154, 653–654.
43 Lindberg, O., de Pierre, J., Rylander, E. and Afzelius, B.A. (1967) J. Cell Biol. 34, 293–310.
44 Nicholls, D.G. and Rial, E. (1989) Methods Enzymol. 174, 85–94.
45 Joel, C.D., Neaves, W.B. and Rabb, J.M. (1967) Biochem. Biophys. Res. Commun. 29, 490–495.
46 Bulychev, A., Kramar, R., Drahota, Z. and Lindberg, O. (1972) Exp. Cell Res. 72, 169–187.
47 Cannon, B., Nicholls, D.G. and Lindberg, O. (1973) in: G.F. Azzone et al. (Eds.) Mechanisms in Bioenergetics, Academic Press, New York, pp. 357–364.
48 Skulachev, V.P. (1991) FEBS Lett. 294, 158–162.
49 Nicholls, D.G. (1976) FEBS Lett. 61, 103–110.
50 Fernández, M., Nicholls, D.G. and Rial, E. (1987) Eur. J. Biochem. 164, 675–680.
51 Aquila, H., Link, T.A. and Klingenberg, M. (1985) EMBO J. 4, 2369–2376.
52 Bouillaud, F., Weissenbach, J. and Ricquier, D. (1986) J. Biol. Chem. 261, 1487–1491.
53 Ridley, R.G., Patel, H.V., Gerber, G.E., Morton, R.C. and Freeman, K.B. (1986) Nucl. Acids Res. 14, 4025–4035.
54 Kozak, L.P., Britton, J.H., Kozak, U.C. and Wells, J.M. (1988) J. Biol. Chem. 263, 12274–12277.
55 Balogh, A.C., Ridley, R.G., Patel, H.V. and Freeman, K.B. (1989) Biochem. Biophys. Res. Commun. 161, 156–161.
56 Casteilla, L., Bouillaud, F., Forest, C. and Ricquier, D. (1989) Nucl. Acids Res. 17, 2131.
57 Cassard, A.M., Bouillaud, F., Mattei, M.G., Hentz, E., Raimbault, S., Thomas, M. and Ricquier, D. (1990) J. Cell Biochem. 43, 255–264.
58 Klaus, S., Casteilla, L., Bouillaud, F. and Ricquier, D. (1991) Int. J. Biochem. 23, 791–801.
59 Rial, E., Muga, A., Valpuesta, J.M., Arrondo, J.-L.R. and Goñi, F.M. (1990) Eur. J. Biochem. 188, 83–89.
60 Heaton, G.M., Wagenvoord, R.J., Kemp, J.A. and Nicholls, D.G. (1978) Eur. J. Biochem. 82, 515–521.
61 Winkler, E. and Klingenberg, M. (1992) Eur. J. Biochem. 203, 295–304.
62 Jezek, P., Houstek, J., Kotyk, A. and Drahota, Z. (1988) Eur. Biophys. J. 16, 101–108.
63 Klingenberg, M. (1988) Biochemistry 27, 781–791.
64 Klingenberg, M. (1986) Methods Enzymol. 125, 618–630.
65 Eckerskorn, C. and Klingenberg, M. (1987) FEBS Lett. 226, 166–170.
66 Strieleman, P. and Shrago, E. (1985) Am. J. Physiol. 248, E699–E705.
67 Nedergaard, J., Bailey, C. and Cannon, B. (1992), submitted.
68 French, R.R. and York, D.A. (1986) Biochem. Soc. Trans. 14, 757–758.

69 French, R.R., Peachey, T.J. and York, D.A. (1986) EBEC Rep. 4, 383.
70 Rial, E. and Nicholls, D.G. (1986) FEBS Lett. 198, 29–32.
71 Chinet, A., Friedli, C., Seydoux, J. and Girardier, L. (1978) in: L. Girardier and J. Seydoux (Eds.) Effectors of Thermogenesis, Experientia Suppl. 32, pp. 25–32.
72 Nicholls, D.G. (1976) Eur. J. Biochem. 62, 223–228.
73 Klingenberg, M. (1984) Biochem. Soc. Trans. 12, 390–393.
74 Nicholls, D.G. (1974) Eur. J. Biochem. 49, 573–583.
75 Jezek, P., Houstek, J. and Drahota, Z. (1988) J. Bioenerg. Biomembranes 20, 603–622.
76 Giovannini, P., Seydoux, J. and Girardier, L. (1988) Pfluegers Arch. 411, 273–277.
77 LaNoue, K.F., Koch, C.D. and Mechitz, R.B. (1982) J. Biol. Chem. 257, 13740–13748.
78 Pettersson, B. (1977) Eur. J. Biochem. 72, 235–240.
79 Nedergaard, J., Cannon, B. and Lindberg, O. (1977) Nature (London) 267, 518–520.
80 Prusiner, S.B., Cannon, B. and Lindberg, O. (1968) Eur. J. Biochem. 6, 15–22.
81 Reed, N. and Fain, J.N. (1968) J. Biol. Chem. 243, 2843–2848.
82 Bukowiecki, L., Follea, N., Paradis, A. and Collet, A. (1980) Am. J. Physiol. 238, E552–E563.
83 Mohell, N. and Dicker, A. (1989) Biochem. J. 261, 401–405.
84 Cunningham, S.A., Wiesinger, H. and Nicholls, D.G. (1986) Eur. J. Biochem. 157, 415–420.
85 Nicholls, D., Cunningham, S. and Wiesinger, H. (1986) Biochem. Soc. Trans. 14, 223–225.
86 Rial, E., Poustie, A. and Nicholls, D.G. (1983) Eur. J. Biochem. 173, 197–203.
87 Cannon, B., Sundin, U. and Romert, L. (1977) FEBS Lett. 74, 43–46.
88 Heaton, G.M. and Nicholls, D.G. (1976) Eur. J. Biochem. 67, 511–517.
89 Rial, E. and Nicholls, D.G. (1984) Biochem. J. 222, 685–693.
90 Barré, H., Nedergaard, J. and Cannon, B. (1986) Comp. Biochem. Physiol. B 85, 343–348.
91 Locke, R.M., Rial, E., Scott, I.D. and Nicholls, D.G. (1982) Eur. J. Biochem. 129, 373–380.
92 Nedergaard, J. and Cannon, B. (1992), submitted.
93 Lardy, H. and Shrago, E. (1990) Annu. Rev. Biochem. 59, 689–710.
94 Gutknecht, J. (1987) Proc. Natl. Acad. Sci. USA 84, 6443–6446.
95 Andreyev, A.Y., Bondareva, T.O., Dedukhova, V.I., Mokhova, E.N., Skulachev, V.P. and Volkov, N.I. (1988) FEBS Lett. 226, 265–269.
96 Kopecky, J., Guerrieri, F., Jezek, P., Drahota, Z. and Houstek, J. (1984) FEBS Lett. 170, 186–190.
97 Donatello, S., Spennetta, T., Strieleman, P., Woldegiorgis, G. and Shrago, E. (1988) Am. J. Physiol. 254, E181–E186.
98 Woldegiorgis, G., Duff, T., Contreras, L., Ruoho, A. and Shrago, E. (1991) FASEB J. 5, A1139.
99 Aquila, H., Link, T.A. and Klingenberg, M. (1987) FEBS Lett. 212, 1–9.
100 Runswick, M.J., Walker, J.E., Bisaccia, F., Iacobazzi, V. and Palmieri, F. (1990) Biochemistry 29, 11033–11040.
101 Zarilli, R., Oates, E.L., McBride, O.W., Lerman, M.I., Chan, J.Y., Santisteban, P., Ursini, M.V., Notkins, A.L. and Kohn, L.D. (1989) Mol. Endocrin. 3, 1498–1508.
102 Wiesenberger, G., Link, T.A., von Ahsen, U., Waldherr, M. and Schweyen, R.J. (1991) J. Mol. Biol. 217, 23–37.
103 Bouillaud, F., Raimbault, S. and Ricquier, D. (1988) Biochem. Biophys. Res. Commun. 157, 783–792.
104 Jacobsson, A., Stadler, U., Glotzer, M.A. and Kozak, L.P. (1985) J. Biol. Chem. 260, 16250–16254.
105 Ceci, J.D., Justice, M.J., Lock, L.F., Jenkins, N.A. and Copeland, N.G. (1990) Genomics 6, 72–79.
106 Boyer, B.B. and Kozak, L.P. (1991) Mol. Cell Biol. 11, 4147–4156.
107 Cannon, B., Hedin, A. and Nedergaard, J. (1982) FEBS Lett. 150, 129–132.
108 Ricquier, D., Barlet, J.P., Garel, J.M., Combes-Georges, M. and Dubois, M.P. (1983) Biochem. J. 210, 859–866.
109 Shinohara, Y., Shima, A., Kamida, M. and Terada, H. (1991) FEBS Lett. 293, 173–174.
110 Loncar, D., Afzelius, B.A. and Cannon, B. (1988) J. Ultrastruct. Mol. Struct. Res. 101, 199–209.
111 Klaus, S., Cassard-Doulcier, A.-M. and Ricquier, D. (1991) J. Cell Biol. 115, 1783–1790.
112 Herron, D., Rehnmark, S., Néchad, M., Cannon, B. and Nedergaard, J. (1992), submitted.

113 Nedergaard, J., Connolly, E. and Cannon, B. (1986) in: P. Trayhurn and D.G. Nicholls (Eds.) Brown Adipose Tissue, Edward Arnold, Paris, pp. 152–213.
114 Obregon, M.J., Jacobsson, A., Kirchgessner, T., Schotz, M.C., Cannon, B. and Nedergaard, J. (1989) Biochem. J. 259, 341–346.
115 Houstek, J., Kopecky, J., Rychter, Z. and Soukup, T. (1988) Biochim. Biophys. Acta 935, 19–25.
116 Sundin, U., Herron, D. and Cannon, B. (1981) Biol. Neonate 38, 141–149.
117 Houstek, J., Janiková, D., Bednár, J., Kopecky, J., Sebastián, J. and Soukup, T. (1990) Biochim. Biophys. Acta 1015, 441–449.
118 Ricquier, D., Mory, G., Bouillaud, F., Thibault, J. and Weissenbach, J. (1984) FEBS Lett. 178, 240–244.
119 Ricquier, D., Bouillaud, F., Toumelin, P., Mory, G., Bazin, R., Arch, J. and Penicaud, L. (1986) J. Biol. Chem. 261, 13905–13910.
120 Jacobsson, A., Mühleisen, M., Cannon, B. and Nedergaard, J. (1992), submitted.
121 Cannon, B., Rehnmark, S., Néchad, M., Herron, D., Jacobsson, A., Kopecky, J., Obregon, M.J. and Nedergaard, J. (1989) in: A. Malan and B. Canguilhem (Eds.) Living in the Cold II, John Libbey Eurotext Ltd., London, pp. 359–366.
122 Rehnmark, S., Kopecky, J., Jacobsson, A., Néchad, M., Herron, D., Nelson, B.D., Obregon, M.J., Nedergaard, J. and Cannon, B. (1989) Exp. Cell Res. 182, 75–83.
123 Rehnmark, S., Néchad, M., Herron, D., Cannon, B. and Nedergaard, J. (1990) J. Biol. Chem. 265, 16464–16471.
124 Herron, D., Rehnmark, S., Néchad, M., Loncar, D., Cannon, B. and Nedergaard, J. (1990) FEBS Lett. 268, 296–300.
125 Kopecky, J., Baudysová, M., Zanotti, F., Janiková, D., Pavelka, S. and Houstek, J. (1990) J. Biol. Chem. 265, 22204–22209.
126 Bronnikov, G., Houstek, J. and Nedergaard, J. (1992) J. Biol. Chem. 267, 2006–2013.
127 Liu, X., Bell, A.W., Freeman, K.B. and Shore, G.C. (1988) J. Cell Biol. 107, 503–509.
128 Freeman, K.B., Chien, S.M., Litchfield, D. and Patel, H.V. (1983) FEBS Lett. 158, 325–330.
129 Klaus, S., Casteilla, L., Bouillaud, F., Raimbault, S. and Ricquier, D. (1990) Biochem. Biophys. Res Commun. 167, 784–789.
130 Liu, X., Freeman, K.B. and Shore, G.C. (1990) J. Biol. Chem. 265, 9–12.
131 Desautels, M., Dulos, R.A. and Mozaffari, B. (1986) Biochem. Cell Biol. 64, 1125–1134.
132 Puigserver, P., Herron, D., Gianotti, M., Palou, A., Cannon, B. and Nedergaard, J. (1992) Biochem. J. 284, 393–398.
133 Lean, M. and James, W. (1986) in: P. Trayhurn and D.G. Nicholls (Eds.) Brown Adipose Tissue, Edward Arnold, Paris, pp. 339–365.
134 Rothwell, N.J. and Stock, M.J. (1986) in: P. Trayhurn and D.G. Nicholls (Eds.) Brown Adipose Tissue, Edward Arnold, Paris, pp. 269–338.
135 Arch, J., Ainsworth, A.T., Cawthorne, M.A., Piercy, V., Sennitt, M.V., Thody, V.E., Wilson, C. and Wilson, S. (1984) Nature 309, 163–165.
136 Cannon, B., Romert, L., Sundin, U. and Barnard, T. (1977) Comp. Biochem. Physiol. B 56, 87–99.
137 Cannon, B. and Vogel, G. (1977) FEBS Lett. 76, 284–289.
138 Svoboda, P., Houstek, J., Kopecky, J. and Drahota, Z. (1981) Biochim. Biophys. Acta 634, 321–330.
139 Houstek, J., Tvrdik, P., Pavelka, S. and Baudysová, M. (1991) FEBS Lett. 294, 191–194.
140 Strieleman, P., Schalinske, K. and Shrago, E. (1985) Biochem. Biophys. Res. Commun. 127, 509–516.
141 Strieleman, P.J., Schalinske, K.L. and Shrago, E. (1985) J. Biol. Chem. 260, 13402–13405.
142 Katiyar, S.S. and Shrago, E. (1989) Proc. Natl. Acad. Sci. USA 86, 2559–2562.
143 Klingenberg, M. and Winkler, E. (1986) Methods Enzymol. 127, 772–779.
144 Kolarov, J., Houstek, J., Kopecky, J. and Kuzela, S. (1982) FEBS Lett. 144, 6–10.
145 Rial, E. and Nicholls, D.G. (1985) Biochem. Soc. Trans. 13, 738–739.
146 Jezek, P. (1987) FEBS Lett. 211, 89–93.
147 Jezek, P. and Drahota, Z. (1989) Eur. J. Biochem. 183, 89–95.
148 Rial, E., Aréchaga, I., Sainz-de-la-Maza, E. and Nicholls, D.G. (1989) Eur. J. Biochem. 182, 187–193.
149 Kopecky, J., Jezek, P., Drahota, Z. and Houstek, J. (1987) Eur. J. Biochem. 164, 687–694.

150 Rial, E. and Nicholls, D.G. (1983) FEBS Lett. 161, 284–288.
151 Nicholls, D.G. (1974) Eur. J. Biochem. 49, 585–593.
152 Ridley, R.G., Patel, H.V., Parfett, C., Olynyk, K.A., Reichling, S. and Freeman, K.B. (1986) Biosci. Rep. 6, 87–94.
153 Sundin, U. and Cannon, B. (1980) Comp. Biochem. Physiol. B 65, 463–471.
154 Gribskov, C.L., Henningfield, M.F., Swick, A.G. and Swick, R.W. (1986) Biochem. J. 233, 743–747.
155 Woldegiorgis, G., Duff, T., Contreras, L., Shrago, E. and Ruoho, A.E. (1989) Biochem. Biophys. Res. Commun. 161, 502–507.
156 Winkler, E. and Klingenberg, M. (1992) Eur. J. Biochem. 207, 135–145 (a copy of this article with significant printing errors corrected can be obtained from Dr. Klingenberg).
157 Ricquier, D., Raimbault, S., Champginy, O., Miroux, B. and Bouillaud, F. (1992) FEBS Lett. 303, 103–106.
158 Jakobs, P., Braun, A., Jezek, P. and Trommer, W.E. (1991) FEBS Lett. 284, 195–198.

L. Ernster (Ed.) *Molecular Mechanisms in Bioenergetics*
© 1992 Elsevier Science Publishers B.V. All rights reserved

CHAPTER 18

Hormonal regulation of cellular energy metabolism*

JAN B. HOEK

Department of Pathology and Cell Biology, Thomas Jefferson University, Philadelphia, PA 19107, USA

Contents

1. Introduction 422
2. Control of oxidative phosphorylation in intact cells 423
 2.1. Distribution of control of oxidative phosphorylation 424
 2.2. Non-phosphorylative energy dissipation 425
 2.3. Control of respiration by ATP consumption 426
 2.4. Control of supply of reducing equivalents 428
 2.5. The need for a 'coupling messenger' 428
3. Effects of hormones on cellular energy metabolism 429
 3.1. Hormones that operate through Ca^{2+} or cAMP 430
  3.1.1. Hormonal signals acting on mitochondria 430
   3.1.1.1. Relationship between cytosolic and mitochondrial Ca^{2+} concentrations 430
   3.1.1.2. cAMP as a mitochondrial messenger 434
   3.1.1.3. Control of mitochondrial adenine nucleotides levels 435
  3.1.2. Mechanisms of activation of respiration by Ca^{2+}-mobilizing hormones 435
   3.1.2.1. Changes in respiratory activity in response to glucagon and Ca^{2+}-mobilizing hormones 435
   3.1.2.2. Ca^{2+} activation of mitochondrial dehydrogenases and substrate oxidation 436
   3.1.2.3. Metabolic consequences of the activation of Ca^{2+}-sensitive dehydrogenases 437
   3.1.2.4. Ca^{2+}-dependent pyrophosphate accumulation and the role of matrix volume changes in the control of mitochondrial energy metabolism 438
   3.1.2.5. Ca^{2+}-control of the ATP synthase 440
   3.1.2.6. Effects of Ca^{2+} on the adenine nucleotide translocator 441
   3.1.2.7. Mg^{2+} as a regulator of mitochondrial function 441
  3.1.3. The place of mitochondrial [Ca^{2+}] changes in the actions of glucagon and Ca^{2+}-mobilizing hormones on cellular energy metabolism 442
 3.2. Thyroid hormones 444
  3.2.1. Mechanisms by which thyroid hormones affect mitochondrial oxidative phosphorylation 444
   3.2.1.1. Synthesis of electron transport enzymes in mitochondria in response to thyroid hormones 444

* This work was supported in part by US Public Health Service grants AA07186, AA07215, and AA08714.

3.2.1.2. Thyroid hormone-induced changes in mitochondrial lipid composition and their role in bioenergetic effects of the hormone 445
3.2.1.3. Short-term actions of thyroid hormones on electron transport activity and oxidative phosphorylation 446
3.2.1.4. Ca^{2+} elevation as a trigger in short-term thyroid hormone actions 446
3.2.2. Thyroid hormone-induced changes in oxidative phosphorylation in isolated mitochondria and intact cells 447
3.2.2.1. Thyroid hormone effects on state 4 respiration: proton leaks or slipping pumps 448
3.2.2.2. Thyroid hormone effects on state 3 respiration: control by electron transport chain and phosphorylation reactions 449
3.2.2.3. Effects of the thyroid hormone state on mitochondrial energy metabolism in intact cells 451
3.3. Other hormones 453
4. Conclusions 453
4.1. How do hormones affect the control of mitochondrial energy metabolism? 453
4.2. Integration of demand and supply: is there a need for a coupling messenger? 454
References 457

1. Introduction

Hormones modulate metabolic flows to integrate the functioning of different cells and tissues and to optimize the use of available resources. One of their important tasks is regulating the distribution of energy utilization to accomodate variations in demand for energy, and regulating the supply of fuel to meet these demands. Organisms can handle large fluctuations in the flux through substrate supply or energy utilization reactions, while maintaining homeostatic control over important metabolic intermediates. The response to an increased need for usable energy is often accomodated without substantially changing the steady-state ATP/ADP ratio, or the mitochondrial proton motive force, Δp; this has been interpreted to indicate the existence of signals that integrate the utilization of energy with the supply of substrates [1]. The mitochondrial and cytosolic NAD redox state are also maintained within relatively narrow limits, despite substantial changes in substrate supply or flux through the electron transport chain. Although the advantages to the organism of maintaining homeostatic control are evident, the mechanisms by which this occurs, or by which it is adjusted to changing metabolic needs in the cell, are not yet well understood.

Hormones are external regulators of cellular metabolism that make use of the internal control mechanisms of oxidative phosphorylation to adapt energy supplying reactions to the needs of the organism. Hence, it is essential to understand the control features of the system of oxidative phosphorylation in order to analyze the conditions under which hormonally induced alterations to the system might affect its output, either through short-term effects, by covalent modification of proteins and allosteric mechanisms, or, over longer terms, by changes in protein and lipid composition of the oxidative phosphorylation machinery.

Our understanding of different modes of action of hormones has expanded dramatically over the past decade. Although the concepts of receptors and second messenger molecules were developed in the late fifties and early sixties, following the discovery of cAMP [2], it took more than 20 years to gain an understanding about other, parallel pathways of signal transduction. Some of these operate through closely related pathways, e.g. the receptor-activated reactions that trigger the hydrolysis of polyphosphoinositides by a specific phospholipase C [3]. Others involve receptor-bound tyrosine kinases or receptor-operated ion channels. Not only is the mode of action of a number of these intracellular signalling processes now being unraveled, it is also evident how pervasive and interactive these regulatory features are at all levels of cellular organization. However, the mitochondria form a cell compartment that has long remained outside of the known regulatory framework of hormonal signalling processes. Only recently has Ca^{2+} been identified as an important mediator for at least some of the hormonal actions at the mitochondrial level [4]. Other mediators are likely to be identified in the future.

The focus of this review is on the question of how these hormonal signals affect mitochondrial energy metabolism, i.e. how certain hormonal messages are transferred to mitochondria and how the hormone-induced changes affect the control of oxidative phosphorylation in the intact cell. Hormonal actions on cellular energy metabolism can be divided into four main categories: (a) those affecting the supply of energy, including the choice of substrates, metabolite transport processes, the TCA cycle and the dehydrogenase activities that transfer reducing equivalents to the respiratory chain; (b) those affecting the electron transport process and the generation of the proton motive force (Δp); (c) those affecting Δp utilization, i.e. phosphorylation of ADP, energy dissipation to generate heat, or ion transport processes to maintain transmembrane ion gradients; and (d) those affecting the transport of ATP to its site of utilization, in particular the adenine nucleotide translocator. Much of what happens in the mitochondria is determined by changes in supply and demand that are set outside of the mitochondrial compartment and that communicate with intramitochondrial processes through transport processes. However, our primary focus is on the reactions associated with oxidative phosphorylation in the mitochondria. These processes are central to much of the cellular metabolism in a wide variety of contexts; equally central are the regulatory mechanisms the cell uses to adjust these processes to changes in demand and supply.

2. *Control of oxidative phosphorylation in intact cells*

Early kinetic studies on isolated mitochondria [5] gave rise to the fundamental concept of the control of respiration by the supply of substrates for phosphorylation, ADP and P_i, reflecting the tight coupling between electron transport, ADP phosphorylation and ATP utilization. The ensuing years have provided a wealth of experimental detail on the molecular basis for these coupled processes. The advent of the chemiosmotic theory [6] gave the (delocalized) proton motive force (Δp) a central place in the coupling of respiration and phosphorylation. More recent models have modified these concepts to offer alternatives to a delocalized proton motive force (e.g. ref. [7]).

Much of this work was based on experiments with isolated mitochondria, where many of the complexities of the control of these processes in intact cells are bypassed. The question has remained how these concepts relate to the control of respiration in the intact cell. In part, the problem is how to analyze the control of oxidative phosphorylation in a cell, where the quantitative analysis of crucial metabolites is much more equivocal, and the controlled disturbance of specific steps more questionable. The development of analytical methods for the determination of the subcellular distribution of metabolites and intracellular gradients of pH, pCa^{2+} and membrane potential has been crucial. More recently, non-invasive studies of energy metabolism in intact cells or tissues using ^{31}P-NMR have been illuminating. In part, a theoretical framework was required to make a systematic approach possible to the analysis of intact cellular systems. The application of the metabolic control theory, developed by Kacser and Burns [8] and others [9], to the control of oxidative phosphorylation has contributed important conceptual insights for intact cellular systems. In recent years, these approaches have also started being applied to the question of hormonal regulation of respiratory activities in intact cells [7,10,11].

2.1. Distribution of control of oxidative phosphorylation

Tager and coworkers [12,13] were the first to apply the quantitative analysis of the contribution of individual steps in the process of oxidative phosphorylation in isolated mitochondria to the control of flux through the system as a whole. Later studies by various other groups expanded their findings [10,11,14–17]. Important messages emerging from these studies were:

(a) Control of flux* through the pathway of oxidative phosphorylation is not usually located at a single site but may be shared to a greater or lesser extent among a number of steps in the pathway. For instance, in mitochondria oxidizing succinate in state 3, control of respiration was found to be distributed between the ATP translocator, proton leak, cytochrome *c* oxidase, cytochrome bc_1 complex and the dicarboxylate carrier [11,12].

(b) The distribution of flux control is not a fixed parameter, but may vary with the metabolic conditions. For instance, in mitochondria in state 4, all [12] or most [17] of the flux control of oxygen uptake was found to be localized in the proton leak pathway; in intermediate states (more likely to reflect the situation in the intact cell), a major fraction of the control was contributed by the ATP regenerating system. These concepts are crucial with respect to hormonal interactions with cellular energy metabolism, since hormones are likely to do just that: shift the balance of control in the system (see e.g. refs. [23,24]).

(c) The distribution of control in a particular step is a reflection of the relative elasticities* of the reactions before and after that step for the common intermediate(s), i.e. of the kinetic sensitivity of the enzymes interacting with the common intermediate [7–10]. For instance, in the studies on isolated mitochondria in intermediate states of respiratory activity [12,14,23,24], the properties of the ATP regenerating system, i.e. its sensitivity to inhibition by the steady-state ATP/ADP ratio relative to that of the adenine nucleotide translocator, were important in determining the relative degree of control exerted by these steps.

For the practical analysis of complex reaction sequences, such as oxidative phosphorylation, it has been useful to group reactions together in supply and demand groups relative to a particular intermediate [1,5,16,23–25]. The distribution of flux control in such pathways over the supply of and demand for the intermediate can then be described by the ratio of the

---

* The flux control coefficient of enzyme E_i for flux J is defined as $C^J_{E_i} = \delta \ln J/\delta \ln v_i$, the relative change in flux through the system for an infinitesimally small change in the activity (or amount) of enzyme E_i, operating at v_i. A metabolite control coefficient describes the relative change in the concentration of metabolite M_j, due to a small change in the activity of enzyme E_i, $C^M_{E_i} = \delta \ln[M_j]/\delta \ln v_i$. The elasticity coefficient is defined as $\varepsilon^{E_i}_{M_j} = (\delta \ln v_i/\delta \ln[M_j])_{M_k, M_l}$, the relative change in the rate v_i of enzyme E_i in response to a small change in concentration of its substrate (or product, or effector) M_j, assuming constant levels of all other relevant metabolites in the pathway. Note that the elasticity of a particular enzyme for an intermediate M_j reflects the kinetic characteristics of that enzyme relative to that intermediate, under the prevailing conditions, and may change markedly, for instance with changes in the concentration of M, or with covalent modification of an enzyme. The elasticity coefficients of enzymes in a sequence are related to the control coefficients through a set of connectivity theorems, which describe the relationships between the local variables and the system variables. See refs. [7–11,18–22] for further details and theoretical discussions of this and other approaches to the quantitative analysis of metabolic control and its application to the analysis of oxidative phosphorylation.

elasticities of the supply and demand sectors, relative to the intermediate of interest [7–10,18–22]*. A process with a relatively high elasticity coefficient would not exert much control over the flux through the system. In a branching pathway, the distribution of flux control over the individual branches is determined by the ratio of the elasticities of the branching reactions. Elasticities also are the operational parameters that describe the ability of a reaction to exert homeostatic control over a particular metabolite [7,15,22]. In particular, reactions with a high elasticity coefficient for metabolite M contribute more to the homeostatic control than reactions with a low elasticity coefficient (i.e., a small change in the concentration of metabolite M would greatly change the rate of that reaction to counteract the change in concentration). The application of control theory to the study of oxidative phosphorylation has contributed important insights and provided a useful framework for analyzing its operation in intact cells and the mechanisms by which hormones can affect it.

The contributions of different sectors of the mitochondrial oxidative phosphorylation machinery to the control of respiration and phosphorylation fluxes in intact hepatocytes oxidizing glucose have recently been analyzed in studies by Brand and coworkers [16,25]. These authors analyzed the relationship between respiration rates, NAD(P) redox states and the membrane potential during titrations with inhibitors and uncouplers of oxidative phosphorylation. These studies attributed about 30% of the control of respiration under these conditions to the processes that generate Δp, i.e., the supply of reducing equivalents through glycolysis and TCA cycle, and the respiratory chain, less than half of which is likely to be due to the respiratory chain itself. Another 20% of the control of respiration reflected non-phosphorylative utilization of Δp. The remainder (50%) of the control over the respiration rates in these studies was due to the phosphorylation system (which included the ATP synthase and the adenine nucleotide translocator, as well as ATP-utilizing reactions). In earlier studies, Tager and coworkers [23] had estimated a flux control contribution of about 26% for the adenine nucleotide translocator to cellular respiration rates, based on a titration of hepatocytes with carboxyatractyloside. Estimates of the control share of the ATP-synthase reaction tend to be low, as this reaction is thought to be close to equilibrium. However, LaNoue et al. [26] found indications that this reaction may be well away from equilibrium under conditions of active respiration. Moreover, the reaction has been found to contribute significantly to the flux control of respiration in isolated mitochondria with elevated Ca^{2+} levels [11]. Nevertheless, in the intact cell a significant proportion of the control of respiration is likely to reside in the ATP-utilizing processes under these conditions. The control of flux through the phosphorylation process in the study of Brown et al. [25] was found to reside predominantly (84%) in the phosphorylation reactions, with the non-phosphorylative Δp utilization (which competes for Δp with ATP synthase) having a negative control coefficient. However, the actual distribution of flux control depends greatly on the specific metabolic setting, in particular on the rate of ATP consumption.

2.2. Non-phosphorylative energy dissipation

The substantial contribution of the use of the proton motive force for other processes than ATP synthesis in the intact cell deserves further comment. The nature and the activity of state 4 respiration has been an issue of considerable debate since the early studies on isolated mito-

---

* This property reflects the connectivity theorem which states that the flux control coefficients of two reactions acting on the same intermediate are inversely proportional to their elastity coefficients.

chondria [5]. Nicholls [27] first emphasized the marked increase in state 4 respiration rates at high mitochondrial membrane potential (>180 mV) and attributed this to a non-linear increase in the membrane proton conductivity, i.e. a proton leak at high membrane potential. Similar phenomena were observed when membrane potentials were generated with a K^+ diffusion potential [28]. The mechanisms involved in this proton leak remain unclear [28–32]. However, as pointed out by Garlid et al. [30], a non-Ohmic increase in proton conductance at high proton motive force is not incompatible with standard models of transmembrane ion transfer and does not necessarily imply variable permeability of the membrane to protons. It is the predicted behavior for ion leaks through a membrane characterized by a single, sharp energy barrier in the membrane (Eyring model) [30]. A non-linear relationship between the membrane potential and the rate of state 4 oxidation would also be obtained with a decreased coupling ('slip') of the mitochondrial proton pumps at high membrane potential [33]. The contribution of 'leak' and 'slip' to the state 4 respiration in mitochondria remains an issue of debate [7,30,31].

In an intact cell, other processes may contribute to the non-phosphorylating respiration, e.g. ion transport processes across the inner membrane through one or more of the recently identified ion channels [34] or the activity of the energy-linked transhydrogenase, processes that would often not contribute significantly under the conditions obtained in isolated mitochondria. The extent to which these processes share in the control of respiration rates in the cell depends on their elasticity coefficient for Δp (relative to that of the phosphorylation reaction and the electron transport chain), parameters that could vary greatly with the prevailing Δp [7,10,15,25]. However, recent studies by Brand and coworkers [16,25] indicated that the predominant contribution to non-phosphorylating respiration in intact hepatocytes is the proton leak pathway. There is evidence that the membrane potential dependence of the non-Ohmic proton conductance (or slip) is affected by the thyroid hormone status of the animal (see below), a process that may contribute to the hormone-induced changes in basal metabolic rate [35].

2.3. Control of respiration by ATP consumption

As pointed out by Balaban [1], the bottom line in terms of the control of oxidative phosphorylation in intact cells remains that the cell should respond to an increased demand for ATP (or Δp) with an activation of respiration. The question is how the system is organized to translate the message of increased demand for usable energy to an integrated response of substrate supply, electron transport, oxidative phosphorylation and ATP transport to the place where it is needed. Recent studies by Chance and coworkers [36–38], Balaban and coworkers [39,40], From and coworkers [41–43] and others [44] have used ^{31}P-NMR to measure phosphorylation states of skeletal muscle, heart, liver, and other tissues, and detected multiple, tissue-specific response patterns to increased workloads under different conditions of substrate supply. The majority of these studies have focussed on skeletal muscle and heart tissue. The presence of creatine kinase in these tissues makes it possible to obtain accurate estimates of the free concentrations of ADP from the creatine kinase equilibrium, and the tissue can be exposed to a controlled workload under well-defined conditions. In skeletal muscle exposed to a variable workload, a hyperbolic relationship exists between the increase in ADP (and P_i) levels and the increased rate of oxygen uptake [36]. Since the steady-state concentration of both ADP and P_i are in the range of the K_m values for their respective translocator systems (estimated to be in the order of 30 μM and 0.8 mM, respectively), it appeared that the rate of oxidation could be kinetically limited by the production of these substrates, i.e. by the rate of ATP utilization. Under these conditions, the elasticity coefficients of the transport systems would be relatively high, which would tend to decrease their flux control coefficients. A similar relationship between the rate of oxygen uptake and the P_i and ADP levels was described by From and coworkers [41–43] in the perfused heart, but the relationship varied with conditions of substrate supply and hormonal status. Chance et al. [37] also obtained more complex relationships in heart and multiple controls were proposed to operate in that tissue.

In tissues containing creatine kinase, such as skeletal muscle and heart, the ATP/ADP ratio is buffered by the phosphocreatine/creatine ratio [45,46]. The activity of creatine kinase is normally well in excess of the capacity of the respiratory chain and the enzyme is poised to respond to a change in this ratio, i.e. it also has a high elasticity coefficient for ATP/ADP promoting homeostasis of the cytosolic ATP/ADP ratio. (Different aspects of this 'energy buffer' function of creatine kinase have recently been reviewed [46].) However, this would not by itself enhance the capacity of the oxidative phosphorylation system to respond to a change in the ATP/ADP ratio. A special role has been attributed to the mitochondrial isoform of creatine kinase, which is located in the intermembrane space [46]. Mitochondrial creatine kinase has been proposed to have selective access to the adenine nucleotides emerging from the matrix on the adenine nucleotide translocator [45,46], thus generating a functional microcompartment for adenine nucleotides in the intermembrane space. Gellerich et al. [47,48] found evidence to support a limited rate of exchange of ADP across the outer membrane not only in isolated heart mitochondria, but also in liver mitochondria, which lack creatine kinase. This may relate to the permeability characteristics of the outer membrane VDAC channel (mitochondrial porin) [46,48,49]. Such a 'dynamic compartmentation' of adenine nucleotides across the outer membrane could significantly affect the control characteristics of respiration in the intact tissue [47,48].

Recent evidence indicates that the octameric form of creatine kinase can bind to both outer and inner membranes [46] and it has been suggested that this structure is functionally involved in the formation of contact sites between the inner and outer membrane, in conjunction with the adenine nucleotide translocator [46,49]. However, the adenine nucleotide translocator is one of the predominant inner membrane proteins, particularly in heart mitochondria, and is active in the inner membrane in other regions than the contact sites [49]. A functional microcompartment for adenine nucleotides associated with mitochondrial creatine kinase in the contact sites may have specific functions associated with that structure, e.g. related to the import of proteins.

The response of the phosphorylation potential to a workload is very different in heart and skeletal muscle, with the latter giving a larger shift in the ATP/phosphocreatine (PCr) ratio (reflecting tissue free ADP levels) for a comparable increase in O_2 uptake [1,36,37,39,40]. Interestingly, continued exercise can shift the response pattern in skeletal muscle to become more similar to that in heart. Also, during neonatal development, the heart changes from a response pattern dominated by ADP control to the adult pattern where a large share of the control appears to be in the supply of substrate [51]. These adaptations are associated with an increased proportion of the mitochondrial isoform of creatine kinase, indicating that this shift of the control of the respiratory rate to the substrate supply may be functionally associated with an improved homeostasis of the ATP/ADP ratio, in which this activity may be involved [46].

The lack of creatine kinase has made tissues like liver and kidney less accessible to the determination of ATP/ADP ratios by ^{31}P-NMR. Tanaka et al. [38] found steady-state levels of free (i.e. NMR-detectable) ADP in liver to be considerably higher (200 μM) than in heart or skeletal muscle, well in excess of the K_m for ADP of the adenine nucleotide translocator; by contrast, the P_i concentration was in the low millimolar range. Tanaka et al. [38] proposed that

P_i rather than ADP supply would be a predominant factor in the control of respiration in the liver compared to muscle tissue [38]. Interestingly, a recent study using transgenic mice that express creatine kinase in the liver reported a considerably lower steady-state ADP level (40–60 μM) [52]; this could indicate that earlier estimates of free ADP in the liver were overestimated, or that the presence of creatine kinase caused a significant shift in the control of respiration in this organ. Such a change could have significant consequences for the characteristics of the ATP utilizing reactions and, hence, for the control distribution of oxidative phosphorylation.

2.4. Control of supply of reducing equivalents

The description of the control of oxygen uptake in intact tissues purely in terms of ATP utilization is clearly incomplete. Conditions of substrate supply and hormonal treatment shift the relationship between ADP and P_i concentrations and myocardial oxygen uptake at different workloads [42]. The same rate of oxygen uptake can be obtained at very different concentrations of cytosolic ADP and P_i, either by providing different substrates for respiration, or by adding insulin or other hormones [37,40,42], presumably because the system operates at different NAD redox states with different hormonal or substrate supply conditions. Katz et al. [40] demonstrated that a wide range of myocardial oxygen consumption could be accomodated in dog hearts in vivo by infusion of phenylephrine, without a signifiant change in the ATP/PCr ratio. Phenylephrine would be expected to increase the mitochondrial $NADH/NAD^+$ ratio by a Ca^{2+}-dependent activation of mitochondrial dehydrogenases (see below) and thereby increase the redox potential difference across the electron transport chain, ΔE_h, that provides the driving force for oxidative phosphorylation. A more complete description of the parameters determining the rate of respiration would therefore take into account both the redox term and the phosphorylation state [7,53].

The concept that a significant control of respiratory activity in heart and other tissues may be exerted by activation of the supply of electrons to the respiratory chain was first strongly promoted by Hansford [54]. More recently, the activation of mitochondrial dehydrogenases by Ca^{2+} has been identified as a critical step in the supply of reducing equivalents to the respiratory chain, in heart and several other tissues (recently reviewed in refs. [1,4,55,56]). At the same time, the Ca^{2+}-dependent activation of dehydrogenase activities could play this role only under conditions where these reactions can influence the overall rate of respiration. The importance of the supply of reducing equivalents to the respiratory chain, and the role of activation of the Ca^{2+}-dependent dehydrogenases in the matrix in regulating that supply, has been demonstrated in liver, heart and other tissues under conditions of limited substrate availability ([55,56]; see below). However, conditions have also been described where changes in mitochondrial Ca^{2+} did not affect respiration rates or the prevailing Δp (e.g. refs. [39,57]). The question to what extent a Ca^{2+}-dependent activation of respiration occurs in an intact cell cannot be answered in a general sense, since it would depend on the sensitivity of the dehydrogenase to regulation by Ca^{2+} and on the flux control of the respiratory activity by the NADH supply under the prevailing conditions.

2.5. The need for a 'coupling messenger'

The need for a mechanism to integrate the rates of respiration, substrate supply and energy utilization by ATP consumption has been stressed by several authors [1,4,58,59], based on observations that often the cytosolic phosphorylation potential does not decrease, despite markedly increased rates of respiration and energy utilization, e.g. with inotropic stimulation

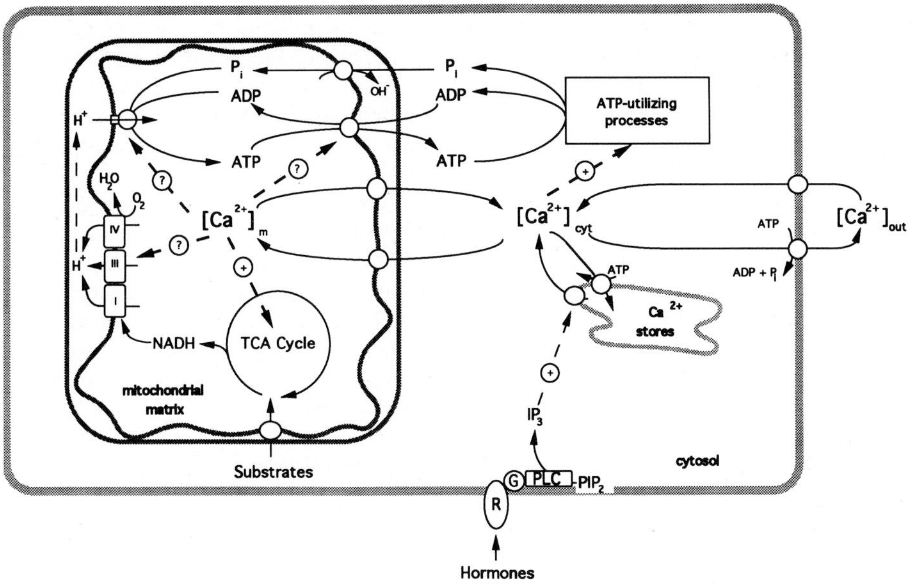

Fig. 1. Schematic diagram of the possible involvement of Ca^{2+} in balancing oxidative phosphorylation and ATP consumption in response to hormonal stimuli. R, hormone receptor; G, G protein; PLC, phospholipase C; I, III and IV, complexes of the electron transport chain. Modified from refs. [1,11].

in heart, or with increased gluconeogenesis or urea synthesis in liver. Balaban [1] proposed that Ca^{2+} could function as such an integrating messenger (Fig. 1). There is evidence that Ca^{2+}-mediated control extends not only to substrate supply, but also to electron transport and ATP synthesis (see below). Additionally, adenine nucleotide transport and matrix volume control may be affected by Ca^{2+}. ATP-utilizing reactions in muscle, liver and other tissues have critical Ca^{2+}-sensitive components. However, as pointed out by From et al. [42], it is difficult to visualize a tightly coupled fine control of two different sets of processes (ATP consumption versus substrate supply and oxidative phosphorylation) exclusively by a third agent such as Ca^{2+}, if there is no mechanism available to exactly balance these rates. This problem is more pronounced if additional Δp-consuming reactions are involved that may not be subject to regulation by Ca^{2+}. If a special 'coupling messenger' exists, hormonal control over the rates of production of Δp and ATP and the utilization of these intermediates should include mechanisms that can account for this integration. If Ca^{2+} is to fulfill this function, the transport mechanisms that interrelate mitochondrial and cytosolic pools of Ca^{2+} are likely to be crucial elements in such mechanisms.

3. Effects of hormones on cellular energy metabolism

In this section, we consider the mechanisms employed by specific hormones to modulate the respiratory activity of the cell and the utilization of the proton motive force, Δp, for ATP synthesis or other purposes. An extensively studied example of such a hormonal effect on respiration is the activation by norepinephrine of the uncoupling protein thermogenin in brown adipose tissue [60]. By relieving the inhibition of thermogenin by ADP or GDP, a proton leak pathway becomes active at very low Δp, greatly favoring the dissipation of Δp at the expense of ATP synthesis, with most of the control of respiration rates shifting to the substrate supply and electron transport chain. The signals employed to achieve this effect are the free fatty acids released by the hormone, which bind to the uncoupling protein and prevent inhibition by ADP and GDP [60].

Usually, the actions of hormones on energy metabolism are more multifaceted and diffuse, and they have given rise to an often equally diffuse body of literature. Our focus is mainly on two classes of hormones, namely (a) those acting through G protein-coupled receptors to activate adenylate cyclase and form cAMP, or to activate phospholipase C and generate Ins-1,4,5-P_3, leading to Ca^{2+} mobilization and (b) the thyroid hormones, which act through nuclear receptors, but which may also employ other, poorly defined signalling mechanisms. These two groups of hormones have been studied in considerable detail, both in isolated mitochondria and in intact cells. Both of these classes of hormones stimulate respiration in intact cells, and stimulatory effects are preserved in isolated mitochondria. However, they appear to represent different modes of interaction with mitochondrial energy metabolism.

3.1. Hormones that operate through Ca^{2+} or cAMP

3.1.1. Hormonal signals acting on mitochondria

A wide variety of hormones and growth factors in almost all mammalian cells operate by causing an elevation of cytosolic Ca^{2+} concentrations ($[Ca^{2+}]_{cyt}$). The Ca^{2+} is released from intracellular stores by Ins-1,4,5-P_3, generated by the receptor-mediated activation of a poly-phosphoinositide-specific phospholipase C; in addition, Ca^{2+} influx from the extracellular medium is usually activated [3]. Ca^{2+} mobilization is only one of several branches of signal transduction pathways activated in response to these hormones, but it is the only one for which there is currently strong evidence that it reaches the mitochondria. All of these hormones set into motion changes in mitochondrial energy metabolism that have been linked to the uptake of Ca^{2+} in the mitochondrial matrix space. Hormones that activate adenylate cyclase and increase cAMP levels (e.g. glucagon and β-adrenergic agonists in liver and other tissues) also affect mitochondrial respiration. The mitochondrial effects of these hormones are, in many respects, similar to those of the Ca^{2+}-mobilizing hormones and it is meaningful to consider these agents together. However, the interrelationship between the cAMP and Ca^{2+} branches of signalling pathways are complex and can vary from one cell-type to another.

A series of recent reviews [4,55,56] has detailed the evidence supporting the involvement of mitochondrial Ca^{2+} uptake in the actions of hormones and other conditions that elevate $[Ca^{2+}]_{cyt}$, and these studies will be shortly summarized below. We will then consider possible implications for the control of energy metabolism in the liver and other tissues.

3.1.1.1. Relationship between cytosolic and mitochondrial Ca^{2+} concentrations. Although mitochondria have a large capacity to actively accumulate Ca^{2+} from the surrounding medium, driven by the proton motive force, this does not lead to a large accumulation of Ca^{2+} in the mitochondrial matrix over most of the physiologically relevant range of cytosolic Ca^{2+} concentrations (0.1–1 μM). This is due to the sigmoidal kinetics of the Ca^{2+}-uptake system, which has a $K_{0.5}$ two orders of magnitude higher than the resting $[Ca^{2+}]_{cyt}$ in most cells [4,61,62]. Ca^{2+} efflux occurs through a Na^+-dependent electroneutral exchange process of much lower capacity and through a poorly characterized Na^+-independent system [4,62]. The relative activity of these two systems varies in different tissues, with heart mitochondria using predominantly the

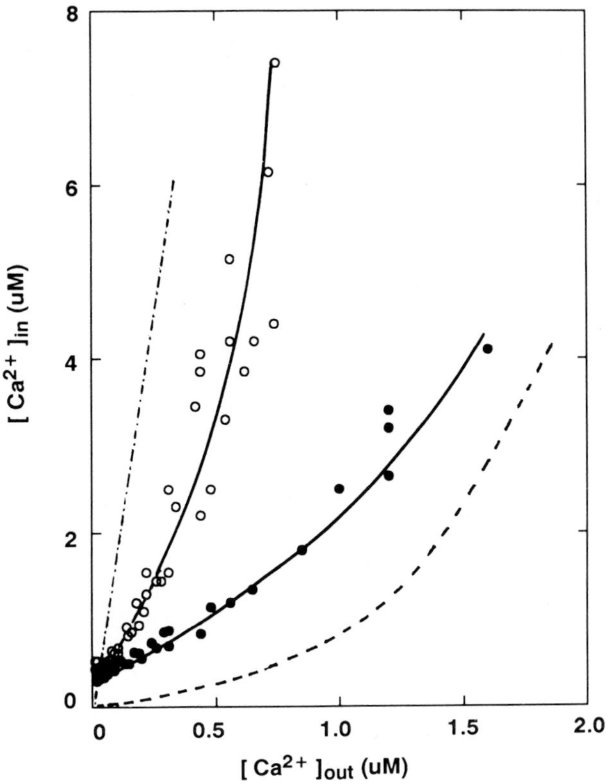

Fig. 2. Steady-state distribution of free [Ca^{2+}] between matrix and extramitochondrial medium in energized mitochondria isolated from liver and heart, in the absence and presence of Na^+ and Mg^{2+}. Dotted lines (---, 10 mM Na^+, 1 mM Mg^{2+}; -·-·-, Na^+ and Mg^{2+}-free), heart mitochondria, data from McCormack et al. [4]. Solid lines (●-●-●, 15 mM Na^+, 1 mM Mg^{2+}; ○-○-○, Na^+ and Mg^{2+}-free) liver mitochondria, unpublished data by Walajtys-Rode, E. and Hoek, J.B. (manuscript in preparation); experimental points were obtained by the simultaneous determination of intra- and extramitochondrial [Ca^{2+}] in a dual excitation spectrofluorometer, using fluor-3-loaded mitochondria incubated in a quin 2-containing medium. Note the much smaller effects of Na^+ and Mg^{2+} on the distribution of Ca^{2+} in liver mitochondria compared to heart mitochondria, reflecting the relatively low activity of the Na^+-dependent Ca^{2+}-efflux pathway in the former.

Na^+-dependent system and liver mitochondria being more dependent on the Na^+-independent pathway [62]. Due to these characteristics, the steady-state Ca^{2+} concentration in the mitochondrial matrix ($[Ca^{2+}]_m$) does not readily reach electrochemical equilibrium with that of the surrounding medium. Recent studies on mitochondria loaded with flurorescent Ca^{2+} indicators, indicate that over most of the physiologically relevant concentration range there is not much of a Ca^{2+} concentration gradient in the steady state, despite the presence of a large electrical potential, providing a driving force for Ca^{2+} uptake. Figure 2 illustrates this steady-state distribution in isolated heart and liver mitochondria, as determined with fluorescent Ca^{2+} indicators. The significant features of this relationship are (a) the marked increase in $[Ca^{2+}]_m$ when [Ca^{2+}]

in the medium increases over 1 μM and (b) the shift in this relationship by the presence of Mg^{2+} (which increases the $K_{0.5}$ for Ca^{2+} uptake from the medium) and Na^+ (which enhances the efflux rate by activating the Na^+-dependent efflux pathway). In liver mitochondria, $[Ca^{2+}]_m > [Ca^{2+}]_{out}$ over most of the concentration range of interest, due to the very slow efflux rate at low $[Ca^{2+}]_m$. However, in heart mitochondria, in the presence of Na^+, the efflux process is more active and the steady-state $[Ca^{2+}]_m$ reaches values even lower than that of the external medium for $[Ca^{2+}]_{out}$ < 0.8 μM. If mitochondria do find themselves in a situation of rapid Ca^{2+} influx (e.g. when extramitochondrial Ca^{2+} levels increase above 1 μM), a high matrix Ca^{2+} level can trigger the opening of a non-specific pore which releases matrix ion contents and temporarily depolarizes the mitochondria [62].

This picture of Ca^{2+} distribution, derived from studies on isolated mitochondria, very recently gained support from an analysis of $[Ca^{2+}]_m$ in suspensions of intact cardiac myocytes [56,63] loaded with fluorescent Ca^{2+} indicators. A substantial fraction of the indicator entering the cells during the loading procedure can enter the mitochondria and other organelles, and this compartmentalized fraction can be used to estimate $[Ca^{2+}]_m$ in the intact cell. Hansford and coworkers [56,63] used a technique in which the cytosolic fluorescence signal was selectively quenched by the controlled uptake of Mn^{2+} in the cell. They found that $[Ca^{2+}]_m$ was significantly less than that of the surrounding cytosolic space, of the order of 50–100 nM in resting cells. Similar findings were obtained by Putney (personal communication) in microscopic imaging studies of single pancreatoma cells, where a pixel-by-pixel analysis identified areas of compartmentalized dye that appeared to be enriched in mitochondria. Although it cannot yet be excluded that $[Ca^{2+}]_m$ is to some extent underestimated, due to the buffering action of the accumulated indicator (intramitochondrial concentrations of the dye may reach 0.3–0.5 mM in these experiments, which could sequester a significant fraction of the matrix Ca^{2+} content) these findings lend strong support to the contention that $[Ca^{2+}]_m$ is in a range of 100 nM or less in resting cells.*

Importantly, in the studies by Miyata et al. [63], when myocytes were paced at increasing frequency, there was a gradual increase in $[Ca^{2+}]_m$ to reach a new steady state, the magnitude of which depended on the frequency of contraction. The inotropic agent epinephrine, which enhances the peak of the Ca^{2+} transient, caused a further increase of $[Ca^{2+}]_m$. These results indicate that, in response spikes of elevated $[Ca^{2+}]$ in the cytosol, mitochondrial Ca^{2+} uptake did occur. A somewhat similar relationship had been predicted earlier by Crompton [65] on the basis of a theoretical analysis of the Ca^{2+}-uptake characteristics in response to contractile Ca^{2+} pulses.

A hepatocyte does not exhibit electrically induced Ca^{2+} transients, but the Ca^{2+} elevation in response to hormonal stimuli is also often oscillatory, in particular at lower levels of the hormones [66–68]. This has become evident from studies on single cells, loaded with the Ca^{2+}-sensitive photoprotein aequorin [67], or with a fluorescent Ca^{2+} indicator, such as fura-2 [68]. An example of a Ca^{2+} oscillation pattern induced by the α_1-adrenergic agonist phenylephrine and detected by fluorescence microsopic imaging in a single hepatocyte is shown in Fig. 3A. Characteristically, the amplitude of the Ca^{2+} peak is invariant with the hormone dose in an individual cell, but the frequency of oscillation is dose-dependent. (Since individual cells oscil-

---

* While this paper was in press, Rizzuto at al. [64] reported a most elegant analysis of $[Ca^{2+}]_m$ in bovine endothelial cells, transfected with cDNA for aequorin, fused with a mitochondrial targeting sequence. The expression of aequorin in the mitochondria enabled the authors to identify a rapid increase in $[Ca^{2+}]_m$ from a basal level < 200 nM, in response to an ATP-induced elevation of $[Ca^{2+}]_{cyt}$.

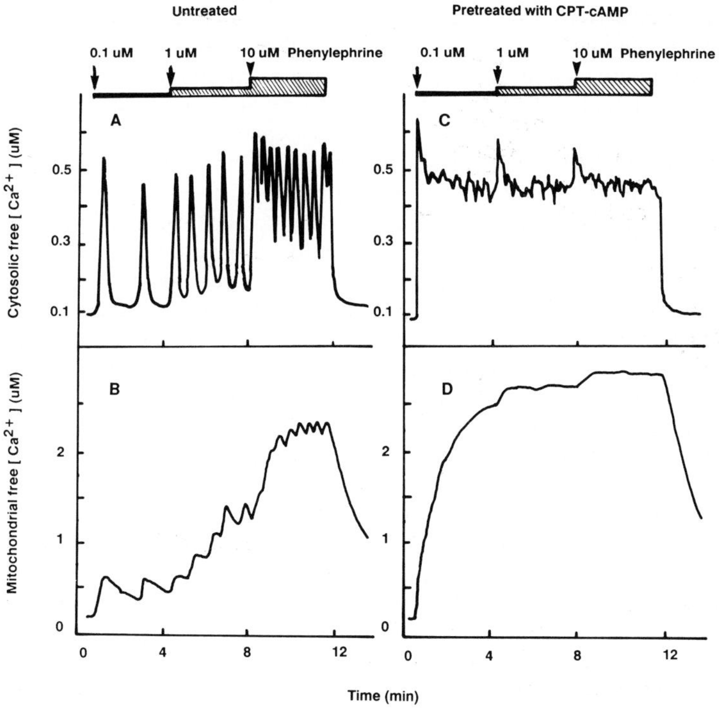

Fig. 3. Intracellular Ca^{2+} oscillations induced by phenylephrine in single hepatocytes. $[Ca^{2+}]_{cyt}$ was measured in individual fura-2-loaded hepatocytes by fluorescence imaging, using a Photonics CCD camera-based detection system, as described by Rooney et al. [68]. (A) Stimulation with increasing concentrations of phenylephrine. Changes in $[Ca^{2+}]_{cyt}$ in a single cell (representative of a population of about 50 cells per field) were followed by ratio imaging at excitation wavelengths of 340 and 380 nm with a time resolution of 300 ms [68]. Under the loading conditions used, the signal derived predominantly from the cytosolic compartment. (B) Mitochondrial $[Ca^{2+}]$ calculated from the net Ca^{2+} flux across the mitochondrial membrane during phenylephrine-induced cytosolic $[Ca^{2+}]$ oscillations, using rate equations for Ca^{2+} influx, and Na^{+}-dependent and -independent Ca^{2+} efflux, from Gunter et al. [62]. (C) After a recovery period the same field was stimulated with phenylephrine according to the same protocol, but cells were pretreated with 2 μM of the permeant cAMP analog CPT-cAMP prior to stimulation with phenylephrine. Images from the same cell were used as shown in (A). (D) Estimated mitochondrial $[Ca^{2+}]$ changes in response to phenylephrine-induced changes in $[Ca^{2+}]_{cyt}$ in the CPT-cAMP-treated cell shown in (C) (Coll, K. and Hoek, J.B., unpublished observations).

late with variable frequencies and out of phase, the net response in a suspension containing a large number of cells does not show evidence of this frequency modulation.) The ability of even low hormone doses to achieve a high $[Ca^{2+}]_{cyt}$ in individual cells is an important feature to explain a Ca^{2+}-mediated action in mitochondria at low hormone concentration. The sigmoidal dependence of mitochondrial Ca^{2+} uptake on $[Ca^{2+}]_{cyt}$ ensures that significant Ca^{2+} influx occurs only during the peak of the Ca^{2+} spikes, when $[Ca^{2+}]_{cyt}$ exceeds 0.4–0.5 μM. The total Ca^{2+} influx (which largely determines the steady-state $[Ca^{2+}]_m$) then reflects a weighted average of the peak $[Ca^{2+}]$ elevation and is a function of the frequency and peak duration. The curve in

Fig. 3B is an attempt to model this response, based on the changes in $[Ca^{2+}]_{cyt}$ in Fig. 3A, using Ca^{2+} transport rate equations derived from studies on isolated mitochondria [62,69]. Since the width of an individual transient of $[Ca^{2+}]_{cyt}$ in hormonally stimulated liver cells is much wider (of the order of 20–30 s or more) than the $[Ca^{2+}]$ spikes associated with contraction of cardiac myocytes, the influx of Ca^{2+} into the mitochondria follows more or less the increase in $[Ca^{2+}]_{cyt}$. However, the decay of the matrix $[Ca^{2+}]$ elevation depends on efflux pathways that are relatively slow in liver mitochondria [62]. Hence, the $[Ca^{2+}]$ changes in the mitochondria appear to be buffered quite effectively and are subject to much less oscillation, i.e., the frequency modulation of the $[Ca^{2+}]$ signal in the cytosolic space is translated to an amplitude signal in the matrix. By the same token, the Ca^{2+} elevation in the matrix will linger for a significant period after the cytosolic signal has decayed. The data of Putney (personal communication) on single pancreatoma cells are in agreement with this interpretation, demonstrating that the region in the cell enriched in mitochondria maintained an elevated $[Ca^{2+}]$ for several minutes after the cytosolic signal had decreased. However, there is evidence that exposure of the liver to glucagon or α_1-adrenergic agonists activates the Na^+-dependent Ca^{2+} efflux pathway [65], possibly enabling the mitochondria to dispose of accumulated Ca^{2+} more rapidly.

3.1.1.2. cAMP as a mitochondrial messenger. Hormones that increase cAMP levels, e.g. glucagon or β-adrenergic agonists in liver, affect mitochondrial metabolism in ways that, in many respects, resemble those of the class of 'Ca^{2+}-mobilizing' agonists. In part, this is due to the fact that these hormones also increase $[Ca^{2+}]_{cyt}$ [70,71], although this effect tends to be small and often transient. In liver, the glucagon-induced increase in $[Ca^{2+}]_{cyt}$ is, at least in part, secondary to the formation of cAMP, since it can largely be mimicked by permeant cAMP analogs [72]. Moreover, activation of protein kinase A potentiates the Ca^{2+} mobilization by other hormones, particularly at low concentrations. Several mechanisms may be involved, one of which is a protein kinase A-mediated enhancement of the response of the Ins-1,4,5-P_3-sensitive Ca^{2+} stores to small elevations of Ins-1,4,5-P_3 [73]. An implication of this synergistic action of cAMP for the Ca^{2+} response to hormones is illustrated in Fig. 3C. The normal oscillatory response to a low concentration of phenylephrine is enhanced to the point of giving a sustained elevation of $[Ca^{2+}]_{cyt}$ in a cell that was pretreated with the permeant cAMP analog CPT-cAMP, but without much change in the peak $[Ca^{2+}]_{cyt}$. Since mitochondrial Ca^{2+} uptake is most effective at the peak of the Ca^{2+} spikes, a marked increase in Ca^{2+} accumulation in the mitochondrial matrix is therefore expected when cells are pretreated to increase cAMP levels. Fig. 3D illustrates the predicted mitochondrial Ca^{2+} accumulation. This prediction is, in fact, supported experimentally, at least in vitro: a large increase in mitochondrial Ca^{2+} level occurs when livers are treated with a combination of glucagon and a phospholipase C-activating hormone, compared to the situation where either one of these is acting on its own [74, but see 75]. It may also explain a long-standing observation that the actions of glucagon on liver mitochondrial function have been much more potent when the hormone treatment was done in vivo, than when isolated cells or perfused livers were treated with glucagon in vitro. In the former system, glucagon probably acts primarily to enhance the response to low levels of catecholamines or other hormones in the circulation.

Until very recently, there was no evidence that cAMP or protein kinase A acts directly on mitochondria, despite intense scrutiny. A cAMP-induced Ca^{2+} release from mitochondria was reported [76], but was not found to be reproducible [77]. However, Romani et al. [78] have now reported evidence that cAMP, in physiologically relevant concentrations (maximal at 50 nM), acts on liver mitochondria to release Mg^{2+}. The effect was observed both in intact and permeabilized hepatocytes and in isolated mitochondria. This action of cAMP was rapid, specific and potent, releasing almost one third of the mitochondrial pool of Mg^{2+} in a matter of minutes. It appeared to be mediated by the adenine nucleotide translocator, as indicated by its sensitivity to the inhibitors carboxyatractyloside and bongkrekic acid. Release of Mg^{2+} was followed by a loss of adenine nucleotides from the mitochondria. Since Mg^{2+} moves against its electrochemical gradient, this efflux would have to be charge compensated. In intact cells, a rise in cAMP levels also caused a loss of Mg^{2+} from the cell [78]. Interestingly, this release of Mg^{2+} was specific for agents that elevated cAMP levels; the opposite response, i.e. an increase in cellular Mg^{2+} levels was found with vasopressin or other hormones that activate phospholipase C. If the intramitochondrial [Mg^{2+}] is significantly and rapidly decreased in response to hormones that elevate cAMP, it could have major implications for the control of Ca^{2+}-dependent processes in the matrix, all of which change their sensitivity to Ca^{2+} with varying Mg^{2+} levels (see below). More work is required to characterize the mechanism and the physiological implications of this interesting effect.

3.1.1.3. Control of mitochondrial adenine nucleotides levels. Apart from promoting the influx of Ca^{2+} into the mitochondrial matrix, a rise in cytosolic Ca^{2+} levels may also affect the distribution of adenine nucleotides across the mitochondrial membrane. Aprille [79] first described an Mg^{2+}-dependent ATP uptake system in liver mitochondria that was independent of the adenine nucleotide translocator and catalyzed the net uptake of Mg^{2+}-ATP, probably in exchange for P_i. Haynes et al. [80] reported that this process is activated by extramitochondrial Ca^{2+} concentrations in the physiologically relevant range of 0.5–1 μM. These observations have been confirmed and extended by Nosek et al. [81]. Interestingly, Ca^{2+} appeared to act on the exterior face of the mitochondria, since the activation was completely insensitive to ruthenium red [80,81]. The uptake of adenine nucleotides by this mechanism is presumably significant in the neonatal development of the mitochondrial capacity for oxidative phosphorylation and pyruvate carboxylation [79]. Although its role in normal, mature mitochondria is more difficult to assess, it could significantly affect the distribution of control between the adenine nucleotide translocator and the intramitochondrial ATP requiring reactions; this would make it a potentially significant messenger to transmit the Ca^{2+}-dependent signal to the matrix, without requiring Ca^{2+} uptake. However, recent mitochondrial studies have indicated other mechanisms by which a net adenine nucleotide transport can occur on the adenine nucleotide translocator (see below). The mechanisms and significance of the changes in matrix adenine nucleotide content in the intact cell remain insufficiently explored.

3.1.2. Mechanisms of activation of respiration by Ca^{2+}-mobilizing hormones

3.1.2.1. Changes in respiratory activity in response to glucagon and Ca^{2+}-mobilizing hormones. An interest in the actions of glucagon on mitochondrial oxidative phosphorylation was stirred early on by studies demonstrating an increased state 3 respiratory activity in liver mitochondria isolated from rats that had been treated with glucagon in vivo [82]. A wide range of changes in the activity of respiratory chain components, ATP synthase, adenine nucleotide translocator, substrate transport processes and other reactions associated with mitochondrial energy conservation were reported, and similar responses were found with other Ca^{2+}-mobilizing hormones (reviewed in refs. [4,59,83]. The relationship to the hormonal responses in the intact cell were not always very systematically explored in these studies. However, several potential target processes for these hormones in intact tissues were found to be stimulated in isolated mitochondria, including pyruvate carboxylase, citrulline synthesis and glutaminase [59,83]. Some of these effects were found to be associated with an increase in the mitochondrial content of adenine nucleotides and Mg^{2+} [59]. However, the mechanism of the activation of electron transport in these mitochondria and its significance remained puzzling and controversial.

Parallel work on perfused liver and isolated hepatocytes demonstrated that glucagon, epinephrine and vasopressin caused an increased rate of O_2 uptake and a transient or more sustained increase in NAD(P)H fluorescence [59,70,83–86]. These processes coincided with an increase in cytosolic Ca^{2+} levels, as detected with fluorescent Ca^{2+} indicators [70,71]. At the same time, the matrix ATP/ADP ratio increased, with no change, or a small decrease in the cytosolic ATP/ADP ratio [87–89]. There was also a net increase in total mitochondrial adenine nucleotides, with a corresponding decrease in the cytosol [87–89], suggesting an uptake of adenine nucleotides into the mitochondria. Measurements of the mitochondrial membrane potential and ΔpH in isolated mitochondria, intact cells or perfused liver did not show consistent evidence of a change with short-term treatment with glucagon or other hormones [90–93]. An activation of the rate of oxidation of glutamine, glutamate and α-ketoglutarate was reported by several groups [92,94–96], indicating that flux through that section of the TCA cycle was activated. Moreover, there appeared to be an increased flux through the malate–aspartate cycle in response to these hormones [93,97–99]. Taken together, these changes were indicative not ony of an increased supply of reducing equivalents to the respiratory chain as the predominant effect of these hormones in the intact cell, due to an activation of the TCA cycle flux, but also of an increase in mitochondrial electron transport activity, and possibly associated with changes in the adenine nucleotide translocator. Mechanistic explanations of these actions have focussed on the effects of Ca^{2+} on mitochondrial processes.

3.1.2.2. Ca^{2+} activation of mitochondrial dehydrogenases and substrate oxidation. An extensive range of studies in heart, liver and other tissues by McCormack and coworkers [4,55], Hansford [56,100] and others, established that an elevation of matrix Ca^{2+} levels within the physiologically relevant range can result in the activation of three mitochondrial dehydrogenases: α-ketoglutarate dehydrogenase (KGDH) and NAD-linked isocitrate dehydrogenase (IDH) are allosterically activated by Ca^{2+}, pyruvate dehydrogenase (PDH) is converted into the active dephospho-form by the Ca^{2+}-dependent activation of the PDH phosphatase. The $K_{0.5}$ for the KGDH and PDH activation is in the range of 0.2–2 μM and that for IDH is slightly higher (5 μM) both in permeabilized mitochondria and in extracts [4]. In addition, mitochondria contain a Ca^{2+}-sensitive pyrophosphatase activity which can lead to a Ca^{2+}-induced accumulation of pyrophosphate [4,59]. A Ca^{2+}-dependent control of the ATP synthase through a Ca^{2+}-sensitive ATPase inhibitor protein has also been suggested [101]. Thus, by a controlled intake of Ca^{2+} from the cytosol, a distinct set of specific mitochondrial proteins can be affected.

A considerable amount of indirect evidence has been provided to support a second messenger role of mitochondrial Ca^{2+} uptake in the hormonal activation of these enzyme activities in intact cells and tissues (see ref. [4,55,56,100] for original references). For instance, hormones that cause Ca^{2+} elevation activate PDH and KGDH in different cell types, and a significant increase in Ca^{2+} content could be detected in subsequently isolated mitochondria. Incubating isolated mitochondria at Ca^{2+} levels similar to those found in the cytosol of stimulated cells caused a corresponding increase in activity of these dehydrogenases, an effect that depended on the uptake of Ca^{2+} in the matrix. When the hormone-induced elevation of cytosolic Ca^{2+} was prevented, e.g. by depleting the cells of Ca^{2+}, there was a corresponding decrease in activation of these dehydrogenases. An inhibitor of mitochondrial Ca^{2+} transport, ruthenium red, was found to prevent the activation of PDH and KGDH associated with an increase in $[Ca^{2+}]_{cyt}$ in perfused heart. (No such inhibition has been observed in liver, where the penetration of ruthenium red into the cell is presumed to be insufficient to achieve inhibition.) The recent demonstration of mitochondrial Ca^{2+} changes in cardiac myocytes [63] and in pancreatoma cells [64] loaded with fluorescent Ca^{2+} indicators has now directly substantiated the conclusion that mitochondrial matrix Ca^{2+} does indeed increase in response to a graded elevation of cytosolic Ca^{2+} levels over the predicted concentration range.

Although the experimental work of McCormack and coworkers [4,55] and of Hansford [56,100] strongly supports a role of Ca^{2+} as a mitochondrial signal in heart exposed to high workloads or inotropic agents, and in liver stimulated with high concentrations of vasopressin or α_1-adrenergic agonists, the role of Ca^{2+} in the stimulation of the liver with glucagon or β-adrenergic agonists is less well established. The Ca^{2+} elevation generated by glucagon in isolated hepatocytes is considerably weaker and more transient than that induced by vasopressin or α_1-adrenergic agonists [70,71]. Moreover, the increase in mitochondrial Ca^{2+} content obtained in mitochondria rapidly isolated from glucagon-stimulated liver cells tends to be small, and this is matched by a much smaller increase in the active form of PDH [102]. Bond et al. [75] studied the subcellular Ca^{2+} distribution by electron probe analysis in rats stimulated in vivo by vasopressin and glucagon, and were unable to identify any significant increase in mitochondrial Ca^{2+} level under these conditions, even when vasopressin and glucagon were added in combination. Instead a marked increase in mitochondrial Mg^{2+} content was observed. This study has been criticized as showing insufficient evidence of stimulation by the hormones [4,55]; however, there was clear evidence that Ca^{2+} release from the endoplasmic reticulum had occurred in the same tissue. It emphasizes the point that mitochondrial Ca^{2+} accumulation in response to these hormones may require relatively strong stimulation and may not be the primary response to low concentrations of the hormones that are likely to be encountered *in vivo*. It would be highly desirable to have more direct quantification of the matrix $[Ca^{2+}]$ increase in intact cells in response to specific hormonal conditions, along the lines of the recent experiments by Hansford and coworkers in myocytes [56,63].

Furthermore, although the enhancement of α-ketoglutarate and pyruvate oxidation by added Ca^{2+} is readily detectable in heart mitochondria, it requires a careful choice of substrate conditions [103]; this is even more true in liver mitochondria [104]. The nutritional state of the animal is an important parameter as well. PDH kinase activity is enhanced during starvation, and this may limit the range of regulation available for the Ca^{2+}-activated PDH phosphatase. PDH kinase is also activated by an increased ATP/ADP ratio and an increased $NADH/NAD^+$ ratio; the same conditions also allosterically inhibit the flux through both the PDH and the KGDH complex and affect the Ca^{2+} sensitivity of these enzymes [55]. Furthermore, the matrix $[Mg^{2+}]$ affects the ability of these processes to respond to Ca^{2+}. Ca^{2+} control of respiration is expected only under conditions where these dehydrogenases are responsive to Ca^{2+} and where this section of the metabolic pathway exerts significant control over the respiratory activity.

Interestingly, recent studies by LaNoue and coworkers [103,105] indicate that conditions can be found where the rate of α-ketoglutarate oxidation is stimulated by Ca^{2+} mobilizing hormones, even when flux through PDH appeared to be completely inhibited, suggesting that the control of the two dehydrogenase complexes can be distinct. This finding could provide an explanation for the long-standing conundrum that glucagon and some other Ca^{2+}-mobilizing hormones in vivo stimulate pyruvate carboxylation and gluconeogenesis in liver, rather than pyruvate oxidation [83]. It also further emphasizes the point that the matrix free Ca^{2+} concentration is only one of multiple factors contributing to the differential control of these enzyme complexes in the intact cell.

3.1.2.3. Metabolic consequences of the activation of Ca^{2+}-sensitive dehydrogenases. Most of the emphasis of the Ca^{2+}-dependent activation of the matrix dehydrogenase activities has been placed on the control of the supply of reducing equivalents to the respiratory chain through the TCA cycle [4,55,100]. However, changes in the dehydrogenase activity will affect metabolite levels in the matrix and thereby alter the distribution of fluxes both in the mitochondrial matrix and, through the operation of metabolite transport systems, in the cytosol. Some recent studies have provided examples of an apparent activation by Ca^{2+} of processes that are by themselves not Ca^{2+} dependent. For instance, a Ca^{2+}-dependent activation of pyruvate carboxylase could be observed in mitochondria from starved rats that had been treated with glucagon in vivo, but not in mitochondria from control rats [106]. This activation was found to be mediated through the Ca^{2+}-dependent activation of PDH, presumably in part by affecting the level of acetyl-CoA or other activators in the matrix. The mitochondria from glucagon-treated rats appeared to be more sensitive to the activation of PDH by Ca^{2+}; the reason for this difference was not evident, but may have involved a change in the matrix ATP/ADP ratio or in the matrix free Mg^{2+} level [106]. Along similar lines, a Ca^{2+}-dependent activation of PEP carboxykinase in guinea-pig mitochondria [107] was attributed to an increased GTP supply due to the Ca^{2+}-dependent stimulation of KGDH.

Activation of KGDH also results in the decrease of matrix α-ketoglutarate levels. Lanoue and coworkers [93,105] established that this change in concentration occurs in intact hepatocytes in a range where it significantly affects the activity of the mitochondrial aspartate transaminase, and thereby the formation of aspartate from oxaloacetate. This reaction is a key step in the malate–aspartate cycle, one of the major mechanisms by which hydrogen shuttling occurs from the cytosol to the mitochondria, the balance of which is essential for gluconeogenesis from lactate. In an intact liver cell, the activity of the shuttling system has to be adjusted to the state of reduction of the gluconeogenic substrates, and its activity can exert significant control over gluconeogenic flux [97]. Evidence that the activity of the shuttle is enhanced by Ca^{2+} has been available [93,97,98], but the site of action has not been unequivocally identified. LaNoue and coworkers [105] propose that the Ca^{2+}-mediated activation of the malate–aspartate cycle is the consequence of the decrease in α-ketoglutarate levels due to the activation of the KGDH reaction. Thus, an increased flux of reducing equivalents to mitochondrial NAD in response to Ca^{2+}-mobilizing hormones may be due in part to the indirect activation of the supply of hydrogen from cytosolic NADH. Other studies [108] indicate that a calmodulin-dependent process may be involved in the malate–aspartate cycle, suggesting that other, as yet unidentified sites of action of Ca^{2+} may also contribute to the control of the malate–aspartate cycle.

Another consequence of the activation of the TCA cycle in liver cells by glucagon is a decrease in the level of succinyl CoA [109]. Quant et al. [109] have demonstrated that these hormone-induced changes in succinyl CoA affect the activity of HMG-CoA synthase in liver mitochondria; this enzyme catalyzes a rate-controlling step in the formation of ketone bodies from acetyl CoA. The control of activity of this enzyme appears to be mediated by a succinyl CoA-dependent inactivation, due to succinylation of the enzyme [109]. Hence, the hormone-mediated changes in the level of this metabolite may affect the flux of acetyl CoA into the ketogenesis pathway.

3.1.2.4. Ca^{2+}-dependent pyrophosphate accumulation and the role of matrix volume changes in the control of mitochondrial energy metabolism. Among the Ca^{2+}-sensitive processes in the mitochondrial matrix, Halestrap and coworkers [4,59] have studied the role of pyrophosphatase. They have developed the intriguing hypothesis that an accumulation of pyrophosphate in response to Ca^{2+}-mediated hormones and glucagon is instrumental in the activation of the respiratory chain and other mitochondrial activities mediated by a change in the matrix volume. The features of this hypothesis are of interest, not only as a regulatory mechanism for mitochondrial respiration, but also because it would qualify as a general signal transduction mechanism to activate mitochondrial processes in response to glucagon and Ca^{2+}-mobilizing hormones [59].

The hypothesis is based on four types of evidence. Firstly, Davidson and Halestrap [110] characterized the Ca^{2+}-inhibition of pyrophosphatase in mitochondrial extracts and found it to be sensitive to Ca^{2+} concentrations in the range of interest for hormonal stimulation. Specifically, at Mg^{2+} concentrations in the physiological range (0.3 mM), half-maximal inhibition required about 4 μM Ca^{2+} [110]. A recent careful kinetic study of the enzyme in toluene-treated mitochondria (which preserves the membrane-bound pyrophosphatase, the predominant form of the enzyme) obtained a similar value of half-maximal inhibition by 12 μM Ca^{2+} at a concentration of 0.4 mM Mg^{2+} [111]. However, at lower Mg^{2+} concentrations the sensitivity to Ca^{2+} increases to values considerably closer to those that activate the matrix dehydrogenases.

Secondly, Davidson and Halestrap [112] demonstrated that a small (0.1–0.4 nmol/mg) but significant increase in pyrophosphate levels in the matrix can be induced by the accumulation of Ca^{2+} or by short-chain fatty acids, e.g. butyrate, either in isolated mitochondria, or by stimulation of intact hepatocytes with glucagon or other hormones (see also ref. [106]). When Ca^{2+} and butyrate were added together, pyrophosphate accumulation increased by a factor of ten or more; this illustrates the substantial capacity of the pyrophosphatase. The source of pyrophosphate is evident with short-chain fatty acids, which are activated in the mitochondria, but not in hormonally stimulated cells in the absence of fatty acids. It is not excluded that the proton motive force directly drives pyrophosphate synthesis, through the membrane-bound enzyme [113]; mitochondrial pyrophosphatase has been suggested to be coupled to transmembrane proton transport, similar to the enzyme from certain microorganisms [113].

Thirdly, over a narrow range of concentrations (up to 0.2 nmol/mg), the pyrophosphate level in isolated mitochondria or in intact hepatocytes correlated with an increase in matrix volume, as estimated by a swelling assay in either system [112]. Halestrap [59] hypothesizes that this swelling represents the opening of a K^+ channel, mediated by the adenine nucleotide translocator [59,114].

Fourthly, in earlier experiments, Halestrap and coworkers [58,115] reported a correlation between small (10–20%) changes in matrix volume and the rate of state 3 oxidation of different substrates, most markedly that of fatty acids. Mitochondrial swelling was also found to stimulate other processes in the matrix that characteristically respond to stimulation by glucagon and Ca^{2+}-mobilizing hormones, e.g. citrulline synthesis, pyruvate carboxylase or glutaminase [59].

Halestrap [59] hypothesizes that the control of matrix volume is affected by Ca^{2+}-mobilizing hormones, as a consequence of the accumulation of pyrophosphate in the mitochondrial matrix due to the Ca^{2+}-mediated inhibition of the pyrophosphatase. However, the determination of these very small matrix volume changes by solute exclusion is not without its problems [116,117]. The volume range reported by Halestrap [58,59] to activate mitochondrial respiration (1.0–1.3 μl/mg protein) in isolated mitochondria is relatively high compared to values for matrix volumes reported by these and other workers (0.7–1.1 μl/mg protein) [58,104,106]. Others have been unable to detect a significant change in matrix volume in isolated mitochondria exposed to low concentrations of Ca^{2+} [104] or isolated from hormone-treated animals [106], despite an increase in respiration. Brown et al. [25] recently reported that vasopressin did not induce mitochondrial swelling in intact hepatocytes (as detected by the change in optical density), although the vasopressin-induced changes in NAD(P)H fluorescence were clearly apparent; presumably, this would indicate that a Ca^{2+}-dependent increase in mitochondrial dehydrogenase activity had occurred. It is conceivable that different incubation conditions are responsible for these differences.

In more recent experiments by the same authors [118,119], the Ca^{2+} induced mitochondrial swelling was found to have an additional component, related to the opening of the Ca^{2+}-

sensitive non-specific pore [62] that is selectively inhibited by cyclosporin A. There is evidence from work by several groups [62] that this process is intimately associated with the adenine nucleotide translocator. Interestingly, in the experiments by Davidson and Halestrap [118], even the vasopressin-induced change in optical density in intact hepatocytes, presumed to reflect mitochondrial swelling, was inhibited by cyclosporin A by about 30%. Thus, under the conditions used, the accumulation of Ca^{2+} in the matrix was sufficiently large to trigger the pore opening in the intact cell. Since the pore opening is a transient and reversible phenomenon, and does not occur simultaneously for all mitochondria in a given population, even under excessive Ca^{2+} loading [62], it is conceivable that cells could maintain a relatively normal function, despite these occurrences. However, it might contribute significantly to an apparent increase in respiratory activity for cells where a significant fraction of the mitochondria is occasionally engaged in such a drastic depolarization. Also, the estimates of mitochondrial volume from the relative penetration of mannitol and sucrose in intact cells [59] become more tenuous in view of mitochondria undergoing these transitions through a transient permeable state.

In the model of Halestrap and Davidson [119], the fraction of the hormone-induced matrix volume changes that is not prevented by cyclosporin A is also proposed to be mediated by the adenine nucleotide translocator. However, in the absence of sensitivity to the characteristic inhibitors of the translocator, the evidence for its involvement in volume control under these conditions is not strong. Several other K^+ channels have recently been identified in the mitochondrial inner membrane by patch-clamp techniques. One of these is inhibited by Mg^{2+} [120], another by ATP [121]. Either may be part of a mitochondrial volume control that involves the operation of a K^+ channel and the Mg^{2+}-inhibited K^+/H^+ exchange process [59,122]. The mechanisms that control the opening of these channels, and the actions of hormones on these processes remain to be characterized. From the data available so far, it is not clear to what extent the hormone-induced changes in matrix volume, and the consequent activation of mitochondrial electron transport, are part of a normal response to hormones or may represent a response to special incubation conditions or an excessive accumulation of Ca^{2+} in the matrix space.

Pyrophosphate accumulation has another consequence that may be of relevance for the control of oxidative phosphorylation in the cell. It has long been known that pyrophosphate can exchange for adenine nucleotides on the adenine nucleotide translocator. Hence, the accumulation of pyrophosphate could contribute to a hormone-induced increase in matrix adenine nucleotides [59]. However, since pyrophosphate has a low affinity for the translocator, it is difficult to visualize an effective competition with the ten- to fiftyfold higher concentrations of adenine nucleotides in the matrix. The Mg^{2+}-linked ATP/P_i exchange transport process [79], discussed above, offers a more feasible alternative mechanism to increase matrix adenine nucleotides in response to Ca^{2+}-mobilizing hormones.

3.1.2.5. Ca^{2+}-control of the ATP synthase. Questions also surround the possible control of ATP synthase by Ca^{2+}. A presumed mechanism of control of the ATP synthase is its inhibition by the inhibitor protein described by Pullman and Monroy [127]. The binding of this inhibitor protein is controlled by the proton motive force, and protects the cell against uncontrolled ATP hydrolysis under conditions of low energization. Yamada and Huzel [101] have identified a Ca^{2+}-dependent ATPase inhibitor protein, which is distinct from the inhibitor protein of Pullman and Monroy [127]. The Ca^{2+}-dependent binding of this protein may provide a control of the ATP-synthase activity in response to varying Ca^{2+} levels over the physiologically relevant range. The Ca^{2+}-dependence of this inhibitor protein indicates an increased inhibition of ATP synthesis over the range of concentrations up to 1 μM, with a concomitant activation of the

ATPase activity [101]. Thus, a Ca^{2+}-dependent increase in oxidative phosphorylation rate is hard to match with these characteristics. The potential role of this protein remains to be further clarified under conditions prevailing in the intact mitochondria.

3.1.2.6. Effects of Ca^{2+} on the adenine nucleotide translocator. Moreno-Sanchez [123] provided interesting data indicating that Ca^{2+} affects the distribution of control of oxidative phosphorylation between the ATP synthase and the adenine nucleotide translocator, and concluded that the adenine nucleotide translocator is a potential site of control by Ca^{2+} levels of physiological interest (up to 1 μM Ca^{2+} in the external medium). However, the nature of this Ca^{2+} effect has been difficult to identify. Recent studies on the cyclosporin-sensitive Ca^{2+}-induced permeability transition [62] demonstrated an involvement of the adenine nucleotide translocator in this process. Compounds such as carboxyatractyloside, which stabilize a conformation of the adenine nucleotide translocator that faces the cytoplasmic face of the mitochondrial inner membrane (C-conformation) promote Ca^{2+}-dependent pore opening, whereas the opposite is true for ADP or bongkrekic acid, which stabilize the matrix-facing orientation of the translocator (M-conformation) [4,124]. Halestrap and Davidson [119] found that Ca^{2+} exerts effects on the heart mitochondrial membrane conformation (as detected in swelling assays) compatible with a Ca^{2+}-induced conformational change of the adenine nucleotide translocator. They propose that Ca^{2+} binds directly to the translocator and promotes its interaction with cyclophilin, a cyclosporin A-binding protein with peptidylprolyl *cis–trans*-isomerase activity which is present in mitochondria from liver and heart [119]. In liver mitochondria, the content of the cyclosporin A binding sites (60–110 pmol/mg protein) [62,119,125] is of the same order as that of the adenine nucleotide translocator (120–300 pmol/mg protein) [125]. This interaction would convert the translocator into a non-specific pore; cyclosporin is proposed to inhibit the Ca^{2+}-induced permeability transition by binding to cyclophilin. McGuinnes et al. [125] also demonstrated that cyclosporin A binding to cyclophilin is decreased by Ca^{2+}; however, these authors distinguished two high-affinity binding sites for cyclosporin A, of which only one, comprising about 10% of the total binding capacity, appeared to be involved in protection against permeabilization.

Recent patch-clamp studies on mitochondrial inner membrane preparations have identified the cyclosporin-sensitive pore with a large conductance (>1 nS) channel [126]. Characteristically, this channel has multiple states of lower conductance; it is, therefore, referred to as the multi-conductance channel [126]. Ca^{2+} affects the opening of this channel, but its effects are complex and may be indirect, e.g. involving the assembly of separate channel elements that together generate the large conductance pore [126]. No direct evidence has been reported yet that identifies the involvement of the adenine nucleotide translocator in this process. Whether the effect of Ca^{2+} is directly on the adenine nucleotide translocator or is an indirect consequence of one or more of the components of the complex that generates the multi-conductance channel, is not clear from the data reported so far. It is also unclear to what extent a possible involvement of the adenine nucleotide translocator in the pore complex is related to the normal function of the translocator of providing ATP to extramitochondrial energy-requiring processes or is a specialized activity, e.g. associated with the formation of the contact sites between the inner and outer mitochondrial membranes or in communication between different mitochondria in a cell. The recent studies have dramatically expanded the scope of potential roles of the adenine nucleotide translocator in diverse mitochondrial functions and it is likely that further work in the near future will throw more light on the interactions of Ca^{2+} with the translocator and its role in mitochondrial oxidative phosphorylation.

3.1.2.7. Mg^{2+} as a regulator of mitochondrial function. Mg^{2+} is only recently emerging as a potentially significant regulator of intracellular events, following the development of NMR and fluorescent methods for the determination of the free Mg^{2+} concentration in cells and tissues [128]. Its free concentration in the cytosol of cells and tissues is in the order of 0.5–1.0 mM [128] and recent measurements in isolated mitochondria indicate that there is not much of a concentration gradient across the inner membrane. Although this implies a large electrochemical gradient, transport of Mg^{2+} has been found to be slow under most conditions [128] and it was not evident how mitochondrial [Mg^{2+}] could vary rapidly in response to external signals, a prerequisite for its role as a regulatory molecule. The recent studies of Romani et al. [78], discussed above, now indicate that matrix [Mg^{2+}] may be significantly and rapidly decreased in response to hormones that elevate cAMP. The implications of a decrease in matrix Mg^{2+} levels for the control of Mg^{2+}- and Ca^{2+}-sensitive enzymes in the mitochondrial matrix could be significant. Mg^{2+} is the primary activator of PDH phosphatase, with Ca^{2+} acting to decrease the K_m for Mg^{2+} [4,55]. Mg^{2+} is the predominant cation complexed to adenine nucleotides in both the mitochondrial matrix and the cytosol. However, the adenine nucleotide translocator, in contrast to most ATP utilizing processes in the cytosol, binds the uncomplexed form of the nucleotides. Hence, the prevailing [Mg^{2+}] may selectively affect the affinity of adenine nucleotides for the translocator relative to those of ATP consuming reactions, and thereby affect the relative control coefficients of these processes. In addition, Mg^{2+}-inhibited K^+ transport in the inner membrane could be activated, possibly providing a mechanistic link to the hormone-induced volume changes that have been described by Halestrap [59]. However, previous reports had observed an increase in mitochondrial Mg^{2+} and adenine nucleotide levels in glucagon-treated liver [75,87–89]; it is conceivable that the action of cAMP is bidirectional and may operate in the intact cell to promote influx of ATP and Mg^{2+}. Clearly, more work is required to clarify the involvement of this system in mitochondrial activation.

3.1.3. *The place of mitochondrial [Ca^{2+}] changes in the actions of glucagon and Ca^{2+}-mobilizing hormones on cellular energy metabolism*

There is now an abundance of evidence, both from isolated mitochondrial studies and from intact cells and tissues, to support the notion that Ca^{2+} can play a significant role in the control of mitochondrial oxidative metabolism, both directly and indirectly. Its primary action at the mitochondrial level appears to be at the level of the mitochondrial Ca^{2+}-sensitive dehydrogenases. However, evidence is gradually accumulating that the dehydrogenases are not always responding in concert to an increase in matrix Ca^{2+} concentration. NAD-IDH has been recognized to require a substantially higher level of Ca^{2+} than what is required for PDH and KGDH activation, and the same may be true of the Ca^{2+}-dependent inhibition of pyrophosphatase. The specific requirements of all of these enzymes would depend on the free Mg^{2+} in the matrix, and the kinetic details that determine the susceptibility of each of these enzymes to Ca^{2+} may be individually different. Furthermore, the range of susceptibility of PDH to activation by its phosphatase will depend on the activity of the other component of the phosphorylation–dephosphorylation cycle, the PDH kinase. This enzyme is subject to various metabolic factors affecting its activity, in particular the state of the adenine and nicotinamide nucleotides [4,55,100], and its activity varies with the nutritional state. Thus, there is considerable room for an adjustment to specific circumstances; the Ca^{2+}-dependent activation of these enzymes has to be interpreted within the context of the functioning cell.

However, even when a significant increase in Ca^{2+}-mediated increase in mitochondrial NAD is recognized as part of the hormonal response, there is considerable evidence that the actions of glucagon, phenylephrine and other Ca^{2+}-mobilizing hormones are more complex than can be described by a shift to a different steady state characterized by a higher activity of substrate supply to the respiratory chain. Several lines of evidence support multiple interactions of these hormones with the reactions of oxidative phosphorylation in the intact cell. The pattern of changes in the reduction levels of NAD and respiratory chain intermediates reported by Quinlan and Halestrap [86], suggested that an initial reduction was followed by a more oxidized state, despite a sustained rate of O_2 uptake. They attribute this change to an activation of the electron transport chain, following swelling of the matrix space [59,86]. Leverve at al. [130] described an equally transient increase in mitochondrial $NADH/NAD^+$ ratio induced by phenylephrine in perifused liver cells, even though O_2 uptake was sustained at a steady rate after an initial increase. This was accompanied by a transient oxidation of cytosolic NAD and a stimulation of gluconeogenesis that appeared to reflect an activation of hydrogen shuttling mechanisms across the mitochondrial membrane, e.g. involving the malate–aspartate cycle. Complex kinetics of the transition between steady states are also suggested by measurements of plasma membrane Ca^{2+} fluxes that accompany hormone-induced activation of O_2 uptake and NAD(P) reduction in perfused liver by several authors [130,131]. The mechanisms underlying these transient changes involved in the transition between different states are still poorly characterized. An activation of intact liver, by glucagon or other hormones leads not only to the mobilization of Ca^{2+} to increase cytosolic levels, but also activates a range of other signal transduction pathways which activate protein kinases in the cytosol, including protein kinase C, Ca^{2+}-calmodulin dependent protein kinase and (for glucagon and β-adrenergic agonists) protein kinase A. These may themselves initiate cascades of further protein kinase-mediated phosphorylation reactions. The combination of these events is likely to induce major shifts in the balance of metabolic fluxes in the cytosol, which will influence the mitochondrial oxidative reactions both by the supply of reducing equivalents and by the ATP utilizing processes. Moreover, the accumulation of Ca^{2+} in the mitochondria may increase over time, due to the oscillatory nature of the Ca^{2+}-mobilizing hormones. This might lead to the graded stimulation of Ca^{2+}-dependent processes over the time required to acquire a new steady-state free Ca^{2+} concentration in the matrix. Furthermore, the activation of Ca^{2+}-dependent enzymes in the mitochondrial matrix would be followed by a readjustment of other metabolic pathways to the new steady state. Secondary hormone-induced processes may also influence the development; these could be reactions initiated by the metabolite changes induced by the initial rapid response, or processes requiring a higher level of Ca^{2+} accumulation in the matrix, e.g. an activation of the repiratory chain through a matrix volume change, a response to an influx of adenine nucleotides, or a change in matrix Mg^{2+} concentration.

These changes in cellular control parameters that occur during the transition phase must have a substantial effect on the balance of fluxes and metabolite concentrations in the new steady state. This is already evident in experiments on isolated mitochondria, as exemplified by the studies of Moreno-Sanchez et al. [132], who analyzed the relationship between the steady-state $NADH/NAD^+$ ratio and the rate of respiration in isolated heart mitochondria exposed to different substrate conditions and treated with low concentrations of Ca^{2+}. The relationship between the NAD redox state and the rate of oxygen uptake was linear when the former was varied by using different substrate concentrations, but the slope was altered when the mitochondria were treated with Ca^{2+}. This finding was interpreted as an indication that Ca^{2+} acts on the system in other ways than by affecting the NAD redox state. The authors suggest that the adenine nucleotide translocator may be a site of interaction. Further careful studies relating the respiratory rate to the redox state and the phosphorylation potential in different hormonally stimulated states are needed. An analysis of the changes in distribution of the flux control over the respiratory rate, in intact cells would further help clarify the physiological implications of these relationships.

3.2. Thyroid hormones

The thyroid hormone state is a significant factor in regulating energy metabolism in mammals and other organisms and in setting the basic metabolic rate [35,133]. In experimental animals, prolonged treatment with thyroid hormone in vivo increases oxygen consumption and heat production in heart, liver, skeletal muscle, and other target tissues. Morphometric studies show an enlargement of the mitochondrial inner membrane surface area [134,135], associated with an increased cardiolipin content, and a higher activity of respiratory enzymes in liver, skeletal muscle and other target tissues of thyroid hormones [136]. Hypothyroidism, induced experimentally either by thyroidectomy or by treatment with deiodinase inhibitors (which decrease the circulating levels of thyroxine (T_4) and triiodothyronine (T_3)), has opposite effects on cellular energy metabolism. It is associated with a decrease in enzyme activities involved in energy conservation and electron transport, including NADH dehydrogenase, cytochromes, ATP synthase [35,133,136,137]. Thyroid hormone treatment in vivo reverses (most of) the effects of hypothyroidism. Corresponding changes in respiratory activity are retained in mitochondria isolated from tissues of hypothyroid or hyperthyroid animals. Apart from these changes in enzyme and lipid composition, which usually develop over several days, thyroid hormones have short-term effects (over a period of minutes to hours) on respiratory activity, well before there is any evidence of significant changes in lipid composition or enzyme content [138]. In recent years, a substantial effort has been devoted to characterize the changes in energy metabolism with altered thyroid states and their implications for the control of mitochondrial energy metabolism in intact cells.

3.2.1. Mechanisms by which thyroid hormones affect mitochondrial oxidative phosphorylation

3.2.1.1. Synthesis of electron transport enzymes in mitochondria in response to thyroid hormones. Protein complexes of the mitochondrial electron transport chain are composed of subunits that are encoded in the nucleus, as well as subunits encoded on mitochondrial DNA [137]. Little is known as yet about the mechanisms responsible for integrating the two transcriptional and translational machineries. Some recent studies have used the thyroid hormone-induced synthesis of mitochondrial electron transport enzymes to gain insight into these processes.

A detailed study of the effects of thyroid hormones on mitochondrial protein synthesis in liver was carried out by Nelson and coworkers [139–142]. When hypothyroid rats were treated with T_3, the synthesis of mRNA for all mitochondrially encoded subunits of electron transport enzymes was enhanced more or less in concert over a period of 1–3 d after T_3 treatment, the time course being dependent on the hormone dose [139,140]. By contrast, the synthesis of mitochondrial 12S and 16S rRNAs, which are also encoded on the mitochondrial genome, was not enhanced under the same conditions. A significant increase in the electron transport chain protein components (e.g. cytochrome *b*, cytochrome oxidase subunits) was often detectable only after up to six days in liver [141]. By contrast, Joste et al. [142] detected mRNA for a limited number of nuclearly encoded mitochondrial proteins that respond early (within 12 h) to T_3 treatment. Among the most pronounced of these are cytochrome c_1 and the Rieske non-heme iron protein. mRNA for these proteins, detected by an in vitro translation assay, increased markedly (more than 15 fold for cytochrome c_1 after 4 d of hormone treatment), following a time course very similar to that of glycerol-3-phosphate dehydrogenase or malic enzyme, two enzymes that have been studied in detail as markers for nuclear effect of the hormone. Nelson and coworkers [137,142] suggest that the cytochrome c_1 may exert subsequent control over the activation of transcription and translation of the mitochondrially encoded polypeptides.

A study with similar scope was reported recently in heart, where thyroid hormone treatment is associated with cardiac hypertrophy [143]. These authors also detected a significant, but non-discriminating increase in all mitochondrially encoded subunits over a period of 12 h to 4 d. The synthesis of mitochondrial proteins encoded in the nucleus was also enhanced over the same time period, and to a greater degree than that of non-mitochondrial proteins, but these proteins were not further characterized. Thus, although thyroid hormone treatment appears to increase the synthesis of mitochondrially encoded subunits of the oxidative phosphorylation machinery, the assembly of functional complexes may be slower and an increase in enzymatic components of the oxidative phosphorylation system may not play a dominant role in much of the short- and medium-term responses to thyroid hormones reported in the literature.

3.2.1.2. Thyroid hormone-induced changes in mitochondrial lipid composition and their role in bioenergetic effects of the hormone. Hoch [136] recently reviewed the evidence that changes in lipid composition of the mitochondrial inner membrane may contribute to some of the functional alterations in respiratory activity and metabolite transport processes with changes in the thyroid hormone state. For instance, Ruggiero et al. [144] reported a substantial (34%) increase in cardiolipin level in liver mitochondria after thyroid hormone treatment, associated with a marked (63%) decrease in the content of 18:2 acyl chains in this lipid. However, the pattern of changes in acyl composition of this and other phospholipids is complex, and varies with the tissue (smaller or no changes in heart cardiolipin were found [136,144]). Paradies and Ruggiero [145] linked the increase in cardiolipin content to an activation of the phosphate translocator in liver mitochondria from hyperthyroid rats. The thyroid hormone treatment caused an increase in V_{max}, with no significant change in K_m, and no change in the total phosphate translocator protein, as indicated by binding of the inhibitor mersalyl [145]. Changes in the activity of the tricarboxylate carrier in hepatic mitochondria also occur [146]. The authors suggest that the lipid matrix in which these proteins operate plays a significant role in determining translocator activity.

Other membrane-associated activities could be similarly affected. For instance, tightly bound cardiolipin is associated with the cytochrome oxidase complex [147] and with the adenine nucleotide translocator [148]. Cardiolipin inhibits the energy-linked transhydrogenase activity in reconstitution studies [149]. The latter activity is decreased in liver and heart mitochondria from hyperthyroid rats and increased in the hypothyroid state [150], even though thyroid hormone treatment caused an increase in the mRNA level for this protein [135]. A possible involvement of cardiolipin in the changes in apparent proton conductivity associated with thyroid hormone treatment has also been suggested [151].

The adaptations in lipid composition may affect the activity of mitochondrial enzymes on a considerably shorter time scale than those requiring protein synthesis. Hoch [136,152] reports significant alterations in the 18:2 acyl content within periods of 1–3 h after thyroxine treatment. How much of a change in specific phospholipid patterns would be required to significantly affect the activity of membrane-associated enzymes of energy metabolism or metabolite transport systems remains to be analyzed. The very complexity of the functional changes in a wide range of membrane-bound enzymes, and the multitude of possible parallel changes in membrane phospholipid composition make it very difficult to establish a convincing causative relationship. On the basis of currently available information, it is hard to go beyond purely correlative speculation in assessing the quantitative contribution of these membrane-composition factors to the control of enzyme activity in intact mitochondria, let alone in an intact cellular system.

3.2.1.3. Short-term actions of thyroid hormones on electron transport activity and oxidative phosphorylation. In addition to the longer-term changes in mitochondrial composition, thyroid hormones also affect mitochondrial energy conservation within a rapid time frame (minutes to hours), well before there is evidence of significant changes in lipid composition or enzyme content [138]. A rapid increase in oxygen uptake in response to short-term treatment with thyroid hormones in vivo has been reported, and this effect has been mimicked in perfused liver [153] and isolated hepatocytes [154]. The mechanism of these rapid alterations in mitochondrial energy conservation in response to thyroid hormones has been a controversial issue for much of the past decade. A direct binding of T_3 to mitochondrial membranes was proposed by Sterling [138] to be responsible for the mitochondrial actions of thyroid hormones. However, direct actions of T_3 on oxidative phosphorylation in isolated mitochondria and submitochondrial particles could not be generally reproduced [154]. The question of whether the rapid mitochondrial actions of T_3 are a direct consequence of the hormone acting on the mitochondria or are secondary to the nuclear actions of T_3, has remained controversial [154].

Recent work by Horst et al. [155] demonstrated that the rapid stimulation of O_2 uptake by T_3 in the perfused liver could be mimicked by 3,5-diiodothyronine (3,5-T_2). This is a minor degradation product of T_3, formed by the action of type I deiodinase, one of the enzyme activities involved in the degradation of T_4 and T_3 in liver and other tissues. Propylthiouracil (PTU), an inhibitor of type I deiodinase completely inhibited the stimulation of O_2 uptake by T_3, but not that by 3,5-T_2, in perfused liver. These authors also indicated that 3,5-T_2 activates O_2 uptake in isolated mitochondria, but experimental data were not provided [155]. 3,5-T_2 was fully active in perfused liver at concentrations of 1–10 pM, presumed to be in a possible physiological range. Importantly, the efficacy of different analogs for the activation of mitochondrial respiration differed markedly from that required for the induction of glycerol-3-phosphate dehydrogenase or malic enzyme, indicating that the requirements for the short-term mitochondrial stimulation are distinct from those required for nuclear activation. Thus, these studies support previous reports that short-term actions of thyroid hormones on oxygen uptake can occur through a direct mechanism that is independent of the actions mediated by T_3 binding to nuclear receptors. However, studies of T_3 binding to mitochondrial constituents may not be relevant for these short-term actions of thyroid hormones on mitochondrial respiration. Instead, the dilemma of the site of action of the thyroid hormone is shifted to a similar question for 3,5-T_2. How 3,5-T_2 exerts its action, whether at the level of the mitochondria or elsewhere in the cell, remains to be established.

3.2.1.4. Ca^{2+} elevation as a trigger in short-term thyroid hormone actions. Shears [156] pointed out interesting similarities in the response of mitochondrial parameters of energy conservation to treatments with thyroid hormone and with glucagon (see also ref. [157]). These analyses were largely based on the characteristics of the mitochondria isolated from animals treated with these hormones and later studies have identified several important differences between the patterns of changes introduced by these two hormones (e.g. the marked change in state 4 respiration characteristic of the response to thyroid hormone is not observed in response to glucagon). Nevertheless, as Ca^{2+} emerged as a mediator of at least some of the actions of glucagon, vasopressin and α_1-adrenergic agonists on mitochondria, the question became relevant to what extent actions of thyroid hormones might be Ca^{2+}-mediated.

Segal [158–160] reported that thyroid hormone acts rapidly (detectable in <30 s) in thymocytes and other cell types to activate an influx of $^{45}Ca^{2+}$ from the extracellular medium. In many instances, this probably reflected an exchange of labeled Ca^{2+}, rather than its net uptake, since the rate of $^{45}Ca^{2+}$ uptake was not affected by decreasing extracellular Ca^{2+} to the μM range [158]. The mechanism of this exchange reaction has not been well characterized. However, in parallel experiments on quin 2-loaded thymocytes a significant, 40–50%, increase in $[Ca^{2+}]_{cyt}$ over the baseline level of about 100 nM was detected (which did require physiological levels of Ca^{2+} in the medium) [159]. Hummerich and Soboll [161] found similar thyroid hormone-induced changes in $[Ca^{2+}]_{cyt}$ in fura-2-loaded hepatocytes; a net efflux of Ca^{2+} was observed in perfused liver, concomitant with an increased O_2 uptake and a stimulation of glucose production. The increase in $[Ca^{2+}]_{cyt}$ was small and required relatively high T_3 levels (a peak of 155 nM $[Ca^{2+}]_{cyt}$ from a basal level of 120 nM was reported in response to 1 μM T_3), but in parallel experiments the response to glucagon and vasopressin was almost equally small [161]. Since fura-2 often is compartmentalized in subcellular organelles during the usual loading procedures, the calibration of this signal may have been problematic.

An increase in $[Ca^{2+}]_{cyt}$ to 150 nM is not likely to induce a substantial mitochondrial uptake of Ca^{2+} in view of the pronounced sigmoidicity for Ca^{2+} uptake (see above). It is unlikely, therefore, that an increase of this size would have the capacity to act directly as a signal to the mitochondrial compartment by a mechanism similar to that indicated for vasopressin or other Ca^{2+}-mobilizing hormones, unless the Ca^{2+} increase in individual cells is much higher than that detected in a cell suspension (e.g. due to cellular heterogeneity in the Ca^{2+} response). However, a small increase in $[Ca^{2+}]_{cyt}$ could have a significant effect on cytosolic processes that indirectly affect the mitochondrial compartment. Mitochondrial glycerol-3-phosphate dehydrogenase, a key enzyme for shuttling reducing equivalents across the mitochondrial membrane, is accessible to cytosolic Ca^{2+}; thyroid hormone treatment has been found to decrease its effective $K_{0.5}$ for Ca^{2+} to 100 nM or less [162]. Segal [158,159] found that the thyroid hormone-induced Ca^{2+} uptake in thymocytes, heart and other tissues was associated with an activation of glucose uptake, which was presumed to be mediated by a Ca^{2+}-calmodulin activation of adenylate cyclase. Thomas et al. [163] reported that, in their experiments, a small change in extramitochondrial Ca^{2+} (up to 0.25 μM) significantly affected the ADP/O ratio for succinate oxidation, but only in mitochondria from hypothyroid rats, and without requiring uptake into the matrix compartment. How such an effect might be achieved is not clear from their data.

In summary, the mechanisms involved in the actions of thyroid hormones on oxidative phoshorylation in vivo remain poorly understood. In the short-term, T_3 or its degradation products may act directly on mitochondria by as yet unidentified mechanisms, or their effects may be indirect, mediated by a small change in $[Ca^{2+}]_{cyt}$, again acting by some uncharacterized process. Longer-term actions of T_3 may be mediated by alterations in mitochondrial enzyme levels and lipid components initiated by its binding to nuclear thyroid hormone receptors.

3.2.2. Thyroid hormone-induced changes in oxidative phosphorylation in isolated mitochondria and intact cells

A decrease in respiratory activity in both state 3 and state 4 is generally observed both in intact cells and in mitochondria isolated from hypothyroid rats, and this can be reversed by overnight treatment with thyroid hormones in vivo, well before a significant increase in electron transport proteins can be detected [136]. The mechanisms underlying these phenomena and their physiological implications have been very much in dispute. In part, this relates to the different procedures used to generate a hyperthyroid state. Despite the evidence that thyroid hormones affect respiration in intact cells and perfused liver in short-term experiments, this protocol has not commonly been found effective in enhancing the oxidative capacity of isolated mitochondria. The efficacy of thyroid hormone treatment in vivo is dose- and time-dependent, and both of these have varied considerably between investigators. The most effective and reproducible procedure has been to induce the hypothyroid state, either by thyroidectomy or by the use of a deiodinase inhibitor, such as PTU; this treatment consistently brings about a decrease in state

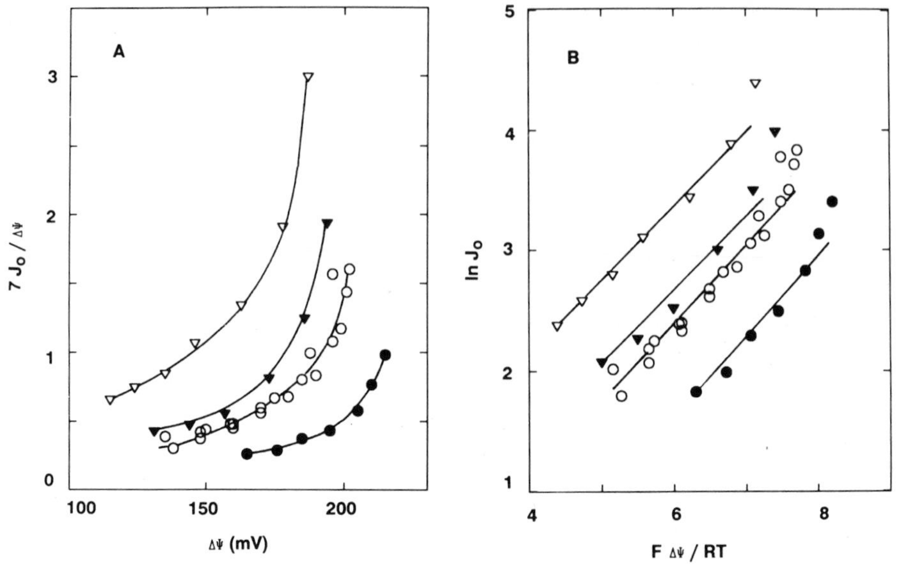

Fig. 4. Changes in mitochondrial proton leak with thryroid hormone status. Data derived from experiments of Hafner et al. [164], using isolated mitochondria oxidizing succinate in state 4, and titrated with malonate. J_O, state 4 respiratory flux; $\Delta\Psi$, membrane potential determined from the distribution of $TPMP^+$. Experimental points were replotted, (A) as the flux/force ratio versus membrane potential, according to Azzone and coworkers (e.g. refs. [33,166]), assuming a proton stoichiometry of 7 H^+/O, or (B) as a log–linear plot of proton flux versus membrane potential, according to Garlid et al. [30]. Symbols: ▽, hyperthyroid, (10 d, 15 μg T_3/100 g body wt.); ▼, hyperthyroid (24 h, 800 μg T_4/100 g body wt.); ●, thyroidectomized (8 wks); ○, corresponding preparations from control animals.

3 and state 4 respiration in subsequently isolated mitochondria, associated with a decreased activity of respiratory chain components. Overnight treatment with T_3 then recovers much of the functional capacity of euthyroid mitochondria. However, this is not a physiologically stabilized situation, since much of the thyroid hormone-induced lipid and protein synthesis, in particular of mitochondrial inner membrane components, is only starting to get underway at that time. Certainly, these mitochondria are not comparable to those from euthyroid animals, and the distribution of control of oxidative phosphorylation may be expected to be substantially different. Moreover, if the hypothyroid state was brought about with PTU, initial actions of thyroid hormones that rely on the degradation products of T_3 [155] may have been suppressed, with unknown consequences for the overall balance of responses of the organism. It is not surprising, therefore, that substantial differences in the characteristics of mitochondrial and cellular performance after thyroid hormone treatment have been observed between different groups.

3.2.2.1. Thyroid hormone effects on state 4 respiration: proton leaks or slipping pumps. The control of state 4 respiration rate is largely determined by the proton leak/slip pathway. Experiments by Brand and coworkers [164,165] demonstrated clearly that the changes in the thyroid hormone state are associated with corresponding alterations in the apparent proton conductance characteristics of the mitochondria, not only in the hyperthyroid state, but also in mitochondria from hypothyroid animals. A recent study by Luvisetto et al. [166] confirmed these observations, but argues that slip in the proton pumps is also compatible with these findings. Figure 4 shows a replot of some of the data obtained by Brand and coworkers [164,165] on preparations from euthyroid, hypothyroid and hyperthyroid animals. Differences are mainly in the region of non-Ohmic increase in proton conductance, i.e. the non-linear increase of the state 4 rate is shifted to a higher membrane potential in mitochondria from hypothyroid rats than in euthyroid controls, while thyroid hormone treatment causes a shift to lower potentials [164,165]. A semi-logarithmic replot of these data should give information on the nature of the change induced by thyroid hormone treatment*. Fig. 4B illustrates that all these relationships are log–linear over most of the membrane potential range of 100–200 mV with a similar slope of approximately 0.6, in accordance with the Eyring model for a proton leak pathway [30] (although a tendency for upward deflection is apparent at the highest membrane potential values). However, a change in the thyroid hormone state was associated with a marked parallel shift, indicating a significantly different intrinsic proton permeability in these preparations. From an extrapolation to $\Delta\Psi = 0$, an apparent proton permeability constant in euthyroid preparations of 3×10^{-4} cm/s is obtained, close to earlier estimates [28–30]; this decreased to 1×10^{-4} cm/s in hypothyroid preparations and to 7×10^{-4} cm/s in mitochondria from hyperthyroid animals. (Note that the results would also be compatible with a change in effective proton concentration at the membrane surface, at constant permeability to protons [30].) The mechanism underlying this shift in apparent proton permeability is not clear, mainly because the control of proton leak itself (or, in alternative models, the control of slip in the proton pumps) is poorly understood. It appears not to be due to an accumulation of phospholipid degradation products [31]. Changes in phospholipid acyl composition may contribute, although there clearly is a role for membrane proteins in this phenomenon [30,31,35]. Interestingly, in Fig. 4, the permeability shift after short-term (24 h) thyroxine treatment was small, with much larger effects occurring in animals treated with thyroid hormone for a prolonged period (10 d). This suggests that the change in proton conductance may depend on the thyroid hormone-induced synthesis of mitochondrial membrane proteins. Since only a limited number of mitochondrial proteins are synthesized in response to thyroid hormone treatment, this observation would argue that specific proteins may play a role in this phenomenon. Brown and Brand [32], in a recent analysis of the factors contributing to the proton conductance, suggest that the leak pathway occurs via the phospholipid bilayer, modified by the presence of proteins. The changes in conductance characteristics with thyroid hormone status were also observed with other cations (K^+ and tetramethylammonium), even though the conductivity for these cations is much lower (by a factor of 10^6 or more) than for protons [167]. (But see Luvisetto et al. [166] for a different finding regarding K^+ permeability.) Brand and coworkers [31,167] argue that this indicates an increase in non-specific cation conductance. This does not exclude the involvement of specific transmembrane proteins, however; for instance, non-specific cation conductance pathways were observed by Dierks et al. [168] when specific metabolite transport systems were reconstituted in lipid bilayers.

3.2.2.2. Thyroid hormone effects on state 3 respiration: control by electron transport chain and phosphorylation reactions. Although the proton leak is a dominant factor controlling respira-

---

* Garlid et al. [30] point out that the flux–force equation for an ion leak at high membrane potential can be approximated by the relationship $\ln J = \ln(PC_1) + \beta F\Delta\Psi/RT$, where P is the permeability constant, C_1 is the ion concentration close to the membrane surface, and β is a parameter describing the nature of the energy barrier ($\beta = \frac{1}{2}$ for a symmetric Eyring model).

tion rates in state 4, the contribution to state 3 electron transport rates is minimal. Yet, the thyroid hormone status markedly affects the state 3 rate of oxidation in isolated mitochondria. Verhoeven et al. [24] compared the state 3 flux control coefficients for succinate oxidation in mitochondria from hypothyroid rats before and after overnight treatment with T_3. The flux control coefficient of cytochrome bc_1 was markedly higher in the hypothyroid preparations than in euthyroid animals, and was decreased significantly by overnight treatment with T_3. The T_3-induced change in flux control coefficient of complex III was accompanied by an increase in the number of antimycin binding sites. These findings may be related to the rapid T_3-induced synthesis of cytochrome c_1 and the Rieske iron sulphur protein of complex III [142]. Studies by Horrum et al. [169] also support the cytochrome bc_1 complex as a target for thyroid hormone-induced changes in electron transport in isolated rat liver mitochondria. Thus, changes in mitochondrial respiratory enzyme activities may have functional implications for the distribution of control of respiratory flux in mitochondria. By contrast, a recent study by Hafner et al. [170] found that changes in the control of state 3 respiration rates in mitochondria from hypothyroid and euthyroid animals were entirely due to Δp utilizing processes, presumably ATP synthesis and transport. A major difference between the data reported by these groups is in the much higher rates of succinate oxidation in the studies of Verhoeven et al. [24] (227 and 338 nmol O/min/mg protein in hypothyroid rats before and after T_3 treatment, respectively), compared to those of Hafner et al. [170] (55 and 67 nmol O/min/mg protein in hypothyroid and euthyroid rats, respectively). The corresponding membrane potentials maintained in these preparations also were markedly different (128 and 129 mV, respectively, in the hypothyroid and T_3-treated preparations in the study of Verhoeven et al. [24], determined from the $TPMP^+$ distribution, and 156 and 166 mV, respectively, in the studies of Hafner et al. [170], using a $TPMP^+$ electrode). The higher flux control coefficient for the substrate supply and electron transport pathways in the studies of Verhoeven et al. [24] may be a reflection of the lower membrane potential, enabling the respiratory chain to operate in a region where its sensitivity to backpressure exerted by Δp is less. The reason for the differences in mitochondrial control characteristics between these studies is not evident in the reaction conditions used, which appeared to have been very similar. Although technical reasons for these differences cannot be excluded, they may also reflect the response of the experimental animals to the details of the treatments given to induce changes in the thyroid state. Thus, the distribution of flux may be highly dependent on even relatively small alterations in experimental conditions.

Hafner et al. [170] reemphasize the phosphorylation system as a potential site of modification by the thyroid hormone state, but do not identify specific sites of action. During previous years, much attention had been devoted to the adenine nucleotide translocator as a possible target for thyroid hormones in mitochondria. Sterling [171] proposed that this protein directly binds T_3, a mechanism he suggested might explain the short-term actions of the hormone on mitochondrial energy metabolism. This interpretation has not been confirmed by others [172], and the evidence in support of a role for the adenine nucleotide translocator was severely criticized by Hafner [154]. However, functional changes in the adenine nucleotide translocator activity were suggested by the findings of Mowbray and Corrigall [173], who measured the kinetics of adenine nucleotide exchange in isolated mitochondria. The hypothyroid state was characterized by a decreased V_{max} and an increase in the K_m for ADP. These changes could be restored to the euthyroid values by a short-term (15 min) T_3 treatment in vivo. To some extent, the changes in the translocator kinetics could be explained by alterations in the adenine nucleotide content of the mitochondria [173]. Changes in both the ATP/ADP ratio and the mitochondrial total adenine nucleotide levels after short-term thyroid hormone treatment were also evident in the intact liver in vivo, in an analysis of the intracellular distribution of adenine nucleotides by the non-aqueous partitioning method [153]. Thyroid hormone treatment caused an increase in the ATP/ADP ratio in the cytosol, with a decrease in this ratio in the mitochondrial matrix. These findings were suggestive of an activation of the translocator by thyroid hormone treatment, although the mechanism of activation was not identified. Very recently, data from Gregory and Berry [174] indicated an increase in carboxyatractyloside binding capacity in hepatocytes from rats that received a prolonged thyroid hormone treatment. By contrast, in experiments designed to determine the distribution of flux control under state 3 conditions, the contribution of the adenine nucleotide translocator was decreased in hypothyroid animals and increased after T_3 treatment [24,175]. This has been used as an argument against a significant role for kinetic changes that increase the capacity of the translocator with thyroid hormone treatment [10,154]. However, the distribution of flux control is a function of the system as a whole and also depends on the characteristics of other components of the system, many of which are affected by thyroid hormone treatment. A further analysis of these components of the system is required for a good understanding of the control of active respiration under conditions of hypothyroidism and thyrotoxicosis.

The effect of thyroid hormone status on the distribution of control of oxidative phosphorylation in intermediate states is more difficult to predict; it would depend on the details of shifts in the contributions of individual components, i.e. the leak pathway, the phosphorylation reactions and adenine nucleotide transport, and the substrate supply and electron transport components. The complex implications of these interactions are well illustrated by the work of Verhoeven et al. [24], who compared mitochondria from hypothyroid animals before and after T_3 treatment during a titration with hexokinase or creatine kinase. Mitochondria from T_3-treated rats maintained a higher extramitochondrial ATP/ADP ratio for the same rate of oxygen uptake, despite their higher proton leak activity, and this was associated with a higher membrane potential. These differences could only partly be reversed by limiting the substrate supply with an inhibitor of succinate transport. The response patterns were not the same when different ADP regenerating systems were used, due to the different elasticity coefficients of these processes. Hence, in intact cells, where the kinetic characteristics of substrate supply and ATP utilizing processes are important controlling variables, the alterations in the response capacity of the mitochondrial oxydative phosphorylation with thyroid hormone state are likely to be a complex function of all the different components of the system. It is not surprising that the changes in the control of mitochondrial respiration in response to thyroid hormone treatment has generated so much diversity of opinion.

3.2.2.3. Effects of the thyroid hormone state on mitochondrial energy metabolism in intact cells. Studies on isolated mitochondria can at best only give a partial picture of the control of oxidative phosphorylation in intact cells and tissues, where the supply of substrates may be more of a limiting factor and where ATP utilizing processes, both inside and outside the mitochondria are likely to be affected by the hormonal treatment. Therefore, a large part of the ability of thyroid hormones to change the control of respiration in the intact tissue by virtue of their actions on mitochondrial electron transport will depend on the ATP-utilizing reactions that are active in the intact cell. Only very recently has this issue started to be addressed in a systematic manner in intact cells and tissues.

As pointed out above, ^{31}P-NMR studies of muscle are particularly suitable to obtain information on the phosphorylation state by non-invasive methods. It is generally believed that the changes in mitochondrial oxidative capacity are a crucial element in the biochemical abnormalities that accompany the myopathies associated with a dysthyroid state, and a few groups have applied this approach to analyze changes in energy metabolism associated with these conditions. In an in vivo study on skeletal muscle in human hypothyroid patients, Argov et al. [176]

found a high value for the P_i/PCr ratio, indicating a low cytosolic steady-state phosphorylation potential in the hypothyroid state. The ratio was elevated further with exercise, as expected, due to a decrease in PCr and an increase in P_i with the workload. Interestingly, in one of these patients, the slope of a measured workrate to the P_i/PCr ratio was about 70% lower than the normal range before treatment; it improved to normal levels in parallel with the recovery of serum T_4 levels during thyroxine therapy. In addition, the rate of recovery of the P_i/PCr ratio after exercise (which reflects the capacity of the tissue to regenerate the phosphorylation potential) was significantly decreased in the hypothyroid patients [176]. By contrast, similar studies in hyperthyroid patients did not detect any abnormalities in the changes in P_i/PCr ratio during exercise, although the rate of recovery in these patients was reported to be abnormally fast. Similar data were obtained with hypothyroid and hyperthyroid rats [176]. These data are compatible with a change in mitochondrial oxidative phosphorylation that impairs the ability to maintain an adequate phosphorylation potential in the hypothyroid state.

In contrast to these findings in skeletal muscle, cardiac muscle of hypothyroid rats showed a small increase in the PCr level in the hypothyroid state compared to euthyroid controls, with no change in ATP levels, but accompanied by a marked decrease in the P_i levels [44,177]. Consequently, there is a significantly lower P_i/PCr ratio, indicating a higher phosphorylation potential [177]. A decrease in heart rate (i.e. a decreased workload), could not account for the change in P_i/PCr ratio, since pacing the hypothyroid heart to the same rate as that obtained in euthyroid controls did not affect the phosphocreatine level and caused only a partial return of P_i levels to control values. These data suggest that the response of the hypothyroid cardiac muscle to a workload is altered in the hypothyroid state, but it is unlikely that a decrease in respiratory capacity can account for these phenomena. There may be additional changes in the control of the adenine nucleotide translocator or the ATP synthase [178] which affect the control of the phosphorylation potential and its response to a workload. Additionally, the effect of mitochondrial creatine kinase on the adenine nucleotide pool in the intermembrane space may affect the response of the adenine nucleotide translocator to a change in the cytosolic ATP/ADP ratio. As discussed above, there is evidence which indicates that the response to metabolic stress in the skeletal muscle is controlled much more by the cytosolic phosphorylation potential than in cardiac muscle, where considerable control may be shifted to the substrate supply and to the adenine nucleotide translocator or the ATP synthase. For an adequate quantitative evaluation of the changes in energy metabolism imposed by the hyperthyroid condition, further information is required, e.g. on the corresponding rates of O_2 uptake, the redox state of mitochondrial NAD or the mitochondrial membrane potential.

Very recently, some of this information has started to be obtained in the studies of Brand and coworkers [151] and Gregory and Berry [174,179] in intact hepatocytes and Soboll et al. [180] in perfused liver. These data have confirmed the persistence of the changes in proton leak across the mitochondrial membrane in response to changes in the thyroid hormone status [151]. The studies of Gregory and Berry [174,179] demonstrate that the expected alterations in mitochondrial energy conservation with cells from hyperthyroid animals appear to result in a significant shift in the mitochondrial capacity to maintain a membrane potential in the intact cell. Similar findings were reported by Soboll et al. [180] in perfused liver. The extent to which that affects the response to increased stress remains to be more carefully evaluated. The limited data available at this point justify the expectation that these analyses will lead to a substantial improvement of our understanding of the changes in energy control in the system.

3.3. *Other hormones*

The Ca^{2+}-mobilizing hormones and thyroid hormones discussed in the preceding sections by no means reflect the complete range of interactions of hormones with mitochondrial energy metabolism and oxidative phosphorylation, but they are by far the best characterized and they may serve in some respects as models for other hormonal actions on mitochondria. Among the hormones having short-term effects on mitochondrial metabolism, insulin has long been known to activate pyruvate dehydrogenase activity by stimulating PDH phosphatase. Insulin has also been noted to stimulate turnover of the TCA cycle [181]. However, the mechanism by which these effects are exerted has remained unidentified. Insulin does not cause an elevation of $[Ca^{2+}]_{cyt}$ levels in the cell. A recent report suggests that even the receptor tyrosine kinase activity, which is required for most other actions of the hormone, is not essential for the PDH activation by insulin [182].

A variety of hormones has longer-term effects on mitochondria, for which the discussion of thyroid hormone actions can serve as a prototype. For instance, hypophysectomy and treatment with growth hormone have opposite effects on liver mitochondrial activity in ways that resemble the effects of thyroid hormones [183]. Hormonal effects on transcription of genes coding for cytochrome oxidase subunits and other components of the oxidative phosphorylation machinery have been noted in several instances, e.g. vitamin D3 was found to increase cytochrome oxidase subunits I and III mRNA and ATP synthase mRNA [184]; Ku et al. [185] found that FSH in Sertoli cells similarly stimulates the transcription and/or stability of mRNA coding for cytochrome oxidase subunit I (encoded on the mitochondrial genome) and subunit Va (encoded on nuclear DNA); glucocorticoids and estrogen may also affect expression of mitochondrial mRNA in different cells [186]. An interesting mechanistic question in all of these systems is the control of the coordinated expression of proteins encoded on mitochondrial and nuclear DNA [137]. In many respects, the thyroid hormone studies may be prototypical of how these actions may result in alterations in the control of oxidative phosphorylation in the affected cells.

Inhibitory effects of glucocorticoids on respiratory activity in isolated mitochondria from liver and other tissues have also been reported, but these direct effects often require high concentrations of steroids [187]. Nevertheless, there are reports that treatment with low doses of glucocorticoids can have beneficial effects in patients suffering from mitochondrial myopathies [188]. The mechanisms by which these actions occur remain to be characterized.

4. *Conclusions*

4.1. *How do hormones affect the control of mitochondrial energy metabolism?*

The studies discussed above identify at least four major sites at which different hormones affect the reactions associated with electron transport and oxidative phosphorylation in different cells and tissues.

(a) The supply of reducing equivalents stands out as a major site of short-term action of the important group of hormones that increase cytosolic Ca^{2+} levels or generate cAMP. A pattern is emerging which suggests differentiation between individual Ca^{2+}-sensitive processes; this could be achieved by their different sensitivity to $[Ca^{2+}]$ or by changes in $[Mg^{2+}]$ or other effectors that affect the range of $[Ca^{2+}]$ to which these enzymes respond. Importantly, the increased supply of NADH is only one aspect of an activation of Ca^{2+}-dependent dehydrogenases: changes in matrix metabolite levels affect a range of other reactions both in the mitochondria and in the cytosolic compartment.

(b) Electron transport activity may be subject to short-term alterations as a consequence of matrix volume changes induced by Ca^{2+}. However, the mechanism and significance of these effects of Ca^{2+}-mobilizing hormones remain to be further characterized, within the context of the permeability transitions that can be associated with the accumulation of Ca^{2+} in the mitochondria. Medium or longer term actions of hormones on electron transport activity are also well documented, the thyroid hormone status being a widely studied example. These effects may reflect changes in membrane lipid composition, as well as changes in the oxidative phosphorylation enzyme complexes. Whether an increased activity of the electron transport components will contribute much to the flux control of cellular respiration is a question that cannot be answered in a general sense; it will depend on the conditions of NADH supply and the rate of energy utilization, as reflected in the Δp.

(c) Particularly interesting is the observed alteration in the proton leak pathway with thyroid hormone status, which has now also been demonstrated in the intact cell. More detailed studies are required to characterize what role this could play in the cell's capacity to respond to changes in workload or other metabolic stress situations. The recent studies of Gregory and Berry [174] and Soboll et al. [180] indicate that the capacity of the cell to maintain a high mitochondrial membrane potential may be significantly affected by the thyroid hormone state. Further analysis of this system should give interesting insights into the internal control features of the system of oxidative phosphorylation and its deregulation in the dysthyroid state.

(d) Recent studies on intact cells have reemphasized the significance of the phosphorylation system and the ATP-utilizing reactions as a possible site of hormonal regulation, and in this regard, the adenine nucleotide translocator retains an important place, by virtue of its position as a mediator between the cytosolic ATP-utilizing processes and the mitochondrial phosphorylation machinery. The role of Ca^{2+} in affecting the function of the adenine nucleotide translocator remains to be clarified, in particular inasfar as this system may be involved in the complex interactions underlying the formation of the cyclosporin-sensitive, Ca^{2+}-activated pore complex and the contact sites between the inner and outer mitochondrial membrane. In cells containing mitochondrial creatine kinase, this enzyme may further contribute to the distribution of control between the ATP-utilizing and ATP-producing parts of the system. To what extent hormonal actions on the phosphorylation system involve these components requires further analysis.

4.2. *Integration of demand and supply: is there a need for a coupling messenger?*

The intriguing question remains of the mechanisms available to the cell to integrate the demand and supply processes, i.e. to match the provision of reducing equivalents with the rate of ATP utilization. The ability of hormonal stimuli to enhance the rate of respiration with little or no detectable change in the cytosolic phosphorylation potential is well documented [1,59]. The question is how tighly controlled and direct a communication mechanism is required in order to accomodate the observed changes. Is there a central communicator, such as Ca^{2+}, which keeps careful tabs on the flux through all individual components and finds a way to adjust these to the desired level? Although there is substantial evidence for interactions of Ca^{2+} with many of the components of the system of oxidative phosphorylation, as well as with some of the ATP-utilizing processes (e.g. muscle contraction, or activation of gluconeogenesis, urea synthesis or ketogenesis in liver), there is no indication of a centralized processing facility that could readjust Ca^{2+} levels to requirements that may be spatially or temporally variable. It is also not clear that such a tightly controlled system is required. A relatively loosely controlled set of changes in the factors that determine the kinetic properties of the phosphorylation reactions may provide a better solution for the cell than a direct system of signals that would maintain demand and supply under tight control, as envisioned in the model of Ca^{2+}-mediated control.

From the perspective of control of energy metabolism, the problem is one of flux control versus homeostatic control. These are related to the elasticity coefficients of the supply and demand processes that affect a particular metabolite (or metabolite pair) of interest, in this case the phosphorylation potential, the proton motive force, Δp, and the $NADH/NAD^+$ ratio. High elasticity coefficients of an enzyme system for a particular metabolite promote homeostasis for that metabolite with respect to small changes in either supply or demand; the distribution of flux control is a function of the ratio of elasticity coefficients of the participating reactions.

In its simplest form, the control of respiratory rate by ATP-consuming reactions in the cytoplasm is dependent on the kinetic characteristics of the ATP-utilizing processes relative to the kinetic characteristics of the P_i transport system and the adenine nucleotide translocator, the latter being responsive mostly to the levels of ADP and ATP in the extramitochondrial milieu. The elasticity coefficient of the adenine nucleotide translocator for a small change in ATP/ADP effectively describes its capacity to transmit a message of the need for ATP to the ATP synthesizing machinery (and hence to the upstream regions of the system). (How effective it is in doing that also depends on the other parameters of the system, i.e. on the capacity of the respiratory chain to meet that demand and on competing reactions in the matrix for ATP or for Δp.) The higher its elasticity coefficient, the more change in respiratory activity can be generated for a small change in ATP/ADP ratio, i.e. the better the homeostatic control.

Its efficacy, of course, also depends on the elasticity of the ATP-utilizing reactions. If such a process causes a large change in ATP/ADP for a small increase in work (i.e. it has a low elasticity coefficient), it will be more effective in transmitting its message to the oxidative phosphorylation system, but it will make it more difficult to maintain homeostatis over the ATP/ADP ratio. A small change in ATP/ADP for a large workload (i.e. a high elasticity of the ATP-utilizing processes) has the opposite effect.

The role of creatine kinase as an 'energy buffer' needs to be considered in this connection. It has been described as a mechanism for maintaining homeostatis of cytosolic ATP/ADP in the face of large energy demands [46]. However, without other adjustments, this would not enable the cell to respond more adequately to a demand for energy, since a similar workload would result in a smaller change in the ATP/ADP ratio, i.e. less of a signal to increase the rate of respiration to respond to the demand. This is exactly how the cell responds, as is illustrated by the recent studies of Brosnan et al. [189] on transgenic mice expressing creatine kinase in the liver, a tissue that does not normally have access to this energy buffering system. The efficacy of the creatine kinase reaction in this system could be adjusted conveniently by feeding different levels of creatine, resulting in variable levels of creatine phosphate in the liver [52,189]. Remarkably, the liver from transgenic animals with different levels of creatine responded to a metabolic stress in the form of a fructose load with an identical pattern of changes in cellular ADP level, irrespective of the presence of the creatine buffer. ATP, however, was maintained at a higher level when the creatine kinase reaction was active [189]. Oxygen uptake could not be measured in these experiments, and the relationship of these effects to the control of cellular respiration rates remains to be characterized. The transgenic mice have only the cytosolic form of the enzyme. It is conceivable that the presence of a mitochondrial creatine kinase isozyme is a crucial component that could alter this pattern. The mitochondrial isozyme, localized as it is in the intermembrane space, has been found to have preferential access to the ATP transported out of the matrix and can maintain a 'dynamic gradient' of ADP across the outer membrane under conditions of active respiration [47,48]. The characteristics of the outer membrane VDAC channel are probably important in imposing a diffusion barrier at low ADP concentrations [46,48,49]. Assuming that phosphocreatine and creatine are more effective in crossing the outer membrane (there is evidence that these molecules have a considerable longer diffusion path than ADP [190]), this system would enable a more effective transfer of the signal of a decreased ATP/ADP ratio to the adenine nucleotide translocator and the phosphorylation reactions. It has been pointed out [46] that the proportion of the mitochondrial isozyme of creatine kinase is increased in heart during neonatal development, when the tissue improves its ability to respond to an increased workload without a substantial drop in the ATP/ADP ratio.

Although this device does not explain the ability of hormones to improve the capacity to increase respiration rates in response to small changes in phosphorylation potential, it does illustrate the principle of how this message could be transferred. If hormones succeed in influencing the degree of homeostasis at the level of the ATP/ADP ratio by increasing the elasticity coefficients of the adenine nucleotide translocator, there is no intrinsic need to utilize a complex messenger system to signal the need for ATP more directly to the NAD supplying processes. There is only a need to ensure that sufficient energy is available, in the form of an adequate redox potential difference, ΔE_h, to generate and maintain the proton motive force. Homeostatic control could be improved, for instance, by enabling the translocator to respond more strongly at the prevailing steady-state ATP/ADP ratio in the cytosol, e.g. by increasing V_{max} or increasing the adenine nucleotide content or the ATP/ADP ratio in the matrix, or possibly by decreasing the free [Mg^{2+}]. As discussed above, there is evidence that cells respond to hormones in part by affecting some of these parameters, but whether they play a role in affecting the homeostatic control over the ATP/ADP ratio by the adenine nucleotide translocator (and other reactions that respond to changes in the ATP/ADP ratio) has not been carefully evaluated.

The intramitochondrial ATP consuming and utilizing processes also have to be considered [11]. Homeostasis of the intramitochondrial ATP/ADP ratio depends not only on the ATP synthase reaction and the adenine nucleotide translocator, but also on the substrate-linked phosphorylation and the ATP-utilizing processes in the matrix. Many of the latter are activated by Ca^{2+}-mobilizing hormones. However, glucagon has been found to cause a decrease in the succinyl CoA level, simultaneously with the activation of α-ketoglutarate oxidation [109]. This indicates that substrate linked phosphorylation is activated. The mechanism of this activation is not well defined. Siess et al. [191] noticed an increased uptake of P_i in mitochondria and cytosol of hepatocytes treated with glucagon, and suggested that the increase in substrate-linked phosphorylation may be associated with this event. The mechanisms underlying the increased P_i uptake were not elucidated. Whatever the mechanism, an activation of substrate-linked phosphorylation would assist in homeostatic control of the system.

If a hormone causes adaptations in the cell that enable it to maintain more homeostasis of cytosolic ATP/ADP, under conditions of increased electron transport flux, a larger share of the flux control over the respiratory rate may be shifted upstream, to the reactions of substrate supply and electron transport. How much of a shift occurs depends on the electron transport activity and the supply of reducing equivalents. A greater influx of reducing equivalents, resulting in a higher $NADH/NAD^+$ ratio, may compensate for changes in the control downstream and the control distribution over the different components of the respiratory chain may not shift very much. The important message, however, is the ability of the system to self-regulate. The need for an internal messenger to signal the demand for energy is not very tight: as long as the potential supply is in place, adjustments that improve ATP/ADP homeostasis can be accomodated with relatively loose controls.

It may be appropriate, in the context of current world events, to compare these two models of control of energy metabolism to the differences between a free-market economy and a centrally planned economy. The centralized model expects carefully orchestrated (and correspondingly bureaucratic) messenger systems to ensure that the demands are met at the supply side and transferred through a transport system that is equally much under the central control, so as to prevent leaks and thefts. In a classical capitalist free market system, the demand is supposed to drive the supply (although the reverse is often possible), mediated through a myriad of local interactions that distribute goods dependent on individual needs and desires relative to the costs involved (reflected in the elasticities of these demands). This system may suffer from substantial swings that could potentially cause serious disruptions. However, a set of carefully placed, but relatively loose regulations can ensure an adequate flow of goods and services, without the need for tight supervision, making use of the free-market interactions rather than strangling them. The hormonal regulation of cellular energy metabolism may be the biological equivalent of this mode of control.

References

1 Balaban, R.S. (1990) Am. J. Physiol. 258, C377–C389.
2 Robison, G.A., Butcher, R.W. and Sutherland, E.W. (1971) Cyclic AMP, Academic Press, New York.
3 Berridge, M.J. and Irvine, R.F. (1989) Nature 341, 197–205.
4 McCormack, J.G., Halestrap, A.P. and Denton, R.M. (1990) Physiol. Rev. 70, 391–425.
5 Chance, B. and Williams, G.R. (1956) Adv. Enzymol. 17, 65–134.
6 Mitchell, P. (1966) Chemiosmotic Coupling and Energy Transduction, Glynn Research Bodmin.
7 Westerhoff, H.V. and van Dam, K. (1987) Thermodynamics and Control of Biological Free-energy Transduction, Elsevier, Amsterdam.
8 Kacser, H. and Burns, J.A. (1973) in: D.D. Davies (Ed.) Rate Control of Biological Processes, Cambridge University Press, Cambridge, pp. 65–109.
9 Heinrich, R. and Rapoport, T.A. (1974) Eur. J. Biochem. 42, 89–95.
10 Brand, M.D. and Murphy, M.P. (1987) Biol. Rev. 62, 141–193.
11 Moreno-Sanchez, R. and Torres-Marquez, M.E. (1992) Int. J. Biochem., 23, 1163–1174.
12 Groen, A.K., Wanders, R.J.A., Westerhoff, H.V., van der Meer, R. and Tager, J.M. (1982) J. Biol. Chem. 257, 2754–2757.
13 Wanders, R.J.A., Groen, A.K., van Roermond, C.W.T. and Tager, J.M. (1984) Eur. J. Biochem. 142, 417–424.
14 Gellerich, F.N., Bohnensack, R. and Kunz, W. (1983) Biochim. Biophys. Acta 722, 381–391.
15 Westerhoff, H.V., Plomp, P.J.A.M., Groen, A.K., Wanders, R.J.A., Bode, J.A. and van Dam, K. (1987) Arch. Biochem. Biophys. 257, 154–169.
16 Brand, M.D. (1990) Biochim. Biophys. Acta 1018, 128–133.
17 Hafner, R.P., Brown, G.C. and Brand, M.D. (1990) Eur. J. Biochem. 188, 313–319.
18 Groen, A.K., van der Meer, R., Westerhoff, H.V., Wanders, R.J.A., Akerboom, T.P.M. and Tager, J.M. (1982) in: H. Sies (Ed.) Control of Metabolic Fluxes, Academic Press, New York, pp. 9–37.
19 Brown, G.C., Hafner, R.P. and Brand, M.D. (1990) Eur. J. Biochem. 188, 321–325.
20 Cornish-Bowden, A. and Cardenas, M.L. (1990) Control of Metabolic Processes, Plenum Press, New York, pp. 195–207.
21 Crabtree, B. and Newsholme, E.A. (1987) Biochem. J. 247, 113–120.
22 Hofmeyr, J.-H.S. and Cornish-Bowden, A. (1991) Eur. J. Biochem. 200, 223–236.
23 Tager, J.M., Groen, A.K., Wanders, R.J.A., Duszynski, J., Westerhoff, H.V. and Vervoorn, R.C.

(1983) in: R.A. Harris and N.W. Cornell (Eds.) Isolation, Characterization, and Use of Hepatocytes, Elsevier, Amsterdam, pp. 313–322.
24 Verhoeven, A.J., Karmer, P., Groen, A.K. and Tager, J.M. (1985) Biochem. J. 226, 183–192.
25 Brown, G.C., Lakin-Thomas, P.L. and Brand, M.D. (1990) Eur. J. Biochem. 192, 355–362.
26 LaNoue, K.F., Jeffries, F.M.H. and Radda, G.K. (1986) Biochemistry 25, 7667–7675.
27 Nicholls, D.G. (1974) Eur. J. Biochem. 50, 306–315.
28 Krishnamoorthy, G. and Hinkle, P.C. (1984) Biochem. 23, 1640–1645.
29 Deamer, D.W. and Nichols, J.W. (1989) J. Membr. Biol. 107, 91–103.
30 Garlid, K.D., Beavis, A.D. and Ratkje, S.K. (1989) Biochim. Biophys. Acta 976, 109–120.
31 Brand, M.D., D'Alessandri, L., Reis, H.M.G.P.V. and Hafner, R.P. (1990) Arch. Biochem. Biophys. 283, 278–284.
32 Brown, G.C. and Brand, M.D. (1991) Biochim. Biophys. Acta 1059, 55–62.
33 Pietrobon, D., Zoratti, M., Azzone, G.F. and Caplan, S.R. (1986) Biochemistry 25, 767–775.
34 Moran, O., Sandri, G., Panfilli, E., Stuhmer, W. and Sorgato, M.C. (1990) J. Biol. Chem. 265, 908–913.
35 Brand, M.D. (1990) J. Theor. Biol. 145, 267–286.
36 Chance, B., Leigh, J.S., Clark, B.J., Maris, J., Kent, J., Nioka, S. and Smith, D. (1985) Proc. Natl. Acad. Sci. USA 82, 8384–8388.
37 Chance, B., Leigh Jr., J.S., Kent, J., McCully, K., Nioka, S., Clark, B.J., Maris, J.M. and Graham, T. (1986) Proc. Natl. Acad. Sci. USA 83, 9458–9462.
38 Tanaka, A., Chance, B. and Quistorff, B. (1989) J. Biol. Chem. 264, 10034–10040.
39 Katz, L.A., Koretsky, A.P. and Balaban, R.S. (1988) Am. J. Physiol. 255, H185–H188.
40 Katz, L.A., Swain, J.A., Portman, M.A. and Balaban, R.S. (1989) Am. J. Physiol. 256, H265–H274.
41 From, A.H.L., Petein, M.A., Michurski, S.P., Zimmer, S.D. and Ugurbill, K. (1986) FEBS Lett. 206, 257–261.
42 From, A.H.L., Zimmer, S.D., Michurski, S.P., Monhanakrishnan, P., Ulstad, V.K., Thoma, W.J. and Ugurbil, K. (1990) Biochemistry 29, 3731–3743.
43 Zimmer, S.D., Ugurbil, K., Michurski, S.P., Mohanakrishnan, P., Ulstad, V.K., Foker, J.E. and From, A.H.L. (1989) J. Biol. Chem. 264, 12402–12411.
44 Taylor, D.J., Styles, P., Matthews, P.M., Arnold, D.L., Gadian, D.G., Bore, P.J. and Radda, G.K. (1986) Magn. Reson. Med. 3, 44–54.
45 Jacobus, W.E. (1985) Annu. Rev. Physiol. 47, 707–725.
46 Wallimann, T., Wyss, M., Brdiczka, D., Nicolay, K. and Eppenberger, H.M. (1992) Biochem. J. 281, 21–40.
47 Gellerich, F.N., Schlame, M., Bohnensack, R. and Kunz, W. (1987) Biochim. Biophys. Acta 890, 117–126.
48 Gellerich, F.N., Bohnensack, R. and Kunz, W. (1989) in: A. Azzi (Ed.) Anion Carriers of Mitochondrial Membranes, Springer, Berlin, pp. 349–359.
49 Brdiczka, D. (1991) Biochim. Biophys. Acta 1071, 291–312.
50 Clark, B.J., Acker, M.A., McCully, K., Subbramanian, H.V., Hammond, R.L., Salmons, S., Chance, B. and Stephenson, L.W. (1988) Am. J. Physiol. 254, C258–C266.
51 Portman, M.A., Heineman, F.W. and Balaban, R.S. (1989) J. Clin. Invest. 83, 456–464.
52 Brosnan, M.J., Chen, L., van Dyke, T.A. and Koretsky, A.P. (1990) J. Biol. Chem. 265, 20849–20855.
53 Van der Meer, R., Akerboom, T.P.M., Groen, A.K. and Tager, J.M. (1978) Eur. J. Biochem. 84, 421–428.
54 Hansford, R.G. (1980) Curr. Top. Bioenerg. 10, 217–278.
55 Denton, R.M. and McCormack, J.G. (1990) Annu. Rev. Physiol. 52, 451–466.
56 Hansford, R.G. (1991) J. Bioenerg. Biomembranes, 23, 823–854.
57 Lakin-Thomas, P.L. and Brand, M.D. (1988) Biochem. J. 256, 167–173.
58 Quinlan, P.T., Thomas, A.P., Armston, A.E. and Halestrap, A.P. (1983) Biochem. J. 214, 395–404.
59 Halestrap, A.P. (1989) Biochim. Biophys. Acta 973, 355–382.
60 Nedergaard, J. and Cannon, B. (1984) in: L. Ernster (Ed.) New Comprehensive Biochemistry, Vol 9, Elsevier, Amsterdam, pp. 291–314.

61 Pietrobon, D., DiVirgillo, F. and Pozzan, T. (1990) Eur. J. Biochem. 193, 599–622.
62 Gunter, T.E. and Pfeiffer, D.R. (1990) Am. J. Physiol. 258, C755–C786.
63 Miyata, H., Silverman, H.S., Sollott, S.J., Lakatta, E.G., Stern, M.D. and Hansford, R.G. (1991) Am. J. Physiol. 261, H1123–H1134.
64 Rizutto, R., Simpson, A.W.M., Brini, M. and Pozzan, T. (1992) Nature 358, 325–327.
65 Crompton, M. (1990) in: F. Bronner (Ed.) Intracellular Calcium Regulation, Wiley, New York, pp. 818–209.
66 Berridge, M.J. and Galione, A. (1988) FASEB J. 2, 3074–3082.
67 Woods, N.M., Cuthbertson, K.S.R. and Cobbold, P.H. (1986) Nature 319, 600–602.
68 Rooney, T.A., Sass, E.J. and Thomas, A.P. (1989) J. Biol. Chem. 264, 17131–17141.
69 Joseph, S.K., Coll, K.E., Cooper, R.H., Marks, J.S. and Williamson, J.R. (1983) J. Biol. Chem. 258, 731–741.
70 Sistare, F.D., Picking, R.A. and Haynes, R.C. (1985) J. Biol. Chem. 260, 12744–12747.
71 Staddon, J.M. and Hansford, R.G. (1987) Biochem. J. 241, 729–735.
72 Staddon, J.M. and Hansford, R.G. (1989) Eur. J. Biochem. 179, 47–52.
73 Burgess, G.M., Bird, St. J.G., Obie, J.F. and Putney, J.W. (1991) J. Biol. Chem. 266, 4772–4781.
74 Altin, J.G. and Bygrave, F.L. (1986) Biochem. J. 238, 653–661.
75 Bond, M., Vadasz, G., Somlyo, A.V. and Somlyo, A.P. (1987) J. Biol. Chem. 262, 15630–15636.
76 Borle, A.B. (1976) J. Membr. Biol. 29, 209–210.
77 Scarpa, A., Malmstrom, K., Chiesi, M. and Carafoli, E. (1976) J. Membr. Biol. 29, 205–208.
78 Romani, A., Dowell, E. and Scarpa, A. (1991) J. Biol. Chem. 256, 24376–24384.
79 Aprille, J.R. (1988) FASEB J. 2, 2547–2556.
80 Haynes, R.C., Picking, R.A. and Zaks, W.J. (1986) J. Biol. Chem. 261, 16121–16125.
81 Nosek, M.T., Dransfield, D.T. and Aprille, J.R. (1990) J. Biol. Chem. 265, 8444–8450.
82 Titheradge, M.A., Binder, S.B., Yamazaki, R.K. and Haynes, R.C. (1978) J. Biol. Chem. 253, 3357–3360.
83 Kraus-Friedmann, N. (1984) Physiol. Rev. 64, 170–259.
84 Sugano, T., Shiota, M., Khono, H., Shimada, M. and Oshino, N. (1980) J. Biochem. (Tokyo) 87, 465–472.
85 Balaban, J.S. and Blum, J.J. (1982) Am. J. Physiol. 242, C172–C177.
86 Quinlan, P.T. and Halestrap, A.P. (1986) Biochem. J. 236, 789–800.
87 Siess, E.A., Brocks, D.G., Lattke, H.K. and Wieland, O.H. (1977) Biochem. J. 166, 225–235.
88 Siess, E.A., Brocks, D.G. and Wieland, O.H. (1978) Biochem. Soc. Trans. 6, 1139–1144.
89 Soboll, S. and Scholz, R. (1986) FEBS Lett. 205, 109–112.
90 Halestrap, A.P. (1978) Biochem. J. 172, 399–405.
91 Hoek, J.B. (1981) Biochem. Soc. Trans. 9, 139P.
92 Strzelecki, T., Thomas, J.A., Koch, C.D. and LaNoue, K.F. (1984) J. Biol. Chem. 259, 4122–4129.
93 Strzelecki, T., Strzelecka, D., Koch, C.D. and LaNoue, K.F. (1988) Arch. Biochem. Biophys. 264, 310–320.
94 Ui, M., Exton, J.H. and Park, C.R. (1973) J. Biol. Chem. 248, 5350–5359.
95 Assimacopoulos-Jeannet, F., McCormack, J.G. and Jeanrenaud, B. (1986) J. Biol. Chem. 261, 8799–8804.
96 Rashed, H.M., Waller, F.M. and Patel, T.B. (1988) J. Biol. Chem. 263, 5700–5706.
97 Leverve, X.M., Verhoeven, A.J., Groen, A.K., Meijer, A.J. and Tager, J.M. (1986) Eur. J. Biochem. 155, 551–556.
98 Sugano, T., Nishimura, K., Sogabe, N., Shiota, M., Oyama, N., Noda, S. and Ohta, M. (1988) Arch. Biochem. Biophys. 264, 144–154.
99 LaNoue, K.F., Strzelecki, T. and Finch, F. (1984) J. Biol. Chem. 259, 4116–4121.
100 Hansford, R.B. (1985) Rev. Physiol. Biochem. Pharmacol. 102, 1–72.
101 Yamada, E.W. and Huzel, N.J. (1988) J. Biol. Chem. 263, 11498–11503.
102 McCcrmack, J.G. (1985) FEBS Lett. 180, 259–264.
103 Wan, B., LaNoue, K.F., Cheung, J.Y. and Scaduto Jr., R.C. (1989) J. Biol. Chem. 264, 13430–13439.
104 Johnston, J.D. and Brand, M.D. (1987) Biochem. J. 245, 217–222.

105 Sterniczuk, A., Hreniuk, S., Scaduto Jr., R.C. and LaNoue, K.F. (1991) Eur. J. Biochem. 196, 151–157.
106 Walajtys-Rode, E., Zapatero, J., Moehren, G. and Hoek, J.B. (1992) J. Biol. Chem. 267, 370–379.
107 Deaciuc, I.V., D. Souza, N.B. and Miller, H.I. (1992) Int. J. Biochem. 24, 129–132.
108 Hamatani, Y., Inoue, M., Kimura, K., Shiota, M., Ohta, M. and Sugano, T. (1991) Am. J. Physiol. 261, E325–E331.
109 Quant, P.A., Tubbs, P.K. and Brand, M.D. (1990) Eur. J. Biochem. 187, 169–174.
110 Davidson, A.M. and Halestrap, A.P. (1989) Biochem. J. 258, 817–821.
111 Dubnova, E.B. and Baykov, A.A. (1992) Arch. Biochem. Biophys. 292, 16–19.
112 Davidson, A.M. and Halestrap, A.P. (1987) Biochem. J. 246, 715–723.
113 Mansurova, S.E. (1989) Biochim. Biophys. Acta 977, 237–247.
114 Davidson, A.M. and Halestrap, A.P. (1988) Biochem. J. 254, 379–384.
115 Halestrap, A.P. and Dunlop, J.L. (1986) Biochem. J. 239, 559–565.
116 Halestrap, A.P. (1988) Biochem. J. 253, 622–623.
117 Cohen, N.S., Cheung, C.-W. and Raijman, L. (1988) Biochem. J. 253, 621–622.
118 Davidson, A.M. and Halestrap, A.P. (1990) Biochem. J. 268, 147–152.
119 Halestrap, A.P. and Davidson, A.M. (1990) Biochem. J. 268, 153–160.
120 Nicolli, A., Redetti, A. and Bernardi, P. (1991) J. Biol. Chem. 266, 9465–9470.
121 Inoue, I., Nagase, H., Kishi, K. and Higuti, T. (1991) Nature 352, 244–247.
122 Garlid, K. (1980) J. Biol. Chem. 255, 11273–11279.
123 Moreno-Sanchez, R. (1985) J. Biol. Chem. 260, 12554–12560.
124 Le Quoc, K. and Le Quoc, D. (1988) Arch. Biochem. Biophys. 265, 249–257.
125 McGuinness, O., Nasser, Y., Costi, A. and Crompton, M. (1990) Eur. J. Biochem. 194, 671–679.
126 Kinnally, K.W., Antonenko, Y.N. and Zorov, D.B. (1992) J. Bioenerg. Biomembranes 24, 99–110.
127 Pullman, M.E. and Monroy, G.C. (1963) J. Biol. Chem. 238, 3762–3769.
128 Murphy, E., Freudenrich, C.C. and Lieberman, M. (1991) Annu. Rev. Physiol. 53, 273–287.
129 Rutter, G.A., Osbaldeston, N.J., McCormack, J.G. and Denton, R.M. (1990) Biochem. J. 271, 627–634.
130 Leverve, X.M., Groen, A.K., Verhoeven, A.J. and Tager, J.M. (1985) FEBS Lett. 181, 43–46.
131 Gonzalez-Manchon, C., Saz, J.-M., Ayuso, M.S. and Parrilla, R. (1988) Arch. Biochem. Biophys. 265, 258–266.
132 Moreno-Sanchez, R., Hogue, B.A. and Hansford, R.G. (1990) Biochem. J. 268, 421–428.
133 Oppenheimer, J.H., Schwartz, H.L., Mariash, C.N., Kinlaw, W.B., Wong, N.C.W. and Freake, H.C. (1987) Endocr. Rev. 8, 288–308.
134 Gustafsson, R., Tata, J.R., Lindberg, O. and Ernster, L. (1965) J. Cell Biol. 26, 555–577.
135 Jakovcic, S., Swift, H.H., Gross, N.J. and Rabinowitz, M. (1978) J. Cell Biol. 77, 887–901.
136 Hoch, F.L. (1988) Prog. Lipid Res. 27, 199–270.
137 Nelson, B.D. (1987) Curr. Top. Bioenerg. 15, 221–272.
138 Sterling, K. (1979) N. Engl. J. Med. 300, 117–123, 173–177.
139 Nelson, B.D., Mutvei, A. and Joste, V. (1984) Arch. Biochem. Biophys. 228, 41–48.
140 Mutvei, A., Kuzela, S. and Nelson, B.D. (1989) Eur. J. Biochem. 180, 235–240.
141 Mutvei, A. and Nelson, B.D. (1988) Arch. Biochem. Biophys. 268, 215–220.
142 Joste, V., Goitom, Z. and Nelson, B.D. (1989) Eur. J. Biochem. 184, 255–260.
143 Leung, A.C.F. and McKee, E.E. (1990) Am. J. Physiol. 258, E511–E518.
144 Ruggiero, F.M., Gnoni, G.V. and Quagliariello, E. (1984) Lipids 22, 148–151.
145 Paradies, G. and Ruggiero, F.M. (1990) Biochim. Biophys. Acta, 1019, 133–136.
146 Paradies, G. and Ruggiero, F.M. (1989) Arch. Biochem. Biophys. 269, 595–602.
147 Robinson, N.C., Stry, F. and Talbert, L. (1980) Biochemistry 19, 3656–3661.
148 Beyer, K. and Klingenberg, M. (1985) Biochemistry 24, 3821–3826.
149 Rydstrom, J., Persson, B. and Tang, H.-L. (1984) in: L. Ernster (Ed.) Bioenergetics, Elsevier, Amsterdam, pp. 207–219.
150 Hoek, J.B. and Rydstrom, J. (1988) Biochem. J. 254, 1–10.
151 Nobes, C.D., Brown, G.C., Olive, P.N. and Brand, M.D. (1990) J. Biol. Chem. 265, 12903–12909.

152 Hoch, F.L. (1977) Arch. Biochem. Biophys. 178, 535–545.
153 Seitz, H.J., Muller, M.J. and Soboll, S. (1985) Biochem. J. 227, 149–153.
154 Hafner, R.P. (1987) FEBS Lett. 224, 251–256.
155 Horst, C., Rokos, H. and Seitz, H.J. (1989) Biochem. J. 261, 945–950.
156 Shears, S.B. (1980) J. Theor. Biol. 82, 1–13.
157 Soboll, S. and Sies, H. (1989) Methods Enzymol. 174, 118–131.
158 Segal, J. (1990) Endocrinol. 126, 2693–2702.
159 Segal, J. (1988) Biochem. 27, 2586–2590.
160 Segal, J. (1984) Endocrinology 115, 160–166.
161 Hummerich, H. and Soboll, S. (1989) Biochem. J. 258, 363–367.
162 Beleznai, Z., Szalay, L. and Jancsik, V. (1988) Eur. J. Biochem. 170, 631–636.
163 Thomas, W.E., Crespo-Armas, A. and Mowbray, J. (1987) Biochem. J. 247, 315–320.
164 Hafner, R.P., Nobes, C.D., McGown, A.D. and Brand, M.D. (1988) Eur. J. Biochem. 178, 511–518.
165 Brand, M.D., Couture, P., Else, P.L., Withers, K.W. and Hulbert, A.J. (1991) Biochem. J. 275, 81–86.
166 Luvisetto, S., Schmehl, I., Conti, E., Intravaia, E. and Azzone, G.F. (1991) FEBS Lett. 291, 17–20.
167 Hafner, R.P., Leake, M.J. and Brand, M.D. (1989) FEBS Lett. 248, 175–178.
168 Dierks, T., Salentin, A., Heberger, C. and Kramer, R. (1990) Biochim. Biophys. Acta 1028, 268–280.
169 Horrum, M.A., Tobin, R.B. and Ecklund, R.E. (1991) Biochem. Biophys. Res. Commun. 178, 73–78.
170 Hafner, R.P., Brown, G.C. and Brand, M.D. (1990) Biochem. J. 265, 731–734.
171 Sterling, K. (1986) Endocrinology 119, 292–295.
172 Rasmussen, U.B., Kohrle, J., Rokos, H. and Hesch, R.-D. (1988) FEBS Lett. 255, 385–390.
173 Mowbray, J. and Corrigall, J. (1984) Eur. J. Biochem. 139, 95–99.
174 Gregory, R.B. and Berry, M.N. (1991) Biochim. Biophys. Acta 1133, 89–94.
175 Holness, M., Crespo-Armas, A. and Mowbray, J. (1984) FEBS Lett. 177, 231–235.
176 Argov, Z., Renshaw, P.F., Boden, B., Winokur, B.A. and Bank, W.J. (1988) J. Clin. Invest. 81, 1695–1701.
177 Sandha, G.S., Steele, R. and Gonnella, N.C. (1991) Magn. Reson. Med. 18, 237–243.
178 Das, A.M. and Harris, D.A. (1990) Biochim. Biophys. Acta 1096, 284–290.
179 Gregory, R.B. and Berry, M.N. (1991) Biochim. Biophys. Acta 1098, 61–67.
180 Soboll, S., Horst, C., Hummerich, H., Schumacher, J.P. and Seitz, H.J. (1992) Biochem. J. 281, 171–173.
181 Mohan, C., Memon, R.A. and Bessman, S.P. (1991) Arch. Biochem. Biophys. 287, 18–23.
182 Gottschalk, W.K. (1991) J. Biol. Chem. 266, 8814–8819.
183 Maddaiah, V.T., Clejan, S., Palekar, A.G. and Collipp, P.J. (1981) Arch. Biochem. Biophys. 210, 666–677.
184 Kessler, M.A., Lamm, L., Jarnagin, K. and DeLuca, H.F. (1986) Arch. Biochem. Biophys. 251, 403–412.
185 Ku, C.Y., Lu, Q., Ussuf, K.K., Weinstock, G.M. and Sanborn, B.M. (1991) Mol. Endocrinol. 5, 1669–1676.
186 Van Italie, C.M. and Dannies, P.S. (1988) Mol. Endocrinol 2, 332–337.
187 Martens, M.E., Peterson, P.L. and Lee, C.P. (1991) Biochim. Biophys. Acta 1058, 152–160.
188 Peterson, P.L., Martens, M.E. and Lee, C.P. (1988) Neurol. Clin. 6, 529–544.
189 Brosnan, M.J., Chen, L., Wheeler, C.E., van Dyke, T.A. and Koretsky, A.P. (1991) Am. J. Physiol. C1191–C1200.
190 Yoshizaku, K., Watari, H. and Radda, G.K. (1990) Biochim. Biophys. Acta 1051, 144–150.
191 Siess, E.A., Kientsch-Engel, R.I., Fahimi, F.M. and Wieland, O.H. (1984) Eur. J. Biochem. 141, 543–548.

L. Ernster (Ed.) *Molecular Mechanisms in Bioenergetics*
© 1992 Elsevier Science Publishers B.V. All rights reserved

CHAPTER 19

The study of bioenergetics in vivo using nuclear magnetic resonance

GEORGE K. RADDA and DORIS J. TAYLOR

MRC Biochemical and Clinical Magnetic Resonance Unit, Department of Biochemistry, University of Oxford, Oxford OX1 3QU, U.K.

Contents

1. Introduction 464
2. The NMR technique 464
 2.1. General 464
 2.2. ^{31}P NMR of skeletal muscle 466
3. Oxygen delivery and consumption 467
4. The relationship between cellular energetics and work 468
5. Control of bioenergetics in vivo 469
6. Creatine kinase in muscle energetics 471
7. ^{13}C NMR and glycogen 472
8. Fatigue 473
 8.1. Non-metabolic fatigue 473
 8.2. ATP 473
 8.3. H^+ and $H_2PO_4^-$ 473
9. Plasticity of skeletal muscle 474
 9.1. Age-related changes 474
  9.1.1. Development 474
  9.1.2. Ageing 475
 9.2. Activity 475
  9.2.1. Training 475
  9.2.2. Denervation 476
 9.3. Disease 476
  9.3.1. Proton handling 476
 9.4. Hormone-induced changes 476
  9.4.1. Thyroid hormones 476
  9.4.2. Insulin 477
  9.4.3. Sex hormones 477
  9.4.4. Skeletal muscle metabolism in heart disease 477
10. Treatment 478
 10.1. Defects in ATP synthesis 478
 10.2. Heart failure 478
Acknowledgements 479
References 479

1. Introduction

Since the first demonstration in 1974 that high-resolution NMR spectra can be obtained from an intact piece of muscle using the phosphorus nucleus for detection [1], the technique has been developed to investigate perfused organs and organs in situ in animals exposed by surgical techniques. With the introduction of the so-called surface coils [2] spectra could be obtained from intact muscle in animals and humans and later from intact organs in humans. The investigation of human organs such as the heart and brain required the introduction of new methods in the NMR measurement to observe the information from a well-defined region or from a series of regions, thus obtaining some form of metabolic image of the system. In this chapter we shall discuss how NMR and in particular, phosphorus [^{31}P] NMR has been used to investigate the bioenergetics of intact muscle. We shall also refer to some investigations in other systems such as the heart and brain. The measurements provide information about the concentrations of specific metabolites. Thus we can investigate the time-course of metabolic changes following stress or intervention. We can also obtain temporal information about fluxes in vivo of some of the main reactions involved in the bioenergetics of muscle. In addition, one can learn something about the control of bioenergetics in vivo and the role of ions, such as hydrogen ions, in such control. The regulation of intracellular pH is also important in this respect. No attempt is made to give a comprehensive review but rather to illustrate with specific examples, the new biochemical information that has been obtained non-invasively that is particularly relevant to our understanding of bioenergetics in vivo.

In the study of humans, the variety of clinical conditions, genetic and metabolic defects provide an additional opportunity to investigate the energetic response of muscle in such cases and to use the information derived from them to enhance our understanding, both of the control processes in the normal state and the abnormalities that arise as a result of the diseased state. Since NMR provides a reproducible, non-invasive and quantitative way of carrying out such measurements, one can investigate human variabilities, development and mutations in a way that could not have been possible in the study of animal models. This adds a new dimension to the role of NMR investigations in bioenergetics.

2. The NMR technique

2.1. General

We shall summarize here only basic aspects and terms directly relevant to the discussion in this chapter. For a more detailed description of both principles and practical aspects of NMR readers are referred to refs. [3] and [4].

This technique exploits the phenomena of charge and spin which together confer magnetic properties on some atomic nuclei. Isotopes used for investigation of biological samples include ^{1}H, ^{13}C, ^{15}N, ^{17}O, ^{19}F, ^{23}Na and ^{87}Rb. ^{31}P, the naturally occurring isotope of phosphorus, has been widely used for the study of bioenergetics in perfused tissues and in intact animals or human subjects. ^{31}P spectra from human muscle in vivo are shown in Fig. 1 and are used here to illustrate the general features of NMR data.

NMR signals are generated as the result of the interaction of the nuclear spins and an externally applied pulse of radio frequency. Each nucleus with a magnetic moment precesses at a rate $\omega_o = \gamma B_o$, where ω_o is expressed in rad/s, γ is the gyromagnetic ratio (a characteristic constant for each nucleus) and B_o is the strength of the magnetic field. The field actually

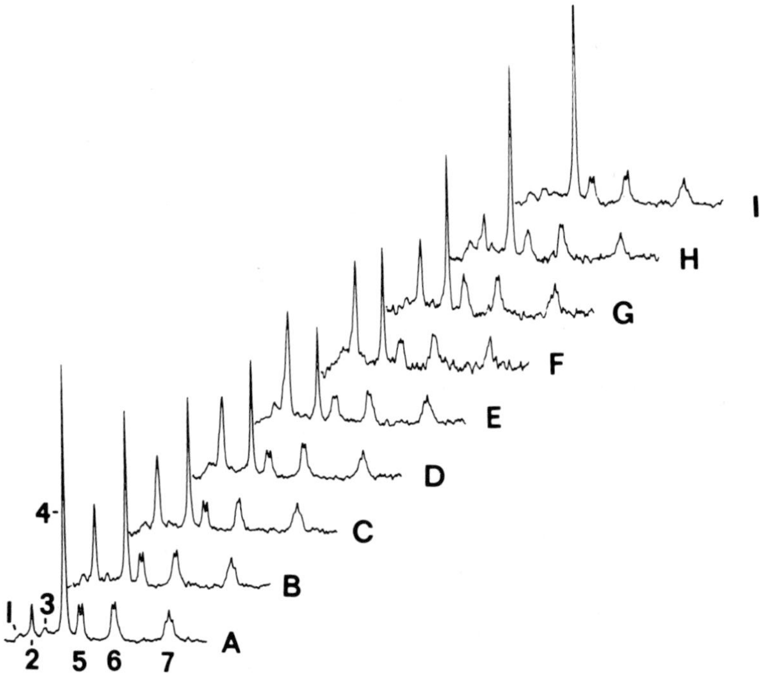

Fig. 1. [31]P spectra from human *gastrocnemius* muscle in vivo. Data were collected using a 6 cm diameter surface coil with an 80 μs pulse length and a 2 s interpulse delay. The 1.9 T magnet (Oxford Magnet Technology) was interfaced to a Bruker spectrometer. On the *x* axis is the chemical shift in parts per million, and on the *y* axis the signal intensity. Peak assignments: (1) phosphomonoesters (PME); (2) P_i; (3) phosphodiesters (PDE); (4) phosphocreatine (PCr); (5) γ phosphate of ATP; (6) α-ATP + NADH and NAD^+; (7) β-ATP. The spectra show muscle (A) at rest, (B–E) during aerobic, dynamic exercise and (F–I) in recovery from exercise. The number of accumulations for each spectrum was (A) 64, (B–E,I) 32, (F,G) 8, (H) 16. The pH_i at rest (A) was 7.03, decreasing to 6.74 at the end of exercise (E).

experienced by a nucleus is modified by the shielding effect of nearby electrons, so that $\omega_o = \gamma B_o(1-\sigma)$, where σ is the so-called shielding constant. The result, illustrated in Fig. 1 for ^{31}P, is that nuclei in different chemical environments rotate at different frequencies. Thus there are several peaks along the *x* (frequency) axis from the phosphate groups in different compounds (e.g. phosphocreatine (PCr) and P_i) or different groups in the same compound (α, β and γ, phosphate groups of ATP).

The peak position (ν_{obs}) is usually calculated relative to the position of a standard (ν_{std}), expressed independently of the magnetic field as 'chemical shift' in parts per million: $[(\nu_{std} - \nu_{obs})/\nu_{std}]10^6$ ppm. In many experiments in vivo, the standard used is one of the constituents of the tissue itself. In ^{31}P spectra this is usually PCr, which is assigned a chemical shift of 0 ppm. The intracellular pH (pH_i) is calculated from the chemical shift difference between PCr and P_i. In the physiological pH range, the chemical shift of PCr is constant, but the position of P_i (pK_a ~6.72) depends on the relative proportions of HPO_4^{2-} and $H_2PO_4^-$. These two phosphate species have a chemical shift difference of ~2.3 ppm, but because they are in fast exchange (~10^9–10^{10} s^{-1}), only a single resonance is observed. Thus, in the experiment shown in Fig. 1, the relative position of the P_i peak was shifted to the right during exercise, indicating a decrease in pH_i from 7.03 (spectrum A) to 6.74 (spectrum E).

The NMR technique depends on inducing transitions between adjacent nuclear orientations, or energy states. The number of possible energy states is restricted to $2I + 1$, where I is the nuclear spin angular momentum quantum number. Thus a nucleus such as ^{31}P with spin $\frac{1}{2}$ will have two possible orientations, with or against the external field, B_0. Transitions between energy states are induced with an oscillating magnetic field, B_1, applied as a radiofrequency pulse (~20–2000 μs) at right angles to the B_0 field. High-to-low and low-to-high transitions take place with equal probability. Nuclei absorb energy at their rotational frequencies, but net absorption of energy depends on a population difference in the different energy states. Increasing the B_0 field strength enhances the sensitivity of the method by increasing the population difference, but in spite of the use of high-field magnets, sensitivity remains low. As a consequence, signals from a series of pulses are often summed before data processing. For example, each spectrum in Fig. 1 is composed of signals from eight or more pulses.

When the radiofrequency pulse is terminated, nuclei relax to their equilibrium positions by two mechanisms with time constants T_1 and T_2. T_1 is the reciprocal first-order rate constant for transfer of nuclei from one energy state to another. In biological samples T_1 may be as long as several seconds. T_2 is the transverse relaxation time, related to the linewidth of a signal (ν in Hz) at half height, $\nu = 1/\pi T_2$. T_2 decreases as molecular motion is restricted, broadening signals so that tightly bound molecules are not detectable. This is the case for ADP in skeletal muscle of which perhaps 90% of the total concentration of 1–2 mM is bound to the myofibrils.

As nuclei relax, an electromotive force is induced in the receiver coil (which may also have been used as the transmitter). This free induction decay, or f.i.d., is amplified, stored and possibly added to other f.i.d.s. before being Fourier transformed to a plot of signal intensity against frequency (Fig. 1). Additional data processing techniques are often used for baseline correction and improving signal-to-noise.

Quantification is carried out by integrating the peak areas and by expressing them as ratios or by relating the peak area to a known concentration of metabolite. When data are collected under appropriate conditions or when the proper corrections are made, peak areas in the spectrum are proportional to concentration. The minimum concentration of any compound which can be detected will depend on a number of factors including sample size, magnet field strength and time available for data accumulation.

Defining the region from which spectra are collected may be as simple as placing the correct size of surface coil over the muscle of interest (Fig. 1). More complex methods have been developed using B_0 and B_1 gradients to encode the nuclear spins according to position in the sample [4,5]. Detailed description of these methods is outside the scope of this chapter. As an example, we shall discuss later how the human heart has been investigated.

2.2. ^{31}P NMR of skeletal muscle

The energetics and metabolism of skeletal muscle is now readily investigated by ^{31}P NMR and also ^{13}C and ^{1}H NMR. Sufficient signal-to-noise is obtained in time intervals of anywhere between a few seconds and a few minutes of signal accumulation depending on the nature of the study and the particular muscle group which is being examined. In general, both in animal and human experiments, surface coils are placed over the muscle group of interest, and that is the main method of spatial selection. Metabolic information can then be recorded at rest, during exercise, and in the period of recovery after exercise (Fig. 1). Thus, the time courses of changes associated with exercise and recovery are recorded. In humans, the two main muscle groups studied are the forearm muscle, *flexor digitorum superficialis* and the *gastrocnemius* muscle in the leg, although other muscle groups have also been examined. In most animal experiments, it is the *gastrocnemius* muscle of the hind leg of the animal that has been investigated. In humans, a variety of different exercise regimens have been used, from dynamic steady state to graded dynamic exercise and ischaemic and isometric exercise. In most animal investigations, stimulation of the sciatic nerve is used to alter muscle energetics. Thus in a typical study shown in Fig. 1, during exercise, one can observe the change in phosphocreatine and the increase in inorganic phosphate and measure, at the same time, intracellular pH and its decrease resulting from glycogenolysis. During the recovery period the rate of phosphocreatine resynthesis provides a measure for the rate of oxidative phosphorylation, as PCr recovery tracks the rate of ATP turnover. Under these conditions, glycogenolysis makes no contribution to ATP synthesis. The characteristic parameters one obtains are the extent of phosphocreatine utilisation and pH change in relation to a particular workload, the rate of PCr resynthesis, the rate of inorganic phosphate recovery and the rate at which pH is restored to its original value after exercise [6].

The study of cardiac muscle in most animal experiments is done in an open-chested model, by placing the surface coil over the heart, while this study in humans, as it will be shown later, requires some kind of spatial selection of the metabolic signal.

3. *Oxygen delivery and consumption*

One of the fundamental questions in the understanding of bioenergetics in vivo is how the demand for ATP utilisation is linked to increased supply of energy via oxidative metabolism and therefore to increased rate of oxygen delivery to the tissue. Oxygen delivery to the mitochondrion within the intact cells has three components: (i) the oxygen carrying capacity of blood, i.e. the nature and amount of haemoglobin present; (ii) the rate of flow of blood through major blood vessels; (iii) capillary density and diffusion from the capillary to the mitochondrion. Despite much study, the mechanisms by which cytosolic reagents control oxygen consumption in mitochondria remain controversial. Studies on frog sartorius muscle [7] or on vascularly isolated muscles in situ [8] have attempted to correlate oxygen consumption in the muscle with measured biochemical parameters such as concentrations of phosphorylated metabolites, lactic acid and pH, obtained from invasive sampling procedures. Recently, several groups have investigated the importance of the oxygen supply as a limiting factor for muscle performance during contractions and recovery from contractions using the perfused rat hind limb model and ^{31}P NMR. Correlations were found between oxygen delivery and oxygen consumption, lactate release, and glucose uptake respectively. ^{31}P NMR showed a correlation between oxygen delivery and the steady-state level of PCr over P_i during the contraction period. The rate of recovery of PCr and P_i after the contraction was also dependent on oxygen delivery [9,10]. A similar preparation was used to determine the relationship of O_2 consumption to changes in cellular bioenergetics during hypoxaemia. The ratio of change in vascular resistance to the corresponding decrease in O_2 transport was taken as an index of vascular autoregulation with hypoxaemia. Different patterns of regulation were observed, in which control of oxygen consumption and the degree of vascular autoregulation contributed to different extents in groups of animals. However, decreases in oxygen consumption were always accompanied by increases in inorganic phosphate and lactate and decreases in PCr, indicating oxygen supply and limitation and anaerobic ATP production. Furthermore, there was no evidence of cellular adaptation to hypoxia by decreasing energy needs under these conditions [11]. Muscle energetics under acidosis during static exercise with calf vasoconstriction was examined recently in humans [12]. Static exercise at 30% of maximal voluntary contraction caused a rise in heart rate, blood pressure, and calf vascular resistance. During a 3 min forearm occlusion after static exercise the heart rate returned to base-line, the increase in blood pressure was attenuated by 30% and calf vascular resistance remained elevated or unchanged. The percentage change in calf vascular resistance was correlated with forearm cellular pH but was only weakly associated with the PCr/P_i ratio. It was concluded that there was an association between forearm cellular acidosis and calf vasoconstriction during the static forearm exercise and large changes in PCr/P_i without concomitant changes in pH were not associated with changes in calf resistance.

The classical clinical condition in which oxygen delivery to the muscle is impaired is peripheral vascular disease, i.e. claudication. Several groups have examined calf muscle metabolism at rest and during exercise during an exercise protocol of plantar flexion of the foot in patients with claudication [13]. During exercise, in the patients with severe claudication, PCr utilisation and intracellular acidosis were greater than in controls and in mild claudicants. After exercise, PCr and P_i recovery rates were slow in the severely effected patients. Although all these parameters measured during exercise did not differentiate the patients with mild disease from normal controls, the rate of recovery of ADP was found to be much slower even in the patients with mild or moderate disease. These observations show that during exercise the patients with impaired blood flow and therefore impaired oxygen availability, depended to a greater extent on glycolytic mechanisms for energy production as demonstrated by a larger drop in intracellular pH. Recovery of PCr at the end of exercise is by a purely oxidative process and can be used as a measure of the mitochondrial oxidative process. It is not surprising that these recovery rates are sensitive in disease to reduced flow and oxygen delivery. Since the calculated ADP recovery depends on both PCr and pH it is therefore a more specific measure of mitochondrial oxidative activity [14]. Many other human diseases produce anaemia and hypoxia, but the situation is often complex in that for example, a reduction in haemoglobin in anaemia causes an increase in cardiac output and therefore compensates for the reduced availability of oxygen by increasing flow. For a summary see ref. [6].

4. *The relationship between cellular energetics and work*

^{31}P NMR is particularly suited for the examination of the relationship between work output and energetics in vivo. This has been achieved in animal models using preferentially sciatic nerve stimulation inducing tetani at different frequencies and with different time intervals. In humans, a variety of experimental protocols have been worked out. Some studies have used the steady-state or graded work rate from a constant load or a Cybex ergometer, while others have used force produced in an isometric contraction. Different muscle groups have been examined in this way, but the two major groups of muscles of study were the forearm muscle (*flexor digitorum superficialis*) and leg muscle (*gastrocnemius*). Chance and coworkers [45] have examined the relationship between reaction velocities (e.g. ATP utilization) and tissue work rates to the concentration of regulatory molecules in skeletal muscle. They examined the forearm muscle and measured values of P_i/PCr and the work by an ergometer in a graded metabolic load situation. They expressed the relationship between work and velocities as a 'transfer function'. The hyperbolic relationship between velocity and P_i/PCr is expressed by the function:

$$\frac{V}{V_{max}} = \frac{1}{1 + 0.6/[P_i]/[PCr]}.$$

The form of this transfer function varies between normal individuals and well-trained athletes and can be used as an indication of exercise performance. This arises because the equation above must be modified to take into account a circulatory adaptation to exercise assuming that the V_{max} is adjusted in accordance with the energy demand: thus $V_{max} = k(P_i/PCr)$. For values of P_i/PCr greater than K_m the transfer function is a straight line of slope k which is a measure of the oxygen delivery to the tissue [15]. Boska studied the ATP cost of force production in the human *gastrocnemius* muscle using ^{31}P NMR [16]. He used an exercise protocol of isometric maximum voluntary contraction and estimated the contributions to ATP production from three different processes. The rate of change of PCr was used to estimate ATP production rates from creatine kinase rates during exercise and from oxidative phosphorylation during the first ten seconds of recovery. The anaerobic glycolysis was estimated from the rate of change of pH from an assumed buffering capacity and hydrogen ion production stoichiometry. The results showed that by the end of 30 s of exercise the total ATP production and ATP cost of force production had stabilised and remained constant until the end of a two-minute period. It was also shown that the ATP cost of force production is lower in the first second than at any other time or possibly that ATP production is underestimated at later time points.

Another experimental study has shown that for submaximal twitch stimulation rates in rat skeletal muscle, the time constants for phosphocreatine changes are independent of work rate and are similar at the onset of stimulation and during recovery afterwards. Furthermore, the relationship between steady-state PCr level and the rate–force product is linear. This is consistent with the model that assumes that muscle oxygen consumption is proportional to the rate–force product for submaximal stimulation rates. It was argued that these results are consistent with a simple first-order electrical analogue model of oxidative metabolism that is applicable to submaximal oxidative rates. The model assumes equilibrium of the creatine kinase reaction, which is modelled as a chemical capacitor with capacitance proportional to the total creatine level and PCr level proportional to the cytosolic free energy of ATP hydrolysis [17].

5. *Control of bioenergetics in vivo*

While the various pathways involved in the energetics of skeletal and cardiac muscle are well mapped out, the way they are controlled in the intact cell is less well understood. NMR studies in vivo have contributed a great deal to our understanding of some of the control functions, particularly in skeletal and cardiac muscle.

In skeletal muscle, mitochondrial oxidation in vivo was shown to depend, in a hyperbolic manner, on the concentration of ADP with a K_m of about 30 μM. This value is very close to that reported by Chance and Williams for isolated mitochondria. Three types of experiments have been done to arrive at this conclusion.

(i) Chance and his colleagues have shown that in a graded exercise protocol in which the work is measured with an ergometer, during performance going from rest (state 4) to exercise (state 3) the transfer function, as defined in the previous section, approximates a rectangular hyperbola (i.e. the plot of work against P_i/PCr is hyperbolic), giving a K_m for ADP of 28 μM. This follows from the fact that if there are no pH changes as is the case in this relatively submaximal protocol used, the P_i/PCr ratio is proportional to the concentration of ADP [15].

(ii) It was shown that the relationship between ADP concentration reached at the end of exercise, and the initial rate of phosphocreatine resynthesis during recovery in human arm muscle, also follows the same relationship, giving a K_m value of 27 mM and a V_{max} for the ADP synthesis rate of 43 mM min^{-1} [18].

(iii) In a different approach ^{31}P NMR magnetisation transfer measurements have been used to measure the flux between ATP inorganic phosphate during steady-state isometric muscle contraction in the rat hind limb in vivo as a function of work [19]. (For a brief description of magnetisation transfer, see below.)

Comparison of the measured ATP synthase fluxes in vivo with those calculated from previous measurements of oxygen consumption or phosphocreatine breakdown during a tetanus, indicated that the measured flux is due predominantly to the activity of the mitochondrial F_1F_0-ATP synthase. The ATP-synthesis rate shows approximately linear dependence with free ADP concentration in muscle up to an ADP concentration of about 90 μM, implying that mitochondrial ATP generation in the muscle is controlled by the free ADP concentration with an apparent K_m of at least 30 μM. The measurements also show a linear dependence of the flux between P_i and ATP with the tension time integral obtained at different pulse frequencies or at a given frequency at different pulse widths of supramaximal sciatic nerve stimulation.

In contrast to skeletal muscle, control of respiration in the heart is dependent upon the substrate that supports oxidative phosphorylation. This field has been reviewed by Heineman and Balaban [20]. There have been a number of reports of studies on the intact, perfused mammalian heart under a variety of different conditions, in which no correlations between tissue concentrations of high-energy phosphates and rates of work of oxidative metabolism could be observed [21–25]. The most notable feature of these studies was the remarkable stability of the phosphorylation potential in the face of varying workloads, suggesting that additional factors may be involved in the control of oxidative phosphorylation in the mammalian heart. Positive ionotropic stimulation by isoprenaline of perfused rat hearts produced no changes in ATP and only transient decreases in PCr and transient increases in ADP and P_i. However, the concentrations of all these metabolites returned to pre-stimulated values in 1 min, whereas cardiac work and O_2 uptake remained elevated. Ruthenium red, an inhibitor of mitochondrial calcium uptake, altered this pattern, in that administration of isoprenaline now caused significant decreases in ATP and also much larger and more prolonged changes in the concentrations of ADP, PCr and P_i. It was proposed that this is consistent with a control that is associated with increases in intramitochondrial calcium, activating the calcium sensitive mitochondrial dehydrogenase, leading to an enhanced NADH reduction [26]. Several whole-animal preparations have been examined by NMR and in dog and sheep heart experiments increase in the cardiac work by electrical or initropic stimulation produced no changes in any phosphate or pH, except at the highest workloads. Interestingly, ^{31}P NMR studies of young lambs show a large decrease in PCr and increases in P_i with increased work in contrast to the adult sheep, indicating that there are changes in the regulation of energy metabolism in the developing heart [27].

The PCr/ATP ratio can now also be measured in the human heart, using special localisation or spectroscopic imaging techniques [28,29]. A one-dimensional metabolic image of the human heart as recorded by the technique of rotating frame imaging is shown in Fig. 2. In normal subjects the PCr/ATP ratio remained constant when the heart was stressed, either by an isometric hand-grip exercise [30] or by a dynamic leg exercise performed in the prone position [31]. The PCr/ATP ratio was found at rest to be 1.5±0.12 and during exercise 1.58±0.14 (SD). Thus, in the normal situation, the heart can adequately regulate its phosphates over a range of

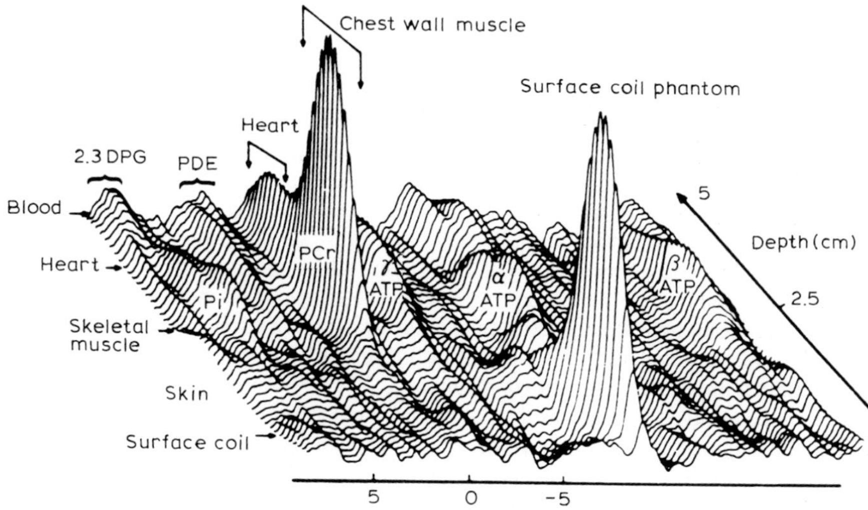

Fig. 2. Stacked plot of spatially resolved spectra. Height corresponds to amount. Position on the horizontal axis identifies the metabolite (cf. Fig. 1). Position on the axis from front to back corresponds to space, starting from the surface coil which contains the phantom and extending through the chest wall and heart muscle to ventricular cavity blood. Each row in the spectrum represents a slice of increasing depth. The PCr peaks labelled 'heart' correspond to the signal from the apical region of the myocardium.

workloads. The results show that free ADP is not the primary regulator of increased ATP synthesis in the normal heart in these situations.

6. *Creatine kinase in muscle energetics*

The role of creatine kinase in skeletal muscle, heart and brain has been debated for some years. Two apparently distinct functions have been proposed for this enzyme: (i) the classical one, in which phosphocreatine acts as an energy reserve, buffering fluctuations in the cytosolic ATP or ATP/ADP ratio, and (ii) the so-called energy shuttle hypothesis, in which creatine kinase bound to the outer side of the inner mitochondrial membrane is functionally coupled to the mitochondrial ATP synthase, providing a special mechanism of transferring energy from newly generated ATP in the mitochondrion to the cytosol (for a review see ref. [32]). One of us has proposed [18] that the role of creatine kinase is to regulate the level of ADP through the equilibrium reaction, ADP being a positive effector of mitochondrial oxidative phosphorylation, and a negative effector of contractility at high concentrations.

One of the major advantages of NMR in cellular studies is that it can provide dynamic information about intracellular events, and give a measure of reaction fluxes catalysed by enzymes whether in the steady state or at equilibrium. The special technique used is called 'magnetisation transfer' [19,33]. These techniques essentially involve a kind of magnetic pulse labelling by irradiating the nucleus in the chemical of interest that makes it possible to follow chemical exchange between two or more species. The study of creatine kinase has provided one of the best examples for the use of this approach in biochemical investigations.

^{31}P NMR magnetisation transfer has been used to measure the flux between phosphocreatine and ADP and, as mentioned earlier, between ATP and inorganic phosphate, during steady-state isometric muscle contraction in the rat hind limb in vivo [19]. Steady-state contraction was obtained by supramaximal sciatic nerve stimulation. Increasing the stimulation pulse width from 10 to 90 ms, at a pulse frequency of 1 Hz, or increasing the frequency of a 10 ms pulse from 0.5 to 2 Hz, did not affect the flux through creatine kinase, suggesting that this enzyme is not acting as a shuttle for phosphocreatine between the mitochondrion and cytoplasm in skeletal muscle in these particular circumstances. It is possible that there is a role for the transport function of PCr (as a spatial buffer) in large muscle cells, establishing the relationship between fibre diameter, mitochondrial content and substrate diffusion [34].

The situation in the perfused heart may well be different, although again there are conditions in which magnetisation transfer measurements on creatine kinase fluxes indicate that the shuttle may not always be an obligatory intermediate in the transport of energy from the mitochondrion to the cytoplasm, involving phosphocreatine as the diffusible substrate [35].

The same kind of fluxes can be measured in humans, both in the muscle and brain, and in the latter, can even be resolved spatially using a combination of the rotating frame experiment (a form of metabolic imaging) with the saturation transfer measurements [33]. We have carried out a detailed study of the creatine kinase activity as a function of its location within the human brain and have shown that the distribution of the phosphorus-containing metabolites (i.e. PCr, ATP and P_i) as well as of pH is independent of the depth on which the slice is taken, i.e. there is no difference between grey and white matter. In contrast, the activity of creatine kinase is two times smaller in the white matter than in the grey matter.

7. *^{13}C NMR and glycogen*

A major new approach to the study of substrate utilisation in bioenergetics is provided by the use of ^{13}C NMR and in particular in the use of ^{13}C-enriched substrates, which can lead to measurements of individual reaction rates or fluxes through specific pathways (for a recent review of ^{13}C NMR see Cerdan and Seelig [36]).

For example, the rate of incorporation of carbon from [1-^{13}C]glucose into the [4-CH_2]- and [3-CH_2]glutamate was measured in the rat brain in vivo [37]. Since glucose carbon is incorporated into glutamate by rapid exchange with α-ketoglutarate, the tricarboxylic acid cycle intermediate, the flux of carbon through this cycle could be determined from the rate of glutamate labelling.

An important new observation for muscle investigations was that, in spite of its high molecular weight, glycogen is fully visible by ^{13}C NMR in organs and tissues [38]. Thus the synthesis and breakdown of glycogen during muscle exercise can now be followed for the first time non-invasively in human muscle. There are many opportunities to study fundamental questions of the regulation of carbohydrate metabolism under a variety of conditions. One example of this was the pioneering study in which the rate of human muscle glycogen formation was measured in normal and diabetic subjects, using infused, isotopically labelled [1-^{13}C]glucose and ^{13}C NMR [39]. The important conclusions from these measurements were that (i) synthesis of muscle glycogen accounted for most of the total body glucose uptake and all of the non-oxidative glucose metabolism and (ii) that in subjects with non-insulin dependent diabetes the rate of glycogen formation was decreased by 60% compared to the rate in normal controls.

8. *Fatigue*

Fatigue can be defined in several different ways. In the investigations we discuss, the term is used to denote the inability to sustain a given force or as the loss of force-generating capacity. The site at which the functional impairment resides may be in the central or peripheral nervous system or be due to changes in the sarcolemma or T tubule membranes or within the muscle fibres themselves. Under what conditions and to what extent metabolic changes cause or contribute to fatigue has been studied extensively by ^{31}P NMR. Muscle metabolite concentrations and pH_i are monitored continuously, so that the changes which occur during and following exercise can be correlated directly with the simultaneously acquired physiological measurements. The involvement of H^+ and the other metabolites which can be measured either directly or indirectly by NMR, can be difficult to establish. This is in large part because creatine, PCr, H^+ and ADP are all related through the creatine kinase equilibrium, so it is difficult to dissociate changes in the concentration of one from changes in the others.

8.1. Non-metabolic fatigue

In some investigations, ^{31}P NMR has been used to rule out metabolic changes as the cause of fatigue. The declining twitch tension and delayed recovery in a study of low intensity exercise in human *adductor pollicis* and *tibialis anterior* were found not to have a metabolic basis and were ascribed to an inhibition in excitation–contraction coupling [40]. In a study of the *latissimus dorsi* of beagles, recovery of the PCr/P_i ratio back toward the resting condition during fatigue at 85 Hz, but not at 25 Hz, led to the conclusion that this high-frequency fatigue was not due to metabolic causes [41]. Three components of fatigue were found during maximal voluntary contraction (MVC) of *adductor pollicis* [42]. One, indicated by a decreased MVC, correlated with the metabolic state of the muscle as observed by NMR, but another was associated with alterations in the muscle membrane, and the third was probably due to impaired excitation–contraction coupling.

8.2. ATP

A decrease in energy supply has not been implicated as a metabolic cause of fatigue. Even if ATP depletion is induced by exhausting exercise, the concentration decreases by about only one third [43], and muscle fatigue associated with metabolic factors can be demonstrated before there is measurable ATP loss. In fact, fatigue probably serves an important purpose in helping to protect muscle by decreasing contractile activity, thereby lowering demand for ATP and preventing excessive changes in energy supply which would inhibit essential energy-requiring processes and threaten cell survival.

8.3. H^+ and $H_2PO_4^-$

Relationships between acidosis and fatigue can be demonstrated convincingly in vitro, but this does not prove that H^+ is a direct inhibitor of contraction in vivo. Wilkie [44] re-analysed previously published NMR data from frog muscle and, using pH_i values and P_i concentrations, showed that the increase in $H_2PO_4^-$ was inversely proportional to the force of contraction. Thus, he suggested, an increase in proton concentration might act by increasing the proportion of P_i in the diprotonated form. Subsequently, other investigators have shown a correlation between fatigue and the concentration of $H_2PO_4^-$ in vivo. A study of metabolite concentration and the fatigue produced by sustained MVC for 4 min in *adductor pollicis* and *tibialis anterior* showed that there was an approximately linear relationship between MVC and the accumulation of H^+ and $H_2PO_4^-$, but the changes in MVC and the concentration of PCr or total P_i were non-linear [45]. In addition, there was a good correspondence between the time course of the recoveries of MVC and $H_2PO_4^-$, but H^+ recovered more slowly. Experiments on wrist flexors showed that in intense, short-term exercise developed force and $H_2PO_4^-$ concentration maintained the same relationship during different exercise protocols, while developed force and pH_i did not [46].

Other results suggest that H^+ and phosphate cannot provide all of the answers. For example, it can be shown that an increase in $[H^+]$ is not essential for development of fatigue, since patients deficient in muscle phosphorylase fatigue easily with no decrease in pH_i [47]. It has been suggested that the very high concentrations of free ADP (~200 μM), which are not achieved in normal muscle, might inhibit contraction in this condition [48]. Some results also call into question a causal relationship between diprotonated phosphate and force development, at least under all conditions. In a study of fast- and slow-twitch cat muscles, the peak tetanic force was measured in the isolated, perfused muscles, and the pH_i of the muscle decreased by tetanic stimulation or with hypercapnea. The results showed that there was no consistent correlation of peak tetanic tension with either pH_i or $H_2PO_4^-$ [49]. In a different investigation, significant changes were induced in pH_i, PCr, P_i, ADP and $H_2PO_4^-$ by voluntary contraction of first *dorsal interosseous* in normal human subjects, but none of them correlated with the decrease in MVC [50].

Relaxation time, which in skinned fibres is slowed by low pH, is another parameter of fatigue that has been studied in intact muscle by NMR. In order to assess the contribution of H^+ to the slowing of relaxation, fatigue was induced in first *dorsal interosseous* by stepwise ischaemic exercise in normal subjects and in one with muscle phosphorylase-deficiency (in which muscle fails to acidify) [51]. The relaxation rate was measured from tetanic contractions applied between the contractions. Slowing of relaxation in the patient could not have been due to changes in H^+ because there was a 50% reduction in rate without a change in pH_i. When circulation was restored, recovery of the relaxation rate was rapid in the phosphorylase deficient muscle, but pH_i remained low and relaxation slow in the normal subjects. It was concluded that there were at least two mechanisms responsible for the slow relaxation, one associated with H^+ accumulation and the other independent of it. Results from isolated cat muscle were also consistent with a negative effect of pH_i on relaxation in both fast and slow muscles and in the rise of tension in the slow-twitch muscle [49].

9. *Plasticity of skeletal muscle*

One of the major features of skeletal muscle is its ability to change and adapt, not just during development, but also throughout adult life. Properties which can change with age, type and degree of exercise and disease state include motor unit size, fibre type and contractile characteristics. Accompanying these changes are alterations in the bioenergetics of the muscle, some of which have been investigated under in vivo conditions by NMR.

9.1. *Age-related changes*

9.1.1. *Development*

Changes in the energetics of developing muscle have been demonstrated in both animals and humans, and the results suggest an increase in phosphorylation potential during post-natal development. Investigations of resting hind limb muscle of normal mice have shown that pH_i and signals from P_i decreased and those from PCr increased dramatically with muscle development after birth [52,53]. These changes were all complete by 30 to 50 d of age, and no further significant changes were noted in animals up to 150 d. In ten boys aged 5 to 15 yr, ATP content increased with age in resting *gastrocnemius* [54]. In the small number of children examined, no significant age-related differences in PCr, P_i or pH_i were seen, but in our own laboratory we found that 11–13 yr old children (n=9) had not reached adult ratios of PCr/P_i or PCr/ATP in finger flexors. These were lower by 11 and 15%, respectively, than in adult muscle, while P_i/ATP and pH_i were at adult levels (unpublished).

In one of the studies on mice [53], age-related changes were compared in normal and dystrophin-deficient animals. It was found that the characteristic pattern seen in older dystrophin-deficient mice (high pH_i, low PCr/P_i and high P/ATP) was not present in the young animals and did not begin to emerge until the mice were about 50 d old. The abnormalities seen in the mice are also those found in Duchenne dystrophy [54,55], in which dystrophin is also absent in muscle. The results from Duchenne boys aged 6–15 years are in general agreement with the findings in the mice: decreases in PCr signal and in PCr/P_i ratio become more prominent with age [54].

9.1.2. Ageing

Although disease and inactivity in old age are likely to affect muscle energetics adversely, the muscles of healthy active elderly subjects may be little affected by the aging process. The pH_i and relative concentrations of ATP, PCr en P_i did not vary in resting finger flexor muscle of six men and six women between the ages of 20 and 80 yr [56]. Changes in PCr and pH_i during aerobic, dynamic exercise were similar in 70–80 yr old and 20–45 yr old subjects. PCr recovery times were not different in the two groups, suggesting that there was no functional impairment of oxidative metabolism in the elderly subjects.

9.2. Activity

9.2.1. Training

Regular stimulation of muscle by exercise training is well known to lead to changes in skeletal muscle. Endurance training, such as long distance running, increases the resistance to fatigue, the proportion of slow contracting fibres (types I and IIa) at the expense of the fast contracting fibres (type IIb), the volume fraction of mitochondria and the activity of oxidative enzymes. Strength or speed training does none of these, but can increase the proportion and size of the type IIb (glycolytic) fibres.

Trained and untrained muscles cannot be differentiated at rest by ^{31}P NMR. Even the severely detrained muscles of forearms recently released from plaster casts showed no significant differences in pH_i and metabolite ratios from those of the contralateral, uncasted arms [57]. Not surprisingly, however, training can be seen to affect the bioenergetics of contracting muscle. Endurance training or chronic electrical stimulation designed to increase the proportion of type I fibres leads to a smaller decrease in PCr (and pH_i) during exercise [58–61]. Such a difference can even be seen without special training in the dominant versus the non-dominant arm [62]. In one study designed to investigate the factors involved in the biochemical changes induced by training [58], muscles were exercised at intensities below the threshold needed to induce cardiovascular adaptations. The more modest changes in PCr and P_i in response to exercise which occurred after training must have been independent of muscle mass, blood flow and $VO_{2\,max}$ because these remained unchanged.

9.2.2. *Denervation*

When the influence of nervous stimulation is removed from muscle there is an alteration in energy status which is at least partially reversible on re-innervation [63–65]. The phosphorylation potential, as judged by the PCr/P_i ratio [64] decreases and pH_i increaes with the degree of clinical involvement. Clinical recovery is accompanied by a return toward normal of the relative concentrations of PCr, P_i and ATP. The pH_i may recover fully in spite of an incomplete return of PCr and P_i to pre-injury levels.

9.3. *Disease*

9.3.1. *Proton handling*

Evidence is beginning to accumulate showing that muscle is able to alter its proton efflux rate. This would be of major importance to muscle energetics because pH_i influences the concentration of ADP through the creatine kinase reaction: the concentrations of H^+ and ADP are inversely related. Efflux can be measured by ^{31}P NMR in vivo from the rate of pH_i recovery after exercise. This rate for normal finger flexor muscle of 0.09 ± 0.02 U/min [66,67] is not invariant. Slow pH_i recovery has been reported in the muscles of hypothyroid patients [68], in normal human muscles depleted of ATP by strenuous exercise [67] and in dystrophin-deficient mouse muscle [69]. Rapid recovery to the pre-exercise pH_i occurs in idiopathic hypertension (mean reate 0.14 ± 0.03 U/min, n=6 [66]) and in mitochondrial myopathy (0.22 ± 0.09 U/min, n=6 [70,71]). In the latter two conditions and in the spontaneously hypertensive rat, protons must also be leaving the muscle fibres more rapidly than normal during exercise, since in both of these conditions the pH decrease during exercise is less than normal relative to the PCr loss. In mitochondrial myopathy such an increase in the ability to handle protons is beneficial to the muscle in two ways. Firstly, it prevents the excessively low pH_i predicted by the large amounts of lactic acid produced by the muscle in this kind of disorder. Secondly, it helps to keep the [ADP] high, thereby stimulating oxidative phosphorylation in a tissue with low mitochondrial capacity.

It has been established in experiments on normal rat hind limb that the majority of exercise-produced H^+ is eliminated from the muscle fibres on the Na^+/H^+ antiporter [72]. The mechanism responsible for the change in muscle proton efflux rate in the spontaneously hypertensive rat (and therefore, probably also in human hypertension) has been shown to be a change in the V_{max} of the antiporter [72]. Since many factors, among them the Na^+ gradient across the sarcolemma and intracellular Ca^{2+} concentration, will affect the activity of the antiporter, it is tempting to speculate that changes in its activity might also be responsible for alterations in H^+ accumulation during exercise and pH_i recovery rates in other conditions. Thus, the decreased Na^+ gradient across the sarcolemma in the mdx mouse muscle and in hypothyroid muscle are consistent with a decreased proton efflux rate, but this has not been confirmed directly.

9.4. *Hormone-induced changes*

The effects of several different hormones on the relationships between glycogenolysis, glycolysis and oxidative phosphorylation in skeletal muscle have been investigated by ^{31}P NMR. In some cases, animal models have been used to look more carefully at abnormalities originally identified in patients.

9.4.1. *Thyroid hormones*

The influence of thyroid hormones on metabolism has long been recognized, and many tissues are affected, including skeletal muscle. Patients with hypothyroidism may complain of fatigue and a decreased ability to exercise. The consequences in vivo to adult skeletal muscle energetics have been evaluated with NMR, and results show that the muscle is clearly abnormal, even at rest. There is an increase in the P_i/ATP ratio, suggesting that the absolute concentration of P_i is also high, and the PCr/ATP ratio is also elevated [73]. As with many other kinds of muscle disease, the exercise intolerance of hypothyroid subjects (and hypothyroid rats) is accompanied by rapid loss of PCr during exercise. Because, in addition, the PCr/P_i ratio recovered slowly after exercise, the abnormal results were interpreted as being due to abnormal mitochondrial metabolism [74,75]. More recently, however, it has been shown that these abnormalities (which are totally reversible with treatment) are more likely to be due to a decrease in glycogenolysis in muscle, which would also limit the substrate available for oxidative phosphorylation [73]. Muscle in hyperthyroidism is also abnormal. In the two studies reported [74,75] no changes were found at rest in patients or rats, but PCr loss and pH_i decrease were greater in patients than in controls during exercise [75], suggesting greater utilisation of ATP. In the other study, recovery in patients and rats was more rapid than normal [74].

9.4.2. *Insulin*

Insulin, or the lack of it, has both long- and short-term effects on the energy status of muscle. Challis et al. [76] found that three weeks after insulin secretion was inhibited by administration of streptozotocin, larger decreases in PCr were needed to maintain isometric twitch tension in rat hind leg muscle. These investigators concluded that chronic lack of insulin led to an increased dependence on glycolytic production of ATP under conditions of low energy demand (1 Hz stimulation) and force failure due to decreased pyruvate utilisation when demand for energy was greater (stimulation at 5 Hz).

In the short term, insulin has been thought to increase the glycolytic rate by increasing pH_i, thereby stimulating phosphofructokinase activity. However, experiments in vivo carried out on normal human subjects have shown, somewhat surprisingly, that administration of physiological amounts of insulin does not raise pH_i, but causes it to decrease slowly by about 0.05 units over 1–2 h [77].

9.4.3. *Sex hormones*

The level of ovarian sex hormones is another factor which may affect energy production. Four weeks after ovariectomy, PCr and P_i changes at >0.05 Hz were greater than in control rat muscle. Recovery rates were also slower than normal. These abnormalities are consistent with a decrease in mitochondrial function, and indeed, a 40% decrease in cytochrome oxidase was found in the hind limb muscle [78].

9.4.4. *Skeletal muscle metabolism in heart disease*

Patients with heart failure develop exercise intolerance, and ^{31}P NMR has been used to investigate the possible contribution of metabolic factors to the muscle symptoms. A high concentration of P_i was reported in resting muscle in one study of forearm muscle [79], and because this is not found in older or inactive subjects, it is in itself strongly suggestive of abnormal bioenergetics. The exericse intolerance is consistent with the rapid acidification and depletion of PCr which have been found in several different muscles [80–84]. These abnormalities cannot be explained by decreased blood flow [81,82,84]. Overall, the results from several laboratories suggest that there is a decrease in the oxidative production of energy in the muscle. This view is supported by a study carried out on rats with surgically caused myocardial infarction. NMR results were similar to those from the patients, and biochemical analysis of muscle confirmed that there was a decrease in oxidative enzyme activity [85]. How much the chronically reduced daily activity levels of heart failure patients are responsible for these changes is under investigation.

10. Treatment

It can be seen from data presented in this chapter that many disorders, particularly those involving muscle, are associated with abnormalities of bioenergetics. NMR therefore has obvious potential for assessing the effects of therapeutic interventions in disease or injury. It can provide an easily repeatable and objective measure for determining the efficacy of treatment and for optimising dosage and determining the time-course of therapy. NMR may also be helpful in determining the site of action of a treatment. Interventions which have been studied in this way include drug treatment, surgical repair, dietary changes and exercise training. A few examples are given in this short section to illustrate the variety and scope of such investigations.

10.1. Defects in ATP synthesis

Most of the published data on treatment of these conditions is on individuals or small groups of subjects, due in part to the limited number of such patients available for study. In one patient, with a severe defect in mitochondrial metabolism at complex III, menadione and ascorbate were administered in an attempt to bypass the affected site and improve electron transport [86]. Within 12 h of beginning treatment, the PCr/P_i ratio in resting muscle increased about two-fold, the low pH_i (6.8) increased to the normal range (7.0) and the kinetics of PCr/P_i during recovery from exercise were markedly improved. These changes in metablism were reflected in an enhanced exercise tolerance and were maintained over a period of months. In an unusual case of defective mitochondrial metabolism secondary to chronic renal phosphate loss, oral phosphate improved symptoms and reversed the kinetic abnormalities in exercise and recovery [87].

Attempts have been made to improve the rate of ATP synthesis in muscle phosphorylase deficiency (McArdle's disease) by relieving the dependence on glycogen as substrate for glycolysis and as the source of pyruvate for mitochondrial oxidation. Infusion of glucose improves exercise kinetics [88,89], but obviously cannot be used for long-term therapy. In one study, amino acids given intravenously did not improve working capacity but increasing dietary protein was found to be beneficial [90], while in another study increasing protein intake had no effect on energetics [88]. Hypothyroid muscle, which may have a milder defect in glycogen metabolism [73], returns to normal kinetics with thyroxine treatment over a period of weeks [73,74].

10.2. Heart failure

NMR has been used in several studies of skeletal muscle in cardiovascular disorders. The drug dobutamine increases cardiac output and total blood limb flow, but when calf muscle of seven patients with heart failure were studied it was concluded that the drug could not have increased local muscle blood flow. The results showed that dobutamine did not alter the pH_i or the slope of the relationship between systemic VO_2 and P_i/PCr during exercise [91].

The effects of training on the metabolism of skeletal muscle in heart failure have been investigated in both animals and man. In the experimental rat model discussed earlier in this chapter, results suggest that training at least partially reverses some of the abnormalities [92]. Training of patients with heart failure may also be beneficial.

Acknowledgements

We are grateful to many colleagues who have contributed to many of the ideas discussed in this article and to the experimental work reported in the cited papers.

Our work was supported by the Medical Research Council, the British Heart Foundation and the Muscular Dystrophy Group of Great Britain.

References

1 Hoult, D.I., Busby, S.J.W., Gadian, D.G., Radda, G.K., Richards, R.E. and Seeley, P.J. (1974) Nature 252, 285–287.
2 Ackerman, J.H., Grove, T.H., Wong, G.G., Gadian, D.G. and Radda, G.K. (1980) Nature 283, 167–170.
3 Andrew, E.R., Bydder, G., Griffiths, J., Iles, R. and Styles, P. (1990) Clinical Magnetic Resonance: Imaging and Spectroscopy, Wiley, New York.
4 Gadian, D.G. (1982) Nuclear Magnetic Resonance and Its Applications to Living Systems, Clarendon Press, Oxford.
5 Styles, P. (1991) NMR Basic Principles and Progress, Springer, Berlin.
6 Radda, G.K., Rajagopalan, B., Taylor, D.J. (1989) in: H.Y. Kressel (Ed.) Mag. Res. Quart., Raven Press 5, 122–150.
7 Mahler, M. (1985) J. Gen. Physiol. 86, 135–165.
8 Connett, R.J. and Honig, C.R. (1989) Am. J. Physiol. 256, R898–R906.
9 Idstrom, J.P., Subramanian, V.H., Chance, B., Schersten, T. and Bylund-Fellenius, A.C. (1986) Fed. Proc. 45, 2937–2941.
10 Idstrom, J.P., Subramanian, V.H., Chance, B., Shersten, T. and Bylund-Fellenius, A.C. (1985) Am. J. Physiol. 248, H40–H48.
11 Gutierrez, G., Pohil, R.J. and Narayana, P. (1989) J. Appl. Physiol. 66, 2117–2123.
12 Sinoway, L., Prophet, S., Gorman, I., Mosher, T., Shenberger, J., Dolecki, M., Briggs, R. and Zelis, R. (1989) J. Appl. Physiol. 61, 429–436.
13 Hands, L.J., Bore, P.J., Galloway, G., Morris, P.J. and Radda, G.K. (1986) Clin. Sci. 71, 283–290.
14 Arnold, D.L., Matthews, P.M. and Radda, G.K. (1984) Magn. Reson. Med. 1, 307–315.
15 Chance, B., Clark, B.J., Niioka, S., Subramanian, H., Morris, J.M., Argov, Z. and Bode, H. (1985) Cicrulation 72 (Suppl IV), 103–110.
16 Boska, M. (1991) NMR Biomed. 4, 173–181.
17 Meyer, R.A. (1988) Am. J. Physiol. C548–C553.
18 Radda, G.K. (1990) Philos. Trans. R. Soc. London Ser. A 333, 515–524.
19 Brindle, K.M., Blackledge, M.J., Challis, R.A.J. and Radda, G.K. (1989) Biochemistry 28, 4887–4893.
20 Heineman, F.W. and Balaban, R.S. (1990) Annu. Rev. Physiol. 52, 523–542.
21 Matthews, P.M., Williams, S.R., Seymour, A.-M.L., Swartz, A., Dube, G., Gadian, D.G. and Radda, G.K. (1982) Biochim. Biophys. Acta 720, 163–171.
22 Allen, D.G., Eisner, D.A., Morris, P.G., Pirolo, J.G. and Smith, G.L. (1986) J. Physiol. (London) 376, 121–141.
23 Balaban, R.S., Kantor, H.L., Katz, L.A. and Briggs, R.W. (1986) Science 232, 1121–1123.
24 From, A.H., Petein, M.A., Michurski, S.P., Zimmer, S.D. and Ugurbil, K. (1986) FEBS Lett. 206, 257–261.
25 Katz, L.A., Koretsky, A.P. and Balaban, R.S. (1988) Am. J. Physiol. 255, H185–H188.

26 Unitt, J.F., McCormack, G., Reid, D., MacLachlan, L.K. and England, P.J. (1989) Biochem. J. 262, 293–301.
27 Portman, M.A., Heineman, F.W. and Balaban, R.S. (1989) J. Clin. Invest. 83, 456–464.
28 Blackledge, M.J., Rajagopalan, B., Oberhaensli, R., Bolas, N., Styles, P. and Radda, G.K. (1987) Proc. Natl. Acad. Sci. USA 84, 4283–4287.
29 Conway, M.A. and Radda, G.K. (1991) Trends Cardiovascular Medicine 1, 300–304.
30 Weiss, R.G., Bottomley, P.A., Hardy, C.J., Gerstenblith, G. (1990) N. Engl. J. Med. 323, 1593–1600.
31 Conway, M.A., Bristow, J.D., Blackledge, M., Rajagopalan, B. and Radda, G.K. (1991) Br. Heart J. 65, 25–30.
32 Bessman, S.P. and Carpenter, C.L. (1985) Annu. Rev. Biochem. 54, 831–862.
33 Cadoux-Hudson, T.A.D., Blackledge, M.J. and Radda, G.K. (1989) FASEB J. 3, 2660–2666.
34 Shoubridge, E.A., Bland, J.L. and Radda, G.K. (1984) Biochim. Biophys. Acta 805, 72–78.
35 Shoubridge, E.A., Jeffry, F.M.H., Keogh, J.M., Radda, G.K. and Seymour, A.-M.L. (1985) Biochim. Biophys. Acta 847, 25–32.
36 Cerdan, S. and Seelig, J. (1990) Annu. Rev. Biophys. Biophys. Chem. 19, 43–67.
37 Fitzpatrick, S.M., Hetherington, H.P., Behar, K.L. and Shulman, R.G. (1990) J. Cereb. Blood Flow Metab. 10, 170–179.
38 Alger, J.R., Sillerud, L.O., Behan, K.L., Gillies, R.J., Shulman, R.G., Gordon, R.E., Shaw, D. and Hanley, P.E. (1981) Science 214, 660–662.
39 Shulman, G.I., Rothman, D.L., Jue, T., Stein, P., De Fronzo, R.A. and Shulman, R.G. (1990) N. Engl. J. Med. 322, 223–228.
40 Moussavi, R.S., Carson, P.J., Boska, M.D., Weiner, M.W. and Miller, R.G. (1989) Neurology 39, 1222–1226.
41 Bridges, C.R., Clark, B.J., Hammond, R.L. and Stephenson, L.W. (1991) Am. J. Physiol. 260, C643–C651.
42 Miller, R.G., Giannini, D., Milner-Brown, H.S., Layzer, R.B., Koretsky, A.P., Hooper, D. and Weiner, M.W. (1987) Muscle Nerve 10, 810–821.
43 Taylor, D.J., Styles, P., Matthews, P.M., Arnold, D.G., Gadian, D.G., Bore, P.J. and Radda, G.K. (1986) Magn. Reson. Med. 3, 44–54.
44 Wilkie, D.R. (1986) Fed. Proc. 45, 2921–2923.
45 Boska, M.D., Moussavi, R.S., Carson, P.J., Weiner, M.W. and Miller, R.G. (1990) Neurology 40, 240–244.
46 Wilson, J.R., McCully, K.K., Mancini, D.M., Boden, B. and Chance, B. (1988) J. Appl. Physiol. 64, 2333–2339.
47 Ross, B.D., Radda, G.K., Gadian, D.G., Rocker, G., Esiri, M. and Falconer-Smith, J. (1981) N. Engl. J. Med. 304, 1338–1342.
48 Radda, G.K. and Taylor, D.J. (1985) Int. Rev. Exp. Pathol. 27, 1–58.
49 Adams, G.R., Fisher, M.J. and Meyer, R.A. (1991) Am. J. Physiol. 260, C805–C812.
50 Newham, D.J. and Cady, E.B. (1990) NMR Biomed. 3, 211.
51 Cady, E.B., Elshove, H., Jones, D.A. and Moll, A. (1989) J. Physiol. (London) 418, 327–337.
52 Heerschap, A., Bergman, A.H., van Vaals, J.J., Wirtz, P., Loermans, H.M. and Veerkamp, J.H. (1988) NMR Biomed. 1, 27–31.
53 Dunn, J.F., Frostick, S., Brown, G. and Radda, G.K. (1991) Biochim. Biophys. Acta 1096, 115–120.
54 Younkin, D.P., Berman, P., Sladky, J., Chee, C., Bank, W. and Chance, B. (1987) Neurology 37, 165–169.
55 Newman, R.J., Bore, P.J., Chan, L., Gadian, D.G., Styles, P., Taylor, D.J. and Radda, G.K. (1982) Br. Med. J. 284, 1072–1074.
56 Taylor, D.J., Crowe, M., Bore, P.J., Styles, P., Arnold, D.L., Radda, G.K. (1984) Gerontology 30, 2–7.
57 Zochodne, D.W., Thompson, R.T., Driedger, A.A., Strong, M.J., Gravelle, D. and Bolton, C.F. (1988) Magn. Reson. Med. 7, 373–383.
58 Minotti, J.R., Johnson, E.C., Hudson, T.L., Zuroske, G., Fukushima, E., Murata, G., Wise, L.E., Chick, T.W. and Icenogle, M.V. (1990) J. Appl. Physiol. 68, 289–294.

59 McCully, K.K., Boden, B.P., Tuchler, M., Fountain, M.R. and Chance, B. (1989) J. Appl. Physiol. 67, 926–932.
60 Clark, B.J., Acker, N.A., McCully, K., Subramanian, H.V., Hammond, R.L., Salmons, S., Chance, B. and Stephenson, L.W. (1988) Am. J. Physiol. 254, C258–C266.
61 Kent-Braun, J.A., McCully, K.K. and Chance, B. (1990) J. Appl. Physiol. 69, 1165–1170.
62 Minotti, J.R., Johnson, E.C., Hudson, T.L., Sibbitt, R.R., Wise, L.E., Fukushima, E. and Icenogle, M.V. (1989) J. Appl. Physiol. 67, 324–329.
63 Chance, B. (1984) Ann. N.Y. Acad. Sci. 428, 318–332.
64 Frostick, S.P., Taylor, D.J., Dolecki, M. and Radda, G.K. (1987) Soc. Mag. Res. Med. (New York) 575.
65 Frostick, S.P., Taylor, D.J., Yonge, R.P. and Radda, G.K. (1986) Soc. Mag. Res. Med. (Montreal) 69.
66 Dudley, C.R., Taylor, D.J., Ng, L.L., Kemp, G.J., Ratcliffe, P.J., Radda, G.K. and Ledingham, J.G. (1990) Clin. Sci. 79, 491–497.
67 Taylor, D.J., Styles, P., Matthews, P.M., Arnold, D.L., Gadian, D.G., Bore, P.J. and Radda, G.K. (1986) Magn. Reson. Med. 3, 44–54.
68 Taylor, D.J., Rajagopalan, B. and Radda, G.K. (1991) Clin. Sci. 81, 27 P.
69 Dunn, J.F., Tracey, I. and Radda, G.K. (1992) J. Neurol. Sci., in press.
70 Kemp, G.J., Taylor, D.J., Dunn, J.F. and Radda, G.K. (1991) Biochem. Soc. Trans. 19, 207.
71 Arnold, D.L., Taylor, D.J. and Radda, G.K. (1985) Ann. Neurol. 18, 189–196.
72 Syme, P.D., Arnolda, L., Green, Y., Aronson, J.K., Graham-Smith, D.G. and Radda, G.K. (1990) J. Hypertens. 8, 1027–1036.
73 Taylor, D.J., Rajagopalan, B. and Radda, G.K. (1992) Eur. J. Clin. Invest., 22, 358–365.
74 Argov, Z., Renshaw, P.F., Boden, B., Winokur, A. and Bank, W.J. (1988) J. Clin. Invest. 81, 1695–1701.
75 Kaminsky, P., Robin-Lherbier, B., Walker, P., Brunotte, F., Escanye, J.M., Klein, M., Forrett, M.C., Robert, J. and Duc, M. (1991) Acta Endocrinol. Copenhagen 124, 271–277.
76 Challis, R.A.J., Vranic, M. and Radda, G.K. (1989) Am. J. Physiol. 256, E129–E137.
77 Taylor, D.J., Coppack, S.W., Cadoux-Hudson, T.A.D., Kemp, G.J., Radda, G.K., Frayn, K.N. and Ng, L.L. (1991) Clin. Sci. 81, 123–128.
78 Roth, Z., Argov, Z., Maris, J., McCully, K.K., Leigh, J.S. and Chance, B. (1989) J. Appl. Physiol. 67, 2060–2065.
79 Massie, B.M., Conway, M., Yonge, R., Frostick, S., Sleight, P., Ledingham, J.J.G., Radda, G.K. and Rajagopalan, B. (1987) Am. J. Cardiol. 60, 309–315.
80 Mancini, D.M., Ferraro, N., Tuchler, M., Chance, B. and Wilson, J.R. (1988) Am. J. Cardiol. 62, 1234–1240.
81 Rajagopalan, B., Conway, M.A., Massie, B. and Radda, G.K. (1988) Am. J. Cardiol. 62, 53E–57E.
82 Weiner, D.H., Fink, L.I., Maris, J., Jones, R.A., Chance, B. and Wilson, J.R. (1986) Circulation 73, 1127–1136.
83 Wilson, J.R., Fink, L., Maris, J., Ferraro, N.P., Power-Vanwart, J., Eleff, S. and Chance, B. (1985) Circulation 71, 57–62.
84 Arnolda, L., Conway, M., Dolecki, M., Sharif, H., Rajagopalan, B., Ledingham, J.G., Sleight, P. and Radda, G.K. (1990) Clin. Sci. 79, 583–589.
85 Arnolda, L., Brosnan, J., Rajagopalan, B. and Radda, G.K. (1991) Am. J. Physiol. H434–H442.
86 Eleff, S., Kennaway, N.G., Buist, N.R.M., Darley-Usmar, V.M., Capaldi, R.A., Bank, W.J. and Chance, B. (1984) Proc. Natl. Acad. Sci. USA 81, 3529–3533.
87 Land, J.M., Taylor, D.J., Kemp, G.J., Radda, G.K. and Rajagopalan, B. (1992) Clin. Sci. 82, 26 P.
88 Argov, Z., Bank, W.J., Maris, J., Chance, B. (1987) Neurology 37, 1720–1724.
89 Lewis, S.F., Haller, R.J., Cook, J.D. and Nunnally, R.L. (1985) J. Appl. Physiol. 59, 1991–1994.
90 Jensen, K.E., Jakobsen, J., Thomsen, C. and Henrickson, O. (1990) Acta Neurol. Scand. 81, 499–503.
91 Mancini, D.M., Schwartz, M., Ferraro, N., Seestedt, R., Chance, B. and Wilson, J.R. (1990) Am. J. Cardiol. 65, 1121–1126.
92 Adamopoulos, S., Brunotte, F., Coats, A.J.S., Unitt, J., Lindsay, D., Rajagopalan, B. and Radda, G.K. (1991) Br. Heart J. 66, 90.

L. Ernster (Ed.) *Molecular Mechanisms in Bioenergetics*
© 1992 Elsevier Science Publishers B.V. All rights reserved

CHAPTER 20

Recent advances on mitochondrial biogenesis

ANNE CHOMYN and GIUSEPPE ATTARDI

Division of Biology, California Institute of Technology, Pasadena, CA 91125, USA

Contents

1. Introduction 483
2. Transcription 483
3. RNA processing 488
   3.1. Intron splicing and intron mobility 488
   3.2. RNA editing 491
4. Regulation of mitochondrial gene expression 493
   4.1. Differential gene expression 493
   4.2. Physiological and developmental control 495
5. Import of proteins into mitochondria 496
6. Assembly of enzyme complexes of the OX–PHOS system 500
7. Perspectives 503
Acknowledgements 504
References 504

1. Introduction

The area of mitochondrial biogenesis has continued to attract the interest of a large number of investigators in the past few years, and considerable progress has been made on several aspects of this complicated process. We plan to review here the most significant information that has come out of the recent studies in this area, and to discuss the new ideas that are emerging and some of the pending issues. For a survey of earlier work on the formation of mitochondria and a more detailed treatment of some aspects of this topic we refer to previous reviews [1–5].

2. Transcription

In vivo and in vitro DNA transcription experiments have shown that the promoters for mtDNA transcription in mammalian cells are located in the gene-free region close to the origin of H-strand synthesis called the D-loop region [6–8] (Fig. 1). There is a single initiation site for L-strand transcription (L in Fig. 1), from which a giant polycistronic transcript originates, that exends over the whole L-strand. This transcript is destined to be processed to yield the mRNA for the ND6 subunit of NADH dehydrogenase and the eight tRNAs encoded in this strand

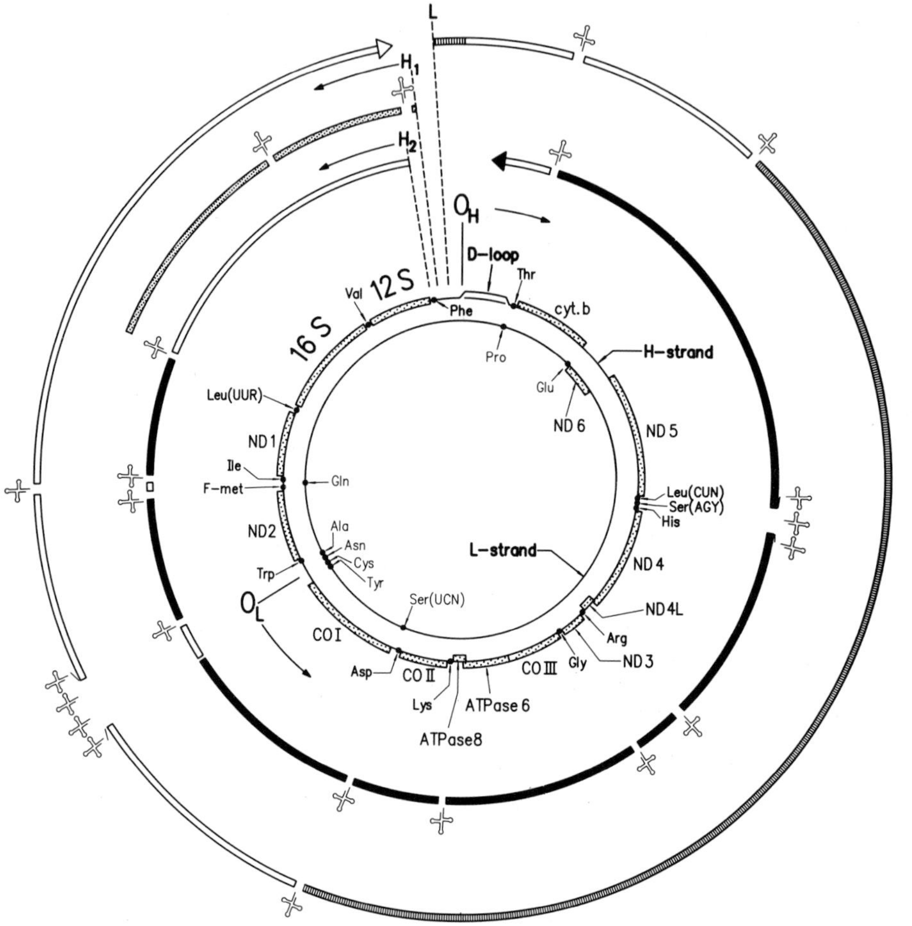

Fig. 1. Genetic and transcription maps of the human mitochondrial genome. The two inner circles show the positions of the two rRNA genes and the reading frames (stippled bars), and the tRNA genes (filled circles). In the outer portion of the diagram, the identified functional RNA species other than tRNAs are represented by stippled bars (rRNA species), filled bars (mRNAs transcribed from the H-strand) or a hatched bar (ND6 mRNA, transcribed from the L-strand). The tRNAs are represented by cloverleaf structures. The open bars represent unstable, presumably nonfunctional by-products. H_1 and H_2: initiation sites of the rDNA and, respectively, whole H-strand transcription unit; L: initiation site of the L-strand transcription unit. COI, COII, COIII: subunits I, II and III of cytochrome *c* oxidase; CYT b: apocytochrome *b*; ATPase 6 and ATPase 8: subunits 6 and 8 of H^+-ATPase; ND1, ND2, ND3, ND4, ND4L, ND5 and ND6: subunits of NADH dehydrogenase; O_H, O_L: origin of H-strand and, respectively, L-strand synthesis (from ref. [9]).

(Fig. 1). There is evidence that the primer(s) for H-strand synthesis also arises from the L-strand promotor [10]. In the D-loop region, there are also two closely located initiation sites for transcription of the H-strand. From the upstream, more active, initiation site (H1 in Fig. 1) proceeds the synthesis of the rRNAs and two tRNAs (tRNA^{Phe} and tRNA^{Val}) in the form of a transcript which terminates at the 16S rRNA/tRNA$^{Leu(UUR)}$ boundary, and which is processed, probably cotranscriptionally, to yield the two rRNA species and the two tRNAs. From the downstream, less active, initiation site, located near the 5′-end of the 12S rRNA gene, proceeds the synthesis of all other tRNAs and the mRNAs encoded in the H-strand, in the form of a giant polycistronic transcript which extends over the entire length of this strand and is destined to be processed to the mature species [11]. Figure 2 shows an experiment in which the di- or tri-phosphate-carrying 5′-ends of the primary transcripts originating from the two H-strand transcription initiation sites in HeLa cells were mapped by capping them in vitro with [α-^{32}P]-GTP and guanylyltransferase and then by S1 protection.

The existence of transcripts extending over the entire length of the L-strand and H-strand implies that mammalian mtDNA is symmetrically transcribed over its entire length, a phenomenon which was discovered about 20 years ago [12,13] and has remained so far unique. As shown in Fig. 1, the primary transcripts contain tRNA sequences interspersed between and butt-joined to rRNA and/or mRNA sequences. This implies the existence of a very precise processing machinery which recognizes the tRNA sequences. A mitochondrial RNase P which cleaves the *E. coli* suppressor tRNA^{Tyr} precursor at the 5′-end of the tRNA sequence, and which is presumably a component of the mitochondrial RNA processing machinery, has been identified in HeLa cells [14].

The development of in vitro transcription systems using submitochondrial fractions [15,16] has allowed the identification and mapping of the L-strand promoter and the rRNA-specific H-strand promoter [7,16]. So far, it has not been possible to obtain in vitro transcription from the downstream H-strand initiation site. The availability of in vitro transcription systems has also permitted the beginning of a dissection of the mammalian mitochondrial transcription machinery. A mitochondrial RNA polymerase activity has been identified and partially characterized [15,16]. Chromatographic fractionation of this activity has permitted the resolution of an intrinsically nonselective or weakly selective RNA polymerase activity and a fraction containing a factor that activates transcription [17], and that was designated mitochondrial transcription factor 1 (mtTF1). Further studies have indicated that this factor, although it binds to sequences upstream of the L-strand promoter and, with much lower affinity, to sequences upstream of the rRNA-specific H-strand promoter [18], has inherently flexible sequence specificity. The recent cloning of the cDNA for mtTF1 [19] has revealed a protein of 204 amino acids exhibiting two domains characteristic of the high mobility group (HMG) proteins, like the functionally homologous yeast mitochondrial HM/ABF2 [19a]. Both human and yeast proteins show a general ability to wrap or condense and unwind DNA in vitro and to bend DNA at specific sequences [19a], and may have mainly a structural and organizational role. It seems possible that another as yet undiscovered factor(s) is the true transcription factor, which confers specificity upon the core RNA polymerase. The isolation of a mitochondrial polymerase activity and mTF1 from both human and mouse mitochondria has permitted an investigation of the species specificity of these two components. The results of this analysis have suggested that the species specificity of mitochondrial transcription in man and mouse resides in the polymerase-containing fraction [20].

A significant advance in our understanding of the mechanisms and regulation of transcription in mammalian cells has been produced by the identification of a protein factor (mTERF) which is involved in termination of transcription of the mitochondrial rRNA genes in human cells [21]. mTERF is a DNA-binding protein, which has been purified by oligonucleotide affinity chromatography, and found to protect a 28 bp region within the tRNA$^{Leu(UUR)}$ gene at a position immediately adjacent and downstream of the positions corresponding to the 3′-ends of the in vivo 16S rRNA gene products (Fig. 3). A tridecamer sequence entirely contained

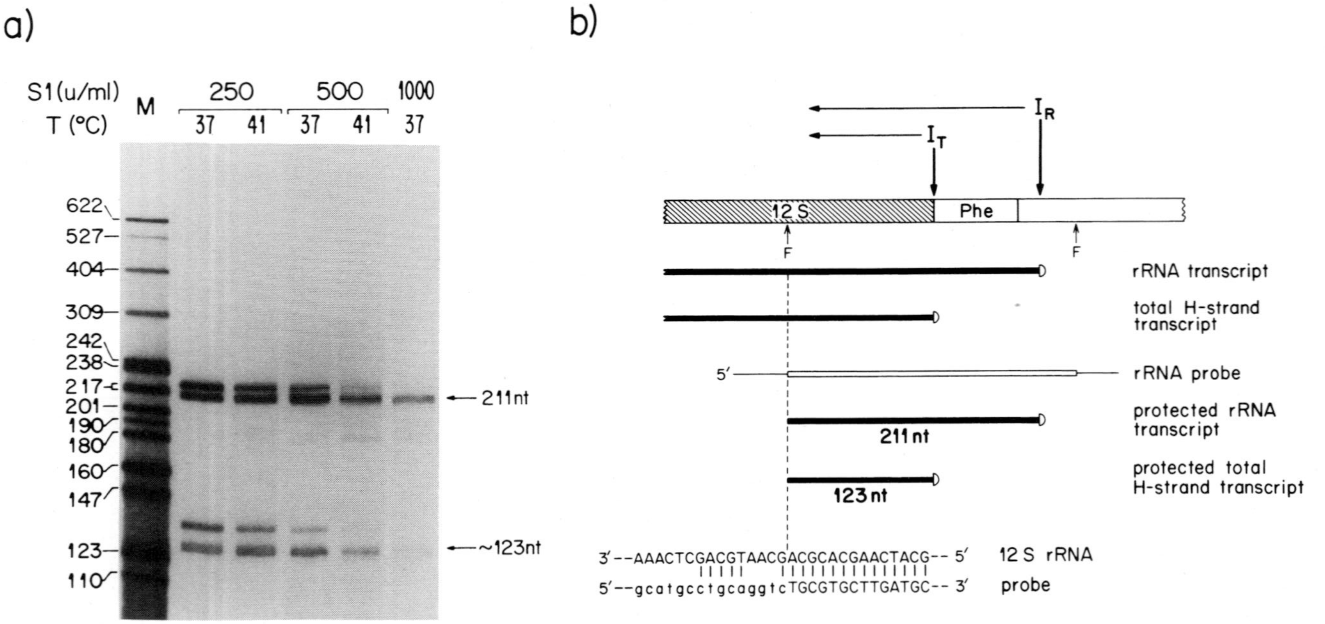

Fig. 2. Demonstration of two in vivo initiation sites for H-strand transcription in human mtDNA. In vivo primary transcripts of HeLa cell mtDNA, starting at the upstrem (I_R) and downstream (I_T) initiation sites for H-strand transcription, were labeled at their 5′-ends with [α-^{32}P]GTP and guanylyltransferase, hybridized with an RNA probe synthesized from a Fnu4HI (F) fragment of mtDNA cloned in pBSKS$^+$ (Stratagene), and digested with the single-strand specific S1 nuclease (panel b); the protected RNA fragments were then analyzed by electrophoresis in an 8% polyacrylamide/8 M urea gel (panel a). The species running slightly behind the expected 211 nt and 123 nt protected RNA bands, which were much less resistant to S1 nuclease than the latter, presumably result from extension of the protected RNA fragments due to a partial fortuitous base complementarity between 12S rRNA and the plasmid sequence adjacent to the rRNA insert (see panel b). M: 3′-end labeled MspI-digested pBR322 DNA marker (unpublished observations by N. Narasimhan and G. Attardi).

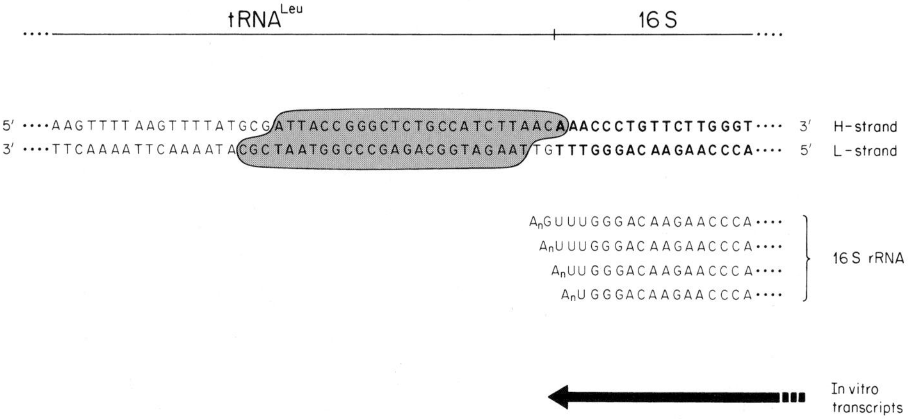

Fig. 3. The mitochondrial termination factor (mTERF) protects an mtDNA region immediately adjacent to the in vivo or in vitro produced 3′-ends of 16S rRNA. The positions of the 3′-ends of in vivo synthesized 16S rRNA are derived from ref. [22]. A_n: oligo(A) stretch, with n varying between 0 and 10. The point of the arrow marks the 3′-end of in vitro transcripts produced in a mitochondrial lysate programmed by a human mtDNA template (16) (from ref. [21]).

within the region protected by mTERF had previously been shown by deletion mutagenesis experiments to be essential for in vitro termination of transcription at the 3′-end of the 16S rRNA gene [23]. Furthermore, in an in vitro transcription system utilizing a mitochondrial lysate [16], it has been shown that mTERF promotes termination at the 16S rRNA/tRNA^{Leu} boundary of the H-strand transcripts starting at the rRNA-specific initiation site [21]. Particularly interesting was the observation that the termination-promoting activity was accompanied by a substantial stimulation of transcription, with indications that this stimulation specifically concerned the transcripts destined to terminate at the 3′-end of the 16S rRNA gene. These observations suggest the possibility that the rDNA transcription termination signal may interact with the initiation sequences, presumably through the intermediary of mTERF and/or other proteins. This interaction may activate the rDNA transcription unit, stimulating at the same time initiation and termination of transcription. mTERF has recently been identified as a 34 kDa protein (A. Daga and G. Attardi, unpublished observations).

In *Saccharomyces cerevisiae*, mitochondrial genes are transcribed either singly or, more often, as polycistronic units from at least 13 promoters, characterized by the conserved nonanucleotide sequence ATATAAGTA, with the underlined A being the site of initiation (for a review, see ref. [1]). Mitochondrial RNA polymerase has been purified from *S. cerevisiae* and shown to consist of two different components: a core RNA polymerase of 145 kDa, which catalyzes RNA synthesis, but does not start transcription correctly from mitochondrial promoters, and a specificity factor of 43 kDa, which confers upon the core polymerase the capacity to recognize correctly the promoters and to produce specific initiation of transcription [24–27]. Neither component is capable of binding specifically to mtDNA, whereas both components together specifically bind to mtDNA promoters [27]. Sequence analysis of the gene encoding the RNA polymerase has revealed significant sequence similarity with the RNA polymerase of *E. coli* bacteriophages T3 and T7 [28].

3. *RNA processing*

3.1. *Intron splicing and intron mobility*

One of the most intricate aspects of RNA processing in fungal and plant mitochondria is represented by the excision of introns from the mRNA precursors. This excision involves the self-splicing properties of some of the introns and the participation of protein factors encoded in the introns themselves (RNA maturases) or in the nucleus. The greatest complexity is exhibited by mitochondrial mRNA processing in *Saccharomyces cerevisiae*. The mitochondrial genes of this organism exhibit introns of both group I and group II, the two types in which introns have been classified on the basis of their characteristic secondary structure forming the active sites for splicing [29–31] (Fig. 4). Several of these introns have an open reading frame, which is in most cases in frame with the upstream exon, and which encodes a protein capable of recognizing specific intron and/or splice site sequences. Two main activities have been found to be associated with these intron-encoded proteins: an activity involved in intron excision (maturase activity) and an activity involved in intron transposition (site-specific DNA endonuclease activity) [32,36–38]. Thus, as shown in Fig. 4, group I introns bI2, bI3 and bI4, in the cytochrome *b* (COB), code for an RNA maturase, and are not mobile [32,39,40]; group I introns aI4α, in the cytochome *c* oxidase subunit I (COX I) gene [33,41], aI5α and possibly aI3

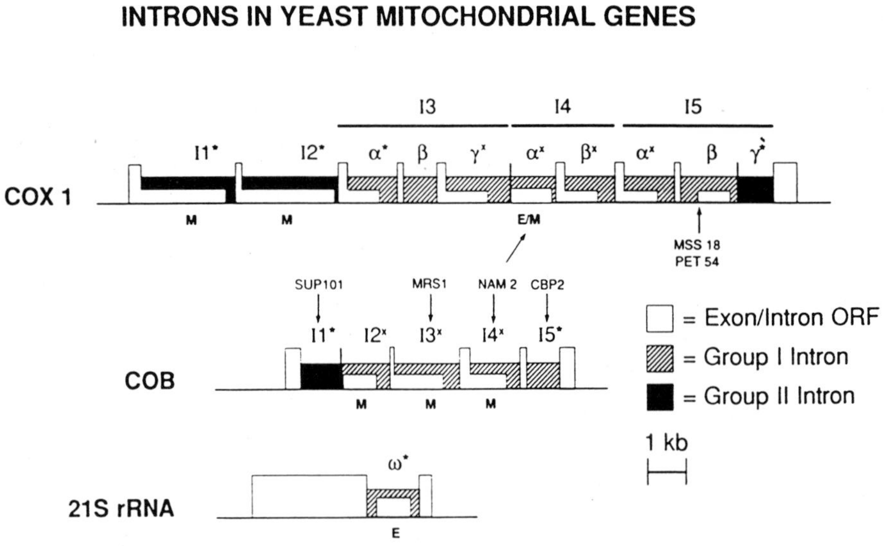

Fig. 4. Schematic representation of the yeast mtDNA introns and of the known proteins required for their splicing and/or transposition. The figure shows the genes for cytochrome *c* oxidase subunit I (COX1), cytochrome *b* (COB) and mitochondrial large (21S) rRNA. The figure also includes COX1 gene introns 3*β*, 3*γ* and 4*β*, recently characterized in mtDNA of *Saccharomyces douglasii* [32] and *Saccharomyces capensis* [33]. Intron designation is placed above each intron, with superscript * indicating presence, and superscript ˣ, absence of self-splicing activity, when known. Letters below introns indicate known activities of intron-encoded proteins: E, endonuclease; M, maturase. ORF: open reading frame. MSS18 and PET54 are nuclear-coded proteins required for splicing COX1-I5*β* [34,35]. See text for details. (Reproduced, with permission, from ref. [36].)

in the same gene (P. Perlman, personal communication; C. Jacq, personal communication), and omega, in the 21S rRNA gene [37,38], encode a DNA endonuclease, and are mobile. There is evidence indicating that the same intron-encoded protein can be involved in both RNA splicing and intron transposition. Thus, the aI4α product, which is closely related to the bI4 maturase [42], has a latent maturase activity, which can be activated either by a single amino acid change (producing the mitochondrial suppressor *mim*2) [42], or by interaction with the product of one of the suppressor alleles of the nuclear NAM2 gene, to compensate for bI4 RNA maturase deficiencies [43]. Therefore, the aI4α product would have both a maturase-like function and a site-specific DNA endonuclease function. Furthermore, there is evidence suggesting that the bI4 maturase can also act on a DNA substrate, since it has been shown to stimulate homologous recombination both in yeast mitochondria [32] and in *E. coli* [44]. Also maturases encoded in group II introns may be involved in intron mobility. In fact, it has been shown that the maturases aI1 and aI2, encoded in the COX I gene (Fig. 4), have structural homology to reverse transcriptases [45]. Furthermore, genetic evidence has indicated that introns aI1 and/or aI2 are necessary for in vivo deletion of introns, suggesting that this deletion may occur by recombination of an intron containing genomic DNA with a reverse transcriptase copy of a spliced transcript produced by maturases aI1 and/or aI2 [46]. However, no direct evidence of transcriptase activity of the aI1 and/or aI2 maturases has been obtained so far.

In addition to the intron-encoded proteins, a growing number of nuclear-encoded protein factors imported into the organelles appear to participate in intron splicing. In several cases, both a maturase and a nuclear-coded protein(s) are required in the RNA splicing of the same intron. With a few exceptions, these nuclear-coded factors have other functions besides that of assisting in the splicing of intervening sequences. This phenomenon is typically illustrated by the two best characterized among such factors, namely the *Neurospora crassa* mitochondrial tyrosyl-tRNA synthetase [47] and the yeast leucyl-tRNA synthetase [43,48]. The *Neurospora* enzyme, encoded in the *cyt-18* gene, is a 67 kDa protein which functions in both mitochondrial protein synthesis and splicing of the intervening sequence in the large rRNA precursor, as well as in the RNA splicing of other group I introns [49]. A missense mutation in the nucleotide-binding fold of the protein produces defects in both protein synthesis and splicing, which can be correlated with deficiencies in the mitochondrial tyrosyl-tRNA synthetase activity and, respectively, in the splicing activity of a soluble fraction which excises in vitro the intervening sequence of the large rRNA precursor [47,50]. Furthermore, second-site mutations clustered near the N-terminus of the protein, in a 63 amino acid span which is absent in both yeast mitochondrial and *E. coli* tyrosyl-tRNA synthetases, restore splicing activity but not the defect in aminoacylation [50]. The observation that the yeast and *E. coli* enzymes lack splicing activity and the analysis of in vitro mutants of the cyt-18 protein synthesized in *E. coli*, which has both splicing and tyrosyl-tRNA synthetase activities [51], have demonstrated that the non-conserved N-terminal domain is essential for the splicing activity, together with other regions of the protein. These observations have suggested that the splicing activity was acquired by the *Neurospora* enzyme relatively recently, after the divergence of *Neurospora* and yeast, as a result of the addition of an idiosyncratic domain not required for animoacylation [50].

Evidence for the involvement of the yeast leucyl-tRNA synthetase in the RNA splicing of the closely related bI4 and aI4α introns was obtained when the nuclear NAM2 gene, carrying suppressor mutations of mitochondrial mutants defective in the bI4 maturase, and therefore unable to splice both bI4 and aI4α intervening sequences, was cloned and found to code for the above synthetase [43,48]. Suppression of the bI4 maturase mutation does not require the bI4 intron, but requires an intact aI4α reading frame [52]. These observations suggest that the NAM2 product interacts with the aI4 product, making it competent for RNA splicing of both bI4 and aI4α. Other experiments have provided evidence supporting a direct function in splicing of the leucyl-tRNA synthetase. In fact, it has been shown that the wild-type gene is required for the function of the product of the universal code equivalent of the bI4 maturase gene carried by a nuclear plasmid, thus excluding an indirect effect involving mitochondrial protein synthesis [53]. It seems very likely that the NAM2 gene product interacts with the bI4 or aI4α maturase forming a complex active in splicing, or that it facilitates somehow the binding of the maturase to the intron in the mRNA precursor.

Among the nuclear-coded factors which have additional functions in the cell, besides their role in assisting in the splicing of introns, there is the MSS16 gene product, which is required for the splicing of some group I and group II introns in the COB and COX I gene, and has homology to the murine translation initiation factor eIF4a and other proteins with helicase activity [54]; the MSS18 gene product, which assists in the first step of the RNA splicing of the intron aI5β, and is also required for other RNA-processing reactions and for the synthesis of several mitochondrial proteins [34]; the PET54 gene, which is required for RNA splicing of the intron aI5β [35], as well as for translation of COX III mRNA [55]; and the NAM1 gene product, which is required for the RNA splicing of group I and II introns and for mitochondrial protein synthesis [56]. Two nuclear-coded factors have apparently only function in splicing, i.e., the CBP2 gene product, which is required for the RNA splicing of intron bI5 in the COB gene [57], and the MRS1 gene product, which is required (together with maturase bI3) for the RNA splicing of intron bI3 [58].

It seems very likely that the protein factors which assist in splicing, whether mtDNA-coded or nuclear coded, have the role of facilitating the correct folding of the intron RNA, thus promoting the RNA-catalyzed splicing reaction. In view of the great diversity of the protein factors, it is reasonable to assume that the mechanism underlying the structural role in splicing of the individual factors may be different, involving, for example, an unwinding activity of the protein, as in the case of the MSS16 gene product, or a recognition of different sequences or structures in the RNA intron. Besides their role in assisting in intron splicing, the mtDNA-coded or nuclear-coded protein factors may be involved in intron mobility by a reversal of the splicing reaction, followed by reverse transcription and recombination with genomic DNA. This pathway has been postulated on theoretical grounds [59,60] as a possible mechanism for intron transposition. Very recently, it has been shown that the *N. crassa* mitochondrial large rRNA intron undergoes reverse splicing in vitro under physiologically relevant conditions in a reaction promoted by the tyrosyl-tRNA synthetase [61].

It has been known for some time that the gene *rps12* in the chloroplast genome of plants and the gene *psaA* in that of *Chlamydomonas* contain exons which are widely separated in the genome and independently transcribed [62–64]. It has been suggested that, in such genes, transplicing of the exons is involved in the formation of the mature mRNAs. Since the sequences flanking these exons have features reminiscent of group II introns, it has been hypothesized that, in the transplicing event, each pair of exons is joined by means of formation of a competent catalytic intron through duplex pairing of the flanking intron sequences. That such reaction is indeed possible has been shown in vitro [65]. Very recently, transplicing has also been observed in plant mitochondria for the *nad1* gene of wheat [66] and *Oenothera* [67]. In both organisms, this gene is fragmented into five exons scattered over a wide region, and, here too, secondary structural interactions between group II-like flanking intron sequences may be involved in the formation of a functional intron.

3.2. RNA editing

A new type of RNA processing, identified in the past few years, is RNA editing, which involves the co-transcriptional or post-transcriptional modification of the nucleotide sequence in the coding region of mRNAs. Isolated examples of RNA editing have been described in nuclear-coded or viral-coded mRNAs, like the tissue-specific post-transcriptional change, due presumably to a chemical modification [68], of a genomic-encoded C to a U, in the mammalian apolipoprotein B mRNA [69,70], the recently reported case of RNA editing in the mRNA for glutamate receptor channel subunits in the brain [71], and the addition of G's in the mRNAs of some viruses, due to stuttering of the RNA polymerase [72,73]. However, the most striking forms of RNA editing have been found in organellar mRNAs of several organisms. Three main types of RNA editing, clearly different in their mechanism, have been described in these organisms: (1) addition of non-genomic-encoded U's or deletion of genomic-encoded U's in mitochondrial mRNAs of trypanosomatids [74]; (2) conversion of genomic-encoded C's to U's at specific sites in mRNAs of plant mitochondria [75–77] and chloroplasts [78]; (3) insertion of C's at specific sites in mitochondrial mRNAs of the slime mold, *Physarum polycephalum* [79]. While the conversion of C to U is probably due to a chemical modification, the mechanism of insertion of C's is totally unknown. In both cases, there is no hint as to the mechanism underlying the site specificity of RNA editing. Much more information, on the contrary, is available today concerning the first type of RNA editing, i.e., addition or deletion of U's in mitochondrial mRNAs of trypanosomatids. The first example of the above phenomenon was the observation that a −1 reading frameshift in the COII gene of *Trypanosoma brucei* or *Crithidia fasciculata* mtDNA was corrected by the post-transcriptional or co-transcriptional insertion of four U's at three sites in the frameshift region [80]. Soon after this finding, other examples of RNA editing were discovered in *T. brucei* and, less frequently, in *Leishmania tarentolae* and *C. fasciculata*. Thus, addition of 34 U's at 13 sites near the 5′-end of the CYTb mRNA in *T. brucei* and similar additions in the CYTb mRNA of *L. tarentolae* and *C. fasciculata* were found to extend the coding sequence and to generate a missing AUG initiator codon [81,82]. In some cases, RNA editing was found to entail also the loss of genome-encoded U's [83]. The term 'cryptogene' was introduced to indicate genes whose transcripts need to be edited in the coding region in order to produce the correct reading frame [74]. Extreme cases of RNA editing have been discovered in which exensive modification of transcripts of unrecognizable cryptogenes has produced mRNAs with identifidable reading frames corresponding to genes previously thought to be absent, like the COIII gene in *T. brucei* [84], or unexpected reading frames, like those in *T. brucei* encoding two subunits of NADH dehydrogenase specified by nuclear genes in other organisms (ND7, ND8 [85,86]), or a reading frame in *L. tarentolae* encoding a protein with sequence similarity to the family of S12 ribosomal proteins [87].

As to the mechanism of RNA editing, the analysis of partially edited transcripts provided evidence indicating that editing proceeds in general in the 3′ to 5′ direction, a conclusion which pointed to a post-transcriptional process [74]. Whether the additional information required for the editing process is contained in a DNA or RNA template has been the object of extensive investigations, as well as speculation, in the past few years [74]. A search for an edited template by direct DNA sequencing, PCR amplification of antisense edited RNAs or Northern hybridization gave negative results [74]. Recently, however, a computer search of the *L. tarentolae* kinetoplast DNA for short sequences complementary to known edited sequences by an approach allowing, besides the classical G–C and A–T base pairs, also the G–T base pairs, has revealed the presence of such sequences both in the maxicircle DNA (the trypanosomatid equivalent of mtDNA), scattered throughout the genome in the spacers between known genes

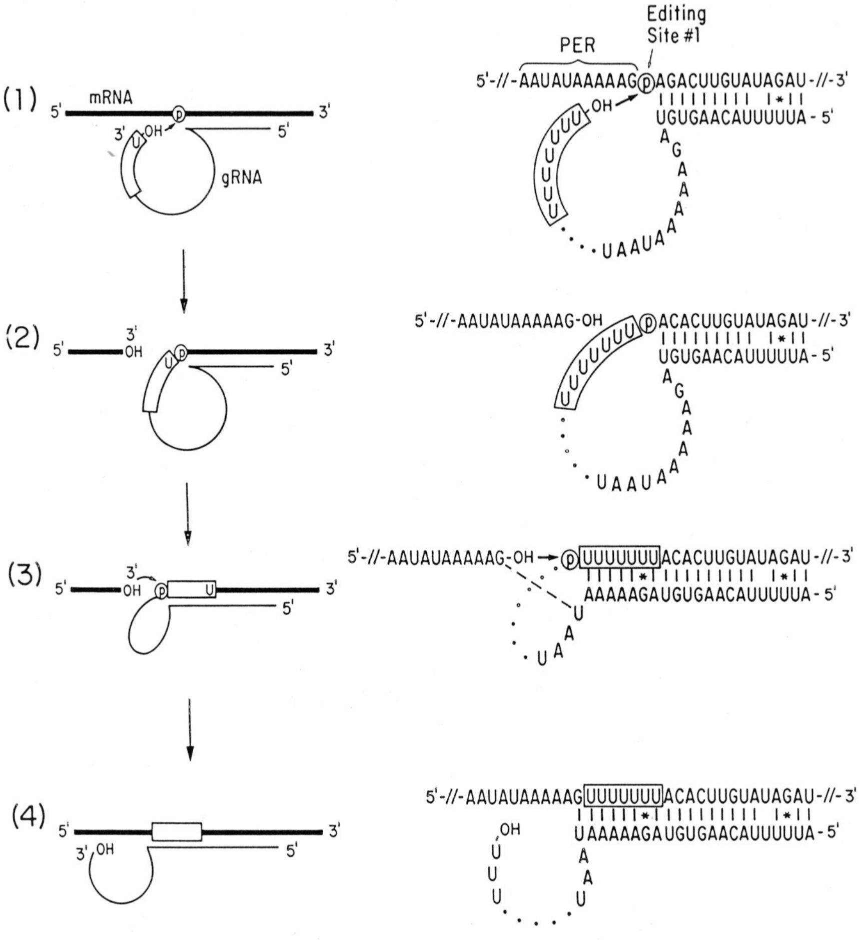

Fig. 5. Transesterification model of RNA editing. The diagram illustrates the transfer of a stretch of U's from the 3′ oligo(U) tail of a gRNA to the first editing site of ND7 mRNA in *L. tarentolae* through two transesterification steps. PER: preedited region. See text for details. (Reproduced, with permission, from ref. [93].)

[88], and within the variable regions of the minicircle DNA [89]. Transcripts of these sequences were soon identified in mitochondria, and designated *guide RNAs* or *gRNAs*. These were characterized for having, at their 3′-end, a nonencoded oligo(U) tail [90], and, at their 5′-end, sequences complementary to regions located immediately downstream of the mRNA segments to be edited. These sequences were named 'anchor' sequences, because they were postulated to base-pair with the complementary regions, and thus to anchor the gRNAs in a position suitable for the editing process. An 'enzyme cascade' model was proposed for this process, which would involve a series of enzymatic steps catalyzed by a multi-enzyme complex [88]. The first step would be the base-pairing of the anchor sequence to the region immediately downstream of the

to-be-edited region. Subsequently, there would be a specific cleavage at the first mismatched base in the mRNA, liberating a 3′-OH group, followed by addition to the 3′-end of one or more U's capable of base-pairing with the guanine or adenine 'guide' nucleotides of the gRNA, and the subsequent ligation of the two ends of the mRNA. The process would then restart at the next mismatched nucleotide. Three enzymatic activities capable, respectively, of cleaving the pre-edited mRNA within the pre-edited region, adding U's to the 3′-end of any RNA molecule (terminal uridyltransferase or TUTase), and ligating two mRNA molecules, have indeed been identified in *L. tarentolae* mitochondria [91,92]. Recently, however, based on the discovery of chimeric gRNA–mRNA molecules with oligo(U) tails covalently linked at RNA editing sites, the model has been modified by the idea that the oligo(U) 3′-tail would be the U donor in a transesterification reaction (Fig. 5) [93]. This would give rise to chimeric gRNA–mRNA molecules attached at the editing site. The oligo(U) tail of the gRNA would base-pair with A or G residues of the gRNA itself, and a second transesterification reaction would occur at the next mismatched base with the 3′-hydroxyl of the 5′ mRNA fragment. In this way, one or more U's will be transferred to the mRNA, the gRNA will be freed to restart the cycle, and the TUTase will regenerate the shortened oligo(U) tail. It has been noted that the involvement of gRNAs in transesterification reactions in RNA editing is formally very similar to the use of internal guide sequences in the self-splicing of group I or group II introns [93].

There is evidence suggesting that RNA editing is subject to developmental regulation in *T. brucei*. In fact, it has been shown that at least some of the edited transcripts are found only in the insect or procyclic stage of the life cycle of this parasite, i.e., in the stage when the organism has a complete respiratory system [81,94]. Since the latter cannot be assembled without the full set of edited mRNAs, it appears that editing may be utilized in trypanosomes as a regulatory mechanism for the activation of the aerobic energy yielding processes. This control may operate at the level of the transcription or processing of gRNAs or at the level of synthesis or activity of one of the several proteins involved in RNA editing.

4. *Regulation of mitochondrial gene expression*

4.1. *Differential gene expression*

In the mammalian mitochondrial genome, both transcriptional and post-transcriptional mechanisms play an important role in the differential regulation of gene expression. The organization of transcription of the H-strand in the form of two overlapping transcription units underlies the mechanism whereby the rRNA species are synthesized at a 15 to 60 times higher rate than the mRNAs encoded in the H-strand [95]. In fact, in vivo labeling studies have shown that the transcripts starting at the upstream initiation site are synthesized at a much higher rate than those starting at the downstream site. Furthermore, the synthesis of the products of the two transcription units in isolated mitochondria show different sensitivity to intercalating drugs, low temperature [96], and Ca^{2+} and Mg^{2+} ions [97], and different ATP requirements [98], pointing to an independent regulation of the two types of transcripts. As already discussed, in addition to being controlled at the level of initiation of transcription, differential expression of the rRNA genes and adjacent tRNA genes relative to the rest of the H-strand-encoded genes is regulated at the level of transcription termination at the 3′-end of the 16S rRNA gene. The role of mTERF in this step is presently being actively investigated.

Apart from the transcriptional control mentioned above, the main mechanisms of differential gene expression in mammalian mitochondria operate on RNA stability and on the efficiency of mRNA translation. Control at the level of RNA stability is clearly seen for the tRNAs [9,99]. The tRNA genes belong to three distinct transcription units, which exhibit widely different rates of transcription. Thus, in HeLa cells, the rDNA transcription unit is transcribed ~25 times more frequently than the whole H-strand transcription unit, while the L-strand transcription unit is transcribed 10–16 times more frequently than the whole H-strand transcription unit [100]. Despite the wide differences in the transcription rates of the tRNA genes belonging to the three transcription units, the levels of the different mitochondrial tRNA species in HeLa cells are relatively similar, ranging between 1.9 and 3.1×10^4 molecules per cell for the 12 tRNAs transcribed from the whole H-strand transcription unit, and between 1.7 and 2.2×10^4 for those transcribed from the L-strand transcription unit. Only the two tRNAs transcribed with the rRNA species ($tRNA^{Phe}$ and $tRNA^{Val}$) are present at levels 2–3 times higher than the other tRNA species encoded in the H-strand. Pulse-labeling experiments have indicated that the uniformity in the steady-state levels of tRNA is not due to differences in the metabolic stability of the *mature* tRNAs, which have half-lives longer than 24 h [101,102]. Indeed, the estimated rate of transcription of the whole H-strand would be barely enough to produce the measured amounts of tRNAs encoded in mtDNA, if the tRNAs were completely stable. These observations imply that the mechanisms producing the uniformity of tRNA levels in HeLa cell mitochondria must operate by eliminating the excess of tRNA transcripts synthesized from the rDNA region and the L-strand transcription unit before they reach the pool of mature tRNAs. It is possible that the tRNA synthetases are present in mitochondria in limiting, approximately equivalent amounts for the different tRNAs, as already shown in *E. coli* [103], and act as stabilizing factors for the newly formed tRNAs. In the case of the tRNAs encoded in the L-strand, it is likely that an additional or alternative mechanism, i.e., the high turnover rate of the L-strand polycistronic transcripts [12,104], plays the main role in reducing the amount of processed tRNAs encoded in this strand, by rapidly degrading those transcripts before any tRNA processing takes place.

Differential metabolic stability also accounts for the much lower steady-state levels of the mRNAs relative to tRNAs encoded in the same transcription unit in the H-strand or the L-strand. In fact, in contrast to the tRNAs which have half lives longer than 24 h, the half-lives of the H-strand-encoded mRNAs vary between 25 and 90 min, while that of the L-strand encoded ND6 mRNA has been estimated to be ~7 min [95]. Also the quite different steady-state amounts of the mRNAs encoded in the H-strand must result from differences in their turnover rates.

Recent evidence has indicated that translational control also plays a significant role in differential gene expression in mammalian mitochondria. Thus, a correlation of the rate of synthesis of different mitochondrial translation products with the steady-state levels of the corresponding mRNAs has revealed that the translation efficiencies of these mRNAs vary over almost one order of magnitude [105]. Nothing is known about the mechanism(s) underlying this variability. It is possible that factors related to the structure of the mRNAs, like degree of secondary structure, accessibility of the initiator codon or affinity of the small ribosomal subunit for the initiator codon and surrounding sequences, determine the efficiency of their translation. It is also possible that the rate of translation of mRNAs coding for subunits of a given complex of the oxidative phosphorylation systems is controlled by the availability of cytoplasmically synthesized subunits of the same complex. An intriguing possibility is that specific translation factors exist which modulate the rates of translation of the individual mRNAs, presumably at the initiation steps, adjusting them to the levels required by the cells under different physiological conditions.

In *S. cerevisiae*, as in mammalian cells, transcriptional regulation plays a significant role in the differential expression of mitochondrial genes. The underlying mechanism appears to be the relative strengths of the multiple promoters, which vary over a 20-fold range, depending on differences in the sequence in and around the consensus promoter [106]. Transcriptional attenuation of the polycistronic units [106] and differential stability of the mRNAs also contribute to variation in the level of expression of the individual genes [3]. It is not known to what extent factors related to the structure of the genes or nuclear-encoded proteins control these attenuation phenomena and the differential turnover rates of the individual mRNAs. However, there is evidence that the product of a nuclear gene, CBP1, affects the stability of the cytochrome *b* mRNA [107], possibly by binding to a region of the 5′-untranslated region and protecting it from endonuclease action [4]. There is also increasing evidence of the role of nuclear-coded proteins in controlling RNA processing, as discussed above, and mRNA translation. The most intriguing aspect of such regulation is the fact that a large number of these proteins appear to be specific for individual mitochondrial genes. Particularly interesting is the existence of nuclearly-encoded mRNA-specific translational activators. Three such activators, PET494, PET54 and PET122, have been shown to be necessary for translation of COX3 mRNA [108]. They act by mediating an interaction between the COX3 mRNA leader and components of the mitochondrial translation system. Two such components have been identified as the small subunit ribosomal proteins PET123 [109] and MRP1 [110]. Also cytochrome *b* mRNA translation depends on specific activators encoded in nuclear genes, in particular, CBS1, CBS2 [111], and possibly CBP6 [112]. A COX2 mRNA translational activator, PET111, which may act similarly to the COX3 and cytochrome *b*-specific activators described above has also been identified [113].

The significance of the control by the nucleus of individual mitochondrial genes specifying subunits belonging to the same complex is unknown. One possibility is that the subunit composition of the various complexes of the oxidative phosphorylation system is less constant and more subtly regulated than so far realized. Another possibility is that the multiplicity of rate-limiting steps that this organization of regulatory events entails allows a greater flexibility of adaptation to the variety of environmental and physiological situations to which a free-living organism is exposed.

4.2. Physiological and developmental control of mitochondrial gene expression

Recent work has revealed that, in differentiated mammalian cell systems an important role in the regulation of mitochondrial gene expression is played by translational control. Thus, an analysis of the pattern of mitochondrial gene expression in isolated quadriceps muscle fibers [9] and brain synaptosomes [114] from rats of different ages has shown that the pattern of mitochondrial protein synthesis differs significantly from that observed in an exponentially growing rat fibroblast line in the relative rate of synthesis of the various translation products. The most remarkable difference was the marginal level or absence of labeling of ND5, one of the mtDNA-encoded subunits of the respiratory chain NADH dehydrogenase. The strong under-representation of ND5 in the rat muscle or brain synaptosome labeling pattern, compared with the rat fibroblast pattern, contrasted with the presence of a comparable level of ND5 mRNA in the three cell types. The implication of these observations is that the reduction or total absence of synthesis of ND5 reflects some form of translational control. Further work will be needed to establish whether the availability of one or more cytoplasmically synthesized subunits of NADH dehydrogenase or some specific translation factor controls the rate of synthesis of ND5. Further evidence of translational control of mitochondrial gene expression in rat muscle was provided by the observation that a sharp decline in the rate of mitochondrial

ysis of ATP [156]. Concomitant with, or prior to its release from mshp70's, the prepiece is clipped from the precursor by the prepiece specific peptidase [153,154,156].

The signal peptidase is a heterodimer consisting of two similarly sized and homologous polypeptides which have been named Mas1p and Mas2p in yeast, and PEP (processing enhancing protein) and MPP (mitochondrial processing peptidase), respectively, in *N. crassa* [157–162]. *N. crassa* PEP is a bifunctional protein, as it has been shown to be identical to the core 1 protein of the cytochrome *c* reductase [163]. The same is not true for the yeast protein, Mas1p.

After release from mhsp70, some, at least, of the precursors go on to bind to hsp60 [156]. The protein hsp60 is a member of another class of heat shock proteins, the 'chaperonins' [164], which serve as catalysts for the assembly of multimeric complexes. The protein hsp60 is a mitochondrial protein highly conserved in *Tetrahymena*, yeast, *Xenopus*, and man, and moreover, is antigenically related to the *E. coli* GroEL protein [165,166]. The yeast, *Tetrahymena*, and *N. crassa* hsp60's form the same kind of cylindrical particles [166,167] that had been described earlier for the *E. coli* GroEL particle [168], namely, that of two stacked rings of seven subunits. The protein hsp60 is the product of the Mif4 gene in yeast, and is essential for the assembly of imported proteins into multimeric complexes [169]. The null mutant of this gene is inviable [170]. Release from hsp60 also requires ATP hydrolysis [123]. For a detailed discussion of the energy requirements for protein import see refs. [3,5,171]. It seems likely that other protein factors in the matrix, besides those discussed above, are involved in the refolding of the imported proteins and their assembly into oligomeric structures.

6. *Assembly of enzyme complexes of the OX–PHOS system*

Work on the assembly of these chimeric complexes has continued at an accelerated pace in the past few years. Most of the progress has been made on the structural characterization of the nuclear-encoded subunits of these complexes. Thus, the sequence of all subunits of cytochrome *c* oxidase, ubiquinol-cytochrome *c* reductase, succinate dehydrogenase and H^+-ATPase from *S. cerevisiae* and mammalian species has been determined either directly or deduced from the cDNA sequence. In several cases, also the nuclear genes have been cloned and their structure characterized. Also in the case of the largest enzyme complex of the respiratory chain, NADH dehydrogenase (complex I), almost all nuclear-coded subunits from the bovine enzyme [171a] and many from the *N. crassa* enzyme [172] have been cloned and sequenced. Reference is made to the other chapters in this book for a detailed description of the information which is presently available on the structure and assembly of the OX–PHOS complexes. Here, only a general discussion will be made of the principles that start emerging about the assembly of these complexes. First, it appears that, in general, the nuclear-encoded subunits of the complexes continue to be synthesized, imported into the organelles, inserted into the inner membrane, and, in some cases, assembled in subcomplexes even in the absence of the synthesis of the mitochondrial-encoded subunits. Under the latter conditions, the half-lives of the nuclear-coded subunits and subcomplexes vary considerably, with some being relatively stable and others very unstable due to degradation by proteases. The inner mitochondrial membrane can be envisaged as harboring various-size pools of parts of each enzyme complex at various stages of assembly, each turning over with a different half-life. The overall assembly of each complex is most probably an ordered process. The individual steps of this process have started being elucidated for various complexes. One of the most intriguing aspects of this assembly is the participation of non-structural components, which are not found in the final oligomeric complex. To such a class of components belong the CBP3 and CBP4 gene products, which are membrane proteins that promote assembly of a stable ubiquinol-cytochrome *c* reductase complex in yeast [173,174], the COX10 and COX11 proteins, which appear to have a similar function in the assembly of the cytochrome *c* oxidase complex also in yeast [175,176], and the ATP10, ATP11 and ATP12 gene products, which are required for the assembly of the yeast F_1-ATPase [177–179]. While some of these proteins may function in the synthesis of prosthetic groups or post-translational modification of the catalytic subunits, it seems very likely that others have a function in directing the assembly of the catalytic and structural subunits as 'scaffolding' proteins, by allowing the proper folding of the subunits and their correct reciprocal interactions in an ordered, sequential process. To this class of 'scaffolding' proteins belongs 'hsp60', a member of the family of 'chaperone' proteins (see above).

A series of comparative biochemical, electron-microscopical and functional investigations have led to a model of the structure of the respiratory chain NADH dehydrogenase (complex I) as consisting of two complementary moieties, which can be distinguished in terms of their evolutionary derivation, genetic control, assembly pathways, arrangement in the membrane, and redox components [172] (Fig. 7). An electron-microscopical analysis has permitted a reconstruction of the tridimensional structure of the enzyme from *N. crassa* as an L-shaped complex containing two arms, a longer, stain-excluding arm, which is membrane-buried, and a more bulky arm which extends perpendicularly to the membrane, protruding into the matrix space [180]. This peripheral moiety contains ~13 nuclear-encoded subunits, including the NADH and FMN binding subunit and the subunits binding the iron–sulfur clusters N-1, N-3 and N-4. The membrane moiety consists of approximately fifteen subunits, including the seven polypeptides encoded in mtDNA; this moiety contains the ubiquinone binding subunit(s) and the subunit(s) binding the high-potential iron–sulfur cluster N-2. It has been proposed that electron flow proceeds from NADH bound to the 51 kDa subunit to FMN, which is thus reduced to $FMNH_2$, and from this sequentially to the cluster N-1 and to the clusters N-3 and N-4 (Fig. 7). It is further assumed, on the basis of the occurrence of a piericidin-A-insensitive ubiquinone reduction site in the small form of NADH dehydrogenase in *N. crassa* (see below), that the electronic connector between the clusters N-3 and N-4 and the high-potential cluster N-2 is an internal ubiquinone. From a putative ubiquinone reduction site in the peripheral moiety electrons are transferred to a hypothetical quinol oxidation site in the membrane moiety, and from this, sequentially, to iron–sulfur cluster N-2 and ubiquinone (Fig. 7).

The original observation underlying the model of Fig. 7 was the discovery of a smaller form of NADH dehydrogenase in *N. crassa* grown in the presence of chloramphenicol, an inhibitor of mitochondrial protein synthesis [181]. This form contains ~13 nuclear-encoded subunits, all apparently identical to nuclear-encoded subunits of the large enzyme and all absent from the hydrophobic portion; furthermore, the small enzyme contains the NADH binding site, FMN, and iron–sulfur clusters N-1, N-3 and N-4, and a low-affinity binding site for ubiquinone, which is insensitive to rotenone or piericidin-A. Subsequent labeling-immunoprecipitation experiments showed that, under normal growth conditions, the small form of NADH dehydrogenase and the membrane moiety, containing all the mtDNA-encoded subunits, as well as nuclear-encoded subunits which have no counterpart in the small enzyme, are independently assembled and then joined to form the whole complex [182]. The hypothesis has been advanced that the small form of the enzyme represents a more ancient enzyme, which was subsequently extended by association with another complex. The idea that the two independent pathways of assembly of the peripheral and the membrane part reflect the evolutionary derivation of these two moieties from two distinct enzyme complexes existing in bacteria, each containing a distinct segment of the electron pathway from NADH to ubiquinone, has received support from the recent demonstration of a striking sequence similarity between the NADH dehydrogenase subunits containing the binding sites for NADH, FMN and the iron–sulfur clusters N-1, N-3 and N-4 and the bacterial NAD^+ hydrogenase [183–186], and between the subunits carrying the electron transfer from the iron–sulfur cluster N-2 to ubiquinone and a bacterial formate hydrogen lyase [172,187]. The association of the two enzyme complexes probably occurred prior to the endosymbiotic event leading to the formation of mitochondria, since rotenone- and piericidin-A-sensitive NADH dehydrogenases containing iron–sulfur clusters similar to the mitochondrial enzyme have been found in present-day bacteria, like *Paracoccus denitrificans* (T.Y. Yagi, personal communication).

The enzymes of the respiratory chain in eukaryotic cells are much more complex than their bacterial counterparts, and contain a large number of subunits which do not participate in electron transport and whose function is largely unknown. It is generally believed that the non-catalytic subunits have the function of modulating the activity of the catalytic subunits in response to developmental or environmental changes. Recent evidence has provided support for this view. Thus, in *S. cerevisiae*, two isoforms of subunit V have been identified, Va and Vb, with the gene COX5a being 25 to 55 times more active than the gene COX5b in cells grown aerobically in glucose, whereas the expression of gene COX5b is favored in cells grown anaerobically or aerobically in the absence of heme [188]. Distinct isoforms of the nuclear-encoded subunit VII of cytochrome *c* oxidase in *Dictyostelium discoideum*, which are expressed with strict dependence on growth conditions [189], and isoforms of several nuclear-encoded subunits of the mammalian enzyme (IV, VIIa, VIIb, VIII), with tissue- or developmental stage-specific expression, have also been described [190]. The mammalian tissue-specific isozymes are

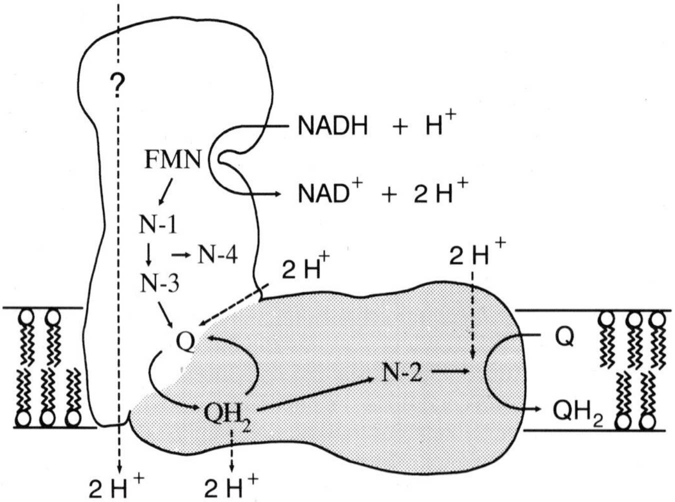

Fig. 7. Schematic drawing illustrating the arrangement of complex I from *N. crassa* and its two moieties with respect to the inner mitochondrial membrane. The vertical arm projecting into the matrix contains the NADH binding site, the FMN, the iron–sulfur clusters N-1, N-3 and N-4 and the putative internal ubiquinone reduction site, while the horizontal arm (shadowed), buried in the membrane, contains the putative internal ubiquinone oxidation site, the cluster N-2 and the rotenone- and piericidin-A-sensitive ubiquinone reduction site. See text for details. (Reproduced, with permission, from ref. [172].)

particularly interesting in connection with the abundant evidence of tissue specificity of defects of the OX–PHOS system in mitochondrial diseases [191,192].

Further complexity in our understanding of the physiological significance of the many additional subunits of the mitochondrial respiratory chain complexes has been provided by the demonstration that a nuclear-encoded subunit of the ubiquinone-cytochrome *c* reductase complex in *N. crassa* is bifunctional, being identical to the processing enhancing protein (PEP), a polypeptide which strongly stimulates the matrix processing peptidase (MPP) (see above) [163]. Therefore, this subunit participates in both electron transport and protein processing. Whether this bifunctionality reflects a still unknown link between precursor presequence cleavage and site of cytochrome *c* reductase activity remains to be determined [193]. A second example of subunits of respiratory chain complexes not being directly involved in respiration or translocation has been provided by the recent observation that a subunit of the peripheral moiety of NADH dehydrogenase from *N. crassa* [194] and from bovine heart [194a] is identical to the mitochondrial acyl-carrier protein, with pantothenate as prosthetic group.

7. *Perspectives*

The past few years have witnessed rapid and exciting developments in several areas of mitochondrial biogenesis, in particular those concerning the unique mode of expression of the mitochondrial genome in various eukaryotic cells and its control by the nucleus, the molecular characterization of the protein import machinery and the assembly of the complexes of the oxidative phosphorylation system. Such studies are not only throwing light on the great variety of mechanisms that operate in the formation of mitochondria in different organisms, but also have profound evolutionary implications. Thus, penetrating studies on RNA editing in trypanosomatids, and RNA splicing and intron transposition in fungi have revealed very intriguing interrelationships between the three processes. It is hoped that such studies will eventually provide insights into ancestral events that took place during the evolution of the 'RNA world'.

In yeast, besides the well-established techniques of nuclear gene disruption and gene replacement, powerful new technologies, like DNA-mediated transformation of mitochondria by the biolistic approach [195,196] and allotopic expression in the nucleus of universal code-equivalents of mitochondrial genes [197] are being successfully used to attack several problems concerning mitochondrial biogenesis. It can be expected that such technologies will soon be applied to mammalian cell systems. Furthermore, the recent development of a generalized approach for repopulation of human mtDNA-less cells with exogenous mitochondria by functional complementation [198] has created a useful tool for genetic manipulation of mammalian mitochondria. This approach is presently being applied [199,200] to the analysis of the ever-increasing variety of mtDNA mutations which have been recognized in the past few years as causes of mitochondrial diseases [191,192]. These studies have opened the possibility of investigating in in vitro systems the effects of specific natural mutations on the expression of human mitochondrial genes. This approach is expected to yield a wealth of information on the functional role of mitochondrial genes in general and on the mechanism and control of their expression. Application of the mitochondria transfer technology should also be useful for investigating the rules that govern selection, segregation and complementation of the mutant mtDNA and of the coexisting wild-type mtDNA, and the role of the nuclear background on the expression of the mtDNA mutations. Finally such an approach may provide a definitive test of the often proposed but still elusive role of mtDNA damage in causing or contributing to senescence.

Acknowledgements

The work from our laboratory described in this paper was supported by the National Institutes of Health Grant GM-11726. We would like to thank many colleagues for sending us reprints and preprints of their recent work and Drs. G. Schatz, L. Simpson and H. Weiss for providing copies of illustrations. We also thank Ms. Lisa Tefo for editorial assistance.

References

1 Tzagoloff, A. and Myers, A.M. (1986) Annu. Rev. Biochem. 55, 249–288.
2 Chomyn, A. and Attardi, G. (1987) Curr. Top. Bioenerg. 15, 295–329.
3 Attardi, G. and Schatz, G. (1988) Annu. Rev. Cell Biol. 4, 289–333.
4 Grivell, L.A. (1989) Eur. J. Biochem. 182, 477–493.
5 Pfanner, N. and Neupert, W. (1990) Annu. Rev. Biochem. 59, 331–353.
6 Montoya, J., Christianson, T., Levens, D., Rabinowitz, M. and Attardi, G. (1982) Proc. Natl. Acad. Sci. USA 79, 7195–7199.
7 Chang, D.D. and Clayton, D.A. (1984) Cell 36, 635–643.
8 Bogenhagen, D.F., Applegate, E.F. and Yoza, B.K. (1984) Cell 36, 1105–1113.
9 Attardi, G., Chomyn, A., King, M.P., Kruse, B., Loguercio Polosa, P. and Narasimhan Murdter, N. (1989) Biochem. Soc. Trans. 18, 509–513.
10 Chang, D.D. and Clayton, D.A. (1985) Proc. Natl. Acad. Sci. USA 82, 351–355.
11 Montoya, J., Gaines, G.L. and Attardi, G. (1983) Cell 34, 151–159.
12 Aloni, Y. and Attardi, G. (1971) Proc. Natl. Acad. Sci. USA 68, 1757–1761.
13 Murphy, W.I., Attardi, B., Tu, C. and Attardi, G. (1975) J. Mol. Biol. 99, 809–814.
14 Doersen, C.-J., Guerrier-Takada, C., Altman, S. and Attardi, G. (1985) J. Biol. Chem. 260, 5942–5949.
15 Walberg, M.W. and Clayton, D.A. (1983) J. Biol. Chem. 258, 1268–1275.
16 Shuey, D.J. and Attardi, G. (1985) J. Biol. Chem. 260, 1952–1958.
17 Fisher, R.P. and Clayton, D.A. (1985) J. Biol. Chem. 260, 11330–11338.
18 Fisher, R.P., Topper, J.N. and Clayton, D.A. (1987) Cell 50, 247–258.
19 Parisi, M.A. and Clayton, D.A. (1991) Science 252, 965–969.
19a Fisher, R.P., Lisowsky, T., Parisi, M.A. and Clayton, D.A. (1992) J. Biol. Chem. 267, 3358–3367.
20 Fisher, R.P., Parisi, M.A. and Clayton, D.A. (1989) Genes Development 3, 2202–2217.
21 Kruse, B., Narasimhan, N. and Attardi, G. (1989) Cell 58, 391–397.
22 Dubin, D.T., Montoya, J., Timko, K.D. and Attardi, G. (1982) J. Mol. Biol. 157, 1–19.
23 Christianson, T.W. and Clayton, D.A. (1988) Mol. Cell. Biol. 8, 4502–4509.
24 Winkley, C.S., Keller, M.J. and Jaehning, J.A. (1985) J. Biol. Chem. 260, 14214–14223.
25 Kelly, J.L. and Lehman, I.R. (1986) J. Biol. Chem. 261, 10340–10347.
26 Schinkel, A.H., Groot-Koerkamp, M.J.A., Touw, E.P.W. and Tabak, H.F. (1987) J. Biol. Chem. 262, 12785–12791.
27 Schinkel, A.H., Groot-Koerkamp, M.J.A. and Tabak, H.F. (1988) EMBO J. 7, 3255–3262.
28 Masters, B.S., Stohl, L.L. and Clayton, D.A. (1987) Cell 51, 89–99.
29 Davies, R.W., Waring, R.B., Ray, J.A., Brown, T.A. and Scazzocchio, C. (1982) Nature 300, 719–724.
30 Michel, F. and Dujon, B. (1983) EMBO J. 2, 33–38.
31 Cech, T.R. and Bass, B.L. (1986) Annu. Rev. Biochem. 55, 599–629.
32 Kotylak, Z., Lazowska, J., Hawthorne, D.C. and Slonimski, P.P. (1985) in: E. Quagliariello, E.C. Slater, F. Palmieri, C. Saccone and A.M. Kroon (Eds.) Achievements and Perspectives in Mitochondrial Research, Elsevier, Amsterdam, pp. 1–20.
33 Wenzlau, J.M., Saldanha, R.J., Butow, R.A. and Perlman, P.S. (1989) Cell 56, 421–430.

34 Séraphin, B., Simon, M. and Faye, G. (1988) EMBO J. 7, 1455–1464.
35 Valencik, M.L., Kloeckner-Gruissem, B., Poyton, R.O. and McEwen, J.E. (1989) EMBO J. 8, 3899–3904.
36 Lambowitz, A.M. and Perlman, P.S. (1990) TIBS 15, 440–444.
37 Perlman, P.S. and Butow, R.A. (1989) Science 246, 1106–1109.
38 Dujon, B. (1989) Gene 82, 91–114.
39 Banroques, J., Delahodde, A. and Jacq, C. (1986) Cell 46, 837–844.
40 Lazowska, J., Claisse, M., Gargouri, A., Kotylak, Z., Spyridakis, A. and Slonimski, P.P. (1989) J. Mol. Biol. 205, 275–289.
41 Delahodde, A., Goguel, V., Becam, A.M., Creusot, F., Perea, J., Banroques, J. and Jacq, C. (1989) Cell 56, 431–441.
42 Dujardin, G., Jacq, C. and Slonimski, P.P. (1982) Nature 298, 628–632.
43 Labouesse, M., Herbert, C.J., Dujardin, G. and Slominski, P.P. (1987) EMBO J. 6, 713–721.
44 Goguel, V., Bailone, A., Devoret, R. and Jacq, C. (1989) Mol. Gen. Genet. 216, 70–74.
45 Michel, F. and Lang, B.F. (1985) Nature 316, 641–642.
46 Levra-Juillet, E., Boulet, A., Séraphin, B., Simon, M. and Faye, G. (1989) Mol. Gen. Genet. 217, 168–171.
47 Akins, R.A. and Lambowitz, A.M. (1987) Cell 50, 331–345.
48 Herbert, C.J., Labouesse, M., Dujardin, G. and Slonimski, P.P. (1988) EMBO J. 7, 473–483.
49 Majumder, A.L., Akins, R.A., Wilkinson, J.G., Kelley, R.L., Snook, A.J. and Lambowitz, A.M. (1989) Mol. Cell Biol. 9, 2089–2104.
50 Cherniack, A.D., Garriga, G., Kittle Jr., J.D., Akins, R.A. and Lambowitz, A.M. (1990) Cell 62, 745–755.
51 Kittle, J.D., Mohr, G., Gianelos, J.A., Wang, H. and Lambowitz, A. (1991) Genes Development 5, 1009–1021.
52 Dujardin, G., Labouesse, M., Netter, P. and Slonimski, P.P. (1983) in: R.J. Schweyen, R. Wolf and F. Kaudewitz (Eds.) Mitochondria 1983, W. de Gruyter, Berlin, pp. 233–250.
53 Herbert, C.J., Asher, E.B., Bousquet, I., Dujardin, G., Groudinsky, O., Kermorgant, M., Labouesse, M. and Slonimski, P.P. (1990) in: E. Quagliariello, S. Papa, F. Palmieri and C. Saccone (Eds.) Structure, Function and Biogenesis of Energy Transfer System, Elsevier, Amsterdam, pp. 201–204.
54 Séraphin, B., Simon, M., Boulet, A. and Faye, G. (1989) Nature 337, 84–87.
55 Costanzo, M.C., Mueller, P.P., Strick, C.A. and Fox, T.D. (1986) Mol. Gen. Genet. 202, 294–301.
56 Ben Asher, E., Groudinsky, O., Dujardin, G., Altamura, N., Kermorgant, M. and Slonimski, P.P. (1989) Mol. Gen. Genet. 215, 517–528.
57 McGraw, P. and Tzagoloff, A. (1983) J. Biol. Chem. 258, 9459–9468.
58 Kreike, J., Schulze, M., Ahne, F. and Lang, B.F. (1987) EMBO J. 6, 2123–2129.
59 Sharp, P.A. (1985) Cell 42, 397–400.
60 Cech, T.R. (1985) Int. Rev. Cytol. 3, 3–22.
61 Mohr, G. and Lambowitz, A.M. (1991) Nature 354, 164–167.
62 Fukuzawa, H., Kohchi, T., Shirai, H., Ohyama, K., Umesono, K., Inokuchi, H. and Ozeki, H. (1986) FEBS Lett. 198, 11–15.
63 Zaita, N., Torazawa, K., Shinozaki, K. and Sugiura, M. (1987) FEBS Lett. 210, 153–156.
64 Hildebrand, M., Hallick, R.B., Passavant, C.W. and Bourque, D.P. (1988) Proc. Natl. Acad. Sci. USA 85, 372–376.
65 Jarrell, K.A., Dietrich, R.C. and Perlman, P.S. (1988) Mol. Cell. Biol. 8, 2361–2366.
66 Chapdelaine, Y. and Bonen, L. (1991) Cell 65, 465–472.
67 Wissinger, B., Schuster, W. and Brennicke, A. (1991) Cell 65, 473–482.
68 Greeve, J., Navaratnam, N. and Scott, J. (1991) Nucl. Acids Res. 19, 3569–3576.
69 Powell, L.M., Wallis, S.C., Pease, R.J., Edwards, Y.H., Knott, T.J. and Scott, J. (1987) Cell 50, 831–840.
70 Chen, S.-H., Habib, G., Yang, C.-Y., Gu, Z.-W., Lee, B.R., Weng, S.-A., Silberman, S.R., Cai, S.-J., Deslypere, J.P., Rosseneu, M., Gotto Jr., A.M., Li, W.-H. and Chan, L. (1987) Science 328, 363–366.
71 Sommer, B., Köhler, M., Sprengel, R. and Seeburg, P.H. (1991) Cell 67, 11–19.

72 Thomas, S.M., Lamb, R.A. and Paterson, R.G. (1988) Cell 54, 891–902.
73 Cattaneo, R., Kaelin, K., Baczko, K. and Billeter, M.A. (1989) Cell 56, 759–764.
74 Simpson, L. and Shaw, J. (1989) Cell 57, 355–366.
75 Covello, P.S. and Gray, M.W. (1989) Nature 341, 662–666.
76 Gualberto, J.M., Lamattina, L., Bonnard, G., Weil, J.-H. and Grienenberger, J.-M. (1989) Nature 341, 660–662.
77 Hiesel, R., Wissinger, B., Schuster, W. and Brennicke, A. (1989) Science 246, 1632–1634.
78 Hoch, B., Maier, R.M., Appel, K., Igloi, G.L. and Kössel, H. (1991) Nature 353, 178–180.
79 Mahendran, R., Spottswood, M.R. and Miller, D.L. (1991) Nature 349, 434–438.
80 Benne, R., van den Burg, J., Brakenhoff, J.P.J., Sloof, P., van Boom, J.H. and Tromp, M.C. (1986) Cell 46, 819–826.
81 Feagin, J.E., Jasmer, D.P. and Stuart, K. (1987) Cell 49, 337–345.
82 Feagin, J.E., Shaw, J.M., Simpson, L. and Stuart, K. (1988) Proc. Natl. Acad. Sci. USA 85, 539–543.
83 Shaw, J.M., Feagin, J.E., Stuart, K. and Simpson, L. (1988) Cell 53, 401–411.
84 Feagin, J.E., Abraham, J.M. and Stuart, K. (1988) Cell 53, 413–422.
85 Koslowsky, D.J., Bhat, G.J., Perrollaz, A.L., Feagin, J.E. and Stuart, K. (1990) Cell 62, 901–911.
86 Souza, A.E., Myler, P.J. and Stuart, K. (1992) Mol. Cell. Biol. 12, 2100–2107.
87 Maslov, D.A., Sturm, N.R., Niner, B.M., Gruszynski, E.S., Peris, M. and Simpson, L. (1992) Mol. Cell. Biol. 12, 56–67.
88 Blum, B., Bakalara, N. and Simpson, L. (1990) Cell 60, 189–198.
89 Sturm, N.R. and Simpson, L. (1990) Cell 61, 879–884.
90 Blum, B. and Simpson, L. (1990) Cell 62, 391–397.
91 Bakalara, N., Simpson, A.M. and Simpson, L. (1989) J. Biol. Chem. 264, 18679–18686.
92 Simpson, L. (1990) Science 250, 512–513.
93 Blum, B., Sturm, N.R., Simpson, A.M. and Simpson, L. (1991) Cell 65, 543–550.
94 Feagin, J.E. and Stuart, K. (1988) Mol. Cell Biol. 8, 1259–1265.
95 Gelfand, R. and Attardi, G. (1981) Mol. Cell. Biol. 1, 497–511.
96 Gaines, G. and Attardi, G. (1984) J. Mol. Biol. 172, 451–466.
97 Gaines, G. and Attardi, G. (1984) Mol. Cell. Biol. 4, 1605–1617.
98 Gaines, G. and Attardi, G. (1987) J. Biol. Chem. 262, 1907–1915.
99 King, M.P. (1977) PhD Thesis, California Institute of Technology, Pasadena, CA.
100 Attardi, G., Cantatore, P., Chomyn, A., Crews, S., Gelfand, R., Merkel, C., Montoya, J. and Ojala, D. (1982) in: P. Slonimski, P. Borst and G. Attardi (Eds.) Mitochondrial Genes, Cold Spring Harbor Laboratory, Cold Spring Harbor, New York, pp. 51–71.
101 Zylber, E. and Penman, S. (1969) J. Mol. Biol. 46, 201–204.
102 Attardi, B. and Attardi, G. (1971) J. Mol. Biol. 55, 231–249.
103 Neidhardt, F.C., Block, P.L., Pederson, S. and Reeh, S. (1977) J. Bacteriol. 129, 378–387.
104 Cantatore, P. and Attardi, G. (1980) Nucl. Acids Res. 8, 2605–2625.
105 Chomyn, A. and Attardi, G. (1987) in: S. Papa, B. Chance and L. Ernster (Eds.) Cytochrome Systems. Molecular Biology and Bioenergetics, Plenum Press, New York, pp. 145–152.
106 Mueller, D.M. and Getz, G.S. (1986) J. Biol. Chem. 261, 11756–11764.
107 Dieckmann, C.L., Koerner, T.J. and Tzagoloff, A. (1984) J. Biol. Chem. 259, 4722–4731.
108 Costanzo, M.C. and Fox, T.D. (1990) Annu. Rev. Genet. 24, 91–113.
109 McMullin, T.W., Haffter, P. and Fox, T.D. (1990) Mol. Cell. Biol. 10, 4590–4595.
110 Haffter, P., McMullin, T.W. and Fox, T.D. (1990) Genetics 127, 319–326.
111 Rödel, G. (1986) Curr. Genet. 11, 41–45.
112 Dieckmann, C.L. and Tzagoloff, A. (1985) J. Biol. Chem. 260, 1513–1520.
113 Poutre, C.G. and Fox, T.D. (1987) Genetics 115, 637–647.
114 Loguercio Polosa, P. and Attardi, G. (1991) J. Biol. Chem. 266, 10011–10017.
115 Glick, B. and Schatz, G. (1991) Annu. Rev. Genet. 25, 21–44.
116 Deshaies, R.J., Koch, B.D., Werner-Washburne, M., Craig, E.A. and Schekman, R. (1988) Nature 332, 800–805.
117 Chirico, W.J., Waters, M.G. and Blobel, G. (1988) Nature 332, 805–810.

118 Zimmerman, R., Sagstetter, M., Lewis, M.J. and Pelham, H.R.B. (1988) EMBO J. 7, 2875–2880.
119 Pelham, H.R.B. (1986) Cell 46, 959–961.
120 Rothman, J.E. (1989) Cell 59, 591–601.
121 Hartl, F.-U., Pfanner, N., Nicholson, D.W. and Neupert, W. (1989) Biochim. Biophys. Acta 988, 1–45.
122 Verner, K. and Schatz, G. (1987) EMBO J. 6, 2449–2456.
123 Ostermann, J., Horwich, A.L., Neupert, W. and Hartl, F.-U. (1989) Nature 341, 125–130.
124 Pfanner, N., Rassow, J., Guiard, B., Söllner, T., Hartl, F.-U. and Neupert, W. (1990) J. Biol. Chem. 265, 16324–16329.
125 Murakami, H., Pain, D. and Blobel, G. (1988) J. Cell Biol. 107, 2051–2057.
126 Murakami, K. and Mori, M. (1990) EMBO J. 9, 3201–3208.
127 Söllner, T., Griffiths, G., Pfaller, R., Pfanner, N. and Neupert, W. (1989) Cell 59, 1061–1071.
128 Söllner, T., Pfaller, R., Griffiths, G., Pfanner, N. and Neupert, W. (1990) Cell 62, 107–115.
129 Steger, H.F., Söllner, T., Kielder, M., Dietmeier, K.A., Pfaller, R., Trülzsch, K.S., Tropschug, M., Neupert, W. and Pfanner, N. (1990) J. Cell Biol. 111, 2353–2363.
130 Pfanner, N., Söllner, T. and Neupert, W. (1991) TIBS 16, 63–67.
131 Hines, V., Brandt, A., Griffiths, G., Horstmann, H., Brütsch, H. and Schatz, G. (1990) EMBO J. 9, 3191–3200.
132 Riezman, H., Hase, T., van Loon, A.P.G.M., Grivell, L.A., Suda, K. and Schatz, G. (1983) EMBO J. 2, 2161–2168.
133 Pain, D., Murakami, H. and Blobel, G. (1990) Nature 347, 444–449.
134 Murakami, H., Blobel, G. and Pain, D. (1990) Nature 347, 488–491.
135 Meyer, D. (1990) Nature 347, 424–425.
136 Phelps, A., Schobert, C.T. and Wohlrab, H. (1991) Biochemistry 30, 248–252.
137 Willey, D.L., Fischer, K., Wachter, E., Link, T.A. and Flügge, U.-I. (1991) Planta 183, 451–461.
138 Flügge, U.-I., Weber, A., Fischer, K., Lottspeich, F., Eckerskorn, C., Waegemann, K. and Soll, J. (1991) Nature 353, 364–367.
139 Joyard, J. and Douce, R. (1988) Nature 333, 306–307.
140 Pain, D., Kanwar, Y.S. and Blobel, G. (1988) Nature 331, 232–237.
141 Pain, D. and Blobel, G. (1988) Nature 333, 307.
142 Vestweber, D., Brunner, J., Baker, A. and Schatz, G. (1989) Nature 341, 205–209.
143 Kiebler, M., Pfaller, R., Söllner, T., Griffiths, G., Horstmann, H., Pfanner, N. and Neupert, W. (1990) Nature 348, 610–616.
144 Baker, K.P., Schaniel, A., Vestweber, D. and Schatz, G. (1990) Nature 348, 605–609.
145 Pfanner, N., Hartl, F.-U. and Neupert, W. (1988) Eur. J. Biochem. 175, 205–212.
146 Schatz, G. (1991) Harvey Lect. 85, 109–126.
147 Hartl, F.-U. and Neupert, W. (1990) Science 247, 930–938.
148 Martin, J., Mahlke, K. and Pfanner, N. (1991) J. Biol. Chem. 266, 18051–18057.
149 Rassow, J., Hartl. F.-U., Guiard, B., Pfanner, N. and Neupert, W. (1990) FEBS Lett. 275, 190–194.
150 Rassow, J., Guiard, B., Wienhues, U., Herzog, V., Hartl, F.-U. and Neupert, W. (1989) J. Cell Biol. 109, 1421–1428.
151 Hwang, S.T., Wachter, C. and Schatz, G. (1991) J. Biol. Chem. 266, 21083–21089.
152 Kang, P.-J., Ostermann, J., Shilling, J., Neupert, W., Craig, E.A. and Pfanner, N. (1990) Nature 348, 137–143.
153 Scherer, P.E., Krieg, U.C., Hwang, S.T., Vestweber, D. and Schatz, G. (1990) EMBO J. 9, 4315–4322.
154 Ostermann, J., Voos, W., Kang, P.-J., Craig, E.A., Neupert, W. and Pfanner, N. (1990) FEBS Lett. 277, 281–284.
155 Neupert, W., Hartl, F.-U., Craig, E.A. and Pfanner, N. (1990) Cell 63, 447–450.
156 Manning-Krieg, U.C., Scherer, P.E. and Schatz, G. (1991) EMBO J. 10, 3273–3280.
157 Hawlitschek, G., Schneider, H., Schmidt, B., Tropschug, M., Hartl, F.-U. and Neupert, W. (1988) Cell 53, 795–806.
158 Witte, C., Jensen, R.E., Yaffe, M.P. and Schatz, G. (1988) EMBO J. 7, 1439–1447.

159 Pollock, R.A., Hartl, F.-U., Cheng, M.Y., Ostermann, J., Horwich, A. and Neupert, W. (1988) EMBO J. 7, 3493–3500.
160 Jensen, R.E. and Yaffe, M.P. (1988) EMBO J. 7, 3863–3871.
161 Schneider, H., Arretz, M., Wachter, E. and Neupert, W. (1990) J. Biol. Chem. 265, 9881–9887.
162 Yang, M., Geli, V., Oppliger, W., Suda, K., James, P. and Schatz, G. (1991) J. Biol. Chem. 266, 6416–6423.
163 Schulte, U., Arretz, M., Schneider, H., Tropschug, M., Wachter, E., Neupert, W. and Weiss, H. (1989) Nature 339, 147–149.
164 Hemmingsen, S.M., Woolford, C., van der Vies, S.M., Tilly, K., Dennis, D.T., Georgopoulos, C.P., Hendrix, R.W. and Ellis, R.J. (1988) Nature 333, 330–334.
165 McMullin, T.W. and Hallberg, R.L. (1987) Mol. Cell Biol. 7, 4414–4423.
166 McMullin, T.W. and Hallberg, R.L. (1988) Mol. Cell Biol. 8, 371–380.
167 Hutchinson, E.G., Tichelaar, W., Hofhaus, G., Weiss, H. and Leonard, K.R. (1989) EMBO J. 8, 1485–1490.
168 Hendrix, R.W. (1979) J. Mol. Biol. 129, 375–392.
169 Cheng, M.Y., Hartl, F.-U., Martin, J., Pollock, R.A., Kalousek, F., Neupert, W., Hallberg, E.M., Hallberg, R.L. and Horwich, A.L. (1989) Nature 337, 620–625.
170 Reading, D.S., Hallberg, R.L. and Myers, A.M. (1989) Nature 337, 655–659.
171 Hwang, S.T. and Schatz, G. (1989) Proc. Natl. Acad. Sci. USA 86, 8432–8436.
171a Walker, J.E. (1992) Q. Rev. Biophys., in press.
172 Weiss, H., Friedrich, T., Hofhaus, G. and Preis, D. (1991) Eur. J. Biochem. 197, 563–576.
173 Wu, M. and Tzagoloff, A. (1989) J. Biol. Chem. 264, 11122–11130.
174 Tzagoloff, A., Capitanio, N., Crivellone, M., Galti, M. and Nobrega, M. (1990) in: Structural and Organizational Aspects of Metabolic Regulation, Alan R. Liss, New York, pp. 71–81.
175 Nobrega, M.P., Nobrega, F.G. and Tzagoloff, A. (1990) J. Biol. Chem. 265, 14220–14226.
176 Tzagoloff, A., Capitanio, N., Nobrega, M.P. and Galti, D. (1990) EMBO J. 9, 2759–2764.
177 Ackerman, S.H. and Tzagoloff, A. (1990) J. Biol. Chem. 265, 9952–9959.
178 Ackerman, S.H. and Tzagoloff, A. (1990) Proc. Natl. Acad. Sci. USA 87, 4986–4990.
179 Bowman, S., Ackerman, S.H., Griffith, D.E. and Tzagoloff, E. (1991) J. Biol. Chem. 266, 7517–7523.
180 Hofhaus, G., Weiss, H. and Leonard, K. (1991) J. Mol. Biol. 221, 1027–1043.
181 Friedrich, T., Hofhaus, G., Ise, W., Nehls, U., Schmitz, B. and Weiss, H. (1989) Eur. J. Biochem. 180, 173–180.
182 Tuschen, G., Sackmann, U., Nehls, U., Haiker, H., Buse, G. and Weiss, H. (1990) J. Mol. Biol. 213, 845–857.
183 Tran-Betcke, A., Warnecke, U., Böcker, C., Zaborosch, C. and Friedrich, B. (1990) J. Bacteriol. 172, 2920–2929.
184 Pilkington, S.J., Skehel, J.M., Gennis, R.B. and Walker, J.E. (1991) Biochemistry 30, 2166–2175.
185 Patel, S.D., Aebersold, R. and Attardi, G. (1991) Proc. Natl. Acad. Sci. USA 88, 4225–4229.
186 Preis, D., Weidner, U., Conzen, C., Azevedo, J.E., Nehls, U., Röhlen, D., van der Pas, J., Sackmann, U., Schneider, R., Werner, S. and Weiss, H. (1991) Biochim. Biophys. Acta 1090, 133–138.
187 Böhm, R., Santer, M. and Böck, A. (1990) Mol. Microbiol. 4, 231–243.
188 Trueblood, C.E., Wright, R.M. and Poyton, R.O. (1988) Mol. Cell Biol. 8, 4537–4540.
189 Bisson, R. and Schiavo, G. (1986) J. Biol. Chem. 261, 4373–4376.
190 Kadenbach, B., Kuhn-Neutwig, L. and Büge, U. (1987) Curr. Top. Bioenerg. 15, 113–161.
191 Shoffner, J.M. and Wallace, D.C. (1990) in: H. Harris and K. Hirschhorn (Eds.) Advances in Human Genetics, Plenum Press, New York, pp. 267–330.
192 Morgan-Hughes, J.A. (1991) in: F.L. Mastaglia and J.N. Walton (Eds.) Skeletal Muscle Pathology, Churchill Livingstone, pp. 367–424.
193 Weiss, H., Leonard, K. and Neupert, W. (1990) TIBS 15, 178–180.
194 Sackmann, U., Zensen, R., Röhlen, D., Jahnke, U. and Weiss, H. (1991) Eur. J. Biochem. 200, 463–469.
194a Runswick, M.J., Fearnley, I.M., Skehel, J.M. and Walker, J.E. (1991) FEBS Lett. 286, 121–124.

195 Johnston, S.A., Anziano, P.Q., Shark, K., Sanford, J.C. and Butow, R.A. (1988) Science 240, 1538–1541.
196 Fox, T.D., Sanford, J.C. and McMullin, T.W. (1988) Proc. Natl. Acad. Sci. USA 85, 7288–7292.
197 Nagley, P. and Devenish, R.J. (1989) Trends Biochem. Sci. 14, 31–35.
198 King, M.P. and Attardi, G. (1989) Science 246, 500–503.
199 Chomyn, A., Meola, G., Bresolin, N., Lai, S.T., Scarlato, G. and Attardi, G. (1991) Mol. Cell Biol. 11, 2236–2244.
200 Hayashi, J.-I., Ohta, S., Kikuchi, A., Takemitsu, M., Goto, Y.-I. and Nonaka, I. (1991) Proc. Natl. Acad. Sci. USA, 88, 10614–10618.

INDEX

acetoacetate, 352, 354
acetogenic bacteria, 50, 66
acetyl CoA, 438
N-acetylglutamate, 363
acetylphosphate, 342
acetylpyridine adenine dinucleotide (AcPyAD), 279
acetylpyridine adenine dinucleotide phosphate (AcPyADP), 277
acidosis, 468
aconitase, 173
acylcarrier protein, 148
adductor pollicis, 473, 474
adenine nucleotide carrier (translocator), *see* ADP/ATP carrier
adenosine-5′-phosphosulphate, 334
adenylate cyclase, 430
adenylation reactions, 332
ADP
– binding to proteins, 268, 269, 402, 430
– concentration, 19, 29, 455
– phosphorylation (*see also* oxidative phosphorylation), 8, 38, 41, 57, 283, 317–334, 423
– regulation of cytochrome *c* oxidase, 252–254
– respiratory control, 268, 269, 402, 430
– transport, *see* ADP/ATP carrier
ADP-ribose, 354
ADP-ribosylation, 352–354
ADP/ATP carrier (adenine nucleotide carrier, ADP/ATP translocator, ADP/ATP transporter), 40, 61, 286–289, 361–378, 389, 392–395, 400, 401, 409–415, 498
ADP/ATP ratio, 31, 362, 377, 422–424, 427, 435, 438, 450, 451, 455, 456, 471
adrenergic agonists, 83, 430, 437, 446
adriamycin, 157
aequorin, 432
age-related changes in skeletal-muscle plasticity, 474, 475
alkalophilic *Bacilli*, 53, 56, 58, 64
alloxan (2,4,5,6-tetraoxypyrimidine), 353, 354
amiloride, 45
amino acid degradation, 360
α-aminoisobutyrate (AIB), 53
aminotransferase, 365
AMP, 19, 402
amytal, 153
anaemia, 468
anaerobic bacteria, 40
antenna chlorophylls, 123, 126
anthracyclines, 157
antimycin A, 201, 202, 209
antiriboflavin antibodies, 178
archaebacteria, 40, 318
ascorbate, 478
aspartate/glutamate carrier, 360–365, 373–378
astrocytes, 52
ATP
– analysis, 334, 365
– binding to proteins, 252, 268, 269, 287, 317–330, 387, 400–406
– concentration, content, 3, 7, 19, 26, 29, 65, 317–330, 350, 452, 475, 476
– control of respiration by ATP consumption, 426
– free energy of hydrolysis, 8, 11, 12, 16, 17, 27–30, 341
– general role in energy transfer, 37, 67, 68, 342, 345
– hydrolysis, utilization, 12, 22, 157, 158, 173, 271, 272, 284, 317–330, 352, 428, 429, 443, 454–456, 467, 468, 473, 477, 493, 496, 500, 501
– regulatory functions, 51, 253, 254, 344, 354, 426, 440–443, 451–455, 497
– synthesis, 2, 22–28, 37–42, 57–68, 113, 122, 200, 241, 242, 273, 283, 287, 295, 317–330, 335, 344, 354, 362, 424–430, 440, 450–455, 467, 470, 471, 478, 497
– transport, *see* ADP/ATP carrier
– turnover, 467
ATP- and PP_i-dependent 6-phosphofructo-1-kinases, 344
atp genes, 11, 285, 292
ATP synthase (H^+-ATP synthase, H^+-ATPase, F_0F_1-ATPase; *see also under individual components and* Na^+-ATP synthase), 7, 8, 18, 25, 29, 31, 39–41, 52, 53, 57, 58, 61, 62, 76, 123, 205, 244, 271, 283–315,

ATP synthase (*cont'd*), 317–348, 425, 435, 436, 440, 441, 452, 458, 470, 471, 498–503
– assembly, 289–297, 308–310
– bacterial ATP synthase, 123, 297–310
– biogenesis, 286–289, 498–503
– chloroplast ATP synthase (CF_0F_1-ATPase), 295–297
– coordination of subunit expression, 307, 308
– gene structure, 285
– mechanism, 320–327
– mitochondrial ATP synthase, 286–295
– structure, 289–291, 295, 296, 298–307, 318–320
– subunit composition, 284–286, 318, 319
ATPase inhibitor protein, 286, 436
ATP/phosphocreatine ratio, 427, 428
ATP:sulphate adenylyltransferase, 334
atractylosides, 361, 362, 389, 434, 441, 450
avidin, 49
2-azido-ATP, 324, 404
azidonaphthoyl-ADP, 324

b-cycle, 207
Ba^{2+} uptake by mitochondria, 351
bacterial chromatophores, 332
bacterial flagella, 39
bacterial photosynthetic reaction centers, 103–120, 139, 188
bacteriochlorophylls, 105, 110–115, 335
bacteriopheophytins, 105–113
bacteriorhodopsin, 16, 38, 40, 75–101, 104, 105, 116, 233, 271
basal metabolic rate, 426
benzenetricarboxylate, 362
binding change mechanism of ATP synthesis, 320–324
biogenisis and import of mitochondrial carriers, 377
biotin, 48, 49
1,6-bisphosphatase, 28
β-blockers, 351
blue cone opsin, 83
blue membrane, 87, 93, 94
blue pigment, 87
bongkrekic acid, 362, 434, 441
brain synaptosomes, 495, 496
branched chain keto acid carrier, 371
branched chain keto acids, 365
brown adipose tissue, 386, 388, 389, 430
brown-fat cells, 388, 406, 414, 415
brown-fat mitochondria, 360, 387–397, 400, 403, 407–410
bupivacaine, 257
tert-butylhydroperoxide, 352–355
BzATP, 324

^{13}C NMR, 472
Ca^{2+}, cytosolic levels, 350, 430, 453
$[Ca^{2+}]_{cyt}$, 432–434, 447, 453
Ca^{2+}-activated PDH phosphatase, 437
Ca^{2+} activation of mitochondrial dehydrogenases, 428, 436–438
Ca^{2+}-ATPase, 41
Ca^{2+} channel, 14
Ca^{2+} channel blockers, 352
Ca^{2+}-control of ATP synthase, 440
Ca^{2+}-dependent activation of mitochondrial dehydrogenases, 428
Ca^{2+}-dependent activation of PEP carboxykinase, 438
Ca^{2+}-dependent ATPase inhibitor protein, 440
Ca^{2+}-dependent dehydrogenases, 453
Ca^{2+}-dependent inhibition of pyrophosphatase, 442
Ca^{2+}-dependent pyrophosphate accumulation, 438
Ca^{2+} distribution, 432
Ca^{2+}/H^+ antiporter, 352–354
Ca^{2+} homeostasis, 349
Ca^{2+} influx from the extracellular medium, 430, 433
Ca^{2+}-mediated activation of the malate–aspartate cycle, 438
Ca^{2+}-mobilizing hormones, 430, 434, 435, 438
Ca^{2+} oscillations, 433
Ca^{2+} release from mitochondria, 351–356
Ca^{2+}-sensitive dehydrogenases, 437
Ca^{2+}-sensitive non-specific pore, 439
Ca^{2+} uptake by mitochondria, 38, 350
calmodulin, 350
Calvin cycle, 122
cAMP, 28, 169, 387, 402, 406, 422, 430, 433–435, 441, 442, 453
cAMP as a mitochondrial messenger, 434
cAMP-induced Ca^{2+} release, 434
cAMP receptor protein, 169–171
canaphite, 342
capsaicin, 154
carboxins, 189
carboxylated biotin, 48

cardiac hypertrophy, 444
cardiolipin, 334, 443–445
cardiovascular disorders, 478
carnitine carrier, 362, 363, 370, 372, 373, 375
β-carotene, 114
carotenoid, 107, 110, 115
CCCP, 46, 56–61, 65
cellular energy state, 10
cellular signal transduction, 13
CF_0F_1-ATPase, *see* chloroplast ATP synthase
cGMP, 402
chemical potential, 9
chemiosmotic systems (coupling hypothesis/theory), 19, 37–73, 113, 157, 423
chlorophyll, 103, 124–126, 132, 135
chlorophyll-binding proteins, 131, 135–138
chloroplasts, 40, 121–143, 199, 200, 295–297
– ATP synthase (CF_0F_1-ATPase), 295–297
– DNA, 131, 135, 136, 285
– envelope, 122, 123, 499
– gene products, 124
– grana stacks, 122
– photosystem I, 40, 104, 115, 123–130, 200
– photosystem II, 40, 103, 115, 131–139, 200
– stroma, 127
– thylakoid lumen, 122
– thylakoid membrane, 122, 123, 127
chondrocalcinosis, 345
citrate carrier, 362, 363, 370, 373
citric acid cycle (Krebs cycle), *see* tricarboxylic acid cycle
citrulline synthesis, 435, 439
Cl^- ATPase, 41
Cl^- permeability, 405, 406, 410
Cl^- pump, 76, 96
Cl^- transport, 394, 395
claudication, 468
coenzyme A (CoA), 363, 404
coenzyme Q (CoQ), *see* ubiquinone
Complex I, *see* NADH:ubiquinone oxidoreductase
Complex II, *see* succinate:quinone oxidoreductase
Complex III, *see* ubiquinol:cytochrome *c* oxidoreductase
Complex IV, *see* cytochrome oxidase
creatine kinase, 426–428, 452, 455, 456, 471
creatine phosphate, *see* phosphocreatine
Cu_A, Cu_B, *see* cytochrome oxidase
α-cyano-3-hydroxy-cinnamate, 362, 364
cyanobacteria, 40, 66, 103, 122–124, 126, 127, 130, 134, 138, 199, 200
Cybex ergometer, 468
cyclic bacterial photoredox chain, 40
cyclophilin, 441
cyclosporin-sensitive Ca^{2+}-activated pore complex, 454
cyclosporins, 352–356, 439, 441
cysteine degradation, 360
cytochrome P-450, 226
cytochrome a, a_3 (*see also* cytochrome oxidase), 220–224, 228, 229, 242–244
cytochrome *b*, 163, 185–190, 202–207, 244, 444, 445
cytochrome *b* anchor polypeptide, 186
cytochrome b_{559}, 114, 131, 132, 135–138
cytochrome b_{562}, 79, 80, 201, 202, 207–209
cytochrome b_{566}, 201, 202, 207, 208
cytochrome b_6/f complex, 123, 132, 199
cytochrome ba_3, 243
cytochrome *bc* (bc_1) complex, *see* ubiquinol:cytochrome *c* reductase
cytochrome *c*, 173, 203, 207, 224, 226, 228, 231, 242–244, 252, 253
cytochrome c', 79, 80
cytochrome *c* oxidase, *see* cytochrome oxidase
cytochrome *c* reductase, *see* ubiquinol:cytochrome *c* reductase
cytochrome c_1, 201–208, 251, 444, 450
cytochrome c_1aa_3, 243
cytochrome $c_1(f)$, 199
cytochrome c_2, 112, 113
cytochrome c_{550}, 243
cytochrome c_{552}, 243
cytochrome c_{553}, 122–124, 128
cytochrome caa_3, 244
cytochrome $c(c_2)$, 199
cytochrome *co*, 243
cytochrome components of bacterial photosynthetic reaction centers, 105–111
cytochrome *d*, 47, 243
cytochrome *o*, 40, 47, 220, 222, 223, 226–228, 231, 233, 243
cytochrome oxidase (cytochrome *c* oxidase, Complex IV; *see also* cytochrome *o*, quinol oxidase)
– binuclear dioxygen reduction site, 221, 230, 233, 242
– biogenesis, assembly, 500–503
– composition of pro- and eukaryotic cytochrome oxidases, 224–228, 233, 242–244

- Cu_A, Cu_B, 218–233, 242
- cytochrome aa_3, 220–224, 228, 229, 242–244
- electron transfer and dioxygen reduction, 228–231
- evolutionary aspects, 244–246
- genes and protein structure of nuclear encoded subunits, 246–251
- isozymes, 242–247

cytochrome oxidase (cytochrome *c* oxidase, Complex IV; *see also* cytochrome *o*, quinol oxidase), 9, 17, 30, 40, 45, 200, 205, 217–239, 241–263, 424, 445, 453, 477, 489, 490, 500–502
- protein framework and membrane topology of redox centers, 224–228
- proton pumping and energy conservation, 231–233
- regulation, 251–257
- structure of metal sites, 218–221
- subunit composition, 224–228, 244, 245

cytochrome reductase, *see* ubiquinol:cytochrome *c* reductase
cytoplasmic petite (*rho*^{0}) mutants of yeast, 292
cytoplasmic PPases, 333, 339, 344
cytosolic hsp70, 287

D-loop region of mitochondrial DNA, 483, 484
D_1 and D_2 protein components of photosystem II, 126, 130–138
DABS, 403
DAN, DAN-ATP, 402–404
deiodinase, 446
deiodinase inhibitors, 444–447
$\Delta\tilde{\mu}_{H^+}$, *see* H^+ gradient, proton-motive force
$\Delta\tilde{\mu}_{Na^+}$, *see* Na^+ gradient
$\Delta\psi$, *see* membrane potential
denervation, 476
dequalinium chloride, 157
diabetes, 472
diadenosine oligophosphate compounds, 324
dicarboxylate carrier, 361, 363, 370, 373, 424
dicyclohexyl carbodiimide (DCCD), 52, 57, 227, 291, 305, 324, 325, 333, 336, 394, 395
diethylammonium actate, 45
diethylammonium (DEA), 46
diethylstilbestrol, 61
dihydrolipoamide acetyltransferase, 48
1,2-dihydroneurosporene, 105
3,5-diiodothyronine, 446
dimethylglycine dehydrogenase, 173
dimethylsulfoxide reductase, 183
diuretics, 351
diuron, 201, 202
divicine, 353
DNA, 14, 28–31
DNA binding proteins, 14, 29
DNA endonuclease, 488, 489
DNA gyrase, 11, 12, 29, 30
DNA-mediated transformation of mitochondria, 504
DNA supercoiling, 11–13, 29, 30
DNA transcription, 483
dorsal interosseous, 474
double-Q cycle, 209
double turnover Q-cycle, 207
DT-diaphorase, 153
DTNB, 403
durosemiquinone, 209
dystrophin-deficient mouse muscle, 476

EGTA, 355
elasticity coefficients, 8, 12, 13, 27, 424–427, 454, 455
electrochemical Na^+ gradient (potential), *see* Na^+ gradient
electrochemical proton gradient (potential), *see* H^+ gradient
electron transport particles (ETP, ETP_H), 155–158
encephalopathy, 159
epinephrine, 435, 3050
ergometer, 469
ethylenebis(oxyethylenenitrilo) tetraacetic acid (EGTA), 354
N-ethylmaleimide (NEM), 52, 337
eubacteria, 40
exercise performance, 469

F_0 (*see also* ATP synthase), 284–286, 292, 325, 394
F_1 (*see also* ATP synthase), 52, 53, 284, 294, 303, 318, 326, 344, 498–503
F_6 (*see also* ATP synthase), 285, 286, 290
F_A, F_B, F_X, iron–sulfur protein components of photosystem I, 126, 127
FAD, 163, 172–179, 269
fatigue, 473, 477
fatty acid oxidation, 360
fatty acid synthesis, 279, 362
FCCP, 407
FeS centers, *see* iron–sulfur centers

Fe^{2+} uptake by mitochondria, 351
ferredoxin, 123–127, 173, 183, 191
ferredoxin–NADP reductase, 124
F_0F_1-ATPase, F_0F_1-ATP synthase, *see* ATP synthase
flagellar motor, 42
flavin-free radical, 178
flexor digitorum superficialis, 467, 468
flow–force relationships, 6, 10, 16
fluorescent Ca^{2+} indicators, 431, 432, 435
fluoroaluminium- and fluoroberyllium-nucleotide diphosphate complexes, 324
FMN, 147, 148, 150, 151, 153, 158, 501, 502
FSBA, 268, 269
FSH, 453
fumarate-reducing complex, 40, 165–172, 184–189
funiculosin, 201, 202

G protein, 429
G protein-coupled receptors, 430
gastrocnemius, 467, 475
GDP, 403–405, 407–410, 430
GDP-binding, 390–392, 395, 396, 401–404, 407
GDP-binding sites, 387, 392
Gibbs free energy, 2, 9, 22
glucagon, 430, 434–439, 442, 446, 456
glucagon-induced increase in $[Ca^{2+}]_{cyt}$, 434
glucocorticoids, 453
gluconeogenesis, 360–362, 428, 437, 438, 442, 454
glutamate carrier, 362, 375
glutaminase, 435, 439
glutamine carrier, 362, 378
glutamine degradation, 360
glutathione, 279, 363
glutathione peroxidase, 279, 352, 354
glutathione reductase, 279, 352, 354
glycerol-3-phosphate dehydrogenase, 444, 446, 447
glycogen, 472
glycogenolysis, 467, 476, 477
glycolysis, 15, 25, 27, 28, 37, 469, 476
Golgi and plasma membrane PPases, 338
Graves disease, 411
green bacteria, 40, 122
growth hormone, 453
GTP, 402, 405, 438
GTP-binding site, 373
guanidines, 351
guide RNAs, 492
gulonolactone oxidase, 173

H^+-ATPase of E_0E_1-type, 40
H^+-ATPase of F_0F_1-type, *see* ATP synthase
H^+-ATPase of vacuolar type, 40
H^+ channels, 85, 278, 284
H^+ conduction, 291
H^+ cycle, 38, 58
H^+ gradient ($\Delta\tilde{\mu}_{H^+}$), *see* also proton-motive force (Δp), 9, 17–19, 23, 27–29, 39–42, 46–48, 52, 58, 62–67, 76, 92–94, 104, 113, 122, 158, 217, 229, 257, 272, 317, 334, 335, 351, 387, 422–426, 429, 430, 450, 455
H^+ leak (leakage), slip, 1–35, 424
H^+-mellobiose symporter, 62
H^+ motor, 40
H^+ permeability, 406, 407
H^+ pore, 284
H^+-PPases, H^+-PP synthases, *see* pyrophosphatases
H^+ pumps, 3, 19, 27, 76, 85–90, 218, 233, 426, 429
H^+-solute symporters, 40
H^+-transhydrogenase, *see* nicotinamide nucleotide transhydrogenase
H^+ translocation, 217, 231, 232, 265, 272, 295, 396
H^+ uniporter, 56
haemoglobin, 91, 467, 468
haems *a* and a_3, *see* cytochrome oxidase
halobacteria, 38, 40, 76
halorhodopsin, 76, 77, 83, 94, 96
heart failure, 478, 479
heat shock proteins hsp60 and hsp70, 288, 294, 295, 496–501
hemerythrin, 79, 80
2-heptyl-4-hydroxyquinoline *n*-oxide, 42
herbicide-resistant PS II mutants, 114
hexamine cobalt, 351
hexokinase, 17, 27
high-resolution NMR, 464
histone proteins, 28
HMG-CoA synthase, 438
H_3O^+, 62
hormonal regulation, 421
hormonal signals acting on mitochondria, 430
hormone-sensitive lipase, 387
HQNO, 45, 59, 201, 202
human DNA, 485
human *gastrocnemius* muscle, 465

human mitochondrial genome, 484
hydrogen peroxide, 229, 241, 279, 280, 352
hydroperoxide-dependent Ca^{2+} release, 354
hydroperoxyeicosatetraenoic acids, 353
hydroxyapatite, 364
hydroxynonenal, 353, 354
hypercapnea, 474
hypophysectomy, 453
hypothalamic area, 386
hypothyroid state, 447
hypothyroidism, 443, 444, 451, 477
hypoxia, 467, 468

idiopathic hypertension, 476
idiopathic Parkinson's disease, 159
import of proteins into mitochondria, 496
import site protein, 288
inorganic polyphosphates, 332, 342
inorganic pyrophosphatases (PPases), 331–348
inorganic pyrophosphate (PP_i), 331–348
Ins-1,4,5-P_3, 430, 434
insulin, 453, 477
insulinase, 209, 210
integrated membrane assembly pathway, 309
intracellular Ca^{2+} concentration, 350, 476
intron mobility, 488
intron splicing, 488
m-iodobenzylguanidine, 354
ionophores, 45
iron–sulfur (FeS) centers (clusters, proteins; *see also under individual proteins*), 104, 107, 112–114, 124–127, 146–158, 163, 164, 173, 178–182, 186, 189, 199, 203, 204, 208, 444, 450, 501
isocitrate dehydrogenase, 402
isocitrate/oxogluttarate shuttle, 360
isozymes of cytochrome oxidase, 242
ISP42, 496

K^+-ATPase, 41
K^+ channels, 440
K^+ gradient, 41
K^+/Na^+ gradients, 42
ketogenesis, 438, 454
α-ketoglutarate dehydrogenase, 436–438, 442
K^+/H^+ exchange, 440
kinase–phosphatase signalling, 14
kinetic control, 4, 5
Krebs cycle, *see* tricarboxylic acid cycle

lactic acidosis, 159
lactose carrier (LacY), 367
lactose permease, 376
lanthanides, 351
lasalocid A, 352
leucyl–tRNA synthetase, 489
Li^+/K^+-ATPase, 51
light-adapted bacteriorhodopsin, 86
light-harvesting antenna complexes, 103, 112, 115, 123
lipid peroxidation, 279
lipogenesis, 360
lysosomes, 40

malate transport system, 367
malate–aspartate cycle (shuttle), 360, 436–438, 443
malic enzyme, 444, 446
malondialdehyde, 279
MAS70 protein, 288
matrix processing peptidase (MPP), 209, 210
matrix protease, 288
maximal voluntary contraction (MVC), 473
membrane lipid composition, 445, 454
membrane potential ($\Delta\psi$), 26, 39, 41, 46, 54, 207, 231, 288, 334, 351–354, 371, 377, 423–425, 446–448, 452–454
membrane receptor, 14
menadione, 353, 354, 478
menaquinone, 179, 180, 185
metabolic control, 7, 8, 18
Metabolic Control Analysis (MCA), 3
metabolic control theory, 423
metabolite carriers in mitochondria (*see also under individual carriers*), 359–384
metabolite control coefficient, 424
methanobacteria, 40, 50, 63, 66
methanogenesis, 59
methanogenesis-linked electron transfer, 63
methanogenesis-linked electron transfer complex, 40
β-methoxyacrylates, 208
N-methyl-4-phenylpyridine, 353
Mg^{2+} as a regulator of mitochondrial function, 441
Mg^{2+}-dependent ATP uptake, 435
Mg^{2+}-dependent ATP uptake carrier, 363
Mg^{2+}-inhibited K^+ transport systems, 442
Mg^{2+} uptake by mitochondria, 351, 352
microbial growth, 19–23, 26
microsomal membrane-bound K^+-stimulated PPase, 336
microsomal H^+-PPase, 336

mitochondria
- ATP synthesis (*see also under* ATP synthase *and* oxidative phosphorylation), 283–348
- biogenesis, 483–509
- Ca^{2+} transport, 349–358, 434, 470
- chemiosmotic systems, 37–41
- contact sites between inner and outer membranes, 427, 454
- control of oxidative phosphorylation, 8, 9, 423–429
- diseases, 159, 433, 476, 503, 504
- DNA, 241, 242, 286, 291, 292, 444, 483–509
- DNA damage causing or contributing to disease and senescence, 504
- DNA transcription, 483–487, 496
- glutathione, 353
- hormonal regulation, 429–453
- metabolite carriers (*see also under individual carriers*), 359–384
- mRNA, RNA polymerase, RNA processing, 486–496
- mutants, DNA-less mitochondria, 504
- nicotinamide nucleotide transhydrogenase, 265–281
- outer membrane proteins, 498
- protein import, 205, 496–500
- protein synthesis, 444, 492–500
- pyrophosphatase, 331–348
- regulation of gene expression, 492–500
- respiratory chain (*see also under individual components*), 145–239
- thermogenesis, 385–419

Mn^+ uptake by mitochondria, 351
MNET, 16, 22, 23, 27
monensin, 45, 49, 56, 57
monoamine oxidases, 154, 173
mosaic non-equilibrium thermodynamics, 14, 16
MPP^+, 146, 154–156, 159
MPP^+ analogs, 155
MPTP, 154, 159
mRNA, 14, 52, 494, 495
mRNA translation, 494
muscle contraction, 454
muscle energetics, 471
muscle glycogen, 472
muscle phosphorylase deficiency (McArdle's disease), 478
MVC, 474
myoglobin, 91
myxalamids, 154
myxothiazol, 202, 203, 208

Na^+-ATPase (Na^+-ATP synthase), 42, 50–52, 57–65
Na^+ channels, 44
Na^+-conducting pathways, 44
Na^+ cycle, 41, 42, 65
Na^+-decarboxylases, 63
Na^+-dependent Ca^{2+} efflux from mitochondria, 351–354, 431–434
Na^+-glucose symporter, 53
Na^+ gradient ($\Delta\tilde{\mu}_{Na^+}$), 38–42, 53–58, 62, 63, 66, 67, 476
Na^+/H^+ antiporter, 41, 45, 46, 54–56
Na^+/K^+-ATPase, 41, 50–54, 57, 61–65
Na^+-metabolite symporters, 61
Na^+-motive decarboxylases, 47
Na^+-motive NADH-mena(ubi)-quinone reductase, 42, 45, 53, 56, 63–65
Na^+-motive oxaloacetate decarboxylase, 42, 49
Na^+-motive quinol oxidase, 63
Na^+-motive terminal oxidase, 45
Na^+-motor, 63
Na^+-proline-symporter, 53
Na^+ pump, 46, 56, 57, 62
Na^+-solute antiporters and symporters, 53, 54, 63
NAD-dependent isocitrate dehydrogenase, 350, 436
NAD^+ glycohydrolase, 353, 354
NAD^+ hydrolysis, 353, 354, 356
$NADH/NAD^+$ ratio, 428, 437, 442, 443, 455, 456
NADH:ubiquinone oxidoreductase (Complex I), 145–162, 173, 244, 444, 483, 491, 501–503
- biosynthesis, 501–503
- diseases related to Complex I deficiency, 158, 159
- energy conservation site, 157, 158
- inhibitors (*see also under individual inhibitors*), 153–157
- substrate specificity, 147
- subunit structure, 147–150, 501–503

$NAD(P)^+$-linked Ca^{2+} efflux, 354
NAD(P) redox states, 425
nicotinamide nucleotide transhydrogenase (energy-linked, energy-transducing), 38–41, 66, 148, 265–281, 352, 353, 426, 445
- AB and BB transhydrogenases, 265–267

- catalytic properties, 269
- energy- and substrate-dependent conformation changes, 275–278
- H^+ translocation, 40, 66
- hydride transfer between NAD(H) and NADP(H), 269
- membrane topology, 269, 270
- nucleotide binding sites, 268, 269
- physiological role, 279
- proton channel, 278
- structure, 267–269

non-cyclic bacterial photoredox chain, 40
non-equilibrium thermodynamics, 10, 16–19, 27
nonactin, 257
norepinephrine, 387, 406, 415
nuclear receptors, 430
nucleosomes, 28

oli2mit mutants, 292
oligomycin, 51, 157, 158, 291
oligomycin sensitivity conferring protein (OSCP), 285, 286, 290, 291, 294, 295
opsins, 83–86
ornithine carrier, 362, 363, 365
osteoporosis, 345
ouabain, 51, 52
outer membrane VDAC channel (mitochondrial porin), 427
ovariectomy, 477
oxaloacetate, 354
oxaloacetate decarboxylase, 48
oxidative phosphorylation, 8, 9, 11–13, 17, 18, 23–27, 30, 38, 62, 64, 212, 283, 317, 360–362, 422–429, 435, 440–448, 451–454, 470, 471, 476, 477, 495, 500–503
oxidative stress, 279
oxoglutarate carrier, 361–363, 365–367, 369, 370, 373, 377, 378
oxoglutarate/malate carrier, 411, 413
oxygenic photosynthesis, 121–143
oxyhaemoglobin, 229
oxymyoglobin, 229

^{31}P NMR, 463–481
P680, P700, 125–135
palmitoyl-CoA, 403, 407, 409, 410
parvalbumin, 350
Pb^{2+} uptake by mitochondria, 351
peptidase, 288
peptidyl-prolyl *cis–trans* isomerase, 356, 441
phenylephrine, 428, 432, 433, 442
phenylmaleimide, 365
pheophytins, 114, 126, 132
phosphate carrier, 18, 39, 286, 361–377, 411, 445, 499
phosphate/ATP ratio, 477
phosphate/phosphocreatine ratio, 451, 452, 469
phosphocreatine, 39, 455, 465–467, 470–477
phosphocreatine/creatine ratio, 427
phosphodiesters (PDE), 465
phosphoenolpyruvate:glucose phosphotransferase, 25
phospholipase C, 422, 429, 430, 434
phospholipid degradation products, 449
phosphomonoesters (PME), 465
phosphorylase, 474
phosphorylation potential, 9, 12, 16–19, 23, 26–30, 360, 427, 443, 452, 455, 456, 470, 474–476
photophosphorylation, 17, 37, 38, 122, 317, 332
photosynthetic bacteria, 40, 199, 212, 285, 296, 334
photosynthetic purple nonsulfur bacteria, 200
photosynthetic reaction centers (*see also under individual components*)
- bacterial, 103–120
- chloroplast photosystem I, 40, 104, 115, 122–130, 200
- chloroplast photosystem II, 40, 103, 115, 120, 130–139

phtalonate, 362
phycobilisomes, 123
phylloquinone (vitamin K_1), 124, 125
piericidin, 146–157
plasma membrane ATPases of plant and fungal cells, 40
plasma membrane Ca^{2+} pump (Ca^{2+}-ATPase), 41, 350
plastocyanin, 122–125, 128, 137, 199
plastocyanin-binding subunit, 127
plastoquinone, 122, 132, 136, 199
polyphosphate, 342
polyphosphoinositide-specific phospholipase, 430
polyphosphoinositides, 422
prebiotic systems, 333
presequence binding factor, 497, 498
processing-enhancing protein, 209, 210, 288, 500

prochlorons, 122
prooxidant-dependent Ca^{2+} release from mitochondria, 352
propylthiouracil, 446–448
protein ADP-ribosylation, 353
protein degradation, 10
protein kinase, 28, 387
protein kinase A, 434, 443
protein kinase C, 443
protein phosphatase, 14
protein phosphorylation, 332
protein transport, 10, 205, 496–500
protonophores, 26, 27, 38, 39, 42, 45, 54
purple bacteria, 40, 103, 104, 113, 114, 122, 132
purple membrane, 76, 85–89, 93, 96
pyridine nucleotide hydrolysis, 353
pyridine nucleotide-linked Ca^{2+} release, 352, 353
pyrophosphatases, 332–345, 438, 439
pyrophosphate, 332–345, 439
pyrophosphate accumulation, 440
pyrophosphate binding enzymes, 339
pyrophosphate synthesis, 335
pyrophosphate:fructose-6-phosphate transferase, 341
pyruvate carboxylase, 435, 437, 439
pyruvate carboxylation, 435, 437
pyruvate carrier, 361–363, 365, 371, 375, 378
pyruvate decarboxylase, 28
pyruvate dehydrogenase, 48, 191, 350, 436, 437, 442
pyruvate dehydrogenase kinase, 437, 442
pyruvate dehydrogenase phosphatase, 436, 442, 452
pyruvate translocator, 394

Q, *see* ubiquinone
Q cycle, *see* ubiquinone cycle
Q radical, *see* ubisemiquinone
Q_A, Q_B, quinone components of
- bacterial photosynthetic reaction centers, 110–114, 188
- chloroplast photosystem II, 126, 132, 138, 139

Q_S, quinone component of succinate:quinone oxidoreductase, 184, 188–190
quadriceps muscle, 495
quantasomes, 138
quantum yield, 90
quinol-binding site, 228
quinol oxidases, 227
quinol:cytochrome *c* (plastocyanin) oxidoreductase, 40, 199
quinol:fumarate reductase, 163, 166–168, 171–184, 187–191
quinol/quinone transhydrogenation, 209

receptor-operated ion channels, 422
redox loop, 231
redox potential, 16, 18
respiratory chain, *see* mitochondria *and individual respiratory chain components*
respiratory control, 18, 30, 256, 257
respiratory control ratios (RCR), 256
retinal, 76, 83, 87, 88, 91–93
retinal binding site, 85
retinal chromophore, 86
retinal isomerase, 87
rhein, 153
rhodanese, 173
rhodopsin, 82–84, 87
rhodoquinone, 185
Rieske iron–sulfur protein, 181, 182, 199, 203, 204, 208, 444, 450
RNA, 10
RNA editing, 491, 504
RNA maturase, 488
RNA metabolism, 10
RNA polymerase, 29, 487
RNA processing, 488, 496
RNA splicing, 504
RNA synthesis, 487
rotenoids, 156
rotenone, 146, 147, 150, 153–155, 157
rRNA, 484
rRNA genes, 493
Rubisco, 122
ruthenium red, 351

sarcolemma, 473, 476
sartorius muscle, 467
Schrödinger's paradox, 2
sdh, 165, 167–171
sensory rhodopsin, 76, 83
sex hormones, 477
signal transduction, 13, 14, 31
singlet oxygen, 134
spermine, 363
Sr^{2+} uptake by mitochondria, 351
steroid biosynthesis, 279
stigmatellin, 202, 208
stoichiometry, 285

streptozotocin, 477
stroke-like episodes, 159
substrate-level phosphorylations, 39, 57, 58, 64, 456
succinate dehydrogenase, 165, 166, 171, 174–179, 183, 187, 188
succinate:menaquinone reductase, 170
succinate:quinone oxidoreductases (*incl.* succinate:ubiquinone oxidoreductase, Complex II; *see also under individual components*), 163–198
– composition, 166, 167
– deficiency, 191
– gene organization, biogenesis, 167–173
– intramolecular electron transfer, 177, 178
– quinone active site, 188–191
– structure, function, 173–190
sugar carriers in bacteria, 367
sulfur metabolism, 360
superoxide radical, 241
sympathetic nervous system, 386

T tubule membranes, 473
targeting sequence, 287, 496
tetrahydromethanopterin, 50
tetramethyl-*p*-phenylenediamine, 46, 230
tetraphenylboron anion (TPB^-), 154
thermodynamic control, 4, 5, 8
thermodynamics of the photocycle, 94
thermogenesis, 256, 386, 406, 407, 415
thermogenin (uncoupling protein), 40, 361, 363, 367, 369, 371–373, 377, 378, 385–419, 429, 430
– gene expression, 388, 411–415
– mechanism, control of activity, 404–410
– mRNA, 414
– reconstitution, 389–391
– structure, 398–400
thermokinetic parameters, 10
thermophilic bacterium PS3, 243, 244, 338
thermoregulatory heat formation, 39
thiamine, 363
thio-NADP, 277
thylakoid membrane, *see* chloroplasts
thyroid hormone, 430, 443–454, 476
thyroid hormone effects on state 3 and state 4 respiration, 448, 449
thyroid hormone-induced changes in mitochondrial lipid composition, 445
thyroid hormone-induced changes in oxidative phosphorylation, 447
thyroid hormone-induced synthesis of mitochondrial electron transport enzymes, 444
thyroid hormone receptors, 447
thyroid hormone regulation of cellular energy metabolism, 443–452
thyroid hormone status, 426, 448
thyroid hormone treatment, 445
thyroidectomy, 444, 447
thyroperoxidase, 228
thyrotoxicosis, 451
thyroxine, 444
tibialis anterior, 473, 474
TNBS, 403
TNM, 403, 404
tonoplast, 56, 336, 337
tonoplast membranes, 332
tonoplast of plant and fungal vacuoles, 40
topoisomerase, 12
tRNA, 494
tRNA genes, 493
training, 475
transmembrane electric (electrochemical) potential ($\Delta\psi$), *see* membrane potential
trialkyltin, 57
tricarboxylic acid cycle (citric acid cycle, Krebs cycle), 15, 168, 173, 360, 423, 425, 436–438, 453, 472
triglyceride droplets, 387
triglycerides, 406
triiodothyronine, 444
troponin C, 350
trypsin, 274
tyrosine kinases, 422, 453
tyrosyl-tRNA synthetase, 489

ubiquinol:cytochrome *c* oxidoreductase (Complex III; *see also under individual components*), 112, 151, 158, 199–216, 231, 500–503
– biogenesis, assembly, 205, 206, 500–503
– components, 199–204
– protonmotive ubiquinone cycle, 151, 158, 206–208, 231
– structure and topography, 204, 205
– ubiquinol/ubiquinone centers *o* and *i*, 208
ubiquinone (coenzyme Q, CoQ, Q), 43, 44, 105, 111, 146, 149–151, 154–158, 180, 185, 188, 199, 204, 207–209, 279
ubiquinone analogue, 204
ubiquinone-binding protein, 204

ubiquinone cycle (proton-motive ubiquinone cycle, Q cycle), 151, 158, 206–208, 231
ubisemiquinone, 188, 189, 207–209
UDP-glucose pyrophosphophorylase, 341
UHDBT, 203
unc (or *atp*) operon, 297, 307
uncoupler-binding protein, 286
uncoupling protein, *see* thermogenin
unsaturated fatty acids, 279
UPD, 402
urea synthesis, 360, 362, 428, 454
urolithiasis, 345
UTP, 402

vacuolar membranes, 318
vacuolar H^+-PPase, 337
vacuolar PPases, 333, 336, 344
valinomycin, 45, 46, 52, 58, 59, 61, 257, 395, 397
vanadate, 52
vasopressin, 434–437, 446, 447
vasopressin-induced changes in NAD(P)H fluorescence, 439
venturicidin, 57, 62
vitamin D3, 453
vitamin K_1 (phylloquinone), 124, 125

xenobiotic hydroxylation, 279